1

Grzimek's ENCYCLOPEDIA OF ETHOLOGY

Grzimek's
ENCYCLOPEDIA
OF ETHOLOGY

Editor-in-Chief

Dr. Dr. h.c. Bernhard Grzimek

Professor, Justus Liebig University of Giessen
Director (Retired), Frankfurt Zoological Garden, Germany
Trustee, Tanzania and Uganda National Parks, East Africa

VAN NOSTRAND REINHOLD COMPANY

New York   Cincinnati   Toronto   London   Melbourne

Van Nostrand Reinhold Company Regional Offices:
New York. Cincinnati Atlanta Dallas San Francisco

Van Nostrand Reinhold Company International Offices.
London Toronto Melbourne

Typography by Santype International Ltd., Salisbury, Great Britain

Printed and bound in Italy by Campi Editore, Foligno, Italy

Published in the United States by Van Nostrand Reinhold Company
450 West 33rd Street, New York, N.Y. 10001

English edition first published in England by Van Nostrand
Reinhold Ltd.

15 14 13 12 11 10 9 8 7 6 5 4 3 2 1

Grzimek's
ENCYCLOPEDIA
OF ETHOLOGY

Edited by:

KLAUS IMMELMANN

ENGLISH EDITION

GENERAL EDITOR:
George M. Narita

PRODUCTION DIRECTOR:
James V. Leone

SCIENTIFIC EDITOR:
Erich Klinghammer

TRANSLATOR:
Monica Baehr

SCIENTIFIC CONSULTANT:
Fritz Walther

ASSISTANT EDITOR:
Peter W. Mehren

EDITORIAL ASSISTANT:
Karen Boikess

INDEX:
Suzanne C. Klinghammer

CONTENTS

PREFACE—K. Immelmann   XI

INTRODUCTION   XIV

1. **HISTORY OF ETHOLOGY**
   by K. Heinroth   1
   G. M. Burghardt   16

2. **GOALS, ORGANIZATION, AND METHODS OF ETHOLOGY**   23
   by J. Lamprecht

3. **STRUCTURE AND FUNCTION OF THE NERVOUS SYSTEM**   40
   by G. Neuweiler

4. **THE STRUCTURE AND FUNCTION OF SENSE ORGANS**   66
   by D. Burkhardt and I. de la Motte

5. **THE VISUAL SENSE**   75
   by D. Burkhardt

6. **THE CHEMICAL SENSE**   98
   by H. Altner

7. **MECHANICAL SENSES**   107
   by K. E. Linsenmair

8. **BEHAVIORAL FUNCTIONS OF HEARING**   129
   The auditory sense, by M. Konishi   129
   The moth ear, by M. Dambach   129
   Hearing in vertebrates, by M. Konishi   132
   The mammalian ear, by H. Wendt   135

9. **THE ELECTRICAL SENSES**   138
   by T. Szabo

10. **ORIENTATION ECOLOGY**   145
    by R. Jander

11. **FORMS AND MECHANISMS OF ORIENTATION IN SPACE**   164
    by H. Schöne

12. BIOLOGICAL CLOCKS   187
   by E. Gwinner

13. MIGRATION AND ORIENTATION IN BIRDS   199
   by K. E. Schmidt-Koenig

14. ECHOLOCATION IN BATS   205
   by H. U. Schnitzler

15. ISSUES AND CONCEPTS IN ETHOLOGY   219
   by G. W. Barlow

16. DISPLACEMENT ACTIVITIES   246
   by P. Sevenster

17. THE INFLUENCE OF SPECIFIC STIMULUS PATTERNS
   OF THE BEHAVIOR OF ANIMALS   258
   by W. Heiligenberg

18. PREDICTABLE PATTERNS IN ANIMAL SIGNALS   275
   by M. Konishi

19. ELECTRICAL STIMULATION OF THE BRAIN—
   A METHOD IN NEURO-ETHOLOGY   282
   by J. D. Delius

20. NEURO-ETHOLOGICAL RESEARCH IN INVERTEBRATES   295
   by M. Dambach

21. HORMONES AND BEHAVIOR   314
   by R. Sossinka

22. LEARNING AND PLAY   325
   by B. Hassenstein

23. NONVERBAL THINKING   345
   by O. Koehler

24. IMPRINTING AS A FORM OF
   EARLY CHILDHOOD LEARNING   362
   by K. Immelmann and C. Meves

25. FORMS, CAUSES AND BIOLOGICAL MEANING
   OF INTRASPECIFIC AGGRESSION   380
   by H.-U. Reyer

26. THE BIOLOGICAL SIGNIFICANCE OF SEXUALITY   419
     by F. Schaller

27. INDIRECT TRANSFER OF SPERMATHOPHORES
     IN THE ANIMAL KINGDOM   437
     by F. Schaller

28. FORMS AND FUNCTIONS OF COURTSHIP   448
     by R. A. Stamm

29. THE NATURE AND FUNCTION OF PAIR-BONDS   469
     by C. J. Erickson

30. PARENTAL BEHAVIOR   487
     by R. Sossinka

31. THE EVOLUTION OF SOCIAL BEHAVIOR IN ANIMALS   495
     by H. Markl

32. CHEMICAL COMMUNICATION IN SOCIAL INSECTS   519
     by B. Hölldobler

33. THE LANGUAGE OF THE BEES   527
     by M. Lindauer

34. GROUP FORMATION IN UNGULATES   540
     by H. Klingel

35. THE ORGANIZATION AND FUNCTION OF
     PRIMATE GROUPS   550
     by H. Kummer

36. SOCIAL STRESS IN MAMMALS: TREE SHREWS   564
     by D. v. Holst

37. THE DEVELOPMENT OF BEHAVIOR   579
     by G. Gottlieb

38. BEHAVIOR GENETIC ANALYSIS   607
     by J. Hirsch

39. ECO-ETHOLOGY—THE ADAPTIVE VALUE OF BEHAVIOR   620
     by P. H. Klopfer

40. THE EVOLUTION OF BEHAVIOR   642
     by W. Wickler

41.   **THE BEHAVIOR OF DOMESTIC ANIMALS**   **654**
by W. Herre and M. Röhrs

42.   **ETHOLOGY AND THE STUDY OF ANIMALS IN CAPTIVITY**   **662**
by H. Hediger

43.   **PHYLOGENETIC ADAPTATIONS IN HUMAN BEHAVIOR**   **670**
by I. Eibl-Eibesfeldt

SUPPLEMENTARY READINGS   **681**

INDEX   **703**

Preface

Ethology today ranks as one of the foremost branches of the biological sciences. For many years this was certainly not the case. On the contrary, it took an amazingly long time before scientists began to study the behavior of animals in earnest. As late as the beginning of this century, a time when many other branches of natural science experienced intense development, the description and interpretation of animal behavior was still left largely to nonprofessionals, whereas the biologists with their more suitable training showed very little interest. We can understand this neglect by looking at the subject itself: Animal behavior may be interpreted in many ways, and man has always had difficulty examining it from a rational and objective viewpoint, explaining it with equal objectivity. Two completely opposing tendencies emerge here: On the one hand, the desire of many people to anthropomorphize the behavior of an animal and to endow it with characteristics that it would certainly not possess, and on the other hand the opposite (and often perhaps unconscious) tendency to see the animal merely as a kind of reflex machine incapable of higher behavior.

Hence ethology experienced a rather late development, but then quickly grew into a comprehensive and autonomous area of research. Today it not only has a firm place within biology as a whole and increasingly in psychology and anthropology, but it has also captured the interest of non-scientists as well. This is easy to understand: First, there was an enormous need to "catch up" on information, and secondly, people began to see more and more that the findings of ethology—if used with the proper caution—could also help us to understand and explain some of the basic facets of human behavior.

The wide-spread acceptance of ethology today may be seen in the large number of books and articles dealing with many important questions and aspects of animal and human behavior. Yet until now the general public has not had ready access to a broad compendium on ethology and its related fields. This volume is designed to meet this need and to further the understanding of the objective examination of animal behavior, un-fettered by emotional prejudices.

However, the large amount of material in this area of science makes it impossible to provide the reader with a complete picture. Thus we were forced from the beginning to be selective. After many deliberations, we decided to work in the following manner: The basic facts and issues of ethology are presented in detail within the more general chapters, providing a general survey of the field. These sections dealing with more specialized topics, however, contain only a few illustrative examples of particular importance and interest, thus allowing a full explication of the broad spectrum of interwoven facets and phenomena. In our opinion, this method of "representation by example," which by now has also been adopted for university instruction and many other kinds of education, yields a greater benefit than any attempt to reiterate as many facts as possible, a method that could hardly deal with a subject in depth. In our selections, we have tried to present those phenomena and vital processes which presumably hold the greatest interest for the reader, and further-more to select those animal groups which display these phenomena in their highest form or whose patterns have been studied in the greatest detail. Thus, the chapter on pair-formation derives its examples primarily from the birds, while that dealing with group formation refers mainly to primates and ungulates. In discussing orientation phenomena the author focused on echolocation in bats, while the chapter on intraspecific com-munication took its examples from the "language" of bees.

Considering the large number of contributions to this volume and the multiplicity of issues discussed, we could not avoid a certain degree of overlap. Some of these examples and arguments serve to provide insight on several questions at the same time. In each case, however, they are pre-sented in a different context, with different emphasis, and with a different purpose in mind. Some of the very early studies that have now become "classic" works of ethology are mentioned in several chapters because they are particularly important in explaining some very different etholog-ical concepts and principles.

An animal's behavior results from its constant interaction with its en-vironment and from the vital processes taking place within the organism itself. We have therefore always tried to emphasize the ecological factors involved, and in the introductory chapters we have also presented the physiological bases of behavior. Today the relationships between behavior and associated physiological processes are of special interest, since in the final analysis we can explain behavior only by understanding how be-havior comes about. Thus even ethologists tend more and more to "look inside the animal" when doing their research.

As with the ecology and evolution volumes which were published with this volume, we have engaged well-known experts to describe and discuss each of their areas within ethology. We are especially happy to have obtained contributions from colleagues in many countries and from diverse ethological "schools." This means that we can present the whole philosophical spectrum of ethology.

We have tried as much as possible to present the subject matter in a simple and easily understandable style. Yet we could not ignore the fact that ethology—like other fields of science—has evolved its own vocabulary. Many technical terms in ethology cannot be adequately replaced or paraphrased especially in those areas extending into physiology. A description of the frequently difficult facts and interrelationships is impossible without using the appropriate technical terms.

Beyond that, any reader with a special interest in this volume will also want to delve into the extended literature, in which the terminology of ethology will naturally be employed without reservation. The reader would surely be disappointed if most of those terms were unfamiliar to him. Thus we do not want to completely eliminate the "technical language" in this volume.

As a compromise, we thought it best to add a glossary of our own. It is at the end of the book and is designed to explain all those technical terms—from ethology and related fields—that the reader needs to know. The introduction to the glossary explains the structure of this material and how to use it.

Special thanks is given to the general editor of this series, Dr. Bernhard Grzimek, and to the publishers for having made possible the publication of this ethological survey. I am very grateful to all the authors for their tremendous effort and for the great care with which these contributions were compiled and written, and to the editorial staff in the publishing firm for their careful editorial and production work. My colleagues and associates working in behavioral physiology at the University of Bielefeld joined in many discussions on the pertinent issues and helped me in many ways. Dr. K. Grossmann, Bielefeld, Germany and Dr. E. Klinghammer, West Lafayette, U.S.A. gave me many valuable suggestions, and H. U. Reyer, Seewiesen, Germany showed unusual willingness and great editorial competence in helping me to select illustrations and in advising me on many technical points. I owe a great deal of thanks to each. I am grateful in a very special way to Herbert Wendt, who in his usual manner exerted every effort to coordinate this volume—with its share of problems in presenting some of the technical content—into a scientific and literary unit.

We all hope this volume will succeed in providing a large circle of readers with the facts, lines of reasoning, and goals of ethology.

K. Immelmann

Bielefeld, Spring of 1974

Introduction

If, like the author of these lines, a person has taken part in the development of a new branch of science for more than half a century, he will know the way in which a truly innovative idea grows to be recognized within the community of scholars. At first, others attempt to prevent any breakthrough by stubbornly ignoring the subject. When silence does not work, an all-out attack is launched. Then—the general opinion shifts often with amazing suddenness—and the new idea becomes a matter of course, its originator forgotten.

Sometimes a new thought initially attracts more followers than it deserves, so that after some time and on the basis of additional findings a reversal occurs that in turn swings to the other extreme, needing further correction. In itself, this back-and-forth movement of opinions and attitudes is a self-regulating process much to be desired. However, much that is true and valid is often thrown out along with those beliefs that needed correcting. When reading Katharina Heinroth's excellent survey of the history of ethology, presented as Chapter 1, we encounter this process over and over again. For example, A. E. Brehm correctly saw that higher animals can have emotions very similar to those in man, but he then went on to assume—without proper justification—that their behavior was guided by reason and morality, even far more so than are man's actions. By contrast, B. Altum correctly deduced that the species-specific drive behaviors of birds occur blindly and without insight into their survival value, yet he did not see that parts of human behavior can discharge in the same way; he then came to support the view of Descartes, that animals are only machines, without a soul. I very much suspect that, in the depths of his nature loving heart, this wonderful old man never really believed it. Many matters of opinion which had been defended acrimoniously became, following the publication of Charles Darwin's discoveries, no more than pseudo-issues, and today Darwin's theory has been accepted by almost everyone.

But widespread scientific consensus does not in any way constitute the ultimate test of truth. It can happen that certain hypotheses dominate both the scientific and public opinion of their time merely by offering a simple and "elegant" explanation, finally gaining such strength that scientific progress is impeded. Only a truly great revolutionary, or several such people, will then be able to disprove this universally accepted theory. For example, one formerly popular theory, known as the reflex theory, states that the nervous system of animals has no other function than to process and respond to stimuli that it receives from the external environment and from the internal organism by way of specialized nerve cells, called sensory receptors. The assumption is that only these bits of information induce an animal to perform behavior patterns necessary for survival of the species. I want to discuss this very notion and the man who was great enough to refute it, Erich von Holst.

The doctrine *animal non agit, agitur* was expressed at an early stage and brought rapid opposition. William McDougall, with a good knowledge of animals, countered it with a far more valid statement: "The healthy animal is up and doing." On the other hand, he was a devout vitalist and claimed "instinct" to be a factor neither requiring nor permitting a causal explanation. The weakness of his vitalistic position gave support to the behaviorists, who strengthened their claim that the reflex and the conditioned response are the only elements of animal behavior which can properly be studied. Sherrington's theory of reflexes gave their notion a more physiological basis, and even today there are a number of people who, unfortunately, remain convinced of this view. I once belonged to their number, stubbornly clinging to the reflex theory long after I should have recognized that it was inadequate for explaining the basic central nervous functions of animals and man. I had learned that the threshold for stimulus combinations releasing the species-typical behavior patterns would decline if that behavior had not been performed for a while; I also knew that this movement pattern, when undischarged, would make the whole organism restless and could perhaps cause it to actively search for releasing stimuli; I had even recognized that this movement, under such conditions of long disuse, could burst forth without any releasing stimuli at all—at least that we know of—and discharge "in vacuo," as we say. It would have been logical to inquire after any other movement processes that also reoccur in cycles without requiring an external stimulus or some other cues. Naturally, the answer to this question is that stimulus-producing mechanisms in the pacemaker system of the heart induce muscle movements displaying just these properties: They arise without external stimulation and are coordinated into a functional working unit without the influence of afferent nerves.

My intellectual autonomy was so much reduced by the controversy between behaviorists and the followers of the vitalistic "purposive psychology" that I thought any divergence from the old stimulus-response theory to be a concession to the latter. Thus, as late as 1935, I had stated that the genetic, homologizable behavior patterns discovered by Whitman and Heinroth (see Chapter 1) were derived solely from the formation of reflex chains, regardless of their tendency for decreasing thresholds and for producing vacuum activities (which, on the other hand, I had described in great detail). A half-hour after I had made this statement in a lecture to fellow scientists, Erich von Holst had succeeded in finally convincing me that the dominant theory of reflexes was fundamentally wrong.

Even today there are some who believe that the central nervous system is nothing more than a vehicle for the processing and transmission of stimuli (information pieces) that originate from outside and are transmitted by the receptor cells, and that its functions in maintaining life and species survival consists only of "reacting" to these influences. This is precisely the viewpoint, fixed into doctrine by the approval of too many people, that I want to discuss; and Erich von Holst was the great revolutionary who refuted it. These lines are dedicated to his memory. The grave criticism he made against the stimulus-response theory was that it provides only for those experiments which necessarily lead to their own confirmation: The experimenter creates constant "controlled conditions" for the nervous system, then causes a change in conditions in order to record the nervous system's response. It is therefore given no chance to demonstrate that it is capable of more than responding to stimuli.

Erich von Holst made a critical experiment designed to give the nervous system of the earthworm just this opportunity. He suspended the animal's abdominal cord—isolated from the body and the supraesophageal ganglion, the "brain"—into Ringer's solution (a saline solution resembling the blood and tissue fluid in its ion composition) and separately recorded the action currents produced by each ganglion pair. In two segments he left the body of the worm intact and connected to the ganglia. It was soon demonstrated that the individual ganglion pairs, connected to each other only through the central nervous system—the abdominal cord—and no longer peripherally, transmit rhythmical and well-coordinated nervous impulses precisely corresponding to the segmental succession of muscle contractions in a normally crawling worm; this was further indicated by the fact that the separate intact muscle segments "crawled along" at the same rhythm.

Using spinal preparations of fish, with their brains removed, von Holst was able to show not only that the nerve impulses for swimming in these animals are produced in the central nervous system, but also that they are coordinated in the same way, that is, without the aid of proprioceptors and afferent nerves. On the spinal cord of a sea horse he demonstrated an effect described by Sherrington with the term "spinal contrast." In a fish (wrasse) given artificial respiration after its spinal cord was cut behind the brain, the central production of stimuli and coordination of the spinal cord resulted in organized swimming movements that continued until the preparation had died off; this is analogous to the crawling movements of the earthworm's abdominal nerve cord. But this does not happen in the sea horse: In the intact animal the dorsal fin—its most important locomotory organ—usually lies tightly folded in a protective groove between the bone plates of the back. Only when the fish starts to swim, which happens only a few times a day, does it completely unfold the dorsal fin. Next, it detaches its prehensile tail from the substrate, and the dorsal fin starts to beat and make waves. As long as the spinal preparation is left on its own, it will never perform these movements; the fin always remains half erect. The normal resting position of the intact sea horse occurs only when a pressure stimulus is applied to the fish's cervical region. If we do this for a certain period of time and then stop, the fin will rise *above* its previous intermediate position, going higher the longer it had been kept in its folded position by the stimulus. If this period was very long, the fin not only unfolds to maximum fullness, but also starts to beat rhythmically for a short time, after which it slowly returns to its intermediate position. For this process, von Holst formulated a hypothesis that also helped to sort out many analogous effects of what we consider a build-up (or damming-up) capacity shown by complex, species-specific behavior sequences. This hypothesis suggests that during latent periods an "action-specific agent"—perhaps a substance like a neurohormone—is accumulated and then consumed when the movement discharges, much like carbon dioxide accumulates in the organism when it stops breathing, to be exhaled as soon as breathing resumes. Von Holst believed that the "half-mast position" of the sea horse's dorsal fin uses up just as much of this agent as is continually produced by the central nervous system. Under normal circumstances, that is, under the influence of the constant central inhibition, this production of impulses is enough to enable the sea horse to swim at any given moment.

This build-up effect is shown by the great majority of all those species-specific movement patterns which, as was discovered by C. O. Whitman and O. Heinroth, can be homologized. Experimental results have demonstrated that the coordination of a large number of these is independent of proprioceptors and of afferent nerve impulses in general. Hence it is certainly a well-founded hypothesis that more complex behavior patterns, what we call fixed action patterns or instinctive movements, are also based on endogenous stimulus production and central coordination in the way that von Holst described. Highly complex systems where a large part consists of this kind of movement patterns have been studied in great detail by Tinbergen, Baerends, and many others. None of their findings have ever been contrary to von Holst's hypothesis.

Based on the findings of Erich von Holst, Paul Weiss, Kenneth D. Roeder, Theodore Bullock, and others, the suppositions of present-day physiologists regarding the functions of the nervous system are very different from those of Sherrington's time. Concomitantly, changes also came about in the views of comparative ethologists. There is probably no such thing as a "purely reactive" cell, no more than there is a complex reactive behavior. When making these statements, however, we must note that "pure reactivity" is very hard to demonstrate in the first place. When a physiologist studying stimulus impulses or membrane potentials is looking for some electrical activity in a cell, and this cell does not oblige him by firing at the time of investigation, he will turn to another one and look down at the first cell as a "lazy neuron." A preparation used in this kind of experiment would not survive long enough to show whether there are impulse-producing neurons that might fire, let us say, only every three days. K. D. Roeder has given convincing evidence that all transitional stages are possible between "stable" and "labile" (i.e., reactive and active) stimulus-producing elements and that, in fact, the same neuron may be one or the other in temporal succession. Roeder also showed how it is advantageous for the species when precisely those cells which are the essence of reactivity, namely, the sensory receptors, transmit endogenously produced impulses in rapid succession. This phenomenon has been demonstrated conclusively in the sensory epithelia of the olfactory, gravitational, and visual senses. The advantage of an active stimulus-producing cell as opposed to a purely reactive one is that the former does not have an actual threshold, which means that a stimulus does not have to surpass a certain minimum intensity to release some kind of response. Instead, even the smallest stimulus effects a change in the sequence of the endogenous impulses. Even in a monosynaptic process, once considered the most primitive kind of central nervous system functioning and where only one nervous relay connection is made to trigger a spinal reflex, the receptor element—the muscle spindle—is found to produce a never-ending stream of rhythmic impulses. Far from representing the element of all nervous phenomena, the monosynaptic spinal reflex proves to be a special instance of a rather complicated feedback control system. When we consider the spontaneity of basic activities, it not only becomes possible to explain the effects of declining threshold, build-up capacity, and vacuum discharge in complex movement patterns, but these processes are now theoretically imperative—as W. Heiligenberg has demonstrated in Chapter 17. In view of the relationship he has shown between stimuli that increase readiness and those that release activity, and the fact that they are interchangeable, it is no longer feasible to differentiate between "spontaneous" and "reactive" behavior patterns and to present them as opposing processes. When seen in the light of a stimulus-response psychology that still clings to the unitary concept of a "reflex," almost all species-typical behavior patterns are "spontaneous"—regardless of the fact that they are more readily performed in the presence of certain stimuli, and that at a biologically suitable moment they can be "reactively" released by external stimuli.

In lower animals, whose functional organization is relatively simple and thus more easily studied in experiments, it has consistently been found that the central nervous motor activity is completely spontaneous. The ordered volleys of stimuli with which the abdominal nervous cord of an earthworm induces crawling movements would produce a continual effect, were they not kept inhibited on an equally constant level by a super-ordinated process. This inhibition, too, results from endogenous production of stimuli, so that one process opposes the other. In itself, the inhibition process is as much an activity as crawling. The higher "center" which controls the release of several movement-coordination patterns, is connected with the external receptors and possesses the vital information about which of the subordinate movement patterns should be disinhibited with a certain set of stimuli. An earthworm's supraesophageal ganglion "knows" when it should allow the movements of crawling, feeding, escape, and so on to discharge.

We have no reason at all to assume that the interplay between the various spontaneously active instances in the nervous system of higher animals takes place in any other way. In humans, for example, we know that what have seemed to be "purely reactive" functions of the perceptual system prove to have an endogenous spontaneity as soon as we eliminate the normal flow of external stimuli: After a short period of stimulus withdrawal, complex hallucinations appear in the most diverse areas of perception. We also know from the work of Sigmund Freud that it takes constant effort to suppress a drive. This knowledge was gained at a time when the scientific world still clung to the reflex theory; just this work would have been enough to establish Freud's scientific reputation.

If we are not to assume the operation of mythical entities, such as some predetermined harmony between organism and environment, then we have no alternative but to hypothesize that all this "innate knowledge," the large store of information accumulated in the course of evolution which manifests itself both on the effector and receptor sides, is already contained in the structure and organization of the central nervous system. When we try to gain an approximate idea of the wealth of information built into the central nervous system of a higher animal species and serving a useful purpose for its survival, we arrive at stupendous amounts even with the roughest of calculations. We need only to think of how much phylogenetically acquired "skill" is available merely to perform relatively simple fixed action patterns such as locomotion, nest-building and getting food, and how much hereditary "knowledge" is contained in all the receptor systems that, in the sense of species' survival, identify important stimulus situations with such precision that they reliably convey to the inhibiting "center" when a certain movement sequence should be "let go." Imagine the wealth of information contained in the mechanisms of spatial orientation, as well as all the other mechanisms that transmit instantaneous or momentary information—all the multiplicity of neural connections which form the basis of what we call insight behavior.

All learning is an adaptive modification of this phylogenetically predetermined basis for behavior. It is a totally unfounded assumption that mechanisms of behavior need or even can be modified this way. In many cases this would even be fatal, because mechanisms like these contain the "teaching apparatus" that directs the learning process into biologically meaningful paths. An "open program" of behavior, which allows ad-aptive modifications to take place throughout the individual's life, requires not less but more structurally stored information acquired in the course of evolution than does the "closed program" of an unalterable behavior sequence, such as a drive activity which an insect or spider performs only once in its life.

Anyone who has taken part in the development of ethology as a fairly independent branch of biological science during many of its decisive years and who knows about the processes—described earlier—whereby new ideas come to influence science and scientists, will find it extremely interesting to observe how thoughts that originally contributed to the rise of this new field may even affect its later development. May the ideas and thoughts of Erich von Holst, which I have described in rather concise form, provide the reader of this book with this kind of attitude.

1 The History of Ethology

By K. Heinroth

Fig. 1-1. Aristotle, Greek philosopher (384–322 B.C.), wrote the most important work in classical times on the history of animals (*Historia animalium*). In searching for the origins of certain behavior patterns in animals, he became a forerunner of modern research on instincts.

Explorations into the realm of animal behavior date back to the classical era. Aristotle, observing that activities of animals were drive-motivated yet highly purposive, had already recognized the central issue, namely, the study of instincts. Through his own observations and those he obtained from the writings of others, Aristotle tried to gain insight into the nature of these drives, which he claimed had the power to change the physical form of organisms. In *The History of Animals* he wrote: "Whenever a hen wins a fight with a rooster, she starts to crow and to imitate cocks by kicking at other hens. With these new drives, she now grows feathers like a rooster." There is some plausibility to this explanation, for we know of several species of fish where individuals which are continually dominant actually do change their sex. Aristotle offers other observations that were more accurate, such as his description of the territorialism displayed by a pair of nesting eagles. Indeed, he saw a purpose in the organization of all living beings. An explanation can be found in his philosophical concepts, or entelechies—a hierarchy of forces working within matter to create form; in turn, these forces correspond to the hierarchy of organization. Aristotle criticized the atomists of his time, representatives of an entirely different approach, because, as he claimed, they blatantly ignored the functional or purposive aspects of animal behavior which are closely tied to drives and are performed without deliberation or virtue.

The atomists viewed all phenomena in terms of causality, of cause-and-effect relationships operating even in the smallest particles. They held that since humans and other animals are of the same matter, their souls must be alike as well. As a result of these formulations and lines of thought, the study of animal behavior took on an uncritical, anthropomorphic character, as exemplified by the works of Epicurus and especially of Plutarch (50–100 A.D.), who believed that like man, animals possessed insight, virtue, rationality, deliberation, justice, and familial love. Aelianus, too, held that animals possessed certain human qualities, interpreting the behavior of nesting birds that pretend to be lame as intelligent action, or the distribution of food by a mother bird as providing her young with a lesson in justice.

The battle between these opposing systems of thought continued through all the following centuries. Many a natural scientist was challenged to offer increasingly more careful descriptions of behavior which might demonstrate the existence of instincts. One of these was William Harvey, a physiologist of the mid-17th Century, who devoted himself to detailed observations of birds during mating and incubation. He discovered that by stroking a solitary female he could induce her to lay an egg. The instinctive behavior that he observed during courtship, mating, and nesting were seen as manifestations of the divine spirit operating within the creature. In 1702 another dissident, Ferdinand Adam von Pernau, launched a written attack on his atomistic and mechanistic contemporaries who, backed by the writings of Descartes, Sorelli, and Claude Perrault, accepted as valid only causal, physical laws, and who claimed that all behavior is essentially reaction to influences from objects in the environment. Von Pernau, who had made a variety of excellent observations, queried cynically: "And is it really true that the atoms and little specks of dust just happened to mix in such a way that the owl, who must do her hunting in silence, will not be heard?" In a manner even more consistent with this line of thought, the clergyman Johann Friedrich Zorn based his studies of birds (1742 and 1743) on the principle of functionality (teleology), strongly emphasizing the way that drive and physical structure would complement each other. An example is the earthlike color of certain birds, corresponding to their drive to squat down, which makes them inconspicuous on the ground. Zorn, by formulating a teleological principle, did greater justice to the facts known at the time than did his opponents who stood upon the principle of causality and who, due to a lack of proper knowledge, could not generate any testable hypotheses.

But it was really the philosophers and not the natural scientists who pondered upon the origin of instincts. The Abbot of Condillac, for instance, wrote in 1755 that instincts first develop on the basis of experience, reflection, and comparison, by means of practice and the combination of concepts with judgment and ingenuity. Once these instincts have matured, there would be no more need for conscious reflection. His ideas were opposed by Hermann S. Reimarus, a careful observer and ingenious thinker who, in 1760, used the spider and the ant lion as examples to demonstrate that animals are born with precisely those abilities at which they are most dexterous and adept, that drives are activated by means of sensory experience and sheer impulse, and that the acting out of a drive is associated with a feeling of pleasure.

Surprisingly enough, it was the theoreticians of enlightened science who wanted no part of this theory of instincts. They suspected some kind

Studies of birds

Fig. 1-2. Even in early times animals were judged by their appearance according to human values and prejudices. The camel, for example, was considered "arrogant" because of the high position of its eyes and nose. The orbital ridges and the drawn-out corners of the golden eagle's mouth attested to its "proud determination."

Fig. 1-3. Charles Darwin, English biologist (1809–1882), produced the theory of natural selection in the fight for survival.

of a masked teleology and this was just the concept they were struggling against. In France, under the leadership of Charles de Roy and his supporters, philosophers were increasingly coming to believe that animal behavior could be explained by intelligent action. Their viewpoint remained prominent far into the 19th Century. In Germany it was followed up by the natural philosopher Oken, who saw a hierarchy of intelligence manifest in animal behavior which is analogous to the gradation of the organism's physical organization. His follower Scheitlin, a professor in St. Gallen, wrote a two-volume treatise, *An Attempt at a Complete Theory of the Mental Life of Animals* (1840), in which he proposed that animals, endowed with feelings, thoughts, and motives like those of children, perform in the spirit of sympathy, pity, pride, love, and hate. These views on animal behavior were popularized by his contemporary, Alfred Edmund Brehm, in his *Brehms Tierleben* (*Brehm's Animal Life*) (1864–1869), along with other quite accurate biological facts. Brehm did not mention drives at all, but instead made liberal use of value judgments expressed in such terms as clever, mean, sly, grateful, malicious, gentle, valiant, and proud.

A clear dispute had now developed between the vitalists, with their teleology and indiscernable force, and the mechanists, who accepted only the natural laws of science as the explanation for the behavior of all animals, including human beings. In the midst of this, in 1859, Charles Darwin published his book *The Origin of Species through Natural Selection*, which revolutionized the existing view of the universe by proposing that all organisms had developed in the course of millions of years from simple forms to increasingly more complex ones, including our own species. This increasing complexity had been accomplished by the process of natural selection: Animal species are engaged in constant competition for survival, and those characteristics best suited for adaptation to existing environmental conditions would perpetuate themselves. Furthermore, the phylogeny of behavioral development would be subject to the same laws as that of morphological changes: all instinctive motor sequences must emerge gradually, developing from many small and useful modifications. Later, Darwin's assumptions regarding the historical development of variations attained their empirical foundation through Mendel's genetic mechanism, mutation, and the modern synthetic theory of evolution (see Ernest Mayr's *Animal Species and Evolution*).

Darwin suggested that we could find clues to the evolution of instincts by examining parallel species or "side lines" of the same origin. At that time, he did not foresee the fundamental value that his theoretical approach would come to hold for a new way of looking at the development of instincts, for the field of ethology. His belief that humans and animals had emerged in accordance with the same evolutionary laws made him an adherent of classical animal psychology: "All have the same sensations and impressions, similar passions, impulses, and emotions, even the more

complex ones like jealousy, suspicion, ambition, gratefulness, generosity
.... They possess the same capacities of imitation, attention, rationality,
choice, memory, imagination, association of ideas, intelligence, though in
different measures."

Darwin's psychological interpretation of the manner in which animal
behavior develops and his materialistic doctrine of the evolution of organisms provoked a sharp reaction from the clergyman and zoologist Bernhard Altum. In 1868 Altum, one of Darwin's most fervent opponents,
produced a highly critical work, *Birds and their Way of Life*, in which he
aimed to discredit Darwin's views with arguments and numerous examples: "Animals do not think or reflect upon things, do not determine
their own goals, and if they nevertheless act purposively, some other
force must have set the purpose for them.... Animals do not know what
they do. Their instinctive acts are rigid, though some are guided in their
adaptability by function.... A nest is nothing but a simple product of
nature... becoming the more complete the stronger is the drive for reproduction.... So-called love for offspring is the drive to feed young
birds with such and such characteristics, screeching, flapping their wings,
and opening their beaks in a particular way; certainly this is not love."
In other words, Altum negates all emotions in animals, demands a rejection of speculative psychology, and urges his contemporaries to return
to the methods of natural observation.

Alfred R. Wallace, who took part in the founding of evolutionary
theory, held an entirely different opinion when it came to the issue of
instinctive development. He stood closer to the philosopher Herbert
Spencer, who saw instincts as primarily intelligent actions which became
habits and then were passed on in the genes. In this way Wallace, in the
realm of psychology, revived the theory of Lamarck which assumed that
acquired characteristics became hereditary—a viewpoint known as
psycho-Lamarckism. Wallace's stand was fiercely opposed by various
scientists, among them the geneticist Weisman. In addition, two zoologists
spoke out against this approach: In Bristol, in 1896, Lloyd Morgan demonstrated in his book, *Habit and Instinct*, that instincts, too, develop phylogenetically by means of selection, modifiable during the individual's
ontogenesis. In the course of development they would combine with
learned behavior through experience and become solidified as instinctive
habits. He also recognized that a number of impulses are associated with
instinctive behavior which hold communicative meaning for conspecifics
and enemies.

The second zoologist who opposed the doctrine of psycho-Lamarckism
gained prominence in North America: In 1898 Charles O. Whitman
summarized his astute observations in the publication of his lectures,
titled *Animal Behavior*. Whitman found support for Darwin's theory even
in the realm of instinctive development through studies on salamanders,
leeches, and pigeons. He claimed that instincts must develop in conjunc-

Alfred Russel Wallace
and psycho-
Lamarckism

Opponents of psycho-
Lamarckism

tion with morphological structures. How else, for example, could the drinking habits of 500 different species of pigeons be identical? His own interpretation of instinctual development is diametrically opposed to that of psycho-Lamarckists: On the evolutionary scale, instincts came first; intelligence itself emerged later from the "ruins" of older instincts. Not until 1918, after Whitman's death, did his student Oscar Riddle publish Whitman's meticulous comparative studies on many species of pigeons. By that time Wallace Craig had further analyzed those instinctive acts which Whitman had discovered in his pigeons. He distinguished between an essentially fixed or rigid final action (consummatory act), which is always innate, and a variable kind of behavior characterized by a restlessness and an apparent striving for the performance of the consummatory act. The latter, termed appetitive behavior would also give clues to the organism's state of readiness or non-readiness for the performance of specific consummatory behavior. In arguments with his countryman Herrick, who, like many others at that time, viewed instincts as complex chain reactions, Craig held that the cyclical nature of many instincts made Herrick's interpretation unsound.

The publications of Whitman and Craig should have marked the birth of a unified field of ethology. Yet in their time they found no echo, no one continued this research, and for years their work remained in oblivion. Perhaps one reason for this lack of response was the observational basis of their studies. More important, however, was the fact that they were overshadowed by the feud between two already dominant theoretical and empirical frameworks, those of the vitalists and the mechanists. The vitalists did lay stress on the functionality of instincts, but they attempted to explain their origin by postulating a life force, a natural force or divine guidance, or a vitalistic imagination. In this way they obstructed any attempts to explore the question of causality. Similarly, the functional psychologists such as McDougall saw the goal itself as the guiding force of animal behavior. A third group in the vitalistic arena was the holistic psychologists who claimed that it is the unity of organic systems which lends direction to purposive behavior.

The mechanistic approaches stood in direct contrast to these vitalistic ones. Their proponents based all behavioral research on the organism's response to external stimuli and confined the causal analysis to reflex activities of the nervous system. One noted scientist with those views was Jacques Loeb, who erected an approach to behavior based on "forced movements" to external stimuli or tropisms. But most mechanists based their views on the work of the great psychologist Ivan Pavlov, who believed that the basic achievement of the nervous system lies in its unconditioned reflexes, and who explained all learning in terms of conditioned reflexes. Within this framework, along with the trial-and-error approach of E. L. Thorndike, American behaviorists continued their studies in animal psychology, gaining insight into the laws of learning by the

Fig. 1-4. The Russian physiologist Ivan Pavlov (1849–1936) formulated the theory of reflexes and so became the founder of psychological learning theory. In 1904 he received the Nobel Prize for medicine.

use of the now-famous mazes and Skinner boxes. On the subject of instincts and in associated discussions they saw a hidden vitalistic meaning and so, in their fanatical empiricism, they rejected the notion of instinctive behavior together. On such a battlefield it was inevitable that the seeds of thought planted by Whitman and Craig withered away.

In turn-of-the-century Europe, too, the search for instincts and their origins failed to gain ground. Although well-trained and enthusiastic biologists tried to bring greater clarity to the issue by observation on spiders and insects (G. and E. Peckham, E. Wasman, Forel, and Escherich), and although field ornithologists, primarily in England, were studying the instincts of birds (E. Selous, H. E. Howard, J. Huxley, and the group working around Kirkman), the causal explanation given for these instincts was, basically, that they are discharged as reflexes or reflex chains by way of innate "cleronomic" pathways (H. E. Ziegler, 1910, 1920).

Fig. 1-5. Oskar Heinroth, German zoologist (1871–1945). In his *Ethology of Anatidae* he wrote: "I had always made it my goal to study, not what is usually called biology, but rather customs and habits. I found that the literature contained virtually nothing on this ethology, so that I entered a fairly uninhabited land."

The fundamental turning point in the study of instincts came only with the efforts and contributions of Oskar Heinroth and his student Konrad Lorenz. Heinroth was well versed in Darwin's theory of evolution as well as in the facts revealed by modern genetics. As assistant director of the Zoological Garden in Berlin, with its large and varied animal population, he had become a very knowledgeable student of animal species. For his psychological studies on animals, Heinroth wisely chose various species of geese and ducks (Anatids). He compared their anatomy, their movements, their calls with all their meanings, and the reproductive and social habits of the entire family. Today we would term this kind of descriptive activity the construction of ethograms. In 1910 and 1911 Heinroth published his work under the title *Contributions to Biology, particularly the Ethology and Psychology of Anatidae*, in short, *Ethology of Anatidae*. Here for the very first time the term "ethology" was employed in the modern sense to denote the comparative study of innate behavioral norms which had emerged in the course of evolution. For Heinroth had discovered what Darwin and Whitman knew, that in addition to reflex movements such as orienting reactions, certain other motor sequences (such as occur during courtship, for example) are performed by all animals of the same species and sex, always with the same movements or postures. Thus he concluded that these patterns must be constant in form by virtue of genetic inheritance. He called them *arteigene Triebhandlungen*, or "species-specific instinctive actions;" these are called fixed action patterns today. Comparing these behaviors with those displayed by related species of Anatids, he would find them again and again with various modifications, typical and constant in form for every species and occurring in the same functional context as homologies. Thus, similar to the manner in which comparative morphology is used to reconstruct the phylogenetic history of species by means of morphological homologies, we are able to gain insight into the evolutionary development of instinctive behavior. We take into account their biological survival value and compare instinctual homologies in their descending order of similarity to closely related species. Konrad Lorenz believed that the discovery of these homologous behavior sequences marked the true birth of a comparative study of behavior. In this way the science of ethology was founded a second time. Heinroth and Whitman performed their pioneering work independently, for they knew nothing of each other. Whitman's studies of pigeons were not even published until 1918.

Heinroth based his research on inductive and comparative methods, and so elevated the study of instinct and with it that of animal behavior from the realm of philosophical endeavor to the disciplines of natural science, of biology. In his publication on the ethology of Anatids he had focused particularly on the phylogenetic development of intra-specific patterns of communication. For example, he derived the introductory pair-bonding movement of head-dipping from the movements of fetching food up out of the water, or the trampling to attract the young of Anatid mothers from their habitual trampling for food in the mud. He discovered that even the mere onset of instinctive actions, which he called intention movements, served to convey information to conspecifics. In the course of evolution these intention movements had developed into a signal language of their own. By describing these phenomena, Heinroth gave direction to the development of ethological methodology. He discerned the signal value of patches and colored parts on the body which some animals may suddenly display, the gesture and posture of defenselessness as a communication of friendship, a ruffling of the feathers as display and an expression of hostility, and the act of folding the feathers as a sign of affection which all conspecifics will understand. These observations played an important part in the further development of ethological research.

At the conclusion of his *Ethology of Anatidae*, Heinroth pointed the way far into the future, writing: "In this work I have placed special emphasis on the patterns of social communication, and here we are suddenly struck with the almost human character of social interaction among these birds which live in society, especially where members of a family, that is, the father, mother, and offspring, enter into such a long-lasting bond as do geese. In their evolution, the sauropod line has acquired emotions, customs, and motives very similar to those which we humans commonly see as beneficial, morally good, and based on rational processes. Further research into the ethology of higher animals—unfortunately still in its pioneering stages—will bring us ever closer to the realization that our behavior toward family and strangers, during courtship, and so on, is in reality influenced by innate factors and far more basic than had been commonly believed." In a letter to Lorenz in 1931, Heinroth went so far as to write that man's nature could be recognized in the animal, a view similar to that expressed in Darwin's at that time much-neglected books, *The Ascent of Man* and *The Expression of the Emotions in Man and Animals*.

Fig. 1-6. Black swans ready to take to flight. O. Heinroth discovered that such indicative movements contain information for conspecifics, and thus he called them "intention movements."

Fig. 1-7. "Positioning the wings, one of the nicest examples of a display posture," wrote O. Heinroth (1910) in his *Ethology of Anatidae*. "As soon as the mute swan wants to drive off a rival, and therefore wants to give itself a frightening appearance, it raises the elbows and wings and draws back its neck."

In following this principle, the field of ethology has continued to grow. The research results of Lorenz and his students, including Irenäus Eibl-Eibesfeldt and Otto Koenig, have borne witness to the amazing accuracy of Heinroth's predictions and his keen insight into the direction that ethology would take.

Heinroth's contemporaries failed to respond to his bid for continued research into the study of instinct from an evolutionary point of view, and thus he pursued this pioneering work alone, with only the help of his wife Magdalena (née Wiebe). The couple aimed to learn more about which instinctive acts are innate in birds and, conversely, which ones are supplemented in their development through experience or through learning processes. Extensive use was made of the so-called *Kasper Hauser*, or "deprivation experiments": The Heinroths raised all central European bird species in isolation, deprived of all influence from their natural parents. The behavior which resulted was then observed and the following questions asked: "How does it come about that in the wild this animal species is equipped to survive? Why does it have a certain appearance or behave in a certain manner? To what extent can we reconstruct their phylogenetic development by observation of their behaviors and characteristics? What we wish to discover here are the details and nuances of the lifestyle of organisms, of their growth and ontogeny, of their molting, of their instinctive acts and mental capabilities. In short, we want to learn about things which have hitherto been scarcely considered." In the years 1925 to 1933, after two decades of immense effort and patience in the raising of vast numbers of individual birds, the Heinroths published their classic four-volume work, *The Birds of Central Europe, Photographed in All Their Stages of Life and Development and Observed in Their Behavior during Ontogeny Beginning with the Egg*. With this valuable and abundant stock of information about both instinctive behavior patterns and learning processes, the Heinroths built a broad inductive basis for further ethological research.

This publication was enthusiastically received by the young medical doctor and zoology student Konrad Lorenz in Vienna. Lorenz recognized the deep significance of Heinroth's research and began a most animated correspondence with him. "And are you aware, Sir," Lorenz wrote, "that you are actually the founder of a whole new field of science, namely that of animal psychology as a branch of biology? That herein lies the profound value of the publication, *Birds of Europe*? That you have developed an approach and a method of research which must be extended and applied to animals of the whole world? I have gained tremendous stimulation from *Birds of Central Europe*. The raven has particularly caught my interest due to your descriptions, and I have therefore acquired several of these birds. Above all, your work has inspired in me great plans for the future." In addition to the Corvidae, Lorenz acquired several species of heron, and before long he turned his attention to the geese and ducks as

▷
Courtship in a pair of great crested grebes. Julian S. Huxley, who first described the finely synchronized courting movements of these animals in 1914, compared them with the courtship of loons and coined the term "ritualization."

objects for extensive research. There can be no doubt that the ten years of intense friendship between Lorenz and Heinroth, with their continuous flow and exchange of ideas and observations, was indeed of primary importance for the further development of ethology. Lorenz himself admits to the great satisfaction and joy which he derived from steady contact with his fatherly friend, and the opportunities he had to discuss his observations and to reinforce their validity and to grow into a keen and cogent observer of animals on the fertile soil of Heinroth's theoretical framework. Heinroth, in turn, recognized that no other biologist but Lorenz would take upon himself the many years of difficult training in the observation and understanding of the most detailed behavior patterns of entire groups of animal species. He saw as well that this student had the talent and educational background needed in biology, philosophy, and psychology to carry on the analysis of stereotyped species-specific, instinctive behavior patterns and their variable components.

It is interesting to note, too, the fact that ethology was not founded within the institutional setting of a university where numerous students could have pursued its research possibilities. Instead, in a curious parallel to the inception of Darwin's theory of evolution, this science was born solely of the private endeavors of Heinroth, director of the aquarium in Berlin, its technical basis created only through the almost superhuman efforts of his wife in the raising of myriad individual birds for observation. Lorenz began his studies in a similar way, for, while working as an assistant at the Institute of Anatomy in Vienna, he kept his own flocks of free-flying birds on his father's grounds in nearby Altenberg. And again, as was Heinroth's good fortune, Lorenz's wife, Margarete, helped to support his endeavors. She accepted work as a physician in a hospital at a time when her husband finally and with great difficulty succeeded in attaining a poorly paid position as private lecturer in animal psychology at the University of Vienna. At last he was able to resign his less creative position as an assistant at the Institute of Anatomy and to devote more time to the study of animal behavior.

In the course of many years during which Lorenz carefully analyzed the behavior displayed by his beloved avian subjects, he perceived that functionally unitary action chains or sequences, consisting of instinctive, innate motor links, alternate or combine with other parts that must be individually learned. He called this phenomenon "instinct-training inter-calation." For every animal species, the exact point at which a gap must be filled by experience and learning is innately determined. In his detailed discussion on *Companions as Factors in the Bird's Environment* (1935), Lorenz emphasized that instinctive actions are fundamentally distinct from all other kinds of behavior. In comparing learned behavior patterns with inborn ones, he showed how the specific sign stimuli needed to release each behavior are basically very different: Learned behavior patterns can only be released by their appropriate stimulus configurations with all the

◁
Top left: O. Heinroth in 1923 with the two European cranes he raised by hand from the egg. For five years he took them for daily walks in the early morning through the Berlin Zoo. Top right: Two experiments by N. Tinbergen on the connection between "sign stimuli" and the behavior released by them. The herring gull chick pecks at the red patch on the bill dummy attached to a shapeless "head" (upper photo). The oystercatcher responds to a "super-normal" sign stimulus as it rolls the huge egg with contrasting speckles into the nest, while ignoring its own egg (left in the photo) and the only slightly larger egg of a herring gull (center in the photo). Below: If, soon after hatching, the young of geese and other precocial birds are presented with a moving object that emits short sounds at regular intervals, they will regard it as their mother and will follow it. Konrad Lorenz, who introduced the term *Prägung* (imprinting) for this phenomenon, uses a microphone to project his own voice through a loud-speaker built into the moveable dummy.

characteristic traits, whereas innate behavior patterns can be released by very few stimuli—often, in fact, only one is necessary. Lorenz drew from the research and terminology of Jakob von Uexküll, who had already demonstrated that the releasing stimuli for instinctive actions were confined to relatively few effective cues in the environment, designating the internal apparatus which filters the releasing stimuli as the "innate releasing schema." Conceptually, he was comparing it to a lock which opened the way to setting the inborn motor behavior patterns in motion by means of a stimulus or a few stimuli, which in turn are appropriately labelled "key stimuli" or, more commonly, "sign stimuli." At the suggestion of Otto Koehler and in conjunction with a special committee, Lorenz later decided upon the term *angeborener Auslösemechanismus* or AAM ("innate releasing mechanism" or IRM) to describe this (somewhat puzzling) stimulus filter.

Fig. 1-8. Konrad Lorenz, Austrian ethologist (born 1903).

Animals with a repertoire of complex instinctive acts must possess many such IRMs for the perception and processing of relevant sign stimuli. In his "bird companion" article, Lorenz drew upon his own research and that of Heinroth to show that among numerous related species the social behavior of animals (for example, courtship) developed phylogenetically by means of selection of those sign stimuli which aided in the survival of the species by being especially prominent and conspicuous to the partner. Hoods or special color spots which serve as signals are good examples of these characteristics. Lorenz termed such sign stimuli "releasers."

One useful strategy in the analysis of IRMs and their innate releasers is the experiment with models, in which single stimuli can be presented in different combinations. Lorenz discovered that in addition to birds, a certain family of fish, the brood-caring cichlids, whose behavior patterns he had come to know in his Institute in Vienna, lend themselves particularly well to this approach as subjects. This experimental work with models opened the door to a broad area of ethological research and attracted a large number of interested scientists. Lorenz found an enthusiastic student in the aspiring zoologist A. Seitz, who in 1940 discovered an important law for the dynamic relationship of individual stimuli to the IRM (the Law of Heterogeneous Summation). In addition, scientists in other countries now took an interest in ethological research, most notable among them the Dutch zoologists Niko Tinbergen and G. P. Baerends, both of whom also led numerous students into the field. As early as 1939, the classical model experiments performed by Tinbergen and Kuehen served to demonstrate that in addition to colors and light-dark contrast relationships, configurational characteristics of stimuli (their relative size and position) play a part. Since that time the investigation of releasers has led to a wide range of insights and discoveries, from an explanation for the phenomenon of mimicry and the development of brood parasitism (J. Nicolai) to an understanding of the process involved in the evolution of complex courtship behaviors.

Fig. 1-9. In a letter to his teacher Oskar Heinroth in 1934, Konrad Lorenz used a great deal of humor in describing his colony of night herons in the Altenberg Park, including with the letter this delightfully funny cartoon.

Heinroth had already made special note of the observation that, before the actual discharge of a species-specific instinctive behavior pattern, the "mood" or disposition of the animal can be discerned by means of the preceding, variable intention movements. Thus Heinroth spoke of the animal being in a mood to fly, to tread, to brood, and so on. In the course of further analysis, Lorenz chanced upon the publications of Whitman and Craig and began to correspond with elderly Craig, who sent Lorenz his works. Lorenz now proceeded to demonstrate the accuracy of Craig's categorical division of behavior into a search element (appetence or appetite) directed at the releasing stimulus, and an innate consummatory action which corresponded to Heinroth's species-specific instinctive behavior. Lorenz accepted Craig's terms "appetitive behavior" for the variable search behavior and "consummatory action" for the rigid, actual instinctive act. In this way he guaranteed that the research of Whitman and Craig would not remain unknown.

Upon further scrutiny of these two kinds of behavior, Lorenz noticed that the consummatory action was more easily released with an increase in the time that had passed since is was last discharged. In other words, the stimulus threshold was being lowered, sometimes to a point where normally inadequate stimuli could bring about their release or even where the consummatory action could be discharged without any stimuli at all, spontaneously or in vacuo. As early as 1911, the Heinroths had described such a vacuum activity in the case of prey-catching behavior of their bee-eater, but they could not offer an explanation. Lorenz too had often observed such "spontaneous" activity with his hand-raised starlings. He deduced from such phenomena that energy was being produced which would accumulate if not released at sufficient intervals.

Phenomena of this kind were not at all compatible with the hitherto popular explanation of instinctive actions in terms of reflex or chains of reflexes, since reflexes themselves are not subject to change in readiness for release nor to vacuum discharge. Thus Lorenz was faced with a dilemma. In this situation of conflict, Erich von Holst's discovery dropped like a bomb (according to Lorenz), to show the way out: Holst had demonstrated that the central nervous system is characterized by a second kind of activity in addition to reflexes. It continuously and automatically generates its own stimuli which, using·neither afferent nor efferent nerves, pass as impulses to release centrally organized, species-specific action patterns (*Bewegungskoordinationen*). Here, then, lies the source of energy which accumulates and determines the specific action readiness or moods. Von Holst's discovery is one of the most important milestones in the history of ethology. Not only does it offer evidence against the vitalistic notion that instincts cannot be explained causally, but also against the mechanistic explanation, which holds that animal and human behavior can be understood scientifically only in terms of reflexes. Von Holst also opened up new research opportunities for the field of psychology, which up to that

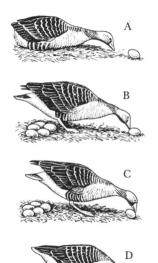

Fig. 1-10. Egg-rolling movement of the greylag goose: After an initial glance, the goose performs several intention movements and then extends its neck fully toward the egg (A). The animal rises with some hesitation, walks toward the egg, and touches it with the tip of its bill (B). It then places the lower bill on top of the egg (C), and with a strange stiffness bends the neck to roll the egg into its nest hollow.

time had not been able to reconcile the principle of cause and effect with psychological phenomena, and thus had failed to develop a truly scientific approach. Most importantly, however, the principle of endogenous central nervous activity led to the formulation of new working hypotheses for the study of instincts.

We may now conceptualize the IRMs as blocks which keep their specific innate motor coordinations—themselves under the continuous pressure of internal drives—locked in, and which can be opened up only by means of specific sign stimuli. To this day one of the most important tasks of ethology is to examine the nature of these drives as well as the selection of IRMs in the instinctive behavior of animals, including human beings.

Fig. 1-11. Erich von Holst, German zoologist (1908–1962).

In addition, the various levels of action readiness as factors in appetitive behavior became a focus for ethological research. The schools headed by Tinbergen and Baerends took considerable interest in this phenomenon. They suggested that these various states of action readiness, which originate from central loci, are organized in a complex hierarchy of systems and subsystems, each like a layer around the other. They also developed elegant yet simple means of experimenting with the behavior of animals in natural or near-natural conditions. They were particularly successful in the use of models to mimic natural sign stimuli.

In the meantime Lorenz and Tinbergen had worked together with great success to produce a comprehensive analysis of all parts of a simple instinctive act, namely, the egg-rolling movement of the greylag goose. Together they performed experiments with egg models in the park adjoining Lorenz's house. When the goose discharged this fixed action pattern, the experimenters noted that the sequence was coupled with taxis movements (directional sideway movements of the bill). This observation provided further inspiration to examine such taxes as they would occur in conjunction with instinctive acts.

A fundamental issue in the history of ethology is the phenomenon of imprinting. Spalding (1873), in raising fifty young chicks from pecked eggs, was probably the first scientist to describe it. His chicks followed him around if they did not hear the call of a mother hen during the first ten days. On the third day they already accepted a human being as an object to follow. Some which had been hooded for the first four days showed panic when their hoods were removed. Spalding was primarily concerned with finding evidence for instinctive movement patterns and dismissed this behavior as no more than the result of a learning process. It was Heinroth (1911) who rediscovered it and saw its real meaning. He demonstrated that a greylag gosling which was hatched in an incubator and then exposed to a human for only a very short period will follow the human and cannot be introduced to a goose family with other young of the same age. But attempts to "smuggle" it in with other geese do meet with success if the gosling is placed into a dark pocket immediately after

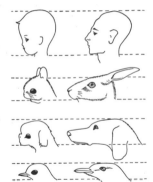

Fig. 1-12. The "baby schema" set up by K. Lorenz. The way the head is proportioned in young organisms (left) makes them look "cute" and releases brood-care behavior. By contrast, the fully developed head shape of adults (right) does not have this effect.

Fig. 1-13. Otto Koehler, German zoologist (1889–1974).

hatching. One could almost get the impression that the newly hatched little creature "looks at one with the express purpose of imprinting the image exactly into the brain...." They "furthermore do not appear to recognize their parents as conspecifics merely on the basis of instinct." The Heinroths frequently observed such cases of bonding even in the realm of their avian charges' sexual behavior. Magdalena Heinroth described the time limitations involved in the song-pattern development of her nightingale.

In 1935 Lorenz coined the *Prägung* term, or "imprinting," to denote this rapid process of learning in a sensitive period. He noted that its characteristic traits included a critical age, rapidity, and seeming irreversibility. His films, which show how young birds follow him when he is in the role of a mother greylag goose and duck, are historical documents which chronicle the growth of research on imprinting phenomena. After the Second World War, scientists in other countries began to show great interest, e.g., Fabricius (1951) in Sweden; O. Ramsay and E. H. Hess (1954) in the U.S.A.; Thorpe (1958) in England; and also the growing ranks of Lorenz's students. Others, to mention only those dealing with sexual imprinting, include Schein (1963); Klinghammer (1964); Immelmann (since 1965); Schutz (since the 1960's). We can recognize the significance of imprinting studies for human behavior when we read, for example, of the results obtained on sensitive periods in rhesus monkeys by M. and H. Harlow in the U.S.A. (1962).

Toward the end of the 1930's, the fledgling science of ethology at last stood on the doorstep of full recognition in the world of science: Lorenz was to build a special research center under the auspices of the *Kaiser-Wilhelm-Gesellschaft zur Förderung der Wissenschaften*, the present Max Planck Gesellschaft. But his plans were blocked by the advent of war. In 1940 Lorenz accepted a call as full professor in comparative psychology at the University of Königsberg, though reluctantly, for although an institute was made available to him there, he could not keep his free-living ducks, geese, and other birds which after all were so important as subjects for his observations. But he was able to set up aquaria for his cichlid fish, on which he had already done a great deal of research. (At the time he jokingly referred to these fish as "the poor man's Anatinae.") Now he began to apply the basic facts discovered during his ethological work on animals in the area of releasers to an examination of the roots of man's (a priori) ethics which are independent of experience. It was at this time, for example, that Lorenz did his work on the now-famous "baby schema."

The Königsberg zoologist Otto Koehler, who worked with Lorenz to make the *Zeitschrift für Tierpsychologie* (*Journal of Ethology and Comparative Psychology*, published by Paul Parey), the first instrument of communication for ethological research, now turned with his students to the large-scale examination of innate, basic mental capacities. Using a forced-choice method he discerned that animals have rather amazing skills to implicitly count and think. After WWII Koehler continued his studies in Freiburg, trying to discover the precursors of human language in animals.

In 1940 the young field of ethology, represented by "the group around Konrad Lorenz," was severely criticized in its epistemological basis by the vitalistically oriented animal psychologist Bierens de Haan, who stated that "we examine instinct but we do not explain it" (*Animal Instincts and their Modifications by Experience*). In the course of formulating his rebuttal (*Inductive and Teleogical Psychology*, 1942), Lorenz became even more conscious of the value which research into evolutionary origins and causes holds for the study of animal and human psychology.

The final years of WWII and the first few years after the war saw a complete cessation of ethological research. It was not until Tinbergen and Lorenz were released from prisoner-of-war and internment camps that they began to work again, with a staff of young colleagues. Not until in 1950, with the help of the Max Planck Society, was Lorenz able to establish a temporary research station in Buldern with a staff of young students for further studies in behavioral physiology. Thus began a new era of research. Successful working hypotheses were now examined in other animal groups as well (amphibians, reptiles, and mammals; e.g., by Eibl-Eibesfeldt). In 1951 Tinbergen published the first textbook of ethology, *The Study of Instinct*, and when Koehler translated it into German, a common terminology was created between English and German. The first International Congresses were organized, initially with a small attendance of twenty to thirty scientists. Today these have blossomed into large congresses attended by hundreds of ethologists from all over the world.

The most explosive development in ethology, however, began in 1956 with the establishment of the Max Planck Institute for Behavioral Physiology in Seewiesen, West Germany. The Institute made possible, at last, the proper keeping of a wide variety of animals, so that the efforts of many scientists could be combined. They exchanged ideas and branched out into diverse directions for which Lorenz himself had paved the way. Jürgen Aschoff heads another nearby Max Planck Institute at Erling-Andechs where he and his colleagues study the effects of "biological clocks" on behavior.

However, ethology is now an internationally recognized science, expecially since the awarding of the Nobel Prize in Medicine and Physiology, in 1973, to Karl von Frisch, Konrad Lorenz, and Niko Tinbergen in recognition of their pioneering efforts. Here we can but briefly note several of the workers and their geographical variation.

By G. M. Burghardt

Ethology in Germany remains strong, and some of the noteworthy contributors not previously mentioned include W. Wickler, who has studied pair-bond formation in many species and who took Lorenz's position in Seewiesen upon the latter's retirement; P. Leyhausen, especially

Fig. 1-14. Nikolaas Tinbergen, Dutch zoologist (born 1907).

known for his work on predatory behavior in cats; and E. Curio, who has focussed on the defense mechanisms of prey against their predators.

Switzerland has produced important ethologists, including H. Hediger, the first person to systematically apply principles of ethology to animals in captivity in zoos, M. Meyer-Holzapfel, a pioneer in mammal ethology; and U. Weidmann, now in England, who has studied releasers in gulls and courtship in ducks.

Scandinavia has contributed through the work of such pioneers as E. Fabricius (imprinting) in Sweden and the physiological studies of H. Ursin in Norway.

France has become a major contributor to ethology, particularly through the studies on insect behavior by R. Michel and others, who produce beautiful films and exquisite analyses. Thus continues a tradition established by H. Fabre in the first part of this century.

In Italy the major ethological center is Parma where G. Mainardi and his colleagues study problems related to imprinting, chemoreception, and imitation.

Holland is not only the birthplace of Niko Tinbergen but also the home of many esteemed ethologists. Here we will only mention two, A. Kortlandt, co-discoverer of displacement activities and pioneer in chimpanzee behavior, and G. Baerends who, through his research and students, has fostered the detailed analytical approach to the study of sign stimuli and the functional organization of behavior.

Great Britain is home to many important ethologists, which is appropriate for the homeland of Darwin. While there are numerous individuals and laboratories, mention must be made of Oxford, where D. McFarland has replaced the retiring Tinbergen; Cambridge, where R. Hinde, who has written one of the major texts in ethology, presides; and Edinburgh, with Aubrey Manning, current Secretary of the International Ethological Congresses.

Much ethology is being pursued outside of western Europe. The Soviet Union and eastern Europe are producing more ethological work. Unfortunately, they are more aware of western studies than we are of their contributions. Hopefully this isolation will soon be reduced. Work by the Japanese on their native macaque has been influential in establishing how traditions spread through primate groups. Israel has produced excellent work on reptiles by H. Mendelson and on birds by A. Zahavi. In Africa, the late R. F. Ewer was a leading mammal ethologist whose books have been deservedly influential. Australia is the home of G. McBride, an influential theoretician on social behavior.

In terms of quantity, the major country for current ethological research is the U.S.A. Slow to see its value, workers in zoology, anthropology, and psychology are now avidly supporting ethology. E. H. Hess and W. S. Verplanck were among the first American psychologists to stress the relevance of ethology to comparative psychology. Other pioneer

Fig. 1-15. Karl von Frisch, Austrian zoologist (born 1886).

American workers in animal behavior and comparative psychology include F. Beach, J. P. Scott, and T. C. Schneirla. To recount the work of D. Griffin on bat echolocation, E. O. Wilson on insect pheromones, as well as the others mentioned above, would be to catalogue the contents of modern texts on animal behavior.

The question arises: Are ethologists today still dealing with the problems and issues of the past or are new concerns demanding attention? Actually both old and new questions are being studied. One of the most enduring is ontogeny. The European ethologists, in bringing back instinct and innate behavior, claimed that animals did many things without having to be trained. When these ideas were introduced to American comparative psychologists, there arose a rather defensive outcry against the notions of instinct and innateness, and the raising of issues long considered dead and buried. Many of these criticisms were based on semantic quibbling, an ignorance of animal behavior outside the sterile environment of the laboratory, and an attitude that evolution and ecology were irrelevant to understanding how and why animals behave. Nonetheless, recent findings indicate that earlier conceptions of the origins of behavior were overly simplified. We now know that many aspects of instinctive behaviors are modifiable in some directions by environmental stimuli and by rewards and punishments. Many behavior patterns are innate in some species and yet may be dependent upon specific kinds of learning in closely related species. The work of P. Marler and W. H. Thorpe on song learning in chaffinches and sparrows demonstrates this nicely.

On the other hand, all "learning" in animals must depend upon the presence of innately given aptitudes. Indeed, the term innate is now used by most ethologists to refer not directly to behavior (except as a convenient shorthand) but to those genetic factors underlying the occurrence of a particular behavior or ability. Americans such as T. C. Schneirla and D. Lehrman advocated an approach to problems of behavioral development that initially appeared in direct opposition to that of the ethologists. Underlying it was the phenomenon that animals are often seen initially to approach areas of weak stimulation (e.g., light) and to avoid areas of strong stimulation. Such responses were thought to be mediated by different parts of the autonomic nervous system, which allowed for the experimental origin of sign stimuli and fixed action patterns. We now know that behavior cannot exist without either genes or experience, and that careful experiments are needed to understand exactly how behavior, sensory processes, and physical structure develop from an undifferentiated fertilized egg. What we actually see and study in an animal is its phenotype, the combined outcome of a complex series of interactions between the genotype and the developing organism's internal and external environment. No single process will explain the development of all behavior in all organisms. Consequently, most of our terms and concepts in ethology—such as instinctive behavior (as distinct from instincts), learning,

▷

Methods used for ethological studies in the wild. The observation of natural behavior in individuals over long periods of time is important for investigating the social organization of an animal species. The tranquilizer gun enables us to capture animals, mark them, and release them unharmed; a shell is filled with an anesthetic drug (top left), then loaded into an air rifle and shot at the animal (top right). On impact it injects the drug into the animal. Next, the shell with its barbed hook is removed (middle left). The scientist now has up to an hour to take measurements and to mark the subject. One method of marking which allows us to track the individual animal more easily is radio telemetry; here a lion is fitted with a collar containing batteries and a radio transmitter (middle right). A receiver to make the transmitter signals audible and a directional antenna enable us to locate the animal at any time, even in dense underbrush (bottom left). Bottom right: A "camouflage tent" is also an aid in observing the natural behavior of animals, which soon get used to it. In this way the scientist can make his observations without being noticed.

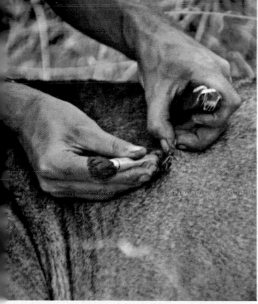

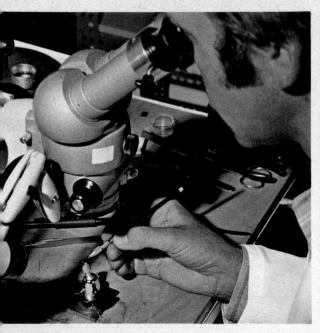
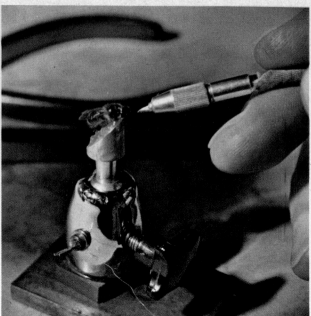

conditioning, experience, displacement, and releaser—must be seen, at least initially, as referring only to descriptive categories which may or may not be underlaid by similar mechanisms in various species.

Our knowledge of the evolution of behavior has likewise increased greatly in recent years. The fact that behavior patterns can be traced and compared between closely related species is now more widely accepted and many examples have accumulated. The analysis of head-bob and pushup displays in lizards, an area pioneered by C. C. Carpenter, indicates the value of using instinctive movements in the understanding of phylogenetic and population differences. We are now aware of the fact that "species characteristic behavior" is a term that cannot be taken too literally, since different populations, and even different individuals, of the same species may show consistent differences as well as similarities to one another. Hence, while the broad features of behavior patterns in a species may be the same in all members, details may differ in different subsets of the species. The origin of these differences may be environmental, as in dialects in white-crowned sparrow song that are due to early experience, or genetic, as in the prey-preference differences in newborn snakes from the same litter.

Another area of active interest is the organization and evolution of social behavior. Ethologists have moved beyond a mere description of interactions between a small number of animals in nature or captivity to a comprehensive look at the entire social organization of a species. Long-term studies of primates, elephants, lions, and wolves to name a few, resemble anthropological studies in following kinship lines, survival rates, and cultural and genetic change. Theories are being advanced about the principles underlying social organization, particularly the evolution of behavior across different groups of animals. Wilson's work on sociobiology is an attempt to develop this knowledge of social behavior and organization and relate it to new understandings of population, genetics, and ecology. Such studies, while innovative and important, carry on and build on the traditions and concepts begun by ethologists many years ago.

Our understanding of the sensory world of animals has expanded remarkably as technical advances in recording and analyzing the signals used by animals in communication, migration, and food finding have increased. For example, the chemical senses are now known to be much more important in many species than was hitherto expected, as in the discovery of the use of pheromones in the courtship behavior of many insects and mammals. Ultrasonic vocalizations are important in the behavior of many animals, but these have only recently been identified. An example is young mice which communicate in this way with their mother.

These studies give an indication of the vitality of animal behavior studies, including studies of humans, and the new directions in which such work is heading. In future years we will see more of a synthesis in the methods, findings, and approaches of students of ethology, physiology,

◁
Methods for making physiological studies of color and movement perception in insects. Above: Under a microscope, bees (to the right, a drone) are prepared for the electrophysiological recording of nervous impulses from direction-specific nerve cells sensitive to movement. Below: These instruments are used to study the reactions of visual receptor cells to pure spectral colors. The color-filter wheel serves to produce the light stimuli. Electronic measuring devices enable us to measure the receptor cell responses (see also Chapter 5).

ecology, genetics, and psychology. Then and only then will the seeds sown throughout the history of ethology noted in this chapter come to flower and fruition. Their value will even be felt by people to whom the mere watching of animals is at best considered a trivial, but perhaps harmless, avocation.

2   Goals, Organization, and Methods of Ethology

Goals, organization, and methods of ethology, by J. Lamprecht

The field of ethology is a part of the natural sciences, specifically biology and its methods are those of all the natural sciences. Like other scientists, the ethologist bases his work on repeatable observations and measurements and tries to establish orderly relationships between them. He deals with the externally observable behavior of organisms, and for this reason cannot examine subjective phenomena such as dreams, thoughts, or feelings. On the other hand, certain manifestations of these phenomena that can be defined objectively are indeed subject to scientific scrutiny.

An ethologist tries to gain insight into the historical development of behavior patterns and to demonstrate the evolutionary processes that would have given rise to such behaviors. He also attempts to understand the mechanisms of behavior well enough to predict accurately—under any given conditions—the behavior of an individual or even a whole society.

The concept of behavior

Ethology focuses on the behavior of organisms. This includes all movements, postures, changes of color, and vocalizations displayed by an individual. These changes can be viewed as more or less short-term deviations from a basic or homeostatic condition, which is never consistent and therefore almost impossible to define. However, changes in the homeostatic condition itself, such as processes of growth or aging, are not part of behavior, although often enough there is some causal relationship.

The behavior of an animal is a continuous, though by no means regular, stream of events from birth to death. It has structure, since it consists of elements related to one another in orderly ways. These elements, called behavior patterns, can be categorized in many different ways, depending on our particular question. If, for example, a scientist wants to know under what circumstances a blackbird performs its song, he may view the entire song as a single element. But if he wants to look at the song's structure, he will have to make finer divisions. In this case he would select, for example, the smallest recognizable song structures as units and then study the way the units are related in time.

Units of behavior patterns are sequences. We can see their pattern only when we consider both the spatial dimension and the temporal one. When describing a behavior, we aim for a precise picture of how all the corresponding muscles (or any larger physiological units such as glands) work together as far as it is overtly expressed.

Vocalizations are usually described according to pitch (frequency), intensity, and temporal sequence, without referring to specific movements of the vocal system. Transcriptions of a sound, such as "Pfumm," "Räbräb," whistling, or howling are now usually replaced by the more precise sound spectrograms, which graphically represent the sequence of sound frequencies.

Depending on the problem at hand, behaviors are labeled and defined according to either their effect (function) or their form. The terms "walking," "flying," or "swimming" tell us little about the form or pattern of a movement. Rather, they denote the medium within which this behavior results in locomotion. Here we emphasize effect—as with feeding activities—which always serve to supply the body with food. This kind of terminology is useful if we want to compare performance in different animal species. If, on the other hand, we are interested in the similarities or differences between organisms—for example, with regard to feeding—it seems more appropriate to define and label a behavior according to its appearance or form. Such a term would then represent an abbreviated description. Examples are "body trembling," "dipping," and "lifting the front leg." Furthermore, terms of everyday language, such as "trot," "gallop," "pick," and "chew," refer primarily to the formal sequence of a movement.

Behavior units can be collected and compiled into an action catalog, the ethogram. The complete ethogram represents a qualitative inventory of all behavior elements contained in the repertoire of an animal species.

Yet we run into problems with even a basic inventory. Animals often behave differently in captivity than they would in the wild. Sometimes whole action sequences are missing in an animal if certain situations (threat of predators, prey defending itself, etc.) never occur. In nature, however, the chances for observing on a continuous basis are often severely limited—for example in a dense jungle environment—and the ethologists will inevitably miss certain details. The only way to eliminate the disadvantages of either condition is to make parallel studies of animals in the wild and in captivity.

Documentation is another aspect of collecting behavioral data. Behavior patterns are recorded visually on film and acoustically on audiotape. In addition, vocalizations can be reproduced by means of the sound spectrograms mentioned above. Unfortunately we have not yet found a way to record olfactory, gustatory, and tactile stimuli.

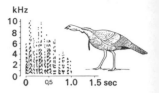

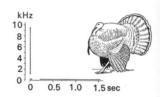

Fig. 2-1. Sound spectrogram of a turkey when "gobbling" (above) and during "pfum" (below). At the right of each spectrogram we see the associated body posture.

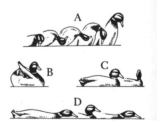

Fig. 2-2. Behaviors may be named according to their function or their form. This figure illustrates behavior patterns from the courtship of a mallard drake. They are labelled according to form: A, grunt-whistle; B, head-up —tail-up; C, displaying back of the head; D, nod-swimming. These movement patterns are stereotypical and therefore easy to recognize.

For the sake of clarity, we try to combine behavior patterns into groups, classifying them according to specific criteria. Of the many possible ways of categorizing, two are most common today: Some behaviors are similar in their effect; other behaviors are associated with a common cause.

In setting up an ethogram for any animal species, behavior patterns are usually organized into so-called functional units, such as food consumption, locomotion, sleep, reproduction, and care of young. Functional units reflect the basic demands made upon the organism by its environment and thus comprise actions effective in one of these areas. The utility of a functional approach is that similar classifications can be employed for very different organisms. In other words, it enables us to catalog the behavior patterns of many animal species according to the same criteria. Unfortunately, however, we often cannot determine the function of an activity with enough certainty to enable us to classify it with one particular unit. In addition, there are always many behaviors, such as those of locomotion, that occur simultaneously in several different areas. We must seriously consider whether locomotion represents a unit just like other functional divisions, since every active change of locale must also be included under such headings as food intake, reproduction, etc. Here we are still faced with unresolved methodological problems.

The second type of classification focuses on activities with some common physiological cause. For example, the behavior of male vertebrates is partly associated with the sex hormone testosterone. If we inject an individual with testosterone and then observe certain actions occurring more frequently or intensively, we may assume that they are facilitated by the presence of this hormone, and we can group them together as "testosterone-dependent" behavior.

But we cannot always manipulate causal factors directly as in that example. In many cases we must rely on indirect evidence. For instance, if we make a continuous recording of all the activities of an animal under the most constant conditions possible, we soon recognize certain temporal associations. Some of these elements often occur in close sequence, while others are almost never observed in this kind of relationship. We may assume—since this had already been demonstrated in many cases—that movements frequently occurring together or in close sequence very probably have a common cause. We may therefore classify them into groups. Instead of naming these groups A, B, or C, scientists usually name them according to the known function of one of their typical behavior patterns. Thus, for example, "testosterone-dependent" behavior, which includes copulation, becomes "male sexual or reproductive behavior," and a category that includes biting rivals is now termed "aggression."

This classification according to common intrasystemic causes is based on something we can measure directly, on the structure of behavior itself. We must emphasize, however, that here—unlike the method of functional organization—units established through observation of one species cannot always be made comparable to those utilized for another. Different animals can accomplish similar goals in different ways, based on differences in their internal organization. Even with this method of classification, certain activities may occur within more than one causal framework. In one instance locomotion occurs together with prey-catching, and in another with copulation when the animal approaches its mate. Despite all their shortcomings, these two forms of classification provide us with a biologically meaningful way to organize behavior and make it easier for scientists to communicate with each other. We should not expect to fit the complex and variable organization of behavior into such simple schemata without some limitations.

On the whole, the above methods of classification have provided a systematic framework for collecting, describing, and organizing research material, giving the ethologist a broad basis for pursuing various kinds of questions. At this point we begin to distinguish several more or less separate areas within the larger discipline. Some of these specialized areas correspond to the ones we know in biological research, such as evolution, genetics, ecology, and physiology. Each field within ethology examines behavior patterns by using basically the same methods as the study of morphological features. However, various new factors play an important part in the study of behavior, such as temporal organization and the role of learning processes. New questions have also come up, giving rise to various branches of research, each with its own special methods.

Every behavior has an effect that normally serves the survival of the species, a history of phylogenetic (evolutionary) development, and, beyond that, physiological factors causing it to appear at that particular moment. These aspects form the basis for three major areas of ethology: the study of adaptive processes, the evolution of behavior, and physiological factors.

In most cases, each behavior will serve the survival of either the individual or the species. Eating provides the body with necessary food, flight prevents physical damage or death, copulation serves the purpose of reproduction, and brood care, the survival of offspring.

Many animals feed, flee, or care for their offspring in very different ways. Yet none of these methods seems superior or inferior, for all of these species have survived, and on the average each of their members produces at least one fertile offspring. But how do we explain that for every animal, its way of moving about (locomotion) is obviously the optimal one? The answer lies in the basic mechanism of evolution—the phylogenetic development of organisms, and their morphological as well as behavioral characteristics. This is based on two processes:

First, the process of mutation creates a range of potentials by varying the genetic substance. Only a few of these constitutional variations end up being successful in the struggle for existence. The rest simply die out

Mutation and selection

after a while. This second process, called selection, is manifested in the environmental demands made upon organisms in their fight for survival.

Nongenetic behavioral traits, or characters, may also vary. Instead of changes in an organism's genetic material, modifications occur in the nervous structures controlling a particular behavior pattern. Here, too, the process of natural selection ensures that only behaviors adapted to the animal's environment actually lead to success. This process takes place in the individual. But when these behaviors are passed on to other individuals by means of tradition, and are constantly retried and tested and perhaps varied in some way, an evolution of learned action patterns will also occur. In contrast to the biological process, this is sometimes referred to as cultural or psychosocial evolution.

While acquired traits are often varied in a directed manner, we have no conclusive evidence that mutation processes in the genetic substance are also directed. Mutations occur randomly, and it is always the process of natural selection that provides the final direction for developmental changes. Thus it is ultimately the animal's environment which determines the kind of behavior necessary for its survival and successful reproduction.

As a result of variation and selection, an organism's genetic substance and memory contain a kind of "knowledge" about the characteristics of its environment. This stored information, which determines and controls the form of an animal's activities and their situationally appropriate discharge, ensures proper adaptation to its natural environment. The feeding movement of a cow is adapted to the properties of grass, and those of the cat, to the characteristics of its prey. Thus feeding behavior looks very different in these two animals, yet for each organism, its own is the ideal way of taking in food.

In order to understand the phylogenetic development of a certain behavior, we must consider the two aspects of function and adaptation. Scientists doing this kind of research are constantly inquiring into the "biological meaning" of a behavior. In other words, what is the effect, the function, the purpose of this behavior? In this context we are certainly not using the word "purpose" to denote any conscious aim of the animal itself. We want to know how this movement serves the animal, rather than the animal's intention in behaving this way. To answer this question, we must first know precisely how the animal behaves in its natural environment, for the question we have to answer is, how would it harm the individual (or the species) if it could not perform this behavior?

Since the behavior being studied has a specific form and occurs in specific situations, our next question is: What environmental factors led to the selection of these special features, as associated with the function of a behavior pattern? Or, more concisely, how is this behavior adaptive?

Adaptations

In practice, we can answer this question only by examining the behavior of different, unrelated organisms displaying similar features in the same environmental circumstances. We may then interpret any similarity between such behavior patterns as functional adaptations to the environment. In this way, we can see that common features in the flying movements of bats, birds, and butterflies evolved as adaptations to locomotion in air.

Questions of function and adaptation are primarily the concern of eco-ethologists, who study the relationships between organisms and their organic and inorganic environment. This requires observation of animals in their natural environments, the demands of which have led to the adaptation of movement patterns.

The second aspect, the evolution of behavior patterns, is the concern of behavioral phylogenetics. (see Chapter 38). The methods stem from phylogenetics, an old branch of biology concerned with the evolution of organ structures and of organisms themselves. In relation to ethology, the central question here is, what was the evolutionary development of a certain behavior? How did it originate, and what were the intermediate steps? The methods for examining these issues are based on the concepts of homology and analogy.

Siblings, and especially homozygous twins, can be amazingly alike. They have inherited a similar genetic constitution from their parents and, in addition, have learned similar things (e.g., the same language). Their common characteristics can be traced back to traits possessed by their father or mother. Such similarities, based on genetic or cultural relationships, are called homologies.

Similarities in the streamlined shape of whales and sharks, however, are based on different factors. These we cannot trace back to common aquatic ancestors with similar morphological features, for the whales are descended from quadrupeds living on land, whereas members of the shark family have always lived in the water. The reason for their similarity is that both groups have adapted to rapid locomotion in the water. Streamlining is ideal for this purpose. Resemblances of this kind, which arose from the independent adaptation of organisms to the demands of their environment, are called analogies or convergences.

The first practical step in describing the development of a behavior is to find a number of similar movements or calls whose correspondents are homologous. In single instances we often have trouble determining whether a similarity has resulted from homology or analogy. This problem is relatively easy to solve when the structures in question are alike even though their function is different. In this way we can say that the pelvis of the whale, strongly involuted and no longer functional, is surely homologous to that of terrestial vertebrates. In the same way, the similarities between the claws and walking legs of a crayfish are homologous. And when a singing lark imitates a shepherd who whistles for his dog, her own whistling call is homologous to that of the shepherd.

However, when organs have similar functions, we have far more difficulty in deciding how much of their structural congruency has resulted

Evolution of behavior patterns

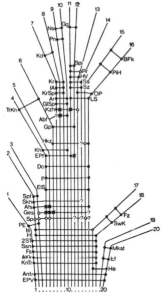

Fig. 2-3. A taxonomy of ducks, drawn up on the basis of behavioral characters. It starts with related characteristics and indicates the relationship between the trait carriers. The vertical and diagonal lines represent individual genera or species. Each horizontal line corresponds to a behavior pattern and links those species (at points where they cross the vertical lines) where it occurs in homologous form.

from parallel adaptation to similar environmental conditions and how much is due to descent from a common ancestor. We do find congruencies among organs and behavior patterns that are hard to explain on the basis of common function alone. The corresponding features in the wings of birds and butterflies are analogous; they have evolved only in response to the demands of flying. By contrast, the numerous congruencies we find in the wings of eagles, partridges, and sparrows must certainly be homologous, for we can argue that if wings with very different structures, such as those of bats and butterflies, are functional for flying, there is no reason to believe that birds so different in their flying methods should otherwise be so remarkably alike in some of their wing characteristics.

With considerations like these in mind, Adolf Remane formulated a number of criteria for establishing a homologous relationship. These allow us to distinguish, with a high degree of certainty, homologous congruencies from analogous ones. The greater the similarity between homologous traits—that is, the greater the extent to which details actually correspond to each other—the more recently their common ancestor can be located on the evolutionary scale and the more closely they are related. Furthermore, experience shows that in most cases a subtrait is older the more widely it is distributed among behavior patterns or organs. With these statements in mind, we may now draw up a taxonomic chart to gain clear understandings of the genetic relationships between certain structures and of their phylogenetic history.

Since early in the history of ethology, scientists have tried to use behavior patterns along with morphological traits for determining relatedness between organisms as a whole. There is a great temptation to use the relatedness of individual traits as a basis for inferring relatedness between the trait carriers themselves. However, we can make this assumption only under specific conditions.

If we compare, for example, different vertebrate skulls and discover homologous concurrences, we may assume a certain degree of relatedness between the animals. A case in point is that of various marine snails found to carry nematocysts in their outer body walls, which they use as defense organs. These nematocysts are remarkably similar to those of coelenterates. Their extreme correspondence in the most detailed structural elements indicates a homology. Yet these snails are not closely related to colenterates. The former merely feed on the latter and incorporate the nematocysts into their own body walls.

Such cases of nongenetic trait transference are more frequent in the realm of behavior. When a Briton and a Chinese speak English together, it tells us nothing about their degree of relationship even though the languages of these two people are homologous. Even if we can show correspondence of traits, we cannot assume that their carriers are related until we know if the basis for this similarity is genetic. The comparable structures would necessarily have had to evolve on the basis of genetic information passed on to the carriers by their parents; in other words, these structures would have to be innate.

The experimental methods used for deciding whether a behavioral trait is innate or acquired will be discussed later. Another source of information on inborn behavioral structures is research done in behavior genetics. This discipline, using the methods common to genetic research, examines the inheritance of individual behavioral, rather than physical, characteristics.

Behavior genetics

In addition to the function and evolution of behavior patterns, physiological causes are the third aspect to consider. To become fully effective in the struggle for species survival, a behavior pattern must occur in the appropriate situation as well as, in most cases, in a particular spatial relation to the environment. Furthermore, it is important that this behavior is temporally synchronized with other activities; and finally, the animal must be able to use it at the proper stage of its individual development.

Physiology of behavior

The internal organization of an organism must allow for behavior patterns to meet these demands. Behavioral physiologists are interested in investigating this internal, physiological organization. They attempt to explain in causal terms—according to the principle of cause and effect—why a particular animal does a particular thing at this very moment in just this way. Researchers in this field regard the living organism almost as a machine that receives stimuli from the outside world and in turn produces movements, postures, and vocalizations, and they want to know how this apparatus works. An animal has sense organs to receive stimuli and a nervous system to process them. This nervous system leads a kind of life of its own, sending nervous impulses to muscles and glands, where the impulses are transformed into overt behavior, which in turn may effect a change in the stimulus situation. This provides many different opportunities where each special area within behavioral physiology can begin its own kind of research.

The simple observation of a dog eating a sausage leads to a number of important questions. All movements occurring in this context have a particular form and require all muscles involved to perform highly complex sequences of contractions organized in space and time. Muscle contractions are released by nervous impulses. Thus, when the animal starts to eat, the nervous system must transmit a well-coordinated pattern of impulses to the musculature. But how and where is such an impulse pattern produced, activated, and transmitted? How are these impulses transformed into muscle movements? These and similar questions are examined by myophysiologists and neurophysiologists. It is possible to stimulate nerve cells electrically by means of tiny electrodes (see Chapter 19). Furthermore, we can record impulses produced by the cells themselves. These and other sophisticated techniques have enabled scientists in this area to make tremendous progress recently.

Stimuli and response

A dog will eat the sausage, which it sees and smells; but it will not

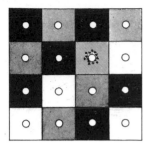

Fig. 2-4. An example of training or conditioning: For a certain period, honeybees were fed with sugar water in a dish on a blue base. In the final test (illustration), the food table is arranged with squares of various shades of grey. One of these squares is blue. None of the dishes contains food. Result: all the trained bees fly to the blue square. This shows that they can discriminate blue as a color.

Fig. 2-5. We can study the effects of sign stimuli with the use of dummies: Here, a herring gull chick begs for food by pecking at the red spot on the adult's bill.

eat furniture or stones, although it obviously perceives them as well because it avoids them, plays with them, and so on. In other words, of the large number of external stimuli perceived by an animal, only a few release a feeding response.

We can find out whether an animal perceives a certain stimulus by conditioning it to respond in a particular way to this stimulus alone. If we cannot condition an animal that is otherwise quite capable of learning, and if at the same time attempts to condition other stimuli do meet with success, we conclude that our subject does not discriminate this stimulus from others presented in the study.

Experimental conditioning is the method we use to determine the functional limits of a sense organ. In this task, the sensory physiologist may record impulses from sensory nerve cells during the presentation of certain stimuli; this can also be done when studying the physiological processes involved when stimuli are received or when examining the relationships between the function and structure of sense organs.

Next, we ask how the nervous system processes incoming electrical excitation from the sense organs. What happens in the brain during the perception of sensory stimuli? Again, these issues are the concern of neurophysiologists.

The fields of sensory physiology, myophysiology, and neurophysiology are occasionally referred to as neighboring disciplines of ethology, because they are related to ethology somewhat in the way chemistry is related to physiology. They examine basic processes underlying overt behavior.

Since a particular response is released by only a small number of the many stimuli which may be perceived, areas associated with sense organs and the central nervous system must also contain a so-called releasing mechanism for that particular behavior. This mechanism acts as a filter, selecting only those stimuli relevant to the behavior. Should such a "sign stimulus" occur, the behavior will be released.

The stimulus-filtering qualities of a releasing mechanism can be analyzed by performing dummy experiments. The dummies or models are representations of sign stimuli more or less similar to natural releasing stimuli. But when an animal responds to such a dummy, its behavior does not have the usual species-preserving effect. The important feature, however, is that dummies can be presented to the animal as often as desired, and always in identical form. This, of course, would not be possible with natural stimuli. For example, we can let an animal listen to a tape recording of a vocalization, using this as an acoustical dummy and perhaps playing it over and over again. Frequently, representations of animals made of cardboard or other materials are used. We may even construct a whole series of models by slightly varying color or shape.

In such an experiment, the animal is presented with different models from fairly large distances and in random order. We then observe whether the subject responds to these in the same way as to the natural stimuli, and if so, to what degree. By experimenting with the releasing effect of different dummies, the experimenter may thus determine the range of stimuli that constitute actual sign stimuli for the behavior under study.

The releasing effect of a stimulus may change. If we beat the dog every time it tries to eat a sausage, it will soon stop reacting to sausages the way it responds to other kinds of food. The dog has undergone a learning process; experience has changed its releasing mechanism for feeding behavior.

In addition to learning the appropriate sequence of an action pattern, animals can also learn the form of a behavior. Foxes as well as monkeys can learn new movement coordinations needed to open the complex lock of a box containing food.

Learning theorists are concerned with the way experience affects behavior and with the conditions that determine what the effect will be. Research on the nature of learning phenomena acquired prominence with Pavlov's famous experiments on dogs, in which he discovered that if a buzzer is sounded just before the presentation of food, a dog will learn that this stimulus signals food. After a while, the animal will salivate even when food is omitted and the buzzer is sounded on its own. The buzzing sound, which originally had no releasing effect, now has become a releasing stimulus.

The suppression of learned material, the process of forgetting, and the learning of movements right up to insightful, intelligent behavior are additional problems to be studied. The study of memory processes has captured the interest of many scientists, foremost among them neurophysiologists, who are searching for the chemo-physiological substrate of stored experience.

The sight of the sausage not only releases feeding behavior in the dog, but also directs these movements. This aspect of behavior, the spatial orientation of movement patterns according to environmental stimuli, is the subject of orientation research (see Chapter 11). By systematically varying the direction and intensity of orienting stimuli and by observing the animal's reactions, scientists in this field try to determine the mechanisms of orientation in space. What happens inside the toad when it catches sight of prey? How do bees and wasps find their way back to feeding or nesting places? And we still do not know how pigeons return to their coop from a totally strange environment over a distance of many kilometers (navigation).

The dog will eat the first and second sausage, but not the fifth or tenth. The cardboard model of a rival will release fighting behavior in male cichlids in some instances but not others. We repeatedly observe that an animal responds differently at different times when presented with the same external situation. Thus, in addition to external releasing stimuli, there must be internal factors influencing the probability of a behavior

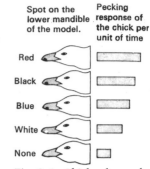

Fig. 2-6. Chicks also peck at cardboard dummies of gull's heads. Out of several models with a differently colored mandibular spot, the one with the red spot, the natural color, releases the strongest pecking.

Pavlov's experiments

Orientation

occurring. A low blood-sugar level, for example, is associated with a particular physiological state where the disposition to eat is strong. In everyday language we say that the animal is hungry.

As mentioned above, certain activities occur repeatedly in close temporal proximity. In this way, for example, seeking out a potential source of food, lying in wait for prey, jumping at it, seizing it and making the kill, and then feeding are all associated activities. As long as there is no sufficient external explanation for this spatial ordering, we must assume that behavior patterns occurring simultaneously or in sequence have a common internal cause. For example, they may all be more likely to occur than others when the blood-sugar level is low; they share, at least in part, a motivational base. The term motivation can refer to the totality of causal factors that increase readiness for action (see Chapter 15). We must emphasize, however, that scientists are still not fully agreed on the usage of this term.

Motivation

The propensity for many behavior patterns is increased by certain hormones. We can raise the probability of such a behavior occurring by injecting the appropriate hormones (see Chapter 21). Conversely, there are cases where blood samples taken from individuals with a natural tendency to perform certain movements more often than others have also shown a higher concentration of the corresponding hormone.

Hormones

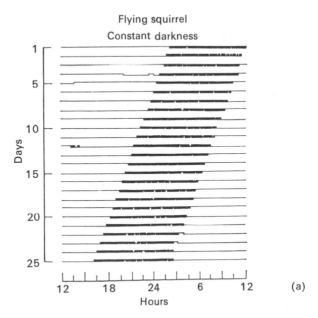

Fig. 2-7. Recording of activity periods of a flying squirrel in constant darkness. These periods (thick black lines) are always approximately the same length. The internal daily cycle, or rhythm, is very regular, but it is somewhat shorter than the external day. This is why the activity period begins a little earlier each day.

Some external stimuli also influence the readiness to perform a certain behavior without releasing the behavior itself. These are called motivating stimuli. For example, virgin female mice can be induced to build a brood nest if we place the newborn of another female in their box. The stimulation produced by the young does not actually release the nest-building behavior. But these stimuli do increase the probability of the virgin mouse performing this behavior when presented with nesting material.

Most animals display a day-night rhythm. Many of their behavior patterns occur only at certain times of the day. The synchronizer ("entraining agent" or *Zeitgeber*) for this daily periodicity is the alternation of light and dark periods, which is dependent on the sun. Some animals also have an annual rhythm. We notice this particularly in migratory birds, which fly south in the fall, return in the spring, and then breed. The tendencies to migrate, to breed, and to perform many other activities fluctuate with the length of the day. When days grow shorter in the fall, these birds get an urge to migrate.

Amazingly enough, some animals with daily or annual rhythms maintain this periodicity even when isolated for a long time from external synchronizers under constant conditions. Evidently they possess an "internal clock" or an "inner calendar" (see Chapter 12) that regulates the temporal organization of behavior independently of external, rhythmically fluctuating stimuli or cues. However, external stimuli may accelerate or decelerate this rhythm. Under natural conditions, they ensure that the cycles correspond to the external year or day.

This knowledge was gained from research in biorhythmics, a subfield of ethology concerned with the physiological and behavioral aspects of rhythmical learning processes (actions recurring evenly and in regular intervals), attempting to find causal explanations for these processes.

When we look at the development of an organism from birth to natural death, a number of special questions arise. In many ways, environmental demands are different for young animals than for adults. Often a whole series of behavior patterns have to supersede one another during an individual's development (ontogeny), so that the young animal is provided with the maximum chances for survival at every stage of life.

Description and causal analysis of behavioral ontogenies is an area of research closely linked with other specializations, such as evolution and physiology. Experimental manipulation of the developmental process, for example, by creating artificial and controlled rearing conditions, has provided us with many new insights into how external factors interact with innate disposition in the individual emergence of behavior patterns.

If a behavior pattern is adapted to the demands of the natural environment, the special traits characterizing this adaptation could have emerged in one of two ways: First, they could have developed in the course of evolution. In this case, the information that determines their presence in the individual will lie in the gene substance. Such behavior elements are called innate. If, however, these traits develop slowly in the course of the individual's life through experience stored in the central nervous system, they are considered learned or acquired.

A cat catches a mouse. Is the skill innate or was this behavior learned?

Internal rhythms

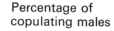

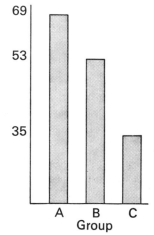

Fig. 2-8. The effect of experience with conspecifics on sexual activity of male rats. After weaning, the males of group **A** were raised in isolation, those of group B, with males and females, and those of group C, only with males. In the final test the animals in group **A** had the greatest percentage of successfully copulating males, while rats in group C had the least.

To answer this question, we have to make further divisions. How does the cat know what a mouse is? The recognition of mice or of other sign stimuli may be innate or learned. It is further possible that an innate, rough "mouse schema" later becomes refined and differentiated through experience. We therefore need to discuss how the releasing mechanism for prey capture acquires its filtering properties. Secondly, how does the cat know how to catch mice? Here we are basically concerned with the form of the movement. And, finally, how does the cat know that it has to catch mice, or that cats fight only with other cats whereas they run away from dogs, etc.? Again, the association of an external stimulus with an appropriate response may be innate or learned. If we want to find out, for example, whether the cat's mouse-catching behavior is innate, we must examine all these aspects even though the behavior pattern appears to be a single unit. By raising an animal in such a way that it cannot obtain the various kinds of experience necessary for learning a particular trait, we may determine which parts of a behavior are innate and which are learned. This procedure of raising an animal under conditions of specific experimental deprivation has been termed a *Kaspar Hauser* experiment after a foundling boy of unknown origin who appeared in Nuremberg in 1828, reputedly having lived in a dark cellar until the age of sixteen without ever seeing another human being. In English it is called a "deprivation experiment."

In order to test whether the flying movements of pigeons are innate, some of these birds were raised in small cages, allowing no wing movements at all. The experimental animals were released when birds of the same age, kept under normal conditions, could fly. The experimental pigeons flew as well as the control animals. Thus, the capacity for flying movements had developed even though the birds had no prior practice.

In any particular case, we usually examine whether the recognition of a particular stimulus must be learned by coupling a stimulus with a specific response. The young of dwarf cichlids raised under artificial conditions until hatching immediately followed the dummy of a female with brood coloration. They did not, however, follow that of a nonbrood-colored female. Thus, they did not have to learn to discriminate a brooding female from other ones. Furthermore, the dummy released the following behavior typical of young fish.

Conclusions from the results of deprivation experiments are possible in only one direction. We consider a behavioral aspect to be innate if it occurs even though the animal has not been exposed to the relevant experience. If, however, the behavioral trait does not appear, we cannot necessarily assume that it must be learned, because we can never know with absolute certainty whether the abnormal conditions under which the animal was raised did not damage it in a more general way, so that even innate behavior elements are eliminated.

Until now, our discussion has focused, for the most part, on the in-

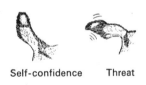

Self-confidence   Threat

Normal position   Uncertainty

Friendly submission   Subordination

Fig. 2-9. Tail signals in wolves, and their meaning. Another wolf, receiving such a signal, "understands" it and responds appropriately.

dividual organism. A number of additional factors emerge as soon as the behavior of an animal is influenced by external stimuli transmitted by other individuals. A male dog attacks strange males that he encounters in "his" backyard; he prefers to raise his leg at locations where other dogs have urinated; if he encounters a female in heat, only force can prevent him from following her. And, finally, dogs learn to come when their master calls.

Here we are dealing with social behavior, with behavior patterns directed toward other individuals that are neither prey nor animals preying upon them, behaviors released or stimulated (motivated) by stimuli coming from others. These behavior patterns are of special significance because they form the basis for living in association with others.

Every form of communal life, also every active withdrawal or dissociation shown by animals and humans, is based on communication, requiring that some information is exchanged between transmitter and receiver. If by coincidence an animal emits a stimulus that is received by another and induces a response, this process may begin the formation of an orderly transmitter-receiver relationship, for if the reaction of the other animal is of advantage to the sender, the sender will attempt to elicit it again the next time. If possible, the sender will even strengthen the earlier stimulus and make it more conspicuous, so that the probability of eliciting the desired behavior will steadily increase.

In this way, simple releasing stimuli have given rise to so-called signals, or "releasers," in the course of evolution, and frequently also during ontogeny. Thus, releasers are sign stimuli specially adapted to elicit a particular behavior in the receiver. Often their effect is enhanced by physical structures and colors that have developed for this purpose. The conspicuous feathers displayed by a courting peacock when it opens its train and the unmistakable songs of many song birds signaling to a conspecific that he is nearing a rival ready to defend his territory are examples of releasers that have emerged through adaptation on the part of the sender. If the receiver, too, gains advantage by responding to such a signal, it may do its part by adapting the releasing mechanism of the response to this stimulus, which will ultimately become the only one to release the behavior in question (adaptation on the part of the receiver).

A relationship that evolved because of adaptation by both the sender and the receiver usually benefits both. There are, however, exceptions: An orchid, which displays a deceptively clever imitation of a bumblebee with part of its petals, transmits to a male bumblebee certain stimuli that the insect usually receives from its female. The male flies to the flower, attempts to mate with the blossom, and thus pollinates the orchid. Such signal imitation is called "mimicry." Here a third part, in this case the flower, becomes a parasite on the established relationship between the female bumblebee as the stimulus transmitter and the male as receiver. At the expense of the male, the orchid profits from his response to the

Fig. 2-10. The conspicuous coloration around the beak and on the throat of many bird nestlings releases feeding in adults. By means of this sender-receiver relationship, one receives food while the other can raise its offspring; both actions aid in the survival of the species.

Fig. 2-11. A songbird is feeding this young cuckoo, which has pushed the songbird's brood out of the nest. The conspicuous throat coloration displayed by the cuckoo's gaping acts as a "supernormal releaser." Instead of breeding again immediately after losing its offspring, the songbird raises the young of another species: an advantage for the sender of the (imitated) signal, a disadvantage for the receiver.

Fig. 2-12. The beeflower *Ophrys apifera* is a model of a female bee. The male pollinates the flower while trying to copulate with it.

Fig. 2-13. Aggressive behavior of two wild boars fighting for a female.

Fig. 2-14. Male jumping spiders wave their brightly colored limbs when courting a female.

imitated signal. This is a rare example where we can clearly see that the receiver may be totally without insight into its actions and that it responds appropriately to an incoming sign stimulus on a completely automatic basis.

Many signals and other sign stimuli are not transmitted continuously. They depend on fluctuating states of action readiness associated with still others in a positive or negative manner. For this reason, the receiver of such stimuli can learn a fair amount about the present states of action readiness in the sender. Thus, sender-receiver relationships already involve a flow of information, or communication, between individuals. This is why communications research, which explores the preconditions and mechanisms of communication, constitutes an important and fundamental part of the whole field of animal sociology.

This area of research examines many different aspects in the relationships between individuals. Hardly any of these factors can be studied thoroughly without some consideration of other features, for the social life of animals is based on a very complex network of processes. Still, we may briefly discuss some of the focal points.

Aggressive behavior (see Chapter 25) between members of a species involves movements of threat and fighting; it may lead to dispersion of individuals over a large area, thus guarding against the overpopulation of any single habitat. Individuals that have been driven off may conquer new habitats, but the battle between males for possession of females generally results in the victorious male having a better chance of winning females and thus siring more offspring. But we don't always understand precisely what advantages aggressive behavior may hold for the species. Our only solution is to make exact comparisons of the number of offspring produced by both more and less aggressive individuals in a closely defined natural habitat.

Research focuses not only on the function of aggression but also on its causes. Why are some individuals in certain situations more aggressive than others? What external and internal causes strengthen or weaken aggressiveness? These issues are examined by behavioral physiologists. Of special importance, too, are mechanisms that inhibit aggression, for they are indispensable in the formation and maintenance of societies whose members are basically aggressive, such as humans.

Males and females of most animal species must come together for at least one purpose, to fertilize eggs. This process is usually preceded by patterns of courtship behavior, which often attract the sexual partner and may even stimulate copulation or ovulation. Among species where eggs and sperm cells must unite in the open, it is especially important that the male and female deposit their sex cells in a precisely synchronized manner.

After the young have developed and are born or hatched, one or both parents or some other adults display active brood-care behavior, depending on the species (see Chapter 30). This contact with older, experienced individuals not only provides the young with protection but also allows them to learn from their elders. Especially in some higher vertebrates, appropriate behavior sequences that have been learned in response to a particular situation may be passed on from one generation to the next (tradition).

In many cases, young and their parents remain together for a long time. But adults, too, may form groups and associations where individuals stay together for life. A number of individuals may simply aggregate when attracted to the same external object, such as a drinking or feeding place. Such aggregations are not necessarily based on social attraction between members (see Chapter 31). By contrast, schools of fish stay together because the individual fish actively seeks to swim with others of its kind. Associations of this type are "anonymous" and at the same time "open." That is, each member can be substituted at any time for any other, without affecting the behavior of the society as a whole. A bee colony is also an anonymous association, but it is no longer open. There is discrimination between colony members and outsiders. All members bear a single scent, characteristic of the particular hive. An individual with a different scent is rejected.

Finally, there are "closed" groups where members know each other individually. Strangers are usually rejected from these "individualized groups," too. This type of society is found with some desert wood lice, with cichlids who live in pairs and groups, and right up the evolutionary scale to monkey troops that comprise numerous individuals (see Chapter Chapter 35).

In many cases we still do not know how a particular association or a group stays together. Similarly, we often know very little about the behavioral mechanisms that cause strangers to be attacked but members to be tolerated. And few studies have been made of the factors limiting the size of a group, for example those which prevent pairs from evolving into larger groups.

In general, groups are not simply aggregations of individuals of the same kind or with equal status. We can detect a definite structure. There are hierarchies or rank orders that differentiate between subordinate and dominant members. Certain individuals function as leaders in a fight against predators, in the search for food, or during migrations and excursions. In some groups there is an actual division of labor for the most divergent activities. Sometimes the "role" of an animal within the group is related to its sex, age, or physical build. Occasionally, individual differences in temperament—as yet hard to define—may determine how roles are distributed.

The description and causal analysis of group structures and of the so-called socialization processes that allow young or initially strange individuals to grow into the structure of a society are important areas of contemporary research in animal sociology. Behavioral physiologists try to

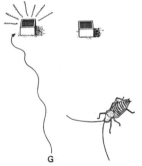

Fig. 2-15. The sexually active female grasshopper goes to the cage containing a singing male. The second cage houses another male who is also ready to breed but who has been muted; the female ignores it. Only the sound attracts the female.

Group mechanisms

Attacker

	A	B	C	D	E	F	G
A							
B	22						
C	8	29					
D	18	11	6				
E	11	21	11	12			
F	30	7	6	21	8		
G	10	12	3	8	15	30	

Attacked

Fig. 2-16. A linear dominance hierarchy, or pecking order, of seven hens living in an enclosure. Records were made of who pecked whom, and how often. The data show that each hen pecks only those beneath it in the hierarchy. A is the highest-ranking animal, G, the lowest.

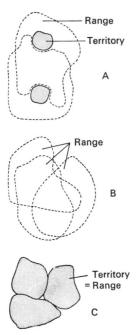

Fig. 2-17. Range and territory: The ranges of different animals very often overlap (A, B), while potential rivals are not allowed to enter the territory (A, C). Some species have a range and defend part of this as a territory (A). Others are not territorial at all (B). Finally, territory and range can be the same (C).

relate the formation, maintenance, and dissolution of groups to the characteristics of their members and to the communication processes within these groups. But we also want to establish an integrated picture of how structures such as hierarchy, role distribution, and so on, evolved. Here, a contributing factor could be found by looking at closely related species to find possible precursors of whatever phenomenon we are investigating.

In rare cases, similarities between group structures are based on homology. Convergent development is much more common; an example is the harem units found in zebras, hamadryas baboons, and domestic chickens—none of which are closely related to each other—where one male forms a permanent group with several females. Research into convergent phenomena, which is concerned with these issues, focuses on such congruencies and examines the environmental conditions that have led to the selection of such similar group structures from very different raw material. Are the advantages of such harem groups (see Chapters 34 and 35) always the same or does their adaptive value differ in each case?

Beyond the group structure itself, we also wish to examine the spatial and behavioral relationships among all conspecifics, even those which do not live in groups. The totality of these relationships represents the social organization of a population or a species. Every animal lives in a spatially limited area, which we call its "range." This may comprise whole continents or simply a few square meters. Sometimes this area, or part of it, is defended against rivals. In such cases we are speaking of a "territory."

In some species of antelope we find not only herds of males or females roaming through a very broad range, but also single males, each of which defends a relatively small area against every other adult buck. The females wander through these plots held by the territorial bucks, and mating will occur only in such an area, with the master of the territory. No buck takes part in reproduction as long as he migrates with others in the herd.

The social organization of a species is not always consistent. Differences in population density, in the nature of a particular habitat, in the quality of food supply, and so on can lead to amazingly divergent forms of organization among populations of the same species; these animals are extremely adaptable in their social behavior. It is one of the most fascinating but at the same time most difficult tasks of ethology to examine the special adaptive value of specific forms of organization under diverse environmental conditions.

Wherever ethologists study social behavior, they have many points of contact with the psychology and sociology of man. We hope that the occasional rivalries still encountered between representatives of these three disciplines will gradually disappear, that ethologists, psychologists, and sociologists are willing to learn from one another, and that they will continue to cooperate in an increasingly fruitful manner.

3 Structure and Function of the Nervous System

Every animal must respond to the environment in which it lives and adapt to continuous change. It requires information not only from within its own organism but also about the nature of its immediate environment and the on-going external changes. These kinds of information are vital, since animals receive energy for maintaining their functions from the environment in the form of organic food. One example: Frogs show a preference for catching flies. If a small dark spot moves within the visual field of a frog's large eyes, the animal will turn toward the object and make a well-aimed thrust with its sticky tongue. On closer examination, we find that this simple prey-catching response consists of a whole chain of information-processing phenomena (Fig. 3-1):

Structure and function of the nervous system, by G. Neuweiler

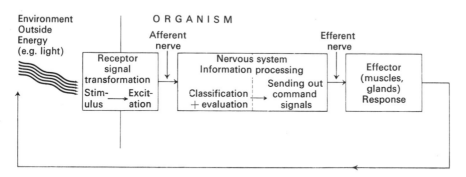

Fig. 3-1. Schematic representation of information flow in the animal organism.

1. The receptor cells in the retina of the frog's eye can be stimulated by the energy of light waves. By means of light rays from the sun, the environment is transformed into a spatially organized stimulus pattern, and the moving fly's image crosses the retina in the form of a small, distinct change in excitation. The information "small dark spot moves within the visual field" is conducted by means of signals—in this case light waves— to the eye, where it is received by the light receptors in the retina. This information is transduced in the receptor cell from the environment's

Sensory cells as information receivers

information carrier—light—into electric signals, the organism's own information carrier.

Information processing

2. Next, the frog must determine the nature of the information received. How large is this moving spot? Is it small enough to be a prey, or large enough to be a predator? How quickly does the spot move, and in what area of the visual field? By using such means of classifying information, and by comparing it with experiences either coded in the genome or acquired by the individual frog, a command signal, or impulse, is created and transmitted to the effector organs, in this case certain muscles. The command signal "catch the prey," resulting from a specific evaluation of the message "small dark spot moves within the visual field," is then carried out by the muscles. This response modifies the originally received information: Either this information now appears at another point in the retina when the frog turns toward the dark spot, or it disappears if the frog has caught its prey. All these processes are included under the general heading of "information processing." This processing is accomplished solely by the nerve cells, whose linkages start to form at the level of the frog's eye but gain prominence in the brain, where they combine into complex neural circuits.

Transportation of information

3. The information receptor—in this case the eye—is located at the periphery of the body, like the muscle or "effector" carrying out the impulse order. By contrast, the highly vulnerable brain, the processor of information, is protectively enclosed in the skull case. Thus, the information must travel from the eye to the brain, and the respective command impulses for the movement of prey catching are then transmitted from the brain to the muscles. For this purpose, the information is coded so that it arrives at its destination with little or no interference from outside influences. The nerve cells accomplish this unobstructed transmission of information just as they transform (or transduce) each message into a suitable information code.

The nervous system of animals is no more than an organ for processing and transmitting information. It receives information from the environment as well as from within the body by means of sensory-receptor cells —which can be regarded as specialized nerve cells—and induces the animal to perform its vital responses. In the following sections we shall demonstrate how the nervous system codes such information, how it conducts the messages in the form of electric signals, and how the nerve cells are able to process this information at their various relay points.

The neuron

The human brain consists of 10 billion nerve cells, interconnected within a structured network of numerous relay stations. These nerve cells are embedded in a mass of glial cells, which are otherwise not directly concerned with the functioning of the nervous system and will therefore be excluded from our discussion. The functional unit of the nervous system is the nerve cell, or neuron; the neuron is defined as a cell that is able to process, generate, and transmit electrical signals.

Like every other cell, the neuron consists of a nucleus and cytoplasm, fully enclosed by a tough cell membrane. The cytoplasm contains different structures essential for the energy-producing metabolic processes. In histological brain preparations, the cell body or soma of a neuron is easily recognized by the nucleus. Various kinds of ramifications project from this soma: These are relatively short, greatly ramified branches, called dendrites. Resting upon them, as well as on the cell membrane, are numerous endings of ramifications from other neurons, through which information signals are received and conducted to the neuron itself. The dendrites therefore carry signals to the neuron, and are thus called afferent fibers. The messages formed on the basis of these incoming signals are transmitted further by another nerve fiber, the axon. This axon originates at the axon hillock of the soma, and leads as a single fiber—in extreme cases several meters long—either to subsequent nerve cells or to muscle or gland cells, transmitting the information processed by its neuron. The axon therefore represents the efferent fiber of the neuron. The widely branched, radiating dendritic ramifications have as their efferent counterpart only one "command cable," the axon; this, however, may again divide at its ending into numerous branches, on the average numbering 100. In this way, the axon is able to transmit the same impulse command simultaneously to several other cells.

When we realize how many different connections are possible with only a dozen building blocks like the neuron, we can see that a system with billions of such units allows for an almost infinite variety of connections. In the face of these overwhelming numbers, it seems completely impossible to explain either the structure or the function of the brain. Yet we already know a substantial amount about the behavior of certain areas of the brain, because the entire central nervous system—in a way similar to our global telephone network—is built up by means of several large connection centers, each in turn controlling numerous "subordinate" neural connection points. These centers are easily recognized in the brain as dense aggregations of neurons. Unfortunately, anatomists of the previous century labeled them as nuclei. There are at least 150 such larger nuclei; since their names rarely tell us the function of these neural centers, beginners in the field have difficulty understanding the structure of the brain. For this reason we have provided a rough outline of the functional organization of the central nervous system of mammals (see Color plate, p. 45), using the human system as our example, and deliberately omitting the above-mentioned anatomical terms.

What we usually call a nerve is actually only one part of any neuron: The thick white "nerves," which run through our body to muscles, glands, or sense organs, consist of hundreds or even thousands of nerve fibers united in a thick bundle. The cell bodies belonging to these fibers are located, with a few exceptions, in the brain or spinal cord—that is, in the central nervous system. Mammals have twelve nerve pairs that run

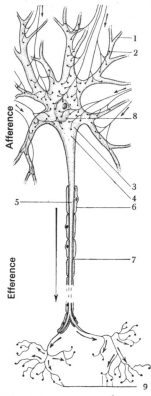

Fig. 3-2. Structure of a nerve cell (neuron). Cell body (3) and dendrites (1) have axon endings (9) of other neurons terminating on them. These form synapses with the neuron in the diagram (2). The arrows indicate the direction of impulse spread. 4. Axon hillock; 5. Axon; 6. Myelin sheath; 7. Node of Ranvier; 8. Cell nucleus.

Sensory and motor nerves

directly to or from the brain, innervating all organs and muscles contained in the head of the organism; there are also thirty nerve pairs that lead from the spinal cord and supply parts of the head, the entire torso, and the extremities (see Color plate, p. 45). Nerve bundles whose fibers conduct messages to the spinal cord and brain are called sensory nerves, and those carrying orders from the central nervous system to muscles and glands are called motor nerves. The nerves of the spinal cord are mixed bundles comprising both sensory and motor fibers. When these enter the spinal cord, however, they separate into the two component parts: The motor fibers emerge from the ventral root, and the sensory fibers lead into the dorsal root. Thus, messages to and from the spinal cord are conducted in the same nerve cable but on two separate "lines," the sensory and the motor nerves.

Just like copper wires in a many-veined telephone cable, dendrites and axons in a nerve cable must be insulated against one another to prevent "interference." This insulating function belongs to companion cells, the so-called Schwann cells, which wrap around the nerve fibers in a dense, many-layered roll filled with an insulation substance of poor electrical conductivity, called myelin. This insulating mass disappears only where two such cell rolls connect to each other on a nerve fiber. These short, myelin-free parts of nerve fibers are called, after their discoverer, the nodes of Ranvier. They play an important part in the transmission of message impulses through the fibers.

The nerves can be compared to cables constructed of many thousands of cords, making up the connection between the periphery—that is, the reception of information through the sense organs and the execution of impulse commands by means of muscles and glands—and the central processing points in the spinal cord and brain. Their function is to transport messages received through the receptor cells quickly and without distortion to the brain and spinal cord, and to carry the command messages originating at the above-mentioned loci back to the muscles and glands.

As we have seen, the nervous system, with its complex relay switches, consists of a single functional building block, the neuron. The manner in which a neuron functions in turn determines the basic functional schema of the brain.

Resting membrane potential of the neuron

But how can an experimenter use a network of billions of nerve cells to derive properties of a single neuron? Fortunately for the scientist, there are a few animal species that have small groups of nerve cells with unusually thick and long axons. Using squid (cephalopods of the genus *Loligo*), we can make preparations of single axon fibers measuring almost 1 mm in diameter. These preparations survive for several hours or even days in a watery saline solution with a composition similar to that of ocean water, maintaining their full functioning capacity. The experimenter can insert a thin measuring probe, an electrode, into the axon preparation. For this purpose we frequently use glass capsules with an extremely thin tip (only 1/1000 mm in diameter), filled with an electrically conductive liquid such as salt solution. This electrode is coupled with a suitable amplifier and connected to an oscillograph, used for measuring and recording electrical potential. The oscillograph has now become a universal biological measuring instrument, because time-dependent electrical biological processes, or those able to be transformed into changes in electrical potential, are represented on its screen in a precise and immediate fashion (Fig. 3-3).

If we insert a probing electrode into an axon preparation and place a second electrode into the surrounding liquid, we discover an electrical potential difference of -60 to -90 millivolts (thousandths of a volt) between the inside of the axon and the outside medium, with the inside potential negative and the outside, positive. This potential difference remains constant in the living axon over long periods of time. It exists not only in the axon but at every point of the neural membrane. The constant difference in potential between the inside of a neuron and the extracellular space (the surrounding medium of the neuron) is called the resting membrane potential; the correct terminology would be "resting membrane potential difference."

If the second electrode is also placed inside the axon rather than into the outside liquid, so that both electrode tips now lie inside the axon, the measuring apparatus will indicate a potential difference of zero. We obtain the same result when both electrodes are placed on the outer side of the axon.

These simple experiments gave us two important results: 1. A state of complete electrical neutrality prevails inside the neuron; no potential difference can be measured anywhere within the cell. The same applies to the outside fluid. 2. By contrast, there is a potential difference of -60 to -90 millivolts between the inside of the neuron and the outside fluid. This is the resting membrane potential.

These two findings—electrical neutrality in the outside medium as well as within the cell, but a potential difference between inside and outside—seem to contradict each other. In order to solve this problem, let us pretend that a neuron is divested of all elements not essential for an understanding of the resting membrane potential. Such a hypothetically simplified neuron consists of a membranous tube filled with water and surrounded by it. This water solution contains protein molecules and salts. When in water, salts break down into ions, atoms and molecules that carry electric charges. Sodium chloride, for example, breaks down into Na^+ ions and Cl^- ions. Proteins dissolved in water also carry a surplus of negative charges; thus they are negatively charged ions, like the Cl^- ion of sodium chloride.

Measurements on countless nerve preparations from different kinds of animals have consistently shown that the inside of a neuron contains a very large number of potassium ions (K^+) and only few sodium ions (Na^+). Since the inside of a neuron is electrically neutral, there must be

▷

The central nervous system of humans. The CNS consists of the brain and spinal cord. Fig. b) is a medial section through the brain, showing the huge extension of the cerebral cortex as compared with the phylogenetically older parts of the brain: diencephalon, mesencephalon, cerebellum, and medulla oblongata. Fig. c) shows the location of the most important association centers on the cerebral cortex. Fig. d): 12 pairs of cranial nerves and 31 pairs of spinal nerves conduct excitation between the CNS and various parts of the body. Fig. e) shows the spinal cord from above. One pair of spinal nerves emerges from each vertebral section; each nerve consists of a bundle of motor- and sensory-nerve fibers, and each supplies a certain section of skin and muscle (see Fig. d). The skin and muscles of the head are serviced primarily by the fifth cranial nerve (I–III). Fig. a) is a schematic cross section of the spinal cord, showing the most important nerve pathways that ascend to the brain, by way of the spinal cord or descend from the brain to the periphery. ●—< is the symbol for neuron: The button represents the cell body, the line is the axon, and the open fork indicates a synapse. Arrows show the direction of information flow.

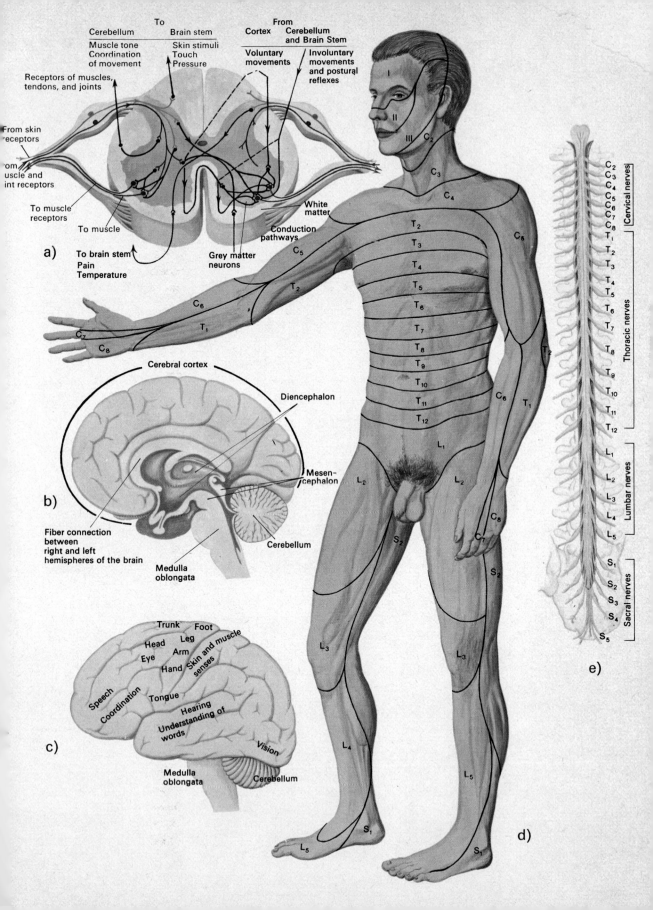

a)

To
Cerebellum   Brain stem
Muscle tone   Skin stimuli
Coordination   Touch
of movement   Pressure

From
Cortex   Cerebellum
and Brain Stem
Voluntary   Involuntary
movements   movements
and postural
reflexes

Receptors of muscles,
tendons, and joints

From skin
receptors

From
muscle and
int receptors

To muscle
receptors

To muscle

To brain stem
Pain
Temperature

White
matter

Conduction
pathways

Grey matter
neurons

b)

Cerebral cortex

Diencephalon

Mesen-
cephalon

Fiber connection
between
right and left
hemispheres of the brain

Cerebellum

Medulla
oblongata

c)

Trunk   Foot
Head   Leg
Eye   Arm
Hand   Skin and muscle
senses

Speech
Coordination   Tongue
Hearing
Understanding of
words

Vision

Medulla
oblongata   Cerebellum

d)

I
II
III   C₂
C₃
C₄
T₂
T₃
C₅
T₄
T₅
T₂
T₆
C₆
T₇
C₇
C₈
T₁
T₇
T₈
T₉
T₁₀
T₁₁
T₁₂
L₁
T₂
C₆
T₁
L₂
L₂
C₈
C₇
S₂
S₂
L₃
L₃
L₄
L₅
S₁
L₅
S₁

e)

C₂
C₃
C₄ } Cervical nerves
C₅
C₆
C₇
C₈
T₁
T₂
T₃
T₄
T₅
T₆
T₇ Thoracic nerves
T₈
T₉
T₁₀
T₁₁
T₁₂
L₁
L₂
L₃ Lumbar nerves
L₄
L₅
S₁
S₂
S₃ Sacral nerves
S₄
S₅

Brain signals

Stimulus

Hy

L

L

Hy

E

Bat

Analyzed brain signals

S

D

C

Hy

Brain signals

◁
Arrangement of instruments for recording electrical signals from the brain of animals after stimulating a sense organ. Example: Recordings of signals from the auditory centers in the brain of an echolocating bat during stimulation of the ears with electronically produced ultrasonic frequencies. Blue: Electronic apparatus for producing the ultrasonic stimulus. This sound is then played to the bat through loudspeakers (L). Red: Electronic apparatus for amplifying and filtering the appropriate electrical signals in the brain. The brain signals are recorded by way of a glass capillary with a tip no larger than 1/1000 mm (E). This electrode E is inserted into the desired location by way of a remote-controlled hydraulic system (Hy). Green: The signals are assessed with a laboratory computer (C) and stored on tape. The computer is given directions with the aid of a distant recording machine (D). At the same time, the result of this evaluation appears on a screen (S) for checking.

Equilibrium potential

as many negative ions as both Na^+ and K^+ ions. These negatively charged ions consist of a large number of protein⁻ ions and a small number of chloride ions (Cl^-).

The external medium, the extracellular fluid surrounding the neuron, also contains an equal number of negative and positive ions. However, while we find many K^+ and few Na^+ ions inside the neuron, there are few K^+ and many Na^+ ions in the outside fluid. Inside the neuron there are many protein and few Cl^- ions: outside we find virtually no protein ions and very many Cl^- ions (Fig. 3-4). The following table summarizes the ion distribution between inside and outside. Bold print shows the high concentration of the respective ions in either the inside or outside solution.

Inside of neuron	Membrane	Outside Solution
K^+	‖	K^+
Protein⁻	‖	
Na^+	‖	**Na^+**
Cl^-	‖	**Cl^-**

At the nerve membrane we have a strong "concentration gradient" for the individual kinds of ions between inside and outside; the direction of this gradient is outward for K^+ and inward for Na^+. According to a general law of nature, these concentration differences should equalize within a very short period if the ions could pass freely through the membrane. The neural membrane, however, is completely impermeable to protein ions, and very difficult for Na^+ ions to pass through. By contrast, the K^+ ions may travel freely through the membrane. Why, then, does the difference in K^+-ion concentration not equalize between the inside and outside? The answer lies in the electrical properties of an ion. We concluded above that there must always be an equal number of positive as well as negative ions in both the inside of the neuron and the outside medium.

If in relation to their drop in concentration level, potassium ions move to the outside of the cell, a negative ion would have to accompany each potassium ion to maintain electrical neutrality. But the negative partners of these potassium ions in the plasma are large protein ions, which could never pass through the small pores in the membrane. When potassium ions flow out, the protein ions are caught on the inner side of the membrane, resulting in the accumulation of a layer of negatively charged particles, which in turn electrostatically draw back the outflowing potassium ions. In this way the membrane becomes charged positively (K^+) on the outside and negatively (protein⁻) on the inside. Thus an electric field is created across the membrane, driving the potassium ions back inside. The potassium flow comes to a halt precisely when the opposing force of the electric field is equal to the force of the difference in concentration that pushes the potassium ions to the outside. When this state of

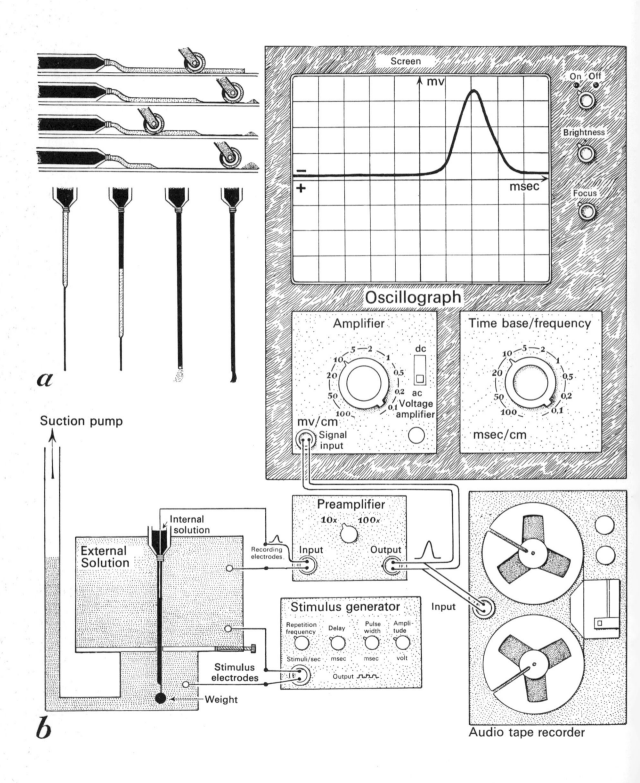

a

b

Suction pump

Screen

mv

msec

On Off

Brightness

Focus

Oscillograph

Amplifier

5 2
10 1
20 0,5
50 0,2
100 0,1

dc

ac

Voltage amplifier

mv/cm

Signal input

Time base/frequency

5 2
10 1
20 0,5
50 0,2
100 0,1

msec/cm

Preamplifier

10x   100x

Input   Output

External Solution

Internal solution

Recording electrodes

Stimulus generator

Repetition frequency   Delay   Pulse width   Ampli-tude

Stimuli/sec   msec   msec   volt

Output

Stimulus electrodes

Weight

Input

Audio tape recorder

equalization between the two opposing forces influencing the ions is reached, some ions may still move back and forth between the inside and outside of the cell. But during any time unit there are as many potassium ions flowing out as in, and thus the net ion flow reaches zero. The force of the electric field, which in a state of equilibrium brings the net outflow to a halt, can be measured as the potential difference across the membrane. It is called the equilibrium potential for potassium. This measures -90 mv for K^+, and corresponds roughly to the actually obtained potential difference at the neural membrane.

We may also determine such equilibrium potentials for the other kinds of ions, depending on the differences in concentration. Their contribution to the combined potential of the membrane is always based on the membrane permeability for this kind of ion. For example, the permeability of the neural membrane in its resting state is twenty times smaller for Na^+ ions than for K^+ ions. Thus, the contribution of Na^+ to the membrane potential is correspondingly smaller.

In principle, the resting membrane potential of -60 to -90 millivolts is a mixed potential arising from the equilibrium potentials of the various kinds of ions. For the most part, however, it corresponds to the equilibrium potential for K^+ ions, because it is difficult for other ions to pass through the resting membrane.

In summary, these are the essential factors in creating the resting membrane potential: 1. There is a difference in concentration for some kinds of ions between the inside of a neuron and the outside medium. 2. The nerve membrane is selectively permeable for certain kinds of ions, while for other kinds it is only semipermeable or impermeable, as for sodium and protein ions. The key to nerve cell functioning will prove to be primarily the selective permeability of the nerve membrane. The example of the mixed membrane potential, brought about by the equilibrium potentials for potassium and sodium, shows how the level of the membrane potential greatly depends on the permeability of the membrane for the different kinds of ions. Should the membrane for some reason suddenly become as permeable for sodium as for potassium, the membrane potential would immediately drop from -75 to -17 millivolts. By varying the ion permeability of the membrane, different potential values can be created on both a short-term and a long-term basis. Later in our discussion we shall see that precisely this kind of change in the properties of a membrane brings about changes in the potential, which in turn make up the signals of the information to be processed.

In its resting state, the neural membrane is largely but not completely impermeable to sodium; that is, during a time unit a relatively very small number of sodium ions will flow to the inside. For every Na^+ ion flowing in, a K^+ ion can be released from its electrical bond to the bulky protein ions inside, and it is then able to move to the outside medium. By means of this slow and constant exchange of inwardly penetrating sodium for

Selective permeability

Fig. 3-3. Experimental method for studying the giant nerve fiber of a cuttlefish. a) The surgically removed and prepared section of axon is tied to an injection canule filled with an ion solution. The axoplasm is squeezed out and the axon membrane tube filled with solution. b) The preparation is suspended in sea water (external solution). The internal solution can be removed with a suction pump and replaced. By way of electrodes, experimenters record electrical changes at the membrane. These changes in potential are then amplified and made visible on the screen of an oscillograph. The audiotape recorder stores the potentials.

The Na^+–K^+ pump

outflowing potassium ions, the differences in concentration should gradually disappear and the membrane potential should drop to zero. However, under normal biological conditions the measurements on the axon preparation of the squid always show the same resting membrane potential over hours and even days. If, on the other hand, we place the preparation into a vacuum chamber, so that the supply of oxygen to the membrane is blocked, the membrane potential soon drops to zero. This experiment demonstrates that an oxygen-consuming mechanism must be at work in the living neural membrane, maintaining the differences in the concentration of potassium and sodium ions so that electrical potentials are created. This high sensitivity to oxygen suggests that the membrane may utilize a "pump" that constantly uses up metabolic energy and forces the inflowing sodium back out, at the same time forcing the potassium (diffusing out in exchange) back in (Fig. 3-4). We do not know how this sodium-potassium pump functions in the membrane. Some experimental results indicate that there is a circular pump, forcing sodium out of the cell and alternately drawing potassium back in.

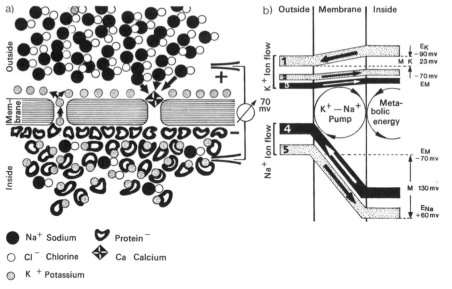

a)

b) Outside | Membrane | Inside

Na⁺ Sodium   Protein⁻
Cl⁻ Chlorine   Ca Calcium
K⁺ Potassium

Fig. 3-4. The resting membrane potential. a) Distribution of ion types at the membrane. Arrows indicate the direction of the ion-concentration gradient. Na ions are prevented from entering by the Ca ions. The outflowing K ions are caught in the external membrane, since they are electrically bound to the protein ions, which in turn are too large to pass through the pores. b) Ion currents (1–5) flowing through the membrane; the volume of current depends on intensity. The incline indicates the electrochemical gradient. Black: active ion flow aided by the energy-consuming K^+/Na^+ pump. Dotted: passive ion flow based on differences in concentration. E_K and E_{Na}: equilibrium potentials for K^+ or Na^+. E_M: membrane potential.

The sodium-potassium pump is the generator, constantly using up energy and thus maintaining the resting membrane potential by creating differences in ion concentration. Ultimately, it is the resting membrane potential that enables the neuron to receive information in the form of electric signals, to process them, and to pass them on. Should this generator break down, the resting membrane potential would disappear and the neuron would no longer be able to function. We can now understand why the brain, of all the organs, is the most sensitive to oxygen, and why it reacts to every circulatory disruption with severe functional disturbances.

Even the tiniest blood clots, which block the blood and therefore the oxygen supply to narrowly restricted areas of the brain, cause neurons to die off very quickly, resulting in a stroke. If larger areas of the brain or vital control centers such as the respiratory center are hit, the stroke quickly leads to death.

Nervous activity

We have already indicated that the neuron could change its membrane potential for a short period of time by varying the ion-permeability of the membrane. According to our previous calculations, the membrane potential would drop to -15 mv if the membrane became more permeable to sodium. Conversely, the membrane potential should rise to -66 mv (the equilibrium potential for Cl^- ions) if the membrane became permeable to Cl^- ions but was no longer permeable to potassium and sodium ions. The electrical signals used by the nervous system to process and transmit information are nothing more than short-term changes in potential, usually only milliseconds in duration, which are incorrectly but conveniently also labeled as potentials. They arise through short-term changes in the ion permeability of the nerve membrane.

When a neuron is active, that is, when it receives, processes, and transmits information, it always deviates at certain points on the membrane from the ion permeability of its resting state, therefore generating changes in potential. These changes are called excitation, and an active neuron is said to be excited. The essential question, the key to an understanding of nervous functioning, is: What determines the selective ion permeability of the membrane, and in what way, so that it leads to short-term changes in potential, that is, in the electrical signals?

Signals must first arise wherever a nerve cell receives information from a receptor cell or another nerve cell. The membranes of the neural body and dendrites receive thousands of axon endings that carry information from other cells. These axon endings lie close to the membrane but never actually merge with it (Fig. 3-5). There is always a small gap of approximately 30–40 hundred-thousandths of a millimeter between the axon ending and the neural membrane. This point of contact is called the synapse, the afferent axon ending, the presynapse, the receptor neuron, the postsynapse, and that part of the membrane directly under the axon ending, the synaptic membrane.

When an information signal arrives at a single synapse, a rapid succession of potential changes—called action potentials—lasting only milliseconds will move over the presynaptic membrane at the axon ending (the presynapse). We will not yet consider how these signals arise and at what point they actually originate. Each time such an action potential (AP) moves over the presynapse, this brief change in potential at the axon ending releases a certain amount of a chemical substance into the synaptic gap (Fig. 3-6). This substance opens the synaptic membrane of the receptor neuron for all kinds of ions—not only for potassium, which can pass through the membrane at any time, but also for sodium and chloride.

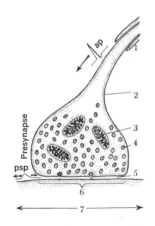

Fig. 3-5. The synapse. The arrival of an action potential (AP) in the presynapse leads to a discharge of transmitter substance into the synaptic gap (5). The transmitter substance alters the ion permeability of the synaptic membrane (6) belonging to the subsequent neuron, so that a postsynaptic potential (psp) arises at this point. 1. Myelin sheath; 2. Axon ending; 3. Mitochondria; 4. Synaptic vesicles; 7. Postsynapse.

The potential of the synaptic membrane flows toward the mixed potential of sodium, potassium, and chloride. The synaptic membrane is depolarized. However, a particular enzyme on the outside of the synaptic membrane quickly breaks down the substance that caused the permeability to change. As the substance disappears, the pores again contract, at the same rate, to become less permeable to sodium and chloride, and the membrane potential returns to its resting state after approximately 10 milliseconds. This depolarization is called the excitatory post-synaptic potential, or EPSP.

The chemical transmitter substance that carried the excitation to the postsynaptic neuron lies tightly packed and stored in small bubbles in the presynapse (Fig. 3-5). Since this transmitter substance is contained only in the axon ending and never in the postsynapse, a chain of neurons can process and conduct information in only one direction, from presynapse to postsynapse.

There are many different transmitter substances in the nervous system, but each nerve cell contains only one of these in its axon ending. The substance examined most thoroughly is acetyl choline. Among other things, it carries excitations from nerves to muscles. Acetyl choline makes the synaptic membrane of muscle fibers permeable to all kinds of ions, thus leading to a depolarization. At the other synaptic membranes, for example, those of the heart muscle fibers, this substance does not effect a depolarization but rather a small hyperpolarization, an increase in membrane potential above the value of the resting state. With the help of acetyl choline, the synaptic membrane of the heart becomes more permeable for potassium ions; all other ions remain unable to pass through the membrane. Thus the change in ion permeability is determined not only by the type of transmitter substance, but also by the nature of the sub-synaptic membrane. In order to achieve a change in membrane permeability, the transmitter substance must always come into contact with this receptor.

The transmission of excitation from nerve to muscle provides a particularly good example for this process: some South American tribesmen use extracts from the bark of certain trees, called curare, as a deadly poison for their arrows. An organism struck with this substance quickly dies from total paralysis of the musculature, including the respiratory muscles. When we examine the effect of curare on an isolated frog-muscle preparation, we discover that after only a few seconds the muscle to which this substance has been applied ceases to contract in response to artificial electrical stimulation of the respective nerve; but if we attach the stimulatory electrodes directly to the muscle itself, it again contracts after every electrical impulse. The muscle therefore remains intact, despite the influence of curare. Obviously what is impeded is the transmission of excitation from nerve to muscle. We have found that the curare molecules occupy the "receptor points" for acetyl choline at the subsynaptic membrane of the nerve muscle synapse, so that the transmitter substance at the subsynaptic membrane is displaced. Since the transmitter substance cannot come into contact with its "receptor," it cannot change the ion permeability of the muscle membrane. In other words, the command signal telling the muscle to contract cannot be transmitted, and the muscle remains slack despite continuous impulses to the nerve responsible for muscle contraction. Thus, contact between the transmitter substance and the receptor point on the subsynaptic membrane is essential for the transmission of excitation.

In addition to acetyl choline, there are many other chemical substances that determine ion permeability at subsynaptic membranes. For example, adrenalin and noradrenalin are excitatory transmitters in the sympathetic nervous system. It is likely that glycine, the simplest of amino acids, and gamma amino butyric acid cause the subsynaptic membrane to become selectively permeable to chloride and potassium ions. Consequently there is a small increase in the value of the resting potential, as hyperpolarization, since the mixed potential for chloride and potassium lies at 78 mv (Cl^-: 66 mv, K^+: —90 mv). Such a hyperpolarizing postsynaptic potential is called the inhibitory postsynaptic potential (IPSP).

We may summarize the chemical control of synaptic ion permeability as follows: Electrical signals (action potentials) moving over the presynaptic membrane effect the release of a transmitter substance, resulting in one of two consequences depending on the type of substance and its receptor point on the postsynapse: a) The membrane potential of the postsynapse is decreased or depolarized and becomes an EPSP (Fig. 3-6) or b) the membrane potential of the postsynapse is increased or hyperpolarized and becomes an IPSP. Every excitation in the nervous system takes the form of a membrane depolarization, and thus every hyperpolarization must be regarded as a decrease in excitation, or as excitatory inhibition. Both changes in potential, either EPSP or IPSP, involve electrical signals that carry information. They arise from electrical signals provided by neural or sensory relay cells. It is the transmitter substance of the presynapse and the receptor point of the postsynapse that determine how incoming information is evaluated: as a command to increase responses (EPSP) or as a signal to decrease responses (IPSP).

Temporal summation of the postsynaptic potentials

We have already mentioned that a short-term depolarization of the presynaptic membrane releases a corresponding amount of transmitter substance by means of an action potential. The greater the depolarization, the more transmitter molecules are released at the same time. The number of molecules in turn determines the number of pores opening up on the subsynaptic membrane, so that the size of the postsynaptic potential depends on the amount of simultaneously released molecules (Fig. 3-7). Since, as a rule, the action potentials of the presynapse have a constant magnitude of approximately 100 mv, the EPSP released by the action potential also remains at a constant level, about 4 mv. After 10 msec, all transmitter molecules previously released have been broken down by the enzyme of the subsynaptic membrane, the ion permeability of the membrane is once more limited to potassium ions, and the potential returns to the value of its resting state. In the 10 msec of a PSP sequence, as many as eight action potentials can follow in succession over the presynapse, each releasing a flood of transmitter molecules into the subsynaptic gap.

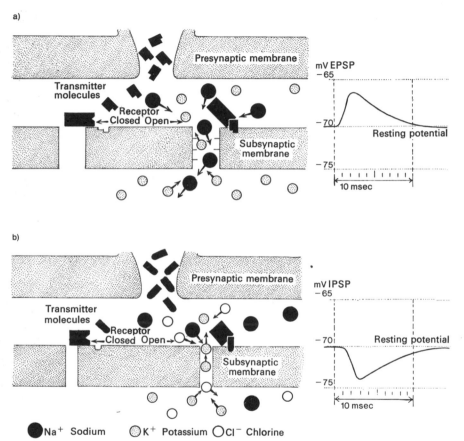

Fig. 3-6. a) Excitatory synapse: The transmitter molecules open the pores of the subsynaptic membrane at the receptor point, so that Na^+ and K^+ ions may flow in and out. The Cl^- ions (not shown) cannot pass through because the inside of the pores is negatively charged. b) Inhibitory synapse: Another kind of transmitter molecule opens a narrow pore at the receptor point, allowing the passage of K^+ and Cl^- ions. The diagrams to the right show the electrical results of ion currents released by the transmitter molecules. EPSP: excitatory postsynaptic potential. IPSP: inhibitory postsynaptic potential. mv: 1/1000 volt; msec: 1/1000 second.

The number of transmitter molecules increases in proportion to the frequency with which the action potentials of the presynapse succeed each other. The molecules in turn open a corresponding number of subsynaptic pores. The following analogy may illustrate this temporal summation: A construction worker shovels sand from a mound onto a mixing tray, and a second worker shovels it from the tray into the mixing machine. If both shovel equally fast, the pile of sand on the tray will always be the same. But if the first worker (action potentials of the presynapse) can shovel faster than the second (breakdown of the transmitter substance by the enzyme of the subsynaptic membrane), the sand pile on the tray (transmitter molecules not broken down, remaining in the synaptic gap and on the subsynaptic membrane) gradually increases. In other words:

Because of this temporal summation, the amplitude of the EPSP can be used to measure the frequency with which the action potentials of the presynapse follow one another. This succession frequency of the presynaptic action potentials encodes information by way of the intensity of excitation. Thus, information is now contained in the amplitude of the EPSP by means of temporal summation at the synapse.

Spatial summation of the postsynaptic potentials

The synaptic potentials spread evenly in all directions, from their point of origin over the entire postsynaptic membrane. With increasing distance from this original point, their amplitude decreases. When more of the numerous synapses of a neuron are active simultaneously or in short succession, the amplitudes of all the synaptic potentials are summated —on their way over the postsynaptic membrane—into a common postsynaptic potential (Fig. 3-7). In this way incoming messages of countless spatially dispersed synapses are summated into a common signal, the combined postsynaptic potential (PSP) of the neuron itself. As a rule, several synapses must activate at the same time in order to release an action potential at the axon hillock. In addition to a temporal summation, this kind of signal processing enables information messages to be summated spatially as well. Thus, for example, messages from the eye and ear, which converge in the mesencephalon over synapses onto a particular neuron, may combine into one unit of information (EPSP). If inhibitory or hyperpolarizing synapses are also active at a neuron, the value of the combined EPSP decreases proportionately. This decrease and suppression of excitations is vital to the functioning of the central nervous system. At every point in time more inhibitory than excitatory processes are active in the brain.

We may demonstrate this with a simple experiment: A frog with its brain removed but its spinal cord intact is still able to perform a whole series of well-coordinated movements, such as hopping, wiping, and so on. If we lightly pinch its foot, the frog will immediately retract the corresponding leg. If we inject this animal with a solution of strychnine and repeat the experiment after 20 minutes, the frog immediately goes into a muscular cramp and dies after a few minutes. Strychnine, a poison extracted from the seed of nux vomica, blocks the transmission of excitation through inhibitory synapses. Because of the strychnine poisoning, the frog loses all its internal inhibitory information signals. The normally harmless pinch stimulus, which in the unpoisoned frog led to a brief and functional reaction—retraction of the leg—now releases a chaotic avalanche of excitation throughout the entire nervous system, brought about by the absense of inhibitory processes that normally suppress excitation. The result is rapid death.

Information-processing signals

The entire postsynaptic membrane may be compared to a data bank: A constant stream of messages flows over hundreds of input channels (synapses). Depending on the type of input channel (inhibitory or excitatory synapse), such messages are evaluated as either positive (EPSP) or negative (IPSP). The input values are recorded (amplitude of the PSP), and multiplied according to the origin of the messages with an evaluation factor: PSPs arising at the dendrite endings must travel further over the membrane than those originating at the synapses of the neurosoma. Thus they will have a lower amplitude on the neural membrane than those arising directly at these points but receiving the same amount of excitatory input. The individual inputs thus evaluated are then summated (spatial and temporal summation of the individual PSPs) in the form of a total calculated from all messages.

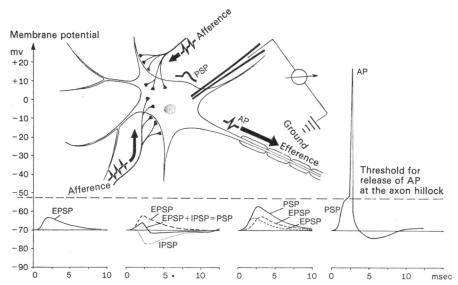

Fig. 3-7. The membrane potential of a nerve cell as a function of the incoming excitations (afferences). The electrode shows the approximate site where potentials are recorded. Arrows indicate the direction of impulse flow. EPSP: excitatory postsynaptic potential. PSP: postsynaptic potential (a mixture of many individual IPSPs and/or EPSPs). IPSP: inhibitory postsynaptic potential. AP: action potential.

Thus, the postsynaptic potentials represent the electrical signals used for calculating or processing information. In this case we have an "amplitude-modulated" code, since the amplitude of the PSP is a measure of the intensity of excitation. As demonstrated above, these amplitudes are easily summated on the postsynaptic membrane in both the spatial and temporal dimensions. Amplitude-modulated electrical signals are created because the selective ion permeability of the synaptic membrane is influenced by chemical factors.

Combined processing of all the information signals passing over the synapses results in an over-all postsynaptic potential that spreads over the membrane of the neurosoma. This combined potential is a depolarizing EPSP if the excitatory signals outweigh the inhibitory ones, and a hyperpolarizing IPSP if the inhibitory signals are stronger. Again, the result of this information processing is read in terms of an amplitude-modulated electrical signal. What now happens to the signal that, as demonstrated with our original example of the frog, ultimately leads to a response such as a leg movement? The information encoded in the combined postsynaptic potential of the neuron must be conducted to another nerve cell

Electrical control of ion permeability

or a muscle fiber, in such a way that the information arrives at its destination without distortion and as quickly as possible.

The action potential

The cable for transporting electrical signals is the axon of a neuron. Since the axon membrane is a continual extension of the membrane of the neurosoma, we might expect that the potential spreads across the axon just like the postsynaptic potentials spread over the neurosoma and the dendrites. But if this were the case, we would find that the amplitude of the signal to be transported decreases with increasing distance from the neurosoma: for one thing, information encoded in the amplitude would be altered; and the length of many axons would prevent virtually all of the signal from arriving at the axon ending. The amplitude-modulated signal must therefore be recoded into a form suitable for transport. This process occurs at the point where the axon originates, at the axon hillock. Ions may pass with somewhat greater ease back and forth through the neural membrane here than at other points. If an excitatory postsynaptic potential arrives at this point, the locus becomes depolarized in proportion to the excitation, as would any other part of the membrane. When this depolarization exceeds a certain value, roughly −50 mv representing the so-called threshold value, the axon hillock membrane suddenly becomes totally permeable to sodium ions for a mere fraction of a millisecond (Fig. 3-7). Since the concentration of sodium ions is much higher in the outside medium than in the inside of the nerve cell, the equilibrium potential of +56 mv lies in counterbalance to the equilibrium potential of potassium ions. The membrane thus undergoes an immediate depolarization of approximately 100 mv in the direction of the equilibrium potential for potassium ions. Two factors then lead to an equally rapid reversal of this depolarization:

1. With a slight delay, a host of potassium ions now follows the sodium ones, in exchange, to the outside medium; in this way they recreate a positive charge on the outside and counterbalance the depolarization.

2. The very same above-threshold depolarization that occurred at the membrane of the axon hillock, and which led to a complete membrane permeability for sodium, now inactivates this permeability; we do not know how this happens.

Due only to the fact that this "sodium brake" is somewhat delayed, a short-term depolarization takes place. The situation is similar to what would happen if the gas and brake pedals of an automobile were coupled. Should we push the common pedal (depolarization) with an above-threshold intensity, the engine would first accelerate and drive the car forward for a few meters (permeability to sodium); but then the simultaneously powered brake, which reacts more slowly, would follow suit, bringing the car to a halt (inactivation of the sodium ions). We do not yet understand the mechanism of this sodium brake, but we could imagine the sudden opening of the membrane for sodium in the following way (Fig. 3-4): With a resting potential of −70 mv, the calcium ions occupy the pores where the sodium ions would otherwise pass, so that the calcium blocks the path for sodium through the membrane. When the membrane is depolarized to its threshold value, the calcium ions move away from the pore entrances, the "corks" are therefore removed, and the sodium ions may pass freely through the membrane. In relation to the total number of sodium ions outside the cell, however, only a few enter; thus, once the permeability process for sodium has been inactivated, the sodium-potassium pump has little trouble in pumping the surplus sodium ions back out and the potassium ions that had traveled into the outside medium back into the cell. This re-establishes the original state of the resting membrane.

The entire procedure—extremely rapid depolarization and an almost equally fast repolarization to restore the membrane to its resting state—is completed in the course of one millisecond. We term this change in potential the action potential (AP). Under chemical control of ion permeability at the membrane of the synapse, the number of transmitter molecules determined the amplitude of change in potential, so that the amplitude itself could vary to any degree. But now we have an all-or-nothing reaction resulting from electrical influence on the ion permeability at the membrane of the axon hillock: Any depolarization, or EPSP-amplitude, that fails to surpass the threshold value will not produce an action potential (Fig. 3-4). On the other hand, any depolarization that, regardless of its amplitude, does exceed the threshold value inevitably leads to a full action potential whose amplitude is always the same. We may compare the release of an action potential to igniting a flash bulb: If we activate the right switch, the bulb is ignited in a preset fashion and then recharged for further use. It makes no difference to the flash itself (action potential) whether we push the button with greater or lesser intensity (amplitude of the releasing EPSP)—the flash lights up as soon as proper contact is made in the switch (threshold value), always with the same appearance and intensity (amplitude and duration of the AP).

Conduction of excitation

The careful reader will immediately protest that our analogy has one flaw: The postsynaptic potential is an amplitude-modulated signal, and the amplitude contains information of considerable interest. If this amplitude-modulated EPSP were transformed into an action potential always showing the same amplitude, the information would be lost. But surely the recoding process cannot be designed to destroy that information!

Frequency of the action potentials

It is the value of the releasing depolarization that determines the number of action potentials created per time unit at the axon hillock, and not size of the amplitude. The succession frequency (number of action potentials per second) of the action potentials is a direct measure of the amplitude of the releasing above-threshold depolarization, that is, of the postsynaptic potential. In other words: The information once contained in the EPSP amplitude has now been recoded into the succession frequency of the action potentials. Usually it is the stimulus intensity that is encoded in the frequency of the AP, for example, the intensity of a light stimulus, the loudness of a sound, the pressure of a tactile stimulus, and so forth. Often there is a logarithmic correlation between stimulus intensity and AP-frequency; for example, the AP-frequency of a particular neuron is doubled when the stimulus intensity increases tenfold.

Fig. 3-8. The conduction of excitation on an axon. An excitatory impulse is just being discharged at the left node of Ranvier (1st AP). Arrows indicate the flow of the current within and outside of the nerve membrane. Although a current is also spreading to the left, an excitation cannot take place in that direction because these nodes of Ranvier have just undergone the process, thus temporarily losing their capacity to be excited. However, the excitation can spread to the right (forward).

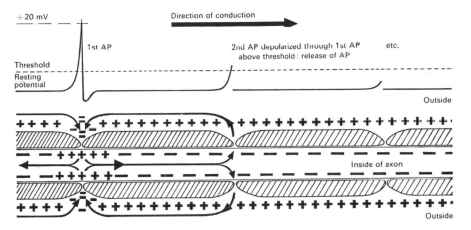

The AP-frequency carrying the information may be conducted in a very precise manner, without any loss at all, even along axons many meters long. This would never happen with the amplitude of a change in potential. The AP-frequency is therefore an ideal code for conducting information.

However, none of the action potentials arising at the axon hillock must get lost on the way, and the distance between these and their predecessors must remain as before. The cables that conduct information, the axons, are embedded close behind the axon hillock in a mass of companion cells with a good isolating substance, the myelin (Fig. 3-2). We have measured an electrical resistance of 100 megohm between the inside and outside across such myelinated axon membranes. This isolating mass is lacking at points along the axon where, in regular intervals, the cells of the myelin sheath make contact with their neighbors. These are the "nodes of Ranvier," and where they occur the axon membrane has a relatively good conducting capacity. If an action potential happens to discharge at the axon hillock, the outside of the membrane at this locus undergoes a short-term negative charge, while the neighboring external parts of the axon membrane that have not yet been excited retain a positive charge. Since the outside of the axon consists of ion solution with good electrical conductivity just like the intracellular space, a number of positive ions will flow away from the nonexcited parts of the membrane to the temporarily negative outer part of the excited locus. A corresponding process will occur inside the membrane, in the axoplasm: At the

Saltatory conduction of excitation

excited membrane section the inner side is positive, while the nonexcited neighboring parts still display a negative charge. The positive ions at the excited locus are attracted by the negative charges of the inner membrane side. This set of conditions does not affect parts of the membrane isolated by the myelin, since virtually no electric current may flow across the rather high resistance there. But at the nodes of Ranvier, where the isolating mass is absent, electrical current will flow from the inside to the outside. Its effect is to depolarize and, once the threshold is exceeded at the nodes, to release an action potential. The same process then occurs at the next node, and so on, until the action potentials have arrived at the axon ending or presynapse.

The amplitude of the action potentials is large enough to ensure with a factor of two or even three that another action potential is released at the node next in line. To return to the analogy of the flash bulb, let us imagine a whole series of flashes: The first one is ignited by a push of the button, then this flash ignites the second one by means of a photosensitive cell, and so on, until the last one is released. If we release the first light (AP at the axon hillock) by activating the button (above-threshold PSP), the remaining lights will flash in rapid succession (action potentials of the nodular chain). The lights will flash in the precise pattern or sequence that the first light was activated. The temporal sequence of the button activations corresponds exactly to the temporal sequence of the flashes. This will continue as long as we press the button; but it must not be activated in intervals shorter than those required for recharging the lights. If we press the button during that interval, the electrical contact will fail to release a flash. The same will occur at the axon membrane, or, more precisely, at the chain of the nodes of Ranvier. If the signals follow one another in a sequence faster than the time needed by the membrane to return to its resting state after a discharge of action potential, the axon will remain inactive. This interval lasts about one millisecond, meaning that 1000 action potentials may pass over the axon in one second. Most axons, however, will handle a maximum frequency of 600 to 800 action potentials per second.

This kind of signal transmission is referred to as a saltatory conduction of excitation, because the excitation (or action potential) seems to "jump" from node to node. In this way we can measure a conduction speed of 80 m/sec at the myelinated axons. For example, an excitation arising in the pressure receptor of a fingertip has been known to reach the spinal cord within a short period of 20 msec. By contrast, the action potential moving along a non-myelinated axon virtually crawls along the entire axon membrane. This latter type of excitatory conduction is pretty well restricted to transmission processes requiring only very short axons.

In summary: Every nerve cell displays a resting potential at its entire membrane that arises both from the different ion concentrations between the inside of the cell and the outside medium and from the selective ion

Elementary functions of the nerve cells

permeability of the membrane itself. The resting potential is maintained by means of an ion pump that uses up energy, and the nerve cell is always at a functioning capacity. A change in the ion permeability of the nerve membrane allows for brief changes in potential, which serve as electrical signals for the information code. When the neuron is depolarized above its threshold due to some prior event, a sequence of action potentials is set in motion at the axon hillock. The succession frequency of these action potentials is a measure of the amplitude of depolarization. The information is contained in part in the AP-frequency, and represents a frequency-modulated information code which is then used to conduct the information to the next neuron without distortion. A transmitter substance is released at the synapse of both neurons to recode the AP-frequency into an amplitude-modulated signal: The amplitude of the postsynaptic potential is a measure of the AP-frequency of the preceding neuron. This amplitude-modulated signal of the postsynapse is particularly suited for processing several bits of information into one new information signal. This capacity is based on the range of possibilities given by both spatial and temporal summation, and on the fact that incoming information signals vary in intensity because of the distance between synapses and their axon hillocks. The combined postsynaptic potential is again transformed at the axon hillock into the message code used for transport, the AP-frequency. The amplitude of the combined PSP is retranslated into a frequency-modulated signal sequence that travels to the synapse where the next neuron begins.

Information is constantly recoded on its way through a nervous system. It becomes a frequency-modulated code for transport (action potentials) and an amplitude-modulated code for information processing (PSP). While this steady process of recoding leads to some time loss, it does allow information to flow with less distortion and makes the nervous system less likely to be interfered with.

If the chain of neurons were to contain an inhibitory one, the AP-sequence would trigger a corresponding inhibitory postsynaptic potential. Not only would this IPSP then fail to release any more action potentials at the axon hillock, but it would also impede the creation of whatever action potentials normally result from other excitatory signals arriving at the synapse. Thus, the activation of an inhibitory neuron suppresses the excitatory potential of those following in the chain. As we have mentioned, these and other kinds of inhibitory processes play a quantitatively greater role in the nervous system than do excitatory ones.

Examples of neural connections

We can activate prey-catching responses in frogs and toads by simulating the released stimulus: Against a light background, we move a dark-colored disc with an optic angle diameter of 0.5–16°—the area of a visual stimulus is measured in relation to the optic angle of the stimulated eye—across the visual field of a hungry frog. The subject responds by turning toward the dark spot, then jumping and snapping at the moving object which he registers as prey. If the disc is larger than 20°, however, the frog turns away and tries to escape into a dark area. He does not respond at all if these discs are presented in the visual field but remain stationary. It is obvious, then, that two stimulus characteristics determine the release of prey-catching behavior: the dark object must be moving and it must be smaller than 20°. Moving objects larger than this are seen as potential predators (birds), based on experience genetically fixed through the generations.

This classification of the stimulus situation, "dark moving object on light background, smaller than 20°," must occur somewhere in the chain of events starting with reception of information in the optic sensory cells of the retina and ending with the command impulse for "prey catching." If we could record these electric signals—such as the action potentials—at a convenient locus in the nervous system under precisely defined stimulus conditions, we should expect to gain some measurable, quantifiable knowledge about this kind of information processing. We can totally immobilize frogs with curare or similar poisons; frogs do not die from this because they can absorb enough oxygen through their moist skin, without lung breathing. In an experiment we place such a paralyzed frog into the center of an evenly lighted hemisphere, then move dark objects across the hemisphere. The objects are varied in size and in speed. All the nervous processes required for the frog's prey-catching behavior occur in its eye and brain, yet it cannot perform any prey-catching or flight movements. The curare poisoning has completely immobilized the frog, even though command impulses for prey-catching behavior are transmitted across the motor nerves. Under these conditions, the experimenter may probe into the frog's brain with a fine metal-filled glass capillary, an electrode with which he can record the electrical signals or action potentials of a single nerve cell for a period of up to several hours. The signals are then made visible on the screen of an oscillograph. A frog's retina contains approximately 1 million receptors and 4 million nerve cells that effect a kind of preliminary processing of electrical signals passed on by the receptors. From the innermost layer of neurons, approximately 500,000 axons bundled together in the optic nerve extend into the visual center of the frog's brain, the roof of the mesencephalon. If we insert the electrode into the frog's mesencephalon and place it directly onto an axon—it is largely coincidence which one of the 500,000 fibers we hit—we can register the electrical activity of these fibers.

Our first discovery is that the retinal nerve cell of the axon we probed responds to light stimuli only from a very narrowly defined area of the frog's visual field (the receptive field). These receptive fields have a diameter of 2–15°, expressed in degrees of the visual field. This means that approximately 40–12000 optic cells are connected to one neuron. Many neurons discharge an AP-sequence only when a dark spot moves across the receptive field; if the spot is stationary, the neuron immediately stops firing.

Receptive field

Many receptive fields are surrounded by a circular inhibitory field. The neuron remains inactive if the moving object is large enough to cover not only the receptive field but also the surrounding inhibitory field. Thus, the neuron signals only visual objects as large as or smaller than the excitatory receptive field. This neuron transmits action-potential sequences only when the stimulus carries two essential characteristics of a prey object. The object must be moving and it must not exceed a certain size. If these properties are met, the sequence of action potential recorded from the neuron should then release prey-catching behavior in the intact frog.

If the AP-sequence discharged by the neuron being measured actually does represent the message "small dark object moves at the respective point in the visual field," we should be able to observe the corresponding prey-catching response even when this AP-sequence is artificially created by means of a stimulus electrode. In Darmstadt, J. P. Ewert carried out such experiments with toads, and he has actually observed that these organisms turn toward the part of the visual field corresponding to the receptive field of the artificially stimulated neuron cluster, as if the toads had sighted a small moving prey object (see Chapter 10). This offers conclusive evidence that the action-potential sequence recorded by the experimenter represents nothing more than the information signal "small dark object moves within a specific area of the visual field."

At the same time, Ewert's experiment illustrates a dilemma for the neurophysiologist. We have precise knowledge of the stimulus situation, the input entering the network where information is processed. By making electrode recordings at the corresponding point in the brain, we also know the product of this information processing, the output in the form of an action-potential sequence. Yet the most interesting phenomenon, the way input stimuli are processed to become output message signals, is still a great mystery. Neurophysiologists have turned this dilemma into a very useful experimental advantage. We do know, after all, the relationship between stimulus characters and output response, which can be determined through such experiments, for example, in what way the neural response is determined by the speed of a dark disc. Furthermore, we know the basic properties of a neuron. As a result, we have been able to develop a theoretical model of a network of nervous connections. For example, we may use a very simple model, well substantiated in numerous experiments, to explain how the visual system of the frog can distinguish the size of a visually detected object (Fig. 3-9).

A group of optic cells connects with several intermediate neurons and then converges upon a single nerve. Optic cells lying in the middle of the visual field are connected with this neuron by way of other neurons with excitatory synapses; they constitute the inner excitatory receptive field. The outer optic cells are connected to the same neuron by means of other neurons with inhibitory synapses. If the optic cells lying in the middle

Stimulation experiments on toads

Models of a network of neural connections

are stimulated, that is, if the stimulus dispersion is small, then only the excitatory synapses are activated and a combined postsynaptic potential results, releasing a response at a single neuron. But if the outer optic cells are stimulated at the same time, that is, if the stimulus dispersion is large, then the inhibitory synapses are activated and the IPSP reduces the amplitude of the combined PSP to such an extent that the threshold can no longer be reached: The neuron becomes inactive. Theoretically, we may distinguish a considerable variety of size stimuli, simply by connecting a corresponding number of optic cells.

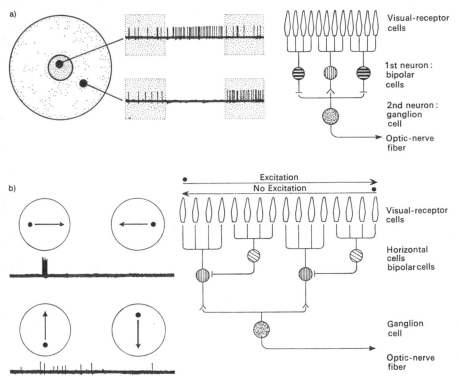

Fig. 3-9. Models of connections in the vertebrate eye. a) To the left, a receptive field of a ganglion cell. If we focus a light onto the center of the field, the associated ganglion cell becomes excited; if the light is focused onto the surrounding field, the cell is inhibited. To the right, the corresponding circuit model of the retina. b) Receptive field of a ganglion cell sensitive to movement. This cell responds to a horizontal movement of the point of light from left to right with an AP sequence, but it does not respond to a movement from right to left. The circuit diagram to the right indicates how connections between different neurons can lead to such a specific response to a stimulus. $<$, excitatory synapse. $\vdash$, inhibitory synapse.

Another model may illustrate how, by means of a relatively simple relay connection, a system of nerves can even classify the direction of a moving object (Fig. 3-9). The optic cells of the retina are connected with a nerve cell whose axon, together with the optic nerve, extends into the brain. This occurs through two different kinds of nerve cells: The bipolar cells, which run vertically in the direction of the brain, and the horizontal cells, which spread and diverge the plane of the retina. Each bipolar cell is connected with two groups of optic cells. One of these is connected directly to the ganglion cell by way of excitatory synapses, the other to the bipolar cell by way of inhibitory synapses in the horizontal plane. If a visual stimulus moves from left to right, the excitatory optic relay cells become stimulated first, and then inhibitory signals follow by way of adjacent optic cells. As a result, the ganglion cell responds to the object moving from left to right with a brief volley of action potentials. But if the visual stimulus moves across the field of optic cells from right to left, the optic cells releasing inhibitory impulses are stimulated first, and the excitatory ones must follow. This inhibition obliterates the excitatory signals, so that the ganglion fails to receive any excitatory input messages and remains inactive. Thus, this ganglion cell will only answer to visually detected objects moving in a certain direction; stimuli traveling over the retina in the opposite direction will not trigger a response. Scientists have discovered neurons that register a specific direction in the central nervous optic pathway of some vertebrates. Cells of this kind have not been found in frogs, however.

Models such as these have predictive qualities, and thus their validity may be tested experimentally. In accordance with experimental results, a model can be altered again and again, until it explains or makes plausible all the experimentally known data. In this way, it becomes highly probable that the nervous network we are studying will actually correspond to the model. Great progress has been made in electron microscopy and in the techniques for marking and dyeing single neurons, now we can even make anatomical examinations of nerve networks with relatively simple structure.

The models we have described are already sufficient to illustrate that information processing in the brain or, as in the example of the frog, in the nervous structure of the eye, is very much determined by connections in the neural network. This network in the central nervous system, in part genetically programed, in part resulting from environmental influences during individual growth, actually determines how we experience our environment. The properties of neurons and the way they are connected provide a biologically meaningful filter for selecting from all the externally impinging units of information only those that are important for the capacity of an animal to respond. In this way, for example, the nerve cells in a frog's eye mainly process parts of the environment that move; the frog probably cannot differentiate anything that is stationary, whereas moving objects immediately trigger excitations and lead to very quick responses.

Almost everything we know today about the function of a neuron and of a nervous system has been discovered through electrophysiological methods, a vital part of neurophysiology for the past twenty years. Explaining the various patterns of connections underlying present-day functional models is a fascinating problem for neuroanatomy and developmental physiology. Developmental physiology also studies how nerve fibers in the embryo and in the first stages of postnatal growth form their ultimate connections in different parts of the nervous system.

Direction-specific neurons

4 The Structure and Function of Sense Organs

The overwhelming majority of all organisms are able to orient themselves actively within their environment. Bacteria and unicellular animals gather in solutions wherever they find a suitable chemical milieu. Plant sprouts grow in the direction of light while roots grow downward. Animals seek places that offer them favorable living conditions in their quest for food, mates, and shelter. Organisms can perform these orienting feats because their cells respond to external stimuli in a systematic way.

Multicellular animals generally possess sensory cells, also called receptor cells, that are highly specialized to respond to certain kinds of stimuli. Often great numbers of receptor cells combine to form, in conjunction with other tissues, sense organs. Many sense organs are located in the head, but whole sensory systems are also distributed in a thin layer over the surface of the body. In addition to sense organs that receive external stimuli, there are others lying inside the body or in the body's outer covering whose function is to monitor the bodily conditions, such as temperature, degree of acidity (pH values), concentration of carbon dioxide, and so on.

Receptor cells are particularly sensitive to what we call their "adequate stimuli." The response of a receptor cell to a stimulus is called "excitation." Once a state of excitation has been triggered, it is transmitted to nerve cells and then conducted to the central nervous system (brain, spinal cord, ventral cord, head ganglia, etc.), where incoming messages are processed and compared with signals from other sense organs as well as with information stored in the brain (memory). The central nervous system may then transmit command signals to "effector organs" such as glands and muscles, causing whatever reactions are adequate in that particular stimulus situation.

Because we can be aware of what our own sense organs are doing, we are tempted to assume that animals have the same experiences. But the zoologist cannot study animals in this way. His statements and conclusions on the sense organs of animals may be based on anatomical findings, on

The structure and function of sense organs, by D. Burkhardt and I. de la Motte

Receptor cells

Excitation

description and measurement of the relationships between stimuli and organismic reactions, or on the recording of stimulus-dependent responses —that is, excitation—coming from receptor cells and nerve fibers. The results of such studies are then summarized in the form of simple drawings, tables, and graphs.

Our proverbial division of the senses into five—which really does not even apply to humans—does not work in zoology. In comparative physiology the sense organs—or, more precisely, the receptor cells—are classified according to their adequate stimuli. Here, a stimulus is defined either in physical or in chemical terms. With this criterion in mind, we may offer the following list of receptors and some of their rather diverse areas of functioning and responding:

Receptors and their functions

1. Light receptors: These respond to light and dark, contrast, form, movement, distance, color, and the direction of light polarization.

2. Mechanoreceptors: These may be very different from one another, responding to the following stimuli: contact with the surface of the body; flow of a medium around the body; sound traveling through air and water; vibration of the ground medium; position, tension, and length of body parts; blood pressure; degree to which hollow organs (e.g., blood vessels, stomach) are full or empty; position of the body in relation to gravity; bodily acceleration, rotation, or acceleration of rotation.

3. Chemoreceptors: These respond to odorous substances in water or air, and to substances which taste sweet, sour, salty, or bitter; carbon dioxide; humidity; osmotic pressure; and degree of acidity (pH).

4. Thermoreceptors: There are cold and heat receptors; some snakes also have special sensory nerve cells that respond to infrared radiation.

5. Electroreceptors: In some fishes, these measure the strength of electrical current on the surface of the body.

6. "Magnetic receptors"—whose existence is still under debate. Scientists have recently demonstrated that certain organisms can orient themselves within magnetic fields. However, we have not discovered any specific sense organs for this.

As a general rule, we can say that it is harder to discover sensory capacities in animals farther removed from man on the taxonomic scale. The greatest difficulty comes in recognizing sensory capacities that we do not possess. For example, it is fairly simple to find a dog's sense of smell, but discovering that insects taste with the "tips of their feet" was far more difficult. Recognizing that insects may also perceive the direction of vibration of polarized light, or that bats orient themselves by means of ultrasonic echolocation, was real pioneer work.

Methods of sensory physiology

The following are some of the methods commonly used in sensory physiology:

In humans, we can study sensations and perceptions directly. For example, we may ask a subject whether a certain light stimulus has the same color as a standard stimulus. In most experiments the subject will be required to establish a balance between two stimuli—such as matching up the intensity of a low tone with that of a higher standard tone presented at the same time. In this way we can obtain quantitative results.

When using animals as subjects, we must work from observations of the animals's responses.

a) Spontaneous choice behavior in certain stimulus conditions: For example, the animal is presented with models of flowers having different colors and light intensities or brightness (shades of gray). To which of these models does it move most often? Here we can test whether a flower-foraging insect can distinguish colors, and which color it prefers.

b) Conditioning experiments: Can a fish be trained to distinguish colors from shades of gray having the same brightness? Conditioning is achieved by rewarding or punishing the animal for the proper response.

c) Behavior in conflict situations: One example is the case of fishes turning their backs toward the sky's source of light, that is, toward the brightest part of the sky. Normally the back would lie at the "top" position, while the sense organ of equilibrium indicates the direction of gravity ("bottom" or below). Under normal conditions, then, a fish uses these bits of information to stabilize its position in the water. In an experiment, we can create abnormal stimulus conditions, using a centrifuge to vary the direction and force of gravity. With a lamp to act as an artificial sun, we can also vary the source of light. By observing the way a fish will compromise to make its adjustments, the experimenter can determine how these units of information coming from the respective sense organs are processed (see also Chapter 7).

d) Orientation toward the stimulus direction: Many organisms direct their movements according to these stimuli, for example, toward light or away from it. While in motion, they may also maintain direction at a certain angle in relation to the source of stimulus. Again, these orienting capacities enable us to observe and draw conclusions about how sense organs work.

The field of sensory physiology, then, began by asking about the kinds of perceptions found in the animal kingdom. In addition, we now study the mechanisms and sequences by which perceptions come to be. Modern electronic techniques have enabled us to discover how receptor and nerve cells respond to stimuli, and how these stimuli are coded. Electronic methods work because when cells are excited they undergo measurable changes in electrical potential. These fluctuations are extremely small (at best one-tenth of a volt, at the least one-millionth of a volt), but they can be amplified and thus become easy to measure and record. While investigating receptor and nerve cells, we can isolate the individual cells through specimen preparation, by direct stimulation, or by placing an electrode so that the changes in potential are finely recorded.

Instead of measuring only the activity of a single cell, we may also record a great number of cells simultaneously to obtain their gross activity

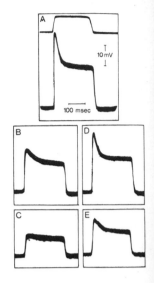

Fig. 4-1. Sometimes receptor cells will respond to changes in the direction of a stimulus in the same way as to changes in stimulus intensity. A to C: Decreasing intensities lead to smaller receptor potentials; D,E: A change in the direction of incoming light also leads to smaller receptor potentials. Results were obtained from a fly's eye.

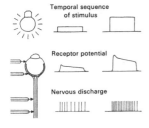

Fig. 4-2. Schematic representation of how stimulus intensity is coded in the sense organs. Stimulus intensity and temporal sequence (above) are reflected in the sequence and amplitude of the receptor potential in the receptor cells. Within subsequent nerve cells, temporal sequence and stimulus intensity are indicated by the frequency and number of impulses.

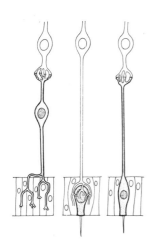

Fig. 4-3. Schematic representation of how receptor cells (dotted) transmit excitation to subsequent nerve cells (blank).

or combined response. While such measurements are technically easier to make, they are more difficult to interpret.

In the future, the methods of biochemistry and molecular biology may help us to discover what basic mechanisms underlie these sensory processes. Scientists working in sensory physiology have always needed to know some basics of biology as well as physics and chemistry to understand sensory processes. Furthermore, biological discoveries are based not only on an understanding of molecular biology and biophysics, but also on a knowledge of organismic structure (morphology and anatomy), of the behavior of organisms (general physiology and ethology), and of developmental history, which includes embryology, taxonomy, and phylogenetic development (evolution).

In comparing the structure of sense organs, we find a number of similar features despite their diverse specializations. One of these common properties helps us to define what, in fact, a sense organ is: An aggregation of receptor cells, of "accessory cells" (stimulus-conducting system) belonging to these receptors, and of nerve cells or sensory neurons. Receptor cells, in turn, have certain features in common. On one side of the cell body, away from the direction in which stimuli are conducted to the central nervous system, we always find cell processes with some peculiar properties. As far as we know, the conversion or "transduction" of stimuli into excitation occurs in the peripheral regions of receptor cells.

We may regard a sensory or receptor cell as an instrument for measuring some physical condition. In many measuring devices, a value may be determined indirectly by way of its effect on some electrical features or characteristics: Photoelectric effects change the resistance of a conductor or induce a photopotential; thermoelectric effects change the potential of a thermoelement or the resistance of a conductor. In other words, it is the change in electric current or in potential that is being measured. Receptor cells perform similar functions: By means of "primary processes"—still poorly understood—an adequate stimulus induces a change in the electrochemical nature of the cell membrane region. Starting here, in these regions, a potential change then spreads through the cell itself. This change in voltage is also called a generator potential. Excitation arising in the periphery of a cell may either be transduced into the typical nervous excitation as soon as it reaches the adjoining cell region or it may be conducted further on to another nerve cell (synapse, see Chapter 3). In both cases, however, the end result is a uniform type of signal that runs along the nerve fibers connecting sense organs with the central nervous system. These signals are the nerve impulses. A single nerve impulse consists of a short-term fluctuation in electrical potential, measuring roughly one-thousandths of a second in duration and about one-tenth of a volt (Chapter 3). The characteristics of each stimulus are conveyed to the central nervous system by variations in the temporal sequence of nerve impulses.

Two questions arise at this point. First, if signals from all receptor cells as well as nerve cells are of the same kind, how does the central nervous system determine whether light hit the eye or sound hit the ear? One hundred and fifty years ago, the physiologist J. von Müller (1801–1858) provided an answer: the "specificity of the nerve tract." Nerves coming from the sense organs terminate, or synapse, in certain areas of the brain. When the message arrives in such an area, it is classified as a specific kind of stimulus, such as light or sound.

"Specificity of the nerve tract"

Secondly, how are inadequate or inappropriate stimuli evaluated if they activate a receptor cell? Examples would be strong pressure stimuli or electric currents. In such cases, the central nervous system may in fact make the wrong evaluation. Getting punched in the eye can make us "see stars" when a sensation of light is created because the signals—even though released by an inappropriate, non-visual stimulus—arrive by way of the optic-tract fibers. In this case, however, there is additional sensory information to help us interpret what is going on: The eyeball's pain and pressure receptors will also respond to the stimulus, something that would not occur with light stimulation alone. Thus we can determine the real cause of this particular sensory experience.

Structural features common to most sense organs include not only tissues forming the stimulus-conducting system in front of the receptor cells; there are also additional nerve cells that process and conduct receptor-cell excitation. These are called the afferent pathways or afferent centers.

People often underestimate the importance of the stimulus-conducting system, which carries stimuli to the receptor cells. The stimulus-conducting system determines which kinds of stimuli should be conducted on to the receptor cells—in other words, which stimuli will be transformed or transduced into the ones utilized by the receptors. For example, vertebrates have three groups of similarly structured receptor cells within their ear labyrinth. The adequate stimulus for all three is a bending or shearing of the hairlike parts of these cells. And yet the receptor cells in the semicircular canal respond to rotational acceleration, those in the static labyrinth respond to the position of the head in relation to gravity, and the receptors in the inner ear respond to sound. These functional dissimilarities are based solely on differences in the structure of the stimulus-conducting systems. Within each system, the various external stimuli are transformed into the same—and for all receptor-cell groups equally effective—utilized stimulus. The overall function of the sense organ as a whole is made possible only through the joint activity of the stimulus-conducting system and the receptor cells.

The stimulus-conducting system

There is one other feature common to many sense organs: They are usually controlled by signals coming from the central nervous system. This means that the message transmitted to the central nervous system depends not only on the actual stimulus but also on whatever command signals are coming the other way, from the CNS. This "efferent control"

Efferent control

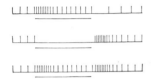

Fig. 4-4. A light stimulus (horizontal line underneath each graph) causes an increase in impulse frequency within certain nerve cells of the eye by its onset (above), in some others by its offset (center), and in yet others by both onset and offset (below).

(efferent meaning that the signal travels from the CNS to the periphery) may take effect within the stimulus-conducting system, the receptor cells, or the afferent (upward-conducting) pathway. Three examples will illustrate these processes:

1. The visual receptor cells in the eye signal exposure to light by means of specific excitation, which is conducted to the brain. In birds and mammals this information leads the central nervous system to send command signals to the muscles of the pupil, by way of certain relay points. As a result, the amount of incoming light is regulated by increasing or decreasing the opening of the pupil. This is similar to the way that a photographer regulates exposure by varying the size of the camera's lens aperture.

2. River crayfish, like other higher-order crabs, have sense organs that measure the deflection of their articulate posterior body end, their "tail." The receptor cells of these organs are connected to nerve fibers that may inhibit receptor excitation. When the tail is deflected, a message is sent to the central nervous system resulting in the activation of inhibitory nerves. This in turn leads to an attenuation of receptor-cell excitation.

When strong stimuli are repeated at a certain rate, they eventually cease to elicit a response—provided that they are not important to the organism. We are startled by a loud noise and will try to find out what caused it. If our neighbor plays the piano every day at the same time, we soon stop noticing it. Yet a closet creaking at this very time would make us jump, indicating that our ears have certainly not lost their sensitivity, nor have our muscles become fatigued. The respective afferent pathway within the central nervous system must have undergone adaptation—that is, it has gotten used to the regular stimulus.

It is as important for an organism to detect the onset of a stimulus as to recognize its cessation. For example, it is just as important to notice when our surroundings are becoming brighter as when they are getting dark. For that reason, many sense organs contain receptor cells that work in an opposite manner. In other words, some receptor cells respond to change in one direction or degree of intensity while at the same time other cells are sensitive to changes in the opposite direction. Should there be no opposing receptor cells, we find in most cases that subsequent nerve cells divide to take over this discriminatory function. Mammalian receptor cells responding to changes in temperature, for example, display one type that signals "getting warmer" and another indicating "getting colder." The signaling is done by increasing the impulse frequency. In the mammalian eye structures, some nerve cells respond with higher impulse frequency when the visual-receptor cells indicate "more light," while other nerve cells respond in the same way to messages from the visual receptors indicating "less light."

Another commonly found functional principle is the merging of impulses from many spatially dispersed receptor cells within a field structure

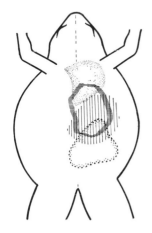

Fig. 4-5. A study of individual nerve fibers in a frog responding to tactile stimuli has shown that these fibers have large receptive fields. The fields overlap each other.

A nerve cell gathers various messages from spatially dispersed receptor cells or from sensory synaptic points. Every stimulus in a given sensory field results in a message transmitted by this nerve cell. The fields of various nerve cells overlap one another, like matting roots from several neighboring plants. The receptive field of a single nerve cell may therefore be very large. When stimuli become displaced within such a receptive field, they do not necessarily trigger a change in the excitatory state of that nerve cell. Yet these spatial stimulus shifts can ultimately lead to the excitation of other nerve cells receiving impulses from neighboring fields. The process of stimulus localization does not arise from a one-to-one stimulus/impulse coordination, but by means of coding processes that transform impulses into a complex pattern of activated nerve cells.

As a final example, we will look briefly at how messages are processed in complex sense organs. Neighboring receptor cells and their subsequent nerve cells may be connected so that the excitation of some will cause inhibition in others, and vice versa (lateral inhibition). One important result of this phenomenon is that contrast is enhanced at the edges. For example, we can take a situation where only the flatly arranged visual receptors on the right half of an eye are strongly illuminated, while those of the left half are exposed to dull light. The receptor cells on the right half will then inhibit each other and respond more weakly to the light than if they were exposed individually. On the left side there is little mutual inhibition, since the illumination is weak.

It is interesting to see what happens at the border between the strongly and weakly illuminated receptors. Those exposed to strong light exert a strong inhibitory influence on the others, whose excitation will thus be even weaker. Conversely, those exposed to dull light will hardly influence their strongly illuminated neighbors. This means that the cells lying in the border area, unlike their counterparts located farther right, do not have neighbors on both sides inhibiting them, so that their excitation is greater. As a result of this lateral inhibition, the contrast of the stimulus pattern is enhanced. To the right of the border line, the already bright area appears even brighter, while to the left the black border area is signaled as being darker.

However, this accentuation of boundaries occurs not only with visual stimulus patterns. In the ear, for example, sound frequency or pitch is represented as a spatial pattern of excitation on the receptor cells of the basilary membrane, that is, on the lower membrane of the organ of Corti (see Chapter 8), where lateral inhibition leads to an increase of contrast between frequencies, enabling the organism to discriminate pitch better.

The greater the sensitivity of a sense organ, the better it is able to signal minute changes in the environment to the central nervous system, hence ultimately effecting responses that contribute to the survival of an individual and, indirectly, of a species. Of course, there are limits to how sensitive a sense organ can be. If it is too sensitive, it will also respond to

Sharpening of contrast

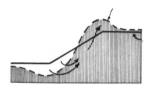

Fig. 4-6. Schematic representation of lateral inhibition. The continuous line represents the spatial distribution of stimulus intensity, while the hatched area indicates the intensity of excitation on the receptor-cell screen.

Thresholds

"noise interference," disturbances that are physically unavoidable and always present, causing unnecessary or useless responses. The human ear, for example, is sensitive enough to respond to a sound stimulus with an energy value of approximately 10–17 watts/second. We cannot imagine such a minute amount of energy. According to the laws of physics, our ear drums are moved back and forth by means of Brownian molecular movements with an energy power of approximately 10–18 watts/second. Thus the sonic information actually utilized is at a level not very much higher than the noise threshold. If our ear were even more sensitive than it is, Brownian movement of molecules would be the cause of continuous noise.

Similarly, the sensitivity of our eye comes very close to the physical limits. A single quantum of light is sufficient to activate a visual-receptor cell. The sensitivity of olfactory receptors, too, may be extremely high. One molecule of a sexual attractant secreted by the glands of the silkworm or silk moth (*Bombyx mori*; see Color plate, p. 96), providing it reaches the olfactory organ of a male, is enough to activate the receptor cells. The reaction of one or a few sensory cells will induce a male to look for a female.

The way sense organs function may be described by means of "characteristic curves." How is the degree of excitation—as indicated, for example, by the size of the generator potential or by the frequency of impulse succession in the nerve fibers—related to stimulus strength?

In many cases there is a logarithmic relationship between strength of stimulus and excitation. In other words, whenever the stimulus is multiplied by a certain factor, excitation increases arithmetically by a constant factor. For example, whenever light intensity increases by ten, the excitation it releases in the nerve fibers of an optic nerve increases by a certain amount. But these regularities apply only within certain limits. In other sense organs excitation is directly dependent on strength of stimulus: Their characteristic curves have a linear shape, meaning that changes in excitation are not many times the changes in stimulus strength; instead, both of these qualities increase in a direct relationship. Finally, there are sense organs where excitation reaches a particularly high or low level with a certain stimulus strength. These would send out warning signals either when normal levels are exceeded or when they are not reached.

The human eye is capable of processing stimuli with a strength of $1:10^{10}$ (a thousand times ten million). On the upper end of the scale, we can see in extremely bright sunlight with a luminosity level of 100,000 lux, while on the lower end we can still recognize objects "in the dark" by starlight illumination of 10^{-5} or 1/100,000 lux. Adaptations of this kind are based on a number of mechanisms: 1. The sensitivity of the receptor cell changes in response to the stimulus strength. 2. Receptor cells with different levels of sensitivity are activated by different degrees of stimulus strength. 3. The stimulus-conducting systems transmit stimuli

Characteristic curves

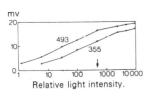

Relative light intensity.

Fig. 4-7. Example of "logarithmic" intensity curves. The excitation of one visual-receptor cell, measured in thousandths of a volt (mv), is shown as a function of the relative light intensity. Numbers beside the curves indicate the wave lengths, in nm, that were used in the experiment.

Adaptation

to the receptor cells either in full or reduced strength. Thus, for example, when the pupil contracts, the retina receives less light.

Common to all receptor cells is the fact that the physical and chemical characteristics of a stimulus are transduced or coded into a temporal succession of nerve impulses. For example, a large amount of light would correspond to a high impulse frequency or large number of impulses in a "visual pathway" leading from the visual-receptor organ to the central nervous system. Similarly, a particularly effective sound frequency would result in a high frequency of impulses being generated in the nerve cells of the auditory tract.

In each case there may be a number of differences in the way stimulus characteristics correspond to patterns of excitation, either in the sensitive receptor cells or in their subsequent nerve cells. Yet there is a process common to all coding or computation systems: the formation of invariants. The individual must recognize color—the subjective experience created by the wave-length composition of a light stimulus—quite apart from the light's intensity. Organisms must respond appropriately to visually perceived movement in their environment, independent of either the brightness or the wave-length consistency of light.

Stimulus changes in the environment—quite apart from changes in direction or intensity—are fundamentally important to the organism's survival. And so, in the course of evolution, we see the common basic principles emerging in the functioning of sense organs, in their capacities, and in the mechanisms whereby messages are coded in their subsequent nerve cells.

5   The Visual Sense

The visual sense, by
D. Burkhardt

Animals with a visual sense are able to perceive and respond to light stimuli by means of visual-sensory cells, or eyes. This does not include non-specific reactions of cells and tissues to light stimulation, as in the case of sunburn. A true visual sense always requires the presence of sensory cells that are specifically sensitive to light. Visual cells—more often called photoreceptors or receptor cells—respond to wave lengths that lie between the red area of the spectrum (750–800 nm; 1nm = one nanometer, equal to 10^{-9} m) and the ultraviolet area (extending below 300 nm). Sensitivity is often greatest at wave lengths between 500 and 550 nm (yellow-green); these approach the maximum wave length of sun rays. The electromagnetic oscillations falling into this range are called light, and their most important source is the sun. Direct sun radiation at the surface of the earth begins at wave lengths above 300 nm, reaches maximum intensity at around 500 nm, and descends far into the area of infrared (heat radiation). The diffuse light spreading across the sky begins at wave lengths of 300 nm, reaches a minor maximum intensity at 360 nm, then a major maximum intensity just below 500 nm and ends at 800 nm (Fig. 5-1). These figures indicate that, as a rule, the spectral sensitivity curves of visual receptor cells are well adapted to the intensity distribution of sunlight.

Other natural sources of light include the moon, stars, and luminous organs of various animals. The wave-length consistency of light emanating from special organs is usually well adapted to the spectral sensitivity of the species's own eye structures. Wherever possible, light coming from artificial sources is made to conform to the consistency of daylight. Every object disperses, reflects, and absorbs impinging light rays differently according to its physical and chemical structure. In this way, secondary light radiation will contain information about the object's position, shape, structure, and movement. Thus, sources of light rays enable animals to orient themselves in their environment. The amount of information to be obtained about the environment by means of eyes would seem to surpass that of any other medium, although there are animals which can-

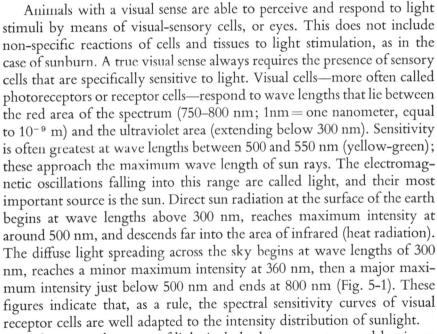

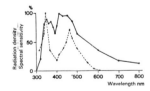

Fig. 5-1. Relative radiation density of blue sky light (extended curve), plotted over wave length. In comparison, the relative spectral sensitivity of the visual receptor cell of a blue-bottle fly is shown (lineated curve).

not perceive visual stimuli and therefore rely on other senses (smell, hearing, touch) for orientation.

If an animal has visual-receptor cells that lie isolated and without accessory organs in certain parts of the body, we term this a diffuse or extraocular visual sense. These receptor cells may lie in the skin, as with earthworms, or in the nervous system, as with lancet fish. Even animals with well-developed eyes may also have an extraocular visual sense. For example, the sixth abdominal ganglion of the decapoda is sensitive to light, and some lower vertebrates have visual-receptor cells that lie in the area of the roof of the diencephalon, forming the so-called parietal eye.

We would speak of an eye structure when receptor cells are combined with accessory mechanisms, including:

1. Cells and tissues that refract light intensively, forming lenses or light conductors and thus increasing light intensity within the eye;

2. Cells and tissues that contain light-absorbing pigments and that shield the receptor cells from light impinging from undesired directions;

3. Cells and tissues that reflect light so as to utilize it more efficiently. Such a layer of reflector cells is called the *tapetum lucidum*, providing the basis for the nocturnal glow emanating from the eyes of certain animals, including cats.

Visual-receptor cells can respond to light only to the extent to which the rays are absorbed. Thus, all of these cells contain light-absorbing substances that undergo physical and chemical changes with the absorption of quantums. The products of this photochemical reaction alter the state of the receptor cell and ultimately trigger the excitation of the optic nerves. Experiments have consistently shown that the light-sensitive substances—the visual pigments—are combinations of retinal (vitamin A-aldehyde) with proteins. Variations in the properties of visual pigments arise because two different forms of the retinal—vitamin A_1-aldehyde and vitamin A_2-aldehyde—can become bound to different proteins.

Only one quantum of light is needed to trigger change in a visual-pigment molecule. The energy absorbed with this light quantum has a value of 10^{-19} watt-seconds (wsec). For the human eye to experience a minimal, just-noticeable sensation of light requires incoming light energy values in the order of 10^{-17} to 10^{-18} wsec, hardly more than 10 light quantums. Since part of this energy is lost due to reflection and dispersion in the eye, we may assume that a single quantum has the power to release excitation within an entire receptor cell, and that the excitation of one receptor cell is enough to produce a response. In the course of a single nerve impulse—the signal of a nerve cell—energy values of 10^{-13} to 10^{-14} wsec are transformed. In this way, receptor cells function not only as highly sensitive detectors, but also as amplifiers: The energy they release is a thousand times greater than that of the triggering agent.

In highly developed eyes, such as those of vertebrates and arthropods, signals transmitted by the receptor cells are filtered either in the eye

Visual-receptor cells

Fig. 5-2.

a. Limpet

b. Ear pond snail

c. Pearly nautilus

structure itself or in the directly adjoining nerve tissue. A large part of these visual-sensory operations may be comprehended only if we go beyond the function of receptor cells and examine other nerve cells taking part in the process of visual perception.

An extraocular visual sense provides the organism with information about the degree of brightness in its surroundings, including changes over time. Among other things, these messages are important for synchronizing activity cycles with the daily alternation of light and dark or with the seasonal changes in the amount of daylight. Rapid changes in brightness may trigger defensive reflexes. Many organisms living on beaches (for example, mussels) close their shell if their source of light is suddenly blocked (shadow reflex). Where receptor cells associated with an extraocular visual sense lie at different points on the body, the animal can perceive or feel out spatial changes in illumination. Tube dwellers, which prefer dark surroundings—for example, earthworms—will immediately retract any protruding tip exposed to light. Many insect larvae detect light or dark areas by swaying their light-sensitive anterior tip back and forth.

Even eyes with a very simple construction, such as those of turbellarians, enable the animal to perceive the direction of a source of light. A small pigment cup shields the few receptor cells from light coming from certain directions. To localize the source of this light, the animal turns its body to the direction of greatest impact.

When a large number of visual-receptor cells are densely grouped, the organism begins to discern a spatial arrangement of light, and it may even recognize rough shapes. Where the layer of receptor cells is arranged like a cup or pit, different areas of the cell screen may be stimulated by light rays coming from different points in the environment. A box-camera eye is formed when the screen of receptor cells contains a bubblelike invagination that has a small hole to allow the passage of light. As a result, a true visual image is projected onto the back of the eye. This image, however, is faint, because the opening is very small. Various mollusks, including the pearly nautilus (Fig. 5-3), have this kind of eye structure. In the case of lens eyes, the opening becomes larger and is occluded by a lens-shaped tissue that causes the light to be highly refracted. The result is a stronger projection of images from the environment onto the screen of photoreceptors. In various classified animal groups we have found closely related species living side by side whose members possess pit-shaped eyes, box-camera eyes, or lens eyes; examples are ring worms, snails, cephalopods, and sea stars. With respect to visual-sense organs, we may therefore assume an independent evolutionary development within different classes of animals.

Highly sophisticated and structurally very similar lens eyes are found in vertebrates and, convergently, in some cephalopods, for example, the octopuses and cuttlefishes. Yet despite their structural similarities, visual

The function of eyes

Fig. 5-3.

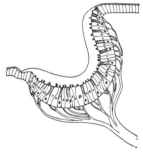

a. Limpet: cup eye

b. Ear pond snail: bubble eye

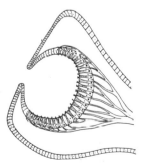

c. Pearly nautilus: box-camera eye

organs found among members of these animal classes evolved in a different way. In vertebrates, the retina has emerged as a bubblelike protrusion of the diencephalon, growing out from the brain rather than being formed peripherally. On the other hand, the retina of cephalopods is formed by a similarly shaped invagination of the epidermis. As a result, receptor cells of vertebrates are turned away from the incoming light (inverse retina), whereas in cephalopods they are directed toward the light (everse retina). Furthermore, the vertebrate retina itself has come to include nerve cells that already process information, but in the cephalopod eye, nerve fibers attached to the receptor cells extend away from this area, and visual data processing begins to take place only in subsequent optic ganglia.

Compound or facet eyes represent a special stage in the evolution of image perception. These are found in arthropods with the exception of spiders, and in various annelids and mussels. Visual perception in facet eyes does not occur by way of a bubblelike invagination of the receptor-cell screen, but rather by protrusion. Tiny lenses that increase the intensity of impinging light are located in front of small clusters of receptor cells, each of which in turn points in a certain direction (Fig. 5-4).

An unusual kind of eye structure has been found in the small sea-dwelling crab *Copelia*. A small cluster of receptor cells lies in the focal plane of a lens, and this cluster can be moved back and forth by means of muscles and suspensory ligaments. This enables the organism to investigate its surroundings, not spatially by means of a screen of photoreceptors, but temporally while using only a few cells.

A single receptor cell conveys only one kind of information to the nervous system: the intensity of an impinging light stimulus. Effectiveness of a stimulus depends not only on its energy value (intensity), but also on other factors such as the wave length. The spectral sensitivity of a cell is basically determined by the absorption curve of its visual pigment. Wave lengths that are well absorbed by the pigment are classified as brighter in stimulus value than those which have the same energy value but are poorly absorbed. When all the receptor cells in an eye have the same spectral sensitivity, the central nervous system cannot decode the wave length of an incoming stimulus message. But when closely adjoining receptor cells have different spectral sensitivities, the wave length consistency of an impinging light stimulus becomes significant. An increase in light intensity triggers a greater excitation in all adjoining receptor cells. On the other hand, a change in the wave length of a light stimulus with equal intensity effects a change in relative degree of excitation within the cluster of receptor cells. The ability to perceive color, then, requires the presence of receptor cells differing in their spectral sensitivity, as well as the operation of a central coding mechanism. Color perception is found among many vertebrates, insects, crabs, and possibly cuttlefishes.

Arthropods and cephalopods are able to record visually the direction with which oscillations of completely or partly polarized light impinge

▷
Visual perception in the compound and lens eye. Above: We will never know how an insect sees its environment. Sub-units of the compound eye divide space into areas subtended by an arc, which are sensitive to violet and white. The area of overlap of both eyes is depicted in red and dark violet. Depth perception is possible only in this area of overlap. Impinging light stimuli (white arrows) are codified by means of complex interconnections (blue arrows) in the central nervous system. The eye, shown as a section in blue, functions as mediator between the environment and the responses produced in the nervous system of insects.
Below: The familiar vertebrate eye projects images from the environment onto the retina by means of its lens. Common features of compound eyes and vertebrate lens eyes: Complex interconnections between the eye and the brain serve to assess and codify messages coming from the eye (section, green arrows).

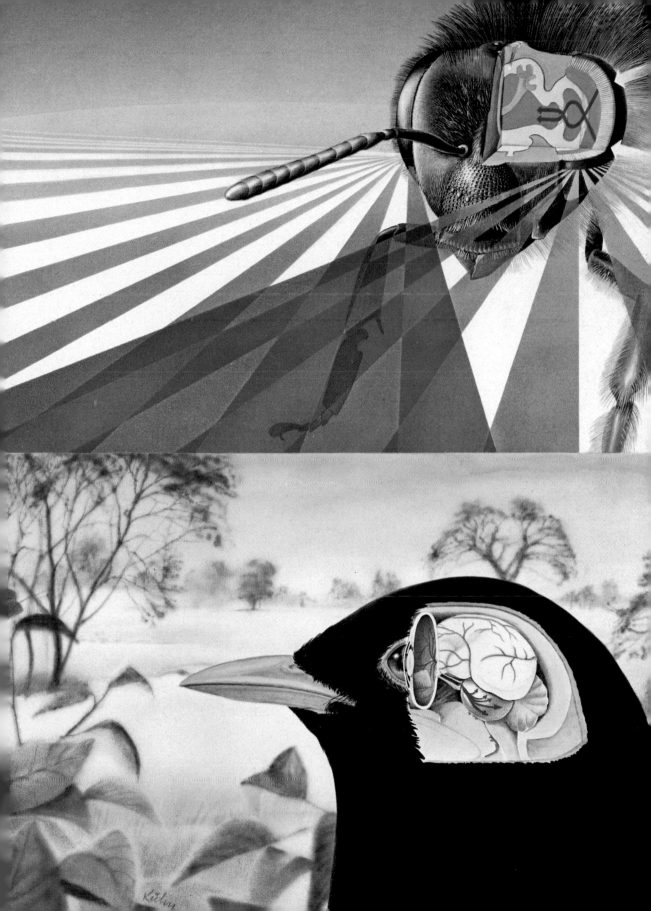

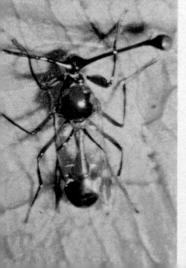

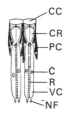

Fig. 5-4. Structure of a compound eye. Above, whole eye; below, section: CC, chintin cornea; C, core; CR, crystalline cone; NF, nerve fiber; PC, pigment cell; R, rhabdome; VC, visual-receptor cell.

Vertebrate eyes

Above: Owl with its nictitating membranes half closed.
Below: Insects have compound eyes which they can project laterally from the head. Left: A stalk-eyed fly; Center: Praying mantis; Right: Gadfly.

upon their eyes. This requires—in a way similar to the perception of color—that single receptor cells respond differently to light variations, in this case differences in oscillation. The light-sensitive parts of receptor cells found in arthropods and cephalopods are shaped like tiny rods (Fig. 5-5) which lie perpendicular to the direction of incoming light. Polarized light oscillating parallel to this finely detailed structure is absorbed far better than that oscillating perpendicular to it; in fact, the absorption is about twice as great. A cluster of receptor cells with different propensities to absorb oscillations coming from different directions will produce an excitatory pattern based on rotations of the plane of polarization.

Many animals perceive not only the shape and location of an object in space, but also its distance. With a pair of eyes, the two visual fields may overlap so that an object within this area is represented at different points in both eyes. Distance may then be calculated from the position of "corresponding" points on the retina. The eye structure of vertebrates has additional mechanisms for distance perception: It allows for the assessment of the position which the eyeball must take to fixate an object and the muscle tension needed for the lens to create a clear image.

Movement perception refers to the ability of animals to record movement in the environment according to the direction and speed of objects. This feat is based on a spatial and temporal displacement of visual patterns on the retina. These animals differentiate between shifts of retinal images which come about through movements of the eyes or the body itself and those created by movement in the environment. The central coding of image shifts also takes account of signal commands to muscles that adjust the direction of sight.

A cross section of the human eye (Fig. 5-6) illustrates the basic structure of vertebrate eyes. Light passes through the cornea, the anterior eye chamber, the crystalline lens, the posterior eye chamber, and the vitreous humor to the retina. Behind the retina lies the chorioid membrane, which absorbs light and has many blood vessels. The iris—ring-shaped in humans but cleft-shaped in cartilaginous fish and predatory animals—becomes the pupil as it encloses the vital eye opening (optic foramen). The strongly pigmented muscle cells of the pupil absorb light impinging outside of the opening. In the rest of the eye, this function of shielding the receptor cells from extraneous light is taken over by pigments of the chorioid membrane and the retina. The pupil's diameter may vary, depending on the amount of impinging light. In higher vertebrates, however, contraction is regulated by the receptor cells, by way of a center in the midbrain (mesencephalon) and of the ciliary ganglion. The human eye can modify the intensity of light through the pupil in a circumference of 1:16. The relative light intensity of vertebrate eyes, measured at a maximum opening of the pupil, lies between 1:0.74 (deep-sea fish) and 1:2.44 (finback); in man, this value reads about 1:2.25. As a comparison, we may note that in a good camera with an open shutter the maximum light intensity is 1:2.

Projection of a visual image takes place through the cornea, the lens, and the anterior eye chamber separating them. The over-all refraction of light within an entire lens is called diopter (dptr). The lens of the human eye with a total refraction of 58 dptr will yield 13 dptr, the cornea, 21 dptr. The cornea, the aqueous humor, and the vitreous humor have a refraction index very close to that of water (n = 1.336; the refraction index of air is approximately 1), whereas the lens has a refraction index in the middle range (n = 1.42). The lens is constructed of layers with different powers of refraction. In amphibious vertebrates the total refraction of light in the eye is based on that of the lens, which in this case yields a higher refraction index (up to n = 1.5); this lens has a more pronounced, almost spherical shape.

By means of accomodation (adaptation of the eye to varying distances), objects at various distances from the eye continue to project clear images on the retina. Humans, like other mammals and like birds and reptiles, accommodate by changing the shape of the lens. In its resting state the lens is flat, accommodated to "far." It becomes more rounded when focusing to create a sharp image of closer objects, through the contraction of the ciliary muscle in the vascular membrane of the eye. The lens is attached to suspensory ligaments that in its resting state will pull and flatten it. When the ciliary muscle begins to contract, the lens escapes the pull of the ligaments and becomes more curved. The power of refraction in a young human may vary by roughly 12 dptr, which is sufficient to focus on objects only 8 cm away. With increasing age, the range of accommodation will decrease, and the nearest point of clear vision moves farther away from the eye. In the case of fishes and amphibious animals, accommodation takes place not through changes in lens shape, but through shifts of the lens to varying distances from the retina. The resting eye of fishes is adjusted for "near," that of amphibious animals, to "far."

Acuity of vision in a vertebrate eye depends primarily on the density of the receptor-cell screen in the retina: The greater the density, the stronger the eye's "grain" or definition. In large eyes, the image on the retina is larger than in small eyes, and the definition is greater even if the density of the receptor-cell screen remains the same. With the definition at its best, the ostrich has 51,000, man has 160,000, and the buzzard has one million visual cells in 1 mm² of retina. Humans are able to distinguish two image points as close to one another as one degree of one-half a minute of arc (1/120°), but the buzzard has a visual acuity of one-quarter of a minute of arc, even though its eyes are smaller than those of man. In most vertebrates, the eye's power of resolution varies within the retina. It is generally very high in a central area of the retina (area centralis), but may also reach a high point in a more lateral area (for example, in the eye of a fish), or in both a central and lateral region (for example, in swallows). Many hoofed animals (ungulates) and birds even possess a bandlike area of high resolution, encompassing the horizon. Some vertebrates have optic

The dioptric apparatus

Fig. 5-5. Schematic diagram of rhabdomes in the eye of a fly. Right: A disc-shaped cross section of a single eye, showing the seven rhabdomes symmetrically arranged and pointing to a common center in the sensory-receptor cell (for emphasis, the rhabdomes are drawn larger, in proportion). Left: Section of a rhabdome. The tiny tubes, visible only under an electron microscope, are perpendicular to the direction of incoming light (see also Fig. 5-10).

Visual acuity

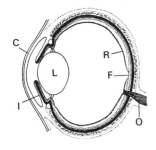

Fig. 5-6. Diagram of a vertebrate eye. C, cornea; I, iris (frontal edge of the optic cup); L, lens; R, retina; O, optic nerve; F, fovea (area of most acute vision).

The visual field

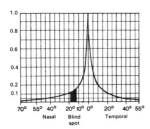

Fig. 5-7. Decline in visual acuity in the human retina, with measurements based on the fovea (at 0°) during light adaptation. Nasal: toward the nose; temporal: toward the temple.

Eye movements

pits (foveae, Fig. 5-7) in the middle of the area centralis, where the arrangement of receptor cells is particularly dense.

Aside from the degree of receptor-cell density, the resolution of the retina also depends on whether each receptor cell has its own connection to the central nervous system or whether large numbers of these cells jointly connect to a single successive nerve cell. In humans every receptor cell in the fovea has its own nerve fiber, but at the periphery of the retina more than 100 receptor cells will connect a single axon. Figure 5-7 shows the decline of visual acuity on the human retina. If a large number of receptor cells are jointly connected to one efferent nerve fiber, a whole area in the retina (the corresponding retinal field) may release excitation in a single nerve fiber. Presumably, the significance of this interconnection is based on two factors:

1. An increase in the eye's sensitivity to light: When illumination is low, the probability that a quantum of light will enter the visual field increases with the range of this field.

2. The photoreceptors and nerve cells forming part of a visual field may be connected so that only certain spatial and temporal patterns within the field will trigger excitation in the efferent nerve fiber.

The visual angle of the eye as a whole varies considerably among vertebrates. In humans, the horizontal range of the visual field of one eye lies at around 160°, in the barn-owl, only 95°, but in the frog, as much as 235°. The extent of the visual field depends on the shape of the pupil and on the focal distance of the refracting (dioptric) apparatus. The size of the retina being equal, eyes with a short focal distance have a larger angle of vision but poorer spatial resolution than those with a more extended focal distance (such as the "telescope eyes" of deep-sea fishes). Binocular vision, too, developed in very different ways among vertebrate species. On the whole, fishes, reptiles, and birds have a very small binocular overlap (between 10 and 60 degrees), whereas amphibious animals and mammals have a large binocular overlap (up to 160°).

In most vertebrates, the eye is attached to six (sometimes seven) eye muscles. Coordination of movement is directed from specific centers in the mesencephalon, by way of three pairs of cranial nerves. The field of vision may be altered through voluntary movements of the eye. In focusing, the eye muscles also adjust so that the fixated object is projected onto the point of most acute vision. Here, the axes of both eyes must converge, coupled with accommodation of the lens. Furthermore, pupil size decreases when relatively close objects are fixated (nearby objects are brighter than more distant ones which reflect the same amount of light.)

In addition to voluntary movements of the eye, two involuntary movements occur:

a) Pacing movements (nystagmus). The eyes move slowly in one direction, then jerk suddenly in the opposite direction. Nystagma are triggered by movements in the environment (optokinetic nystagmus, as in the perception of a moving train) and through the organ regulating the body's balance (labyrinthic nystagmus, as when the individual turns about on a rotating chair while keeping his eyes closed; these eye movements can be felt through the eyelid). Similar pacing movements of the eye may be observed in a person reading one line and then switching to the next.

b) Involuntary fluttering movements. More refined methods of measurement have enabled us to detect tiny ocular fluttering movements. Experiments have shown that, at least in humans, these eye movements are extremely important for maintaining contrast and acute vision. There are several components: a tremor with amplitudes of up to 0.5 minutes or arc and frequencies of up to 150 hz, irregular jerking movements lasting 0.03 to 5 sec with amplitudes of 50 minutes and arc ("flicks"), and slow drifts with speeds of about one second on arc per second which occur between flicks. These involuntary eye movements are very important for normal vision. If in an experiment we stabilize the image of the retina, this image will soon start to fade, and even color sensations begin to disappear. Instead of the normal value of 2:100, the sensation of contrast goes down to 3:1. Presumably, these involuntary eye movements serve to prevent excessive receptor cell adaptation.

Fishes generally lack eyelids, but all four-footed animals (quadrupeds) have at least two, with a rich supply of glands on the inner side to keep the eyeball moist and to remove foreign bodies. Reptiles, birds, and mammals have a third eyelid, the nictitating membrane (see Color plate, p. 80), which is drawn from the side of the nose to the outside, across the eye. This membrane, situated underneath the other eyelids, is transparent. All terrestrial vertebrates have tear (lacrimal) glands that secrete a thin and saline lacrimal fluid to cleanse the eyeball. Excess fluid is drained from the corner of the inner eyelid by way of the nasolacrimal duct to the nasal cavity.

The back of the eye consists of the retina and the adjoining choroid membrane. This membrane contains numerous blood vessels that supply the retina with oxygen. Birds have almost no retinal blood vessels, but in the lower part of their eyes we find a strongly folded structure (pecten) extending into the vitreous humor and containing an ample supply of blood vessels. We can only speculate about their exact function. Among other things, moving objects may be more easily recognized on the grid-shaped shadow which the pecten projects onto the retina.

The retina consists of supporting cells, nerve cells, and receptor cells. The receptor cells point away from the lens. Appositioned at the front of the receptor cells are the transparent nerve cells, through which light must pass on its way to the receptor cells. This "inverse" construction results from the fact that during its embryonal development the retina is formed as a protrusion of the diencephalon. The nerve cells form a bundle, leave the eyeball at a certain point through the retina, and become the optic

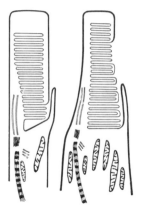

Fig. 5-8. Diagram of the visual receptor cells of a vertebrate eye, based on photographs made with an electron microscope. Right, a rod; left, a cone; only segments of the cell are shown. The visual pigments are a vital part of the receptor-cell membrane; they are folded to look like a roll of coins, and they occupy a large part of the receptor cell.

The retina

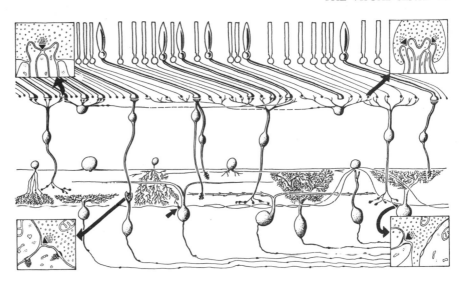

Fig. 5-9. Basic diagram of the way in which receptor cells in a mammalian retina may be "wired" to subsequent nerve cells. Upper layer of visual-receptor cells (photoreceptors): Rods (angular) and cones (lanceolate). Middle layer: "Bipolar" nerve cells; these may connect with several receptor cells (left and right) or with only one (left from the center). Bottom layer: The bipolar nerve cells are connected with the brain by way of other ganglion cells. In addition to these interconnections leading from receptor cells over bipolar and ganglion cells to the brain, there are other nerve cells (behind the receptor cells and in front of the ganglion cells) which provide lateral cross-connections. Inserts at the four corners illustrate the appearance of points of contact between cells, as seen under an electron microscope.

nerve. At this particular point, there are no receptor cells, and it is therefore termed the "blind spot" (optic disc).

In most vertebrates, there are two kinds of visual receptor cells in the retina: slender rods and slightly more flattened cones. When there are no cones, for example in deep-sea fishes, we speak of a rod or rhabdome retina; if the rods are missing, as in diurnal reptiles, we refer to a cone retina. In the human retina there are about sixteen times as many rods as there are cones. The number of photoreceptors in the human eye is about 100 million. Many findings support our assumption that the rods are particularly sensitive to light, enabling the organism to see in twilight (photopic vision), whereas the cones are relatively light-insensitive and thus function in daylight vision (scotopic vision). Since humans are color-blind during twilight vision, and since full color vision occurs only in the fovea which contains cones almost exclusively, we may speculate that the cones provide for the perception of color. There are at least three kinds of cones with different spectral sensitivities (absorption spectra). In fishes, we can observe that in twilight the rods will protrude from the pigment at the back of the eye, and during the day, the cones will protrude.

Both kinds of receptor cells are divided into three sections: outer segment, inner segment, and nerve process of fiber (Fig. 5-8). In the outer segment, folds in the cell membrane create numerous flat structures, which are piled one on top of another as in a roll of coins. The light-sensitive substances are nested within this part of the receptor cell. Visual pigments apparently constitute the main part of the membrane's protein components.

The excitatory processes triggered by incoming light are conducted across the inner segment and nerve fiber of the receptor cell to adjoining nerve cells, called bipolars. In the fovea, each bipolar receives the excitation of only one receptor cell, but at the periphery one bipolar receives excitations from numerous photoreceptor neurons. While still in the retina, excitatory impulses of the bipolar cells are transmitted to subsequent nerve cells. The axons of these ganglion cells then combine into the optic nerve (*nervus opticus*), which leads away from the eye. In lower vertebrates, the optic nerves of both eyes cross each other completely and synapse in the mesencephalon. The situation is quite different in mammals, where the crossing-over is only partial and where part of the receptor-cell fibers remain on the side of the brain which contains the respective eye. Here, the axons of the receptor cells synapse in the geniculate body of the diencephalon. Their excitation is transmitted to another nerve cell (neuron) and ultimately arrives at the cerebrum. A few nerve fibers extend from the optic nerve into the mesencephalon, where they terminate in regions controlling the extrinsic eye muscles, the iris muscles, and the ciliary muscles.

Additional nerve cells (horizontal cells and amacrines; Fig. 5-9) establish lateral cross-connections within the retina, leading us to assume that messages from the receptor-cell screen are processed within the retina. We have made interesting discoveries about these processes by way of electrophysiological investigations.

For instance, studies on the retina of the cat have shown that the excitation of the ganglion cell is always released from a large area in the retina. By exploring the retinal area with tiny spots of light, we discover that it has a diameter of several millimeters. One ganglion cell is connected to roughly 100 visual receptor cells. These receptor neurons are not tightly packed, but spread loosely across the entire receptor area of the ganglion cell. The receptor areas of neighboring ganglion cells overlap and intersperse with one another. When their field is exposed to light, the ganglion cells may respond with a signal sequence following the beginning of light stimulation, with a sequence at the end of the stimulation, or with a volley both at the beginning and at the end of the stimulus. In responding to stimulation at the center of their visual field, some ganglion cells will produce an *on*-discharge and some an *off*-discharge. In either case, the ganglion cell will respond with the opposite kind of discharge when the periphery of its visual field is exposed to light. Finally, the *on-off* signal is released within an intermediate zone.

The size of the visual field depends on the adaptation of the eye (see below), the illumination of the background, and the intensity of the light. The visual field increases with stimulus intensity as well as with adaptation to dark, but it decreases with a rise in background illumination. The response of ganglion cells in the frog's retina and of nerve cells in the mammalian geniculate body of the diencephalon is far more specific with regard to stimulus patterns in the retina. Some of these nerve cells produce excitations only when small objects of a certain shape enter the corresponding retinal field. Other nerve cells respond exclusively to a specific direction of movement, and still others signal only a decrease in brightness (dimming detectors), regardless of the absolute intensity of the light. The retina of goldfish—fishes have good color vision—contains ganglion cells that produce an *on*-response if the stimulus light has certain wave lengths, and a light-*off*-response if the wave lengths are those of the complementary colors. This coding mechanism probably plays an important role in the color vision of vertebrates.

The human eye can process light stimuli covering an exceptionally wide range of intensity. The energy values which lie between the threshold values at maximum dark adaptation on one extreme and the pain or blinding threshold on the other, lie in a relationship of $1:10^{10}$ (10^{-6} lux$:10^5$ lux).

By way of adaptation processes, the threshold for noticeable light stimuli varies by a factor of 100,000. There are a number of mechanisms involved:

1. The change in the size of the pupil increases the intensity of incoming light by sixteen.

2. The thresholds of rods, sensitive to weaker intensities, are a thousand times lower than those of cones, which are activated at higher intensities.

3. When light intensities are lower, the receptor areas of the retinal ganglion cells are larger than with stronger intensities. Thus, any impinging light quanta are more likely to release excitation in the ganglion cell.

4. When light intensity is high, part of the visual pigment in the receptor cells becomes ineffective. Due to decomposition (or breakdown), the visual pigment may not be immediately available for new quantum absorption.

Facet or compound eyes

Like the camera eyes of cephalopods and vertebrates, the compound eyes of crabs and insects are highly efficient. The earliest arthropods, trilobites which existed in the Paleozoic era (Cambrian to 600–200 million years ago), already possessed compound or facet eyes. The basic structure of such an eye is shown in Figure 5-4. The receptor cells are usually long and slender—in the blow fly, for example, they have a diameter of 7 micrometers (mm) and a length of 250 micrometers; they are found in groups, usually numbering eight. The axes of these clusters diverge by a few degrees. Several hundred (in fruit flies) to a maximum of 20,000 groups (in dragon flies) make up the retina. Each cluster of receptor cells lies wrapped in a layer of pigment cells and is shielded from light coming from other directions. There is a visual apparatus made up of a small lens (corneal lens) and an adjoining crystalline cone in front of every cluster.

The lens is formed by a secretion from a layer of cells (hypodermis) lying under the skin. The crystalline cells that internally secrete a powerful light-refracting substance (euconic eyes, as in the bee), or of an extracellular secretion from the crystalline cone cells that are directed toward the lens (pseudoconic eyes, as in the common fly). Sometimes the crystalline cells lie behind the corneal lens without additional specialized structures (aconic eyes, as in beetles). In many species, such as fireflies, a long, light-refracting fiber (*processus corneae*) runs from the crystalline cone to the receptor cells (in this case located further down). Of the pigment cells, the primary ones directly envelop the crystalline cone. These are joined by a second layer, the accessory pigment cells, which in most cases also covers the receptor cells. A third group, the retinal pigment cells, lies at the basis of the receptor cells. The retina is sealed by a basal membrane, which allows the passage of nerve processes from the receptor cells on their way to successive ganglia forming the visual centers. Individual clusters of receptor cells with their auxiliary structures are known as ommatidia.

Detailed studies have indicated that the focal distance of the dioptric apparatus (lens and crystalline cone) pointing toward the retina is just right for allowing parallel rays to combine (focus) onto the tip of the receptor cells. Thus, infinitely distant objects are clearly focused in this plane. On an absolute scale, the focal distances are quite narrow (in the blowfly, approximately 75 μm), and therefore the depth of focus in facet eyes is large even though there are no accommodation mechanisms. Since roughly speaking, objects which lie 100 times as far as the focal distance approach infinity, then the area of focus for facet eyes begins at a few millimeters distance from the animal. The organism's own limbs thus can extend into the area of focal depth.

Focal distance of facet eyes

Measured by the focal distance, the lenses have a relatively large opening (orientational value approximately 1:2). At least in some cases, the high refractive power of the dioptric apparatus results from the fact that the cornea has a lamellated structure with layers having different refractive powers. The highest conclusive refraction indices found run at $n = 1.55$ (water beetles). The crystalline cone may have an optic homogeneity (as in flies) or a cylindrical layering around the longitudinal axis (lens cylinders, found in fireflies). Such a lens cylinder acts as an accumulator with high refractive power.

Depending on the position of receptor and pigment cells, facet eyes are called either appositioned or superpositioned eyes. In appositioned eyes the receptor cells lie close to the dioptric apparatus and, through pigments, are optically totally isolated from their neighboring ommatidia. Light reaches the receptor cells only by means of the dioptric apparatus associated with each ommatidium. In superpositioned eyes the receptor cells are located farther in the eye, and in darkness the area between the tips of the crystalline cone and the receptor cells has no pigments at all. Here, light can also be refracted on its way to the receptor-cell cluster by the dioptric apparatus of neighboring ommatidia, resulting in a more efficient use of incoming light. During light adaptation, the pigments spread farther from the crystalline cone and the receptor cells, so that the ommatidia become optically isolated, as in the appositioned eye. Most crabs and, among insects, fireflies, click beetles, and moths, have superpositioned eyes.

Appositioned and superpositioned eyes

In addition to the alternating exposure of protective pigments to light and dark in the superpositioned eye, compound eyes may vary light intensity by means of other adaptive mechanisms. In meal beetles there is a pigment layer that constricts like a pupil when exposed to bright light. Flies have small bubbles (vesicles) that attach to the receptor cells (the rhabdomeres) during periods of weak light; the refraction index of these bubbles is smaller than that of the receptor cells. This facilitates the conduction of light in the rhabdome, because the combined angle of refraction is increased. The bubbles disappear during strong illumination, replaced by absorbing grains of pigment. Finally, some insect eyes (for example, those of backswimmers and the tsetse flies) in which receptor cells undergo shifts depending on the light conditions.

Eyes of moths

Some animals, including moths, are able to increase the intensity of light by means of a reflecting tapetum. This tapetum consists either of reflecting pigment cells or of air-filled tracheal bubbles at the base of the receptor cells. In various moths, reflection of light at the cornea is greatly reduced ("compensated vision"). By means of cone-shaped elevations with a diameter of about 200 mm, each of a similar height and distance from one another, the medium refraction index does not increase suddenly from air to lens, but instead changes gradually. However, since reflection occurs as a result of sudden change in the refraction index, these corneal surfaces in fact reflect very little light. We are still not sure about the biological significance of this mechanism, since the resulting use of light energy improves only slightly, by about five percent. Perhaps this process creates some kind of "camouflage," since the "compensation" reduces brightness on the surface of the eye.

Optic resolution

According to J. Müller's (1801–1859) theory of mosaic vision, the dioptric apparatus of each individual ommatidium produces an image point. Thus, an upright image of the environment is formed within the whole of the retina, made up of as many image points as there are ommatidia. The resolution of the facet eye, therefore, depends on the number of ommatidia within a particular area. There is a strong interrelationship between the physiological resolution and the angle of inclination of the axes of neighboring ommatidia: The narrower the angle of inclination, the better the resolution, which at its best lies at one degree.

Rhabdomeres

As in vertebrates, the receptor cells display a light-sensitive structure, called a rhabdomere, an offshoot from the actual cell body. This extends from the middle of the ommatidium along the length of the receptor cell. In most insects and in crabs the rhabdomeres of an ommatidium lie very close to one another and form a solid visual rod (rhabdome). The rhabdomeres have a higher refraction index than the surrounding receptor cells or the axial cavity of the ommatidium (if present). In this way, they act as a light conductor; at their peripheries, light which impinges from within a narrow field around the visual axis becomes completely reflected and thus is conducted along the entire rhabdomere. The upper end of the rhabdomere is the light-absorbing cross section in the image plane of the dioptric apparatus. Since a rhabdome's diameter is very small compared to its focal distance, the whole system works like a camera with a telephoto lens and a very small angle of projection. The angle of opening, whereby light rays pass into an ommatidium, measures only a few degrees; it is adapted to the anatomical angle of inclination of the ommatidia. Since in most arthropods—with the exception of flies and beetles—the rhabdomeres are fused into one visual rod, they act as a single light conductor. Thus we cannot schematize the detailed structures inside an ommatidium; where the spatial resolution is concerned, ommatidia work as a single physiological unit. Visual acuity of these facet eyes, one degree at its best, is approximately one hundred times weaker than that of vertebrates.

There are two areas, however, where ommatidia do not function as a unit. A receptor-cell cluster distinguishes between different oscillation directions and between different wave lengths. On the whole, then, every ommatidium transmits to the central nervous system the momentary brightness of a perceived image point as well as its direction of oscillation and its wave-length composition. Insects are a special case, since their individual rhabdomeres are separate. Here, the visual axes of the receptor cells of an ommatidium point in different directions, diverging by only a few degrees of arc. But neighboring ommatidia would in turn have receptor cells oriented in the same directions as those of the first ommatidium. In joining up with subsequent nerve cells, these receptor cells pointing in the same direction are transformed into a functional unit. Because of this linkage, the ultimate effect is the same as in insects with fused rhabdomeres.

The visual acuity of vertebrate eyes declines at the periphery. Similarly, in insects the ommatidia at the eye's lateral periphery have a larger angle of opening, so that the resolution is reduced. But in the case of insects, the change in acuity within the visual field of an entire eye is never as great as in the eyes of vertebrates. Thus there is a more equal comparison between these two kinds of eye structures when considering performance at the periphery. The whole visual field of a compound eye can be very large. In the common fly, one eye covers more than half of the insect's surrounding space, so that it may simultaneously view all of its surroundings except for a tiny space at the back, which is blocked by its own body. The visual fields of both eyes overlap and form a binocular field above, beneath, and in front of the head. This binocular visual field may extend over 90° or more in the praying mantis, dragonfly, and other insects with a good visual capacity (see Color plate, p. 79).

Visual field of compound eyes

Many insects are able to compensate for a poor spatial resolution (at least in part) with a very high temporal resolution. Bees, flies, and dragonflies may perceive separately up to more than 200 light flashes per seconds (humans perceive only 20–40). Thus, while in the vertebrate eye the flow of spatial information is high, it seems that at least some insects can cope with a rapid flow of information. The net flow of information transmitted

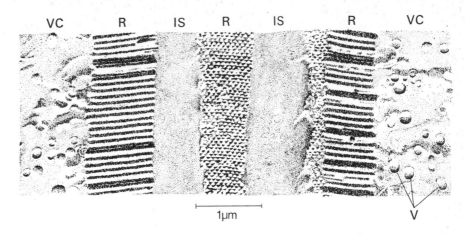

Fig. 5-10. Electron-microscopic photograph of rhabdomes in the eye of the blue-bottle fly. A special method of preparation (frozen cauterized section) emphasizes the details of spatial structure. VC, visual-receptor cell; R, rhabdomere; IS, intercellular space; V, vesicles (bubbles inside the cell).

by the eye to the central nervous system, then, is only a little less than that of vertebrate eyes.

The rhabdomeres of the receptor cells contain the visual pigments. As in vertebrate eyes, they are a combination of the retinal with proteins. The electron microscope shows that rhabdomeres have a fine, regular structure consisting of tube-shaped protrusions of the receptor-cell membrane (Fig. 5-10). These tubules have a diameter of 50 nonameters and are positioned perpendicular to the axis of the receptor cell. The tubules of a receptor cell, numbering roughly 100 thousand, are about 1 μm long, and they lie stacked in parallel formation close on top of one another. As a result, the rhabdomeres show a striped formation when sectioned perpendicularly to the visual axis. It is this special orientation of ultrastructures which enables arthropods to perceive the oscillation direction of polarized light. A receptor cell absorbs polarized light most effectively when that light vibrates parallel to the orientation of the rhabdomeres. Light with a vibration direction perpendicular to the rhabdomere orientation is absorbed only half as much. Therefore, if light intensity remains constant, excitation of a visual-receptor cell depends on the oscillation direction of polarized light.

Since the individual receptor cells in an ommatidium are oriented differently, the distribution of excitation within the receptor-cell cluster of an ommatidium will depend on the oscillation direction of light. These variations enable the nervous system to assess the location of light polarization. Under natural conditions, polarized light results from the scattering of light across a blue sky and from reflection on the surface of water. Since the direction of polarization of the light of the blue sky depends on the position of the sun, the perception of polarization enables animals to orient themselves toward the sun by means of a blue sky, as with a compass, even when the sun is not visible.

In all known cases, the spectral sensitivity of insects goes far into the ultraviolet range, up to approximately 300 nm. Spectral sensitivity how-

Polarized vision

ever, is shorter for the visual spectrum (butterflies possibly excepted) in the area of red, terminating above 600 nm. The sensitivity to ultraviolet is based on the fact that—in contrast to vertebrates—the dioptric apparatus does not absorb ultraviolet light above 300 nm. Some insects have photoreceptors specialized for ultraviolet, with a maximum absorption of about 350 nm (bees, St. Mark's fly, Syrphidae, and backswimmers). Other species have receptor cells with a maximum peak in the visible spectrum and another peak in the area of ultraviolet (flies, dragonflies).

It has been demonstrated that many insects perceive color. Particularly detailed studies have been made on the honeybee. As in man, color vision in the bee is based on the perception of three primary colors. For humans, these primary colors are red, green, and blue, whereas in the bee they are "yellow," "blue" and ultraviolet (UV)—the quotation marks are to remind the reader that the names given to these spectral colors are based on our own color perception. A mixing of spectral light rays of these basic colors produces all of the bee's color perceptions: "yellow" + "blue" = "green," "blue" + UV = bee's "violet," UV + "yellow" = bee's "purple." As with humans, the combination of pure spectral colors from both ends of the visible spectrum makes purple, a mixed color not contained in the spectrum itself.

Thus, again in the same way as with humans, we may construct a circular color chart for the bee. The three primary colors lie at three different points on the chart, and in between are the mixed colors obtained by combining two primary colors. As we do, the bee perceives the light resulting from the spectral combination of sunlight as colorless, or "white.' If one basic color is filtered out of this mixed light, the remaining light appears in the color which lies across from the filtered color on the disc: the complementary color. Conversely, an appropriate mixing of complementary colors results in a "colorless" appearance. By varying the combination of the three primary colors, every color perception can be made to appear less complete or less saturated. The more two complementary colors are mixed in the appropriate ratio, the more the corresponding primary color appears in the proportion found in sunlight.

It is the shift in the visual spectrum which differentiates the bee's world of colors from that of man. For instance, purely red objects, such as red dahlias and roses, appear black to the bee. But if red petals reflect ultraviolet, as do those of the corn poppy, the bee will perceive these as purely UV. Yellow petals which appear similar to us, such as those of the related species charlock and rapeseed, have different appearances for the bee. The first flower reflects a lot of UV, and the second reflects only a little: some of the other yellow flowers do not reflect UV at all. Thus, the bee sees charlock as "yellow," whereas rapeseed appear as the bee's "purple." The petals of some species of spurge, which to us appear almost as green as grass, look very different from grass to the bee. In addition to the green light which we perceive, grass also reflects UV which for the bee is the

Color vision in insects

▷
Olfactory bristles (Sensilla basiconica) on the antennae of the burying beetle *Necrophorus vespilloides*. Each receptor cell (blue) is encased in three spiral cortical cells. The dendrite of the receptor cell projects into the hollow bristle and splits into several ramifications. Stimulus-substance molecules reach the membrane of the dendrite ramifications through the pores and the pore basin and tubuli lying underneath (further enlarged section, top left).

A bee's world of colors

▷▷
Pit cones (Sensilla coeloconica) on the antenna of *Locusta migratoria*. Top view of a segment of an antenna (top left; length ca. 500 μm); top view (top right) and section (bottom left) of a pit cone; screen- or irradiation-electron-microscopic photographs (demarcation length 1 μm). Bottom right: Part of the instrumentation for the recording of potentials form olfactory bristles: A. Spray-olfactometer; B. Experimental animal (the cockroach *Periplaneta americana*); C. Electrode fasteners.

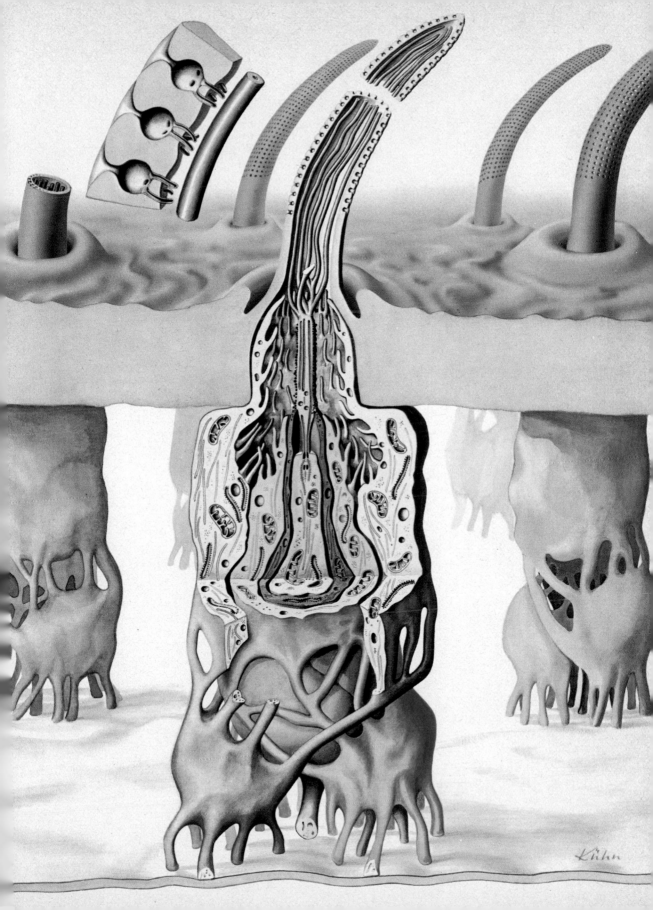
Kühn

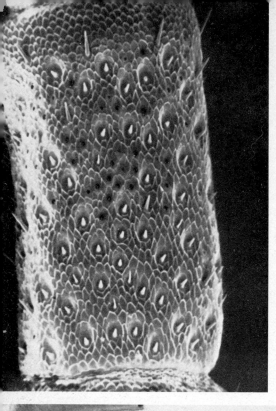

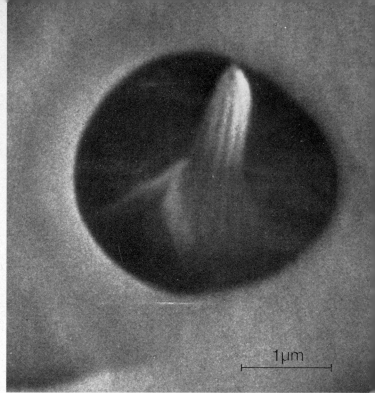

1μm

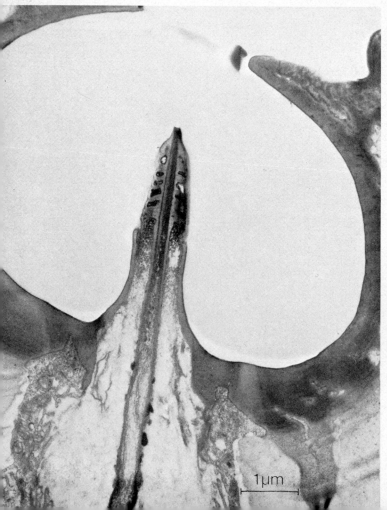

1μm

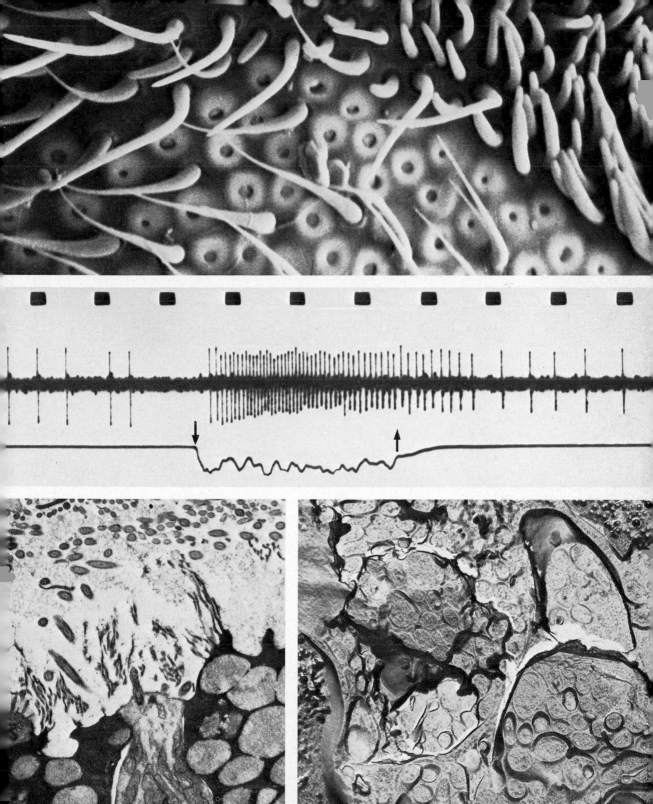
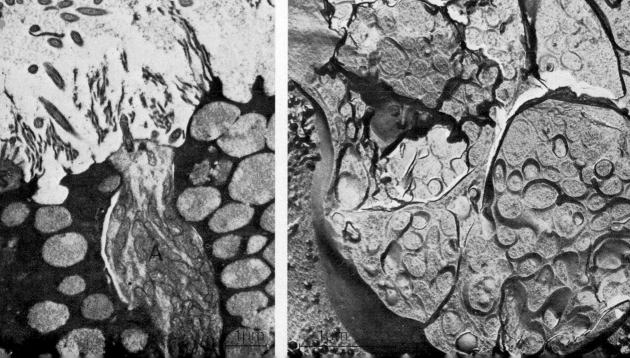

◁

Above: Front view of a male Viennese eyed hawk-moth (*Saturnia pyri*). The pinnate antennae carry thousands of olfactory bristles.
Below: Female mulberry silk moth or silkworm (*Bombyx mori*) sitting on a cocoon, in a luring posture: The pheromone glands protrude from the end of the dorsum (abdomen). These glands appear as well-rounded yellow sacs on both sides of the dorsal tip.

◁◁

Above: Screen-electron-microscopic photograph of coelospheric and basiconic olfactory sensilla on the antenna of the burying beetle *Necrophorus vespilloides*.
Center: Response of a Sensillum coelophaericum on the antenna of the burying beetle to the smell of carrion. The impulse frequency (center track) increases markedly during stimulation (lower track, arrows). Time recording (upper track): demarcation intervals 100 msec.
Bottom left: Section of the olfactory mucous membrane of a blindworm (*Anguis fragilis*). An olfactory cell (A) is surrounded by supporting cells, within which are embedded round mucous particles. Electron-microscopic photograph; demarcation length 1 μm.
Bottom right: Section of the olfactory nerve of a grass frog (*Rana temporaria*). The bundled axons of the olfactory cells, covered by a sheath, can be seen in relief. Electron-microscopic photograph of a frozen section; demarcation length 1 μm.

complementary color of green. And so the bee will perceive grass as gray or "colorless." From this background, those petals which reflect only green light will contrast with the rest and appear as pure bee's "green."

Many studies have been made on the capacities of insect eyes to perceive form and movement. In general, the form perception of insects is well developed, but the mechanisms are different from those of vertebrate eyes. Bees assess figures according to their arrangement of parts and their contours. These insects distinguish quite well between figures which we would perceive as similar, but they do not differentiate between other shapes which seem different to us. Ants are able to distinguish two vertical beams that diverge by 0.5 degrees of arc from a continuous beam, although the physiological vertex angle of an ommatidium is only about 10 degrees. Obviously, there are larger groups of ommatidia which work as a functional unit. Very small, almost insignificant changes in the visual pattern taking place within these functional groups produce noticeable changes in the pattern of excitation.

Studies of movement perception in insects have focused primarily on the orientation of flight. An insect placed in the center of a revolving cylinder with alternating black and white stripes will try to turn with the cylinder, so that its visual surroundings remain as constant as possible (optomoter response). Under natural conditions, this response would allow the insect to maintain a stable course of locomotion or flight in the face of outside disturbances, such as air vortices. Furthermore, pattern displacements in the retina of compound eyes enable insects to control their altitude and speed of flight.

We may also use electrophysiological methods to examine visual-data processing in the central nervous system. As in the visual pathway of vertebrates, numerous kinds of neurons have been discovered which become active in response to very different stimulus situations. There are neurons with large visual fields, combining messages from numerous sensory cells. Other neurons have small visual fields. Some receive messages from one eye, and others are triggered by impulses from both eyes. Still other kinds of neurons undergo short-term activity in response to changes in brightness (dimming detectors), and some discharge continuously under constant exposure to light. To complement neurons which transmit more signals under "brighter" conditions, there are those which produce more impulses when the environment becomes "darker" (dark detectors). Finally, there are neurons which respond to certain directions of movement in the visual field (changing contrast detectors).

Yet, despite considerable knowledge about a great many details, there is still a great deal that we do not understand about the visual sense. At this point, we simply do not have the basis for analytically relating the behavioral responses of animals in certain situations to the functioning of single neurons, of entire neural systems, or of the central nervous system as a whole.

6 The Chemical Sense

As human beings, we habitually divide the chemical senses into a *sense of taste* (gustatory) and a *sense of smell* (olfactory). The organs of taste lie on the tongue, those of smell in the nose. In the category of taste substances we include salts, acids, sugar and other sweet-tasting substances, and bitter ones such as alkaloids. By contrast, there is an immense variety of odorous substances, belonging to greatly diverse classes of chemical compounds. These include fatty acids, alcohols, aldehydes, ketones, and esters, as well as terpenes and various aromatic combinations, to name only a few categories. As a rule, taste substances reach the sensory (or receptor) cells in a water solution, while odorous substances are carried to the olfactory cells by an air current, as vapor or scattered as mist (aerosols). Even with amphibious animals, we differentiate between smell and taste according to the stimulus substances. Here, of course, the odorous substances are also contained in the water.

Our own experiences provide relatively little information about the capacity and biological meaning of chemical sense organs. Our language, itself, is misleading: Delight with the "pleasant flavor" of a good wine would refer less to sensations of taste than to those of smell. Odorous substances pass through the interior nasal aperture, the choane which lies in the mouth cavity, to the olfactory cells in the nose. On the whole, the olfactory sense plays a secondary role in humans. Certainly we sometimes examine food for odors, and several reflexes are triggered by olfactory stimuli. The saying "I can't stand the smell of him" indicates that the olfactory sense plays some part in human relationships, yet this phrase is meant figuratively—at least in our way of life. As shown in the chapter on chemical communication, however, the sense of smell is vitally important for a multitude of animal species in the context of intra-specific communication. Many creatures dwell in a world of odors hardly imaginable to us.

We can apply this division of taste and smell quite freely to other animals, even to invertebrates such as insects. On the other hand, annelida, mollusks, and crayfish are generally said to have only a chemical sense,

The chemical sense, by H. Altner

Fig. 6-1. The club-shaped antenna of the burying beetle *Necrophorus vespilloides* has eleven segments. The last three segments project olfactory bristles (Sensilla basiconica; black area), and the last segment has additional coleospheric sensilla (dotted area). For a better general view, half of the tenth segment has been omitted.

without further differentiation. But these groups of animals have been examined less than have the insects and vertebrates, to which we will direct our focus.

Insects taste and smell mostly with chemically sensitive bristles or seta (so-called *bristle sensilla*). These bristles are hollow, and they always have the same basic structure (Fig. 6-2). Beneath their base, there are one or more receptor cells whose stimulus-receiving processes (dendrites) project into the hollow space. As a rule, taste bristles have only one opening, at the tip, but olfactory bristles are generally perforated by numerous pores. When the molecules of a stimulus substance reach the surface of a bristle, they accumulate and quickly slide—probably by means of surface diffusion—through these pores into the hollow space. In this way they travel to the dendritic membrane, where they exert specific influences. Taste substances may pass through the terminal openings.

We have now traced the path of a stimulus substance to the receptor cells. The location of dendrites within a stiff bristle guarantees that these vulnerable structures are protected from dehydration and mechanical damage. Electron-microscopic studies of these sensory bristles have revealed that in most cases the pores are not simply perforations of the bristle wall; they are quite complicated structures, especially in the case of olfactory bristles. One common type of pore, found for example in the sensilla of the burying beetle's antenna (insect species of the genus *Necrophorus*; Fig. 6-1), first leads through an opening into a tiny globular basin measuring 50 billionths of a meter (nanometers) in diameter. This is attached to very delicate tubules (*pore tubuli*) extending into the liquidinous inner cavity of the bristles containing the dendrites. These tubules are about 120 nanometers long, with a diameter of approximately 10 nm. They are formed as parts of the cuticular armor by special epidermal cells. Taste bristles have a more simple structure; they are open at the tip, allowing the taste substances to reach the dendrites of the receptor cells. Most taste bristles, however, are also organs for the mechanical sense: They are connected at their base with the surrounding cuticle by way of a flexible articular membrane, and they have an additional, mechanically sensitive receptor cell.

Insects carry their olfactory bristles mainly on the antennae, where thousands of these bristles may stand closely packed together (Fig. 6-2). Here, they are also frequently embedded in the cuticula, where they are called pit cones (*Sensilla coeloconica*) or pit bulbs (*Sensilla coelospherica*). Very thin, hairlike, and elongated sensory bristles are called *Sensilla trichodea*, whereas the more compact ones are called *Sensilla basiconica*. In addition, the honeybee has pore plates (*Sensillae placodae*), roundish, plate-shaped, slitted areas. Beneath these apertures, in turn, we find the end parts of dendrites belonging to olfactory-receptor cells. Taste bristles in insects, of course, are located at the complex structures of the mouth; but that is not the only place. The bottom of the fly's tarsal joints carries a thick, furry layer of bristles containing numerous taste bristles. Flies test the surface on which they dart around for its taste qualities. When its "soles" touch a sugary solution, a fly quickly extends its proboscis and sucks in the substance. If we gently pour a saline solution in the fly's direction, it will immediately retract its proboscis.

The nose usually has a homogeneous olfactory mucous membrane (Fig. 6-4) containing the tightly packed olfactory cells. Two nerve processes extend from each of the round cell bodies. The stimulus-receiving dendrites climb to the surface. Here, we find the origination of cilia or microvilli, embedded in the layer of mucilage that covers the olfactory epithelium. The second kind of process, the axons, extend downward, bundle together into the olfactory nerve, and lead directly into the brain. In the salamanders, the olfactory cells are joined together in groups of budlike cells. These buds are separated from one another by supporting cells. The olfactory mucous membrane, with its unified structure, also contains supporting cells, whose somata (bodies) lie above the olfactory cells. Their irregular processes project between the olfactory cells, while the dendrites of the receptor cells project to the surface between the cell bodies of the supporting cells.

The same structure occurs in the Jacobson's organ, found in most reptiles as well as in many mammals as a special organ of smell that is either separate from the nose or connected to it only by a narrow passage. This organ is especially well developed in snakes. It seems that during their rapid flicking movements these animals catch olfactory-substance molecules with the tip of their split tongue and gather them into the organ's hollow space. It is believed that in artyodactyla (even-toed animals) the acts of flehmen (lip curl) and a so-called yawn combined with smelling function in the ventilation of this organ (Fig. 6-6).

The gustatory-receptor cells are usually joined into buds. In humans one such bud contains about 40 to 60 cells and measures approximately 40 millionths of a meter (micrometers) in diameter. The buds can be seen only with a microscope. They are embedded in the mucous membrane, which covers foldlike elevations on the tongue. We differentiate between several such elevations (papillae), three of which, the fungi, foliate, and vallate papillae, carry taste buds. The buds, in turn, contain supporting and sensory cells. The sensory cells have a dendritic process, carrying microvilli that terminate within a small indentation. But the receptor cells do not have axons. Instead, they adjoin to the processes of nerve cells, so-called afferent fibers, where they form points of contact (synapses) for the transmission of excitation. Each afferent fiber branches out in various directions. The branches make contact with a number of receptor cells, and they also radiate into various buds and papillae. As a result, each afferent fiber receives messages from numerous receptor cells lying within a large area on the tongue. In this way, each fiber is connected with a receptive field. The receptive fields of individual fibers overlap.

Taste buds

▷
Fig. 6-2. Chemically sensitive sensilla in insects. Top to bottom: A. Sensillum coeloconicum; B. S. basiconicum; C. S. trichodeum; D. S. coelosphaericum; E. Pore plate.

We have seen that, despite many variations, all chemically sensitive organs share a basic structure: The receptor cells are positioned in such a way that stimulus-substance molecules from the surrounding medium can reach the special areas of the cell membrane. Understandably enough, we now encounter tremendous difficulties in trying to explain the way in which incoming stimulus-substance molecules interact with membrane molecules. Using modern electrophysiological techniques, we can record the responses of single receptor cells during stimulation. Insects are especially good subjects for experimentation. Electrodes are mostly made of finely sharpened tungsten wires. Using these, we record the signals transmitted by the cells (the action potentials). We can read on the screen of a connected oscillograph that, as a rule, the frequency of these action potentials markedly increases with stimulation. In this way, we can at least examine the response transmitted by the receptor cells to the central nervous system. Of course, the information gained from these response recordings is relevant only when we apply precisely defined stimuli. For this purpose, experimenters have designed suitable olfactometers (instruments for the production of specific concentrations of odorous substance-air mixtures). Clouds made up of specific kinds of molecules, controlled in concentration and duration, are blown across a receptor organ. The purity of an odorous substance can be tested with a gas chromatograph, and the quality of an olfactometer can be determined with the help of radioactively marked substances. Yet this procedure gives us only an indirect picture of the molecular interactions taking place at the membrane; there has already been a long chain of events preceeding the rise of action potentials—and these potentials represent information about the situation at the receptor surface which has already been translated (encoded) into the language of the central nervous system.

Our next question is: How does an organism recognize which stimulus substances—or even which combination of substances—are being presented, and in what concentration, since its receptor cells can only transmit the same kind of action potentials in varying frequency? We know from our own experience—and the same applies to other organisms—that we can distinguish between thousands of odorous substances. Evidently we are dealing with receptor-cell specificity. The simplest case would be that each receptor cell reacts to only one specific substance, so that the response is unequivocal. Researchers have in fact, discovered such "specialists" among the receptor cells. While receptor cells reacting with less selectivity are more common, even these kinds of elements can portray a clear-cut stimulus situation.

One example of cell specialization is found in the pheromone receptors of the mulberry silk moth or silkworm (*Bombyx mori*). Pheromones are chemical substances that attract conspecifics of the opposite sex. Males of the mulberry silk moth respond with a typical whirring motion or humming activity of the wings to the pheromone stimulus produced by

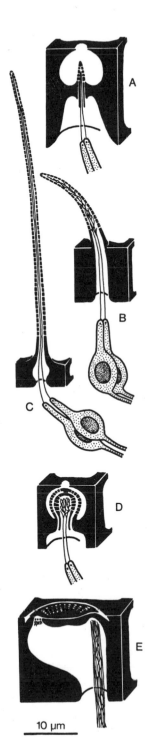

10 μm

special glands in the female. In 1959, A. Butenandt and his students discovered the structure of this pheromone. Two years later it even became possible to synthesize it. Through detailed investigations, the scientists working with D. Schneider analyzed the functioning of receptors on the antenna. They discovered that the silk moth has very specific receptor cells which respond to the pheromone called bombykol, reacting much less or not at all to other pheromones. By experimentally modifying the chemical structure of the pheromone molecule, it was discovered that even minimal changes will reduce the effect to a thousandth of that of the natural pheromone. It is hardly possible to "fool" a system at such a high level of specific adaptation. The response of the receptor cell to a substance of vital biological importance is specific and—as we have indicated—highly sensitive. Even very similar substances must have a much stronger concentration to release the same response.

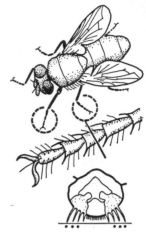

Fig. 6-3. Taste bristles on the tarsal joints of the fly *Phormia terraenovae* (below: Cross section of the third tarsal joint; the taste bristles are indicated by dots).

Based on these observations, we can turn to another direction for some very interesting experiments. If sex pheromones are found within a group of animal species, we may ask whether each species has its own species-specific pheromones (and corresponding specific receptors). In an experiment using a series of cross-tests, E. Priesner exposed the receptors of each of 104 species of emperor moth with preparations of pheromones belonging to every single species. In each case, he only needed to record and measure the summated responses of numerous receptors (so-called electro-antennograms), a procedure technically easier than the recordings of single discharges. Looking at a total of 1900 species combinations, Priesner found that 104 species could be relegated to nineteen response categories. Within a response category, all the animals behaved, as if their pheromones were identical. Furthermore, the species belonging to a response category were directly related to one another. These findings not only suggest a totally new method for rechecking the taxonomic system of classifying animal groups, created primarily on the basis of morphological characteristics; they also provide us with an answer about the evolutionary significance of the pheromone system. In the present case it was discovered that, contrary to previous assumptions, closely related species are not usually isolated from one another by different pheromones, but rather by their way of life, including differences in daily or seasonal activity, environmental preference, and so on.

But how do animals codify stimulus situations requiring not just one special receptor for a particularly significant sign odor, but rather the ability to distinguish among a large number of attracting odors—for example, in the search for food? Again, experiments on the olfactory sets of insects have provided an answer. These subjects are especially well suited for long-term experiments using extensive batteries of odorous substances. Thus, we have found that, among others, ants (*Lasius fuliginosus*), honeybees (*Apis mellificia*), and cockroaches (*Periplanata americana*) carry receptors that react to a greater number of substances by increasing

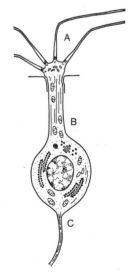

Fig. 6-4. Single olfactory cells from the olfactory mucous membrane of a vertebrate (simplified): A. Cilia; B. Dendrite; C. Axon.

the frequency of action potentials (spike frequencies). But these responses are not random. A more detailed analysis has shown that there are several receptors with clear differences in receptivity. Every type is characterized by a specific response spectrum, in other words, by a list of stimulus substances that will trigger certain impulses. The spectra may overlap, but we know of substances to which receptors belonging to corresponding response categories are especially sensitive.

We can label the response categories according to these stimulus substances. Among the olfactory bristles on the antenna of a cockroach, for example, we can distinguish between "groups" sensitive to pentane, octane, butyric acid, and formic acid. We must note, however, that odor combinations effective under natural conditions, such as food odors, do not correspond simply to the spectrum of one response category. Animals apparently differentiate and categorize odor stimuli from the environment in a way that is far more complicated than the simple excitation of a response group. Most likely, various groups are activated in different intensities, giving rise to complex but characteristic excitation patterns in the subsequent nervous relay stations. This coding system may be compared to a large board strewn with numerous colored lights: Every light represents a receptor cell, and every color represents a response category. The "lights" will glow with different intensities according to changes in the level of excitation, that is, to alterations of impulse frequency. We can see that in this way a tremendous variety of light patterns may be created, each corresponding to a specific stimulus situation.

We have no reason to assume that the olfactory organs of vertebrates utilize a different coding system. There are still only a few cases of single olfactory-cell recordings, but all of these experiments indicate that even vertebrates have receptors with different specra of response.

We know very little about how the impulse patterns coming from the receptor cells are processed by the secondary relay stations of the central nervous system. Even in the case of insects, where the structure and function of sensory organs are relatively well understood, studies on the projections of the olfactory and gustatory pathways are only just beginning. More detailed studies have been made on the olfactory pathway of vertebrates, but even here our knowledge about the various relay connections and the flow of information is very incomplete. Yet we are able to state three basic principles: First, impulses are conducted from a great many receptors to a far smaller number of subsequent nerve cells. The color plate on page 95 shows a small bundle of approximately 130 olfactory cell ramifications that extend into the olfactory lobe of the brain. In this area, however, thousands of these axons will synapse together on the excitatory receptors of one large nerve cell. Second, there are complicated switch connections that will inhibit any large inflow of excitation by means of negative feedback. Third, even higher centers can modify the inflow of excitation.

Fig. 6-5. Distribution of taste buds on the body surface of the sheatfish *Ictalurus natalis*. Each dot represents 500 buds.

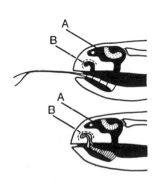

Fig. 6-6. Tongue-flicking in snakes: Longitudinal section of the nose (A) and the Jacobson's organ (B).

In the case of the gustatory sense, where we may classify sensations into such basic qualities as sweet, sour, salty, and bitter, the problem of codification is by no means easier than with the sense of smell. We have found areas on the human tongue that are particularly sensitive to single qualities. Bitter sensations occur mainly on the base of the tongue, sweet sensations, on the tip, and sour and salty ones, on the lateral margin. But in each case the receptivity is preferential and not exclusive. We have said that the taste organs of vertebrates contain numerous afferent fibers, and that each of these fibers is connected with numerous receptor cells extending from different papillae. We have established by means of electrophysiological methods that—with a few exceptions—neither single receptor cells nor single afferent fibers are specifically responsive to substances from only one of the four taste categories. As a rule, cells and fibers will respond to substances of different qualities. We have, however, made some interesting discoveries by comparing the reactions of different fibers to the same kind of stimulation, meaning to the same concentration of a taste substance solution. Every fiber will display a specific pattern of response, a so-called taste profile. Thus, in a way similar to the action of olfactory receptor cells, taste receptors will produce within their entire system of fibers certain patterns of excitation that will reflect the characteristics of the stimulus situation (types of substances and their concentration).

Based on observations of "specialists," we can begin to speculate about what kind of interactions must take place between the stimulus-substance molecule and the molecules in the membrane of a receptor cell. These interactions are also called the primary process. This is beginning of a long chain of events that ultimately give rises to transmitted action potentials. Remember that even minimal changes in the bombykol molecule will reduce its effectiveness many times. This indicates that the interactants in the primary process—the stimulus substance molecule and the responding molecule in the receptor membrane, which is also called the acceptor molecule, or, simply, the acceptor—must allow for a certain "fit." They specifically complement one another. The acceptor molecule determines whether a stimulus substance molecule will be able to "lock in" and release the resulting processes or whether it will "rebound." By systematically varying the stimulus-substance molecules, we can trace the characteristics that determine a molecule's effect. Such experiments have been done on the pit cones of the antennae of migratory locusts (*Locusta migratoria*) (see Color plate, p. 94), among others. The results lead us to conclude that there are two significant molecular traits: Dispersive power and dipolar momentum. A dispersion is a mixture in which one substance is finely distributed, but not dissolved, within another. In a case like this, every particle of one substance is enveloped by the other substance. A dipolar arrangement involves the juxtaposition of two equally strong but opposite electrical charges. While it may be assumed that both molecular

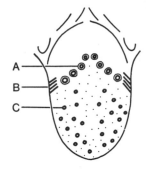

Fig. 6-7. Distribution of different kinds of papillae on the human tongue: A. Vallate papilla; B. Foliate papilla; C. Fungi papilla.

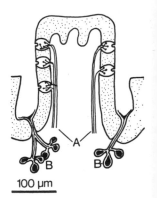

Fig. 6-8. Cross section of a foliate papilla (diagrammatic) with five tast buds; A. Afferent fibers; B. Rinsing glands.

traits are important for attachment to the acceptor, it is likely that the dipolar forces are responsible primarily for those changes in the acceptor necessary for the production of subsequent reactions.

It requires an extremely careful planning of experimental method to provide a step-by-step explanation of those chemoreceptive elementary processes that occur at the molecular level. There are, however, very simple observations which—naturally with some basis of scientific knowledge—may lead us to the same conclusions. For example: From the Indian plant *Gymnema sylvestre* we can extract gymnema acid. This substance has the strange quality of blocking the sensation of sweetness in humans. If we let a subject chew gymnema leaves or wash his tongue in this acid, and then apply sweet-tasting substances to the tongue, these substances will produce no sensations at all. Sugar will taste like "sand." How can we explain this effect which—depending on the concentration of the substance—lasts for several minutes? Evidently, the molecules of the gymnema acid have combined with the acceptors for sweet substances. But, in contrast to the ordinary kind of interaction, these molecules have formed a more persistent bond with these acceptor molecules, blocking the triggering of the response.

The gymnema experiment

Moreover, a number of observations indicate that the taste seta of insects may be particularly well suited for further research into the primary process occurring at the gustatory-receptor cells. The taste bristles of flies contain specialized receptor cells that respond exclusively to certain kinds of sugar. Using biochemical methods, we can show that the tarsal joints which hold the insect's taste bristles, will display a particularly high degree of activity involving an enzyme that splits sugar molecules. These experiments indicate that this enzyme represents the acceptor. In this case, the primary process would be a so-called enzyme-substrate reaction. The characteristics of this kind of reaction are well understood, which means that continued research along these lines should be fruitful indeed.

We have now pretty well established that the primary process at the chemical receptor cells is an interaction between stimulus-substance molecules and acceptor molecules contained in the membranes of the receptor cell. Following the completion of this primary process, there is a string of further reactions. These give rise to action potentials, which are transmitted to the central nervous system and provide information about the respective change in the external situation. But just how sensitive are the taste and olfactory cells in their capacity as molecule detectors? How many molecules of a stimulus substance must impinge upon the receptor-cell membrane and make their way to the acceptors before taking part in the production of a transmitted signal?

We all know of the olfactory acuity of many organisms—for example, the sensitive nose of dogs—and therefore suspect that very small amounts of an odorous substance will be enough to activate a single receptor cell. Measurements have shown that dogs are able to detect butyric acid that

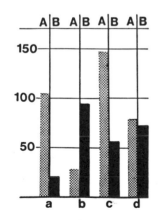

Fig. 6-9. Responses of olfactory cells on the antenna of a cockroach. Comparison of excitation (impulse/sec) from receptors of the pentane group (A, dotted) with those of the octanol group (B, black) during stimulation with pentane (a), octanol (b), banana (c), and lemon (d).

has been diluted as much as $1:5.10^{16}$ since they will respond behaviorally even to this dilution. After all, it must be noted that, in this concentration, one milliliter of air still contains about 6000 odorous-substance molecules. Fishes have especially sensitive noses. In their spawning migrations from the sea into the rivers, salmon (*Oncorhynchus, Salmo*) are able to relocate precisely that tributary where they hatched and grew. Obviously, their olfactory sense plays a decisive role in helping them to find their native waters, a fact that has been demonstrated by plugging their noses. Minnows (*Phoxinus laevis*) are shoal fish with a highly sensitive warning system. If, for instance, one minnow is injured by a predatory fish, the injured skin area secretes a warning substance that alarms the other members of the shoal, who will smell the substance and quickly swim away. Some well-designed laboratory experiments have shown that river eels (*Anguilla anguilla*) can recognize a test substance (B-pnehyl ethyl alcohol) diluted as much as $1:3.10^{18}$. This value means that one milliliter of water will contain only about 1800 odorous-substance molecules.

Measurements and calculations of single receptor-cell receptivity have been obtained for the mulberry silk moth or silkworm. As mentioned earlier, males of this species will respond to a pheronome stimulus with a typical whirring motion of the wings. In an experiment, this organism is exposed to a brief air current that is passed over a filter paper soaked with droplets of the pheromone. Results have shown that even as little as 3.10^{-6} micrograms (three-billionths of a milligram) of bombykol are sufficient to release the whirring response. This value, then, represents the threshold.

We also know that roughly 25,000 bombykol receptors lie on the antenna of the silk moth. These receptors may be responsible for creating a response in the threshold area. More precise statements can be made with the help of electrophysiological studies and with concentration measurements made by marking the bombykol radioactively. Results of these experiments have shown that under the stimulus conditions named above, only about 600 molecules per second will strike the antenna, and of these only about 300 will reach the 25,000 bombykol receptors. If we also consider that the whirring reaction begins as early as 0.86 seconds after stimulus onset, we arrive at the conclusion that roughly 200 nerve impulses are enough to produce a reaction; that is, 200 receptor cells each produce one action potential, generally as a result of a single impinging molecule. Thus, one pheromone molecule is enough to release an action potential at a receptor cell. And it takes only about one percent of the available pheromone receptors, each sending out only one impulse, to trigger the whirring activity of the male silk moth.

Olfactory acuity

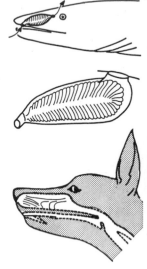

Fig. 6-10. In Teleostae, for example the eel (*Anguilla anguilla*), the nasal grooves (nostrils) lie at the side of the head (above); the olfactory mucous membrane is tightly folded (center). Arrows indicate the direction of the water current. In higher vertebrates (below), the nasal cavity is connected with the mouth cavity.

7 Mechanical Senses

Mechanical senses, by
K. E. Linsenmair

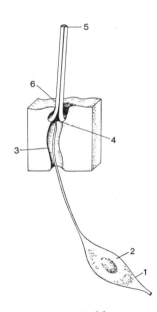

Fig. 7-1. Typical hair
sensilla (from the cervical
joint of the honeybee), as
seen under an electron
microscope. 1. Axon;
2. Receptor cell; 3. Cuticula canal; 4. Place of
attachment for the nerve
process (dendrite) near the
articulation; 5. Cut bristle;
6. Mechanical force giving
the receptor cell a specific
direction.

Anyone who has ever tried to approach a chirping great green bush
cricket (*Tettigonia viridissima*) without being noticed will know how difficult it is: The insect usually stops singing when we are still several meters
away. Only if we have the patience to stay absolutely quiet until the
chirping resumes can we sneak closer step by step. Notably the distance
at which the cricket stops for the first time depends very much on whether
the observer moves lightly or heavily, but to a far smaller degree on
whether he is talking or not, approaches with or against the wind, or is
hidden rather than out in the open. The cricket responds to shock vibrations in the ground. Since these vibrations are very much attenuated by
the ground, this creature must have very acute sense organs to register
vibrations; that is, it has highly sensitive vibratory perception.

A few species among the barbels, gobies, characins, and other fishes,
have become adapted to living in totally dark caves. Some of these, such
as the Mexican blind characin, *Anoptichthys jordani*, have completely lost
the ability to see, and are even born blind. Yet they behave in such a way
that the observer would not even notice their blindness. These fish manage
to avoid even small obstacles and to accurately locate and catch prey
animals. During courtship and spawning, when they are very active, both
partners achieve a precise synchronization of responses. If we spray a blind
characin with a fine stream of water, it whirls around quickly, snapping
at it just as if it were prey. In an experiment we can easily demonstrate
that the fish reacts not only to extremely light water currents in its
vicinity, but also to the pressure waves created both by objects approaching it and by its own body approaching another one. Its tactile receptors
enable this fish to perceive pressure stimuli from a relatively great distance.

Except for our sense of hearing, the signals transmitted by our own
mechanical senses have little influence on our conscious experiences, which
are dominated by visual and auditory impressions. Because of that, we
tend to underestimate the significance of the biological role that mechani-

cal senses play both in humans and in other organisms. We regard these senses as more primitive or of a "lower" order.

There are many kinds of mechanical stimuli impinging upon the organism: gentle, aperiodic touch stimuli, regular (that is, periodic) vibrations, persistent pressure either localized or coming evenly from all directions, traction, gravity, linear as well as rotary acceleration, and so on. If a creature responds differently to different mechanical stimuli, we may assume that it can distinguish between them. For that it must possess mechanically sensitive receptor cells (mechanoreceptors), each responding primarily to one kind of mechanical stimulation.

Mechanoreceptors

The basis for this selectivity varies. In photoreceptors and chemoreceptors, the key factor lies in the chemical make-up of those membranous areas where the primary processes take place (see Chapter 6). In mechanoreceptors, all activating stimuli ultimately effect either a distension or a contraction of the stimulus-receptive structures. Selective responsiveness, or stimulus selectivity, occurs primarily in so-called accessory structures, mainly those of the corresponding stimulus-receiving and stimulus-conducting system.

Let us take a look at the hair sensilla of arthropods (articulate animals). The utilized stimulus, which activates the receptor cell, is always the same: it causes a bending or shearing of the sensory bristle. Yet, in spite of this uniformity, these sensilla respond to very different input stimuli, the responses depending on both the mechanical properties and the physical location of the stimulus-conducting hairs. One that is long, thin, and articulated over a flexible membrane will be susceptible to even a light waft of air. By contrast, a thick and firmly implanted bristle will not bend to the required extent unless hit by far greater pressures, such as a direct physical force. Once the displacement is great enough, however, the input stimulus (pressure) is converted into the proper or adequate stimulus utilized by the receptor (shearing). Should an easily displaced hair not extend beyond the surface of the body, but rather stand—for example, as a filiform (threadlike) hair—within the enclosed statocyst (equilibrium organ) of a crab, it will be articulated only by movements of the surrounding statocyst fluid. The corresponding receptor will therefore respond only to accelerations (which lead to movement of the statocyst fluid relative to the membrane wall), but not to stimuli consisting of flow in the outside medium. In many cases, then, it is primarily or exclusively the accessory structures that allow a specialization within the mechanosensitive receptor cells, performing the essential tasks of filtering, conducting, and sometimes even transforming (or transducing) the mechanical stimuli.

We do not know the primary way in which mechanical stimuli become effective. It is uncertain whether the distortion of the membrane leads directly to structural changes which then alter its permeability to ions or whether certain processes are set in motion that lead—only on a secondary

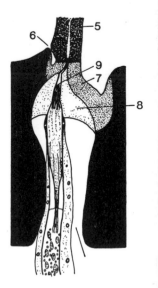

Fig. 7-2. Articular region or the receptor cell in Fig. 7-1, as seen under an electron microscope.

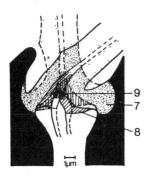

Fig. 7-3. Distortions at the base of the hair when the bristle is bent. The folding articular membrane (7) puts pressure on the cap (8), causing the tip of the dendrite (9) to press in; (6) mechanical force; 7 and 8 consist of resilin, a material with extremely elastic qualities. See also Fig. 7-2.

Fig. 7-4. Stimulus-testing hair with flexible bristle.

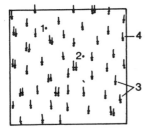

Fig. 7-5. Section (2 cc) of the top part of a human forearm, showing the distribution of pressure points (3) as derived from stimulus-hair experiments. 1 and 2 are the only pressure points not lying at the base of a hair (4).

level—to changes in the electrical activities of the membrane, creating a receptor potential (see Chapter 4).

There is no such thing as "the" mechanical sense. The fifth of the classical senses, the tactile sense, is not purely mechanical in its function, since the tactile impressions that we perceive as unitary also contains signals from hot, cold, and pain receptors. We cannot make any absolute statement on the actual number of mechanical senses, nor on what criteria should be used to classify them. Natural science does not offer a clear definition of the term "sense," and the concepts used to define modal and qualitative properties in other senses cannot be applied to mechanical senses.

But if we do not want to discard the idea of a "mechanical sense," we will have to define what it is really supposed to mean. Our main interests here are to know which mechanical stimuli organisms can perceive and how these particular sensory capacities function in the interactions between the animals and their animate and inanimate environment. We are thus inclined to classify these sense modalities according the physical nature of their effective stimuli. Accordingly, an organism possesses a certain mechanical sense if it is able to recieve a physically defined mechanical stimulus by means of specialized receptors, process the stimulus, and respond to it with some kind of behavior.

In this chapter we will deal only with those sensory capacities where receptor cells are activated by external stimuli (exteroreceptors). The organism also has other sensory cells, called proprioreceptors, that signal impulses generated within its own body. It must be emphasized that especially in the case of mechanical senses the exteroreceptors cannot always be functionally separated from the proprioreceptors—in other words, the same receptor cell frequently functions in both ways.

The tactile receptors are stimulated when they come into direct passive or active contact with outside objects. Occasionally they may also receive stimuli coming through indirect contact, such as pressure stimuli through the compression of the surrounding medium. Usually the stimuli impinge irregularly over time; that is, they are aperiodical. When there is a gradual transition from one kind of stimulus to another, we can distinguish between contact and local-pressure stimuli by their intensity and duration. We do know of special tactile organs, such as the combs of scorpions (see Color plate, p. 117) or the hammer-shaped organs of the spider *Galeodes* (see Color plate, p. 117), but nowhere in the animal world is the tactile sense itself restricted in concentrated form to a single sense organ. As a rule, any part of the body surface that is constantly or even intermittently exposed to the outside environment will be sensitive to touch stimuli. This lays the groundwork for a biologically very important feature of the tactile sense: its spatial resolving power or "definition," which, like the resolution of the visual receptors (see Chapter 5), allows for a discrimination of detail. It enables the animal to localize the source of stimulation and is therefore prerequisite to appropriate reaction. The quality of this spatial definition depends not only on the screen density of surface receptors but also on the precision with which excitatory impulses are transmitted to the brain (central representation). The more individual conduction pathways there are between the peripheral receptors and the central neurons of associated projection-field areas, the more detailed will be the representation of body surface in the brain. One way to measure this degree of precision is the "simultaneous space threshold." It indicates the minimum distance at which two narrowly defined tactile stimuli (e.g., the two points of compass dividers) have to be applied simultaneously so that they just manage to elicit two separate signals. In humans the smallest space thresholds are found on the tip of the tongue, the lips, and the fingertips, so that these are the areas with the most detailed spatial definition.

We can use a stimulus-testing hair to create a narrowly defined pressure stimulus with its maximum force set at a certain value. Beyond that, pressure applied to the bristle will merely cause it to bend without increasing force on the stimulated surface. By moving this instrument about on our skin, we discover well-defined, sensitive "pressure points" whose density changes within different areas of the skin in a characteristic way. At least in humans, we should have little difficulty in determining which receptors indicate pressure and which signal contact stimuli. The method is to make microscopic histological examinations of those areas of the skin shown in the testing-hair experiment to be sensitive to touch. Yet it turns out that this microscopic preparation contains such a variety of possible receptors that morphological investigations alone do not allow us to associate certain receptors with certain functions in any clear manner.

Using morphological criteria, we can divide the skin receptors of vertebrates roughly into two groups: nerve endings with special terminal structures ("end organs") and so-called free nerve endings. The latter include pain and thermoreceptors as well as mechanosensitive receptors. Our best knowledge is about mechanically activated free nerve endings that form networks around the shaft of normal hairs (vibrissae).

These hairs are the major touch receptors in the pilose (or hairy) skin of mammals. The long ends of the hairs themselves extend evenly across the surface of the skin, forming a two-pronged lever pivoting within the horny, rigid upper layer of the skin. The long arm of the lever serves as an amplifier, transmitting even very small mechanical stimuli. These impulses then lead to a minor local alternation of compression and distension of the tissues surrounding the shaft and root of the hair. It seems that only the bulging movement constitutes the adequate stimulus. These hair networks respond in a purely phasic manner, and prolonged stimulation causes excitation to drop back to the zero point. Therefore these hair receptors signal only the locus and time of stimulation, as well as rapid

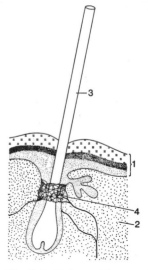

Fig. 7-6. Hair sensitive to tactile stimuli. 1. Epidermis with a horny layer on top, the layer consisting of dead cells; 2. Dermis (true skin); 3. Cut hair, consisting of a shaft that extends over the surface of the skin and acts as a long lever arm, and a root that acts as a short lever arm; 4. Network of nerves with branching fibers; the nerves are activated by distension.

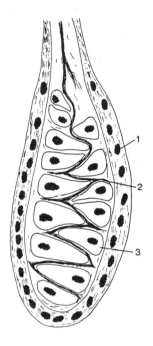

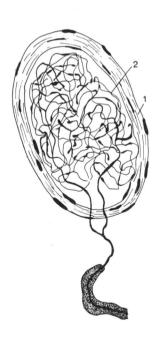

Fig. 7-7. Above: Meissner's corpuscles. Below: Genital bodies. 1. Connective-tissue covering; 2. Mechano-sensitive nerve endings; 3. Tactile receptor cell.

changes in the temporal stimulus pattern. In other words, they may indicate where, when, and how quickly the hair is bent over, but they cannot signal a prolonged and even displacement resulting from, say, pressure or pull.

All vertebrates have mechanosensitive end organs in their skin; these occur in the greatest number where parts of the body are naked and exposed: on fingertips, the snout of earth-grubbing or burrowing mammals (e.g., pigs and moles), lips, the tip of the tongue, and (in animals such as woodcocks) even on the tip of the bill. In general, all these end organs have quite a similar structure: Cell bodies lying in the spinal ganglion are attached to nerve fibers that extend to the skin, sometimes reaching a length of several meters. The end section of this ramification is exposed; that is, there is no myelin sheath for electrical insulation. The end organ is encapsulated by connective tissue and thus is isolated from neighboring tissues. Such end organs, which are probably mechanoreceptors, include among other types the Herbst's corpuscles of birds, Merkel's disks, which are supposedly pressure receptors with a slow adaptation period, Meissner's corpuscles, the Pacinian corpuscles, and the touch-sensitive genital organs of mammals.

Tactile receptors that lie beneath the surface of the skin will receive external mechanical stimuli only where the skin is able to conduct these impulses. In arthropods, with their rigid exoskeleton, most of the surface is unsuited to this purpose. They have therefore adapted by extending the stimulus-conducting system of their tactile receptors, specific hairs structured according to their function, largely above the surface. Again, the adequate stimulus is a bending of the hair.

One of the basic functions of the tactile sense is to indicate the borderline between the body and its external environment. This dividing line may be shifted away from the direct surface of the body into extended space. For example, the stimulus-conducting systems as well as those parts of the body containing large numbers of touch receptors may be lengthened.

The tactile hairs in the snout area of many mammals are much longer and stiffer than the rest of their hair (see Color plate, p. 118). The hair follicles of these receptors signal the speed of displacement as well as the duration of a more prolonged bending action. By removing these vibrissae, we can demonstrate how important they are for spatial orientation: Cats become markedly unsure of their movements in the dark; they keep bumping into obstacles and have great difficulty in finding and assessing narrow passages. Rats that have learned to run a very complicated maze quickly and without errors are still able to do so even when their eyes are covered and their sense of smell is eliminated, but they make many errors when we cut off their vibrissae.

Another way to extend tactile space, as mentioned above, is to lengthen certain body parts containing touch receptors. This phenomenon is very marked in cave-dwelling insects and crabs. Their antennae, which are often completely covered by tactile hairs, frequently reach a length many times that of their body.

The lateral-line organ found in some fishes and amphibians also serves to extend tactile space by responding to pressure waves as well as local water currents and vibrations. In fishes, this lateral-line organ consists of a system of canals at the head and at the sides of the rump, opening to the external environment by way of pores. Receptor cells standing in clusters on the bottom of the canals have hairlike processes that extend into a gelatinous, dome-shaped vault (cupola). The adequate stimulus for the hair cells of the lateral-line organ is the bending-over of this cupola by water currents within the canal. The receptor cells react to specific directions: Displacement in one direction leads to excitation, and in the opposite direction, to inhibition. In amphibians with a lateral-line system, as well as in some fishes and fish larvae, the receptor cells lie at the surface of the skin so that their associated cupolas project freely into the water. Because of the spatial distribution of both the free-standing sensory hillocks (neuromasts) and those contained in the canals, and also because of the directional specificity of their receptor cells, these organisms are able to localize precisely the direction and therefore—if close enough—the source of stimulation.

Tactile stimuli play a very important part in releasing and controlling a large number of behavior patterns. In this chapter, we are forced to restrict ourselves briefly to a few points:

1. In many cases, tactile stimuli trigger protective behavior. For example, a cockroach runs off immediately when touched lightly with a hair. A snail receiving a blow will retreat into its shell. Sleepers (*Vireosa*) exploit the protective behavior of other creatures for their own advantage: When danger threatens they will flee into the open shells of living giant clams (*Tridacna gigas*) and cause them to close by quickly providing a touch sitmulus. In this way the fishes are shielded from all attacks. Some insects "play dead" when touched: They will fold their limbs to the body, let themselves fall (see Color plate, p. 118), stay motionless for some time, and so escape their predators that respond to movement. In humans, too, important protective reflexes are triggered with the stimulation of tactile receptors, such as the closing of the eyelid when a light mechanical stimulus is applied to the cornea, or the coughing reflex in response to mechanical stimulation of the mucous membranes in the trachea and bronchial tubes. Some crabs and arachnid species, when subjected to a stronger stimulus, will throw off one of their legs—which then frequently jerks around on its own, much like a lizard's tail after it has been broken off. If such a discarded but active limb succeeds in drawing the attention of the predator, its "self-mutilating" owner may have time to find safety. Animals capable of defending themselves often utilize their defensive weapons when subjected to a strong tactile stimulus. For example, many bedbugs,

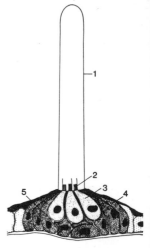

Fig. 7-8. Sensory hillock (neuromast) standing free on the surface of the body. 1. Cupola; 2. Cilia of the hair cells (several stereocilia, one kinocilium per receptor cell); 3. Receptor cell; 4. Supporting cell; 5. Nerve fiber.

Fig. 7-9. The lateral-line organ in fishes consists of canals that open to the outside by means of pores.

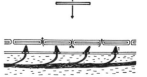

Fig. 7-10. Within the canals of the lateral-line organ, a current of water (arrows) is created through impounding pressure as soon as an object approaches the fish. This causes the gelatinous cupolas with their hair-cell cilia to bend.

MECHANICAL SENSES 113

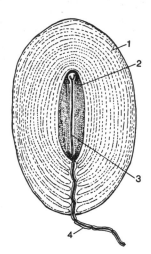

Fig. 7-11. Pacinian corpuscle. 1. Outer layers (lamellae); 2. Inner layers; 3. Exposed tip of dendrite; 4. Nerve fiber covered by the electrically insulating myelin sheath.

Vibratory sense

beetles, and ants squirt their defensive secretions in the direction of such stimulation.

2. Scorpions use neither their visual nor their olfactory sense to find food, especially with small arthropods. But if the touch receptors of scorpions—located mainly on their pincer legs, or pedipalpi—are activated by a colliding prey animal, these predators will seize upon it quickly. If we touch a blind and hungry clawed frog somewhere on its body surface, it immediately turns its mouth to that spot, snapping at the source of stimulation.

3. In many animals, especially insects, the act of copulation requires that the individual attach its genital organs in some complicated way. These genital organs, in turn, are monitored by mechanoreceptors. Even with an indirect transfer of spermatophores, tactile receptors often play a very important role (see Chapter 27). For example: At the meeting of a male and a female scorpion, both of which are ready to mate, the male seizes the female with his pincers and walks backward, pulling her along and feeling the ground with his combs (see Color plate, p. 117). He will deposit his spermatophores only when he has found a firm base; then he pulls the female across it. Now the female locates the spermatophores with her combs and, after feeling them, pushes them into her genital opening, located directly in front of the combs.

4. Tactile stimuli often also influence the way an animal moves. In an experiment we may carefully glue a thin thread to the back of a fly and then suspend the creature in the air, letting it hang freely. It will not fly as long as its tarsals (foot joints) are in contact with an object. All it needs is a small paper ball to hold with its legs. As soon as we remove this "hold" the animal begins to fly.

5. Infant monkeys raised without being allowed contact with conspecifics will show severely disturbed social behavior as adults, even if all other conditions are optimal (see Chapter 25). Young rhesus monkeys that could hear and smell their mother and other conspecifics, but could not touch them, were as disturbed in their behavior after a few months as youngsters kept in total isolation over the same period of time. Infant monkeys, like human children, need the occasional safety and comfort that only direct contact with conspecifics will provide. This comfort contact, mediated by the tactile receptors, is essential for normal development.

The vibratory sense is activated by rhythmical changes in pressure or by the movement of a substrate or an object in contact with the receptors, or by such changes in a medium surrounding the stimulus object. It is unusual for the energy from mechanical vibration to be directly transmitted: this happens when the transmitter and receiver are in direct contact with each other. In most cases, there is a stimulus-conducting medium between the two that responds to mechanical stimulation with a spectrum of sine waves. Here, the surface waves are biologically very significant, arising and propagating only at the borders of conducting media such as between water and air or ground and air. Rhythmical currents in the surrounding medium, such as might be produced by insects beating their wings, can also activate the corresponding receptors and sense organs; but we will not describe these further here. In each case the investigator will have to examine what properties of an oscillation will activate the vibratory-sense organs, be they pressure, speed, acceleration, or direction of the medium particles. The lower limit of adequate frequency stimulation is generally set arbitrarily at 10 Hz, which means 10 oscillations per second. Periodic changes in pressure that have a lower frequency are then considered as tactile stimuli. The upper frequency limit (a few hundred to several thousand Hz) will depend on the capacity of the sense organ itself and on the attenuating and oscillating properties of surrounding tissue areas and body parts.

If we let an object vibrate on our skin at different frequencies, sensations will change: At 60 Hz we will experience a "fluttering of the skin" that we can locate quite accurately at higher frequencies this accuracy diminishes, and we begin to feel as if the layers far below the surface are vibrating along with the rest of the skin. Our skin has a strong attenuating effect on vibrations with a low frequency, so that these will activate only surface receptors lying close to the point of stimulation, probably Meissner's corpuscles.

Vibration frequencies of 100–300 Hz undergo the smallest degree of attenuation and are therefore conducted to the farthest points, also traveling deep into the organism's body and across the bones. Here, the receptors are the Pacinian corpuscles, the only end organs whose morphology and physiology have been examined in greater detail. These corpuscles are relatively large (up to 4 mm long and 2 mm thick), and therefore easy to locate and to surgically isolate in a specimen preparation —as with, for example, the Pacinian corpuscles lying in the mesenteries of a cat's intestine.

Vibratory receptors usually have a high temporal definition, or resolution. As with the touch-sensitive hair networks, continual stimulation (such as constant pressure) causes excitation to decline rapidly to zero. In the case of Pacinian corpuscles, we know the cause of this phenomenon: It is not any property of the mechanosensitive membrane that is responsible, but rather certain structural qualities of the stimulus-conducting bodies—namely, the layered form of the casings. If we exert a constant pressure on an individual Pacinian corpuscle, it will emit a single one response, at the onset of stimulation. This impulse can be recorded. When we carefully remove the layers, however, the exposed nerve ending will respond to this constant stimulation with a prolonged generator potential that is a function of the pressure force. In the intact Pacinian corpuscle, the distension created by the pressure stimulus (an impression of 0.5 to 1/1000 mm is enough) releases a single pressure wave that is then transmitted to the sensitive nerve ending. During prolonged pressure the

Frequencies of vibration

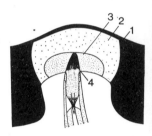

Fig. 7-12. Campaniform sensillum of a honeybee. 1. Chitin cuticula; 2. Cupola; 3. Cap (2 and 3 consist of resilin); 4. Tip of dendrite, becoming distorted when the cap forces a distension of the elastic cupola.

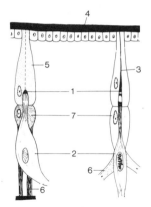

Fig. 7-13. Scolopidium as seen under an electron microscope. The tip (1) of the receptor cell (2) is connected either by way of a long end fiber (3) with the cuticula (4), or by means of a long cap cell with a fibrillary inner structure (5). Ligaments (6) originating from a cortical cell (7) run to the second point of attachment.

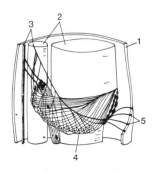

Fig. 7-14. Subgenual organ of a cockroach (extended transverse to the longitudinal axis of the tibia, and attached to the tibia's wall). 1. Cuticula; 2. Trachae; 3. Nerve fibers; 4. Subgenual organ, consisting of three separate organs; 5. Campaniform sensilla (these are activated by low-frequency vibrations, 30–100 Hz).

layers, or lamellae, become distended and so catch or absorb the continuing inflow of mechanical energy much in the way a pile of feathers would. Thus, the Pacinian corpuscles owe their capacity as vibration receptors solely to the filtering properties of the stimulus-receiving and stimulus-conducting system.

The vibratory sense of arthropods can be extremely sensitive, responding to waves in the surface of substrates or water. Their receptor organs consist of campaniform sensilla or scolopidium. These scolopidia, in turn consisting of one receptor cell and a few accessory cells, generally form clusters within a sense organ and are spread between two rigid skeletal parts or between a stiff section of the cuticula and a membrane. These, like the campaniform sensilla, respond to bulging and therefore register the movements of structures relative to one another.

If seen as a measuring device, the subgenual organ found in the leg of a cockroach—just under the "knee" formed by femur and tibia—can be classified as an acceleration recorder. When amputated and examined as a specimen, this organ will respond within its optimal frequency range (1–2 kHz) to substrate vibrations as small as 2.1^{-10} cm (by comparison, the diameter of a hydrogen atom is two to the power of ten times larger, 10^{-8} cm). But the receptor cells do not respond directly to these very small stimuli. Instead, the stimulus has to be amplified in the leg, and we are uncertain about which mechanisms govern this process. Furthermore, we have yet to discover the smallest vibrations that an intact animal can perceive. When a cockroach's leg is forced to carry part of the body weight, its subgenual organ becomes less sensitive by about two to the power of ten—probably because the resonating impulses are attenuated.

The biological significance of perceiving vibrations varies among animals. Approaching predators may be detected by the vibrations they produce, so that even in the dark a bird, for example, will fly off a branch on which it has been sleeping if it is disturbed by low-frequency vibrations that could be coming from a climbing predator. Here, the receptors in question are probably Herbst's corpuscles in the bird's legs. A spider's vibratory sense helps it to locate victims struggling in its web (see Color plate, p. 118). When an animal moves about on the ground, it inevitably produces shock vibrations in the substrate and thus may be detected by predators. Some crabs living on land are experts in this method of hunting. Several species of ghost crabs (*Ocypode*; see Color plate, p. 370) hunt for insects at night along suitable strips of beaches. If during this activity period we drop a pebble on the sand from a height of 1.5 m, a ghost crab will be likely to run right to this spot, at a very great speed, even from a distance of more than 1 m. The crab is able to locate the object without seeing it, and will try to grab it.

Vibratory signals play an extremely important role in the communication system of many animal species, especially arthropods. Male spiders pluck at the strings of a female's web during courtship. Land crabs use different methods to produce vibratory signals for courtship and conflict with conspecifics: They may beat the ground with their pincers, pummel it with their legs, or stridulate with special organs. In the ghost crab (*Ocypode ceratophthalmus*), conspecifics are able to register these signals from as far away as 7.5 m. In most cases, however the strong attenuating effect of the substrate, as well as the low capacity of these sense organs, is enough to restrict such communication to a range of centimeters or decimeters; the intensity of vibrations such as those produced by arthropods is just too low to travel farther. They are really effective only when conspecifics live close to one another.

Ideal conditions for this kind of communication exist in the hive and nest structures of insects that live in societies. It is here, in fact, that organisms use vibatory signals to a markedly great extent. In many termite species the members will beat on the ground or substrate with their body if disturbed within the nest or within the tunnels that lead to food. This induces others in the colony to make a quick retreat into the deeper areas of the nest. Bees use vibratory signals, in addition to a temporal coding of information, in their waggle dance (see Chapter 33) to indicate the distance to a food source. Leaf-cutting ants build very deep structures into the ground where part of the earth is quite loose, so that individuals often get buried after a rainfall or as a result of being "raided" by a predator. They will then send out very weak stridulation signals (only about one hundred-billionths of a watt). But as others in the colony pass over that spot, they can pick up these signals if the covering layer of earth is no thicker than 5 cm, and they will rescue the buried ants. Hungry wasp larvae use their mandibles to scratch on the walls of the cells, whereupon the adult workers become excited and feed the larvae.

The frequency spectrum of vibratory signals not directly transmitted to the receiving animal will depend on the physical properties of the conducting substrate. This means that the factor of frequency cannot be very important to the signal value of the vibrations. Our conclusion is supported by all the experimental findings to date, which show that a conspecific reacts primarily to the information contained in the temporal sequence of vibratory signals.

Whirligig beetles (Gyrinidae) usually aggregate on the surface of water. When something disturbs them, they all start swimming much faster and whirl about in total disorder. Yet the "group" stays together. The result of this activity is that a predator cannot possibly focus on one beetle. Even at their greatest speed, these insects do not collide with one another. In the dark, or when blinded in an experiment, they still have no trouble avoiding obstacles that project over the surface of the water. But if we cut off their antennae, the whirligig beetles will collide with one another and bump into the glass walls of an aqaurium even in bright daylight and with full vision. The second segment of their antenna (the pedicellus) normally lies on top of the water, while the flagellum reaches

▷
Above: During locomotion, scorpions (here *Androctonus australis*) constantly feel out the ground with their combs. Bottom left: These combs (here belonging to *Pandinus imperator*) are modified appendages of the anterior end. They are attached directly under the genital opening, which is closed by a lid. Bottom right: The hammer-shaped organs (of *Galeodes*), located at the underside of the basal segment of the last pair of cursorial (walking) legs, are mechanical sense organs of Solifugae. We know very little of their biological function.

How whirligig beetles avoid objects

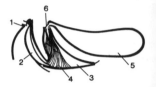

Fig. 7-15. Antenna of a whirligig beetle. 1. Point of attachment to the head; 2. Stem; 3. Pedicellus; 4. Organ of Johnston; 5. Flagellum; 6. Pedicellus-flagellum articulation.

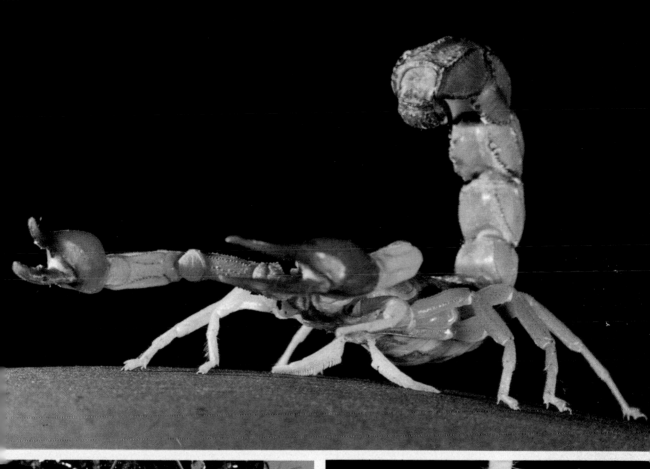

up into the air. When these two parts oscillate against each other—for example, when a surface wave causes a rise-and-fall movement in the pedicellus—this will be registered by a scolopidium organ, the organ of Johnston. Due to the large number of scolopidia and their particular arrangement, the beetles can figure out the exact direction of oscillation.

A slow-swimming beetle produces surface waves in the range of 5–20 Hz, with a maximum at 10 Hz. Waves of this frequency are only slightly attenuated at the surface of water, and they spread out at a rate of about 24 cm/sec. Unless disturbed, a beetle swims slowly, and the low-frequency waves this produces are then reflected off of objects above the surface. When these waves come back to the beetle, it registers them with the organ of Johnston as an echo effect. If, however, the beetle swims very fast, it overtakes its own waves and so, of course, there is no echo effect. But since the insect can perceive a rise in water level (the meniscus) roughly 1–1.5 cm away from an object, it is able to avoid objects even when swimming at its greatest speed.

Water striders (*Gerris*), which live on the surfae of water, feed largely on insects that accidentally fall into the water. Backswimmers (*Notonecta*), which move about underneath the surface, do the same. Both of these kinds of insects are very sensitive to surface waves, such as might be produced by a struggling prey animal. Backswimmers are pressed against the surface by upward currents in the water, and the scolopidial organs located in their tarsals will signal any displacement of the end claws. Water striders are pressed onto the surface from above by their own body weight, and we assume that a similar mechanism is at work with them.

Continuous movements of water or air cause a bending-over of hair and other stimulus-conducting systems that project over the surface of the body and are easily articulated. Under normal conditions, these currents involve more or less turbulent movements, and their force may vary within short periods of time; this means that the elastic stimulus-conducting structures will be displaced unevenly over certain lengths of time, undergoing arythmical oscillations only.

Water and air currents

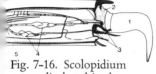

Fig. 7-16. Scolopidium organ displayed in the dissected first tarsal joint of a backswimmer. By movement of the claws (1), the end sections of the dendrites (3) extending into the claw joint (2) can be stretched and therefore activated. Receptor cells; 5. Nerve fiber.

Many aquatic animals orient against the flow of water. If they are completely immersed, they perceive the direction of flow only by sight. Thus, a minnow blinded in an experiment shows no preference in the way it moves or stands relative to the current. But as soon as contact is made with some solid object, the fish starts to adjust its orientation against the current. As yet we do not know what mechanoreceptors are used to indicate regular water currents. Experiments where such receptors have been removed or incapacitated have shown that this kind of orientation is possible even without the lateral-line system. Similarly, we have little understanding of the receptors used to indicate the direction of current by animals living at the bottom of fast-flowing bodies of water, including crabs, insect larvae, snails, planaria, and others that at times move against the flow of water (positive rheotactic response). Biologically, this behavior is important because otherwise such animals would simply drift with the current all the time, piling up at various points downstream.

The larvae of the caddis flies (*Hydropsyche*) are best suited to fast-flowing bodies of water. Here, with the use of a "spinning finger" in their mouth cavity, they construct intricate webs for trapping prey, spun as evenly as any spider web. These larvae always position their webs across the stream of water. If in an experiment we change the direction of flow, they will abandon their old web and spin a new one that is transverse to the current. If we then apply some laquer to a row of morphologically different sensory bristles on the animals's heads; thus incapacitating these structures, our subjects will no longer be able to build their webs according to the direction of flow.

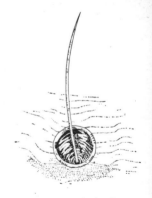

Fig. 7-17. Trichobothrium of a scorpion.

A large number of terrestrial animals orient against currents of air. In part, this would represent no more than simple spatial orientation, allowing animals such as scorpions to move along a straight line. In many cases, however, wind currents also enable members of certain species to locate a source of olfactory stimulation. If such a source is very far away, the concentration of olfactory molecules becomes too small to allow the animal any comparison with other concentrations. At such levels, olfactory molecules only have the effect of triggering an orienting response directed against the wind. In other words, the animal will move against the wind until it finds the source of stimulation, or it will systematically search for the smell as soon as the concentration suddenly decreases—a sign that it has passed the odor source.

Humans need special aids to detect the direction of very light wind. By contrast, scorpions have receptors that are (called trichobothria) sensitive to these currents, enabling them to register the direction of air puffs moving no faster than 2 cm/sec. Because of their structure, these trichobotheria tend toward a particular plane of vibration. With a constant force of wind current, their angle of displacement varies with the changing direction of flow: This angle is greatest when the plane of vibration coincides with the direction of flow, then decreases when the two factors move apart, reaching zero when the wind current approaches at a right angle to the plane of vibration—assuming that the stimulus strength is of the usual order of magnitude. Another property makes the trichobothria into excellent "weather vanes." Within their plane of vibration they respond specifically to the direction of flow. Depolarization occurs only when the trichobothria are bent to one side (called the plus side). Displacement to the other side (minus side) never leads to any conductible excitation. Scorpions have their trichobothria only on the pincer legs, whereas many other arachnid species have them on all walking legs (plantigrade limbs). In each case, the individual hairs have different planes of vibration and plus sides relative to one another. As a result, whatever the direction of wind, it will be represented by a unique and characteristic pattern of excitation.

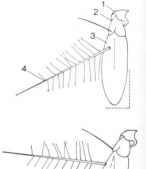

Fig. 7-18. Above: Resting antenna of a fly (Muscidae). Below: position during an approaching current. 1. Scapus; pedicellus (2) and funiculus (3) are displaced together a little to the outside; the arista (4) works as a wind sail and rotates the funiculus about its longitudinal axis relative to the pedicellus.

Without visual control, a flying insect cannot determine the direction of an air current that has a constant force. The insect does however, control its own speed relative to that of the surrounding air. One way we can demonstrate this phenomenon is by suspending a fly (Muscidae) by its back in a wind tunnel. If we blow air at it with a constant force, the insect will spread its wings to a certain width for beating. In accordance with the species-specific wing-beat frequencies, this stroke amplitude of the wings will then determine the speed of flight. When we increase the speed of the approaching air current, two things happen: The insect's antennae are turned outward, and the stroke amplitude decreases. It is no accident that these two processes occur together, because the way that the wind hits the antenna causes the amplitude to change. When we artificially and evenly displace the fly's antennae ourselves, its wing-stroke amplitude will change even when the speed of approaching air remains the same. Whether in reality we change the speed of air flow or simply make it "appear" so by directly displacing the subject's antennae, the effect will be the same: The insect responds as though its own speed of flight has inadvertently increased. The insect's feedback control system that keeps flight at a constant speed makes the animal slow down.

Normal flight speeds are still subject to visual control. A flying insect recognizes its own speed relative to the ground below because of the speed and direction at which optic patterns seem to pass by. When the insect's wing-stroke amplitude remains the same, this flow of ground patterns increases in reaction to a tail wind and decreases with a head wind. The fly now acts to maintain a certain speed of ground flow: It increases its own speed when flying into a head wind, at the same time turning its antennae inward. Although the head wind now approaches with greater force, the fly's antennae are not displaced any more than before. Therefore the sense organ continues to operate within its most favorable range. If we cut off the fly's antennae, or block them by gluing the joints, its speed-regulating feedback system no longer works because it needs to be triggered by the sensory processes described above. The sense organs in the antennae are the organs of Johnston.

Gravity (corresponding to an object's weight) is the force of attraction that the earth constantly exerts on all material objects within its field. This gravitational pull operates everywhere on the surface of the earth with only minimal variations, in approximately the same direction, roughly toward the center of the earth. The vector by which gravity exerts its force on a living organism depends on that animal's position in space, which in turn is measured by its body axes. This means that any organism able to determine the gravity vector at a given moment will also be able to control its position in space consistently, without needing to perceive any other external stimuli.

The principle by which the factor of gravity is evaluated or measured is basically the same everywhere. A body that is specifically heavier than

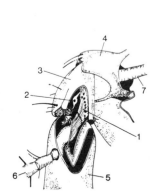

Fig. 7-19. The organ of Johnston (1), measuring the rotary vibrations of the funiculus relative to the pedicellus, and the sensillum campaniforme (2), measuring the long-term displacement of the funiculus relative to the pedicellus. Both are shown in the dissected preparation of a pedicellus (3) belonging to a fly's antenna; 4. Stem; 5. Funiculus; 6. Arista; 7. Nerve fiber of the antenna.

Fig. 7-20. Statocyst of the clam *Pecten* with the statolith lying free on top of the sensory hairs.

the surrounding material will lie on the sensory bristles of mechanosensitive receptor cells. The so-called statolith organs are the most numerous among the gravitational sense organs in the animal kingdom. In many cases these structures evolved autonomously in different animal phyla, and, despite numerous variations in detail, they are built in basically much the same way: Stimulus-conducting bristles of the sensory epithelium—in vertebrates the cilia of the hair cells—extend into a liquid-filled bubble which normally is enclosed. A statolith may articulate freely within this bubble, or vesicle, or it may be firmly connected with the bristles (or cilia) by means of a gelatinous mass. In mammals the statoliths consist of calcium-carbonate crystals; they are eliminated by cells in the statocyst wall. In decapods they are grains of sand from the outside environment, glued together by glandular secretions. They are lost everytime the animal sheds its skin, and are then replaced by new grains of sand.

Ctenophora possess only one gravitational sense organ, but medusas often have more than two. The great majority of organisms with statocysts have bilateral organs, usually lying symmetrically to the longitudinal axis of the body.

What are the functions of these statolith organs, and how do they work? If we anesthetize a fish, it will usually turn on its side or back. This shows that in its normal position the fish maintains a precarious balance, keeping upright only by constantly moving its fins. Control of this normal posture occurs through the use of its eyes as well as the sense organs of equilibrium. The fish tries to adjust itself so that both eyes receive the greatest amount of light possible from above, morphologically speaking. At the same time it strives to attain a position in which both statoliths stimulate their respective receptor cells in symmetrical fashion, signaling "belly down." Since, as a rule, the greatest amount of light in the water comes from above, the direction of the surface, both kinds of stimuli work toward the same end and serve to stabilize the fish's normal position. However, we can easily make these sources of stimulation compete with each other by making the light come from the side, keeping all other conditions constant. The fish will then turn more toward the source of light, choosing a compromise position in which its vertical axis deviates from direction indicated by either sense organ. An increase in light stimulation will cause the animal to tilt even farther in that direction, while a decrease allows gravity to exert a greater influence. Thus, in the latter case, the fish would tilt its horizontal axis more to the upright position, more parallel to the bottom.

If we compare the statolith organs of decapods such as shore crabs or river crayfish with those of vertebrates, we are struck by their remarkable functional correspondence despite the fact that they certainly evolved autonomously. In decapods as in fishes, spontaneous activity of the sensory epithelium is modulated by the statolith's shearing effect (activity increases when the animal tilts over to one side, and decreases with a tilting

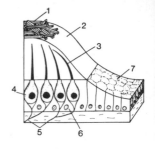

Fig. 7-21. Section of a mammalian statolith organ. The statolith consists of numerous calcium-carbonate crystals, (1) resting firmly on the gelatin of the macula (2); cilia (3) of the receptor cell (4); 5. Nerve fibers; 6. Supporting cells; 7. Epithelium cells.

Fig. 7-22. Statocyst of Ctenophora. 1. Statolith; 2. Sensory hair.

movement to the other). We find statocysts with a similar structure in mollusks, including cephalopods such as octopuses (Octopodidae); here, too, the adequate stimulus for the receptor cells is a shearing effect. In octopuses, however, the intensity of a response (e.g., eye movements) does not depend on the strength of the shearing stimulus but simply on the direction of force by which the statolith produces its shearing effect on the sensory epithelium.

In tilting over from the normal position, vertebrates, cephalopods, and crabs perform compensatory eye movements. Within certain anatomical limits, these movements serve to keep the visual field at a constant angle, especially the upright. But they stop as soon as we remove their statocysts. An intact octopus adjusts its eyes so that the slit-shaped pupils lie roughly horizontally even when the animal deviates from its normal posture. It is easy to train an octopus to discriminate a pattern made up of horizontal stripes from one with vertical stripes. It loses this ability, however, when we remove its sense organs of equilibrium.

Compensatory eye movements

Fig.7-23. Response of a fish when the mechanical field intensity (F) is doubled. Note the compromise position between light and gravity. The direction of incidence of the light (L) in the fish's eye (angle β) remains constant. Left: adjustment with normal field intensity (1 g) and light coming from the side. Right: Adjustment after the field intensity is doubled (2 g). The fish positions its vertical axis relative to the vector of the field intensity (angle α) so that the shearing force (length of the vector S) acting on the statoliths remains the same. The pressure (D) that the statolith (St) exerts on the sensory epithelium changes considerably in both cases. This demonstrates that the shearing force is the adequate stimulus of the receptor.

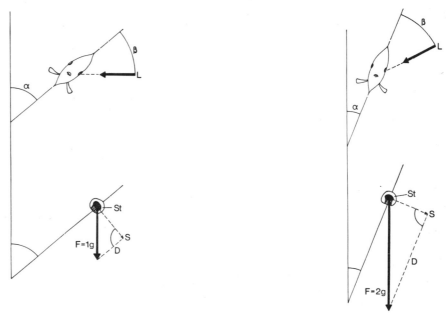

In most animal species the sense organs of gravity serve not only to keep the individual in its normal posture, but also to maintain any particular angle in relation to the pull of gravity. This directional orientation to gravity can be found most often in insects. Bees, for example, use their waggle dance to transpose the compass bearing from hive to food source into the field of gravity, and their accuracy is astounding (see Chapter 10); in other words, they translate the angle of the sun's position from a horizontal to a vertical bearing. Some beetles that scoop out cave holes in the ground accomplish this task by maintaining a certain angle to the force of gravity. Insects, with the exception of a few aquatic species, do not have statolith organs. It took many years to locate their gravity receptors; it was only in the 1960s that we learned that they are concentrated fields of bristles on different body joints. These bristles are activated by movements of the connected body parts, as well as by the relative displacement of these parts through the influence of gravity. The actual stimulation, again, is produced by a shearing effect. Thus, like many other mechanoreceptors, these gravity detectors also function as proprioreceptors and exteroreceptors. In bees and ants the bristle areas located on the joints of the trunk play a more important part than those on leg joints. By contrast, in all beetles that we know of, the entire body functions as a statolith. Any displacements of the body activate bristle areas on the leg joints, and perhaps also the scolopidial organs located in the legs.

The adequate forms of stimuli for these organs are rotary and linear accelerations as well as, of course, their corresponding decelerations. In cephalopods, decapods, and vertebrates the receptors of gravity and acceleration are combined into one complex sense organ. The functional principle is the same in all these animal groups: The statocysts or, in vertebrates, the semicircular canals of the equilibrium organ in the ear (labyrinth) are filled with a fluid. During acceleration, this fluid's inertia causes it to stay behind the movement of the membrane walls, which are firmly connected with the system as a whole. Thus the fluid actually moves against the direction of the membrane walls. When the animal slows down or decelerates, the fluid continues to move in the same direction as before. In both instances the relative movement of this fluid causes a bending of the stimulus-receiving structures. These may consist of individual long and easily activated sensory bristles (for example, the filiform hairs in the statocysts of decapods) or of cupola organs sensitive to different directional movements. The latter structures resemble the cupolas of a lateral-line organ, which in the course of evolution gave rise to the organ of equilibrium found in vertebrates.

Most vertebrates—including humans—have three semicircular canals in their middle ear, stacked at three different levels in a roughly vertical arrangement. Because of the way these structures are arranged, at least one cupola organ will always be activated whatever the direction of rotary acceleration. In addition to the semicircular canals, the statolith organs also respond to different degrees of acceleration.

By way of the highly responsive vestibular organs in vertebrates (that is, the semicircular canals with the statolith organs), complex compensatory movements (postural reactions or adjustments) may be triggered in the animal, for example when the animal is falling. Normal reflexes that occur during locomotion are also largely under the control of the vestibular system, especially when they occur during swimming, leaping, or flying in three-dimensional space. Certain rather unpleasant reactions of the vegetative nervous system, such as sea sickness or airsickness, are the result of excessive stimulation of the equilibrium organ.

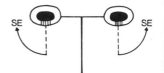

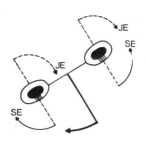

Fig. 7-24. Above: In a resting state, the rotary excitations are neutralized. These excitations would result from spontaneous excitation (SE) of the sensory epithelium; they are always directed to the counter side. Below: When deviating from the normal position, the spontaneous excitation in a statocyst (here to the left) is amplified by the excitation (JE) induced by the shearing stimulus. Spontaneous excitation would then be reduced in the other statocyst. Thus, both induced excitations work to create the same effect.

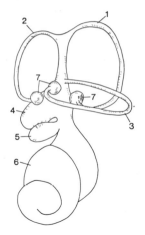

Fig. 7-25. Vestibular system of a mammal. 1 and 2. Vertical and 3. Horizontal semicircular canal; 4. Utriculus and 5. Sacculus (sense organs of equilibrium); 6. Cochlea with the organ of Corti; 7. Ampoules with cupola organs.

Flies and other Diptera have a completely different kind of sense inertia. Their hind wings are strongly involuted and have been transformed into organs called halteres, for controlling flight. If we cut off a fly's halteres, the insect will reel about in the air and then quickly plunge to the ground, even if all the other sense organs involved in flying (distension receptors at the base of the fore wings, eyes, antennae) are intact. It has not been fully explained precisely how the halteres control flight. However, we do know certain parts of this function: The halteres vibrate with the normal wing-beat frequency, but alternately with the strokes (antiphasic) and in a vertical direction, perpendicular to the animal's longitudinal axis. These halteres, which, depending on the species, vibrate at 200–1000 Hz, react like a gyroscope: As soon as the fly turns so that its halteres are rotated out of their plane of vibration, the moments of inertia act to maintain the organs in this plane. These forces wield their influence perpendicularly to the plane of haltere vibration, thus generating rotational moments that activate sense organs at the base of the halteres. The corresponding receptor cells (campaniform sensilla) then mediate the control of the animal's wing beat (especially amplitude and angle of attack). In addition, results of elimination experiments have indicated that the halteres, despite their small mass, exert a direct mechanical stabilizing influence quite independent of the receptor cells: They seem to attenuate the movements of the thorax and with it the wings.

The air or water pressure acting upon the organism evenly from all sides constitutes the adequate stimulus for the sense of pressure. We can easily condition minnows to an increase or decrease in water pressure. These fish learn to differentiate degrees of pressure varying only by an equivalent of 1 cm in the height of a water column. How is this possible? Fishes are heavier than water. To maintain their position in the open water they must either swim continuously—apparently as is done by many sharks, many deepsea fishes, and most bony fishes living at depths of 1000–2000 m; or they must be able to change their specific density. This change is achieved by means of an air bladder. It contains mainly oxygen, which is much lighter than the surrounding water medium and therefore makes the animal more buoyant. When this air bladder, or swim bladder, reaches a certain volume, the effect of buoyancy just matches that of passive sinking, and the fish is able to stay on its level in the water without actively swimming up. The air bladder works according to the general laws governing gaseous substances. When the fish swims upward, pressure decreases and thus volume increases, causing the air bladder to expand. Only rarely is this distension picked up and signaled by receptors in the bladder wall; instead it is conducted to the inner ear by way of the Weber's ossicles, which also serve to conduct sound stimuli. In fishes whose swim bladder is connected to the intestine (i.e., by structures called physostoma), such a signal may trigger an impulse to spit out air. In fishes where this connection is lacking, the same

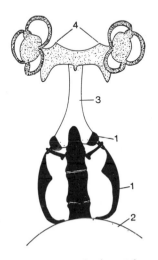

Fig. 7-26. Weber's ossicles (1) will conduct changes in pressure within the swim bladder (2) over the perilymph cavity (3) to the inner ear (4).

kind of signal may lead to a resorption of gas in certain areas of the air bladder.

Obviously, a fish could obtain information about the relative depth of its position at any given moment only by refraining from actively changing the contents of its swim bladder during ascent or descent. Such an indicator source would be especially useful for fishes that normally keep to certain depths but may change levels temporarily, for example while hunting. We do not know to what extent they use this method, but we may safely assume that the capacity to perceive pressure is a primary means of controlling the amount of air in the swim bladder. Again we have an example of how exteroceptive functions may be achieved by means of proprioceptive processes.

We have not been able to demonstrate conclusively that any particular sense organs serve only to perceive hydrostatic pressure. Similar gaps exist in our knowledge of how organisms perceive changes in air pressure. We do know that such changes influence many organisms in some more-or-less nonspecific manner. But scientists have not yet discovered whether this perceptual capacity plays some fundamental biological role and whether there are special sense organs that serve to indicate the intensity of air pressure.

▷
The anatomical structure of a vertebrate ear, using the bullfrog as an example. 1. Medulla oblongata (hindbrain); 2. Labyrinth; 3. Skull bone; 4. Columella (rudimentary form of the malleus and the incus in humans); 5. Window; 6. Tongue bone; 7. Upper jaw; 9. Auditory nerve; 10. Eardrum. Top right: Sound spectrogram of the bullfrog's call. It shows the distribution of acoustic spectral energy in the various frequencies.

8 Behavioral Functions of Hearing

The auditory sense,
by M. Konishi

Many animals use sound signals to recognize conspecifics, including mates, territorial rivals, and offspring. Others hunt by hearing, and some escape their enemies by detecting their sounds. In order to serve these biological functions, various anatomical and physiological adaptations for sound reception and signal processing evolved. A few examples suffice to illustrate this point.

Male mosquitos use their antennae to hear the wing noise of females

Male mosquitos—such as *Ades egypti*—perceive the wing noise of females of their species by means of their bushy antennae. These antennae vibrate synchronically with the sound waves. They have a large surface area for collecting sound. There are nerve cells in the base of each antenna that are sensors to register its vibration. Any acoustical system has a resonance frequence with which the system vibrates at the greatest amplitude. Male mosquitos are precisely "tuned" to the wing noise of females; they hear neither their own wing noise nor that of other males in the swarm. According to studies by Tischner and Schief, the resonance frequency of male antennae is around 350–360 Hz, close to the frequency of the female wing noise, which lies at an average of 385 Hz. By contrast, the frequency of the male wing noise is higher than that of the female by about 150 Hz, so that it is not perceived. It is possible to show that the female wing noise is the major cue used by the male to recognize the opposite sex of his species. Sounding a tuning fork which produces a pitch between 300 and 350 Hz attracts sexually mature males of *Ades egypti*. If both antennae are prevented from moving by fixing their bases with glue, the mosquito fails to recognize females.

A sexually immature male does not chase females because his antennae cannot collect sound due to the non-erect orientation of their hairlike structures. It is particularly interesting that, as the frequency of female wing beats increases with temperature, so does the resonance frequency of the male antenna.

Mosquito antennae are one example of a simple auditory organ. Another highly remarkable acoustic-warning system is found in certain

The moth ear, by
M. Dambach

◁
Infrared photograph of a barn owl flying in the dark. This photograph shows a mid-flight course correction. The owl starts by flying toward one loudspeaker (300 msec), then turns to the other side as the sound signal is shifted to a second speaker located in that direction.

nocturnal moths. Its function is to detect the ultrasonic calls of bats, one of the moth's greatest enemies. This detection system helps the moth to evade its predator. Roeder spent many years studying the "moth's ear." As we know, bats orient themselves by means of ultrasonic calls. Their echoes enable them to determine the direction, distance, and shape of objects in their environment (see Chapter 14). On their nightly forays they also locate moths, which can escape their predators only if they hear the bats' sounding calls early enough.

Many moths have auditory organs of the type classified as tympanic organs. In noctuid moths each side of the anterior surface of the thorax carries thin circular membranes which function as eardrums, or tympanic membranes. Behind each membrane lies an air sac, and crossing through this is the sense organ, attached to the inner side of the tympanic membrane with a fine tissue strand.

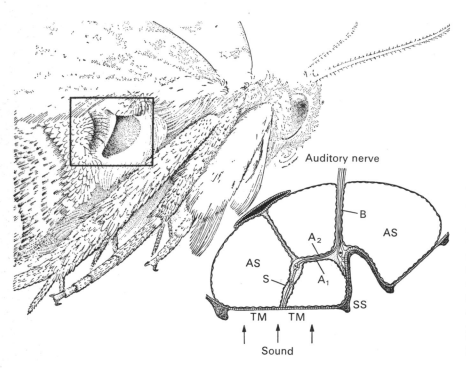

Fig. 8-1. Diagram of the auditory organ of a noctuid moth. The tympanic membrane (TM) forms the outer boundary of an air-filled cavity, or air sac (AS). A fine tissue strand, the acoustic sensillum (S), which is attached to the tympanic membrane and suspended in the air sac, contains both auditory receptors, neurons A_1 and A_2. The nerve fibers B, A_1, and A_2 come together in the auditory nerve. SS = Skeletal support.

The sense organ, called a chordotonal organ, contains only two receptor cells, each with one efferent axon (A_1 and A_2 in Fig. 8-1). Near the auditory receptor cells lies a third neuron (axon B), but this has nothing to do with the actual hearing process. Thus, an auditory nerve contains three nerve fibers, but only two of these are auditory fibers.

An electrophysiological analysis seemed the ideal method for studying the capacity of this system, because the smaller the number of neurons involved, the easier the technique. By using silver electrodes we can record impulses from the auditory nerve and match them with the individual receptor cells. Large impulses independent of acoustic phenomena belong to the B-neuron. When no sounds are produced, the A-neurons will also show a certain spontaneity. But even the rattling of a bunch of keys, which contains many overtones in the ultrasonic range of frequencies, causes strong electrical discharges. If we test different intensities of pure tones measuring, for example, 40 kHz, initially only the A_1-cell will respond. With increasing intensity the number of impulses will also rise systematically. With higher sound intensities we also obtain a response from the A_2-cell; this neuron is less sensitive than A_1 by a factor of 10. Once it has started to respond, its discharge frequency will also increase with sound intensity.

The coding of sound intensity

This, however, is not the only way of coding sound intensity. Increased intensity also results in a shorter latency period. This interval is the time between stimulus onset and the first impulse coming from the recording site. If the tympanic organs are stimulated with tones of different frequency, they will respond with impulses to sounds ranging from 3 kHz right up to 150 kHz. The greatest sensitivity lies between 50 and 70 kHz. Hence, these receptors cover the frequency range of the localization calls of insect-hunting bats, i.e., 20-100 kHz. (In comparison, the range of hearing in humans is 16 to 20 kHz). Since both A-neurons show the same responses to sound frequencies, we must conclude that moths cannot discriminate pitch.

Other findings were made when the moth research was set up in a yard with flying bats, and recordings were obtained from both tympanic organs simultaneously. Our first surprise came with the long range of receptivity: The moth's organ can register a bat from as far as 35 to 50 m. If we make parallel recordings of impulses from both tympanic nerves, we see that the side toward the bat responds sooner and with greater intensity than the side away from the bat. This is related to the way the actual receptors function and respond, as we described above, and results from the fact that these auditory organs function like directional microphones. As the bat comes closer, this difference between left and right decreases, so that both impulse responses end up being very strong and almost simultaneous. This means that the moth can determine the direction of its enemy best when the bat is still at a fair distance, and hardly at all once it has come close. This alone would indicate that the moth uses different methods of evading the predator, depending on how close it is to the bat. This was confirmed while studying the flight paths of these moths, which we can observe and record with spotlights and an open camera aperture: When the source of ultrasonic frequencies is some distance away, the moth flies directly away from the danger zone; should the bat suddenly appear above the moth, it responds—seemingly at random, but with success because the bat cannot predict its path—by letting itself drop, by diving along a circuitous course, or by flying in tight loops.

Methods used to escape

The ears of these insects are tuned to specific frequency ranges, but they cannot discriminate between frequencies within the tuned range. Although a recent physiological study by A. Michelsen indicates the existence of neurons in the auditory nerve of the locust that respond to different tonal frequencies, the vertebrates show, both behaviorally and neurophysiologically, the ability to discriminate tonal frequencies. Studies by Fay and Tavolga indicate that fishes can discriminate frequencies quite well.

Frogs are among the most vocal animals. Their songs are highly species specific, each species having a unique song. M. T. Littlejohn has provided evidence that female frogs recognize males of their own species by their song. We have seen that the mosquito and the noctuid moth use simple acoustic cues for biological purposes. The ability to discriminate tonal frequencies introduces a new dimension to the use of acoustic signals. The bullfrog is named for its bull-like voice, which characterizes the male. When a male bullfrog hears this call, he responds by calling back; he does not respond to the calls of other frog species. R. R. Capranica used this evoked vocal response to determine which acoustic parameters of the call are used by the bullfrog to define its species specificity. A complex sound such as an animal call contains many frequencies. When we analyze the frequency components of the bullfrog's call by audiospectrograph—used to measure and record sounds—we find two amplitude peaks: one in the low-frequency range, 200–300 Hz, the other in a higher region, 1400–1500 Hz, separated by a relatively low-energy band of 500–700 Hz (see Color plate, p. 127). Tests with filtered and electronically synthesized calls show that the male frog responds best when both the low-frequency and the high-frequency peaks are present. However, the response is greatly reduced if sound energy in the middle-frequency region exceeds the low-frequency peak.

L. S. Frishkopf and M. H. Goldstein made a neurophysiological study of the bullfrog's auditory nerve. In general, the response of a vertebrate auditory neuron is a function of sound intensity and frequency. Each auditory neuron has a frequency of that neuron. Different neurons have different optimum frequencies. When neurons in the bullfrog's auditory nerve are classified according to their best frequencies, the resultant histogram matches the distribution of sound energy in the call; the largest numbers of neurons occur in the frequency ranges corresponding to the two dominant frequency bands mentioned above (Fig. 8-2). It is interesting to know that these two groups of neurons innervate two separate sensory organs in the frog's inner ear.

As mentioned above, the frog's response declines when sound energy in the middle-frequency region exceeds the low-frequency peak. Frishkopf and Goldstein discovered that the response of neurons sensitive to lower frequencies (below 700 Hz) could be inhibited by the addition of a second tone which, when presented alone, has little stimulating effect. Inhibition is graded. The most effective inhibitory frequencies range from

Hearing in vertebrates
by M. Konishi

The singing of frogs

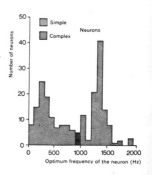

Fig. 8-2.

500 to 700 Hz, and their amplitudes must exceed the excitatoy tone by at least 6 decibels for total inhibition. These physiological findings on the distribution of optimal frequencies and inhibition correlate remarkably well with the behavioral results obtained by Capranica. More recent studies by J. J. Loftus-Hills indicate a good match between auditory neural frequency responses and the species songs in several frog species.

Bird songs

The bird ear does not seem to be as specialized as the frog ear. Each species of songbird has a unique song, and although there is no experimental evidence for its role in attracting females, its function in territorial advertisement has been amply demonstrated. The type of specialization observed in the frog ear is only partially applicable to birds. Two basic parameters of hearing are tonal frequency and intensity. These are related to each other in hearing. The lowest audible sound level is called the auditory threshold, which varies with frequency. A plot of thresholds for different frequencies is called an audiogram. The audiogram of a bird can be constructed either by determining the thresholds of auditory neurons at their optimum frequencies or by behavioral conditioning methods. [Konishi] compared the audiograms and sound spectrograms of vocalizations in several species of songbirds and found the following correlations between them. The highest audible frequency of a species is usually close to the highest frequency in its vocalizations. The frequency range of maximum sensitivity seems to be correlated with that of vocal frequencies. However, all songbirds studied so far can hear low frequencies which do not occur in their vocalizations. In birds, the recognition of one's species's song and calls is only one of the functions of hearing. The ability to hear low frequencies may be related to the detection of predators.

Hearing and prey-catching

Some animals, such as the noctuid moth, use acoustic cues to evade predation, while some, like bats, use sound to locate prey. Nocturnal animals often have an acute sense of hearing. Their ears may be highly specialized for specific purposes. Owls are well-known nocturnal hunters, although some owls hunt only at dawn and dusk, R. S. Payne demonstrated the ability of the barn owl (*Tyto alba*) to catch mice in total darkness by means of hearing alone (see Color plate, p. 129). The barn owl uses the rustling sound of prey to locate them. Owls do not have structure homologous to the external ear of mammals, the pinna, but some of them have a fold of skin extending from the forehead above the eye and along the orbit behind the ear to the base of the lower mandible. In the barn owl the skin fold itself is not so prominent, but it carries a tall curved wall of densely packed feathers that encircles each half of the face. The left and right halves meet along the midline of the face, where the feather walls from the two sides form a pointed ridge. On each side of the face, the curved wall looks like a trough with a paraboloid inner surface. The entire facial structure makes up the well-known heart-shaped outline of the owl's face, called the facial disc. This seems to play an important role in the acoustical location of prey by the barn owl. It amplifies sound and makes the ear directional. If the facial-disc feathers are removed, the barn owl makes large errors in sound location.

The barn owl can hear extremely faint noises. Figure 8-3 compares the audiograms of humans, the cat, and the barn owl. Note that the cat and the owl have similar auditory sensitivities up to about 7 kHz, beyond which the cat continues to be sensitive while the owl's sensitivity starts declining sharply. Both animals are much more sensitive than man in the frequency range from about 500 Hz to 10 kHz. Above 12 kHz, however, humans are more sensitive than the barn owl.

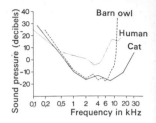

Fig. 8–3.

The rustling noises of a mouse contain a wide range of frequencies. My [Konishi] experiments show that the barn owl needs only a small portion of the mouse noises. Noises containing frequencies between 6 and 9 kHz are essential and sufficient for accurate location of prey. Rustling noises are produced by moving mice. Can the barn owl catch a moving mouse in the dark? In order to analyze this behavior, barn owls were trained to strike hidden loudspeakers emitting electronically generated noises in a dark room. These experiments clearly demonstrated the owl's ability to make mid-flight course corrections based on auditory signals. The owls made larger errors when the signal stopped upon take-off than when it continued until landing.

The most crucial test of the owl's ability to make mid-flight corrections involves the use of two loudspeakers; the signal shifts from one speaker to the other during the owl's flight. The plate on page 128 shows an owl changing its flight direction as the signals shift from one speaker to another. Notice the direction of the owl's face; the owl turns its face toward the new target position before orienting its body. In this series of photographs, infrared flashes were delivered at a constant interval of 250 msec. In the second and third exposures the owl's locations are closer together than in the others. This is because the owl reduces its flight speed as soon as it hears a shift in the target position. When the owl has to make a large course correction, it comes to a sudden halt in midair and hovers before advancing toward the new target position.

These experiments also show that the barn owl needs a very small amount of noise for mid-flight course corrections. Hearing faint and brief noises in flight and therefore correcting the flight course must be a difficult feat: How does the owl manage to do this when its own wing noises might mask the signal? Owls are known to fly much more quietly than other birds. The barn owl's flight noises are faint, and they also lack high-frequency components; most of their energy is concentrated below 1 kHz, with little energy above 3 kHz. This suggests that the owl's wing noises would not intefere with the detection of acoustic cues for mid-flight correction, since the useful cues are noises between 6 and 9 kHz. The lack of high frequencies is also advantageous for the owl because small rodents capable of hearing high frequencies cannot hear and thereby locate the approaching owl.

In many animals there are needs for parents to communicate with their young. Sound signals can play important roles in this communication. In precocial birds, chicks can walk and feed for themselves shortly after hatching. A considerable amount of vocal-auditory interactions is known to occur between chicks and parents before and after hatching (see Chapter 37). In guillemots (*Uria aalge*) B. Tschanz found that chicks in the egg can learn to recognize individual characteristics of the calls of their parents. The Carolina wood duck (*Aix sponsa*) uses tree holes for nesting. Newly hatched ducklings jump many meters onto the ground or water when their mother calls them from under the nest. According to G. Gottlieb, incubator-hatched ducklings of the mallard duck (*Anas playtyrhynchos*) approach a loudspeaker emitting the assembly call of the wild mallard duck in preference to a speaker producing the assembly call of other species.

This ability of ducklings to discriminate between the signals of its own and alien species appears as early as five days before hatching. Neurophysiological studies by me [Konishi] show that auditory neurons in the cochlear nucleus of mallard duck embryos begin responding to low-frequency tonal stimuli as early as eight days before hatching. Vocal-auditory interactions between the mother and the embryo begin two or three days before hatching. By this time, the embryo's auditory system can register the entire frequency range of the maternal brooding and assembly calls. Thus, the development of auditory neural responses is timed to prepare the embryo for the onset of prenatal and postnatal mother–young communication.

With this small selection of examples, we have tried to demonstrate how hearing functions in the behavior of animals. Other information is provided in the chapters dealing with orientation, including Chapters 14 and 20.

The mammalian ear, by H. Wendt

Nowhere in the behavior of mammals does the vocal apparatus and the emission of sounds—with some rather amazing exceptions including bats, dolphins, and primates—play as large a role as in frogs and birds. But the mammalian ear has developed into an extremely complex organ, whose means for receiving, conducting, and transducing sounds have taken on a far more complicated structure than those of other vertebrates. Most amphibians, reptiles, and birds still lack an external ear; these organisms have only one auditory ossicle in their middle ear for conducting sound, the columella, and the inner part of the ear, which transforms the sounds, is still small and uncoiled.

In mammals, as seen from outside, we particularly notice the ear conch, called the auricle or pinna. In some groups the pinna is quite prominent and may also be very mobile. Special muscles enable these animals to turn the pinna to the direction of a sound; in humans these muscles have become very much stunted. The development of such a "sound funnel," however, is not always a sign of unusually good hearing. For example, bats, which because of their echo-orientation virtually live in a world of sounds (see Chapter 14), usually have gigantic ears, while dolphins have no pinnae at all even though they have a similar system of echo-location. Apart from these exceptions, however, the occurrence of large erectile ears is a consistent indication that the auditory sense plays an essential role for the mammalian species concerned. Many of these species can perceive the finest tones, too soft or too high for our ears.

The ear drum or tympanic membrane, which closes off the middle ear against the external ear canal, receives the sound waves from the outside air. Three small bones (auditory ossicles) inside the middle ear conduct the vibrations to the internal ear. These ossicles, located in the upper part of the tympanic cavity and connected to each other, have an interesting history of development. Only one of these, the stapes (stirrup), corresponds to the auditory ossicle (columella) of reptiles and birds; the other two, the incus (anvil) and the malleus (hammer), are the remnants of the quadrate and the articulate bones, two bones that in reptiles form the articulate connection between the mandible and the skull (primary temporo-maxillary joint). In the phyletic group of mammals a new articulation developed (secondary temporo-maxillary joint), while the remnants of the quadrate and articulate bones shifted into the tympanic cavity of the middle ear. During the growth of every mammalian embryo, the joint between the malleus and the incus starts out as a temporo-maxillary joint and later moves into the middle ear.

The internal ear of mammals is coiled, much like a snail shell. This labyrinth consists of a system of tubes closed within themselves and communicating with each other. The labyrinth is filled with a clear fluid, the endolymph, which has little protein. The cochlear nerves, which take part in the hearing process, and the vestibular nerves, which serve to maintain equilibrium terminate in its epithelium. The actual auditory sense organ, which contains the endings of the chochlear nerves is called the organ of Corti. Here the incoming vibrations are evaluated and distributed to the countless receptor cells—the hair cells or auditory cells—which then transduce the vibrations into nervous excitation and conduct these impulses to the brain. In the brain they are "interpreted" according to their number and the location of the stimulated receptor cells at various sound intensities, kinds of pitch, and tonal qualities.

It is remarkable that every mammal possesses the three auditory ossicles, the malleus, incus, and stapes; none of these have ever become involuted. This means that, in addition to receiving and amplifying sound waves, they must also have another function, especially since reptiles and birds, which can hear very well, have only one such ossicle. There is a theory—as yet unproven—that mammals have a "spatial" (stereoscopic) perception of sounds, just as they can perceive visual images spatially. In any case, we may consider the external ears of mammals as directional aerials, discriminating as much as possible the nature and intensity of sound.

Malleus, incus, and stapes

By means of their auditory apparatus, mammals are able to hunt by hearing or to perceive the sounds of their predators. As demonstrated by the external ear of many murids and lagomorphs, these small mammals are capable of excellent hearing; mice even perceive frequencies of up to 100,000 kHz. The prominent ears of many species of deer and antelopes and the pinnae of other ungulates also serve primarily in the avoidance of predators. By contrast, predators such as the fennec fox, the big-eared fox, and many species of cats can use their extremely sensitive hearing to localize even the smallest of sounds for catching their prey.

In the same way, the large membranous ears of some prosimians, including bushbabies and tarsiers serve primarily in the perception of prey and perhaps also in vocal communication. With the onset of darkness, these nocturnal animals wake up, unfold their pinnae, pick up sounds, and respond to even the faintest buzzing of insects by turning their ears in the direction of the sound. Tarsiers are able to turn one pinna to the front and the other to the back, a feat probably not possible in any other mammal.

Intraspecific communication

In prosimians of medium to large size, the lemurs, sifakas, and indris, we find the practice—already described in birds—of marking off territories with the aid of calls. In its most narrow sense, this is a form of intraspecific communication like what we find in some other social mammals, including wolves. Such "auditory walls" are also created by very different monkey species, of which we shall mention only the particularly loud and vocal howler monkeys and gibbons. When male gorillas make booming sounds by beating on their chest, or when male chimpanzees beat on fallen tree trunks, it can also be interpreted as a signal for other groups that this place or territory is occupied.

This means that some species of mammals, especially primates (see Chapter 35), also use their vocal system and auditory sense for intraspecific communication. In humans, however, verbal language evolved as a completely new technique, which is not genetically determined as are most types of sounds in animals, but which must be learned. By means of this communicative system we can pass our experiences and insights on to future generations, creating a view of the universe that other organisms will never attain.

9 The Electrical Senses

Most animals use a wide variety of organs to obtain an image of their environment. The capacity of sense organs goes beyond merely indicating the presence or absence of physical stimuli; they can provide the animal with precise information about its surroundings by transmitting messages about these various stimuli. Thus, an eye will function not only to identify a source of light, but, more importantly, it enables the animal to discern shapes and colors as well as movements within the environment; the eye distinguishes between different light intensities and wave lengths. In addition to light stimuli (Chapter 5), we are familiar with chemical (Chapter 6) and mechanical ones (including sound waves; Chapter 7).

Humans do not have proper sense organs for evaluating electrical stimuli. But with the aid of measuring instruments, people discovered the phenomenon of electricity and began to study the nature of electric as well as magnetic stimuli. Then the following questions arose: 1. Can electric stimuli be placed in a separate category—that is, do they represent a different stimulus modality—providing organisms with information about their environment in a totally different way? 2. Are there organisms with specific sense organs receptive only to this kind of stimulation?

Although some biologists and physicists had speculated that some animals can orient themselves by the earth's magnetism and others by sources of electricity, only a few scientists had looked for a discrete sense responding only to electric stimuli. Years went by and no evidence for such an exclusive sensory system was offered. In the past few decades, however, more intensive investigations on so-called "electric fishes" were undertaken, and interest in the possible functions of electric discharges generated by these kinds of fishes was renewed. We should note that fishes, more than most other kinds of animals, live in an environment extremely well suited for transmitting and conducting electricity.

Scientists have assumed that the ocean-dwelling electric ray (family Torpedinidae), the South American electric eel (*Electrophorus electricus*), and the African electric catfish or thunderfish (*Malapterurus electricus*) use

The electrical senses, by T. Szabo

Strongly electric fishes

their strong discharges (50 to 600 volts) for either predation or defense. In fact, recent experiments have demonstrated that the high-frequency discharges generated by these large fish during predation are strong enough to paralyze their victims.

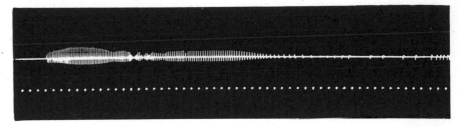

Fig. 9-1. Oscillogram of the discharge volley of an electric ray during predation.

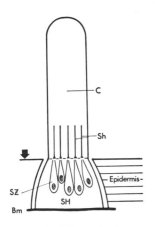

Fig. 9-2. Schematization of a mechanical receptor belonging to the lateral-line system and carrying sensory cilia (above), and of a specific electroreceptor (below). Bm, basal membrane; C, flexible cupula; Sh. sensory hair or cilium; SH, supporting cell hillock; SZ, sensory or receptor cell. Arrow indicates the surface of the skin.

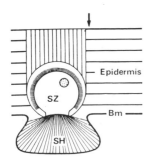

On the other hand, there are also what we call mildly electric fishes (for example, the gymnotids and mormyrids) with electric discharges only 0.1 to several volts strong—obviously much too weak for using against other creatures. Humans cannot feel these discharges, but in 1951 H. W. Lissmann used amplifiers to make them visible on an oscillograph screen and audible by means of loudspeakers, enabling him to observe the constant electrical activity of these animals. He wondered if they used their discharges for spatial orientation or for localizing objects in the water. Another idea also occurred to him: Perhaps this electrical transmission activity enabled conspecifics to communicate? In either case, the fishes would require sense organs to respond to the species-specific type of electric impulses and impulse sequences.

Nineteenth Century anatomists had in fact already discovered and described special sense organs that lay in the skin of mormyrids. Similar ones are found in the skin of gymnotid eels and gymnarchids (*Gymnachus niloticus*), which belong to a different family. These sense organs are spread over the entire body of the fish, but are concentrated in the head region. In some species we have found up to eighty such organs per square millimeter. Similar to the lateral-line organs found on all fishes (Chapter 7), these specific sense organs possess secondary receptor cells, but without sensory cilia. On the other hand, the membrane surface of these receptor cells is densely covered with a rather long, ultrafine, and striated (brushlike) border (microvilli). The receptor cell lies in a cavity within the epidermis (outer skin), and its basal plasma membrane is attached onto a knob of supporting cells lying on the basal membrane of the epidermis. Each organ is serviced by a single nerve fiber, which relays the excitatory state of the receptor cell to the brain in a coded form. Unlike ciliate cells, these specific sense organs in the skin of fishes do not contact any efferent nerve fibers extending from the central nervous system, and so their activity is not influenced by the brain.

The afferent nerve fibers leading back from the specific sense organs to the central nervous system arrive at the brain together with the nerves of the lateral-line organ. Both synapse in the lateral region of the cerebellum (lobus lateralis of the rhombencephalon). We therefore conclude that these electrical sense organs in the skin are a specialized part of the lateral-line organ.

We find this kind of sensory system not only in mildly electric fishes (mormyrids, gymnotids, and gymnarchids), but also in the electric eel, which in addition to its main organ has a mildly electrical one. The latter organ is sensitive to electric stimuli, especially those generated by the fish itself. In order to using this sensory system for orientation, the fish must feel out its surroundings by means of this electric field. How is this done?

With every discharge generated by the electrical organ, the fish surrounds itself with a momentary electric field (duration 0.1 to 10 milliseconds, depending on the type of fish). These impulses will affect all susceptible elements within the electric field. Of course, it is primarily the animal's own sense organs that are sensitive to these discharges: The organism stimulates itself.

Every receptor measures the local stimulus current, or rather the intensity of the field. This value is then coded and conducted over the sensory fiber into the central nervous system. If the electric field now becomes distorted by encountering an object with a conductivity different from that of water, a change is effected in the intensity of the stimulus current within the receptors. In this way the fish's electroreceptors can provide information about changes in its electric field, allowing it to localize obstructions within the environment. The animal obtains this information through active participation, always generating a new electric field. For that reason we use the term active electrical localization (Fig. 9-8).

The intensity of a stimulus following a discharge can be coded by means of "characteristic numbers," or parameters, of the receptor-cell re-

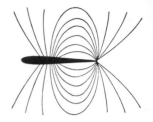

Fig. 9-3. Lines of the current generated within the electric field of a mildly electric fish.

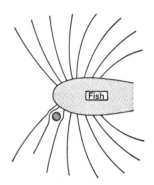

Fig. 9-4. Distortion within the electric field generated by a mildly electric fish (dotted). This distortion is caused by a nonconducting object (lined).

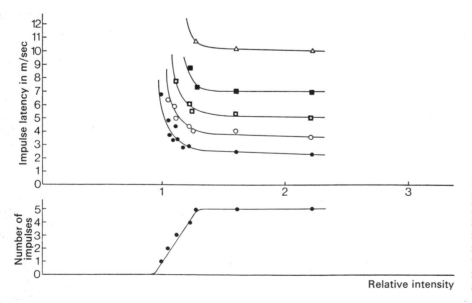

Fig. 9-5. Changes in the number (below) and latency (above) of nerve impulses in the afferent nerve fiber of an electrical receptor. These changes were brought about by an increase in the intensity of artificial electric stimuli.

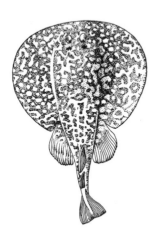

Fig. 9-6. Some species of strongly electric fishes: Above, the African electric catfish, or thunderfish (*Malapterurus electricus*); middle, the South American electric eel (*Electrophorus electricus*); below, the sea-dwelling electric ray (*Torpedo*).

sponse. Thus, changes in stimulus intensity may cause a change in the frequency of impulses or in the time interval between signal and response (latency period). An increase in stimulus intensity results in a greater number of sensory impulses and a shorter latency of individual impulses. Therefore, the first parameter encodes the stimulus intensity by number (a relatively crude method), while the second allows for a finer distinction between two levels of intensity. Typically, the mildly electric fishes with a low rate of discharge (up to 60 Hz) encode stimulus intensity by means of varying both the frequency and latency of nerve impulses. In fishes that generate higher rates of discharge (up to 1600 Hz in members of the sternarchids), this method of coding is no longer feasible because the intervals between discharges are too short. We know that only one or, at the most, two nerve impulses are produced per discharge pulse in the sensory fiber. Stimulus intensity at the receptor is shown by the probability of a discharge pulse releasing or not releasing a nerve impulse.

This is the method of probability coding. But we know of another form of coding for those gymnotid eels that display regular high-frequency discharges. Here, the appropriate receptors respond to discharge pulses with a single nerve impulse. The temporal displacement (phase relation or phase displacement) of this impulse, as related to the preceding discharge pulse, varies in proportion to the stimulus intensity.

Whatever the coding mechanism used in active electrical localization, the electroreceptors utilize the stimulus parameter "intensity" to detect a distortion of their own electric field. In this way the animal perceives the presence of a conspecific or another object. Of course, a single receptor built into this system cannot pass on information about whether an object is moving, and in what direction. Instead, the fish must combine and compute the responses of many primary fibers at different levels in the central nervous system. Certain neurons in the rhombencephalon (lobus lateralis) assess changes in stimulus intensity as coming from a moving object. The fish then perceives the movements of an object, including a conspecific, by computing the changes in its rate of primary impulses within the sensory system. It is therefore plausible that this primary activity can serve as a precise indicator for direction of movement, since opposite directions will be associated with either an activity increase or a decrease.

An object's dimensions are computed within higher-order neurons in the lobus lateralis. Complicated mathematical laws are at work in such processes (summation of derived functions), because different receptor activities (including those indicating the movement of an object) are combined.

These are the mechanisms by which mildly electric fishes use active electrical localization to investigate their immediate surroundings. They can therefore localize organisms or other objects and perceive their dimensions, movement, and direction of movement. The mechanism involves coding electrical stimulus intensities at the receptor level.

With this electrical sensory system, information gained about the immediate environment will be as detailed and accurate as that obtained through the auditory or visual senses. But the range of the electrical system is only a few centimeters all around the animal. Although the intensity of the electric field generated by a 10-cm-long *Gnathonemus* is such that it affects a conspecific 20 or even 30 cm away, distortion caused by an obstructing object will change the electric field at the surface of the skin only when the fish and the object are no more than 4–5 cm apart. Members of *Gnathonemus* (a mormyrid species) were trained to distinguish between conducting objects (metal pieces or live conspecifics) and nonconducting ones (plastic objects). These subjects, however, could no longer perform their task when the distance between themselves and the test objects was greater than 5 cm.

The receptors involved with electrical localization are special lateral-line organs. There are two kinds of electroreceptors, and both have phasic properties, which means that they are sensitive to stimulus change; therefore they will be sensitive to bioelectric stimuli or short electric discharges. These receptors have a relatively high stimulus threshold and respond only within a small range of stimulus intensities.

We may compare this active localization system used by mildly electric fishes to the sonarlike system by which many bats and dolphins orient themselves—at least insofar as all of these animals send out as well as receive the localization signals. As with sonar and radar, however, bats analyze information by evaluating the time interval between transmitting and receiving their supersonic orienting signal (echo-location). By contrast, mildly electric fishes analyze the fluctuations in intensity of their orienting signal (the electric discharge) which result from changes in the environment. In other words, bats localize objects by supersonic auditory signals, and fishes localize objects by means of direct electrical self-stimulation.

The second kind of electrical localization is referred to as passive. Here, the fish makes use of a wide variety of external electric stimuli occurring in its natural habitat. Biologists have assumed that electrically sensitive fishes can detect the direction of currents in seas or rivers by means of direct electric currents arising through the movement of large masses of water. In this way the animals can orient their migrations according to the water currents. Similarly, they utilize the earth's magnetic field by means of the induction of currents. The fish itself is a conducting object, producing an electric current by crossing the magnetic field during its travels east or west. The intensity of such a current corresponds to a potential voltage gradient of 0.2 millionths of a volt to the centimeter (μv/cm). Interestingly, the properties of electroreceptors found in all cartilaginous ocean fishes are geared to just this kind of electric stimulus (direct current), enabling these animals to perceive even very small drops in voltage.

Direct currents found in nature do not always derive from inanimate,

Fig. 9-7. Some species belonging to the three families of mildly electric fishes:

Gymnarchus

MORMYRIDS

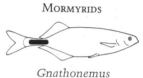

Gnathonemus

Hyperopisus

Mormyrus

GYMNOTID EELS

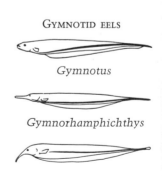

Gymnotus

Gymnorhamphichthys

In the mormyrids, unlike with the other families, the electric organs (black) are located in one concentrated area.

purely physical sources; they could also be generated by a living organism, such as a resting fish. For example, we know that sharks use their electro-receptors to detect prey hiding in the sand: The victim is out of sight but remains "electrically visible."

Far more interesting than direct-current phenomena, however, are those in which an alternating current is generated by the electrical organs of certain fishes. We mentioned earlier that strongly electric fish, such as electric rays and catfish, use their discharges not only as weapons when hunting but also as calling signals. For example, an electric eel sending off volleys of discharges will attract conspecifics. Impulses constantly gene-rated by African mormyrids and by Central and South American gym-notid eels are used not only to localize objects but also to communicate with conspecifics.

When placed in isolation, such a fish will emit the same transmitter rhythm for hours and even days. But when we add a conspecific to the tank, we find that the frequency of electrical discharges undergoes a radical change as soon as the two individuals approach to within 20–30 cm from each other. This, in fact, is the electrically perceptible distance. The "territory holder" in his hiding place will stop generating discharges as soon as the "intruder's" frequency increases. In responding to certain stimulus patterns, test animals were found to emit a tremendous variety of discharge patterns (increase in frequency, erratically changing fre-quency, discharge with constant frequency, and discharge synchronization occurring between two individuals). It seems that a mutual relationship of action and reaction develops between two conspecifics the moment they come within "calling distance" of each other (that is, within the threshold distance for electrical interaction).

However, a variety of factors influence the formation and nature of such an action-reaction relationship. A typical stimulus pattern does not always trigger the "appropriate" or corresponding response. In the mor-myrids, for example, a typical increase in discharge frequency signals a readiness to fight, yet this does not necessarily cause the rival to move away. This means that social influences are at play in the mutual discharge activity of two individuals, and that this activity is controlled by the animal as part of its hierarchy of motor behaviors. Now, as we have seen, the discharge activity of one fish can influence electrical behavior in a con-specific in certain ways. This interaction may therefore be defined as a way in which two individuals exchange information; we can say that mildly electric fishes communicate by means of electric signals.

At least some of the discharge patterns generated by these organisms do function as information signals. A mormyrid, for example, emits a typical discharge pattern when feeding. This can stimulate a conspecific who will then start to perform motor and electrical searching movements typical of food intake. The electrical activity of the first fish therefore triggered behavior in the second, much like a chemical food stimulus will

Fig. 9-8. Schematic diagram of passive electric-ally active conspecifics. Dotted area: spatial boundaries of active electrical localization (see also Fig. 9-9).

release certain responses in an animal. This means that discharge patterns of conspecifics have informational value for the fish.

It seems, then, that most of the electrical feedback processes in these mildly electric fishes are based on changes in the number of discharges. The special properties of electroreceptors associated with this sensory mechanism do, in fact, enable the organism to perceive clearly the impulse frequency of external stimuli within its entire range of sensitivity. Furthermore, when compared with the other type of electroreceptors, these organs have a relatively low stimulus threshold and are therefore specially built to detect electric signals coming from greater distances.

In summary, we may conclude that some animals do have a genuine electrical sensory system. Two functions have been described: 1. localizing animate as well as inanimate objects, and 2. localizing both physical and biological sources of electric stimuli—among other things enabling the organism to communicate with conspecifics. Up to this time, however, we have found such an electrical sensory system only among fishes.

Fig. 9-9. Schematic diagram of active electrical localization of conspecifics or other objects. Thick arrows: self-stimulation, and how this is modified by obstruction in the form of a nonconducting plastic plate (white rectangle).

10 Orientation Ecology

Orientation and the spatio-temporal structure of the biosphere, by Rudolf Jander

Life, as we know it, takes place in a shell around the earth, the biosphere. Physical living conditions within this shell fluctuate greatly from place to place and from time to time. The complexity of this spatio-temporal pattern is greatly amplified by biological diversity. Each of the several million living species of organisms is specifically adapted to only a subset of the totality of conditions permitting life. Due to such ecological restrictions, as well as geographical barriers, every species occurs only in a spatially limited section of the biosphere; its habitat. Typically, habitats of different species overlap to various degrees. What holds for the whole biosphere also holds for habitats: biotic factors, that is other species, and abiotic factors like temperature, humidity, and so on, are all unevenly distributed in space and time. This general situation requires from every individual organism, to seek and find its localized optimal combinations of living conditions, in order to survive and prosper. This requirement is met either by chance, as in the passive random dispersal of seed in higher plants, or by active, self-controlled movements in space. Active maintenance or change of position is orientation behavior.

Orientation in this sense occurs in all mobile organisms, comprising such diverse forms as bacteria, microscopic plants, and most animals, including man. Roughly a century of increasingly intense research on the orientation of living organisms has resulted in an overwhelming quantity of knowledge, recorded in many thousands of publications. These vast numbers of facts and theories can be looked at and organized from many points of view. Traditionally scientific writers preferred one of three; they either emphasized the various mechanistic aspects of orientation as in the following Chapter 11, they focused on one particular group of species as for example in Chapter 13, or they classified the respective knowledge in terms of the sensory modalities involved, like light orientation, mechanical orientation and others. It is only recently that we became fully aware that the greater number of generalizations, and hence predictions, are possible by organizing our knowledge about orientation in ecological terms. In other words, orientation is to be understood as one aspect of the adaptive interactions between individuals and their environments. Incomparably more details are known about the ecology of orientation than about it's mechanisms.

The newly defined area of orientation ecology is best framed conceptually by spatial (geometric) descriptions of the possible movements of mobile organisms, by the various sources for spatial information and finally by the spatio-temporal distribution of biologically relevant factors in the individual environments. These three aspects will be discussed in the order just mentioned.

In order to describe exactly the positions and movements of individual organisms in space it is necessary to refer to the three main body axes, the longitudinal, the transversal and the dorso-ventral axes, as is described in Fig. 10–1. A body may actively or passively rotate about any one of these three axes, and we discriminate correspondingly between roll (longitudinal axis) pitch (transversal axis) and yaw (dorso-ventral axis), as indicated by circular arrows in the illustration. In addition to these three types of angular displacement there is another set of three possible linear displacements, as indicated by straight double arrows in the illustration, forward-backward (longitudinal), left-right (lateral), and up-down (dorso-ventral). Thus, for a complete description of an animal's movement in space it is necessary and sufficient to account for the magnitude and the sign of all six geometric types of displacement. Of these six movements, the control of roll and pitch is most important for positional orientation, and the control of longitudinal displacement and yaw for orientation from place to place.

Pure descriptions of movements in space are scientifically unsatisfactory unless they are accompanied by some causal, evolutionary or ecological explanations. The latter entail the understanding of orientation as a form of behavioral adaptation, learned or evolved, to the spatio-temporal pattern in which biologically relevant environmental factors (such as food, predators and so on) occur. This statement gives rise to the fundamental question of what are the orientation strategies and means that organisms use for increasing their adaptiveness; how do they manage to be at the right place at the right time? To approach this central question of orientation ecology, it is expedient to classify all biotic and abiotic environmental factors into two broad categories, resources, that are all the factors that promote well-being, and the opposite, stress-sources that are all the anti-biotic factors. These two key terms allow the definition of an important third one, orientation-fitness, which is the ability of living beings to minimize their distances from resources and to maximize their distances from stress-sources.

Orientation, self-controlled movements in space, requires in every instance spatial information or knowledge. Therefore, sources of information have to be available for every act of orientation. All sources of

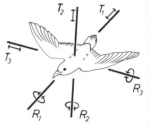

Fig. 10–1. The six degrees of freedom for movement in space of a freely moving animal. Double headed arrows indicate three possibilities for rotation and three possibilities for translatory movements with respect to the three main body axes.

Orientation and the sources of spatial information

spatial information for the control of orientation fall into four, and only four mutually exclusive categories. *Immediate environmental information*; *learned information*, that is past environmental information stored in the memory of an individual; *innate information*, that is environmental information accumulated through the process of evolution and stored genetically and finally, *random information* controlling arbitrary spatial movements. These four sources of spatial information play different characteristic roles in the control of spatial orientation depending on the ecological context.

Immediate environmental information instructs about directions and distances in the environment of an organism by means of sensory perception. A hungry toad, for instance, turns around toward a worm it had visually discovered (directional orientation), approaches the worm up to striking distance (distance orientation), and then snaps at its prey. For everyone watching wildlife while walking, it is a common experience to see birds or mammals escape away only after being approached too closely, less than the flight distance. It is not very difficult to recognize in these two and countless other examples how orientation behavior is controlled by instantaneously available environmental information about directions and distances. A common aspect of many such examples can be used to establish a general rule: If the where and when of a resource or stress-source is highly unpredictable, fitness increasing orientation is best achieved by means of instantaneous sensory control. This aspect of spatial orientation is so obviously important that most writers in the past mistakenly considered all spatial orientation as exclusively controlled by immediate environmental spatial information.

Even though immediate sensory spatial information is necessary for controlling orientation behavior in the instances mentioned above, it is definitely not sufficient for controlling all spatial aspects of the orientation response discussed above. With the direction to a resource or a stress-source given by sensory information additional spatial information is necessary for selecting the proper orientation movements out of all the possible movements (Fig. 10–2). The information for deciding between the various possible responses has to be available inside the animal. Such internal spatial knowledge can only be either learned or inherited, in

Fig. 10–2. Interaction of extrinsic and intrinsic directional information. Both the frog and the fly see an object on their right side. This is a source of exrinsic directional information (dashed arrows). The response to the directional stimulus is a left hand turn by the fly but a right hand turn by the frog (drawn out arrows). The difference in the two turning responses is controlled by intrinsic directional information.

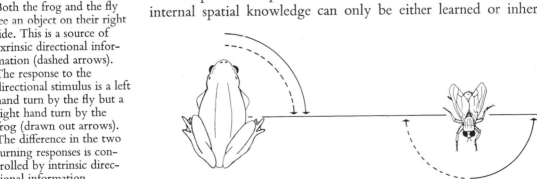

other words, be memory or instinctive information. Young domestic chicks, frogs and toads, for instance, are known to turn towards grains or small moving objects, respectively, even without prior spatial experience with such a situation. So the spatial information necessary for them to select the proper orientation response is innate or instinctive. A newly born quokka, a kangaroo, crawls upward on the belly of its mother rather than in any other possible direction and thus finds her pouch. The European garden warbler (*Sylvia borin*), even when raised in captivity, knows innately to head to the SW for the fall migration by properly responding to the earth's magnetic field, and many young vertebrates even human babies are known to avoid crawling over a steep cliff even without prior experience of falling down. The common aspect of such examples can be cast into a rule that states that relatively simple, highly predictable spatial information of biological value tends to be provided by inherited experience or instinct. The respective orientation movements, however, cannot simply be called "instincts" because only part of the spatial knowledge used in their execution is provided by an instinctive information source.

Learned information is the second type of information, besides ininstinctive information, that can be used for the intrinsic spatial control of orientation behavior. Whereas immediate sensory information tends to be used in highly unpredictable situations and innate information in highly predictable contexts, the use of learned spatial information is biologically meaningful in ecological contexts that fall between these two extremes, provided the respective organism has enough brain capacity to cope with a particular orientation problem. Spatial situations that are most important for orientation learning are those that stay relatively constant over the whole or part of an individual life time, but are different from individual to individual. This very set of conditions is met, for instance, by the layout of individual home-ranges, in which the spatial constellation between various resources and stress-sources, such as nesting sites, water holes, food patches, or activity ranges of predators, tends to be fairly constant within an intermediate time range, but is generally very different from place to place, and over long periods of time. Home-range or topographic orientation will be discussed below in a separate section.

Partial or complete ignorance about the location of resources and stress-sources is not a sufficient condition for an animal to forego oriented movements. Instead newly generated or randomly guided spatial movements are generally biologically more meaningful. The result is "trial and error localization" of resources or stress-sources. If the places thus located are memorized for the purpose of later using this spatial information, the whole process is labelled "spatial trial and error learning."

Ignorance is not the only reason for resorting to sources of random information in the control of orientation. Apart from the fact all movements by necessity are conducted with a certain randomness or error, there is one group of situations where random movements, that is highly unpredictable movements, are advantageous. This is when an animal is chased, say by a predator or a conspecific in a territorial or other dispute. The difficulty of approaching and catching a prey is certainly greater the more unpredictable the directions and speeds of an escape course are.

The six ecological categories of spatial orientation

Having discussed the ecological and orientational relevance of the four basic sources of behavioral information attention is now focused on the spatio-temporal structure of the environment. Such structural properties will be used to characterize six fundamental ecological types of spatial orientation. These six types will be defined first, and then discussed in more detail with examples.

First, a mobile organism may already be at the proper place and its body axes may be properly aligned. In such a situation orientational movements quickly become necessary in order to counteract outside disturbances. These movements that serve to maintain a certain position in space are called *positional orientation.*

Second, once the organism is about to change its place and position, the spatial patterning of resources and stress-sources becomes an important factor. Such sources are either concentrated on local spots, and spatially restricted patches or they extend in at least one dimension. In the former two cases such sources are considered as biologically relevant objects and all spatial orientation with respect to such objects will be called *object orientation.*

Third, there are among the spatially more extended resources and stress-sources those that exhibit a rather characteristic vertical or horizontal distribution or gradation. The former aspect is referred to as stratification or layering and the latter as zonation. According to these two spatial attributes it is convenient and meaningful to discriminate between *strato-orientation* and *zonal orientation.*

Fourth, the spatial structure within the daily activity range of an animal may have more complex and unique spatial attributes that do not change too frequently over time. If within such a framework the location and constellations between various resources and stress-sources are learned, then, an orientation being controlled by such learned spatial information, is called *topographic* or *home range orientation.*

Finally, animals may leave their normal daily activity range for the purpose of dispersing or long distance migrations. Spatial orientation utilized during such far reaching movements is classified as *geographic orientation.*

Positional orientation

Virtually all freely moving multicellular animals ranging from coelenterates and flatworms to arthropods and vertebrates exhibit species characteristic body positions in space which they maintain by positional orientation. Positions are usually related to two environmental factors, the direction of gravity and the position of the substrate. In weighty terrestrial animals the direction of gravity is more important; they are mostly seen in upright position, even on slanted substrates. With decreasing size positional orientation with respect to the substrate gains in importance over positional orientation relative to gravity as house flies demonstrate by walking equally easily on the ceiling as on the floor. Smaller vertebrates, insects, and many other terrestrial animals, frequently exhibit a positional orientation that is simultaneously controlled by the direction of gravity and the direction of the substrate. For instance, locusts as woodpeckers mostly keep their heads upward while facing a vertical substrate.

Once the solid substrate is left for free flight or swimming, one might first expect that position is only maintained with reference to the direction of gravity. However, it has been shown for many aquatic animals like shrimps and a variety of other crustacea as well as for many species of fish that in addition to gravity, the average direction of the overhead light is also an important sensory reference for maintaining positional stability against passive tumbling that is rolling and pitching (see illustration page). This type of visual positional orientation is referred to as "dorsal light orientation," or in the few instances that a species normally swims upside down, as brine shrimps (*Artemia salina*) do, as "ventral light orientation." More details about the mechanisms of positional light and gravity orientation are found in Chapter 11.

The positional orientation of animals in flight is only poorly studied. Nevertheless, some interesting specific adaptions have been discovered. Flying birds, for instance, face the problem of having their main gravity sense organs located in the head which they want to turn quickly for the purpose of visually scanning their three-dimensional environment. This, no doubt, renders the maintenance of body stability with respect to gravity a difficult problem. It has been solved, as shown for domestic pigeons, by having another set of less specialized gravity sense organs in the intestines that are used for maintaining flight position.

In flying insects the role of gravity for maintaining body position is still poorly investigated. We do know, however, that at least some insects, like locusts and dragonflies, utilize a dorsal light orientation. Migratory locusts also refer to the visually detected line of the horizon by matching the position of their horizontal body plane to the position of the horizon. Flies finally, as all dipterans (crane flies, mosquitoes, midges, deer flies etc.) command an unique means of positional orientation. Their second pair of wings has been transformed beyond easy recognition into small drumstick shaped structures, the halteres, that rapidly oscillate during flight so that any initiation of tumbling during flight is quickly detected mechanically by batteries of small sensory structures at the bases of these organs. After amputation of their halteres house flies, for instance, are incapable of maintaining their flight position and helplessly tumble down to the ground.

For swimming organisms type and importance of positional orienta-

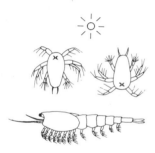

Fig. 10–3. Change in the predominant light orientation and position in space during the development of a crustacean, for instance some shrimps. The microscopic nauplius larvae prefer orientation toward or away from the light whereas the adults prefer the dorsal light orientation. The larvae are greatly magnified in size.

tation is a function of overall body size. Large to moderately small animals typically prefer a position in which the longitudinal body axis is kept horizontal. As we approach the range of millimeter-sizes and below there is an increasing tendency to maintain the longitudinal body axis vertical as is seen in various species of water-fleas of the genus *Daphnia*. Apparently, size itself is not the decisive criterium for the preference of an upright swimming position, but locomotory capabilities since large but relatively sluggish jellyfish also demonstrate this preference for the upright posture.

Since many fair sized aquatic animals start their individual lives as tiny larvae, examples for the above rule for preferred positions can even be found in succeeding developmental stages. The microscopic nauplius larvae, typical for many species of crustacea, are only known to show orientational reactions to the directions of light and gravity that keep them in vertical positions whereas later developmental stages and the adult forms typically prefer the transversal orientation to these two directional stimuli (Fig. 10–3). This changeover of preferred positional orientation coincides with the morphological development of the paired lateral compound eyes instead of, or in addition to, the primitive unpaired median eye of the nauplium larva.

Animals not only exhibit positional orientation by maintaining their angular body position around all or some of their three body axes: positional orientation manifests itself also as the ability to counteract linear or translatory displacement. This is particularly important for swimming and flying animals that may passively be carried away in any direction by currents of the substrate. Thus, any fish in running water has to swim constantly against the stream in order to maintain its position. This is achieved by detecting passive displacement visually. Therefore the fish has to see at least some sections of the surrounding solid substrate. Down stream drift can, of course, also be prevented by some mechanical interactions with solid substrate. Flying insects, except for very weak flyers like plant lice (aphids), are generally known to compensate passive wind drift by heading appropriately into the wind direction. Birds compensate similarly.

Maintaining the body position relative to the environmental space serves a variety of functions. Most important among those is counteracting the tilting and displacing effects of gravity in conjunction with the need to place locomotory organs on the substrate or to have wings produce lift that compensates for falling. Sinking in water is not a problem for locomotor orientation in fish that can control the volume of their swim bladder and thus are in almost neutral equilibrium with their environments. Animals without swim bladders, like sharks or shrimps and most other aquatic invertebrates, are all typically heavier than the water volume they replace and therefore they have to position themselves in such a way, while swimming, that a sufficient component of their swimming force is directed downward in order to counteract passive sinking. A series of other less general biological needs are met by positional orientation in various contexts. Camouflage is an example. Many freely swimming animals have a dark dorsal and light ventral surface that renders them less conspicuous against the respective backgrounds, provided they maintain the proper position in space. The best solution for this problem of camouflage has been discovered by a pelagic squid (*Abraliopsis* spec.) which produces on its downwardly kept ventral surface light of an intensity just enough to make his silhouette disappear when viewed against the bright upward background. Another example is positional orientation in the service of moulting. Larger insects in the vegetation typically hang downward when shedding their cuticle thereby utilizing the support of the gravitational pull to slip out. Finally, the important role of positional orientation in behavioral temperature regulation is mentioned. Many insects, reptiles, birds and mammals at suboptimal temperatures are known to expose a large body surface to heat radiation by placing themselves in a position perpendicular to the sunlight. Conversely, if it is too warm, and no other ready solution at hand, like moving to a cooler place, body position is changed from perpendicular to parallel with the sunlight, thereby minimizing heat exposure.

In microscopic, swimming, oligocellular to unicellular organisms positional orientation decreases in importance with reduction of size. Typically, all such small organisms; bacteria, protozoa, unicellular algae, small cellular colonies of algae and very small larvae of animals, like the trochophora larvae of many marine annelid worms, propel themselves by means of rotating or undulating flagella and cilia thereby describing swimming courses that are spirals. In other words, they do not possess a defined, and well oriented ventral side since their bodies continously rotate during swimming. The only rudimentary form of positional orientation that might be found in these microscopic forms is some tendency to swim upward in order to compensate passive sinking. For virtually all freely swimming bacteria even this is not necessary because sinking speed is extremely slow due to their minute body size.

Positional orientation is mainly a concomitant of animal mobility but its major function is approaching resources and avoiding stress-sources. This will now be discussed.

Object orientation is the ability to find and approach pointlike or spot-shaped resources and avoid similarly concentrated stress-sources. Object orientation is the only form of spatial orientation, and the only form of behavior, occuring in all freely moving organisms ranging the enormous span from bacteria to man. Despite countless specific adaptations of object orientation in various contexts, some general principles can be formulated that hold for the great majority of organisms with object orientation. In the sequel, first these general principles will be formulated, then examples for specific adaptations will be given.

Object orientation

A complete sequence of object orientation can be broken down into two distinct phases, search and approach, which are first discussed for resource localization. During the first phase an organism has no precise knowledge concerning the whereabouts of a required resource. Orientation movements in such situations that increase the probability for finding the location of an object are called *searching movements*. Once a resource is precisely located but is till some distance away, the second, or *approach phase* follows which is typically a movement straight towards the goal.

With respect to stress-sources the orientation strategy is largely the converse of resource orientation. Given that a stress-source, say a predator, has been precisely located, then it is frequently meaningful to avoid this place by moving straight away. However, active search for a stress-source may not be advantageous, especially in the context of predator avoidance. On the other hand, it is meaningful for the holder of a territory, say a fish or a bird, to patrol its range in search for potential conspecific intruders that can normally be driven out, once being spotted, without too much risk involved.

Typically, the search phase of object orientation can be broken down into two subphases, *ranging* and *local search*. During ranging the searching organism has virtually no knowledge about the places where resources, or perhaps stress-sources, are located. For instance, as one watches a housefly ranging on the breakfast table or a thrush ranging on the lawn, they both move along fairly straight lines, thereby covering much ground and hence, increasing the likelihood of finding patches of food objects, sweets or earthworms respectively. Straightline locomotion, however, usually does not continue over the whole ranging phase. Ranging animals typically turn onto new directions after periods of straight locomotion. Such turns are either spontaneous or occur at the edge of the search area. The effects of such occasional turns are always the same, the ranging animal remains within its spatially restricted search area, habitat, or home range.

During ranging the searcher may encounter an indicator of an object's proximity. For instance, a stimulus like a diffuse odor may indicate the closeness but not the precise location of food or a predator; also a resource object may be found that is too small to completely satisfy a need. Such proximity indicators elicit local search that differs from ranging by its restriction to a much smaller spatial area (Fig. 10–4). Local search is usually executed as a series of relatively tight alternating turns resulting in a convoluted search path. Local search, as ranging, serves to take the animals to places where they can localize and then directly approach resource objects. In the instance of finding a small, non-satiating food item the locally searching animal acts in a biologically meaningful way because resources in natural and artificial human environments usually occur in patches.

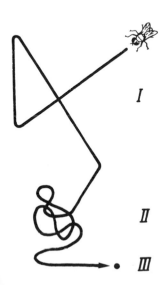

Fig. 10–4. The three fundamental components of object orientation, ranging (I), local search (II) and approach (III).

It is a remarkable fact that the three fundamental components of object orientation, ranging, local search, and approach or avoidance, are already present in the most primitive type of behavior known; that of motile bacteria. Take, for instance, the human colon bacterium, *Escherichia coli*, an abundant inhabitant of our intestines. For this minute organism (1/500 of a millimeter) a local concentration of a sugar (e.g. lactose) constitutes a resource object and a local concentration of an acid a stress-source. During ranging such a bacterium maintains a straight swimming course in any direction of its three-dimensional aquatic environment by continuously rotating around its longitudinal axis and propelling itself with cilary motion. As in the ranging of animals, the straight course is occasionally interrupted by brief changes of direction.

Food objects are surrounded by decreasing gradients of molecular concentrations. Once the ranging bacterium enters such a gradient of sugar it could swim either up or down the gradient. It is capable of discrimination between these two possibilities by comparing stimulus intensities sequentially. If such sensory information indicates movement down the gradient this is taken as a message of resource proximity; the bacterium starts tumbling, in other words, engages in local search. If the sensory message indicates movement up the gradient of sugar concentration, perhaps after some tumbling, this is interpreted by the bacterium as the successful sensory localization of the resource object which is then approached by straight swimming.

Needless to say, object orientation does not always follow the complete sequence of its fundamental components as described above. A ranging animal may suddenly spot a resource and approach it without intervening local search. A sit-and-wait hunter, for instance, does not search and sometimes need not even approach its prey. Finally, some animals in certain life stages not only completely lose object orientation but any form of orientation such as the filter feeding sedentary oyster that moves about and orients only at its early, dispersing larval stage. Also, many parasites of animals and plants give up locomotion and orientation for prolonged periods of time after the host has been located either during an early active larval stage as in many parasitic worms or still earlier during the maternal egg laying phase, as in flies with parasitic larvae. The female bot-fly of the New World tropics so locates a mammalian host, even man, deposits an egg on the skin where the larva grows and develops in the subcutaneous tissue.

Much more interesting than the degeneration of object orientation is its diverse elaboration in countless species and situations. The following few examples can only give a feeble impression of an immense, still little organized research area within orientation ecology. Diverse perceptual mechanisms supplement or substitute for primitive chemo-location of stress-sources and resources shown by *E. coli*. Vision is of prime importance in birds and many other animals; small bats use acoustic echo-location as described in Chapter 14 of this book; and certain fish rely on electrical senses as detailed in Chapter 9. The search phase, too, is subject to many specific modifications, depending on the constraints of individual environments and the degree to which partial knowledge about object localization is utilized. This is illustrated by the following examples.

Partial knowledge can render search movements more efficient than just random locomotion. Thus, larvae of lacewings (Genus *Chrysopa*) like many other insects that eat plant lice (aphids) concentrate search efforts on the tips of plant shoots where their prey tends to concentrate. Unilateral or mutual knowledge of the opposite sex's whereabouts facilitates mate finding; in most species of bees searching males patrol from flower to flower where they are most likely to find females, and in many species of ants the swarming sexuals tend to converge at defined rendezvous places, the tops of tall outstanding objects like towers, big trees, etc. The common feature of those latter examples can best be described as two-step object orientation. First, a larger object is searched for and approached, then on or near this object a similar orientation procedure is repeated with respect to the ultimate object.

Cropping strategies are another way of elaborating the search phase of object orientation. A cropping animal has to locate many food objects in order to satisfy his needs. Those objects are often sedentary and easily detectable so that the searched area need not be revisited until the resource has renewed. Efficient cropping implies that each place is visited only once at a particular time and revisited only after a specific resource renewal time. Furthermore, because of the energy costs ensuing and exposure to potential dangers such as predation, the overall length of a cropping path should be minimized. Under a variety of environmental constraints various cropping paths have been discovered. Some characteristic types will be described.

"Outline tracing" is the characteristic strategy of ants foraging on more or less horizontal branches and leaves (Fig. 10–5). The search is ruled by either one of two decision sequences that are merely mirror images of each other. It suffices to present one of the two. While walking on a branch the ant turns right at every intersection. Upon reaching dead ends on the tips of leaves or shoot terminals the ant makes a left-handed turnabout. On the surface of the leaves the ant tends to stay near the right-hand edges, and thereby covers larger leaf areas by undulating movements. In the alternate decision sequence another individual, or the same individual at some later period of time, is simply exchanging sides in the decision process so that it traces the outline of the branch on its left side rather than on the right as in the previous example. It is obvious from this account that the objective of the cropping search is met efficiently and elegantly; every leaf of the branch is visited only once and the overall path length is as short as possible.

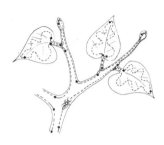

Fig. 10–5. "Outline tracing" in an ant searching for food. The ant keeps to one side and by this strategy visits all parts of the branch just once.

On the two-dimensional continuous surface of the ground other cropping strategies have to be employed. Two major, not sharply separated, categories can be discriminated, depending on whether the respective animal is nomadic or occupies a more or less permanent refugium to which it returns. For many nomads "meander cropping" has been described. An aquatic snail grazing a patch of algae or a fly licking a layer of sugar may first turn left, clearing a strip, then double back to the right for the next strip and so on as sketched in Fig. 10–6I. This strategy also meets the stated efficiency criteria for cropping.

For an animal with permanent resting and hiding places (refugia) meander-cropping is hard to reconcile with the need for returning to the starting point. They employ various kinds of radial cropping (Fig. 10–6) Fiddler crabs as many other ground animals with a burrow may either follow straight radial feeding paths so that the return trip simply retraces the outgoing path or they employ the more sophisticated radial looping which appears more efficient because unexploited places are continuously encountered during the whole roundtrip. A final example is "radial sweeping" in which the animal goes out in one direction from its burrow and keeps foraging on one side of its path while going to and fro along the radius thus harvesting a gradually increasing sector which finally closes to a complete circle (Fig. 10–6IV). This type of cropping is characteristic, for instance, for the intertidal crabs of the genus *Scopimera* (Ocypodidae) relatives of the above mentioned fiddler crabs (Genus *Uca*). These crabs live on somewhat muddy sandy tropical beaches where they sift organic material out of the sand. They disappear in their burrows during high tide and manage to "sweep" approximately a full circle during the period that the sand is dry. A similar cropping strategy has been described for the North American, grain harvesting, desert ants of the genus *Pogonomyrmex*.

It is noteworthy that meander cropping and various types of radial cropping are one of the few forms of complex behavior whose existence is geologically documented by so-called trace fossils present since the Cambrium, about a quarter billion years ago.

Object orientation is the most ancient, most general, and most direct type of spatial orientation. In addition, all animals also have more indirect ways of approaching resources and avoiding stress-sources. One of these involves the proper use of habitat stratification by strato–orientation.

Strato-orientation, self-controlled movements across the vertical layers of a habitat, is widespread among mobile organisms from minute unicellular forms to mammals as large as monkeys and apes. The ecological importance of such vertical movements relative to movements in horizontal directions increases with a decrease in body size of the orienting animal. This rule is made plausible by two facts. First, the smaller the size of an organism the easier it is for it to move vertically due to an improving ratio between locomotory power and body weight. Second,

Strato–orientation

the smaller an organism the smaller, in general, is its average speed of locomotion. This puts a premium on vertical movements because vital physical variables like temperature, humidity, light intensity, usually change much faster in the vertical than in the horizontal direction.

Obtaining the sensory information for the directional control of vertical movements is no great problem for most animals because they have sense organs for detecting the direction of the earth's powerful gravitational pull. Light is frequently used as an additional source of directional information; animals that tend to go downward usually also tend to turn away from the direction of light and the opposite holds for animals moving upward.

The most spectacular, and best studied, of all movements controlled by strato-orientation are the vertical migrations of aquatic zooplankton, mostly small crustaceans but also many other freely swimming forms, jellyfish, various worms (Nematoda, Chaetognatha, Annelida), snails, insect larvae, fish larvae and so on. Each day, both in still, fresh water and in all the oceans, billions of animals migrate upward for the night and downward again for the daytime. The amplitudes of these daily vertical migrations range from a few centimeters to several hundred meters.

How are these movements controlled? The up-down directions, as is typical for all strato-orientation, are usually given by the directions of light and gravity. The specific problem for understanding the control of the daily vertical migrations of zoo-plankton is their initiation and termination. Some releasing and terminating factors have to be responsible for the timing and the amplitude control of these migrations. Observations in the field and the laboratory showed that change in light intensity is the major but not the only factor responsible for the required control.

One of the most instructive natural experiments is provided by a total solar eclipse. The moment the daylight disappears the planktonic organisms migrate upward and then downward again as the eclipse terminates. It is easy to elicit similar responses to changing light intensities in the artificial environment of an aquarium. It is now definitely known from evidence of this sort accumulated over a period of almost a century that increasing light intensities release migratory movements in zoo-plankton away from the light source and downward and that decreasing light intensities have the reverse effect. In other words, planktonic animals have a photo-preferendum, an optimal light intensity toward which they move; a higher than optimal intensity initiates downward migration and a lower than optimal intensity releases the upward migration. This relationship is still a simplified description of the major control system, in reality the overall control of vertical plankton migration is much more complex and not yet understood in all its finer details.

The complement to the use of environmental information for the control of daily vertical migrations is internal control by a physiological

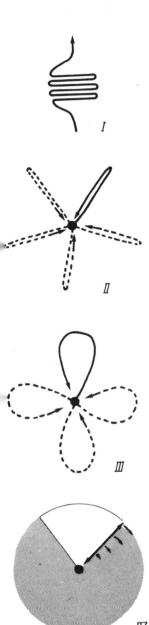

Fig. 10–6. Various cropping paths of ground living animals. Meander cropping (I), radiating U-returns (II), radial looping (III) and radial sweeping (IV). Black dots indicate the locations of burrows and arrows the directions of locomotion.

clock. This has been demonstrated in animals that burrow into the ground after their early morning downward migration, as in the pink shrimp (*Penaeus duorarum*). It is conceivable that a buried animal has difficulties in detecting changing light intensities above ground and if this is so, an intrinsic timing system telling it when to come up again has definite biological advantages. More about the nature of such physiological clocks is found in Chapter 12.

One may wonder what the biological benefit is for such an elaborate and energetically expensive behavior as vertical migrations of the zoo-plankton. A definitive answer cannot be given yet, but the most plausible is the interpretation that the daily downward migration is an escape from visually hunting predators and the nocturnal upward migration then serves feeding in the upper photic layers that are rich in algae. Strato-orientation as preventative predator avoidance also occurs on land. Baboons in Africa, for instance, typically forage on the ground during daytime and sleep high on cliffs or in trees at night.

The spatio-temporal pattern of the abiotic environment is responsible for countless adaptive types of strato-orientation. Virtually all mobile aquatic organisms move upward whenever there is a shortage of oxygen and, conversely, if the water temperature is too high or if a saltwater animal accidentally is surrounded by freshwater, downward migration is initiated. Given normal stratification of abiotic factors these responses all tend to move animals to less stressful environments.

In terrestrial habitats strato-orientation is a vital component in the humidity control systems of many small animals, especially those living in the soil. Whenever the topsoil dries out most soil inhabitants, earth-worms, mites, insects (to mention only a few) migrate downward. This behavior can be used for collecting purposes. A device for this is the "Berlesi-Funnel," a vertically suspended funnel with a wire mesh on top. A soil sample is placed on top of this mesh and gently dried out with an overhead lamp. The downwardly migrating soil inhabitants fall into the funnel and from there into a collecting vessel underneath.

The short-term avoidance of stress-sources and the direct approach of resources are not the only adaptive functions of strato-orientation. It also aids in passive dispersal. Most bottom living marine invertebrates, from sponges to starfish and crabs, have small dispersal larvae. Typical of such larvae is an initial upward migration into higher aquatic strata where currents carry them some horizontal distance whereupon they migrate downward again in order to settle somewhere on the substrate. Similarly on land the rather slowly flying winged plant lice (aphids) fly up into the air to be carried by air currents some distance, hopefully finding a new host plant. Similarly, in the Fall, young spiderlings of several species climb up in the vegetation, and spin threads of silk out into the wind until an aerial raft results to which they can cling and then passively float as gossamer to some undetermined destination.

Though strato-orientation allows organisms to adjust their position in space vertically, zonal orientation is used for adjustments of position in the horizontal plane.

Zonal orientation

Zonation or horizontal, and subhorizontal, differentiation of habitats is brought about primarily by sloping ground levels, by local differences in the soil or by various random events like tree-falls in forests. Animals with zonal orientation take advantage of changing living conditions across zones. The most drastic zonal change takes place where land and water join. *Cross shoreline orientation* is one of the most conspicuous and also the best studied form of zonal orientation. We can discriminate between three major types of zonal orientation by the complexity of the mechanisms involved, rudimentary, primary and secondary.

An example for rudimentary zonal orientation is seen in clams of the genus *Donax*, that live on sandy marine beaches in the zone where wave action produces high concentrations of suspended organic material which these mollusks filter out as food. This nutritious zone, however, constantly shifts back and forth with the incoming and outgoing tides. Therefore the clam must resort to some form of zonal orientation in order to stay in its optimal feeding zone. This cyclic zonal migration is achieved by a remarkably simple mechanism. It is mainly a matter of proper timing. When the tide moves in or out the clam leaves the sand in which it is normally buried and is passively carried in the proper direction by the tidal current whereupon it digs itself down into the sand again. This mechanism is rudimentary in comparison with some others. In more complex primary zonal orientation animals are utilizing sensory information directly connected with the zonal transitions. Near shore-lines primary directional information is provided by the direction of the sloping ground, by the usually greater horizontal light intensity towards the water due to the higher elevation of land and, perhaps, by the sound and other mechanical stimuli produced by wave action.

Resorting to primary zonal orientation is necessary in all those cases where prior learning of the zonal directions is impossible. Aquatic reptiles typically lay their eggs into self-dug holes on land. Whereas the babies of the Nile crocodile are carried toward the water in their mother's mouth, the hatchlings of all species of sea turtles are not that fortunate. They have to find their own way in two stages. After hatching from the egg buried in the sand of some warm beach they first have to dig themselves up to the surface, an activity which requires strato-orientation. Then on the surface they turn into the direction of greatest horizontal brightness and rush to the sea at maximal speed in order to minimize exposure to terrestrial predators.

Many invertebrates are known to utilize slope direction in primary zonal orientation. The beach-hopper, *Talitrus saltator*, an amphipod crustacean, tends to go downslope on dry sand and upslope on wet sand thus ending up in either case in the beach zone of optimal humidity.

Primary zonal orientation is frequently supplemented by secondary zonal orientation; some form of compass orientation which typically utilizes directional, celestial cues (usually the sun), but some arthropods also utilize the polarization pattern of the blue sky and in some instances also the moon at night.

An animal with secondary zonal orientation like the North American cricket frog or the above mentioned beach hopper "knows" the local compass direction across the zones of its home area. If the celestial compass cue like the sun is referred to, its daily movement across the sky is to be taken into account. A beach facing South in Europe or North America is thereby approached by keeping the sun to the left in the early morning, in front at noon and to the right in the late afternoon and at the proper angles all other hours in between. Secondary zonal orientation normally implies prior individual learning by trial and error after first responding to cues for primary zonal orientation. Provided the zones have a high degree of permanence like continental coastlines evolutionary "learning" is conceivable and the beach hopper *Talitrus saltator* has in fact been shown to know innately or instinctively the compass direction perpendicular to his native beach.

Zonal orientation not only serves the relocation of an optimal zone that is shifting but an animal may also move between zones in order to approach resources or avoid stress-sources. A basking frog takes an escape jump into the water when it notices the approach of a potential predator and the alligator like other crocodiles is known to perform regular daily migrations, spending the day mainly basking on land and the night in the water. It is not yet known what kind of orientational cues crocodiles use in order to control the spatial aspects of their rhythmic zonal migrations.

Zonal orientation between terrestrial zones is common, but less frequently analysed. Butterflies may move into the forest clearing for foraging on flowers, but fly back into the forest once the temperature becomes too high. Various species of deer frequently come out of the forest at night for foraging and hide inside it during daytime. The role played by primary and secondary orientation between terrestrial zones is best studied in the European cockchafer, a melolonthid beetle.

After overwintering in the soil as a young beetle, the cockchafer leaves the ground at an open area in May. All the beetles then fly towards the forest, first being guided by large dark silhouettes on the horizon, a form of primary zonal orientation. After periods of foraging, egg maturation and mating the female cockchafers shuttle several times back and forth between the forest and the open field where they originally left the ground. In the open field the eggs are laid in the soil. These shuttle flights take place along the same axis as the original first flight to the forest. However, all these subsequent flights are guided by some still unknown form of secondary orientation. This is substantiated by displacing females which then maintain their original flight directions irrespective of where dark silhouettes on the horizon are located.

Strato-orientation and zonal orientation derive their importance from vertical and horizontal distribution patterns of vital environmental factors that are constant over time. If such constancy is reduced but still spans the lifetime of an individual topographic orientation becomes important.

Topographic orientation

Topographic orientation, by definition, always requires some form of learning. In its simplest form, U-turn orientation, the animal leaves an object, its burrow or a source of food, travels a short straight distance and then returns by the same path. It has been shown for some spiders and fiddler crabs that such a simple short distance return orientation can be achieved without reference to external cues; the necessary spatial information is intrinsically provided. The animal somehow records the length of its path away from the object to which it wants to return. At the end of its path the animal turns 180 degrees and then reproduces the same amount of locomotion as originally recorded. The change in the direction of locomotion at the far end of the U-turn trip is particularly simple and elegantly solved for crabs. They prefer sideways walking and without turning their body around they simply switch from progressing towards one side, say left, to progressing toward the other side, say right.

Topographic orientation is complex and well developed in virtually all vertebrates as judged from the ease and determination with which they move from place to place in their respective home ranges. Yet, very little is known about the sensory cues used and the respective mechanisms of orientation. Best studied is the topographic orientation of digger wasps (Sphecidae), ants and honey bees. The basic structure of the orientation system of all these insects is rather similar as far as our present knowledge goes. The topographic orientation system of the honey bee has three components, orientation by landmarks, distance orientation and directional orientation. The latter two components provide polar coordinates by which the spatial location of any resource with respect to the hive can be determined. Honey bees also communicate their knowledge about the spatial coordinates of resource objects to other bees as discussed in Chapter 33.

The mechanism of distance orientation in honey bees is not yet fully understood, but it is certainly an intrinsic form of orientation similar to the one discussed above for the primitive U-turn trip. Honey bees measure and reproduce flight distances somehow by the amount of effort or energy spent during flight or walking. If a honey bee or an ant that is about to return to its nest is displaced into another environment without much disturbance they fly or walk in the same direction and the same distance as they would have in their familiar environment and then start searching for their nest. This outcome is evidence that both these insects are capable of controlling their distance of locomotion without reference to specific topographic features of their environments.

The directional orientation employed by bees, ants, and wasps as components of their topographic orientation is a celestial compass orientation. They maintain the bearing of their course by proper responses to the changing positions (azimuths) of the sun and the associated light polarization pattern of the blue sky. This component of the topographic orientation system is functionally similar to secondary zonal orientation as discussed above.

Landmarks are of prime importance for bees, wasps and many species of ants for pin-pointing the precise location of resources within the home range; resources like nest entrances, water holes, or abundant sources of food to which they frequently return. Distance and directional orientation allows only approximate localization of places. During exploration flights bees and wasps learn a constellation of landmarks, plants, rocks, holes and so on, with respect to the various resource objects, attending to finer and finer details as they approach their goals. Landmarks, however, are not only utilized for the purpose of such short distance approach orientation. Large conspicuous landmarks like trees, buildings, streets, shorelines are also components of long distance orientation which may cover stretches of up to about 10 kilometers in honey bees. Such landmark orientation is of particular importance whenever heavy cloud cover prevents celestial orientation.

For animals walking on the ground like ants, termites and mammals (including man) a common additional component of topographic orientation is guide-line orientation, the following of paths, edges, ridges, grooves, tunnels and odor trails that either pre-exist or are constructed for this purpose.

Seasonal and other changes may render a home range temporarily uninhabitable. This requires that an animal leave its home range for long distance travel guided by geographic orientation.

All animals that migrate far away from their daily home range across vast geographic areas can do so only with some sort of geographic orientation. Usually the timing of such migrations is correlated with seasonal changes in living conditions. The mechanisms of timing are discussed in Chapter 12 on bio-clocks. The best known are the migrations of birds, but some mammals also migrate (bats, reindeer and whales) as do many other species (butterflies and other insects, sea turtles and various species of fish like salmon and eel).

Best known among the butterflies is the monarch of North America whose northern populations migrate south-westerly from September into November to overwinter around the Mexican Gulf Coast and in Southern California (Fig. 10–7). In late February the north-eastern return migration to the breeding areas begins where two or more generations are produced before the fall migration starts again. Four types of geo-

Geographic orientation

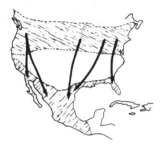

Fig. 10–7. Fall migration of the monarch butterfly from the northern breeding area to the southern overwintering range. The Spring migration takes place in approximately the opposite direction.

graphic orientation mechanisms can be distinguished. The more complex these mechanisms are, the less we understand their nature. Controlled drift is the most primitive type of geographic orientation and comparable in its structure with the zonal drift orientation of the clam *Donax*. The migratory locusts of the subtropical to tropical Old World are the best studied examples for controlled geographic drift orientation. After the build-up of large concentrated local populations, desert locusts (*Schistocerca gregaria*) as well as the migratory locusts (*Locusta migratoria*) form gigantic swarms that let themselves be carried by the prevailing trade winds that take them—hopefully—to some area with high precipitation and hence abundance of plant food. The timing of this swarming and migration behavior is less determined by seasonal rhythmic changes of climate than by high local population densities. This fact is important for monitoring swarm formation so that efficient control measures can be organized.

To-and-fro sun orientation represents the next higher level of geographic orientation. As yet it is only known for some European butterflies, especially the two common cabbage whites (*Pieris brassicae* and *Pieris rapae*). In England these butterflies simply fly toward the sun in the warm noon hours in the fall and similarly fly directly away from the sun in Spring. Thereby they cover some 100 to 200 kilometers between their breeding and their overwintering areas. The directional orientation thus achieved is not true compass orientation because these migrants change their flight direction in step with the directional change of the sun. How migrating monarch butterflies maintain their direction of flight is still unknown.

The last two forms of geographic orientation, vector orientation and navigation, are best known for birds and will be described in a separate Chapter (13). Geographic vector orientation has a functional structure comparable to topographic vector orientation as described for the honey bee. The migrating bird maintains a given compass bearing utilizing a sun compass, a star-compass or a magnetic compass and then keeps flying some predetermined distance. Finally, true navigation or goal directed geographic orientation, is the amazing ability of birds to directly migrate from any place to a distant goal, such as a nesting site or overwintering area. The homing of artificially displaced pigeons demonstrates such navigational abilities.

11 Forms and Mechanisms of Orientation in Space

Three kinds of information are required to specify an object or a state of affairs: place, time, and quality. Accordingly, we distinguish three kinds of orientation: in space, in time, and regarding the nature of things and events. Spatial orientation involves the ability of animals to determine their position in space and to use this position as a basis for assessing changes in the environment resulting from the animal's own movements. In short, we are speaking of the capacity to organize behavior within space as well as in relation to spatial features.

Suppose that we are sitting in front of an aquarium which contains angel fish. These fish prefer to swim around freely, in an upright position (back to the top), avoiding the plants in their tank. One fish may turn toward a worm at the bottom of the aquarium, fixate the prey, and snap at it. All these behavior patterns are regulated in relation to some stimuli: Swimming in the open water is guided with respect to the plants, and the fishes' upright position is maintained in relation to gravity and light; or the behavior is directed toward an object, like feeding behavior toward the worm. Every time we observe animals—including humans—we can see how movements and behavior patterns are oriented, how they stand in relation to spatial reference points. Elements of behavior that bring about and control these relations are called orientation processes.

To begin with, we shall organize the many forms of orientation according to three basic criteria, as formulated in the following questions: What is being oriented (i.e., what object is being used for orientation)? What is the nature of the spatial relationship (geometry)? What is the meaning of this orientation for the animal, what does it accomplish within the realm of behavior (function)? Next, we will examine the orientation processes and how they originate (physiology).

The object being oriented may be the animal as a whole. However, certain parts of the body may be oriented on their own, by individual movement. A hen pecks at a kernel with its head and beak; we reach for an apple with our hand. Furthermore, external objects can also be used for

Forms and mechanisms of orientation in space, by H. Schöne

What do we mean by spatial orientation?

The object being oriented

orientation. In building a nest, birds direct each stalk toward the nest structure, but orient the nest itself in relation to gravity so that the hollow always faces the top. People also orient a painting by gravity, making sure that the picture hangs "straight."

We can cite two elementary geometric processes that form the basis for all spatial changes: rotation (turning) and change of place (translation). A rotation is the same as a directional change, for example, alteration in the direction of a body axis. Here, the angle of directional change may be measured by the original position or with reference to a fixed point in space. The orientation process involved is called rotary, angular, or directional orientation. On the other hand, when the adjustment is to a fixed reference point—like bees orienting to the sun—we speak of compass orientation. This process is comparable to that of a compass needle oriented to the earth's magnetic field.

When an animal changes its position or place, it may orient itself in two ways: in relation to distance and/or to the direction of locomotion (rotary orientation). The directional control of locomotion is called course orientation. A course that is oriented in relation to some external direction is called a compass course. There are two methods of course orientation: Many animals orient their body axis to the course direction, but others are able to maintain a constant body position even while changing course.

When an animal that is changing position uses both kinds of orientation—direction and distance—we speak of vector orientation. A vector (see Chapter 13) is a geometric quantity determined by direction and length. In this way, bees can orient themselves by distance and direction when flying to a food source: Direction is gauged by means of the solar compass course (see Chapter 10), while distance is measured according to the energy expended during the trip. We can speak of navigation when a certain vector orientation presupposes that the animal knows both its present position and that of the goal (using place coordinates). We assume that homing carrier pigeons, for example, determine the spot where they are released by means of certain (as yet unknown) place coordinates. By using this information as well as the solar compass, these birds fly straight to their homing point, again using place coordinates.

Rotary orientation implies that the animal has a normal or preferred position. Many animals keep their back uppermost, especially when resting, but also during locomotion. The term "uppermost" may be relative to gravity, to light, or to the ground on which the animal stands or walks; its ventral side then points in the direction of the ground. Organisms frequently use all three reference criteria. We can observe this in cases where two or three directions no longer coincide, causing the animal to take up a compromise position—for example, between the force of gravity and the influence of light coming from the side, or between gravity, and a tilted substrate.

Basic geometric processes

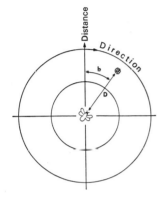

Fig. 11-1. Geometric features of orientation. A bee flies at an angle (b) to the sun. After covering a distance (D) it arrives at a food source. Depending on how we want to describe this process, we can view the bee's adjustment to the sun as directional, rotary, or angular orientation, its flight orientation (change of position) as compass course orientation, its monitoring of range as distance orientation, and the entire process (control of range and bearing) as vector orientation.

How orientation functions

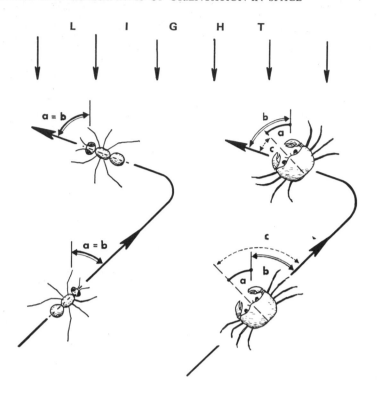

L I G H T

Fig. 11-2. Two types of course orientation: Insects, such as ants (left), always move with their head foremost, orienting their longitudinal body axis in the direction of their course. By contrast, crabs (right) maintain a constant body posture in relation to their spatial environment; nevertheless, their locomotion is oriented to the reference stimulus (course orientation with spatially constant body position).

Animals that must keep their heavy bodies in balance while perched on relatively narrow bases formed by their legs are faced with special problems in maintaining their normal position. Examples are terrestial animals with long, upright legs. Gravitational force exerts a pull on their body mass, making it difficult to balance. Maintaining this balance requires the animal to measure its body position, and for this it has mechanical sense organs sensitive to the force of gravity, such as sense organs of equilibrium, or just receptors that record the direction in which the animal's body weight is resting on the substrate. Visual organs register position relative to optic reference points within the environment. By using these measurements, organisms perform their movements to maintain equilibrium.

Orienting acts, involving a change in position may lead the animal to a certain goal. Such goals differ greatly in area as well as distance. Sand fleas and shore-dwelling spiders (Pyczogonidae), for example, move back to shore if placed on solid land or in water. Their movement is guided vertically to the main stretch of their home beach, keeping a sun compass course. Thus, the goal of this orienting movement is the beach zone, a relatively narrow strip but virtually unbounded in one dimension. Many planktonic animals orient their vertical movements toward a certain stratum of water—an area virtually unbounded in two dimensions.

In other cases, the goal region is bounded in all directions. Migratory birds find their winter homes in the fall, and return to their habitual breed-

Fig. 11-3. Orientation with certain objects. Carnivores, for example, orient their prey-catching movements mostly according to direction and distance of the prey: A chameleon uses its tongue to strike at a mantis (above). A toad strikes at an insect with its folding tongue (middle). A leopard leaps at a gazelle (below).

ing areas in the spring. Often enough these birds head for the same territories, or even the same nesting sites, that they occupied before. We may assume that the animals know their place coordinates, finding direction by familiar landmarks once they are in the vicinity of their goal. This capacity is an important factor in many kinds of place orientation. Many animals live in fixed places—caves, nests, or shelters—which provide them with a refuge. In finding their way back to them, organisms often utilize landmarks and spatial patterns. Examples are the orientation processes of sand wasps and bees.

All these forms of goal orientation require the animal to move over certain distances. By contrast, object orientation requires only relatively short movements or brief acts. These include striking and grasping movements performed with certain body parts, aimed at hitting particular objects (Fig. 11-3). Chameleons, toads, and frogs strike at prey with their tongue, while dragonfly larvae use parts of their oral apparatus; mantids strike with their grabbing legs. Many mammals seize or grasp pieces of food with their forelegs. Birds peck, striking at objects with their head and beak. Even the prey-catching leap of carnivores, the striking blow of raptors, or the forward swoop of predatory fishes upon their victim represent this kind of orienting behavior. In many of these acts, the animal judges both the direction and the distance of the object.

How do animals orient?

An orientation process is part of a behavior, and therefore subject to the same physiological limitations as the behavior itself. Thus, a particular orienting act also depends on the very processes that release the behavior. Ethologically speaking, a releasing factor in the broadest sense is anything that contributes directly to the occurrence of a behavior pattern. More narrowly defined, releasers are specific stimuli acting by way of the sense organs (sign stimuli, see Chapter 15) and leading to the discharge of a certain behavior.

External and internal conditions

Releasing stimuli are different from the orienting or reference stimuli, which are relevant only to those elements of a behavior concerned with orientation or spatial relationships. N. Tinbergen illustrates this difference between releasing and orienting stimuli with the behavior of nestling birds. The adults landing on their nest create shock vibrations, releasing upward-stretching movements of the head and neck as well as gaping in the young. Orientation of these movements is then guided by stimuli originating on the head of the parent birds, and the young turn their gaping mouths to these stimulus patterns. On the other hand, behavior patterns may arise spontaneously, occurring without external stimulation. Here, the effective stimuli would be endogenous (internal) "releasers."

Physiological state

Ethologists refer to a specific physiological state leading to a certain behavior as the motivational element of a mood. This refers primarily to processes within the central nervous system or the brain. A given physiological state (disposition) will also determine the orienting behavior. For example, an organism may be in the mood to seek out a food source by taking a certain orienting direction, like bees heading for their goal at a certain angle to the sun. The physiological disposition changes when a bee has taken in enough food, causing the insect to fly off in another direction. The causes of such changes in orienting disposition have been examined in butterfly caterpillars. Toward the end of their caterpillar stage, and prior to pupation, these organisms cease to move in the direction of light; they reverse their tendency, now seeking the dark. Experiments have shown that this reversal is determined by an increase of the molting hormone ecdysone in the blood.

The direction an animal seeks to attain therefore depends on its endogenous disposition. The special target value of the orientation mechanism, specifying a certain direction, is also called the *Sollwert*. This word may be literally translated from the German to mean "the value that should be attained"; another way to describe it would be "equilibrium position." There is no direct translation, and our use of words, as pointed out by other ethologists, depends on the context. The concept itself, however, is extremely important. In most cases, the Sollwert are determined before the action begins.

Target value (Sollwert)

Fig. 11-4. Fishes are trained to swim toward the sun (south) at noon. The dots in our diagram indicate the bearing toward an artificial "sun" (the angle between the swimming direction and the light, shown on the left scale) when observed after training at different times of the day. The curve indicates position of the real sun in the sky at the same times. The concurrence of dots with the curve demonstrates that the fishes's "internal clock" continuously readjusts the Sollwert for their compass course. In this way, the daily "movement" of the sun remains even, and the animals can maintain a fixed geographical bearing.

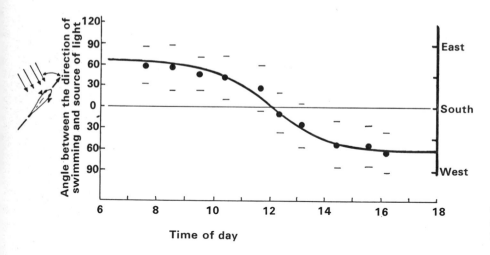

One important determinant of Sollwert is time. Spatial orientation is frequently coupled with a mechanism for temporal (or successive) orientation, with an "internal clock." This temporal mechanism determines, for example, the orientation of bees and fishes toward the sun. The Sollwert disposition for swimming direction in water-beetle larvae may be influenced by both internal and external factors (Fig. 11-5). Another example is that of the animals that habitually seek out moist places: Dry air causes them to turn away from the light and toward the dark.

The stimuli influencing the nature of the Sollwert are again different from the orienting stimuli. External stimuli such as shade and humidity do affect the target value of the orientation; but they do not constitute the actual orienting stimuli, which are provided by the light. Thus, two factors determine orientation: the Sollwert disposition, and the reference points in the animal's spatial environment, which are the reference or orienting values.

We see, then, that orientation as a process of organizing behavior in spatial terms requires spatial reference values. If these values are picked up and recorded by the sense organs, we speak of reference or orienting stimuli.

Orienting stimuli must be separated from other kinds of stimulation. The orientation mechanism has to "recognize" the stimulus as a reference value. A prey object, for example, has certain characteristics that trigger a response in the orientation system of the predator. The characteristic values of an orienting stimulus may be determined with experiments using models. A praying mantis, for example, has a preference for turning toward objects with many glittering surfaces, resembling insect wings. This turning movement, however, also depends on the spatial characteristics of the stimulus. The orientation mechanism has to determine the spatial features of this stimulus in relation to the animal itself—the stimulus must be localized.

Recognition and localization

Fig. 11-5. Shrimp larvae orient toward an up or down direction by considering light. A: Breathing position and various swimming postures; a shadow (external factor) can induce the animals to swim downward, while the physiological factor of breathing (internal) causes upward movement. All orientation is gauged with respect to the source of light; this is demonstrated by B: The entire set of orientation behaviors is reversed when light is made to come from below.

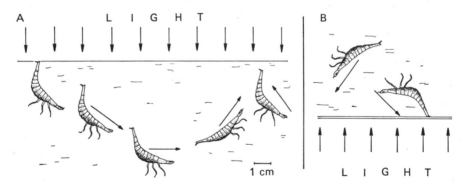

Reference values

The reference values may lie outside of the animal; for example, they may be gravity, light, or various goal objects. But we do know of other means of orientation without any reference to external stimuli. Vertebrates have semicircular canals in their inner ear that serve as rotary sense organs to control turning movements or to allow the animal to move in a straight line by avoiding rotations. Both angular and linear movements may also be oriented by means of proprioception (see Chapter 7). Proprioceptors are sense organs that respond to movements of body parts relative to each other, giving the animal information about its own movements. But this kind of information need not be obtained "the long way around" through proprioceptors. In principle, it is already present when the central nervous system gives a command signal for the appropriate movement. Thus, the animal could note the directional information "to

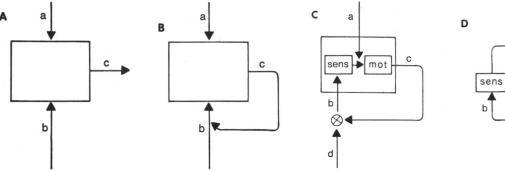
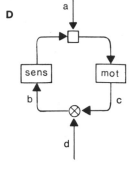

the left" which is derived from a central command for a left turn—the information would be stored in the memory. In this way, the orientation system would be informed about its new direction of locomotion.

Orientation according to internal reference values may also be called idiothetic, in contrast to allothetic orientation, which occurs in relation to external stimuli. An animal orienting idiothetically has information about only the orientation of its body in relation to the previous direction. Here, the ultimate deciding factor in spatial orientation would be the animal's original position.

An orienting movement arises from the interaction of two functional values or control factors: first, the central physiological state, which sets up a certain Sollwert (target value) for orientation; and second, the reference value, affecting the mechanism by way of, for example, sense organs. We often represent these effects and their functional connections in block diagrams. Control factors (effects) are then symbolized by arrows, and connections, by circles or blocks. These block diagrams say nothing about either the material structure of the mechanism or its functioning processes. In other words, the diagrams do not show how a sensory apparatus becomes activated, how the excitation is transmitted, how these stimuli are processed in the brain, or how the motor organs function. A block diagram indicates only the connection and transmission of control factors. For example, it may show how a change in the stimulus direction or the source of stimulation affects the orienting movement, given a certain Sollwert. A more detailed explanation of block diagrams is provided in Figures 11-6 and 11-7. Here, we describe the mechanism of controlling an orienting act, such as the adjustment and maintenance of a certain direction by constant measurement of the directional value and, if necessary, correction of any deviations (feedback control system).

We have emphasized the importance of the central disposition that determines target value in the process of orientation. But ethologists studying orientation phenomena have been making this emphasis only during the past two decades, since E. von Holst and H. Mittelstaedt—parallel to similar ideas entertained by R. W. Sperry—proposed their "reafference principle." According to this principle, a self-initiated change

Fig. 11-6. Block diagrams showing the effective connections in the functional organization of control systems. A: The effects of the Sollwert (a) and stimulus direction in relation to the animal (b) are the input quantities, which are evaluated (box)

Orientation as a control system

The reafference principle

and which then lead to the adjustment of the orientation movement (output quantity (c). B: This quantity (c) in turn influences orientation to the stimulus (b). C: The evaluation block has been subdivided into an input box (b: sens, which corresponds to the sensory system measuring orientation) and an output box (c: mot, the motor system for adjusting orientation). With the effect of sensory system on motor system, the target value (a) is calculated into the process. The reafference from (c) to (b) enables feedback to take place: (b) is continuously being measured by senses, and the measurement value compared with (a). Discrepancies in the comparison (deviation of stimulus direction from the target value, or Sollwert) lead to orienting movements (c), aligning (b) back with (a). Redrawing the block diagram C, we obtain D, which is the typical diagram of a feedback control system; (b) may be interfered with by extrinsic factors and deviate from (a), whereupon the difference between (a) and (b) is corrected and the old target direction reestablished. For reorientation, the animal has to change (a); again the difference between (a) and (b) is corrected, but now the animal takes up a new orientation value.

Orienting responses

in direction involves the following events (Fig. 11-7): In response to a dispositional change, a "higher center" induces a change of state in a lower one, which in turn sends a "command" to the motor apparatus (motor efference) and therefore causes a directional change. This then leads to a stimulation of the sense organs, whose excitation (afference) is transmitted to the center. Afferent impulses indicating a change in direction normally release orienting reactions that serve to re-establish the previous position. This does not happen, however, if the animal has initiated the change itself, in other words, when the central physiological state was altered prior to the change. The response-triggering effect of afferent impulses will then be inhibited by the central nervous alteration. The assumption was that something is diverted from the motor efference; this was called the "efference copy" by von Holst and Mittelstaedt, and "diverted excitation" by Sperry. The feedback afference is then compared with this efference copy, and the two compensate each other. This means that the impulses have opposite signs. The sensory stimulation produced by the organism's self-initiated movement was called "reafference."

If we do assume that a higher center sets up an efference copy in a lower one, we may explain in a simple way how, for example, people orient their eye movements to look in a certain direction. In other cases we can assume that the afferent impulse is evaluated directly with the efferent one, something like in the feedback system of Figure 11-7(C): Here, the circuit model of the reafference principle may be transformed into a feedback control-system diagram.

Controlled orientation—that is, adjustment to the source of stimulation regulated by sensory input—is frequently found in phenomena of long-term directional orientation such as course orientation. By contrast, short-term orienting acts, such as the rapid prey-catching movement of toads or the sudden strike of mantids, are often not regulated but merely guided. Regulation means feedback control in a closed-loop system. Guiding or directing effects, by contrast, are involved in the kind of orientation processes where only one orienting measurement is made prior to the act. On the basis of this measurement, the animal performs its orienting movement without any further control through the sense organs. This kind of open-loop causal sequence, occurring without feedback control, may also be termed simply a chain of events.

All orienting movements, whether controlled or merely directed, mean that the animal strives to attain the direction corresponding to its central physiological state (Sollwert), and that it corrects for any deviations from this target value. This process of compensation is called an orienting response. The important factor is therefore the deviation from a Sollwert, or the difference between orientational value and target value. It makes no difference whether the deviation was caused by some external interference or by a change in the Sollwert resulting from redispositioning in the higher center. In either case, the input from the receptors will indicate that the orientational value of the stimulus is different from that of the Sollwert. The system then responds to this deviation with an orienting response.

The greater the deviation from the target direction, the more intensively the animal will attempt to regain that direction. The strength of these efforts, or the turning tendency, have been measured experimentally in a number of ways. For example, we have recorded the energy expended by a prawn in attempting to gain its normal position in relation to light by turning its dorsal side toward the light source. In this experiment, the animal was required to turn counter to a small spring, whose tension then indicated the strength of the turning tendency. Another method involved two stimuli made to act in opposing ways. Fishes, for example, take up a compromise position between gravity and a lateral light source. While the experimenter alters the direction of light, he can derive the relative effects that these two reference stimuli have on the subject's turning tendency from the fishes's orientation at varying angles of intersection.

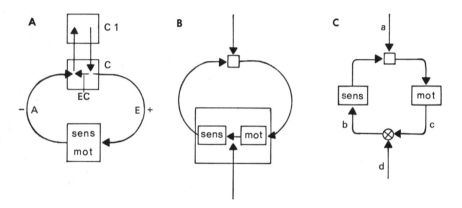

In many similar studies we have found a peak in the strength of the turning tendency when the animal deviated from its target direction by about 90 deg (Fig. 11-8). Experiments were made, for example, on larvae of *Dytiscus* and *Acilius* water beetles. We used the radius of rotation as a measure of the subject's turning tendency. If we couple the animal with the stimulus direction, it will continue to rotate endlessly about its longitudinal axis, turning in circles whose diameters correspond to the strength of the turning tendency. We observed what happened when these animals were induced to strive for different target directions. The different results also told us something about the mechanism involved in orientational change. Let us pursue this example more closely:

The compound eyes of a water-beetle larva were covered so that only the stemmata pointing in the intended direction were left free (Fig. 11-9). When we then placed the larva into a glass container exposed to diffuse light, it was able to see the light incidence only from a limited angle. If this angle now deviates from the animal's Sollwert, the larva displays a continuous turning tendency: It swims in circles, and the shape of these

Turning tendency

Fig. 11-7. A: Reafference principle. A change in the physiological state in the higher center (C 1) leads to an efferent command signal (E) in the lower center (C), resulting in an orienting movement and thus a stimulation of the sense organs (both represented in the sens/mot block). The sensory afference (A) in C encounters an "offshoot" of the efferent command, the efference copy (EC), which neutralizes the effect of the afferent impulse. B: Figurative section inserted between the reafference principle and the feedback system. C: Feedback control system (see also Fig. 11-6).

Change in direction

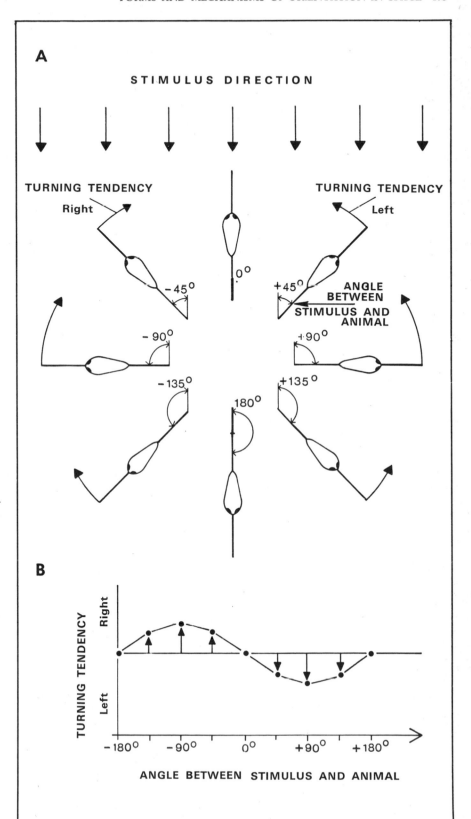

Fig. 11-8. We speak of a turning tendency when the animal strives to correct a deviation from the Soll-wert by turning back toward the target position. The strength of this tendency depends on the amount of deviation. A: An animal is forced to change from its orientation toward the stimulus (0°; target position), successively adopting positions between +45° and −45°. Its turning tendency is strongest in the 90° positions. B: If we superimpose the various strengths of turning tendency over the angles between the stimulus direction and the animal's position, we obtain a sine-like curve.

circular movements will depend on difference between the angle of light incidence and the target direction.

The target direction itself may be modified by experimental factors. When scared away from the water surface, larvae with intact vision swim diagonally downward, away from an overhead source of light. If we then force them to stay underwater, they cannot breathe, with the result that the animals swim back up as soon as possible, in the direction of the light source. In the ones that could see the light only from a certain angle, these

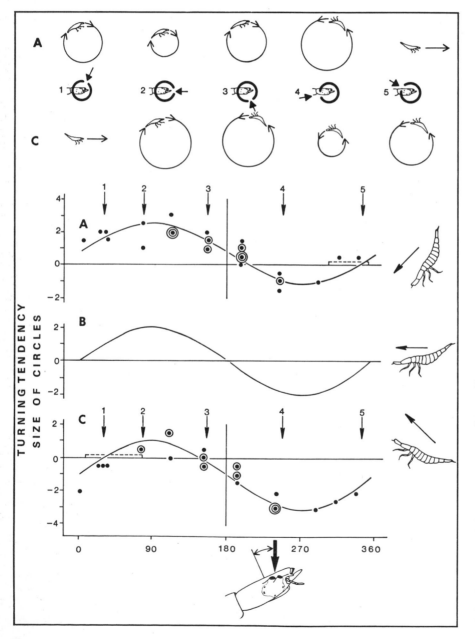

Fig. 11-9. Change in orientation in shrimp larvae. Row A: Circular movements of larvae after being startled (Sollwert = diagonally away from the light). In each case, the animal was "coupled" with a particular direction of light incidence (1–5). Row C: Circles performed by the same larvae when they want to breathe (Sollwert = diagonally toward the light). Explanation of diagrams: Horizontal scales: direction of light incidence; vertical scales: circular movements (− dorsal circles; + ventral circles; 4 = 2 − 4 cm∅; 3 = 4 − 7 cm ∅; 1 = more than 11 cm ∅; o = straight ahead). A: Values obtained from larvae moving away from the light. B: Curve obtained for animals swimming horizontally (derived from values for A and C). C: Values obtained for larvae with Sollwert "diagonally away from the light." The values for 1–5 correspond to the above examples. The turning-tendency curves (e.g., for A and C) are horizontally opposite to each other. This means that for one change in orientation, all turning tendencies are altered in the same way and with the same intensity by the amount it takes to equalize the turning tendency at zero (= swim-

two methods of treatment led to different orienting movements, because each treatment set up a different Sollwert and therefore created different deviations of angle of light incidence from the target value. For example, if we scare a larva that is able to see light only from the ventral side, it will swim in circles with the ventral side turned inward. But after a while it must breathe, at which time the larva swims in the opposite angular direction, with the dorsal side turned inward. Two other kinds of circular movement (Fig. 11-9) result when we couple animals with different angles of light incidence. The overall finding of these studies is that in conjunction with any orientational change (change in target value), the turning tendency curves are shifted vertically in relation to each other—i.e., all turning tendencies change in the same way. We therefore conclude that an orientational change is made when all turning tendencies are altered in the same direction and with the same intensity.

This alteration mechanism is only one of many that are possible. Another mechanism could function so that, for example, the turning tendency curves must shift horizontally rather than vertically. This would mean that with a change in Sollwert the corresponding set of turning tendencies is reoriented to different sources of light without any change in intensity.

Taxes and kineses

The central processes that determine Sollwert are difficult to study experimentally. This is why they are often excluded in studies of orientation mechanisms, where the central Sollwert disposition is simply taken as constant. The orienting movement is then related only to the stimulus, and the act is seen as a response to stimulation. Such orienting responses are also called taxes (from *taxis* = arrangement, position). In 1919, A. Kühn developed a classification system for taxes, later extended by G. Fraenkel and D. L. Gunn, that is still commonly used. For a start, the concepts contained in this system permit a shorthand characterization of the properties of simple orienting movements. For example, the terms "phototactic," "geotactic," and "chemotactic" indicate the nature of corresponding stimuli (light, gravity, and chemical stimulus, respectively). The terms "positive," "negative," and "transversal" refer to the geometrical arrangement, or direction, that the animal takes up in relation to the stimulus (turning toward, turning away, and transverse orientation, respectively). A caterpillar is positively phototactic, for example, when crawling toward a light source. These qualifying terms do not carry any further meaning, however—in contrast to the key concepts like phototaxis, tropotaxis, and telotaxis, which distinguish the different physiological mechanisms involved in the various taxes.

Phobotaxis (from *phobos* = fear, fright) refers to orientation within a gradient of stimulation where the animal displays "fright reactions" or startle responses (stimulus gradient = sequence of stimuli with decreasing intensity). For example, *Paramecia* landing in water with a high concentration of carbon dioxide will stop, turn, and continue swimming. If they

ming straight ahead) with the target direction of light. In this manner, a change from A (away from light) to C (toward light) would mean that narrow ventral circles become wider ones, wide ventral circles change to straight-ahead swimming movements (1: target direction), straight swimming movements become wide dorsal circles (5), and so on.

again meet with a high carbon-dioxide concentration, the process will be repeated. These turns are random, i.e., not oriented in any relation to the stimulus, and the *Paramecia* escape the danger area only by trial and error. Fraenkel and Gunn describe similar mechanisms when talking of kineses.

Kinesis (from *kineo* = to move) refers to changes in form of locomotion, based on a change in intensity of stimulation but not oriented in relation to this stimulus. For example, animals will run faster (orthokinesis), or turn more frequently and to a greater degree (klinokinesis). Although such turns are not spatially oriented with respect to the stimulus, these mechanisms ultimately lead the animal in the direction of a particular stimulus gradient. Turbellaria make fewer turns and therefore move over greater distances in a linear direction as their surroundings become darker. In this way they move from light to dark spaces.

Contrary to the kineses, the orientation mechanisms of tropotaxis and telotaxis are directly oriented to the source of stimulation. In both cases, the animal attains a symmetrical balance with respect to the stimulus: it orients toward or away from it. Kühn describes the tropotactic mechanism (from *tropē* = turn) in the following way: The excitations in the sense organs of the right and left sides of the body flow into corresponding right and left centers, which in turn exert a more-or-less direct influence on the movement musculature. A central imbalance of excitation releases turning movements. For instance, if an animal is standing diagonally to the stimulus direction, those sense organs facing the stimulus will be activated more strongly. This means that the corresponding lateral motor centers will also be "charged" more strongly than those of the other side. As a result, the animal makes a turning movement in the direction of stimulation, stopping when the excitation on both sides is equal. We must emphasize, however, that this central balance of excitation is only figurative; we still do not know what processes take place in the central nervous system to evaluate receptor input from both sides, causing the animal to turn.

Telotaxis works in a completely different way (from *telos* = goal): The differences in excitation between the two sides are not important; the major factor is locus of stimulation in the receptor organ. This excitation in the activated receptor triggers a turning movement with a certain value, leading the animal toward the source of stimulation so that the frontally oriented parts of the sense organs (fixating parts) are directly exposed to the stimulus.

Contrary to the many diagonal or transverse adjustments that we observe in animals, those that are symmetrical immediately strike us as orientation processes because their relationship to the stimulus is so obvious. This is why the mechanisms of symmetrical orientation (tropotaxis and telotaxis) have attracted more attention than others.

Kühn assumed that the other taxes involved a mechanism derived from tropotaxis, called menotaxis (from *meno* = to stay, remain). Here, the animal tries to maintain a certain "disequilibrium" of excitation be-

Tropotaxis and telotaxis

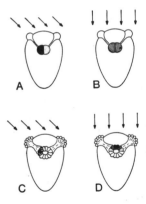

Fig. 11-10. Directional localization with two measuring units (A, B) and with screens (C, D). A: Stimulus incidence from the side activates the left measuring unit and therefore causes a greater excitation in the left center than in the right (blacker marking). The size of the central nervous difference in excitation provides information about the source of stimulation. B: Symmetrical stimulus incidence activates both measuring units, and therefore both centers,

tween the two sides of its body. But we could also picture the diagonal and transverse orientations from a telotactic point of view: The animal fixates the stimulus with a corresponding lateral-fixating part belonging to the receptor organ.

We may summarize the basic elements of these two mechanisms in the following way: In tropotaxis, the response of the mechanism depends on a comparison of two streams of excitation, transmitted by some kind of bilaterally symmetrical (paired) receptor organs to the nervous center. By contrast, telotaxis involves a response to the locus of stimulation on the animal's body.

Localization of stimuli

It frequently happens that the stimulus-recording processes through the sense organs and the movements of orientation do not occur in discrete succession, first one and then the other, but instead run parallel or alternate in small stages. In this way, they influence each other. Gaining access to spatial information, in one form or another, is called localization. Sensory systems have four properties particularly important to the localization process:

1. Sensitivity to differences in stimulus intensity. This would be expressed by the fact that stronger stimuli activate the receptor to a greater degree than weaker ones. For example, greater light intensity would stimulate a photoreceptor more, or a higher concentration of odorous substance would lead to greater excitation in a chemoreceptor.

2. Sensitivity to differences in direction of stimulus incidence. Here, the excitation of a receptor changes with the angle at which it is exposed to the receptor.

3. Sensitivity to time differences in stimulation. This is an important property both for the receptors and for their adjoining central evaluation systems. These systems assess the temporal discrepancies between excitation in two receptors.

4. Special kinds of receptor properties are the positional order, or local signs, and the order of body orientation. These contain information about the body area or part of a receptor organ that is activated, and/or about the direction of a stimulus in relation to the body. These data are essential to any orienting movement. We can cite a simple example: If the right antenna of a bee is more highly stimulated by a scent than is the left, it releases a turning movement to the right—even when some inquisitive experimenter crosses both antennae over, pointing the right one to the bee's left side (see Chapter 10). The turning direction "to the right" is determined by the positional order (or value) "right" associated with the more strongly activated antenna. In this case, the body-orientational order is firmly anchored within the network of central nervous fibers by means of receptor connections. In many other cases, however, the order of body orientation is a variable to be recorded separately—for example, when the animal is in a position to change receptor orientation with respect to its body, because the receptor is located on a mobile appendage.

equally. Screens: Stimulus incidence from the side (C) and from the front (D) maximally activate those parts of the screen turned toward the stimulus, therefore causing the greatest excitation in the associated parts of the central screen. Information about direction is derived from the body-orientational order of the screen sections.

In classifying the various principles of localization, we basically refer to the body-orientational values. Receptors with the same orientational order provide the system with the same units of information about the direction of stimulus incidence. They will therefore have the same effects on orientation. Conceptually, we combine these receptors and call them a directional or orientational measuring unit. According to the number of effective measuring units involved, we can distinguish three types of localization: Type A= localization with one measuring unit; Type B= localization with two measuring units; Type C= localization with many such units, or screens.

A mechanism may work with a measuring unit responding only to differences in intensity. Changes in stimulus intensity then lead to positional changes with a resulting orientation within stimulus gradients. This involves the kinesis mechanisms described earlier. On the other hand, if the measuring unit is sensitive to direction, the mechanism will respond to changes in stimulus incidence. In such cases, resulting movements will be directly oriented, as with flagellates moving in light coming from a certain direction.

Localization with one measuring unit

There are three kinds of localization using two measuring units: a) assessing time differences between excitation in two meauring units; b) comparing the degree of excitation in two measuring units sensitive to stimulus intensity; c) comparing the degree of excitation in two measuring units sensitive to direction.

Localization with two measuring units

For stimulus localization (a), the significant factors are temporal differences of stimuli impinging on two points of the body. In this way, for example, the sound localization used by mammals involves taking into consideration the different intervals needed for sound to reach one ear as opposed to the other.

Localization (b) and (c) work by assessing differences in intensity of the two receptor excitations. In this respect, we may compare these kinds of localization with the tropotactic mechanism. On the other hand, tropotaxis does not discriminate between (b) and (c).

Mechanism (b) responds only to discrepancies in stimulus intensity. The animal may therefore determine the direction of stimulus gradients but not the direction of stimulus source. Examples are insects, such as bees that follow a scent gradient with the use of chemoreceptors located on both their antennae.

The measuring units of (c) are sensitive to the direction of stimulation. We can represent the relationship between the source of stimulation and the intensity of excitation in the form of characteristic curves of directional excitation (directional characteristics). For example, a stimulus coming from a source ahead of the animal will activate both measuring units (right and left) equally, while stimulation from other sources will be uneven. From this difference in stimulus intensity, the animal learns about the exact directional order between right and left (Fig. 11-10). In addition to echo-location in mammals (see Chapter 8), so far as it works with intensity differences, we can cite mechanisms in crabs where the position of the body is recorded in relation to gravity: When a river crayfish tilts its body to the side, the countermovement of its eyes—an indication of posture provided by the mechanism—will depend on the difference in excitation between the right and left statocysts (sense organ of equilibrium).

In localization with many measuring units (screens), the animal gains information primarily from the local signs (order of body orientation) of the maximally stimulated units (Fig. 11-10). The lens eyes of vertebrates, the compound eyes of arthropods, and the statocysts of snails function according to this principle. Telotaxis also requires the use of screen mechanisms.

In Chapters 4–9, the authors described how, in addition to the more familiar senses of vision, hearing, smell, and touch, there are other senses that are important to many organisms: those of gravity, heat, current perception, and electricity. We shall add a few comments dealing specifically with the principles of localization.

Visual perception

Almost all visual receptor organs that we know about display a directional characteristic. In the simplest case, this characteristic arises through a pigment disc located beside a sensory spot (a single measuring unit). *Euglena* has such an eyespot, and while swimming it circles spirally in a frontal direction. When light shines from a lateral direction, the animal's sensory spot is alternately stimulated and blocked by the pigment. This leads to changes in the flagella's movement and as a result the animal orients toward the direction of light stimulation. Most higher animals have facette eyes: We can find lens eyes with optic surfaces consisting of many individual receptors in, for example, vertebrates, spiders, and cephalopods. The compound eyes of insects and crabs also consist of many individual eye structures.

Heat perception

Pit vipers (Crotalidae) can localize prey by perceiving heat radiation emanating from the victim (Fig. 11-11). These animals have two pits with a small opening, each containing a heat-receptor organ, at the front of the head. These organs are sensitive to even very weak heat radiation, such as that coming from mice and other small, warm-blooded animals. We have not yet determined whether the heat-receptor mechanism of pith eyes utilizes two measuring units sensitive to stimulus direction, or whether there are whole screens.

Perception of electrical stimuli

Chapter 9 gave a detailed description on the electric stimuli that mormyrids and gymnotids use for their orientation. Here, we will only emphasize that the electric impulses discharged by such fishes build up an electrostatic field, in turn recorded by many receptors distributed over the animal's entire body. Objects approaching the fish, or being approached by it cause alterations in the electric field and stimulate the receptors. In this way, the fish's localization mechanism gains information about the position of external objects.

Chapter 6 includes descriptions of how fishes and bees, for example, discriminate intensities of odors by using two chemoreceptive organs. This allows for orientation within a scent gradient. Woodlice (isopods) use only one measuring unit, sensitive to intensity differences, when orienting with respect to relative air humidity. The dryer the air, the more these animals move about; conversely, increased humidity decreases locomotory activity. As a result, they aggregate in most places because that is where movement ceases.

Media that move or flow, such as wind or water currents, serve to

Chemical senses

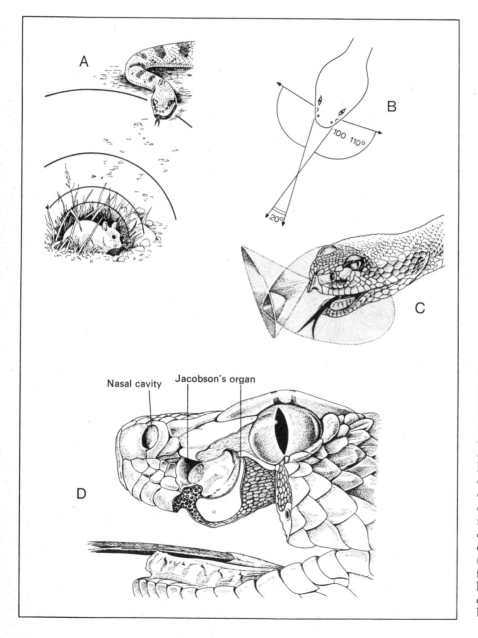

Fig. 11-11. Pit vipers detect their prey (warm-blooded small mammals) by the prey's heat radiation (A). By means of their pit organs (D), these reptiles can register very weak heat radiation at a distance of several decimeters. The openings of the pits are directed diagonally to the front and outside, and the cone-shaped receptor parts overlap in the middle (B, C). This probably facilitates the process of localization: The animal can "fixate" its prey with both organs.

carry scent stimuli from one place to another. An animal orienting toward a source of odor frequently responds not only to the chemical stimulation, but also to the mechanical one provided by the current. Sometimes the chemical stimulus only serves to trigger a response, inducing the animal to orient itself in relation to the current by means of its current receptors.

Orientation by tracking

In cases where an animal seeks and finds a goal by following a whole trail (or chain) of odor sources, we speak of orientation by tracking. For instance, dogs continuously sniffing on the ground take in odorous substances, using their noses. Similarly, snakes follow the trail of some prey animal they have bitten, collecting scent samples with the tip of the tongue. These chemical samples are then analyzed by a special olfactory organ (Jacobson's organ) inside the mouth.

Gravity receptors

Gravitational stimuli do not affect the sense organs of equilibrium directly (Fig. 11-12), but instead influence these receptors through objects stimulating the sensory hairs (cilia) through mass or weight (see Chapter 7). In vertebrates and crabs, excitation in the sensory cilia occurs through the lateral action of statoliths as they "shear" the hairs by their weight.

Fig. 11-12. Sense organs of equilibrium in the river crayfish. A: Crayfish; arrows point to section shown in B. B: Head section from above; the pointed ramification (rostrum) has been removed; between the eyes are the base segments of the first antennae with the triangular openings of the statocysts, normally closed by bristles (left; removed at right). Right, an indication of statocyst contours. Arrows point to the line of intersection for C. C: Statocysts; view from the front. Right: curved row of sensory hairs; left: statolith mass, baked in with the cilia and consisting of many small granules.

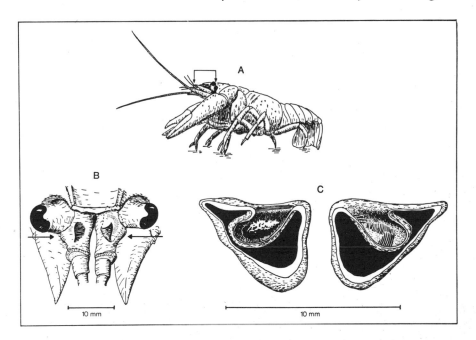

Different degrees of tilting in one and the same body plane (for example, to the side), are discriminated by using two principles of unit measuring; on the other hand, different planes of tilting (for example, sideways or forward) are recorded on the basis of receptor elements arranged like a screen within the sense organ. The spheroid statocyst chambers of some snails and mussels contain loose pebbles lying on a "carpet" of hairs (Chapter 7). These act as statoliths, pressing down on that part of the screen positioned downward and thus in the direction of gravity. Water bugs have gravitational sense organs where bubbles of air are connected with the sensory cilia, acting, we might say, as "reverse statoliths."

Chapter 8 described the variety of ways animals may orient themselves in relation to sound. We will briefly summarize the different means of localization based on auditory processes:

Fishes, such as minnows, are able to follow the intensity gradient in a field of sound, using a kinesis mechanism with only one measuring unit that is sensitive to intensity. The sound localization used by birds and mammals is based on a combination of several principles, but they use only two measuring units; these are sensitive to direction. The mechanism of these animals evaluates differences in time taken by a sound to reach both ears, as well as intensity differences. In addition, the animal assesses tonal discrepancies resulting from the fact that longer sound waves curve more easily around the head than do shorter ones; a sound composed of notes with different frequencies will sound higher to the ear pointing in the direction of the sound source than to the one pointing away from it. Female crickets apparently localize singing males by measuring time differences, using two measuring units, the auditory organs in their two forelegs (see Color plate, p. 248). Mosquitos carry auditory receptor organs in their antennae (Chapter 7); they record the stimulus-directional value by means of the screen principle.

The use of echo-orientation (see Chapter 14) enables animals to use sound waves to localize objects that do not emit sounds. The passive objects are made into sources of sound by a secondary process, because they reflect the sound (or supersonant) produced by the localizing animal— e.g., a bat or a dolphin.

Mechanical vibrations of a medium, such as water, whose frequency lies far below that of sound are called vibratory stimuli. These are recorded by special receptors. In this way, the backswimmer (*Notonecta*; a waterbug) localizes prey by recording the waves created on the surface of water by the prey animal's movements. Here, orientation is based on measuring the time difference between stimuli reaching two units. These measuring apparatus are located on the extended natatorial (swimming) legs of the backswimmers.

Stimulation caused by moving currents (current stimuli) may originate from two completely different sources, with a corespondingly different orienting effect:

1. Animals locomoting through water or air create a current that flows along their bodies. The individual is then able to monitor its own direction and speed by gauging the direction and force of the current.

2. The medium (water or air) flows in space; its course and intensity changes in relation to some reference system, such as a shore or bank, a river bed, or the ground below a wind. Animals can record this current in space only if they are in touch with the reference system—like a lobster sitting on the bottom of the ocean. This animal's pincers have a layer of

Auditory perception

Perception of vibrations

Perception of currents

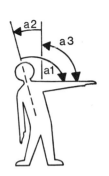

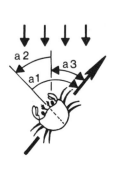

sensory hairs that respond to the movements of water. The cilia have different arrangements, so that one group may register water currents in one direction while other groups respond to a different one. All these groups of cilia together form a screen with which the animal can discriminate different directions of current.

While the lobster is in direct contact with the reference system, an animal moving freely in space has to keep in touch over a distance. To this end, fishes and birds use their visual sense. Seeing the ground, they are able to orient themselves within a current (flowing water, wind) by sight. An animal having no contact with the ground cannot obtain information about the movement of media in space; it would be a part of the medium, carried along without noticing it. No animal could fly against the wind without some visual contact with a fixed object or substrate. Therefore, two kinds of spatial information are required for orientation within a current: visual data (or some other kind) on the relationship of the animal to the reference system, and mechanical information about the flow of current in relation to the animal.

A number of systems require several kinds of spatial information. A person hanging a picture on the wall, or a bird constructing its nest with the hollow facing upward each utilizes several reference values. The person is registering the position of his body in relation to gravity, and the angle of the picture in relation to his body as perceived visually. By using these two measurement values, the person can orient the picture in relation to gravity.

In this example, two external (allothetic) pieces of spatial data are recorded and computed. By contrast, Figure 11-13 goes on to show how one external and one internal (idiothetic) cue may be involved: holding an arm horizontally, orienting to a course with the body position constant, and the gravity orientation of spiny lobsters.

The capacity and performance of combined orientation systems depend on the precision of incoming spatial information, which in turn depends on the quality of the "measuring instruments." Discrepancies and errors in orientation may be illustrated by one case of spatial orientation in humans. In an experiment, we reduce the picture on the wall to a

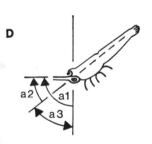

Fig. 11-13. In combined orientation systems, spatial information is gained by computing two spatial reference values. In (A), both quantities originate outside the body (allothetic), while in (B)–(D) one originates within the organism (idiothetic). A: To obtain information about the arrangement of a picture (angle a3), the person's mechanism has to compute the reference values of body to gravity (a2) and picture to body (a1). B: The position of the arm in relation to gravity (a3) derives from the relational values of body-gravity (a2) and arm-body (a1). C: The bearing of the crab in relation to the sun (a3) derives from calculation of the angle body-sun (a2) along with the angle between the direction of leg movement and the body (a1). D: The spiny lobster computes its position to gravity (a3) from the reference values "angle of statocysts (antennae) to gravity" and "antennae to body."

horizontal, dimly shining line and gaze at it in the dark so that we are un-affected by other visual stimuli. We perceive the line as horizontal as long as we look at it with our head straight up or bent slightly to the side. But once we tilt our head or body further down, the line appears to slant. Psychologists call this perceptual phenomenon the Aubert phenomenon or the Müller phenomenon, depending on the direction in which the line seems to slant—either to the side of our tilt (Müller) or to the other side (Aubert).

When the observer is asked to subjectively adjust the line to the ver-tical position, he will increase the angle between himself and the line in the first case (Müller) and decrease the angle in the second (Aubert). This subjectively vertical line now deviates from the real vertical position. The person's "miscalculation" can be explained by the functioning of the human sense organs of equilibrium. These apparatus lie in the labyrinth, a part of the middle ear; when the body is tilted to a relatively large de-gree, they no longer indicate the "correct" position of the body. For example, when the person bends to the right at 60°, the gravitational sense organs signal only about 50°. By contrast, the eyes register the actual change of 60° of the vertical line relative to the body. From these two measurements, the perceptual apparatus (Fig. 11-14) computes a difference of 10° between the line and the vertical position; that is, the observer per-ceives the line as 10° off the vertical after he has tilted his body. In this way he gets the impression that while he is bending over the line changes position in space.

We can observe this phenomenon with ourselves, by gazing at a ver-tical object—such as a free-hanging lamp—and tilting our body far to one side. The background and surroundings should have as little visual structuring as possible, to reduce visual cues. Now the lamp seems to hang "crooked." What this means, however, is that we perceive the vertical position differently from before. The spatial setting therefore did not re-main unaltered during our change in posture—in other words, we ex-perience a disturbance in spatial constancy.

Performance of combined orientation systems

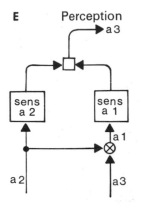

Fig. 11-14. Block control diagram for Figure 11-13: In sens a2 and sens a1, the angle sizes a2 and a1 are recorded and transmitted as measurement values (output quantities). These measurement values are combined and computed as information about a3 (percept a3). Applying only to (A) and (D): the quantity a2 splits up, affecting not only sens a2 but also sens a1; when in (A) the position of the body changes, this also causes an alteration in the angle between body and picture (a1; correspond-ingly also D). In addition to a2, a3 and a1 also exert an influence (that is, a2 and a3 overlap in the effects on a1).

▷
Section of a large flock of brent geese (*Branta bernicla*), as may be observed when these birds rest at the sea coast in winter.

12 Biological Clocks

Biological clocks,
by E. Gwinner

◁

Above: Wild geese in the
familiar wedge-shaped or
"V" formation, sometimes
also taking the shape of a
"one" or a diagonal line.
Many larger birds such as
cranes, gulls, and swans
also prefer this formation
during short (non-migra-
tory) flights. Below: At
certain times of the year,
sandpipers resting from
their flight form partic-
ularly concentrated and
large flocks, for example,
on the coastal areas of the
North Sea. During low
tide the animals disperse
into the shallows to forage
for food, while during
high tide they congregate
on a few areas of coastal
land.

Almost all animals and plants live in an environment whose properties
are not constant over time, but instead show considerable periodic fluctuations. In this way, most organisms are exposed to often rather drastic
changes resulting from the rhythms of night and day or summer and winter. In addition, there are tidal and lunar fluctuations that exert their
greatest influence on organisms inhabiting the tidal zones. Thus, organisms
may be subject to daily rhythms, seasonal cycles, and, in coastal areas, also
to the turns of the tide and the phases of the moon.

Animals and plants have adapted in a variety of ways to these four
kinds of environmental rhythms. For example, most terrestial reptiles inhabiting the temperate zones are active only in the summer months, and
even then only around midday. This is because these animals are poikilothermal (cold-blooded) and need high external temperatures for optimum
activity levels. Many bird species reproduce mostly in the spring and
summer because at any other time of the year there would not be enough
food to raise the young. Many invertebrate animals living in the tidal
zones will confine their predatory activities either to low or high tide—
depending on whether they naturally hunt on land or in the water.

There are countless examples to illustrate how biological activities are
distributed over time to reflect the rhythmic fluctuations of environmental
conditions. Adapting to these rhythms often involves the animal's entire
physiological and morphological organization. One of the first questions
that biologists try to answer is whether such periodicity is determined
only by external circumstances (exogenous) or whether it comes about
through factors within the organism itself (endogenous). Scientists originally believed that these biological rhythms were directly related to changes
in external conditions. But this happens only in some of the cases. Many
organisms have an endogenic periodicity. Experiments have shown that
many animals maintain their biorhythms even when the relevant conditions are held constant.

We have illustrated some typical cases of different biorhythms in a number of species (Fig. 12–1). The example of a daily rhyhm is offered by the activity of bees. In experiments, these insects can be trained to visit a food dish at a certain time of the day. Their periodic searching or foraging activity is performed independent of regular daily fluctuations in external stimuli. Our graph shows that the bees remember the specific time to which they were conditioned to visit the food source even when the daily periodic time markers are absent. In other words, these animals work according to an endogenous daily periodic clock. We must note, however, that under constant conditions this entire cycle does not correspond to exactly twenty-four hours. It was noticed that the bees' search activity came a little earlier each day, so that the length of this period is obviously somewhat shorter than twenty-four hours. We therefore speak of an "approximate daily rhythm," known as a circadian rhythm (*circa* = approximate, *dies* = day). This term indicates that the periodic length does not quite correspond to the length of the day. With the aid of their internal circadian clock, bees can remember various times of the day and therefore visit certain flowers only when these flowers release nectar and pollen.

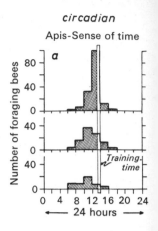

Biological tidal rhythms, too, may be endogenously programed in an organism, and again they will diverge from environmental periods when conditions are kept constant. In this case we speak of a circatidal periodicity, analogous to a circadian one. An example is the swimming activity of a small crab, *Synchellidium*, which inhabits the tidal zone along California's coast. Under normal circumstances, these crabs bury themselves deeply in the sand when the tide is low. When the beach becomes flooded, they emerge to forage for food in the water. However, this tidal activity rhythm does not depend directly on the water level, nor on any other environmental stimulus conditions; instead, it is maintained even when conditions are kept completely constant. But now these activity phases begin increasingly later relative to the normal time of high tide, so that one period is actually a little longer than 12.4 hours. *Synchellidium* profits from this circatidal periodicity because it can start digging itself out long before the coming of the tide, thereby being ready to swim in search of food when the beach becomes flooded.

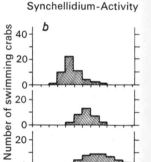

Some marine organisms are known to have an endogenous lunar rhythm. Once again, with environmental conditions held constant this period proves to be somewhat shorter than the external rhythm—that is, it is circalunar. An example here is the lunar-periodic swarming activity of *Platynereis*, a marine member of Annelida. These annelids normally inhabit the bottom of the sea, but once a month, when the moon is in a particular phase, they display their genital organs at the ocean surface, and reproduce. The people on the Fiji Islands know when these animals swarm; they consider them delicacies, and they catch them with nets to eat at feasts. *Platynereis* perform their swarming activity during the days around the new moon, but this periodicity is independent of external

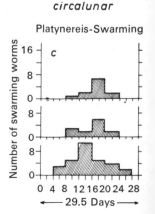

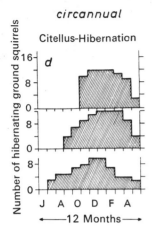

circannual

Citellus-Hibernation

d

Number of hibernating ground squirrels

16—
8—
0—
8—
0—
8—
0—

J   A   O   D   F   A

←——12 Months——→

Fig. 12-1. Examples illustrating a circadian rhythm (a), a circatidal rhythm (b), a circalunar rhythm (c), and a circannual rhythm (d). The length of the abscissa represents a full daily, tidal, lunar, or annual cycle. The frequency distribution of active (a–c) or hibernating animals (d) during three successive periods are recorded one below the other. We can see that the period of the endogenous rhythm deviates somewhat from that of the environmental cycle: The endogeneous period is longer than the environmental one in the case of circatidal periodicity, and shorter than the environmental period in the case of circadian, circalunar, and circannual periodicity.

periodic stimuli. When experimental conditions are constant, these annelids will swarm at intervals of about thirty days. In this case, however, the endogenous rhythmicity only incidentally corresponds to circalunar and circatidal rhythms which are otherwise irrelevant to the real purpose. These periods—in contrast to our previous examples—do not represent any adaptation to particularly favorable or unfavorable conditions in the environment. Instead, this rhythm synchronizes the swarming activity of *Platynereis* so that the chances for fertilization are at their greatest. We might say that these animals have "agreed" to meet at the ocean surface during a certain lunar phase in order to reproduce.

Finally, there are annual rhythms that occur endogenously, within the organism itself, and again they do not correspond precisely to environmental periodic changes. These periods are therefore circannual. The American thirteen-lined ground squirrel generally begins to hibernate in October and emerges from its sleep during the following April. This rhythm of hibernation is independent of annual changes in external conditions, since it can be observed even when the animals are kept in an environment of constant temperature and regular twelve-hour periods of "daylight." This independent, endogenous circannual periodicity allows the animals to awaken at the proper time. Still buried deep in the ground, they would otherwise take far too long to notice the beginning of spring. Correspondingly, this circannual biorhythmic behavior ensures that organisms begin to develop large reserves of fat in the fall, essential as a source of energy during hibernation.

While we know relatively few organisms with endogenous tidal, lunar, and annual rhythms, internal daily rhythms occur so frequently that we may consider them a basic feature of life itself. These internal daily periods control a multitude of physiological and ethological processes, in one-celled organisms as well as in the highest vertebrate species. It is because of these rhythms, for example, that Paramecia pair up and reproduce only at certain times of the day, that insects requiring warm temperatures become active only around midday, and that the majority of songbirds do their most intensive singing in the early hours of the morning.

Humans, too have many daily periodic functions that are based on some endogenous physiological periodicity. Figure 12-2 illustrates the walking-sleeping cycle of an experimental subject who stayed in an isolated room for seventeen days without a clock. As we can see, this person still maintained a clear periodicity, although his hours of wakefulness shifted to a somewhat later time each day. The subjective day of this experimental subject was therefore a little longer than 24 hours—an average in fact, of 25.9 hours. Similarly, under constant conditions the waking-sleeping cycle of animals diverges somewhat from the 24-hour day (Fig. 12-5).

Circadian rhythms allow the organism to adapt its behavior in the best way possible to the course of a day's external events. This endogenous

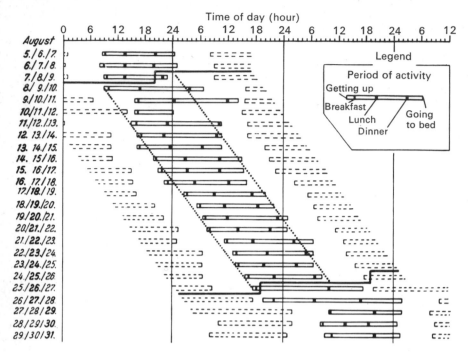

Fig. 12-2. Endogenous daily periodicity in a human subject isolated in an experimental room with constant conditions. Successive days in the experiment are entered one below the other. Horizontal bars: waking periods; black demarcations in the bars: mealtimes. The first three and the last four rows show the subject's behavior under normal conditions with a clock and outside the isolation room, before and after the experiment. Dates in bold face to the left of the graph each mark the beginning of a waking period. As we can see, waking periods begin later on each succeeding day.

periodicity determines, for example, that many insects hatch from their pupae only in the early morning or evening hours. This is vital because when these otherwise fully developed individuals emerge, their skin is not yet completely hardened (developed into a protective armor), which means that the insects would dry up very quickly. Thus it is essential that they hatch at a time of day when humidity is high and external temperatures are low.

In some parasites, the daily pattern of activities is adapted to the daily rhythm of their host animals. For example, the larvae of the elephantiasis-producing *Wuchereria bancrofti*, which normally live in the lungs of humans, travel at a very definite hour into the blood vessels lying in the skin —at the very time that mosquitoes—which carry the nematodes to other people—begin to bite. The swarming periods of the worm larvae differ somewhat from one area to another, but they always correspond exactly to the activity phases of mosquitoes. In Africa, for example, the larvae are transmitted by a nocturnally active mosquito, and here they travel to the peripheral blood vessels at night; but on Samoa and the Fiji Islands the carrier is diurnally active, and here the nematode larvae are found in the peripheral blood vessels only during the day.

In a number of organisms, circadian rhythms fulfill important and very special functions. Some animals utilize a circadian clock to orient themselves to the sun. In this way, certain species of migratory birds find their route to their winter quarters (see Chapter 13). Many animals and plants "measure" the length of the day by means of a circadian periodicity,

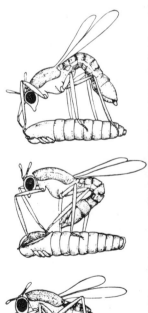

Fig. 12-3. In the mosquito *Clunio*, the male helps the female emerge from the pupa.

and this tells them what season it is. Their basic procedure is similar to ours when we try to determine whether the day is a long one in summer or a short one in winter: These animals find out by means of their internal clock whether the day is still light at a certain hour. If the answer is yes—for instance, at 6:00 p.m.—then it must be summer, if no, then it is still winter. This principle is very much a simplification of the highly complicated processes that actually occur, but it does illustrate the basic idea.

As a result of continuously assessing the length of days, many flowers blossom only at a certain time of the year. Many of the stages that insects have to undergo in their development are controlled by the length of daylight. In numerous species of birds breeding in the temperate zones, a circadian clock helps to measure the length of days and thus determines when the animals reach sexual maturity, when they perform their courtship behaviors, when they molt, and when they migrate.

In some species of birds, however, annual periodicity cannot be determined exclusively by seasonal changes in the length of days. This applies particularly to migratory birds that spend the winter at the equator, where each day is approximately equally long. Here, the animals show an endogenous annual rhythm ensuring that molting and homeward migration take place at their proper times. Figure 12-4 gives the results of an experiment in which willow warblers were subjected to a constant twelve-hour day for

Fig. 12-4. Annual periodicity of migratory restlessness, body weight, and molting of six willow warblers. The two birds whose behavior is represented at the top lived during the entire experimental period under natural conditions of light in their native breeding sites in southern Germany. The four other birds were subject to the same conditions, but at the end of September they were transferred to an isolation room (LD 12:12). Shaded area: changes in intensity of migratory restlessness. This activity occurs at night in many nocturnally migrating birds who are forced to sit in a cage during their normal migration periods. Unbroken curve: changes in body weight. Black bars: molting of the retrices and remiges (wings and tail) of the birds's plumage. Shaded bars: molting of the body feathers. Arrows: date of hatching from the egg.

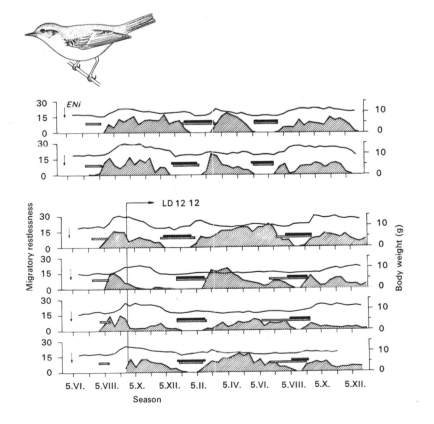

a period of fifteen months. It shows that these birds experience their migratory restlessness and undergo their molting twice a year, even when external conditions remain constant. The experimental subjects therefore showed the same behavior as their conspecifics who are exposed to natural lighting conditions and external seasonal fluctuations. Under constant experimental conditions, this periodicity is maintained for at least three years. This may be compared to the ground squirrels mentioned before, whose circannual periodicity makes them wake up with the onset of spring even though they receive no relevant or reliable cues about outside circumstances during their hibernation.

Endogenous rhythms frequently ensure not only that an animal's behavior is adapted to temporal fluctuations in the environment, but also that individual members of a species meet at certain times to perform certain activities, such as reproduction. We have already illustrated this point with the circalunar swarming activity of *Platynereis*. One very good example, where two endogenous rhythms work together allowing the animal to adapt its own behavior to periodic fluctuations in the environment as well as synchronizing with conspecifics, is the mosquito *Clunio*, whose hatching behavior corresponds to daily and lunar rhythms. This species of Diptera is one of the few insects that extend their activities to the sea. Their larvae inhabit the tidal zone, specifically the low-lying stretches of land that are almost always flooded. Their habitat is free of water only twice a month, shortly after full moon and shortly after new moon at the time of the spring low-level mark—and even then for only a few narrowly determined hours. *Clunio* females can lay their eggs in the ground only at these points in the daily and lunar cycles. Since they live only a few hours as fully matured insects, the time when they hatch from the pupa must fit in with the rhythm of the tides.

In fact, these mosquitoes hatch a short time before the spring low-level mark is reached. The males, which have wings, emerge a little earlier than the wingless females. They then help the females to hatch, and they carry them around in their mating flight until the tide has reached its lowest level. Next, the males deposit the females close to the water line; the females then lay their eggs, dying a short time later. All this happens within a few hours and only once every fifteen days. Researchers have found two internal clocks working to ensure that these insects hatch at the proper time: a circadian clock to determine the right time of day, and a circalunar (more precisely, circasemilunar) clock to pinpoint the proper phases in the cycle of the moon.

In our discussion we have already emphasized one important property of all four biorhythms: Under constant external conditions, their period generally deviates somewhat from the period of the environmental rhythm—these internal clocks go either fast or slow. For example, if a number of starlings are kept under the same constant experimental conditions, each animal shows a slightly different circadian periodicity in its

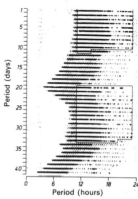

Fig. 12-5. Entraining the circadian activity rhythm of two chaffinches by means of a 24-hour light-dark alternation. The recording lines made on successive days are glued one beneath the other. Both birds were exposed to a 24-hour light-dark alternation with 12 hours of light (white) and 12 hours of darkness (shaded) between days 1 and 11 and between days 20 and 33.

activity cycles. These and many other results have demonstrated that circadian periodicity is indeed endogenous to the animal, not determined by some outside daily-periodic factors. After all, if the latter were the case, we would expect each animal to show a daily biological periodicity of exactly twenty-four hours, without any individual variation. Similar results have been obtained in measuring circatidal, circalunar, and circannual rhythms, and we may therefore conclude that these, too, are endogenous.

At first scientists believed that the daily biological rhythms were imprinted into the animal in the course of early development, determined by the periodicity of their environment. Today we know for certain that this is not the case. K. Hoffman, raising lizards from when they were still in their eggs, created three different cycles of light and dark: One group was incubated under the influence of a 24-hour day, with 12 hours of light and 12 hours of darkness; a second group was subjected to an 18-hour day, with 9 hours of light and 9 hours of darkness; and a third group was incubated in a 36-hour day, with 18 hours of light and 18 hours of darkness. After hatching they were placed under constant conditions of light and temperature. The result was that all members of the three groups showed a circadian activity rhythm whose period diverged only slightly from a 24-hour day.

Again, the outcome of this experiment indicates that circadian periodicity is innate. This has been conclusively demonstrated in plants, too. Hybrids that were the offspring of bean plants with different daily leaf movements under the same constant conditions, showed a period lying half-way between those of the parents. In the fly *Drosophila*, mutants have recently been produced in the laboratory with circadian activity-rhythm periods of 19 or 29 hours. Both of these mutants probably result from changes in the same gene, localized on the X chromosome. Although similar genetic-control experiments have not yet been carried out with tidal, lunar, and annual rhythms, it is probable that these, too, are innate; we have observed them developing even in animals that were placed under constant environmental conditions before they had a chance to experience one complete external cycle.

One of the most puzzling factors of circadian rhythms is that they are largely independent of outside temperature. While almost all chemical and physiological processes take place at twice to three times their normal speed when the temperature rises ten degrees, the period of a circadian rhythm rarely changes by more than ten percent under the same warmer conditions. This applies to warm-blooded vertebrates as much as to insects or single-celled animals. Naturally a change in temperature always has a certain influence, if only very slight; it is interesting that a temperature rise causes processes to speed up in some animals and to slow down slightly in others. It is likely that internal clocks have some compensational mechanism whose workings are not absolutely precise, but which overcompensates at one time and undercompensates at another. We have some

During this time their activity cycles were synchronized with the 24-hour light-dark cycle. The rest of the time these birds lived in constant darkness during which time their activity periods deviated from the 24-hour cycle. In the bird whose record is shown on the left, the activity cycle was shorter than 24 hours, while in the bird to the right it was longer.

evidence that tidal and annual rhythms also allow for this kind of temperature adjustment.

This high degree of independence from changes in temperature is extremely important in a biological sense. It ensures that circadian rhythms can meet their function as biological clocks. The "sense of time" inherent in bees would be useless if their circadian clock were to go faster or slower by several hours every time the bees were subjected to changes in their body temperature. Similarly, many poikilothermal animals that—comparable to migratory birds—utilize both the sun and a circadian clock for orientation would not be able to calculate properly the apparent movement and ecliptic of the sun if their endogenous clock were as strongly influenced by temperature as are most of their other physiological processes. Consider that if a circadian period were to accelerate or decelerate by only one hour, the animal would diverge from its course of migration by approximately fifteen degrees.

In many-celled organisms, different daily-periodic processes are frequently controlled by different circadian clocks. This could be determined by studying isolated organs or tissues belonging to the same animal. For example, both the isolated eye and a particular paired posterior sensory ganglion (the parieto-visceral ganglion) belonging to sea hares (*Aplysia*; a sea snail), show a continual daily periodicity, maintained over many cycles despite constant external conditions. This periodicity applies to the snails's frequency of spontaneous discharge. In a similar way, the isolated adrenal tissue of rats continues to show a daily periodicity in hormone secretion at the same time that the donor animal passes through various other circadian rhythms.

Even in intact organisms we are occasionally able to observe different processes controlled by different circadian rhythms. People who have stayed in an isolated underground experimental room without being kept informed of the time have, at various points, showed a circadian activity rhythm whose period diverged from that of their circadian temperature rhythm; the two periods did not always coincide with each other. It is probable that the time-memory mechanism in bees is based on at least two circadian rhythms—one of these can be blocked by deep carbon-dioxide anaesthesia while the other remains unaffected.

Single-celled animals may also possess more than one circadian clock. If we remove the nucleus from the unicellular alga species *Acetabularia*, the individual organism concerned will continue to show a daily rhythm of photosynthesis despite constant light conditions. This means that one internal clock is to be found in the cell plasma. However, there must be a second clock located in the nucleus. We know this from experiments where denucleated algae were implanted with the nucleus of another alga species whose rhythm of photosynthetic processes stood in a different phase relationship to the natural day. It was found that the receiving plant adopted the phase position of the cell from which this nucleus originated.

▷

Indian false vampire bat (*Megaderma lyra*). During experiments testing place recognition, this bat always flew through the same square in the mesh created by horizontal and vertical wires.

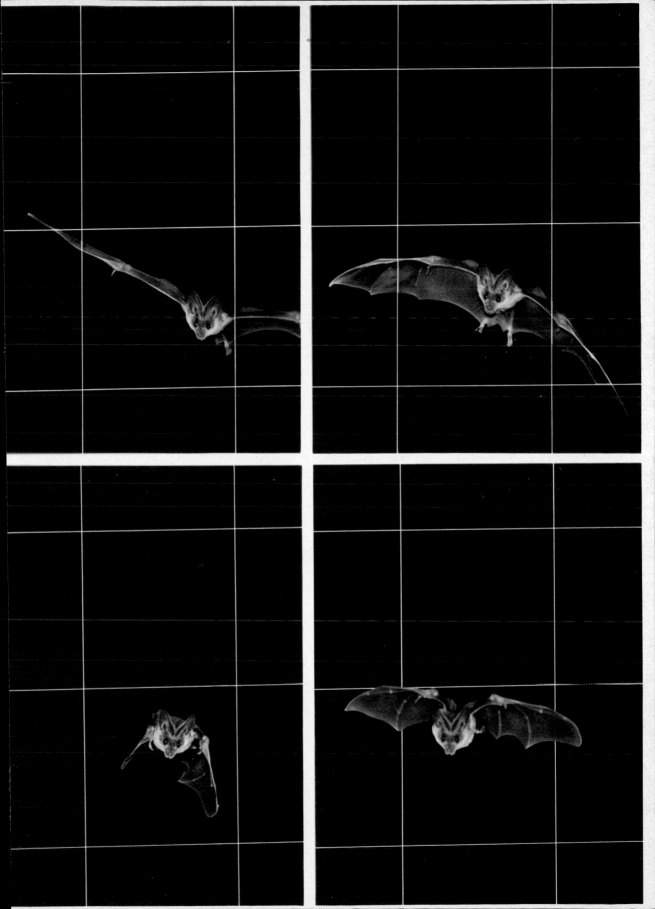

The rhythm arising in the nucleus is therefore paramount over the plasma rhythm, forcing its phase position and probably also its period on the denucleated host. We do not know how this coupling is made.

Furthermore, we do not know the mechanisms by which many-celled animals synchronize their different circadian rhythms. In at least a few cases the nervous system definitely performs an integrating function. Cockroaches and crickets lose their circadian activity rhythms when we block the connection between certain parts of the brain, known as Lobi optici, and the rest of the brain. In mammals, too, there are important integration channels located in the central nervous system. We eliminate the circadian activity rhythm of rats by destroying certain parts of the Nucleus suprachiasmaticus (a brain nucleus in the diencephalon). Birds become permanently active when we remove their pineal body, and the daily periodicity of their body temperature also disappears. Perhaps there are certain areas of the central nervous system, such as the ones described above, which contain a central clock for synchronizing the various circadian rhythms.

As we have seen, under artificially constant conditions the period of endogenous daily, tidal, lunar, and annual rhythms deviate a little from the periodicity of respective environmental rhythms. Under natural conditions, however, they correspond exactly. This means that the endogenous rhythms normally become synchronized (entrained) with the periodicity of the environment by means of periodically fluctuating external forces. Such environmental stimuli or cues which synchronize endogenous rhythms are known as entraining agents or as synchronizers (*Zeitgeber*). Their function is to correct the biological periodicity at least once in every periodic cycle by precisely that amount which the endogenous period has been either lacking or adding.

The most prominent entraining agent of circadian rhythms is the daily illumination cycle. We have discovered this, for example, in experiments where "free-running" circadian rhythms were entrained by a regular 24-hour light-dark cycle. In addition, we can also synchronize the endogenous daily periodicity of many organisms with cyclical changes in temperature. Some birds will utilize their species-typical song, when we play it back on tape in a regular daily cycle, as a synchronizer for their circadian rhythms.

Only in a few cases do we know anything about the external periodic factors that synchronize the other three "circa"-rhythms. One important entraining agent of the circatidal activity rhythm shown by the water flea *Synchellidium*, the small crab discussed earlier, is the fluctuation of water movement—in turn coupled with the ebb and flow of the tides. A number of circalunar rhythms are entrained by means of lunar-periodic changes in nightly illumination. This may be observed, for example, in the previously mentioned periodic swarming activity of *Platynereis* and hatching periodicity of *Clunio*. The length of daylight within the seasonal cycle is probably an important synchronizer for circannual rhythms. This

Above: Greater horseshoe bat (*Rhinolophus ferrumequinum*) during the cruising phase. The animal beams its localization sounds through the nose precisely in the direction of our camera. Its nostrils are embedded in a nose piece shaped like a horseshoe. Bottom left: Greater horseshoe bat. The nose piece with its ridge and lancet probably increases the directional effect during emission of the long cf-FM sounds. Bottom right: Linne's false vampire bat (*Vampyrum spectrum*). It beams FM sounds through the nostrils, which are surrounded by a nose piece.

latter point, however, has been conclusively demonstrated only with one species, the Sika deer (*Cervus nippon*).

When a biological rhythm is precisely entrained with external fluctuations, it not only corresponds to the synchronizer's period but also shows a characteristic phase relationship with it. Here, the exact phase position of the endogenous rhythm is determined in part by the properties of the entraining agent and in part by the "response properties" of the biological periodicity. The fact that diurnal species are active during the day and nocturnal animals during the night arises from the diverging response properties of their circadian activity cycles. Since the various circadian rhythms belonging to the same animal respond differently to a particular synchronizer, they adopt different phase relationships to this monitor and therefore also to one another.

These characteristic phase relationships of different rhythms to one another are probably very important if the organism is to function properly. For example, the periodicity of body temperature is synchronized with the animal's activity rhythms in such a way that high temperatures correspond to high levels of activity. If in an experiment we eliminate all the synchronizers, the normal phase relations begin to change, and in some cases different rhythms may be completely dysphasic. The consequences can be very bad. If tomato plants are kept under continuous illumination, they become "sick," but they remain healthy if allowed to grow in a 24-hour temperature cycle that acts as a synchronizer. Similarly, small fruit flies have a shorter lifespan under conditions of continuous light than they do when subjected to a 24-hour light-dark cycle. The fact that people do not feel very well for several days after an overseas flight may be due in part to similar factors: The various circadian rhythms require different lengths of time to become resynchronized with the new, shifted entraining agent. Thus, the normal phase relationships of these rhythms are temporarily out of balance. We should take these disturbances seriously: Flies, for example, have a shorter life span when we force them to "fly" a few thousand kilometers to the east or to the west every couple of weeks just by shifting their external light-dark cycle one way or the other.

13 Migration and Orientation in Birds

Migration and
orientation in birds,
by K. Schmidt-
Koenig

One of the most remarkable phenomena in nature is the migration of animals, especially that of birds. Year after year we can observe how the migratory birds in our part of the world fly away to their winter quarters in autumn and return in the spring. One day they are gone; another day they are back. Anyone watching the sky when bird migrations are fully under way will also see numerous species flying past, individually or in flocks, or he may hear the calls of nocturnal travelers after dark.

An inquisitive person, not content merely watching them fly for a short time, will accompany the birds in his thoughts, inevitably asking a number of interesting questions: Where do the birds fly in the fall? How do they find their way? How did bird migrations come about in the course of evolution? What causes the birds to embark upon their long journey with such regularity? Do birds also migrate in the Southern Hemisphere? There are many other questions. In this chapter we shall focus on the way birds navigate—often half-way across the globe.

Everybody knows that migratory birds fly south in the fall. But this "southern" direction extends from south-east to south-west. Furthermore, it is not enough to know the direction; these birds must also know the distance to their goal, in other words, how far they have to fly. Northern European birds that spend the winter in central or even southern Africa have to accomplish a far greater feat than birds flying from northern Europe to Spain or northern Africa.

For over fifty years, ornithologists have been placing rings on birds to find out about the migratory routes and winter homes of many species. It seems that the record for long distance is held by the arctic terns (*Sterna paradisaea*). They breed in the northern to arctic latitudes, for example in northern Canada, flying east across the Atlantic in the fall changing course to the south off the coast of England, and finally spending the winter far in the south reaching Antarctica. The migratory route of the arctic terns is one of many examples showing that, contrary to popular beliefs, bird migration does not take place in a simple north-south direction. Eastern

Fig. 13-1. Breeding areas (black dots) and migratory routes of arctic terns.

and western routes are also observed, and, in fact, straightforward north-south flights are a rarity. The migratory course of the bar-tailed godwit (*Limosa lapponica*), which is native to Alaska, offers another very impressive example. This bird flies south-west across Japan and the Malay Archipelago, crosses the equator, and then flies south-east over eastern Australia to New Zealand.

Birds native to the Southern Hemisphere also migrate. Their migrations however, take place in the Southern climatic fall (which concurs with spring in northern latitudes), and they orient toward the north. Thus one species may fly from the southern tip of South America to central South America, or across the equator due north to Central America or North America. A very good example of the kind of migratory route and navigation displayed by the birds is seen in the slender-billed shearwater (*Puffinus tenuirostris*). It breeds on several islands in the Bass Straight off the Australia's eastern coast and on Tasmania. Once a year it flies in a large clockwise loop around the Pacific, crossing the equator on two occasions; in this way it is exposed to a variety of changes in length of day, apparent path of the sun, climatic conditions, and so on.

Modern technological methods of radar and radio tracking have largely superceded our earlier means of observing birds with binoculars. Today, individual birds can be spotted on their migratory route, and their course and behavior can be determined over longer distances with a great deal of precision. Miniature transmitters, weighing 2–3 g, have been developed in the U.S.A. Easily carried by small birds, these transmitters send a signal that may then be located by means of stationary direction-finding aerials or by vehicles and airplanes. The routes of a great many migratory birds have been tracked with this method, sometimes over a distance of several hundred kilometers. (Figure 13-3 shows the nocturnal route of a thrush flying over Illinois.) Almost all birds tracked in this way flew a fairly straight line once their course was determined. Since even this method has its limitations, however, researchers are working on a largely automatic method of tracking birds on their migrations, with the aid of satellites.

With the use of radar tracking instruments, we have gained further knowledge of what happens to birds on their migration. In contrast to the flight-monitoring radar instruments used earlier, this new device detects individual birds and automatically "keeps an eye" on them. When such a bird passes the instrument at a greater or lesser angle, the exact range, bearing, and altitude as well as ongoing changes in these values can be recorded or printed out, and the animal's course of flight relative to the ground can be plotted on a chart. Furthermore, the bird's wing-beat frequency can be determined; this identifies or at least indicates the species and ensures that the same individual is still being tracked. Finally, the exact migratory routes of individual birds as well as other bits of information can be recorded from the natural course of events during the flight as they occur within the range of the radar instrument.

Fig. 13-2. Breeding area (shaded) and migratory route of slender-billed shearwater in the Pacific region. Black dots indicate the areas where the birds were recaptured.

Fig. 13-3. Flight path of a thrush (black line, with clock times), plotted from a car across Illinois and leading beyond Lake Michigan.

We hope to gain some particularly interesting results from experimental manipulation of the natural phenomena. The following are some attempts presently being made; Experimenters catch a bird resting on its migratory route and glue several small magnets to it so that any magnetic orientation will be disrupted; or the experimenters create a shift in the bird's internal clock, changing its sun-compass orientation or other orientation mechanisms that depend on time. Next, the bird is placed in a cardboard box, attached to a balloon, and floated up and released in the vicinity of the radar station which then tracks the bird once it has reached a predetermined height. The radar now follows the animal on its flight, registering the changes that occur in orientation behavior as a result of experimental manipulation. This puts us on the threshold of a great deal of new and basically different information.

At this point we can already make one important statement: When radar sets (that were less suitable) were first being used to study bird migration at the beginning of the 1950s, researchers came to the conclusion that migratory birds became disoriented whenever the sky was overcast. But today we know that these birds—at least as indicated by radar—fly a straight course even under a heavy cloud cover. There are indications, in fact, that they can maintain their course even when they are inside the clouds. This leads us to conclude that eyes—in other words, optic orientation to stars or to landmarks—probably play a far less important role than was previously assumed; certainly visual orientations is not the only method used.

There are some aspects of bird migration that we can study in the laboratory, where the animals live in cages or aviaries. Here we can conduct conditioning experiments or we can perform tests on birds in the state of intense activity characteristic of migratory restlessness (*Zugunruhe*). One of the results found with conditioning experiments was that birds use the sun as a compass—not unlike an experienced hiker who knows that the sun is to be found south-east at 9 a.m., south at 12 noon, and west at 6 p.m. And just like the woodsman or scout who only needs a wristwatch to find the desired direction, birds have a "timesense" or time compensated mechanism, an "internal clock" that allows them to calculate the position of the sun at any time of day and to use it for their compass orientation (see Chapter 12).

In experimenting with birds that are ready to migrate, researchers take advantage of the fact that even with the best care, these animals will become restless, jumping and fluttering about in their cage as the time of migration approaches. The activity of such a bird can be recorded in a number of ways. Figure 13-4 illustrates a rather ingenious method. The bird sits on an ink pad located inside a "funnel" or cone covered with blotting paper. While attempting to leave the cone in an upward flight direction, the bird stains the blotting paper with its footprints. The experimenter then assesses the amount of ink staining on the paper. Under cer-

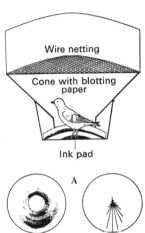

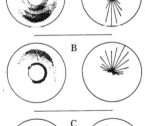

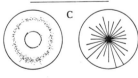

Fig. 13-4. Above: Recording apparatus for small birds during periods of migratory restlessness. Below: Analyzing the patterns of their footprints. In (A), this bird was oriented toward SSE, in (B), toward NNE, and in (C), no preferred direction was evident. Left, footprints; right, analysis in vector diagrams.

tain conditions, activity during migratory restlessness becomes directional, usually corresponding to the species-typical direction of migration. Experiments with robins and warblers have shown that these birds also use the earth's magnetic field as a compass. When this magnetic field was blocked out, the birds became disoriented. Furthermore, experimenters could predict the direction in which the robins tried to fly by changing the magnetic field. We do not know how these birds perceive and evaluate the magnetic field, and no magnetic sense organs have been discovered or recognized yet.

Birds may also use the stars for orientation; but their method seems to be different from what was assumed fifteen to twenty years ago with the hypothesis of a star-navigation system. We have known for some time that the last few weeks before flying off are extremely important to the orientation capacity of young migratory birds. Results of recent studies on indigo buntings (*Passerina cyanea*) have shown that a young bird just out of the nest will watch the stars intermittently determining which part of the starry sky undergoes the least rotation. In the northern latitudes this turns out to be the North Star, or rather, the stars closest to it. This is the section of the sky that birds will use as their reference point for migration —i.e., they will fly mainly to the south away from the North Star. Experiments carried out in a planetarium served to demonstrate that this is the case. Young indigo buntings were left free to watch an artificial starry sky that did not rotate about the North Star but instead was made to turn about another star, belonging to the southern constellation Orion. The animals then started off in a direction opposite to their normal, species-specific one, north away from Orion.

All laboratory experiments have shown that birds may choose their flight direction with the aid of "compasses" using the sun, the stars, or the earth's magnetic field. This explains some important aspects of bird migration, but there are still other problems to be solved. Navigation requires more than a compass, because the animal must not only determine the proper direction; it must also find its goal at the end of the journey. We can imagine one simple and rather limited way of finding a goal if some information about the migratory distance is added to that about the direction. A goal can be found from a starting point if the animal knows both the range and bearing. Many studies both in the field and in laboratories have indicated that in nature some animals may indeed utilize this kind of "vector navigation." A vector is a quantity defined by both direction (or bearing) and distance (or range). Numerous experiments have been done in which migratory birds were intercepted in their fall migrations, ringed, transported to another place located diagonally to their original route, and then released. These studies have produced a number of common results, and as an example we may cite the behavior of Baltic starlings.

Figure 13-5 shows the route of starlings native to the Baltic Sea: Their flight to the wintering grounds (the South of England and Ireland and the north of France) takes them across the Netherlands, where for several years thousands of these birds have been intercepted each fall, ringed, and then transported to Switzerland, where they were released. In these experiments, older birds that had traveled their natural migratory route at least once were able to navigate to the species-specific winter quarters; they could compensate or correct for the distance and direction by which they had been displaced. But young birds flying the route for the first time traveled parallel to their species-typical route and spent the winter in a completely new area. We can see that even on their first migration, these young birds possessed information about the direction they had to follow, and they had the capacity for maintaining this bearing on the basis of solar, stellar, or magnetic compasses.

We are now beginning to understand something about the way birds "know" what distance they have to fly to reach their goal. In laboratory experiments, people measured the amount of activity expended by the restless birds in their special cages during the usual time of migration. The result was that amount of activity was proportional to the distance of the respective, species-specific migration routs: Birds with a long route (e.g., willow warblers normally flying from Europe to southern Africa) expended proportionally more activity than individuals with a normally shorter route (e.g., wood warblers, from Europe to central Africa), and these in turn were more active than birds with an even shorter range of migration (e.g., chiff chaffs, from Europe to northern Africa).

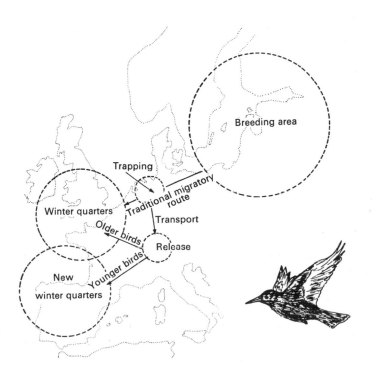

Fig. 13-5. Results of experiments with starlings native to the Baltic region.

Birds might therefore know the distance they would have to fly by the amount of activity expended during their migration. This level is preprogramed, and once it is reached the birds stop migrating. Activity level alone would certainly be only a crude distance indicator, and we have no idea how it might work in detail. Yet this principle would make a form of vector navigation plausible, albeit of limited use to young birds on their first migration. It does not allow the birds to correct any diversions from their course, experimentally or naturally (as may occur during a storm, for example).

The method of vector navigation is still the only one supported by results of both field and laboratory studies. All other navigational hypotheses, such as solar, stellar, or intertial navigation, have not yet been substantiated by experiments. Thus our present state of knowledge encompasses only a small number of the problems we are faced with when studying the migratory phenomena of birds and other animals. A great deal of further research is needed. But as new and ever-more-suitable technological aids such as radar, radio tracking, and so on become available, we can hope to gradually uncover the mysteries of bird migration.

14   Echolocation in Bats

Echolocation in bats,
by H. U. Schnitzler

Bats can avoid obstacles and find food even in total darkness. This puzzling ability has fascinated scientists for a long time. In Italy Lazzro Spallanzani was the first to attempt an explanation of how bats find their way in the dark. Around 1793 he demonstrated that blinded bats could detect obstacles as skilfully as intact animals. During further experiments, he eliminated the senses of touch, hearing, smell, and taste, but again found no effects on their flight performance. Since, in these first experiments, eliminating the five known senses made no difference, Spallanzani began to suspect that bats used an unknown sixth sense for orientation; but he was uncertain, and he challenged other scientists to make the same tests. In Geneva, Ludwig Jurine answered the call and made experiments of his own. In contrast to Spallanzani, he found that bats could no longer perceive obstacles once their ears were fully blocked.

Discovering echolocation

When Spallanzani heard of these results he performed some new experiments, inserting and gluing brass tubes into the auditory canals of his subjects. As a result, he found that intact as well as blinded animals collided with obstacles when these tubes were blocked, while the sightless bats had no more difficulty than the sighted ones if their tubes were left open. In another experiment, Spallanzani caught fifty-two bats in the bell tower of the cathedral of Pavia, removed their eyes, and released them. Four days later he managed to recapture three of these bats at the same place, early in the morning. Their stomachs contained as many insect remains as did those of sighted bats he had examined. Spallanzani then concluded that bats catch insects and avoid obstacles in the dark by using their auditory sense. This brought him very close to solving the problem.

In France, however, Georges Cuvier ignored these discoveries. Though he had never done experiments of his own, he claimed that bats detect obstacles with a highly developed sense of touch. Despite the fact that Spallanzani had already disproved this hypothesis, Cuvier, a prominent scientist prevailed in his opinion, and he determined what scientists believed on this subject right into the 20th Century. Official dogma did not

even change when R. Rollinat and E. Trouessart (France) in 1900 and
W. L. Hahn (U.S.A.) in 1908 performed more tests and reasserted that the
sense of hearing played a very important role in the orientation of bats.

It was not until 1920 that H. Hartridge (England) opposed Cuvier's
hypothesis by suggesting that bats emit high-frequency sounds, above
the range of human hearing, using echos for orientation. In 1938, D. R.
Griffin (U.S.A.) used instruments developed by G. W. Pierce to demonstrate that bats do, in fact, emit ultrasonic sounds with a frequency of more
than 20 kHz. Humans can hear sounds only up to 20,000 cycles per second
(20 kHz). On the other hand, the sounds emitted by bats lie between
20,000 and more than 100,000 cycles per second. This, however, was a
later discovery.

In the following three years, Griffin and R. Galambos were able to
show without a doubt that bats utilize the reflected echos of their ultrasonic calls in the detection of obstacles and to navigate in their environment. In 1943, without knowing about these studies, S. Dijkgraaf (Holland) obtained the same results using only ethological experiments.
Griffin called this kind of orientation echolocation.

After the second World War, scientists both in Europe and the U.S.A.
resumed the study of echolocation in bats, using methods from ethology
and neurophysiology. We shall summarize the most significant findings,
especially in the area of ethology.

The smaller insectivorous bats (Microchiroptera) and the flying foxes
(Megachiroptera) are suborders of Chiroptera. At night, flying foxes hunting for food orient by means of their well-developed nocturnal eyes. An
exception are the rousette bats, (*Rousettus*). E. Kulzer demonstrated that
these megachiropterans make smacking sounds with their tongue when it
is very dark, using the echos for orientation. But this method of echolocation is quite primitive and very different from that of microchirop-
terans.

On the other hand, all of these smaller bats constantly emit localization
sounds in flight as well as in a hanging or sitting positions when they explore their surroundings. This sound emission is independent of the
brightness level in their environment. Comparative studies on the localization sounds of different bat species have uncovered both large differences
and certain correspondences in the frequency, duration, and intensity of
these calls.

In all bats there is a coordination between sound-pulse emission and
breathing rhythm. During localization, these animals emit either one pulse
per inspiration or a whole group of sounds (Figs. 14-1 and 14-2). When
flying, they emit either a single pulse or a group of sounds for every beat
of the wing. Thus, in correspondence to the interrelationship between
sound pulse emission and breathing rhythm we can deduce that flying bats
draw one breath per wing beat.

The number of sound pulses within a group will depend on the orient-

Orientation with
ultrasonic sounds

Localization behavior

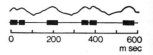

Fig. 14-1. Breathing
rhythm and sound
succession in the greater
horseshoe bat.

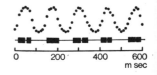

Fig. 14-2. Wing-beat
rhythm and sound
succession in the greater
horseshoe bat.

ing situation. Undisturbed bats hanging or sitting usually emit four to eight sound groups per second, with one or two pulses per group. However, the number of sounds in a sound group can increase dramatically when the animal localizes and homes in on a particular object. In the Mediterranean horseshoe bat (*Rhinolophus euryale*), for example, the frequency may rise up to ten pulses per sound group.

A useful description of localization behavior in flying bats is provided by the way Griffin categorized this activity during insect-catching. He divided it up into the cruising, closing in, and fielding or final phases. Other scientists described these as the searching, approach, and terminal phases.

When bats are cruising, they fly around in a space without obstacles either to reach a certain goal or to search for prey. Animals we have studied, including a variety of species, typically performed seven to twelve wing beats and inspirations per second while cruising in the laboratory, each time emitting a single group or a group with two sounds. In this orienting situation the bats always emitted their longest sounds. For example, the greater horseshoe bat (*Rhinolophus ferrumequinum*)—a species with very long localization calls—achieved a pulse duration of 50–65 msec while cruising, whereas the big brown bat (*Eptesicus fuscus*) only reached 2–4 msec and Seba's short-tailed bat (*Carollia perspicillata*) showed a maximum sound duration of only 1–2 msec.

The closing-in phase begins when bats make their first response to a localized object, such as an insect, an obstacle, or a landing spot. At this point the number of cycles per sound group starts to increase. This orientation phase will cover four to six groups of sound pulses when the animal is flying in a room with familiar obstacles or landing at a familiar spot. But if an obstacle or prey object should suddenly appear, the closing-in phase becomes much shorter and covers only one to three groups. As the bat nears the object, its pulse rate in the successive groups will increase and the sound duration will decrease.

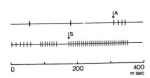

Fig. 14-3. Sound succession in a big brown bat while pursuing a pebble thrown in the air.

In their final phase of orienting activity and just before seizing the prey, landing, or passing an obstacle, the bats emit a group of sounds with a high rate of short pulses. This final set typically has a longer duration than the preceding ones. Here, the animals make more wing beats per sound—pulse group, and for all of the bats the duration of the sounds as such reach the shortest value. For instance, the sounds emitted by the greater horseshoe bat decrease to a minimum of 10 msec, and those of the big brown bat, to a minimum of 0.3 msec. These sounds follow one another in rapid succession. The smallest interval between pulses is 11 msec in the greater horseshoe bat and as little as 4 msec in the big brown bat. Figure 14-3 shows the sequence of sounds in a big brown bat observed by Griffin as it pursued a pebble thrown into the air, while Fig. 14-4 shows the sequence recorded in a greater horseshoe bat flying through an obstacle of vertical wires. In both graphs, A represents the transition from the

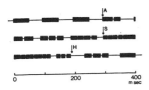

Fig. 14-4. Sound succession in a horseshoe bat while flying through a wire obstacle.

cruising to the closing-in phase, while S indicates the beginning of the final phase with many short sounds. At H, the horseshoe bat passed its obstacle and reverted back to cruising.

Although sound patterns may be similar in comparable orienting situations, more detailed examinations of orienting sounds in different bat families and species reveal considerable differences in frequency progression duration, and intensity. Not all bats emit their orienting calls through the open mouth. Quite a few beam the sounds through their nostrils, which are then usually surrounded by a nose piece. When looking at the frequency progression of pulses, we can distinguish two major groups of localization calls. In frequency-modulated calls (FM sounds), the frequency drops from high to low. These FM sounds are always short, lasting no more than 5 msec even during cruising.

In the localization sounds of some species, the first harmonic (keynote) is by far the most prominent. In the FM-sound emission of the little brown bat (*Myotis lucifugus*) during cruising, for example, the keynote decreases from 100 kHz by more than an octave to 40 kHz (Fig. 14-5). In other species, localization sounds mainly contain upper harmonics (overtones). For instance, the localization calls of Linne's false vampire bat (*Vampyrum spectrum*) during cruising will display only upper harmonics between 65 and 115 kHz, belonging to a first harmonic that falls from 35 to 20 kHz. In the Indian false vampire (*Megaderma lyra*), we recognize primarily the fourth and fifth harmonics belonging to a tonic that slowly descends from 20 to 16 kHz.

We know of some species with FM sounds where, in addition to altering pulse duration, the frequency progression of localization sounds is also changed in different orienting situations. For example, the little brown bat emitted short pulses of sound in its final orienting phase, falling from 50 to 30 kHz with a duration of 0.8 msec and from 40 to 22 kHz with a duration of 0.4 msec. By contrast, during cruising it emitted localization sounds lasting about 3 msec and descending from 100 to 40 kHz.

In contrast to FM sounds, some species of bats emit constant-frequency sounds with a frequency-modulated end section (cf-FM sounds). Examples are horseshoe bats, Old World leaf-nose bats, mustache bats, and a few other families. These localization calls are usually longer than FM sounds, lasting more than 5 msec in many species during cruising.

Horseshoe (rhinolophid) bats emit typical cf-FM sounds. The part with constant frequency usually takes up nine-tenths of the entire pulse in these species. In the short frequency-modulated end section, frequency falls by about 12-16 kHz. The frequency of the cf section varies among horseshoe bats. In large horseshoe bats occupying a hanging position, this part lies between 81 and 84.2 kHz. It corresponds to the second harmonic, belonging to a less prominent first harmonic at half the frequency.

Figure 14-6 gives a table of localization sounds in different bat families. In addition to frequency progression, it also shows whether the calls are

Fig. 14-5. Frequency progression in the localization calls of some bat species:

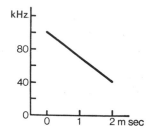

Little brown bat.

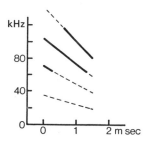

Spear-nosed bat.

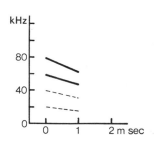

Malayan false vampire bat.

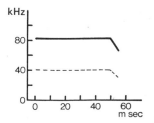

Greater horseshoe bat.

Fig. 14–6

Family	Source of food	Sounds beamed through/ Sound intensity/ Duration	Frequency progression (cf = constant frequency) (FM = frequency modulated)
Emballonuridae Sac-winged bats	Insects	Mouth high mostly short	Mostly slightly descending FM sounds some species also cf sounds
Noctilionidae Bulldog bats	Fishes Insects	Mouth high long	cf-FM sounds
Nycteridae Slit-faced bats Megadermatidae Large-winged bats	Arthropods Insects Small vertebrates	Nose low short	Slightly descending FM sounds several upper harmonics
Rhinolophidae Horseshoe bats	Insects	Nose high long	cf-FM sounds second harmonic emphasized
Hipposideridae Old World leaf-nosed bats	Insects	Nose high long and short	cf-FM sounds second harmonic emphasized
Desmodontidae Vampire bats Phyllostomatidae Big-eared leaf-nosed bats	Blood Fruit Nectar Pollen Small vertebrates	Nose low short	FM sounds several upper harmonics
Chilonycterinae Mustache bats	Insects	Mouth high long and short	cf-FM sounds second harmonic emphasized
Natalidae Funnel-eared bats	Insects	Mouth high short	FM sounds second harmonic emphasized
Vespertilionidae Vespertilionid bats	Insects	Mouth high short	FM sounds mostly the first harmonic emphasized sometimes also the second and third harmonics
Molossidae Free-tailed bats	Insects	Mouth high long and short	Some species FM sounds (*Tadarida*) others cf-FM sounds (*Molossus*)

emitted through the mouth or nose, what is the intensity and duration of the sounds, and what is the main source of food for each family.

Localization sounds are designated as "long" if they last more than 5 msec during cruising. Data on sound intensity are usually not very precise because it is difficult to measure the pressure waves. Therefore, we have attempted to give a more general view in Fig. 14-6 by designating as "high" any sound-pressure waves above 100 db re 0.0002 dyn/cm² when measured at approximately 10 cm in front of the bats, in other words, if the pressure waves had a value higher than 20 dyn/cm². Even this rough division supports Griffin's finding that bats which hunt small objects (insects) emit loud localization sounds, while species pursuing relatively large objects (fruits and small vertebrates) need only soft localization calls. This is why he called the latter species "whispering bats."

When a bat is flying toward a stationary object, the animal's flight movements create so-called Doppler effects whose magnitude will depend on the flying speed. One such effect arises during sound emission and another when the animal receives its echo. For this reason, the echo frequency created by flying bats is always higher than the emission frequency. With flight speed at 4.1 m/sec, the greater horseshoe bat can already achieve a Doppler shift—a difference between emission and echo frequency—of 2 kHz.

Doppler-effect compensation

By making careful measurements of the cf sound frequencies contained in the localization calls of flying horseshoe bats and mustache bats (*Chilonycteris rubiginosa*). I was able to show that these animals compensate the Doppler effects created by their flight speed simply by lowering their emission frequency the same amount as the degree of Doppler shift. As a result, echo frequency is held at a nearly constant level within a small range of about 200 Hz, lying just above the frequency of the cf sections contained in localization sounds that these animals emit while still in a hanging position.

For example, a greater horseshoe bat that emits a cf frequency of 83 kHz ∓100 Hz while roosting, then flies toward a stationary obstacle with a cruising speed of 4.5 m/sec, will emit localization sounds with a cf frequency between 80.85 and 81.05 kHz and then perceive their echoes in accordance with the Doppler shift at 4.5 m/sec with a cf frequency between 83.05 and 83.25 kHz. During the slower flight speeds in the closing-in and final phases, the bat will lower its emission frequency less, again holding the echo frequencies within the range of 83.05 to 83.25 kHz.

Measurements made in a wind tunnel with a greater horseshoe bat flying against the wind have shown that, in such cases, the animal will make its Doppler-effect compensation correspond with the air speed above the ground, not in accordance with its higher speed of flight in the air. Furthermore, a bat flying in a gas mixture of helium and oxygen will continue to compensate the occurring Doppler effects, and even a roosting animal in localizing a swinging pendulum will try to compensate in this

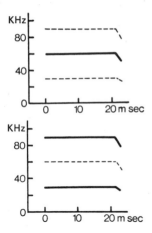

Fig. 14.7. Localization calls of a mustache bat in air (above) and in a mixture of helium and oxygen.

way. We therefore conclude that these animals have a feedback control system which compares the frequency of the cf sound sections contained in its echoes with some target value of frequency. If the echo frequency deviates from the target value, the bat will continue to alter its emission frequency until the echo frequency has reached the desired value.

Griffin and Novick eliminated the laryngeal apparatus and associated nerve fibers in bats of various species, and demonstrated that special membranes in the larynx are tightened by muscle action during sound production. D. Pye proposed that, similar to human vocalization, a base vibration with harmonics is produced in the larynx of bats, and that the throat while conducting these sounds acts as an acoustic filter by letting some harmonics pass but not others. In experiments with a helium-oxygen gas mixture, which changes the nature of the acoustic filter due to transmission of higher sound frequencies, I found support for Pye's hypothesis for horseshoe bats, mustache bats and short-tail leaf-nose bats. The acoustic filter of the mustache bat (*Chilonycteris rubignosa*; Fig. 14-7) in normal air is adjusted so that only the second harmonic can pass through, suppressing the first and third harmonics. In a helium-oxygen mixture, the nature of this air-passage filter changes to such an extent that now the second harmonic is blocked, while the first and third harmonics pass through. In the vespertilionid bats *Myotis lucifugus* and *Eptesicus fuscus*, on the other hand, the sounds emitted are the same whether the animal is exposed to air or a mixture of helium and oxygen. The throat and mouth of these species do not seem to have any filtering qualities, so that the sound pulses beamed out are the same as those produced in the larynx.

In bats that emit sounds through the open mouth, the wave length is always smaller than three times the mouth opening. For this reason we can assume that interference alone would cause a strong directional effect, intensifying with decreasing wave lengths or with increasing frequency. A further influence on directional effect probably comes from the megaphonelike structure of the animal's open mouth.

Where sonar sounds are emitted through the nose, interferences occur between the pulses beamed through both nostrils. F. P. Möhres demonstrated for the horseshoe bats and Pye for the Old World leaf-nosed bats that the distance between the nostrils corresponds to about half the wavelength of the cf frequency. These interferences therefore lead to amplification of sound-pressure waves ahead and attenuation to the sides. In greater horseshoe bats, the resulting directional effect is so strong that even at an angle of 20.5° the pressure waves are only half as strong to the sides as to the front. As yet, no one has examined whether the nose pieces surrounding the nostrils of some bats—which in part are very complicated—have any additional compounding effect on the direction of sound.

Localization capacities

When looking for food or detecting obstacles, all microchiropteran bats use echolocation. Their navigational feats are astounding at times, especially those of insect-catching bats. In the wild, E. Gould observed free-flying bats, *Myotis lucifugus*, that hunted a different insect on the average of every three seconds and, according to his estimation, caught about half of these. Griffin and F. A. Webster came up with similar figures while observing bats hunting fruit flies in the laboratory. One fielding maneuver with the typical sound patterns of the closing-in and final phases lasted no longer than half a second, and in one case, two successful pursuits were observed during this short period. When placed in a dark experimental room, these bats hunted with abated speed, as evidenced by their weight increase. Nor were they distracted by loud noises designed to drown out the whirring sounds of flying insects. Only when the experimenters produced very loud noises within the frequency range of the bat's localization sounds did the bats stop hunting. This shows that they catch insects by echolocation alone.

Fig. 14-8. Catching a fruit fly (F); *Myotis lucifugus.*

Bats do not always seize their insect prey directly with the mouth. Films made by Webster showed that vespertilionid bats more often caught flying insects in a pocket, or apron, formed by their interfemoral membrane stretched out between their hind legs. While still in flight, these animals would reach into the apron with their head, seize the prey, and eat it. Frequently, a bat may also field an insect with its wings and then transfer it to the mouth. Their accuracy in pursuing and catching flying prey indicates that bats can localize their object with a great deal of precision.

A very different set of localization problems confront the fish-eating bulldog bats (*Noctilio*). Toward evening, these animals fly in long and narrow stretches back and forth over the water, at certain intervals thrusting their feet into the water and seizing fish swimming close to the surface. Here the bats use sharp claws pointing forward from their hind feet. R. Suthers discovered that these bats do not localize the entire fish directly under water, but rather detect either water movements caused by the prey or simply the fish's body parts projecting above the surface.

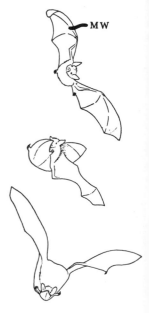

Few investigations have been made on echolocation in the "whispering bats" during food hunting, because their very soft localization sounds are difficult to record. With true vampries (*Desmodus rotundus*), which cut into the skin of vertebrates with their knife-sharp teeth and then lick up the blood, we are nonetheless reasonably certain that olfactory and perhaps even visual stimuli play an important part in finding prey. We may assume as much for fruit and nectar eaters.

One method for quantitatively recording the localization feats of bats is the use of obstacles for experimentation. Subjects are trained to fly through obstacles made of vertical and horizontal wires, and we then record their number of passage flights made without touching the wires. A number of scientists who have made these experiments with vespertilionid bats, leaf-nosed bats and bulldog bats report that animals cleared the obstacles with better than random success even with wires as thin as 0.12–0.2 mm. The performance of large-winged bats and horseshoe bats

Fig. 14-9. Fielding a mealworm (MW) thrown in the air: The prey is caught with the interfemoral membrane; *Myotis lucifugus.*

Fig. 14-10. Fielding a noctuid moth with the tip of the wing; (*Rhinolophus ferrumequinum*).

was even more impressive. They detected wires as thin as 0.08 in diameter with a high degree of certainty. By the sound pattern emitted by a Mediterranean horseshoe bat, I was able to observe a response in three of eight recorded cases to wires only 0.05 mm thick. In all the animals studied, the closing-in phase began around 2 m in front of the object when the obstacles were large or the wires were thick. With decreasing thicknesses of wires, this phase would begin at correspondingly decreasing distances to the obstacle.

Obstacle studies with bats that have been experimentally treated in some way provide more information about how echolocation functions. The experiments of Spallanzani and Jurine had already shown that blinded bats navigate as surely as sighted animals, while blocking their auditory passages prevents them from detecting obstacles. Even if we tightly close just one ear, the bats perform no better than chance; this was found in all observed species. If we lightly close one ear so that echo perception is decreased by about 20 db, a greater horseshoe bat will accomplish 30 percent fewer flights through vertical wires without touching them. An interesting observation was that the subjects regained their normal localization skills when the second ear was also lightly blocked. This shows that both ears must be at the same level of receptivity if the bat is to localize obstacles accurately.

N. Suga examined the effects on localization when he ablated the auditory center in the animal's brains. He found that bats continued to avoid obstacles without the auditory center in the cerebrum and even without the dorsal part of their posterior corpora quadrigemina. Only the elimination of the latter's ventral parts prevented the bats from avoiding the wires.

With obstacle experiments, scientists could also examine whether bats are distracted by noises in the range of ultrasonic frequencies. Studies by Griffin, J. J. G. McCue, and F. A. Webster have shown that bats will stop flying around only when the noise level has reached a very high pitch, within the area of localization sound frequencies. If they are nevertheless forced to fly, the animals perform poorly. In these experiments, subjects were able to detect echoes with intensities less than those of the interference noises, provided that the echo and the noise came from different directions. This feat is possible because the bats's ears are highly directional in their receptivity and because echoes are evaluated from stimuli registered by both ears. Such a tolerance for disturbances would explain how the bats can congregate in large numbers in tropical caves without interfering with one another.

Role of vision in localization

The question of how bats use their eyes in orientation was also studied in obstacle experiments. J. Chase used leaf-nosed bats as subjects, more specifically spear-nosed bats (*Phyllostomus hastatus*) and Seba's short-tailed bats (*Carolia perspicillata*). She blocked both ears of her animals and noted that those who could see avoided white strips of cloth 30 cm wide somewhat better than did the blinded subjects. On the other hand, sighted bats with blocked ears oriented far more poorly than blind animals with intact hearing. Eyes therefore play only a subordinate role in the detection of these very large obstacles, and they are probably no use at all in perceiving thin wires. However, T. C. Williams and J. Williams found that visual stimuli do play a more important part in orientation over longer distances. Spear-nosed bats were able to find their way home from as far away as 60 km, providing that their eyes were not covered. This indicates that these animals orient themselves by landmarks, or visual reference points, when flying over greater distances. They could not possibly use echo-location during such flights, because the range of sound waves is very limited, especially under tropical conditions.

Bats utilize their system of echolocation not only for localizing objects but also for discriminating among different objects. Griffin and Webster showed that trained bats were able to catch mealworms that were thrown into the air together with plastic chips. Their most skilled mouse-eared bat caught 98 percent of all the mealworms and avoided touching 85–90 percent of the plastic chips thrown into the air at the same time. A. Konstantinov and his co-workers taught bats to discriminate between cubes, spheres, and cylinders of the same volume, and between rectangular, round, and triangular discs. Furthermore, subjects were able to discriminate a rectangular disc from rectangles with jagged edges and from concave and convex rectangles, and even to distinguish between aluminum rectangle pieces and wooden ones of the same shape and size.

J. A. Simmons studied how bats performed when they had to discriminate distances. He conditioned big brown bats so that they always received food whenever they turned toward the closer of two triangles presented simultaneously. As the difference between distances of the two triangles to his subjects decreased, the bats made more and more errors in discrimination, starting at 3 cm. When the difference was reduced to zero, the subjects made the expected random choices, i.e., their error frequency was 50 percent. When Simmons compared these errors made at the different discrepancy levels with the error probability curve deriving from signal theory, he came to the conclusion that the orientation feats of these big brown bats approximate the best performance theoretically possible.

Bats that are brought into an unfamiliar room will fly in apparently random circles. Starting at the ceiling, they spiral downward slowly and finally, soar underneath tables and chairs, getting to know every corner of the room. After a while, they will remember and recognize certain areas quite accurately and fly straight in that direction—for example, the door of the cage, a preferred roosting place, and food and water places. This orientation with the use of memory is so greatly relied upon that greater horseshoe bats crashed at full speed into a glass pane closing off a normally open cage entrance, even though the animals are constantly using their echolocation. Another example is their attempt to take up a

Discriminating
distances

hanging position at a spot in the air usually occupied by a landing post. For that reason it is very important to repeatedly change the position of obstacles in an experiment; otherwise, the bats could avoid these obstacles by using only their memory.

Place recognition

G. Neuweiler examined the capacity of the Indian false vampire bats (*Megaderma lyra*) to remember certain places—i.e., their place-recognition ability. His subjects had to fly through an obstacle made of vertical and horizontal wires, forming a mesh of square openings measuring 14 cm² (see Color plate, p. 195). After a short time, each animal restricted its flights to a few of these squares, maintaining this preference for weeks. The more consistent its preference, the better it was able to pass through the squares without touching the wires. Even when the mesh was shifted just slightly, their orientation grew less accurate. After a six-week rest period, the bats still showed good recognition of the obstacles, but after twelve weeks they hardly remembered the positions at all. Few investigations have been done on the interaction between place recognition and echolocation. We do know that place recognition is at times the dominant factor, because bats have been observed to collide with easily identifiable obstacles placed into a normally empty flight path.

Hearing

Bats receive echoes by "catching" them with their external ears, gigantic structures in some species. The sounds are transmitted over the auditory canal to the tympanic membrane, causing it to vibrate. These vibrations are then conducted by the auditory ossicles to the cochlea, where mechanical impulses are transduced into nervous excitation. The auditory nerve conducts these impulses to the auditory pathway, which consists of various auditory brain centers where nervous impulses are processed.

The external ear structures of various bat species are very different in size, shape, and mobility. In the members of many families, a prominent ear lid, or tragus, projects in front of the external ear and the opening to the auditory canal. Its function has not been defined. The external ears effect a directional amplification, so that only echoes from a narrowly defined area around the ear will be clearly received. In some families, members can move their external ear structures while localizing. In the horseshoe bats, these ear movements are associated with the emission of localization calls.

Many neurophysiological studies have been made on the auditory capacities of several bat species. To summarize, we can say that all bats can hear best when sounds are within the frequency range covered by their own localization calls. In bats with FM sounds, the acoustic frequency range is therefore rather large. On the other hand, bats that emit cf-FM sounds have a particularly low threshold in the area of the cf frequencies. The horseshoe bats, which as we know keep the cf parts of their echoes within a small frequency range through Doppler-effect compensation, show an extremely narrow threshold minimum just in this area of target

Fig. 14-11. Head of a long-eared bat (*Plecotus auritus*). A very prominent ear lid (tragus) projects in front of the ear opening.

frequency. It is precisely this synchronization of the auditory system with target frequency that we would expect in a feedback control system if echo frequency is to be held constant. Another way that the auditory system of bats is specifically adapted to the demands of echolocation is the capacity to discriminate sound phenomena following in rapid succession, gauging sound phenomena following in rapid succession, gauging the direction of sound stimuli with both ears, and recognizing echoes even in the midst of noise interferences.

We can see by the way bats orient that echolocation allows for the detection of objects, noting distance and direction. In addition, bats can recognize movement and discriminate objects according to size, shape, and quality. Perhaps they can even identify prey animals according to individual qualities. Here are the most plausible hypotheses, according to present scientific knowledge, that researchers have suggested in order to explain these capacities.

Hypothetical explanations of the echolocation phenomena

In measuring distance, the most likely explanation is the one H. Hartridge published in 1945. According to this hypothesis, bats gauge the elapsed time between the emission of their pulse and the return of the echo, the traveling time of the sound waves. This would then be a measure of the distance from the object. In animals with FM sounds, this involves no problem at all, because their localization sounds are always kept at such a short wave length that they do not overlap with the returning echoes. In bats with long cf-FM sounds, however, there is the problem that the latter parts of the emitted calls with cf vibrations do overlap with the echoes. We assume that these bats do not use the cf parts of their pulses as measuring points for traveling time, but rather utilize the FM parts of emitted sounds and returning echoes, just like other species producing only FM sounds.

Measuring distance

By contracting the muscles of their middle ear during the emission of FM sounds, or of the FM parts of cf-FM sounds, bats hear their own localization sounds at a weaker intensity than they had when the bats produced them, and so their reception of the weak echoes returning shortly after emission will be clearer. This process is also facilitated by a special evaluation procedure in the nervous system.

Bats need to use both ears for determining the direction of an object. It is highly probable that differences in intensity between echoes picked up on both sides are considered when these animals gauge the direction of the echoes. According to A.D. Grinnell, bats with FM sounds, which do not usually move their ears when orienting, determine the intensity discrepancies for the various frequencies contained in the echoes. By contrast, bats with cf-FM sounds are constantly moving their ears during orientation, and they possibly evaluate the intensity differences of cf parts within their pulses at different ear positions.

Gauging direction

The capacity to recognize movement is very good, especially in bats with cf-FM sounds. The Doppler effects arising through the bat's own

Detection of movement

movement and that of a localized object lead to a difference in frequency between the cf vibrations contained in emitted sounds and their echoes. The bat uses this Doppler shift as a direct measure for the relative speed between itself and the object.

This process holds an added advantage for horseshoe bats, which use Doppler-effect compensation to maintain the cf frequencies of their echoes at a constant and most favorable level. These animals are able to assess the frequency and amplitude modulations additionally produced within the cf parts of their echoes when a localized object moves on its own. The bats will therefore be able to detect, for example, an insect's beating wings. Possibly they use this information to identify various prey objects.

The FM sounds are not well suited for measuring Doppler shifts and recognizing movement in this way, because they are very short and they undergo rapid frequency changes. In this case, animals probably measure the relative speed between themselves and a localized object only by gauging the variations that come about when repeatedly and successively measuring the distance to the object.

Discriminating objects according to size, shape, and quality

For discriminating objects according to size, shape, and quality, all the relevant information is contained within the frequency spectrum of the echoes. In an experiment, bats are required to determine the pressure-wave amplitudes of various frequencies within the echoes, then compare this spectrum with the values learned for certain objects. Griffin demonstrated that bats are indeed able to evaluate this kind of information. As mentioned earlier, he managed to teach a mouse-eared bat to discriminate between mealworms and plastic chips, then showed how the frequency spectra of echoes reflected from these two objects were clearly different. But we still do not know whether cruising bats really use this capacity to recognize prey. Since some of these animals also pursue pebbles thrown into the air, it seems that movement alone already provides the organism with an important clue to the presence of prey.

At present, researchers studying the phenomenon of echolocation measure bats's orientation behavior under precisely defined conditions using methods of behavioral physiology, then compare their results with the theoretically possible limits of these capacities as calculated from the laws of signal theory. To these comparisons are added anatomical investigations and neurophysiological recordings in the brain's auditory centers. In this way, scientists are trying to understand how bats can accomplish all these amazing orientation feats with a brain not even 1 cc in size.

Echolocation in other animals

In the 1950s it was found that certain other animals also orient themselves by means of echolocation. For example, toothed whales (Odontoceti) have a good system for echolocation; the most detailed studies to date have been made on dolphins. These animals emit exceedingly short clicking sounds (maximum duration approximately 0.2 msec) with a band spread of up to 150 kHz, and then utilize their echoes to detect obstacles and objects when visibility is poor. Experimenters demonstrated that dolphins use this method by taking two trained animals and blocking their eyes with rubber suction cups. Nearing a feeding place, these dolphins would emit their localization calls in increasingly rapid succession. These calls were not diffuse, but were oriented in a specific direction. One dolphin with blocked vision even discriminated gelatine capsules—used in a test to distract the animal—from pieces of meat of the same size, using echolocation. When the experimenters hung a grid with 36 metal bars 25 m apart from each other into the dolphin compound and continuously changed the spatial arrangement, the animals sped through and around the bars unabatedly and without touching them even when it was completely dark.

Clicking localization sounds audible to humans are also emitted by two species of birds which nest in caves. These are the oilbird (*Steatornis caripensis*) and the swiftlets (*Collocalia*), which use their sounds to orient within the dark caves. Some blind people also use a rudimentary kind of echolocation: They listen for the echoes produced by the rattling sounds of their steps, the tapping of their cane, or a snapping sound made with their fingers, in this way detecting large obstacles in their path.

15   Issues and Concepts in Ethology

Issues and Concepts
in Ethology, by
George W. Barlow

Introduction

In the past, ethologists have studied a spectrum of behavior ranging from feeding through anti-predator behavior and modification of the habitat. The main thrust, however, has involved social behavior, especially in reproduction. Thus, much attention has been devoted to the (at times) spectacular behavior evoked when two rival males meet, or when courtship occurs between animals of the opposite sex. Likewise, much attention has been given to parental behavior. The trend in the last two decades, however, has been toward a broader statement of the field, with a more balanced inquiry into the manifold expressions of animal behavior. That which sets ethology apart, in the final analysis, is that the questions have been inspired by observations on naturally occurring behavior.

It is difficult to say when ethology began. Perhaps it was with Aristotle, for he wrote some excellent descriptions of animal behavior; unheralded observers before him probably also did so. Most of us, however, mark the beginning of ethology with the writing by Charles Darwin of *The Expression of the Emotions in Man and Animals.* But the field progressed from there in an inchoate fashion until after the beginning of this century. Then Konrad Lorenz and Niko Tinbergen pulled the loose threads together into a coherent fabric. It is this theoretical framework that for several years has been called ethology.

The synthetic labors of Lorenz and Tinbergen were not without difficulties. With relatively little information at hand, but an abundance of personal experiences, they formulated a number of concepts to deal with the phenomena as they saw them. You should appreciate the difficulties inherent in formulating concepts. To define an idea, a concept, is to put boundaries around it. Hence, discrete categories are formed. Nature has little understanding of categories. Its diverse phenomena commonly lie on a continuum which we divide up for convenience. Consider the growth stages of human beings. We recognize infancy, childhood, adolescence, and so forth, but also we realize that each

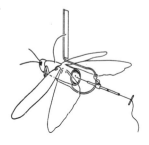

Fig. 15-1. Insects can tolerate rather severe operative procedures. Shown here is front part of the body of a tethered grasshopper in flight. This method enables the investigator to stimulate and record from nerves that participate in the coordination of flight movements. With this procedure D. Wilson was able to show for the first time that central control plays an important role in locomotion.

grades insensibly into the next. While we cannot say exactly when, for instance, an adolescent becomes an adult, we find the concept useful.

Forming a concept means the creation of an ideal. It has been epitomized in the history of biology in the concept of the type specimen of a species. One individual is held representative of the entire species. Modern taxonomists, however, study a series of individuals deriving average values and estimates of variability. No one specimen ever fits precisely this statistical description, which is thus also an ideal for the species but one embracing the species's range of variation. This more dynamic approach does not damage the species concept. Rather, it illuminates the nature of the concept and focuses on the processes of speciation.

The type concept is evident in controversies in ethology. The tendency is to pick one example of behavior and to claim it as an ideal. This is difficult to avoid, especially at the level of a single chapter in a book.

In spite of all this, the formation of concepts or of categories is indispensable to clear thinking. Concepts provide the framework for organizing new information to make comparisons. The new relationships discovered in this way lead to modification or replacement of the old concepts, and hence to more accurate formulations.

A major goal of science, then, is to disprove existing concepts and to replace them with better ones. Thus it often seems a marvel to the lay reader, and even to many young scientists, that the great men in science were actually so great. After all, everyone tries to replace their ideas, and often someone succeeds. It is, however, actually a tribute to the greatness of the pioneers that their ideas were so clearly stated and stimulating that others were inspired to test them. In ethology, this is exemplified by the theories of Lorenz and Tinbergen. Their synthetic powers and keenness of conception produced the broad theoretical framework and also the specific ideas that have inspired succeeding scientists. In what follows, I shall be reviewing many of the original concepts of Lorenz and of Tinbergen, and some of the debate that has surrounded them.

The study of animal behavior has experienced, at times, unfortunately acrimonious arguments between the proponents of the different schools. This is, in part, because the sociological implications of behavioral theory are so evident in comparison to other areas in biology. The various investigators have brought to behavioral research different histories, cultures, ideologies, and personalities and consequently have often asked quite different questions about the same behavioral phenomena. (Recall the parable of the blind men trying to describe an elephant by touch: one grasped the tail, another a tusk, a third a leg, and so on; each described a very different creature.) With such divergent viewpoints, arguments are to be expected. Too often the question has been, "Who is wrong?" It should be, "How can the differences be resolved?" This is illustrated in our first issue, whether locomotion is centrally or peripherally controlled.

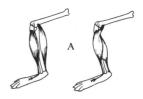

Fig. 15-2. P. Weiss demonstrated in an experiment with rats that a transposition of tendons of flexor and extensor muscles results in a disruption of function which cannot be compensated for completely by learning. Left—the normal position of muscles, right—the transposition.

Fig. 15-3. A hermit crab *Calcinus laevimanus* examines the house of a snail *Tegula funebralis* to see if it is suitable as a home.

The argument of central versus peripheral control has been a recurrent theme in the study of behavior. I have chosen locomotion to introduce the issue because the behavior involved is relatively simple and the resolution of the problem is reasonably adequate. I shall not then treat other comparably simple types of behavior but, having made the essential point, I shall move on to more complex aspects of behavior that have occupied the attention of ethologists.

Locomotion

It has been argued back and forth whether the coordination of locomotion is from patterned output of the central nervous system or whether it depends on a reflexive chain of events regulated by input from the sensory system. These opposing views often are linked with positions in the nature-nurture problem. The proponents of central control have been associated more with the innate argument, whereas those of peripheral control have been aligned with learning.

The evidence for peripheral control comes from experiments such as the following. The brain of a toad was surgically destroyed, leaving only the spinal cord intact. The toad was then placed so its legs hung over a revolving drum that turned away from the toad. If one leg was stretched out it was reflexively withdrawn while the revolving drum slowly stretched out the other leg. So started, the toad continued to step alternately as each leg was stretched. No central patterning was needed for this roughly coordinated pattern of stepping. Each movement was considered part of a reflexive chain. Some sensory stimulation fired a reflex arc producing a movement. That movement, in turn, triggered further reflexive arcs that kept the animal stepping.

The first experiment clearly demonstrating the overriding central control of locomotion involved the flight of locusts, as done by Donald Wilson. Sensory nerves run from the wing muscles to large clusters of nerve cells (essentially a central nervous system) in the locust. These cells, in turn, send commands to the muscles to produce the flight movements. Cutting the sensory nerves should destroy the coordination of flight if the control is peripheral. Wilson cut the sensory nerves and abolished flight. Then he artificially stimulated the cut sensory nerves running to the central nervous system. Even random stimulation—that is, stimulation completely lacking in pattern—resulted in coordinated output and hence normal flight movements. Thus the role of the sensory nerves here is not to control the coordination, but merely to provide some excitation. The faster the sensory nerves fire, the faster the locust flies, but the pattern of coordination itself is controlled by the organized output of the locust's central nervous system. Sensory input, on the other hand, is important in regulating balance.

An experiment that has been informative in higher animals involves the muscles that control flexing and extending. These are made to work with opposite action by surgically cutting the tendons in the legs and reversing the attachment of the muscles. When Paul Weiss did this to

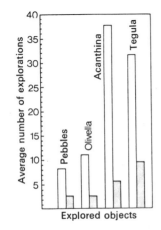

Fig. 15-4. *Ordinate:* Average number of explorations. *Abscissa:* Explored objects. *Above bars in histogram from left to right:* Pebbles, Olivella, Acanthina, Tegula. The frequency with which the hermit crab *Coenibita perlatus* investigates the houses of the snail species *Olivella, Acanthina,* and *Tgulea* depends on which house the crab inhabits at the time. If the crab lives in a non-preferred house, like that of *Olivella* (white bars) it will spend more time searching for a suitable shell than if it inhabits a shell of the preferred species *Tegula funebralis.* (After the work of E. Reese.)

a rat, it extended the legs when it should have flexed, and flexed when it should have extended. It never learned to do otherwise. Thus the control here is central and cannot be unlearned.

The more advanced the mammal, the more learning can be involved. Some human beings, for instance, have had their leg muscles surgically reversed out of necessity, whether due to illness or serious accidents. These people can and do learn to flex when they should extend and vice versa. But there remains a central organization of the walking movements which is not subject to learning, thus, in moments of panic they revert to the original coordination and are unable to walk. Learned movements can be used, however, for noncritical activities.

Even this cursory review should make clear that there is merit in the arguments both for peripheral and for central control of locomotion. Sensory feedback is obviously important in the refinement of locomotory movements, as when an insect walks over rough terrain where some legs dip low and others are high, depending on the obstacles. Clearly, too, local reflexes are an important part of locomotion; after all, the spinal cord of vertebrate animals can be thought of as an extension of the central nervous system. Finally, central coordination, producing patterned output, is probably common to all animals. In fact, central patterning is apparent even in the more complicated movements that we often think of as displays, as will be discussed shortly.

The distinction between appetitive and consummatory behavior has been an issue of considerable historical importance in ethology, although the distinction is seldom discussed in current writings. Originally there was confusion among the students of animal behavior because behavior seemed so continuously variable. But to deal scientifically with any phenomenon, it is necessary to break it into units. Then it becomes feasible to make quantitative models to guide the neurophysiologists in their quest, on the one hand, and, on the other, to compare species for evolutionary studies.

Wallace Craig proposed, and was followed by Lorenz, that behavior be divided into two types. These he called *appetitive* and *consummatory behavior*.

Appetitive behavior is relatively unstructured but appears to be goal-directed. Take, for example, the plight of a hermit crab that has been removed from its shell. Ernst Reese has shown how such a hermit crab will restlessly move about until it has found a suitable shell. If you give it a shell that is inadequate, it will continue to move about until it finds one of the proper size and shape. Likewise, a male dog will strive to reach a female in heat. Such a male will try to run around a fence, dig under it, climb it, or bite its way through it. It shows a wide range of adaptive but variable behavior. The idea may be extended to include the notion of avoidance. Thus the shell-less hermit crab could be considered to be trying to avoid being naked, that is, showing aversive behavior.

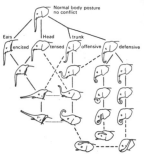

Fig. 15-5. The various combinations in which the positions of head, ears and trunk of an African elephant show a variety of expressions. Certain elements can be readily recognized. The heads are arranged according to a hypothetical, motivational heirarchy.

Appetitive and consummatory behavior

Consummatory behavior usually follows appetitive behavior and consists of one or more relatively stereotyped behavior patterns. The occurrence of consummatory behavior is said to reduce the probability that it will soon recur. While a male dog may spend hours trying to get to a female, the frantic searching behavior stops shortly after copulation. Thus copulation here is called consummatory behavior because it is followed by the cessation of appetitive behavior.

There is, however, no rigid separation of appetitive from consummatory behavior, as Robert Hinde has pointed out. Many events which seem to be consummatory (in German, Endhandlung) actually are appetitive to some succeeding behavior. Thus, in the dog, climbing the fence is appetitive to being with the female, but being with the female is appetitive to the first step in a long series of behavioral interactions leading to copulation. Furthermore, it may not be the performance of the behavior that is consummatory, but the situation in which the animal finds itself. In the hermit crab it is not the act of climbing into the shell that is consummatory, but rather the stimulation the hermit crab receives from the shell. For this reason, it is better to speak of the consummatory situation than of the consummatory act, since situation embraces both an act and the sensory situation.

Modal action patterns (M.A.P.)

In ethological studies the kind of behavior that has been treated as unitary is most easily assigned to consummatory behavior. The difficulty here is that there is a spectrum of behavior. It ranges from that which is so stereotyped that it is nearly the same every time it is done, and hence highly recognizable, to that which varies in a number of ways, often grading into another unit of behavior. Among the less variable and hence more recognizable actions are those concerned with bodily care, as when a dog scratches its ear or shakes water out of its fur, or when a housefly cleans its wing. In social interactions, easily recognized behavioral events include the courtship displays of ducks and fruit flies, the raised-claws defensive threat of a crab, and the precise song of the chaffinch.

It is possible, however, to point to many examples of behavior that, while recognizable, are also highly variable. Contrast, for instance, the changing performance of cats and dogs when licking their bodies as opposed to the more stereotyped scratching of the ear. Other easily noted examples of more variable displays are seen in encounters between two stallions, or in the quivering and associated behavior shown by courting cichlid fishes. And who has tried to predict the next syllable in the complex and highly variable song of the European blackbird or the American robin? Yet there is sufficient pattern in each instance to recognize the action.

These motor patterns have been given a variety of names, such as innate motor pattern, fixed action pattern, and instinctive movement. I use the term *Modal Action Pattern (M.A.P.)* because it emphasizes that the behavior is recognizable in terms of its usual appearance. The other terms often are burdened by explanatory concepts, such as whether the behavior is instinctive, triggered in a particular fashion, or highly rigid in its performance. I suspect that M.A.P.s are indeed heritable and centrally patterned. But the evidence suggests they are also sensitive to environmental feedback and, depending on the species, to some degree of modification during development.

The original proponents of the concept of modal action pattern certainly recognized that there is considerable variation. In fact, an important contribution of Lorenz and of Tinbergen was to point out that one of the major sources of variation lies in the directionality, the *taxis*, of the performance. This idea was beautifully elaborated in their classic study on how the graylag goose retrieves an egg that has been taken from its nest. The goose puts its bill on the far side of the egg and then gently rolls it back into the nest. The bill weaves from side to side, compensating for the departures that would otherwise be made from the straight path by the eccentric egg. If a small cylinder is substituted for the egg, the weaving disappears. The lateral motion, therefore, is not an integral part of the M.A.P., but rather a response to the wobbling of the egg. Remarkably, if the egg or cylinder is removed after the retrieval has commenced, the goose completes the movement, tucking the bill under her breast.

This phenomenon reveals an important principle that applies to many M.A.P.s. The orientation of the response, here the directing of the head toward the egg and the weaving of the bill, can be separated from the "core" action pattern, the retrieving of the egg. (Note the parallel to the discussion of the central control of locomotion and the role of sensory feedback.)

Many modal action patterns, however, can be recognized only by their orientation, that is, their taxic component. For instance, when the male stickleback displays before a rival male by standing on its head, no special pattern of coordination is observed, just the relative orientation of the male to the substrate. This raises an important theoretical question. It would appear that in this case the taxic component alone has been exploited in the evolution of a display. Thus it is the taxis that is stereotyped, not the ordinary swimming movements involved.

M.A.P.s are often used as signals in the communication of animals. There are two conflicting needs an animal faces in evolving a signal. One is to communicate quickly, economically, and without ambiguity. When this need arises, as in the rapid identification and mating between species such as sticklebacks, evolution will favor the elaboration of simple and relatively fixed signals, that is, stereotyped M.A.P.s, color patterns, or chemicals. This process of simplification and emphasis of signals is sometimes called ritualization. In contrast, when species stay together for some time, as in pairs of cichlid fish that have a prolonged court-

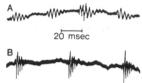

Fig. 15-6. Sounds produced by animals are the result of coordinated nerve and muscle activity, as is the case with all movements. Since vocalizations can be objectively recorded, and are thus free from subjective interpretations, they are an excellent source of information for the analysis of modal action patterns. The figure shows the sound spectograms of courting fruit fly males: A—*Drosophila pseudoobscura*; B—*D. similis*. The large spikes each represent one sound burst. The larger the interval between bursts, the lower the frequency.

Ritualization

ship before spawning, the problem to overcome is one of monotony. To hold one another's attention, and to create the arousal necessary for spawning, such species should not do the same thing the same way all the time. They should introduce variation to break the monotony. Selection here will favor some degree of variability, which accounts for those signals that are both recognizable and variable; their sequence will also vary.

Releasers

As long as man has thought about such things, it has been obvious that most of an animal's behavior is in response to stimulation. When ethologists made their original contribution here, the analysis of stimulus and response was already at a fairly high level in psychology and psychiatry. In his landmark "Kumpan" article, Konrad Lorenz developed the idea that species simultaneously evolve special stimuli and the capacity to respond to them. Lorenz called such stimuli *releasers*, and the mechanism that enabled the species to recognize these releasers, the *innate releasing mechanism*. I will consider first the concept of releaser, then its counterpart, the releasing mechanism.

The original examples of releasers struck the fancy of many readers. Lorenz described, among others, the special signal whereby nestling birds of some species void their feces without soiling the nest. When the parent is at the edge of the nest, the baby bird points its cloaca skyward, then opens the feathers around it revealing an inner ring of bright white feathers, literally making a bullseye out of the cloacal opening. The parent bird attends closely, staring at the white ring. The baby bird extrudes its feces, neatly wrapped in a mucous lining. The parent immediately picks up the fecal sack and flies away, dropping it a safe distance from the nest.

Fig. 15-7. In the flicker the sexes recognize each other by the mustache-like spot below the bill, which is only present in the male.

Another early example was provided by G. K. Noble. He noticed that the male flicker has a patch of black under its beak, resembling a mustache; the female does not. He painted a "mustache" on a mated female and released her. When she returned to the nest, the male attacked her as though she were a rival male.

Since these original observations, there has been a rapid growth in number of examples and in our understanding of the range of relationships. There has been little adjustment by modern workers to this growth, however. Some new terms have arisen, but many of them have been employed with little precision and often with overlapping meaning. It is possible, nonetheless, to point to a simple and relatively unambiguous usage.

Fig. 15-8. The lowered neck of the ostrich on the left indicates a submissive posture which is a social signal. Right: normal posture.

All stimuli having special importance to a species may be termed *sign stimuli*. One would count here both the stimuli whereby the predator recognizes its prey, as when a shrike pounces on a grasshopper, and those stimuli that are more conventionally called releasers.

The decision to call a sign stimulus a releaser is made when there appears to have been an evolutionary coaction to enhance communication. This can involve even interspecific signals, as when a flower evolves a special signal (color, odor, shape) to attract and guide its insect or bird pollinator.

Similarly, those emphatic signals that serve in the recognition and coordination of activities between males and females, and sometimes their offspring, in the same species are included here, although some workers would classify these into a subgroup of *social releasers*. A social releaser, then, is usually a signal that communicates such information as species, age, and sexual identity as well as the readiness to carry out behavior important to the survival of that species, such as fighting, mating, and escaping.

Sometimes another category of signals is recognized in order to set apart a very important type of behavior, the active inhibition of attacking in a dominant animal. This type of behavior is widespread among animal species, although it is not universal; it is best developed among animals that live in groups and that have dangerous weapons, as in wolves. The subordinate, when threatened by the dominant, performs *appeasement behavior*.

In general, there are two types of appeasement. In one, the subordinant behaves *antithetically* to the behavior of the dominant animal. That is, if the dominant erects its hair, stands tall, and shows its teeth in threat, the subordinant sleeks its hair, lowers itself, and keeps its mouth relatively closed while turning its head away, in effect, removing its weapons from view.

The other tactic used by the subordinant is to lower the hostility of the dominant through *diversion*. The subordinant stimulates behavior in the dominant that is incompatible with aggression. In highly social animals this often means an attempt to evoke parental behavior. Thus an adult but subordinant dog will roll on its back and invite grooming, as would a puppy to its mother. Likewise, an adult dog will nudge the corner of the dominant dog's mouth, the way a puppy asks for food. In song birds, the female of a pair begs for food from the male.

Sexual or grooming behavior is also used to divert the dominant animal. In many primates, the baboons being noteworthy, even males present their hindquarters to the dominant male, who mounts and may deliver a cursory pelvic thrust. This act, however, is usually seen at times other than those involving overt aggression.

An insight into the behavioral mechanism of releasers was afforded by the discovery by Alfred Seitz of what has been called *heterogeneous summation*. To explain what is meant, it is necessary first to describe how responses are evaluated, since responses to the same stimuli may vary from time to time. For one, the response to the stimulus may or may not occur. For another, the form of the response may be more or less complete, sometimes called more or less intense. Thus, the response to a stimulus may be evaluated either by the numbers of times it occurs or by the intensity of the response itself, or both. Using this type of approach,

Fig. 15-9. The posture of the wolf on the right indicates submission to the higher-ranking animal on the left who does not attack.

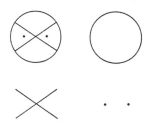

Fig. 15-10. Human infants that were shown a schematic face and its separate parts (top) reacted with varying degrees of head-turning. The Table shows the intensity with which infants of different ages react.

Age (weeks)	Circle	Cross	Dots	Sum of Parts	Total Face
8	40.00	14.66	15.00	69.66	69.00
12	67.00	13.00	21.00	101.00	100.50
16	55.00	15.00	31.00	101.00	114.33
20	23.00	10.00	16.00	49.00	106.00

Seitz discovered that a few major stimuli alone, such as color pattern and shape, can elicit partial courtship or aggressive responses from a male cichlid fish. Taken together, these stimuli result in the complete response.

Daisy Leong put this phenomenon on a solid quantitative basis in her experiments on yet another cichlid fish. She discovered that two aspects of the color pattern of a male influence the level of attack. She counted the number of attacks by a male on some juvenile fish kept in the same tank with it before and then after the presentation of a dummy that was variously colored. If the dummy had a black eye bar but no orange patch on its side (the orange patch characterizes a territorial male), the probability of attacking the juveniles increased by 2.79 bites per minute. In contrast, if the dummy was marked with an orange patch on its side but with no black eye bar, the probability of attacking decreased by 1.77 bites per minute. The difference in the attack rate between the two models is $+1.02$ bites per minute $(2.79 - 1.77)$. The critical test, therefore, was to present a dummy that carried both the eye bar and the orange patch. When this was done, the change in the rate of attacking was $+1.08$ bites per minute, or almost precisely the same as predicted from the tests with the different stimuli alone.

The releaser concept, as extended through the notion of heterogeneous summation, is often compared to the concept of *Gestalt perception*, which holds that an organism does not recognize a pattern by its separate parts, but only by the integrated whole. Thus in the releaser concept the whole is equal to the sum of the parts, whereas in the Gestalt concept the whole is greater than the sum of its parts.

The difference stands out in an experiment done by T. G. R. Bower on very young human infants. Each element of an abstract face evoked a trained response, head turning, in some amount. When the responses to the different components were added up, they equaled the response to the face as a whole. Thus the concept of heterogeneous summation is applicable. When the infants reached an age of about 20 weeks, however, there appeared to be a shift to Gestalt perception. Then the sum of the responses to the elements was much less than the response to the whole.

A comparable shift, in time, from heterogenous summation to Gestalt perception has been reported by Ulli Weidman in gulls. After surveying the literature, Peter Klopfer concluded that there is an evolutionary trend from heterogeneous summation in lower animals to a greater reliance on Gestalt perception in the more advanced species. Thus the change in ontogeny parallels the change in phylogeny, suggesting a similarity in mechanisms.

Often when analyzing the response to the parts of a complex configuration it is found that the response is not a simple yes or no. Rather, the strength of the response is often proportional to some regularly changing attribute of the stimulus. For example, Gerard Baerends found that gulls show preferential responses when retrieving their eggs. Preference depends on degree of speckling, darkness of the egg, shade of color, and so on. That the response is proportional to the magnitude of the stimulus is particularly clear with regard to the size of the egg. The larger the egg the more it is preferred. Thus gulls, and also oyster catchers, when given a choice, will retrieve and attempt to incubate huge eggs, including those they can barely straddle. (Obviously these eggs are much larger then they themselves can lay, and such a predicament would never present itself in nature.) The term *supernormal releaser* has been applied to stimulus objects such as these. They are exaggerations of the normal stimuli, and they evoke responses that are preferential over the normal stimuli. This phenomenon is of interest because it provides some clues to the physiological mechanisms underlying the preferential responses, the so-called *releasing mechanism.*

It is difficult to deal with the concept of the releasing mechanism because it is based on an implied physiological process and is consequently not open to direct observation. There, the concept is more hypothetical than that for the releaser or sign stimulus.

The original formulation of the concept of releasing mechanism stated *innate* releasing mechanism. Now, however, it is clear that experience may play a role in the selectivity of the releasing mechanism, as has been demonstrated by Wolfgang Schleidt in his analysis of the way turkeys respond to models of predators and other objects. He demonstrated that it is not just the shape of the model that evokes alarm. Novelty is extremely important. Thus if a falcon model becomes familiar it no longer evokes a predator response (the turkey is said to have *habituated* to that model). Yet a simple disk-shaped model, if unfamiliar will still elicit alarm.

Another instructive example is that of the gull chick. Shortly after hatching, it obtains food by pecking at the tip of its parent's bill. With no instructions or example, it pecks preferentially at a model of its parent's bill, as seen from below—a thin object with a red tip. Jack Hailman has shown, however, that the chick soon learns the details of the parent's head. In fact, it is possible, though it takes more effort, to teach a chick to peck at heads of species quite different from its own.

Still another example is that of the Siamese fighting fish (see Color plate, p. 589). While it ordinarily gives threat displays only toward other Siamese fighting fish, it can be taught to threaten when a light goes on. It learns that the onset of the light is followed by the appearance of a rival male. Through experience, the light has been substituted for another fish. This response to the light is soon lost, however, if the fish experiences many presentations of the light that are not followed by the appearance of another rival male.

The theme that emerges from a study of releasing mechanisms at this level is that most animals are capable of responding to the appropriate releaser without instruction or example. But, as will be discussed further on, experience can and often does round out or even alter the association between stimulus and response, especially in higher animals.

Search image

The more recent interest in experience as the agent influencing responsiveness has come from the concept of *search image*. As developed by Luke Tinbergen and subsequent workers, this idea applies primarily to sign stimuli and the factors determining to which such stimulus an animal will respond at a given moment. The basic idea came from observations on the feeding behavior of small birds, such as the great tit. If one has been having success in finding a particular kind of prey, it attends more to that prey than to other types. It has formed an image of that prey, so to speak, and therefore it detects it more readily than other types. As the prey becomes less abundant, the bird "resets" its search image because other prey will have been more frequently encountered.

The physiological basis of releasing mechanisms

Before leaving the topic of the releasing mechanism altogether, it is appropriate to say a few words about what is now known of its physiological basis. As in locomotion, historically there have been differences of opinion about the location of the mechanism. One school contended that it is central, and the other that it is peripheral. There is merit to both points of view, and for different reasons, as we shall see.

Perhaps the most extreme cases of peripheral perception are found among insects. A good example is the way males of certain mosquitoes detect the humming sound of the female. The sound sensor of the male is a small hairlike projection from its body; it is literally a tuning fork. It vibrates only to the frequency of the female's hum. These mechanical vibrations excite a sensory nerve whose sole message is that a female of the correct species is present.

Some aspects of hearing in moths present a next higher stage of complexity. Certain noctuid moths respond to the calls of bats that are hunting them by taking evasive action. Kenneth Roeder has demonstrated that there is a continual processing of the information contained in the bat calls as the neural message moves up further and further into the nervous system. In effect, the releasing mechanism is stretched out in the moth's nervous system.

A parallel exists among frogs. When alarmed, they jump toward the color blue. The retina of their eye is selectively responsive to blue, and it communicates this information to the central nervous system. The releasing mechanism involves the nerves of the retina and those just behind it.

The response to food objects is comparable. The frog has what has been called a bug detector in its retina. Objects that are shaped and moved like small insects evoke a special response in the retina; this information is encoded in the first two steps of neural transmission by nerve cells intimately associated with the retina.

Still more complex is the way the ears of the frog discriminate between the calls of frogs of different species and ages. As the impulses pass up the major auditory nerve toward the brain they are progressively analyzed. The information about frequencies and pulse duration is thus extracted before the message reaches the brain.

Remarkably, quite similar processes go on in the sensory systems of mammals. They eyes of cats and of rabbits are capable of rather independent processing of information. But there seems to be continual extracting of information and encoding it in simpler form all the way up the optic pathway to and including the brain.

Our present state of knowledge suggests then that the ability of animals to respond particularly to certain objects in their environment is a process that begins in the receptors themselves. Since any organism is continually bombarded by an enormous number of stimuli, at least some decisions must be made peripherally as to which will be passed on and which ignored. In most animals, selecting out aspects of the stimuli, often termed *filtering*, continues up through the sensory pathways of the nervous system and on into the brain, where further evaluation takes place. Thus the releasing mechanism is not a single mechanism but a composite of continual processes. Even the more complex releasers will have been at least partially encoded before reaching the central nervous system. While the concept of a releasing mechanism played an important historical role, and may still be a useful concept in comparative studies, it is no longer a guiding concept in the search for physiological mechanisms.

Stimulus filtering

The final assembling and evaluation of sensory input in the central nervous system will depend on the motivational state of the animal. These states are the subject of the next section.

I have been discussing behavior as though it consists only of overt acts—the output—and stimulation that evokes them—the input. While discussing the nature of stimuli, however, I brought in some elements of integration, that is, what stimulus causes what response.

Motivation

Now I want to consider more complex aspects of integration by taking up the topic of motivation. By motivation, I mean fluctuations in the physiological state of an animal, however caused, that result in the animal responding in different ways to the same stimulus at different times. At the level of discussion here, you may substitute such terms as drive, mood, or motivation.

Traditionally, ethologists have started the study of motivation by analyzing the functional relationships between the different M.A.P.s. Three approaches have been used. The first is to describe the *context* in which the behavior occurs. For example, the male European robin attacks a dummy with a red breast *if* that dummy is in its territory during the breeding season. However, the male does not attack its red-breasted rival when the male is out of its own territory, or even when in its territory at certain other times. Furthermore, a female robin does not attack a red-breasted dummy. Whether attacking occurs depends on the bird's relation to the physical situation, its endocrinological state, and the presence of appropriate stimulation.

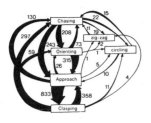

Fig. 15-11. Modal movement patterns can also be grouped according to their similarity. Shown here are the behavior patterns of courtship in a minnow *Cyprinidon rubrofluviatilis*. The numbers indicate the frequency with which one behavior follows another. The thickness of lines corresponds to these numbers.

The second approach is to analyze the behavioral events to see if they fall into groups. This is done today through a statistical evaluation in which M.A.P.s are judged to be related if they occur closely in time. This may be determined by measuring whether they occur in the same short time interval, or whether they tend to follow one another.

The third technique is to group the modal action patterns according to resemblance. To illustrate, most of the M.A.P.s used in threat display in a given species will have several elements in common. In a dog, this may include a deep-throated vocalization, erect hair, bared teeth, erect ears, and so on. A number of threat displays may contain one or more of these elements.

A practicing ethologist utilizes all of these factors, even if more intuitively than statistically, together with descriptions of the M.A.P.s and the context in which they occur. This material is synthesized into an *ethogram*, which provides a framework for further studies.

It is inherent in such a process, at least initially, that the resulting ethogram is somewhat static. Nonetheless, it provides a point of reference so that the changing relationships can better be detected and dealt with. An ethogram is also an important reference for comparative studies and for the analysis of communication (which will not be treated here).

The study of animal behavior has witnessed the development of a number of formal models of motivation. Most of these have emerged from psychological theories of learning. Ethology has not contributed greatly in the way of formal models, but it has produced a stimulating and rather unique view of the problem of motivation.

The first and best-known development in ethology was that of Konrad Lorenz's "psychohydraulic" model. He proposed that an area in the central nervous system steadily produces, or accumulates from elsewhere, excitation in some form. The analogy is of a fluidlike substance dripping into a receptacle. Lorenz has emphasized that his model is purely heuristic, and that he has no knowledge of, or belief in, a substance responsible for the changes.

This idea can be traced back to the writings of the psychoanalysts Breuer and Freud; and in their writings they credit Charles Darwin and yet earlier scientists with the basic concept. Breuer and Freud claimed that when a state of excitement finds no expression through overt behavior, the resulting physiological state is damaging to the organism.

Lorenz refined this idea in two significant ways. First, he proposed that excitation steadily builds up even in the absence of obvious external stimulation. Second, the excitation was said to be specific to certain functionally related groups of M.A.P.s. The performance of these M.A.P.S, therefore, would "use up" the excitation specific to them. The hypothetical excitation received the name of *specific action potential* or *reaction specific energy* to emphasize its specificity.

A puzzling phenomenon that led Lorenz to postulate specific action potential was what he called *vacuum behavior*. He first noted it in a captive starling. The bird in question had had no opportunity to capture live insects for some time. Lorenz saw the bird apparently hallucinate a fly and carry out normal prey-capture, eating, and bill-cleaning movements. The conclusion was that the longer the bird was denied the opportunity to capture live prey, the more specific action potential it accumulated, until the complete behavior broke out in the absence of adequate stimulation, that is, in vacuo.

One of the difficulties is that vacuum behavior is so rarely reported. In my own now fairly extensive experiences I have only seen one unequivocal example: a young but mature female cichlid fish (*Etroplus suratensis*) was noticed fanning "imaginary" eggs. She had never bred before, and had had no opportunity to see eggs. Yet during the course of two days she kept fanning at different places on a wall, indicating that there was no particular suboptimal stimulus there, such as egglike objects. Nonetheless, this does not preclude the possibility that certain otherwise unsatisfactory stimuli might have triggered the behavior, such as the wall itself, for it is well known that the more highly motivated an animal is, the less discerning it becomes. This, in fact, is a central tenet in Lorenz's development of the concept of vacuum behavior.

The important point is that when an animal has no opportunity to express certain types of behavior that are important to it, its readiness to do so is apt to change, whether it becomes more responsive or less. Examples are easy to find. For instance, when obvious tissue needs are involved, as in feeding and eating, an animal becomes more responsive, and also less discriminating, the longer it is deprived. More germane to the psychohydraulic model, however, is the increase in readiness to respond sexually as a consequence of deprivation. Yet there is now some evidence to suggest that if sexual deprivation proceeds too long, animals become less sexually motivated.

It is when we turn to aggression that the application of this idea becomes so controversial. The notion of the need to be aggressive, to vent the built-up energy, leads logically to the postulation of appetence for aggression: the animal seeks a fight. Applied to human conduct, this line of thinking produces the conclusion that war is inevitable unless harmless outlets for aggression are provided or physiological intervention is employed. In light of the importance of this hypothesis, there has been little experimentation to test it. Most recently, Anne Rasa has presented evidence for appetence for aggression in the highly territorial juvenile of one of the coral-reef damsel fishes. And others have reported similar findings in territorial species of birds. The difficulty is in the failure to integrate the effect of other processes that reduce the probability of overt aggression, particularly inhibitory effects of previous fights. Not uncommonly, animals show appetence for aggression but fail to fight upon

Fig. 15-12. Vacuum activity of a willow warbler; the bird is "attacking" an imaginary opponent.

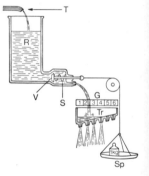

Fig. 15-13. Lorenz's "psycho-hydraulic" instinct model: The input (T) denotes the action-specific energy which is collected in the container (R). The releasing mechanism (illustrated by the valve V) is kept closed by a spring (S): which represents the inhibition emanating from higher nerve centers. The weight (Sp) denotes the stimuli which activate the releasing mechanism. The heavier the weight (stimuli) the fuller the container (motivation), the stronger the reaction whose intensity can be read on a scale (G) above the pan (Tr).

meeting the object of their aggression. Thus the problem of the reality of appetence for aggression, and its consequences, remains one of the most important and unanswered problems in biology.

The phenomenon underlying appetence for aggression is that of the change in aggressiveness when an animal has no opportunity to be aggressive. The most commonly performed experiment to test this idea is to hold the animal in isolation. Later it is exposed to another member of its species to see whether it has become more aggressive. The difficulty is that the studies by different scientists are not in agreement. Some experiments have shown that animals become more aggressive, others that they become less aggressive, and still others have revealed no differences. It has been demonstrated by the same experimenter, furthermore, that mice of one genetic strain become more aggressive in isolation while those of another strain do not change. The animal's age also makes a vast difference, as the work of J. Paul Scott has shown. Walter Heiligenberg, while still Lorenz's student, discovered that the longer a cichlid fish was held in isolation, the *less* aggressive it became.

One source of confusion in these studies is that little attention was given to the complex nature of the aggressive behavior under investigation. Most workers measure but one aspect of aggressive behavior. Eric Courchesne and I studied the effects of isolation on aggression in hermit crabs. The introductory threat displays showed no effect whatsoever. However, the actual fighting behavior increased the longer the crabs were isolated. Furthermore, the more damaging the particular act of fighting, the longer in isolation it took to increase in responsiveness. Thus the danger of extrapolating from studies on animal behavior to that of human beings should be obvious. Until we know better, we must take care in talking about all actions involved in aggression as though they were the same.

Evolutionary considerations

There is also an evolutionary consideration. As with any other type of behavior, aggression is adaptive; and it will be differently adaptive in different species. The key factor is always a vital or limiting resource, such as food, a mate, progeny, or shelter, in the evolutionary competition for genetic survival. Aggression is therefore more likely in those species in which, and when, aggression produces an advantage. It is adaptive for a marsh wren, for instance, to fight to exclude competitors from its feeding territory. In contrast, it is impossible for a swallow to exclude trophic competitors from the immense air space in which it feeds. Yet, when breeding, the same swallow is highly territorial over its nest, which is both an essential and defensible resource for which it must compete.

On the other hand, aggression is potentially maladaptive. Aggressive behavior, even when highly ritualized, is done at a significant cost in metabolic energy; and the heart of the evolutionary struggle is the maximizing of energy intake while minimizing its loss. Another factor is that animals are enormously preoccupied with one another during aggressive encounters, and they are consequently less attentive to the approach of a predator. Perhaps more important is the risk of serious injury and subsequent infection if overt fighting breaks out; even the winner faces this ominous prospect.

In the competition for evolutionary success, therefore, each species must continuously juggle the gains and losses that result from aggressive behavior. In some this means an almost total loss of aggression, as in the schooling fishes of the open seas. In most species, however, it means that aggression occurs only when and where necessary, and in a manner that minimizes injury, time involved, and the expenditure of energy. Any model of aggression, such as the psychohydraulic model, must be consonant with these larger evolutionary considerations.

There is one final difficulty with the psychohydraulic model, leaving aside the evolutionary condiderations and also the physiological difficulties that will not be explored here. In Lorenz's view, the critical factor is the actual performance of the behavior since it must, theoretically, use or release the specific action potential. Other ethologists, most notably Robert Hinde, claim it is equally plausible that it is a change in stimulation that causes the change in responsiveness. So far, there has been no resolution of this difference. However, I anticipate a period in which the parable of the blind men palpating the elephant will apply.

The other important development in motivational theory in ethology has been the *tripartite model*. It has come largely from Dutch and British ethologists. Since ethologists have not been greatly concerned with such behavior as feeding or body care, the focus has been on social displays.

The tripartite model seeks to explain social behavior in terms of the interaction of three major motivational axes: fear, aggression, and sex. The model accounts for displays as resulting from mixtures of those elements. The displays seen when two gulls approach one another, for example, reflect a conflict between fleeing, attacking, and mating. A male approaching a rival male is on the one hand stimulated to attack the rival, but on the other frightened of it because it represents potential injury. Thus it may position its beak to peck at the rival, while turning its body to the side to be in a position to retreat. Similarly, when the male and female of a pair come together after a separation, each shows signs of being both slightly afraid of and aggressive toward the other, for each strongly resembles the neighboring rival gulls; at the same time, each is sexually motivated because they recognize one another as being of the opposite sex and their mate. As Tinbergen has pointed out, even in a well-mated pair there is always some element of fear and aggression in their interactions.

The tripartite model of motivation has not been of great service in discovering the physiological mechanisms underlying displays. Such a formulation may even be a hindrance. It is becoming increasingly apparent that fear, sex, and aggression are not unitary drives but rather are complex

Fig. 15-14. Three different behavior patterns during attack-flight conflicts:

Fish
Re-direction: This cichlid directs its behavior to a substitute object; it bites the sand instead of its opponent.

Fig. 15-15.
Roosters
Displacement activity: In fighting cocks pecking on the ground (as when eating) is an indicator of conflicting motivations.

Fig. 15-16.
Baboon
This baboon vacillates between attack and flight. The exposed teeth indicate threat to the opponent, while his body indicates a readiness to flee.

manifestations of a variety of subsystems. Recall the isolated and highly aggressive hermit crabs in which different elements of aggressive behavior responded differently to the effects of isolation.

In spite of these reservations about the usefulness of the tripartite model to physiology, it has been of value in understanding the origin and therefore the evolution of displays. This comes in part from a better understanding of the behavior that occurs when an animal finds itself in conflict. A modest vocabulary has appeared to describe some of these manifestations. *Ambivalance* and *superposition*, which have slightly different meanings, are terms applied when an animal shows two types of behavior simultaneously. Another expression, *redirection*, refers to those instances when an animal is stimulated to do a particular type of behavior but is at the same time inhibited and thus directs that behavior to an inappropriate object. An excellent example, and one commonly observed, is when one animal is stimulated to attack, but is inhibited from doing so, perhaps because the other animal is larger than it is. So it turns and attacks some inanimate object or a smaller individual. (There are parallels here in the psychoanalytic literature dealing with defense mechanisms.)

The type of behavior resulting from conflict that has received the most attention is that of *displacement behavior* (see Chapter 16). Often when an animal is in strong conflict it expresses behavior that is unexpected in that the behavior performed is inappropriate to the stimulation; the behavior is also not in the appropriate functional context. An often-cited example is that of two roosters threatening one another; one or both may then stop and peck at nonexistent grain. This has been called displacement eating.

Several motivational models have been proposed to account for displacement behavior. Most of these, but not all, make use of some surplus-excitation hypothesis. Charles Darwin probably gave the first clear description of what we now call displacement behavior, and he explained it in terms that are compatible with the psychohydraulic model.

Displacement behavior still remains one of the most intriguing types of behavior to study in the evolution of displays and their physiological mechanisms. One of the difficulties in tackling the problem, however, has been that there are doubtless may different types of behavior collectively called displacement behavior. Some contemporary writers seek to avoid the difficulty by inventing new names such as irrelevant behavior or stereotyped behavior.

No other issue in the study of behavior has so aroused the emotions and so stimulated bitter and at times personal arguments between scientists as has the nature-nurture issue. The more extreme of the nurture advocates have accused ethologists of having an excessive preoccupation with instinct, for one, and of studying a nonexistent phenomenon, for another. This is in spite of the fact that ethologists have made some of the most original contributions to the study of the role of experience in the development of behavior.

In the attacking and counterattacking, each side has tended to ignore much of what the other has been saying, focusing on those parts of the opponent's argument they felt to be unsound. Unfortunately, the problem often extends beyond the realm of the scientific. Social, political, and even theological experiences of the scientists in question seem to have influenced their formulations of scientific hypotheses. A quote from Bertrand Russell is particularly appropriate here, even though he wrote it in reference to the opposing views of different theories of learning: "One may say broadly that all the animals that have been carefully observed have behaved so as to confirm the philosophy in which the observer believed before his observations began." In offering this quotation, I intend no self-righteous posture. Rather, I hope only to sensitize you to the problem that exists not just for scientists, but for me as the writer and for you as the reader.

The nature-nurture issue has more often been termed the instinct-learning argument. Further on I will shift away from instinct-learning to genetic-environmental. But before doing so, a few words about the concepts of instinct and of learning are in order. One of the major difficulties in dealing with the word instinct is that its usage has varied so greatly between writers. Probably the best general formulation of the concept is that of William James, who wrote in 1872 that instinct is "... the faculty of acting in such a way as to produce certain ends, without foresight of the ends, and without previous education or performance." That remains a good postulation, but it is not one that is easily testable. To most investigators, instinct has become virtually synonymous with some drive or motivational state in which the animal finds itself, and the consequences of that state. In ethology, it has meant largely the modal action patterns one observes, and the releasing mechanisms controlling them. In a sense, this is simply a more externalized view of motivation.

You should be aware, also, that there is an appreciable difference between the use of the terms *instinct* and *instinctive*. Many scientists shy away from the word instinct, since there is little agreement about precisely what an instinct is. Many scientists, however, use the term instinctive to refer to any aspect of behavior that appears to be inherited.

The concept of learning, in contrast to instinct, suffers by virtue of being too narrowly defined in most instances. In its broadest usage, however, it means any change in behavior as a consequence of experience. In practice, in contrast, it is usually precisely delimited by the relationship between stimulus and response, and the assocaition that develops between them.

A major problem has been in the basic postulation, and the loose way in which the arguments have been stated. It will help, for the moment, to lay out a simple scheme to make a few points:

NATURE: Instinctive = Maturation = Genome
NURTURE: Learned   = Experience = Environment

▷
Many behavior patterns are variable, while others are displayed in a stereo-typical manner, always looking the same. These more of less invariable behaviors are called modal movement sequences. They include many body-care behaviors, such as the shaking motion with which a polar bear removes water from its coat (above). Most animals have different behavior patterns to groom various parts of the body, but each of these behaviors can be recognized by its stereotypical sequence. Center left: A graylag goose scratches its head with its toes because this area cannot be reached with the bill, its tool for grooming most other parts of the plumage. Center right: Ibexes and many other horned animals use the tips of their horns for body care. Below: A Thomson's gazelle scratches its ear with its hind foot, because its tongue and horns cannot reach that part of the body.

The instinct con-cept

The learning con-cept

These three pairs of roughly dichotomous terms have often been applied in rather different ways in the literature. Since they have somewhat different meanings, this has been a source of some confusion. For instance, one writer may contrast genome with learned, while another may play instinctive against environmental. Perhaps I can make a little clearer how these terms should be used. Since I have already discussed instinctive and learned, I will pass on to the other two pairs.

The genome

Genome refers to the genetic material an organism starts with, hence the genome is that which is inherited. *Maturation* refers to the interaction that ensues between the genome and the tissues that appear during development under the direction of that genome.

When the term *environment* is used, there is immediately a difficulty. In the strict sense, it refers to the physical environment in which the genome finds itself, and is thus clearly separable. However, the immediate environment in which the genome operates, the extragenetic contents of the cells and their membranes, is a product of its own action. Thus the proponent of the nature argument accepts the genetically elaborated environment, the embryonic tissues, as being genetic. The proponent of the nurture argument, however, would claim that it is instead a part of the organism's environment. Both are correct, of course. But this has been a growing source of confusion as the controversy has moved from the behavior of the adult organism to the development of behavior during embryonic stages. Gilbert Gottlieb is a case in point. He has done a series of fascinating experiments on the embryogenesis of behavior in birds. He believes that his results are strong documentation of the nature side of the argument, whereas most ethologists wholeheartedly accept them as bolstering the nurture side.

The concept of "experience"

The same dilemma arises when discussing *experience*. Broadly seen, any environmental influence may constitute experience, including experience with one's own body. It is in fact difficult to draw the line between the experience an embryo has with its own tissues, as opposed to those it has with the external environment provided by its mother, whether in the uterus or in an egg.

It should be apparent from the foregoing that much of the disputation in the nature-nurture argument has arisen from the biases inherent in the investigator at the outset of the investigation. The person interested in genetical aspects is more apt to look both at animals where this phenomenon is more easily detected and at processes that make the action of the genome evident. Similarly, a person interested in the role of environment in the development of behavior will seek animals and aspects of behavior that will make his point. To show that the phenomena lie on a continuum, I will develop a discussion moving from clearly genetical to clearly environmental contributions to the elaboration of behavior.

As Daniel Lehrman has recently pointed out, the most commonly accepted meaning of inherited behavior is *interindividual*, that is, popula-

◁

Examples of "Batesian mimicry": These harmless imitators (Syrphidae, top right; *Necydalis* major L. center left; hornet moth, center right; crane flies, bottom left; cocoon of the magpie moth, bottom right) are protected by the fact that they bear the same conspicuous body coloration as the wasp, (top left) which used its sting as a weapon.

tional. The distribution of a given character, whether behavior or other-wise, may be predicted in the offspring by noting its distribution in the parent generation. Such a prediction can often be made in spite of differences due to the environment. Consequently, the focus is on those traits that characterize a species. Examples would be the bowing and cooing of a pigeon, the urination posture of a male dog, or the smile given together with a quick up-down movement of the eyebrows in human beings.

It follows that variation between individuals should be greater than that within a given individual. This has been shown to hold in comparisons within and between individuals of birds singing their songs, and of ducks and lizards performing their complex displays.

If these differences are heritable and therefore genetic, one should be able to breed for such behavioral differences. This has been done. Well-known genetic strains based on behavioral differences are known among pigeons, such as rollers, tumblers, and homers; in dogs bred for different tasks; in horses with different gaits; and in rats that solve mazes quickly or slowly. Aubry Manning has even done an artificial "natural selection" experiment on fruit flies in which he was able to change the speed with which the flies mate.

An extension of this approach is to hybridize different species and to observe the consequences. In most cases, the behavior of the hybrids is intermediate between those of the parent species.

While behavior is obviously heritable, no claim is made from such studies as the foregoing that it is immune to the effects of the environment. If one looks closely, the impact of the environment is usually detectable.

It is the second approach, the intraindividual one, that has given rise to so much controversy. (This might also be termed the developmental approach.) In discussing it, I wish to leave aside a consideration of embryological processes because of the difficulties involved in resolving what is meant by experience independent of the genome.

In attempting to demonstrate intraindividual genetic influence, scientists have tried to show that there is no obvious way that learning could be involved. That is, there is no tutor, example, or opportunity for trial-and-error learning. This is not to say that experience plays no role whatsoever. Illustrating this point requires going back to interindividual, in fact interspecific, comparisons. While an instructor is needed in some species, as in the learning of song in certain birds, it may not be needed for the same task in yet another species. Thus the degree to which an individual relies on experience as opposed to maturation in the development of its behaviors depends on the species under consideration. Examples of behavioral phenomena that lead to this type of thinking come from observations in nature in which there is no apparent opportunity for the animal to learn the complex behavior that it develops. Many cases are known among insects where, following a long larval life, the animal metamorphoses into a completely different organism that immediately seeks a member of its species of the opposite sex, engages in elaborate courtship, copulates, and dies. Examples among higher animals are not lacking. Otto Heinroth filmed the European cuckoo which immediately upon hatching engages in a complex motor act, throwing the eggs and nestlings of its foster parents's species out of the nest. Moreover, this cuckoo experiences an alien species for parents, yet overcomes that experience to mate with its own species as an adult.

Some of the most persuasive experiments done on the independence from experience have been those aimed at demonstrating that some behavior may develop even in the face of contrary experience. Eckard Hess showed this in the development of the ability of a chick to peck accurately at a seed. He put goggles on his chicks. One group had plain glass; the other had lenses that shifted the field of vision a known amount to the right or to the left of the seed. During the first few days of life, in both groups of chicks, the pecks were widely scattered about. In the course of a few days the scatter shrank to a narrow cluster. The chicks with plain glass were now getting the seed most of the time. Those with displaced vision were now pecking to the right or to the left of the seed in a comparably narrow cluster, but missing it all the time. This shows that the ability to peck more accurately develops independent of the experience of accurately pecking a piece of grain.

C. Richard Dawkins also used chicks to demonstrate the resistance to environment of the early feeding behavior of chicks. He discovered, first, that chicks prefer to peck at seeds that are illuminated from above, that is, shaded underneath, as would occur in nature. Even when the chicks were raised in an environment in which the light came from below so that all seeds were shaded on top, when given a choice they still preferred seeds that appeared to be lighted from above.

Another class of experiments on the development of behavior is the so-called isolation experiment. In this, the animal is not presented with interfering experience. Rather it is kept away from members of its own species—usually in a rather bleak environment—to preclude learning from its fellows. Several such studies have shown that many animals are capable of carrying out normal behavior as adults. This is true of the three-spined stickleback in its mating behavior, and of the production of adult song in bluebirds and roosters.

However, some animals, even lower ones, have inadequate behavior when raised in isolation. For example, isolated male platyfish were shown by Evelyn Shaw to be relatively incompetent in inseminating their females. Jaap Kruijt found that jungle fowl roosters, reared in isolation, took to attacking their own tails. Female rhesus monkeys that were reared in isolation on artificial mothers were unable as adults either to copulate normally or to care adequately for their babies; this deficit was prevented in a subsequent experiment by Harry Harlow by permitting the young female to play with other young.

Fig. 15-17. E. Hess was able to demonstrate that young chicks can peck accurately small seeds, and that this ability is not a function of learning, but of maturation. Dark-hatched chicks were given hoods containing prisms, and nail heads embedded in clay were presented to them. The experimental chicks' (1) prisms displace the visual field by about 7 degrees, while the controls (2) could see normally. After four days the pecks were closer together for both groups (3 and 4), although the experimental groups never hit the nail.

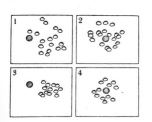

It seems generally true that the more advanced the species the more disruptive is the effect of being raised in isolation. It is as though being raised in such an austere environment is an effect in itself, rather than merely the removal of information from which an animal might learn.

There are other problems with the isolation experiment, as illustrated in the oft-cited observation by Konrad Lorenz on hand-raised European shrikes. In the wild, these birds capture sizable prey such as grasshoppers and small rodents. Since their feet are too weak to hold the prey while they dismember it, they generally impale it on a thorn or wedge it in a crotch of a bush or tree. The hand-raised shrikes were not allowed to have experience with wedging or impaling. As grown birds they were unable to do either. Lorenz drew from this that experience with thorns or crotches is indispensable for the development of impaling or wedging. Later his close associate Gustav Kramer repeated the experiment, but he fed his shrikes a superior food. These shrikes, without benefit of experience, successfully impaled. Lorenz concluded that his birds had had the ability to impale but were not in good enough health for the behavior to come to the surface.

Thus the difficulty with doing an isolation experiment is that any time an animal fails to perform the behavior, one can still claim that if the organism had only been healthier, it could have done it. Therefore the experiment can never prove that experience is necessary.

There is a further difficulty with the isolation experiment as a means of demonstrating the independence of behavioral development from experience with the environment. Proof in this experiment rests upon there being no difference between the animal raised in isolation and the one raised under more normal circumstances. Logically, it is impossible to prove that two things are the same. A meaningful scientific experiment is one that reveals differences, not one that shows no effect of the treatment.

Despite this, the isolation experiment has proved persuasive. It has shown that many animals have the capacity to develop remarkably complex behavior in the absence of any obvious example or instruction, but without precluding less obvious environmental effects. To make the point, imagine how startled a scientist would be if he were to do the unspeakable experiment of rearing a human infant in isolation and then discover that it could, say at age two, say a few sentences in its native tongue.

It is becoming increasingly obvious that for most animal species nature and nurture act in an inseparable fashion. To turn the argument about, one could say that learning is instinctive, or just as well that instincts are learned. By way of comparison, the classical approach to learning has seen little consideration of the nature of the unconditioned response. Yet it is patently the consummatory behavior, the M.A.P., of the ethologist. Likewise, the unconditioned stimulus would be the sign stimulus

Fig. 15-18. In some behavior patterns experience plays hardly any role. This applied to the behavior pattern of a newly hatched, still blind European cuckoo who ejects the eggs or young of its foster parents by lifting them onto his back and pushing them out.

Fig. 15-19. Raising animals in isolation from others leads to abnormal behavior patterns. In jungle fowl it leads to abnormal mating behavior: They may attack squatting females (A, B), peck her on the head instead of mounting her (C), mount from the wrong side (E) or show mating patterns while holding a feather in their beak (D).

or releaser. In fact, there has been remarkably little concern with showing how an animal learns the unconditioned response or stimulus. Rather, the attention has been the association formed between some arbitrary new stimulus, the conditioned stimulus, and the consummatory behavior. Recall the example of the Siamese fighting fish that learned to give its threat displays when a light goes on, in anticipation of the appearance of the appropriate releaser. Here again we see an important source of confusion between the opposing schools. Most ethologists have reference to the development, the maturation, of the *form* or coordination of the behavior and its releaser when they use terms like innate and instinctive. On the other hand, most psychologists take these as given, and are interested in studying how associations are formed between these pre-existing pieces of behavior. There are, in all fairness, exceptions in both camps.

It is now becoming more apparent that even admittedly learned associations may have a genetic basis. Thus the Russian psychologists have been writing of late about how the ease of conditioning depends upon the "naturalness" of the conditioning stimulus. This has been known to ethologists for some time. The social insects are capable of rapid and reversible learning when it relates to finding food, as has been so beautifully demonstrated for honey bees by Otto von Frisch. Niko Tinbergen has reported that a female wasp can learn the appearance of her nest entrance in only one circling flight above it; one-trial learning is difficult to demonstrate even in mammals. Furthermore, chicks show color and shape preferences for the objects they peck at for food. They learn to peck at these much more quickly than they do at nonpreferred colors and shapes. Nonetheless, they can learn to peck at the nonpreferred patterns, but it takes more time and effort.

An obvious parallel is in the learning of speech by human beings. If we use the criterion that all members of the species show the trait, then speech is obviously genetic—if you will, instinctive—in human beings. Just as obviously, we learn to speak a particular language. It is difficult if not impossible here to separate maturation cleanly from experience.

One of the most effective ways an organism has of incorporating experience into its development is to have *sensitive periods* during its maturation when it is more attuned to certain kinds of environmental influences than to others. These "time windows" seem to be largely genetically regulated, that is, maturational. Nonetheless, it has been shown that through manipulation the termination of a sensitive period can sometimes be extended. The most widely known and spectacular example of sensitive periods involves the phenomenon called *imprinting*, which will be covered elsewhere in this volume.

Seen in a broader framework, many species develop behavior at a particular point in their life if they have the appropriate experience.

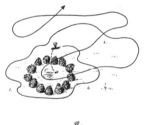

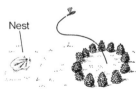

Nest

Fig. 15–20 Cues that are important in the lives of animals are often quickly learned. The digger wasp (*Philantus triangulum*) needs only one circling flight around her nest entrance to remember its location (top). Object surrounding the nest serve as cues, as N. Tinbergen was able to demonstrate when he displaced pine cones. When returning, the wasp searches in the center of the displaced ring.

But after they have passed that stage they seem unable to develop the behavior even when given the right experience. Eberhard Curio has done experiments illustrating this principle, on the development of bathing in the bullfinch. This would seem at first to be a slow process of learning in which the young bird becomes ever more skilled in the complex movements involved in bathing as it grows older. By withholding from selected young birds the opportunity to bathe, he discovered that after a certain age they no longer could develop the bathing response; remarkably, however, one bath just before the end of this sensitive period sufficed to produce the full response. Thus both maturational and experiential changes were occurring in parallel.

My favorite example of the complex interaction of the genome and the environment is that of the development of the snail-smashing behavior in the songthrush, which has been studied by Clifford Henty. In the normal behavior the adult first finds a snail and then carries it to a so-called anvil, usually a rock. There it holds the snail by its foot and smashes it against the rock by means of a rhythmic sideways swinging motion of the head. There is apparently a long period of learning in the young birds as they perfect this behavior. As in the bullfinch, the response fails to appear if the critical period is passed. Thus the songthrush learns how to smash a snail through a process of trial-and-error at an early stage.

Henty went on to a most revealing experiment. He knew that the European blackbird, a close relative of the songthrush and therefore morphologically very similar, relishes smashed snails even though blackbirds apparently do not know how to break them open themselves. He observed that young blackbirds were never able to learn to open the snails through trial-and-error. He then attempted to teach them how to do it by offering them at first smashed snails, then snails that were progressively less broken up. Under no circumstances were the blackbirds ever able to develop the ability to smash a snail.

The difference between the two species in their capacity to learn how to open the snail is therefore genetic. One could say that *learning* to smash snails in the songthrush is *instinctive*, or vice versa.

It should be clear from the foregoing that all behavior involves both genetic and environmental influences. Depending on the species taken as the point of comparison, first one or the other factor may appear to be paramount. It is legitimate for analytical reasons to consider on the one hand the role of the environment, and on the other the contribution of genetics to the development and control of behavior. This in no way denies the interaction of the two factors.

It would be difficult and perhaps presumptuous of me to say just where ethology is headed today. But it is obvious to any observer, I believe, that the clearest trend is toward a broader framework and a movement away from the original preoccupation with limited aspects of social behavior. One major thrust is toward the interaction of behavioral

Time-dependent developmental processes

Where is ethology heading?

and ecological theory, particularly as expressed in the intensive field work of late, and the growing awareness of a mathematical theory of ecology. At the same time, we are witnessing the development of a more precise and experimental attitude in the study of behavioral mechanisms. This is producing a more rewarding working relationship with neurophysiologists and endocrinologists. The genetics of behavior is also becoming a field in its own right. As the ethologists and other students of animal behavior have discovered evermore their common ground, a rapprochement has developed. The growing concern of ethologists with human behavior has helped to erase the boundary between ethology and comparative psychology. Now it is often difficult to say whether a psychologist is to be counted as an ethologist, or whether a given ethologist is to be considered a neurophysiologist or another an ecologist. In short, ethology is becoming an even-more-integral part of the main stream of biology than it was before.

16   Displacement Activities

▷
Examples of releasing stimuli. Visual: The red throat coloration of gaping jaws elicits feeding behavior in the parents (top left). Auditory: The croaking of male frogs attracts females for mating (top right). Tactile: Smooth, wet surfaces induce clinging in male toads (bottom left). Chemical: Water beetles are attracted by chemical secretions from the prey animal (here, a newt; bottom right).

A person who has been observing one animal species for a period of time will begin to feel that he can predict an individual's behavior fairly accurately. He will then try to test and, if necessary, correct his predictions by making further observations. At some point this procedure allows him to form a rough idea of the factors and conditions determining the animal's behavior. If it has not eaten for a while, it will probably look for food; if threatend by a predator, it will probably flee, and so on. To fulfill these functions, an animal performs a complex series of behaviors which appear significant and appropriate to the situation at hand. But this may apply to individual behaviors as well: Most of the time, we can associate each part of a complex activity with the adaptive function of the behavior pattern as a whole. An example is the response of a territorial animal to its neighbors: Depending on where they meet, the movements released will be those of attack (in the core of the territory), escape (outside the territory), or a mixture of the two (at the periphery). When an animal performs a combination of behaviors derived from the first two categories, it may alternate movements at a high rate, express them as incomplete "intention" movements, or display "ambivalent threatening movements" containing both attack and flight components (see Chapter 25).

Displacement activities, by P. Sevenster

Ambivalent threatening movements: attack and flight

On the other hand, among all these meaningful and adaptive behavior patterns the animal frequently also displays some that do not seem relevant to the situation at all, thus arising unexpectedly in the eyes of the observer. A. Portielje, for example, described in 1936 how a male swan responded to an intruding rival by showing both attack and escape behaviors. Suddenly the swan picked up twigs and straw from the ground and arranged them to the side in a species-typical nest-building activity. Portielje concluded that the tendency to attack was blocked, or thwarted, by an equally strong desire to flee, and this caused the swan to discharge the tendencies through the unexpected and irrelevant nest-building behavior. Other observers had also remarked upon this kind of unexpected activity. Starlings occasionally interrupt their agressive encounters to preen themselves just

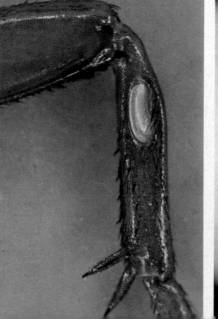
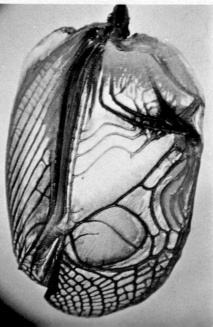

like after a bath. Domestic cocks stop in the middle of a fight to peck at the ground, as if looking for food.

Observations like these were at first recorded individually for each species. In 1940, A. Kortlandt and N. Tinbergen tried to establish a common category for such unexpected movements, which may occur not only during fighting but also during courtship and nest-relief ceremonies, in general occur appearing whenever a situationally appropriate behavior is obstructed. This category is referred to in English as displacement activities; in German, *Übersprüngbewegungen*. Tinbergen defines it as any movements belonging to another instinct than the initially activated instincts (or instinct). The word "instinct" today probably corresponds more to what we have labeled a behavior system (functional group of behaviors). The definition itself implies two assumptions: 1. Every movement can be assigned to a behavior system. 2. In every situation it is possible to determine which behavior systems are activated and which are not.

Both assumptions, however, are subject to a number of criticisms. In order to say something about the causes of behavior patterns, detailed examinations must be made. This applies in particular to behavior patterns that arise unexpectedly, in other words, where the causes are not easily discerned. A definition of such behaviors in terms of causes would therefore be inadequate. This is why others have tried, instead, to define them in terms of function, i.e., with reference to their biological consequences. A displacement activity would then be defined by way of its ineffectiveness and irrelevance to the situation. After all, the fighting cocks peck without eating anything, and the threatening swan arranges straws without really nesting. But again we would argue that observers can only determine whether a movement is ineffective (in the biological sense) and therefore a "displaced" activity after examining it very closely. Furthermore, a number of so-called displacement activities have been shown to be biologically effective in a specific way. In the case mentioned above, it is entirely plausible that the male swan's nest-building behavior intimidates his rival, contributing to his territorial defense. Perhaps we can say the same for the food-pecking behavior of fighting cocks. Ethologists have also strongly suggested that displacement activities may help an animal in conflict to discharge its thwarted behavior, thus re-establishing its neurophysiological state of equilibrium. This process would indeed be biologically significant. We therefore suggest that the criterion of "ineffectiveness" cannot be an important one in defining displacement activities.

A more useful definition is one based on descriptive criteria. We can relate movements to others occurring before and after. In this way, they are characterized in terms of the behavioral context or of the behavior sequences where they arise exclusively or most frequently. Using these criteria, we can even calculate the probabilities of changeover between various behavior elements, a precise mathematical aid in classifying behaviors. With or without mathematical refinement, however, this general

Tinbergen's definition of displacement activity

◁

Above: Male house cricket (*Acheta domesticus*) attached to an experimental apparatus and holding on to a "walking ball" made of cork. We can elicit species-specific chirping —performed with the stridulatory apparatus of the wings—by electrical stimulation of the brain, administered through the fine wire seen top left. Below, from left to right: The auditory organs of the cricket lie in the front tibia and are equipped with tympanic membranes. The forewings of male crickets have special structures for producing and beaming off their sound pulses. Lastly, a dissected specimen of the mollusk *Aplysia*. The drawn-out section of intestine is surrounded at its lower end by a wreath of orange-yellow nerve ganglia, which send out ramifications of milk-white nerve fibers.

Deriving displacement activities from their behavioral context

procedure frequently enables us to discriminate two sets of sequences containing the behavior in question. The following is a good example of what we mean: Male sticklebacks position themselves in front of their nest entrance and perform rhythmical movements with their pectoral fin to provide the nest with fresh, oxygen-rich water. This ventilation behavior aerates the eggs, a vital factor in their development. The movement itself is called "fanning," and lasts for a few seconds to several minutes. Lack of oxygen and a surplus of carbon dioxide in the nest encourage this behavior, which increases with the age and number of eggs. The fanning activity is therefore determined by what is required to care for the eggs. So much are its form, orientation, and determining factors adapted to the vital function of aeration that we could easily assume the movement to have evolved for this very purpose. At this point, it is important to note that brood fanning is almost never followed by what we shall describe as "gluing" behavior.

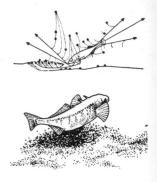

Fig. 16-1. Male stickleback during fanning (above) and gluing (below). The arrows and dotted lines indicate water currents created by the fanning movements. These currents can be made visible by dropping colored crystals (black dots) into the water. Both fanning and gluing may be performed as normal movements or as displacement activities.

Exactly the same fanning movement occurs during courtship. As long as the female fails to follow the male, he will vacillate back and forth between her and the nest, making brief but very frequent fanning movements at the nest even when it contains no eggs. Ethologists have long been using this fanning behavior as a classic example of displacement activity; their main reason was probably its ineffectiveness (irrelevance) in the absence of eggs. Fanning during courtship is very often followed by "gluing," where the male passes over the nest with its belly and secretes a sticky substance.

We can see, then, that fanning occurs in two very distinct behavior sequences, and it matters little for any further scrutiny which of the two we call displacement fanning. We could make a division on the basis of functional or phylogenetic criteria: For which behavior sequence does this movement seem to have been developed? Where does it serve its primary purpose, as judged by the adaptive function of its form? Whichever sequence we choose, it provides the "normal" setting for the movement, while the other context would make it a "displacement" activity. In our example, fanning during brood care (behavior sequence without gluing), where its aerating function is clear, would be considered as normal, while fanning during courtship (behavior sequence with gluing) would be labeled as displacement fanning. More generally, a displacement activity is any movement occurring outside of the behavior sequence to which—according to our knowledge—it is primarily adapted. The advantage of this description is that it fits in nicely with our original idea: The occurrence of such behavior would be unexpected; in other words, it would seem out of place or out of its original context. We can retain the initially causal meaning of the term *Übersprung* (litterally translated as "sparking-over") and add a new interpretation, namely, that the movement "changes over" (*Übergang*) to a behavior sequence where it no longer serves its primary function.

On the other hand, even this definition is hypothetical, because we seldom (if ever) really see the full adaptive value of a certain behavior, let alone trace its phylogenetic history. We simply cannot give a definition that has no drawbacks at all. Our circumscription is designed only to indicate the problem, not to solve or explain it. As we shall see in the following discussion, there are many possible explanations of what we call displacement activities, and they, in turn, probably encompass a number of distinctive phenomena.

Ethologists have been actively concerned with the causal question of how displacement activities come about. The oldest hypothesis, suggested by Kortlandt and Tinbergen, is associated with the word *Übersprung* itself, literally meaning "sparking-over." It should be noted, however, that for various reasons ethologists have come to accept the term "displacement activity" in English, rather than "sparking-over activity." According to this formulation, the excitation of activated but at the same time obstructed behavior systems sparks over to other behavior systems, so that instead of the obstructed behavior, a movement from this other system is elicited— i.e., the displacement activity. This hypothetical process corresponds to the psychological concept of catharsis: If the behavior that would fit the situation is thwarted, the aroused animal has to do something else. However, what it does for a substitute is by no means arbitrary; rather, the animal performs in a way typical for its species and the situation: Fighting starlings preen themselves, fighting sticklebacks dig, and courting sticklebacks fan. This would mean that the excitation sparks over only to certain specified pathways in the nervous system.

The sparking-over hypothesis rests mainly on the assumption that in a particular situation there is nothing to trigger the movement within its own behavior system, so that the energy has to be drawn from other systems. We have, in fact, some impressive examples to support this idea: If a person approaches the nest of an arctic skua, the bird may break into intensive bathing behavior on dry gorund. During this activity, these birds not only shake themselves in a characteristic way, with their wings half spread, but they also perform the deep scooping movements of the head and breast which during real bathing serves to throw water on their back. No water or loose sand need be available. When the observer moves away a little, the bird immediately stops its bathing movements; renewed approach will make it start again, until it finally flies up and swoops at its enemy in attack.

Cases like this virtually force any observer to accept the initial sparking-over hypothesis. The intensity of the bathing movements does not seem to be directly caused by anything in the situation; instead, it corresponds to the intensity with which the animal attacks. This would suggest that the tendencies to attack and flee are in equilibrium when the observer reaches a certain distance, so that the bird remains rooted to the spot and instead vents its arousal in bathing behavior, though there are neither

Fig. 16-2. This arctic skua feels threatened by a person who has approached up to a certain distance from its nest. In conflict between attacking or fleeing, the bird performs intensive bathing movements although the ground is dry.

water, dust, nor soiled feathers to provide any reason for doing this. Despite all these findings, however, numerous objections arose to the concept of sparking-over. First of all, the idea that undischarged excitation is somehow dammed up and must find an outlet is not necessarily compatible with neurophysiological viewpoints. The idea of an action-specific excitation that sparks over into a strange behavior system is closely tied to the so-called "energy model" of behavior and to the psychological concept of catharsis. Although not entirely impossible, this concept of specific excitation is at least questionable. The present-day tendency is to assume that any specific excitation is almost always combined with a more general arousal encompassing various groups of behavior. This concept, supported by the experiments of J. Delius and J. Fentress, does show a connection with the old hypothesis, but it dispenses with both the energy model and the notion of catharsis.

Fig. 16-3. Experimental method for conditioning a cock to respond to certain pieces of plastic (for example, green triangles). Only when it pecks at these (above) does the animal get a food reward (below). F. Feekes was able to show that during a fight these conditioned cocks preferred the food-associated pieces in their displacement pecking.

Secondly, a number of experiments have shown that the same factors influencing the occurence of any particular movement in its normal functional context could also give rise to the corresponding displacement activity. For instance, in certain conflict situations, terns showed more displacement preening when it rained than when the weather was dry (J. J. A. van Iersel and A. Bol, 1958). Similarly, in a conflict situation chaffinches with wet plumage preened more than dry control animals (C. H. F. Rowell, 1961). During courtship, sticklebacks did more displacement fanning when their nest was flooded with water saturated with carbon dioxide than when the water was normal. These displacement activities are obviously facilitated by the presence of such external factors. Furthermore, certain intrasystemic factors may also have an influence, as shown by the experiments of J. Kruijt and F. Feekes: During a fight, hungry cocks pecked on the ground more and also swallowed more than others who had just been fed. Feekes also discovered that the hungry animals pecked more at food grains than at undigestible pebbles or grains of sand. In short, these cocks behaved as they would during normal food-foraging activity. In a convincing experiment, Feekes then offered the animals small pieces of plastic of varying shapes and color: blue crosses, red rectangles, and green triangles. Cocks that had learned, for example, to associate green triangles with a food reward also preferred these chips in their displacement pecking during a fight. Results like these lead us to conclude that any movement occurring as displacement activity is determined by the same causal factors which influence it in its normal behavioral setting. This view, however, makes the concept of an excitation sparking over seem at the least superfluous.

Some studies on situations leading to displacement activities also provide information about what behavior systems are involved in the conflict. While making their observations of a colony of terns, van Iersel and Bol frequently triggered alarm reactions by waving from their tent. The frightened animals would return after a short time and remain for a few

The conflict situation

seconds in an alarm posture with varying degrees of intensity. The experimenters distinguished four different intensity grades, each showing a certain degree of flight tendency. The distance at which the terns landed from their nest was seen as a measure of their tendency to return to the nest and resume their incubation activity. After landing, some of the birds preened in a hurried and frantic manner before going back to their nest. The likelihood of these preening movements occurring or not depended on the proportionate strength of the two tendencies (flight and return to the nest), as expressed in the alarm posture and the landing distance to the nest. The probability of preening was highest when the most intense alarm posture was combined with the smallest distance from the nest, or when the weakest alarm posture came together with greatest distance from the nest.

These findings indicate that for a displacement activity to occur, a particular and well-defined state has to be present in the animal. Only when the two tendencies of flight and return are equal—i.e., both strong, both weak, or both of intermediate value—will preening take place. When these conflicting tendencies are unequal, the birds do not show some other behavior but simply stand there, in the same alarm posture, but at a different landing distance. This is why we cannot assume, as some authors do (e.g., B. R. J. Andrew, 1956), that these animals just preen when they have nothing else to do.

C. Rowell came up with similar results in his simple experiments with chaffinches. The cages in which these birds were housed contained a few perches and a food dish. Sometimes, when an animal approached the dish, the experimenter fired a flash bulb. This caused the chaffinch to fly in a hopping motion back and forth in the cage, showing an obvious conflict between approach and avoidance. While hopping, it would intermittently stop and rest briefly on a perch, occasionally making a few preening movements. In the turning back, or when the bird flew back to the perch from which it came, preening occurred much more often than during stop-over pauses, where the bird continued in the same direction after stopping. Rowell makes the following point: In the case of turning back, the bird most certainly passes a point where approach and avoidance are in equilibrium; but in the case of a stop-over pause, this is not at all certain, and more likely the tendency to continue hopping in the same direction is stronger here. As with the terns, aside from their greater likelihood to preen, there does not seem to be any difference in behavior between stop-over pauses and turning-back pauses.

Again, these experiments with chaffinches demonstrate that any displacement activity would arise from a very specific internal state, and not just because the tendency for some other behavior pattern is lacking.

These investigations also revealed that, under different circumstances, each of the behavior systems involved in the conflict producing a certain displacement activity will inhibit the very same behavior within its nor-

Turning-back vs. stop-over pauses

mal setting. For example, preening is inhibited by attack as well as by flight tendencies. Since we have already demonstrated that a sparking-over model is unnecessary, the following hypothesis now suggests itself: When the behavior systems involved in the conflict are in equilibrium, they no longer inhibit the system where the displacement activity has its primary function. As a result, the displacement activity that is influenced by whatever factors are present will be disinhibited. I was able to demonstrate in my own experiments on displacement fanning in sticklebacks (1960 and 1961) that a removal of inhibition really does take place. But instead of describing this rather complicated example in detail, I will discuss the general principle of the "disinhibition hypothesis" by means of another movement occurring during courtship in sticklebacks, namely, the "gluing" behavior.

As mentioned before, we are quite certain that this gluing behavior evolved in the service of nest-building, since it very often occurs after the fish has carried material to the nesting site, and then almost always after this material has been deposited in its proper place. The material itself is glued together when the fish secretes a sticky substance and performs its gluing movements. We can calculate the speed of gluing from the duration of this movement and the distance traveled.

The disinhibition hypothesis

This gluing activity also occurs during courtship. At that time, however, it is almost never preceded by the carrying of nest material; instead, very often it is preceded by fanning movements. Thus, according to phylogenetic criteria, gluing behavior during courtship can be seen as a displacement activity, just like courtship fanning. We can now compare the duration, speed, and frequency of displacement gluing with the same values during a previous period of normal gluing. We then find that during courtship the frequency is always higher and the duration shorter than in a normal context; and the frequency and duration of the gluing movement during courtship is independent of its frequency and duration in the preceding normal behavior sequence. On the other hand, there is a connection between the two situations in terms of the speed of movement: If a fish performs its nest gluing slowly during its normal behavior sequence, it will also be slow in its courtship gluing. That is precisely what we would predict according to the disinhibition hypothesis: Frequency and duration of the disinhibited movement would, after all, be determined by the behavior systems involved in the conflict, while within a period of disinhibition the speed would be influenced by whatever factors are present that normally determine the movement.

Although the disinhibition hypothesis does provide a good explanation for most results obtained in related experiments, we must caution that its key premise is still open to doubt. If chaffinches that have been sprayed with water do more displacement preening, should we then assume that the displacement preening of dry control animals is also caused by some stimulation of the skin? Generally speaking, if certain factors associated with the movement are shown to have an influence on the displacement activity does it mean that this displacement behavior is activated only by such factors? Feekes has examined this question with detailed studies on displacement pecking in fighting cocks. Results led her to reject the disinhibition hypothesis for that particular case in favor of another explanation. Her reasoning is most interesting and deserves discussion.

Activation through a variety of factors

Feekes presented both her cockerels with inedible plastic pieces of different color and shape. The animals pecked at them with particular eagerness while fighting. It was found that hunger increased pecking behavior only when the chips had been previously coupled with a food reward. Without this prior conditioning, the pecking was in no way associated with hunger and therefore could not be regarded as disinhibited food-getting behavior. In some way, the pecking itself is obviously caused by the fighting, and especially by factors associated with a state of internal conflict because it occurs most frequently when the two cocks are about equally strong. Kruijt had already discovered, in 1964 that cocks which had frequently been induced to fight by stimulation with a stick of a particular color also preferred pecking at grains of this same color during those fights. Kruijt then suggested that ground-pecking may be regarded as a redirected act of aggression, where aggression against the stick was inhibited by the conflict and instead discharged at the grains.

However, Feekes saw various reasons for suggesting different lines of explanation. First of all, pecking activity was only vaguely associated with the nature of the fight, provided that this conflict did not stop too soon and the opponents were fairly equal in strength. Given these conditions, the cocks will peck in the most diverse circumstances and more or less independent of what the opponent is doing. Therefore the factors giving rise to ground-pecking independent of hunger are apparently not bound as closely to a precise internal state of equilibrium as van Iersel and Bol had suggested when describing the conditions for displacement preening in their terns.

Secondly, Feekes discovered that neither fighting cocks nor isolated cocks pecked at the ground if it was completely bare. When she used this method to prevent the animals from pecking, they were less aggressive but also more alarmed and ready to flee. This applies, however, only to cases where according to previous observations pecking would be expected if the ground had been covered with sand. The absence of pecking opportunities therefore seemed to result in a motivational change. The change itself could not have been caused by the clean-swept floor, because the behavioral changes did not arise in situations where pecking on a sand-covered floor would not have been expected either. Furthermore, the greatest changes occurred in cases where the animals would normally have pecked the most. The impression was that ground-pecking reduced escape tendencies and led to an increase in aggressive behavior. Feekes wondered if in fact the animals had learned to control their own state by pecking, where a frightened cock might use this behavior to calm down, so to speak (instrumental response). To see if this explanation was at least plausible, Feekes devised another experiment. She trained a cock to peck at small cross-shaped pieces of blue plastic by rewarding this behavior with the withdrawal or removal of its rival. In a subsequent test (as yet not entirely conclusive, however), the cock preferred blue crosses to red rectangles. It therefore seems that Feekes's idea, although still very much a speculative one, is nonetheless worthy of note. She also suggested the possibility that chickens might learn at a very early age to associate food-pecking near the mother hen with a feeling of security and relaxation, so that in later life the reverse might happen: Pecking behavior is used to create a state of de-arousal. This would be comparable to the pacifying effect of thumb-sucking in small children.

In addition to the sparking-over and disinhibition hypotheses, other explanations have been suggested. Though none of these has been experimentally tested, there are two that may be regarded as highly plausible. The first one states that the posture adopted by an animal within a certain behavior sequence may give rise to a movement also belonging to another behavior sequence where the same posture is displayed (Tinbergen, 1940, 1952; Lind, 1959). Thus, a bird turning its head in anticipation of flight would switch to preening its wings because the head-turning movement is also a part of this second behavior.

Another possible interpretation, which would apply mainly to preening and other body-care activities occurring in a displaced context, was described in detail by D. Morris in 1956. He stressed the physiological changes observed during conflict, such as increased blood circulation in the skin, constriction of blood vessels, erection of hair or feathers, and so on. These kinds of peripheral effects may then lead to preening, scratching, and other body-care activities. Regulation of temperature may also be influenced by conflict situations, possibly leading to activities such as panting, drinking, or shaking. However, experimental data on this interpretation are virtually nonexistent, and probably hard to obtain in the first place.

Furthermore, it is highly likely that a combination of factors, as indicated by the various hypotheses, would provide us with valid explanations. For example, peripheral effects may be combined with the disinhibition hypothesis and seen as facilitating factors. We also suggested an association of the disinhibition hypothesis with that of a sparking-over. We may therefore assume that not all displacement activities are based on the same kind of mechanism; more detailed analysis may reveal that these activities are generally caused by a variety of factors.

We have already remarked briefly on the possible meaning of displacement activities. But we still know very little about it, at least experimentally. Many of these movements could simply be regarded as inevitable "by-products" of particular conflict situations. However, most authors

The function of displacement activities

Fig. 16-4. Many displacement behaviors have taken on a secondary function as signals. The frequently almost vertical threat posture of male sticklebacks (below) probably developed from displacement digging (above). In both these movements the dorsal spines are erect, contrasting with nest-digging activity.

tend to assume that their function is to regulate or eliminate a conflict; this viewpoint is expressed in concepts such as "abreaction" and "change in motivation" to mention a few. The basic premise is that an animal would enter an unfavorable motivational state if it could not perform the displacement activity. This interpretation has been tested experimentally, not only by Feekes, but also by K. J. Wilz (1970). He found that male sticklebacks remained aggressive toward females and could not lead them to the nest if, in an experiment, the males were prevented from performing certain nesting activities (e.g., gluing). This indicates that the nesting behaviors may be necessary to resolve the conflict in favor of courtship.

As a final remark, we would like to point to the signal function of many displacement reactions. Since these activities are frequently highly indicative of a certain motivational state, it is hardly surprising that they evolved into signals for social communication. Tinbergen (1952) examined such developments in detail, including the way movements are modified to serve as signals: Their signal effectiveness is increased in the process of ritualization—for example, by becoming more and more conspicuous or more uniform in the course of evolution, and perhaps always occurring with the same intensity. Finally, they may also become increasingly independent of their original causes (for example, a state of conflict) until ultimately they arise solely from special factors in the social situation, or even from a completely new set of internal factors.

We have derived these possible evolutionary developments from observations on animal species living today. Related species can display qualitatively similar (i.e., homologous) displacement movements that differ in form, degree of expression, and causal factors. By taking these differences into account, we can easily arrange the movements occurring across species into a certain order, representing different stages of evolution.

17 The Influence of Specific Stimulus Patterns on the Behavior of Animals

The way animals behave is as typical of their species as many of the anatomical features traditionally used to classify them. A large number of bird and insect species, for example, can be recognized and distinguished just as well by their calls as by their physical traits. In some cases, animals that seemed to belong to the same species because they showed no obvious physical differences were later reclassified on the basis of differences in behavior. Certain species-specific behavior patterns, such as the chirping of crickets, seem to occur spontaneously, that is, in the absence of demonstrated external stimuli. But the majority of such behaviors is elicited by certain "stimulus patterns," and is frequently oriented to particular elements within these patterns. In this way, many song birds respond to approaching predators with typical alarm calls; a male stickleback in its territory will attack red-bellied male conspecifics and perform zigzag dances to court gravid females. In their own holes, male crickets attack chirping conspecifics but court intruders that do not make any sounds.

While simple reflexes such as our pupillary and knee-jerk reflexes are almost certain to occur with the corresponding external stimulus, no such clear-cut stimulus-response relationship seems to exist for behavior patterns like those described above. The presentation of a particular stimulus pattern does not always release a corresponding behavior pattern; the "readiness" of an animal to respond in this way depends very much on environmental conditions and the animal's overall physiological state. On the other hand, such behavior patterns sometimes occur even in the absence of releasing stimuli that would normally have to be present. In 1937, K. Lorenz reported, for example, how a starling hunted nonexistent flies. A cricket alone in its hole may respond to the chirping of an unseen neighbor by performing head thrusts and leaps of attack "in vacuo"—as if fighting with an opponent. These behavior patterns also differ from ordinary reflexes in that a certain stimulus pattern may elicit different responses depending on the animal's state, or conversely that a certain behavior can be released by very different stimulus patterns. For instance, a cichlid fish

The influence of specific stimulus patterns on the behavior of animals, by W. Heiligenberg in its territory will attack any intruding male conspecific with characteristic threatening movements, while outside of its territory the fish will probably flee from the same opponent. Begging movements of young songbirds, as another example, may be triggered by such diverse stimuli as shaking at the edge of the nest and the call of a parent.

In contrast to simple stimuli and their reflexes, then, the relationship between stimulus patterns and associated species-characteristic behavioral responses is not rigidly determined but rather is statistical. We shall describe a few examples to illustrate the properties of such stimulus patterns and the means by which they elicit a response.

Many behavior patterns can be elicited almost as strongly with dummies as with their natural counterparts. For example, a territorial stickleback will attack even very rough models of a fish, as long as the model's underside is red like the belly of an actual rival. By contrast, the fish will hardly take notice of good imitation sticklebacks, however realistic, if these do not sport a red belly. The trait "red underside" is therefore an essential element of the rival's stimulus pattern if a territorial stickleback is to respond aggressively. According to N. Tinbergen (1951), details of shape and surface texture are of little or no importance. Similarly, the zigzag dance—a courtship ceremony with which a territorial male stickleback greets a female ready to spawn—can be triggered with fish models displaying a swollen belly, and especially when this model is made to point the front end upward like a real female. Such models are far more effective in releasing the zigzag dance in males than are imitation females with a realistic appearance but no swollen belly.

The water beetle *Dytiscus* catches minute animals that it hunts down by their scent rather than by their shape or movements. We know this because if one of these beetles is presented with a tadpole behind the glass of a test tube, so that it can see the prey but not smell it, there is no reaction. But when the beetle is allowed to smell meat, it will start hunting inside the cloud of odor and blindly grab at any objects it accidentally touches. Although this animal has well-developed compound eyes, it hunts only by responding to chemical and tactile stimuli (this was also discovered by Tinbergen), not to optic ones.

The grayling butterfly *Eumenis semele* pursues paper shreds just like it does females, provided that they stand out sufficiently from the background and flutter closely enough to the way the insect itself does. But the color of the shreds is unimportant, despite the fact that these butterflies do perceive colors quite well—as we can see by their preference for blue and yellow when looking for food.

Young turkey hens without brooding experience will show maternal behavior toward their chicks (gathering them under the wings, defending them) only if they can hear the young chirping; they also respond just as maternally to chirping models. If, on the other hand, we offer only mute models of chicks to the hens, they will show aggressive behavior

Essential vs. nonessential features of a stimulus pattern

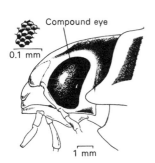

Compound eye

0.1 mm

1 mm

Fig. 17-1. The water beetle *Dytiscus marginalis* has well-developed compound eyes (above). Yet its hunting behavior is not triggered by visual stimuli but rather by chemical cues, such as the smell of meat (below).

instead of maternal responses. Turkey hens whose auditory organs have been removed in the first days of their life and whose social, sexual, and defensive behavior is hardly different from that of normal conspecifics do incubate their eggs, but they then kill the hatching chicks, whose chirping they cannot hear. These findings were made by W. M. Schleidt (1964).

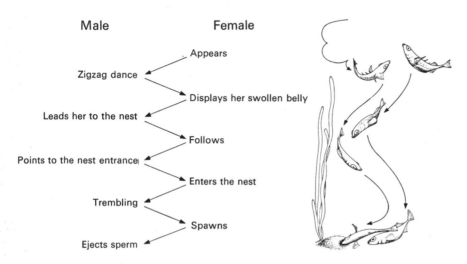

Male   Female

Appears

Zigzag dance

Displays her swollen belly

Leads her to the nest

Follows

Points to the nest entrance

Enters the nest

Trembling

Spawns

Ejects sperm

Fig. 17-2. Reproductive behavior of the three-spined stickleback. After performing a zigzag dance in front of the gravid female (top right), the male leads her to the nest (bottom right) and points to the nest entrance. The female is characterized by her swollen belly. This figure is a schematic representation of the chain responses in stickleback mating.

In some cases a particular situation will evoke a whole chain of activities in the animal. Here, different stimulus properties elicit the different response elements in the chain. For instance, females of the wasp *Philanthus triangulum* hunt for bees by searching from flower to flower. Initially, the wasp responds to every small, moving object by whirring and howering in the air at about 10 cm leeward to the object. If it then catches the scent of a bee, it will make a sudden strike; otherwise it flies away. Thus, while females of *Philanthus* detect potential prey (destined to provide reserves for its brood) by visual means, the rest of the prey-catching sequence is determined by scent. According to Tinbergen, the stinging action which follows upon the wasp's sudden leap and which paralyzes its victim seems to be elicited by a further stimulus trait, perhaps physical contact; in any case, just movement and a bee scent emanating from simple models have turned out to be inadequate for evoking this stinging action.

Our examples show that only certain parts of a stimulus pattern are required to release a corresponding behavior, while other elements—even if the animal is actually able to perceive them—are of little or no significance. The fact that individual elements within a stimulus pattern are enough to trigger an associated behavior in a specific way has led ethologists to formulate a simple notion: The combination of such stimulus elements fits like a key into a lock, and the opening of this lock corresponds to the occurrence of associated behavior patterns. In 1935, K. Lorenz consequently labeled these stimulus components as "key stimuli" (*Schlüs-*

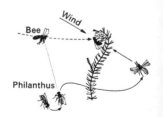

Fig. 17-3. Response chain of the hunting wasp *Philanthus triangulum*, a species of Hymenoptera. This animal is first alerted by the visual stimulus of a bee flying toward a flower. The wasp then moves into a position leeward of the feeding bee in order to catch its scent.

selreize) or, to use a more accepted term in English, "sign stimuli." Together with J. J. von Uexküll (1934), he also called whatever provided a fitting for the "key" within the animal the "releasing schema." Tinbergen later spoke of a "releasing mechanism."

Animals frequently respond appropriately to a particular stimulus pattern without ever having experienced that stimulus before. F. M. Barraud (1961) found that twelve-day-old great titmice who were raised by hand and never exposed to conspecifics would immediately crouch and remain still the first time they heard a parental alarm call. Other, comparable sounds did not evoke such a response. R. A. Hinde (1954) observed how thirty-day-old chaffinches mobbed an owl in a typical way, even though they had never seen one before. Ethologists therefore considered these reactions "innate," meaning that animals respond with the appropriate repertoire to a certain stimulus pattern without ever having encountered it before. Lorenz (1935) spoke of an "innate releasing schema," and Tinbergen (1951), of an "innate releasing mechanism" (IRM).

For a time this notion was misunderstood to imply that an innate behavioral response could develop in the absense of any individual experience whatever, completely separate from all environmental influences. This interpretation, however, is meaningless and even absurd, because a developing organism is constantly interacting with some environment; its genes then determine both the norm and the range of variation with which the animal responds to environmental demands. A developing creature is always exposed to external influences, no matter how hard an experimenter may try to isolate it from its surroundings, and therefore such an animal will always gather some kind of experiences during its growth. Furthermore, it was found that, for example, the eyes of a growing kitten have to be exposed to certain structured stimuli in order to develop their capacity for Gestalt perception, or not to lose the functions present at birth (D. H. Hubel and T. N. Wiesel, 1963). The way an ethologist tries to answer the question of whether a certain behavioral response is innate (in the sense we have described) or whether it has to be learned by individual experience with the corresponding stimulus pattern is to raise the animal under the most natural conditions possible so that it will not be damaged in any way. But the experimenter has to keep his subject from experiencing any stimulus patterns associated with the releasing mechanism under study. If during its first exposure to the stimulus the animal does not respond with the appropriate behavior, there is still no proof that this reaction must be learned, because experimenters cannot guarantee that developmental defects are ever really avoided. In addition, animals raised under isolated or deprived conditions may show very different degrees of readiness for various behavior patterns when compared with conspecifics growing up in a natural environment. In this way, an animal raised by hand may become so tame that it hardly

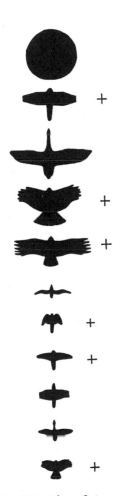

Fig. 17-4. These flying models were presented to gallinaceous birds familiar with flying ducks and geese but less familiar with birds of prey. The animals therefore crouched only when sighting short-necked models, which resembled birds of prey (+), and displayed no "fear" of long-necked models.

ever displays flight reactions—regardless of the stimulus—and responds to predator stimuli only when its keeper has managed to "frighten" it in some way, inducing an appropriate state of flight readiness.

Basically, there are two ways in which a certain stimulus pattern and its corresponding behavioral response may become "fitted" to each other: Either an animal has been "wired" from the start in such a way that it automatically produces the appropriate response when a particular external stimulus is presented; or it has a learning mechanism by which, under certain environmental conditions, it chooses those responses leading to a desired effect. Thus, an animal may be equipped with a mechanism for predator recognition which becomes activated and then "innately" causes the animal to flee or hide. Another possibility would be that an organism is equipped to remember any external situations where danger threatened, and thereafter to avoid these situations. The first case would then imply that the animal is already structured to associate "enemy traits" with "flight," while the second example represents an association of "being seized (danger)" with "avoidance," with a particular learning mechanism enabling the organism to anticipate the danger of being seized or attacked. The first of these two "solutions" is more of an advantage if the predators almost always hit their mark. Both solutions, however, may be used by the same animal. In the technical language of ethology, the first kind of mechanism comes about in the course of phylogenetic development, while the second arises through the animal's ontogenetic (individual) growth.

More detailed investigations of releasing mechanisms have shown that innate elements can be modified or in part even replaced by individual experience. We have already described inexperienced turkey hens that acted maternally toward their chicks only when they could hear the chirping of the young. But within a few days these hens learn to recognize visual traits in their chicks, and eventually they accept even mute dummies and dead chicks that they would have attacked without prior individual experience. Gulls learn to know their own young. Tinbergen observed that strange young are then attacked and perhaps killed. On the other hand, these gulls never learn to discriminate their own eggs from those of other pairs, no matter how different the eggs might be.

Turkeys that have never experienced flying objects will crouch and hide if they see such an object pass across the sky at a certain speed. Their reaction is not only to predators, but also to airplanes and to harmless large birds such as geese, storks, and cranes. But in time turkeys no longer respond this way to flying objects that appear frequently and never turn out to be threatening (Schleidt, 1961). As a result of this learning mechanism, gallinaceous birds, which are daily exposed to the sight of flying geese and ducks, will no longer react even to flying models that have a similar appearance of long neck and short tail. Yet the same models trigger fleeing responses when made to pass over in reverse, resembling predators which have short necks and long tails.

Fig. 17-5. The gallinaceous birds crouched in fear of this model only when it was moved across the sky with its short "neck" to the front (+).

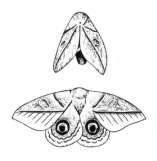

Fig. 17-6. We find this "eye spot" in many insects. By suddenly displaying this pattern, as the South American moth *Automeris memusae* is doing, it scares off predators.

Gull chicks peck at the tip of their parents's bills, causing the adult birds to regurgitate food (Tinbergen, 1951). As demonstrated by J. P. Hailman, inexperienced chicks prefer long and thin objects, especially if these are held to point downward and moved around horizontally. During this pecking activity the chicks aimed at a patch resembling the one their parents would have on the tip of the bill, although the position of this patch on the model as a whole was unimportant. It was only necessary to present the patch at eye level. Furthermore, the properties of the "gull's head" to which the bill model was attached were also insignificant. On the other hand, chicks that had lived with their parents for a while showed a preference for models resembling the head and bill of adult conspecifics, whose bill patch was located on the normal spot.

All these examples give us the impression that animals are born with certain releasing mechanisms to help them survive (preferences for particular stimulus patterns), but that these innate mechanisms can be modified through individual experience to suit the conditions of the organism's environment. Schleidt (1962) has called them IRM's (innate releasing mechanisms) modified by experience.

Certain behavior patterns, such as the courtship movements of ducks, are conspicuous by their stereotypic nature and their strange displays of sound, form, and color patterns. Many of these so-called courtship, threat, and attracting movements seem to derive from simple, everyday behaviors, such as preening, food foraging, and intention movements of flying (see Chapter 15). Ethologists presume that these more basic movements developed in the course of evolution into "ritualized" behaviors that serve as signals for social communication (Tinbergen, 1952). This concept was formed in comparing the behavior patterns of anatomically related animal species. Investigators found that different species "ritualized" common movement patterns to different degrees. Derived from a common basis, such behaviors came to diverge in different ways from the original movement so that some varied a great deal and others remained fairly close, with all modifications arising through stereotypical ritualization and the parallel development of additional conspicuous traits. In turn, these behavior patterns are believed to have acquired different degrees of signal value, becoming "releasers" (Lorenz, 1935) and functioning in the context of pair-formation, delineating a territory against rivals, parent-offspring communication, and other social situations. It is quite possible that such releasers attract the attention of other animals because of their conspicuousness, and through their specific nature play a role in the recognition of conspecifics. Correspondingly, these social releasers would help to keep different species apart.

Fig. 17-7. The melba finch (left), belonging to the weaverfinches, is host to the whydah (right). The whydah's nestlings have the same throat coloration as the weaverfinch young. We assume that this is why the former are fed by weaverfinch parents.

Releasers of this kind not only have a social function within species, but also serve to communicate between different species of animals. Various butterflies carry conspicuous eyelike markings (eyespots) on their hind wings, suddenly displaying these when threatened. Some will add rocking movements to the display. A. D. Blest (1957) was able to show that such eyespots scare off predators. Syrphidae bear the conspicuous black-and-yellow ring markings of wasps on their posterior end—and birds that have had unpleasant experiences when hunting wasps will henceforth avoid even the similarly marked but otherwise harmless Syrphidae. The nestlings of some weaverfinches have very conspicuous throat coloration; they stretch their throats towards the parents when begging for food. But the same marking is also found in the throat of nestling whydahs, who—according to J. Nicolai—are raised as nest parasites by weaverfinches. All these examples show that eye-catching and impressive stimulus patterns and gestures are used to elicit certain behavior patterns. The very fact that such stimulus patterns are mimicked by other species proves their effectiveness.

We mentioned earlier that certain behaviors can be released by models combining the necessary elements of the normal stimulus pattern. We also found that these models are more likely to release the behavior, the more completely and prominently they display the relevant stimulus components. When one of these components is weaker, it can be compensated for by increasing the strength of another. A. Seitz (1940/43) observed, for example, that threatening and fighting behavior could be evoked in cichlids by means of fish models, and with greater intensity the more these models were endowed with traits representing "display coloration of a territorial male," "fin-spreading of a threatening rival," and "lateral tail beats of a fighting fish." Models showing a combination of the first and second stimulus characters were about as effective as others combining only the second and third. Seitz postulated that each of these three traits possessed a certain releasing value, and that the effect of the model as a whole resulted from a kind of sum total of the releasing values of its components.

This phenomenon or Law of Stimulus Summation—as Seitz called it—also showed up in various later studies. For instance, the courtship approach of male grayling butterflies—mentioned earlier—is more likely to be elicited by models of females, the greater their contrast with the background, the more lively their fluttering movements, and the closer they seem to the male (as judged by size). A deficiency in one of these characters can be compensated for by increasing one of the others correspondingly. In this way the effectiveness (or releasing value) of a model can be retained if the weakening of one character is balanced out by a sufficient increase in the strength of another. If we present all three characters in their optimum effectiveness (i.e., in exaggerated form), the model releases courtship pursuit in the male with a greater probability than if the stimulus traits were all copied precisely from corresponding natural objects. The "supernormal" effect of these models was demonstrated in a number of other cases:

Fig. 17-8. A hen responds less to the visual than to the auditory cues (chirping sounds) of chicks "calling for help." In the above experiment, a chick has been placed under a sound-proof bell dome. In the set-up below, the hen cannot see the chick but can hear its distress calls.

Fig. 17-9. Ringed plovers prefer a model egg (left) to their own less-contrasting eggs.

Fig. 17-10. The oyster catcher prefers a much larger egg than its own to its own (front) and to a gull's egg (left).

Birds incubating on the ground roll eggs lying outside their nest back into the nest hollow. If we offer an incubating bird two different eggs at the edge of the nest, and repeat this test several times, we can count the number of times the bird chooses to roll either one into its nest. This provides us with a measure of the bird's preference and therefore of the model's effectiveness relative to each other. In this kind of choice experiment, a ringed plover showed preference for a white egg with large black speckles rather than for its own egg, which has dark-brown spots on a light-brown base, creating less of a contrast. Oyster catchers prefer larger eggs to smaller ones, and they will try to roll an imitation egg into their nest even when the egg's size is so exaggerated that rolling becomes almost impossible. The superlarge models are preferred to normal eggs.

Stimulus characters of an egg: size, base color, and speckling

G. P. Baerends (1962) used this relative preference for rolling various egg models into the nest hollow to make a quantitative study of the Law of Stimulus Summation—or Heterogeneous Summation, as it is now often termed—in herring gulls. Characters of major importance were found to be size, speckling, and the base color of the egg. A larger egg was preferred to a smaller one whose properties were otherwise the same, a speckled one was chosen over one without spots, and an egg with a greenish base coloration was preferred to a similar one with a brownish coloration. However, the lack of speckling or of speckling or of the more effective greenish coloration could be compensated for by an increase in egg size, and here the amount of necessary increase also served as a measure for the effectiveness of the stimulus trait that was lacking. The experimenter then found that neither the releasing value of the speckling nor that of base coloration—when measured this way—depended on the initial size of the egg. Thus, speckling and color have consistent values independent of size. But this mutual independence could not have been assumed from the start, because the effects of speckling and color each depend on the other: On a greenish base, speckling is more effective than on a brownish base. It is possible, however, that the significant stimulus factor is not whether an egg has spots, but how such speckling contrasts with the base, and that greenish coloration offers a greater contrast to the spots than does a brownish color. This could mean that the effects of all three stimulus characters—size, base color, and degree of speckling contrast—are really independent of each other.

This example shows that any attempt to break down a stimulus pattern into its elements is arbitrary, at least at the beginning, and that our categories, which are based on human perception and language, may be artificial and do not necessarily apply to the sensory system of whatever animal we are studying. These categories may then fail to describe adequately the phenomena under investigation. It is therefore quite possible that mathematical formulations of some Law of Heterogeneous Summation will appear simple enough if our choice of stimulus parameters and trait categories happens to be convenient.

Fig. 17-11. Many birds "mob" predators, attacking in a group by making loud calls or by flying in a thick swarm above the enemy and periodically swooping down on it.

There is another aspect to the Law of Heterogeneous Summation, when we observe two different stimulus patterns releasing certain behaviors. Brooding flycatchers will mob red-backed butcherbirds as well as owls. In a number of experiments using models, E. Curio (1963, 1967, 1972) showed that when feathers and eyes were added to the model of an owl, its effect was considerably increased. Similarly, the upright model of a butcherbird had greater releasing value when its profile was presented, displaying the horizontal eye stripe. But while the stimulus traits of an owl and those of a butcherbird in themselves increase the effectiveness of either model, a combination of owl and butcherbird traits in single models has no such effects. In other words, a butcherbird model does not become more effective when owl eyes are added, and neither is the value of an owl dummy increased by the addition of a horizontal eye stripe. Curio therefore concluded that these flycatchers have different recognition schemata; that is, there is one releasing mechanism for owls and one for butcherbirds. The effectiveness of either stimulus pattern will increase with the number of stimulus characters belonging to that schema, whereas inappropriate characters decrease rather than increase the model's chances of releasing behavior. This hypothesis gains further support when seen in the context of "habituation" phenomena, discussed later.

For a long time, ethologists had believed that the Law of Stimulus Summation (or Law of Heterogeneous Summation) could be applied solely to "innate releasing mechanisms," while animals would only respond to learned stimulus configurations if these were presented complete. However, the studies of Baerends and his associates (1965), in which pigtailed macaques were conditioned to respond to artificial stimulus patterns, demonstrated that even such learned configurations can be recognized with increasing facility when approaching completeness: Different gradients evoke different degrees of response, and even here the absence of some stimulus characters can be compensated for by an increase in others. T. G. R. Bower (1966) "conditioned" human babies to recognize a stimulus pattern composed of three components. He then found that the sum of responses to each of the three components presented in isolation was equal to the number of responses to the pattern as a whole with infants up to fifteen weeks old. In older babies, however, the value of this sum was less than that of the complete pattern, meaning that the Gestalt of the entire stimulus could no longer be replaced evenly by successive presentations of its elements.

As we mentioned at the beginning of this chapter, an animal's responsiveness toward a certain stimulus pattern depends on various external conditions as well as its own physiological state. Thus a male stickleback will attack red-bellied conspecifics and corresponding models only when it has formed a territory; flycatchers mob butcherbirds and owls more readily and more intensively during their own breeding season. In addition, however, a certain behavior can be elicited with increasingly de-

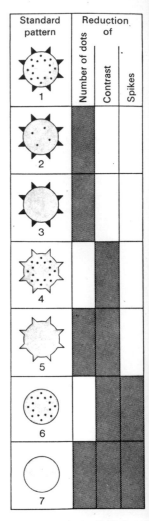

Fig. 17-12. A pig-tailed macaque was trained to look for food under the stimulus pattern 1. The experimenter then presented him with all possible combinations each of two of the patterns on the left and he preferred them in this sequence, from top to bottom. Experiments with reduced traits (shading) showed that these characters supplement each other (Law of Heterogenous Summation). But they do have different values. For example, a decrease in contrast reduces the overall effect of the pattern more

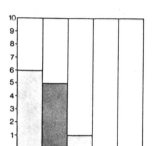

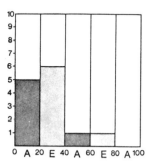

Fig. 17-13. Nestlings of various species gape in response to shaking of the nest (E) and to certain sounds (A). If we use two-second intervals and each time produce ten shakings followed by ten sounds, the animals react with about equal frequency in both series. The same results are obtained when the series are reversed. Only with continued presentation of these alternating series do the subjects begin to habituate. The fact that no significant waning of response was observed from the first to the second series shows that the animal habituates separately to shaking and sound stimuli.

ficient stimulus patterns, the greater the animal's readiness to respond. Conversely, if responsiveness is low enough, the behavior in question will not be released even if the appropriate stimulus pattern has an optimum combination of stimulus traits. While studying the courtship behavior of male guppies, Baerends and his co-workers found that females could more easily evoke such responses the larger their physical size. The degree of a male's courtship readiness could be gauged by its body coloration. With greater receptiveness, the male responded with certain courtship behaviors to correspondingly smaller—i.e., more "deficient"—females, while conversely the male needed increasingly larger females to elicit these behaviors when courtship readiness decreased. We can therefore say that the strength of an external stimulus and the degree of an animal's response readiness will compensate each other in a way similar to the Stimulus Summation phenomenon, where different strengths of stimulus characters balance out each other's effects within a stimulus configuration.

In order to study the effectiveness of a stimulus pattern, experimenters choose different measurement criteria depending on the nature of the respective behavioral responses. For example, relative frequency of positive responses to a certain stimulus pattern may be used as a measure where behaviors normally occur singly and in greater intervals. But if a response is easily elicited more than once and in shorter intervals, we can use the number of reactions within a certain period of time after presentation of the stimulus. Still another measure is the total duration of behaviors usually displayed for some time. The effects of a stimulus pattern may therefore be measured in a number of ways, and the observer has to decide what criteria will be meaningful and provide him with the necessary standard of accuracy. He must also find an appropriate measure for the animal's responsiveness, such as the quality and degree of some coloration pattern in fishes, or the strength of response to a standardized experimental stimulus.

After choosing a proper system of measurement and running a sufficient number of tests, the experimenter will find certain quantitative relations between the stimulus and the releasing effect, between the effects of various characters within a stimulus pattern, and between the effect of a releasing stimulus and the animal's present degree of response readiness. For example, Baerends et al. concluded that the product of stimulus strength (as provided by the female) and the male's courtship readiness must approach a minimum value in order to produce a certain courtship movement. According to Curio, the strength of response in flycatchers to predators and corresponding models increases—as measured by the number of mobbing calls per minute—during the breeding season by an additive amount, which is constant for all stimulus patterns presented. Furthermore, the effect of an owl model presented to Darwin finches —again measured by the number of mobbing calls per minute—could be formulated as a weighted sum derived from the individual effects of

than does a decrease in the number of dots.

parts of an owl model presented in isolation. Such mathematical representations facilitate the quantitative description of observed regularities between stimuli and responses. We must caution, however, that these formulations are based on arbitrary measurement categories, and are therefore not meant to imply that identical mathematical processes are actually operating within the organism.

If we present a certain stimulus pattern repeatedly and in sufficiently short intervals, the animal will respond with ever-decreasing intensity and may finally cease to react altogether. We call this process "habituation to a stimulus." Various experiments have shown that habituation cannot be the result of either physical fatigue or sensory adaptation to the stimulus, but that it is rather something in between. H. F. P. Prechtl, for example, showed that the gaping of young chaffinches could be elicited both by shaking of the nest and by imitations of parental calls. If we release this begging behavior several times with the same stimulus, the nestlings's response will wane. But they will start begging again if the other stimulus is presented. This means that the muscles used for the begging movements cannot have become fatigued. Neither is habituation simply an "exhaustion" of the corresponding receptor cells which mediate the perception of a stimulus: A repetitive sound such as the ticking of a clock will no longer be heard after a while, but we do take notice as soon as this familiar ticking stops.

Habituation

This notion has been supported by other findings. Again, Curio was able to show that flycatchers will habituate to either of the owl or butcherbird stimulus patterns, and that this process does not even require the performance of mobbing reactions. After being exposed to a less effective owl model, which they ignored most of the time, these birds turned out to respond hardly at all to another, normally effective model. But if the experimenter presented them with a model looking more like a butcherbird, the same normally effective owl model would elicit the usual response when offered afterward. Similar responses were observed when the roles of owl and butcherbird models were reversed. We therefore conclude that these flycatchers will habituate to the releasing stimuli associated with either the owl schema or the butcherbird schema even when the stimulus configurations are offered in such incomplete form that the birds do not even respond with a behavior.

In a number of different cases, behaviors were elicited by electrical stimulation of certain brain areas, just like with presentation of the appropriate stimulus pattern. E. von Holst and U. von Saint Paul were able to trigger a certain behavior pattern by stimulating either of two different regions in the brain. After repeated stimulation of one of these fields, responses decreased; but the same behavior could still be elicited by stimulating the other field. These two fields in the brain could therefore be compared to different stimulus inputs, to which the animal habituated separately.

Once habituation has occurred, the animal's response readiness will recover if the stimulus is not presented for a while. The speed of recovery depends on the behavior in question. For example, once the prey-catching behavior of a mantis has been elicited "to exhaustion"—i.e., the insect has completely habituated to the stimulus—we can observe some response recurring after several minutes; but the animal will not be fully responsive again until several hours later. Similarly, the prey-catching response of toads requires from several hours up to a day to recover, while the mobbing reaction of chaffinches to owls takes several days, or even weeks, and even then the degree of readiness will be less than that of experimentally naïve animals. Habituation processes that remain effective for that long actually become more like "learning": We might say that the chaffinches learn that the owl model is harmless, and eventually they will hardly take notice of it. The same process can be seen in turkey hens escaping from objects that look like flying predators: If these objects appear often enough and prove to be harmless, the hens cease to respond.

Most likely, habituation represents a central process much like learning, serving to eliminate from the animal's perception any stimulus patterns that are present all the time or repeated too often, sparing the organism from responding needlessly. Valuable as this mechanism may be for the animal, it sometimes interferes greatly with experimental procedure, since the ethologist working experimentally is forced to repeat certain stimulus patterns many times in order to make any statements of general validity about their effects on behavior. These experiments therefore require that stimuli are presented in time intervals large enough to allow for recovery from habituation.

Certain stimulus patterns not only trigger corresponding behaviors, but apart from that will also change the readiness of an animal to show these behaviors in the first place. This effect is less obvious than the releasing one, and probably because of that it has not been investigated as much. A male cichlid in its territory attacks other tank dwellers fairly often. But if it is presented briefly with a male conspecific that it can see and possibly approach, but not fight for any length of time, it will then attack others in the tank more frequently than before. We can observe this increment in attack readiness even if the presentation of the rival is so brief that our fish cannot engage in fighting at all. Thus the appearance of a male conspecific not only releases fighting responses as usual; it also increases general attack readiness in the territorial fish. We can examine this phenomenon in the following way:

In order to find out what changes occur in attack readiness as a result of extrasystemic changes, we have to measure this tendency to attack both before and after the presentation of the stimulus and then compare these values. If we put an adult male cichlid into an aquarium together with young fish of any cichlid species, we observe that soon after establishing a territory the adult fish will occasionally rush upon and snap at one of

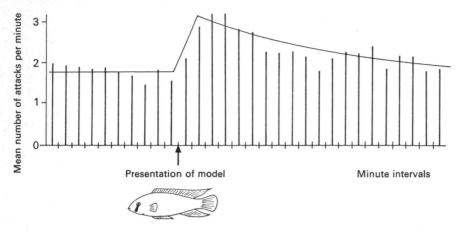

Presentation of model   Minute intervals

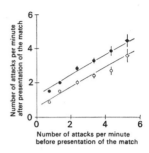

Fig. 17-14. Presentation of a fish model with a black eye stripe leads to increased attacks on standard test fish. Left: This increment in the rate of attack declines in an exponential curve (half-life period approximately 3 minutes) after presentation of the model. Right: A comparison of attack frequencies before and after presentation of the model shows that the mean rate of attack is higher by a constant amount shortly after presentation (dots) than would have been expected without the stimulus (circles). This increment in the rate of attack corresponds to the vertical distance between the two regression lines.

the younger fish. These fry usually escape and rarely suffer serious bites. But the more frequently the territorial fish attacks male conspecifics, the more it also rushes at young fish in its tank. We can therefore take the number of attacks per time unit directed only at fish fry to serve as a measure of attack readiness in the territorial fish. The advantage in setting up experiments with young fish is that they never engage the territorial fish in long and exhausting fights as an adult male would; otherwise our subject could end up losing and, as a result, change all its subsequent behavior. The young fish therefore represent a constant, randomly distributed, and rather weak attack stimulus which enables us to measure the territorial animal's aggressive tendency without too much interference.

In order to standardize the experiment even more, we can dispense with a live male conspecific—whose behavior may differ from one test to the next—and instead briefly present a model made to look like a male fish. Since the fry are supposed to represent the same kind of test stimulus before and after presentation of the model, their own behavior must not be influenced and so they cannot be allowed to see the model. For that purpose, the experimenter can sever their optic nerve, making them blind for a few weeks until the nerve regenerates.

If we use these experimental conditions and offer the territorial animal a model of a conspecific for periods of 30 seconds each, our subject will at first watch the model closely but show very little direct threatening or fighting behavior. However, results of a number of these tests show that, on the average, the territorial fish attacks the fry in its tank more frequently after presentation of the stimulus. But in the minutes following, it gradually falls back to its pre-presentation rate of attack. This decline in response can be expressed by means of an exponential curve with a half-life period of a few minutes. That is, the increase in the rate of attack declines after this time to half value. If we then compare the rate of attack after presentation of the model with that expected in the absence of stimulation and on the basis of immediately preceding activity, we find a mean increment with a constant value, D. This tells us that the mean rate of attack

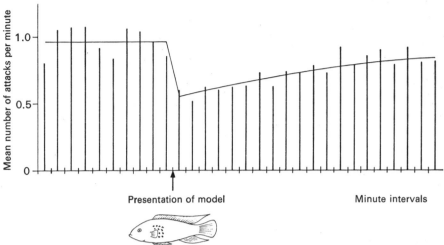

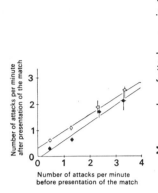

Fig. 17-15. Presentation of a fish model with orange spots above the pectoral fin leads to decreased attacks on test fish. Right: This decline in the rate of attack decays in an exponential curve (half-life period approximately 11 minutes) after presentation of the model. Left: A comparison of attack frequencies before and after presentation of the model shows that the mean rate of attack is lower by a constant value shortly after presentation (dots) than would have been expected without the stimulus (circles). The size of this decline corresponds to the vertical distance between the two regression lines.

after presentation of the model is higher than that expected without presentation by the constant value D, independent of initial value. This increment D, then, provides us with a measure for the model's influence on the tendency to increase a response.

By using this method, we can test the effectiveness of various models. Another particularly useful feature is that such models can be made to display individual traits of a conspecific one at a time, and we can then compare these single effects with that of a model combining all the characters. In this way, C. -Y. Leong computed that a black eye-stripe pattern like that of the cichlid *Haplochromis burtoni*, when displayed by itself on a model, increased the rate of attack by a constant value, D_1, while an area of orange spots above the pectoral fin had the surprising effect of decreasing the rate of attack by a smaller amount, D_2. But a model combining both characters raised the rate of attack by a value of $D_1 - D_2$. In other words, the "arousing" effect of the eye-stripe pattern was superimposed on the "inhibiting" influence of the orange spots. Furthermore, it was found that the inhibiting effect of the orange spots decayed about three times as slowly after presentation of the model as did the arousing (or facilitative) effect of the eye stripe. Thus the effects of the two characters differ not only in how they change the rate of attack (D_1 or D_2), but also in their decay periods.

In looking for the relevant properties of the eye-stripe pattern, experimenters found that this eye stripe had a stronger incremental effect on the rate of attack in *Haplochromis burtoni*, the more it approached a position parallel to the model's front profile, which corresponds roughly to its natural position in real fish. Conversely, when the eye stripe was perpendicular to the model's front profile, it had no effect. This applies to models both in a horizontal position and when presented vertically, head down. Consequently, the orientation of the stripe is significant only in relation to the rest of the body, not to gravity.

On the one hand, we found that the animal's attack increment, caused by seeing the eye-stripe pattern, decayed within a few minutes after presentation of the model; on the other hand, a more detailed analysis showed that a very small part of this increment persisted over several days. We can demonstrate this in the following way:

Once we isolate a territorial male in a tank with young fish for several weeks, its rate of attack will decrease to nearly zero. If we now present the male with an eye-striped model for ten consecutive days at an interval of 15 minutes, its rate of attack will slowly increase. At the end of this period of continuous stimulation, we find that attack responses wane again, but very slowly, with a half-life period of about seven days until the previous level is reached. Thus the eye-stripe pattern causes two kinds of arousal, one very strong and decaying within a few minutes, the other very weak but decaying over a period of several days. This latter change cannot be demonstrated within single tests because it is too faint, but due to a slow rate of decay its effects can be seen over repeated trials.

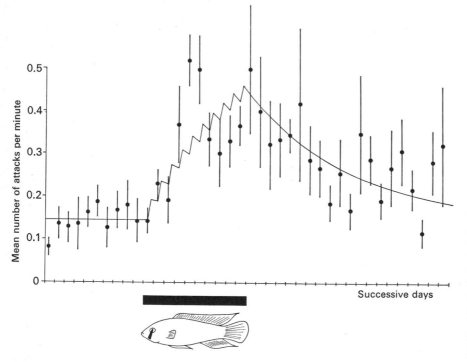

Fig. 17-16. The low rate of attack displayed by an animal kept only with test fish for several weeks can be raised by a frequent presentation of a fish model with a black eye stripe. This rate of attack subsequently declines and falls back to its initial level with a half–life period of approximately 7 days. The thick black line underneath the abscissa indicates ten days during which the model was presented from morning to night for 30 seconds every 15 minutes. The superimposed curve represents a theoretical process of excitation.

When an animal is presented with a certain stimulus pattern often enough and in sufficiently short intervals, its response to this pattern will wane. This habituation to a releasing stimulus, a process we discussed earlier, may be masked in the early stages of a series of trials by an opposing process which causes a temporary build-up of response strength, until habituation finally takes over. This means that an animal may at first hardly respond to a repetitive flight stimulus of medium strength. But if we continue with repetitive stimulation, the animal begins to show more direct fleeing and hiding responses, until eventually it comes to ignore the stimulus again. This initial increase in response strength may be interpreted as a greater readiness to respond, which was caused by presentation of the stimulus but could not be sustained in the face of increasing habituation to the very same stimulus. When running this kind of experiment, then, the degree and duration of an initial response increment that we may observe during a series of tests will depend on how the responsiveness increases and on the accompanying habituation process.

In an attempt to understand how an animal could profit from being influenced by certain stimulus patterns in its tendency to respond, we may look at *Haplochromis burtoni* and venture the following interpretation: Territorial males live in colonies where each one has its own tightly drawn-up territory. These animals bear the eye stripe as well as the orange spots above their pectoral fin, and they engage in frequent fights at the territorial boundaries. Males approaching adulthood and looking for a territory also have an eye stripe, but no orange spots. In this way a young male who intrudes into a colony of established conspecifics will considerably raise their aggressive tendency, and the territorial fish then increase their attacks on the intruder and drive him away. Conversely, however, the sedentary males raise each other's aggressive tendency to a lesser degree because of their orange spots, so that a certain "truce" is maintained.

A second question concerns the appearance of both a short-term and a long-term increment in attack readiness, caused by the eye-stripe pattern in the males. We could speculate that a male, noticing its rival's approach because of the eye stripe, undergoes a rapid increase in attack readiness that prepares it better for the coming fight. The additional weak but far more persistent increase in the tendency to attack serves to maintain this readiness at a higher level during frequent encounters with rivals, again providing the male with a better position for fighting. We must caution, however, that interpretations of this kind are not satisfactory until the animals' behavior is observed in their natural environment, and many other external factors are considered.

Crickets chirp spontaneously, without any external stimulation that we know of. But if we play back such a chirping sound to a singing cricket, its own chirping calls will increase for a short time. The degree of increment depends solely on the "quality" of our stimulus sound, and not on the cricket's previous rate of chirping. Most of this additional sound-pulse rate declines again after a few seconds, but a small remnant can be detected even a few minutes later. This weak but slowly decaying part of the chirping increment will lead to a gradual rise in the mean chirping rate if we play our stimulus sounds to the cricket often enough. The mechanism involved in this process therefore corresponds very much to the influence of an eye-stripe pattern on attack readiness in cichlids, except that the processes associated with chirping are more rapid, and this

Relatedness of the releasing and responsiveness-increasing function of a stimulus pattern behavior is not directed at an outside stimulus. Crickets may also be silent for several minutes, even for hours. If we then play chirping sounds to them, the crickets frequently will start singing again. Should these animals have been generally mute instead of just for short periods, it would mean that chirping is elicited by a particular outside stimulus. It is probably just another step from increasing the responsiveness to actually releasing the behavior, and it is therefore possible that both of these stimulus effects are no more than two different ways of looking at a common process.

We could imagine that a certain behavior pattern will occur whenever a corresponding hypothetical and physiological quantity X passes a minimum value, i.e., a critical threshold, X_0, and that this event becomes more likely the higher the mean base level of this quantity X. Spontaneous behavior patterns such as the chirping of crickets would then be characterized by the fact that their corresponding X value generally reaches the critical threshold X_0 without the influence of an external stimulus. By contrast, environmentally elicited (nonspontaneous) behavior patterns would require that their X value be increased by some appropriate outside stimulus in order to pass the critical threshold. According to this hypothesis, the nature of the stimulus effect—whether increasing responsiveness or increasing releasing behavior—would depend on how the value X increases over time as a result of stimulation. A strong and short-term increment would mean that the behavior immediately follows stimulation, and we would than speak of this behavior as being elicited by the stimulus. On the other hand, a persistent and slowly declining increment would have the consequence that future stimuli have a better chance of raising the already increased X value above its threshold, thus releasing the behavior; here we would speak of an increase in the tendency to respond. In the experiments we described with cichlids, the young fish serving as test stimuli would occasionally raise the X value of aggressive behavior above its threshold, with short-term and relatively transient effects. But the model of a conspecific would raise the X value on a more permanent basis, so that the stimulation coming from younger fish has a greater likelihood of releasing aggression. In other words, fish fry represent a releasing stimulus while models of adult fish have more the effect of increasing responsiveness.

This model complies with the statistical stimulus-response relationship we described at the beginning. So-called "vacuum activities"—the spontaneous emission of behavior patterns normally dependent on external stimulation—comes less as a surprise when seen in the context of this model, and would now seem very much expected. In order for this completely spontaneous behavior to occur, it is only necessary that the associated X value rises by a sufficient amount without the corresponding input of an exteroceptive stimulus. The probability of this happening is very small, but certainly it must be considered.

Spontaneous vs. non-spontaneous behavior patterns

18   Predictable Patterns in Animal Signals

Predictable patterns in
animal signals, by
M. Konishi

The elaborate displays, gorgeous plumage, and musical songs of birds have always attracted people's attention. Darwin, in his *Origin of Species*, explained how such traits evolved, writing: "I can see no good reason to doubt that female birds, by selecting, during thousands of generations, the most melodious or beautiful males, according to their standard of beauty, might produce a marked effect." Darwin called this process sexual selection. Subsequent arguments by critics against assigning the sense of beauty to animals hampered the study of the roles of these traits. Ethology, which Konrad Lorenz defines as a Darwinian approach to behavior, re-opened the study of animal signals and communication. Systematic studies of animal signals helped discover some general patterns in them. One of the most conspicuous features of animal signals is their diversity. Songs of birds, frogs, and insects are species-specific; that is, each species has a unique song. For species living in the same environment the rule of species-specificity has not been violated. Some signals, such as sex pheromones of certain moths, seem to be less specific. In such cases mate recognition is facilitated by a combination of sensory cues and ecological factors such as habitat selection and synchronization of activity rhythms. The diversity of signals evolved as a result of natural selection for species-specific signals, which are mostly used to ensure reproductive isolation.

Consider an ecological community in which many species use signals in various forms of energy for species-recognition and for other types of intraspecific communication. Such a community may be said to have a signal environment. This can be subdivided according to the forms of energy and sensory modalities used: thus there are acoustic, visual, and chemical signal environments. Just as each member of a community must occupy a different ecological niche, so each of the species utilizing a common modality must use a different signal code or set of codes if confusion is to be avoided. Therefore, the composition of the signal environment is an important factor affecting the characteristics of signals.

Selection for species-specificity inevitably encourages the evolution of signals that are uniform among all members of the particular species. The uniformity of signals is one of the prerequisites for species-recognition. There are occasions in which the recognition of individuals is advantageous. Songbirds defending territories learn to recognize one another by song so as to avoid wasting energy in further dispute after the initial phase of settlement. Individual recognition requires characteristics which vary from member to member. Therefore, there can be two opposing forces operating on signaling systems, one favoring uniformity, the other variability, as pointed out by P. Marler.

This apparent conflict can be easily resolved if animals use different signals for the two purposes. However, many animals seem to adopt a more economical solution. That is, they use one signal to achieve both aims. For example, certain aspects of bird songs are extremely uniform within a species, while individual birds differ from one another in other features of song. It is similar to the fact that all humans have five fingers on each hand but each finger has a unique fingerprint. Since the opposing forces are constantly operating, there ought to be a point of the best compromise. This point should shift according to the relative strength of the two forces. In a signal environment where species-recognition is not difficult, signals can be more variable. Marler discovered an example of this situation among island populations of a bird species. In the absence of closely related species on the islands, the songs of these populations were more variable than those of their continental counterparts which lived with closely related species.

It is generally agreed that signaling behaviors were "derived" from behaviors which originally did not have communicatory functions. The anatomical and physiological mechanism that controls behavior tends to be conservative. The derived signaling behavior often preserves some of the basic characteristics of the original behavior. It implies that the nature of a derived signaling behavior can be partly predicted from that of its evolutionary precursor. For example, the temporal patterns of sound emission code all information for cricket species, not only species-identity but also other messages for intraspecific communication. Some cricket species have a continuous trill-like song. According to W. Kutsch and D. R. Bentley, the neurophysiological mechanism of singing in these species is identical to that of flight. Other crickets sing more complex songs involving chirps which are produced by periodically holding the wings open or closed. This periodic act resembles the ventilatory movement. It is possible that the flight and ventilatory rhythms are combined to produce the complex songs. It is not a coincidence that all cricket acoustic signals are periodic. Even the most complex ones can be shown to be combinations of two or three basic periodic patterns.

Communication involves the production and reception of signals. It is likely that the two aspects evolved hand in hand. Therefore, the study of sense organs is expected to help detect patterns in animal signals. For

An environment made up of signals

Species-specificity of signals

example, most insect sounds are noisy, lacking the musical qualities of bird songs, although there are exceptions. The noisy quality is due to the lack of orderly arrangements of frequencies. This suggests that the frequency structure of insect sounds is not used in coding messages. The insect ear cannot perform frequency analysis of complex sound although they may be "tuned" to particular ranges of frequencies (see Chapter 4). This condition makes signaling by amplitude modulation of sound imperative among species that produce sounds of similar frequency ranges.

Origin of signals

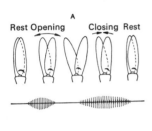

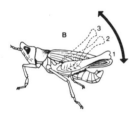

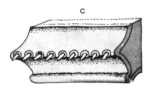

Fig. 18-1. Insects produce sound with the aid of their chitinous skeleton. Bush crickets and crickets, for example, produce chirping sounds by periodically opening and closing their forward wings (A; below: the oscillogram of these sounds). Grasshoppers (B) rub a rough surface (C), located on the inside of their upper hind legs, against an enlarged blood vessel of the wingcover of the forward wing.

Such natural wonders as the beautiful colorations of fish, birds, butterflies, and flowers tend to make us feel that random assortments of sounds, colors, and chemicals were selected to produce signals. In other words, the physical and chemical structures of signals emerged irrespective of their functions, their internal designs being accidental by-products of natural selection. However, animal signals are various forms of physical energy and they must be transmitted from one animal to another. There are physical laws that govern the transmission of energy. Therefore, animal signals must evolve within the constraints of these laws. They must be obeyed before biological rules such as species-specificity are followed.

The effects of physical laws on animal signals are clearly seen in the selection of signals under different environmental conditions. Except for animals with their own sources of light, signaling by visual cues uses reflected light. Since sound and chemical cues can be transmitted in the dark, nocturnal animals use vocalizations and odors to communicate. Light of visible wave lengths does not bend around corners, whereas sound and odor can. Communication by these means is expected to prevail in environments rich in visual obstructions, such as forests. Deposition of chemical signals is a unique feature impossible for either visual or acoustic cues. An animal which has a large territory or which patrols its territory infrequently can advertise its presence more profitably by scent marking rather than by emphemeral visual and auditory signals.

Since sound and light travel much faster than odors, vocal and visual signals are more effective than olfactory cues when rapid communication is necessary. Also, unlike sound and light, odors are not fit for signaling by modulating the temporal pattern of emission.

E. O. Wilson and W. H. Bossert made more specific predictions about the design features of chemical signals on physiochemical bases. This should serve as a model for the analysis of other types of communication. They predicted that the number of carbon atoms in pheromones should lie between 5 and 20, with a corresponding range in molecular weights from 80 to 300. The need for signal diversity cannot be met by less than 5 carbon atoms since the number of possible molecular configurations will be too small. The number of possible molecular species increases almost exponentially as the carbon number augments. It is also known that larger molecules in a homologous series of substances have greater stimulatory effects. The maximum number of carbon atoms (20) is determined by other functional constraints. Locatability is an important factor in determining the features of signals. Many chemical signals must become airborne in order for their sources to be located. The volatility of chemical substances decreases with increasing molecular weight. Also, large molecules would require more energy for synthesis and more space for storage. Available facts substantiate these expectations.

As Wilson and Bossert further predicted, the molecular size of sexual attractants tends to be larger than that of other chemical signals, since more complex signals are needed to code species distinctiveness. This rule seems to apply to acoustic signals; mating and territorial calls of birds, frogs, and insects tend to be longer and more elaborate than other sounds such as alarm calls. If mixtures and combinations of different substances can be used in coding, elaborate systems of signals can be developed. Some animals seem to use such methods. Wilson points out that the use of these methods should be restricted to communication over short distances, since the differential rates of diffusion alter the composition of a signal over longer intervals of time.

Since the transmission of sound waves is also subject to physical constraints, some logical predictions can be made as to the design features of acoustic signals. The intensity of sound decreases as it radiates from its source; it attenuates by 6 db for each doubling of distance. Sounds of all frequencies are subject to this rule. It is obvious that animals should produce louder signals to communicate over longer distances. A more significant factor which affects the evolution of acoustic signals is frequency-dependent losses of sound energy. The viscosity of the air causes a greater loss of energy for higher frequencies than for lower ones. Other factors being equal, long-range communication should use low-frequency signals.

Sound can be bounced back by objects in its path, and the effects of reflection become greater as the diameters of the objects become larger than the wave length of the sound. For example, tree trunks can impede the transmission of high-frequency sound. Therefore, animal's living in a dense forest are expected to use acoustic signals of long wave lengths or low frequencies. Recent work by E. S. Morton seems to support this prediction. He measured the rates of sound propagation in dense tropical forests and open grassland in Panama in an attempt to correlate the acoustic properties of different habitats and bird sounds. In the lower strata of the forest habitat, sound attenuation increases rapidly as frequency exceeds 2500 Hz. The songs of birds living in lower levels of the forest tend to be low-pitched. The attenuation of higher frequencies is even more pronounced in grassland than in forest. This is perhaps due to a number of factors such as absorption by vegetation, temperature gradients, and wind. However, the songs of grassland birds tend to contain higher dominant frequencies. Thus, the simple prediction made above does not hold here.

Morton explains this discrepancy by assuming that grassland species

In darkness and densely covered habitat, chemical and acoustic signals are better suited in communication than are visual signals.

Fig. 18-2. A wild rabbit makes the boundaries of its territory with odor substances secreted by its sub-maxillary glands.

Fig. 18-3. In its rather large odor glands (D), the female silk moth produces odorous substances which attract males. This chemical signal cannot be smelled by humans.

use the temporal pattern of the songs, which is not directly subject to the rule of frequency-dependent attenuation. As already pointed out, the distance between the signaler and the receiver is an important factor in the evolution of sound signals. The differences in bird sounds described by Morton may be partly related to variation in signaling distances among different species. If grassland species communicate over shorter distances than forest species, the former can afford to use higher frequencies than the latter. Whenever permissible, higher frequencies are favored because they are easier to locate than low frequencies.

Animals which are capable of discriminating sound frequencies can use combinations of frequencies to convey different messages. The mating calls of frogs are species-specific and their frequency spectra are an essential or important cue for species recognition. Here, too, the signaling distance and the acoustic properties of its habitat are important factors affecting the mode of signaling.

If different frequency components of a signal attenuate at different rates, the receiver at a long distance registers a frequency spectrum different from that produced by the sender. Therefore, this method of coding is expected to be employed more frequently among animals which communicate over short ranges, such as frogs which gather for breeding. Acoustic signals with wide frequency spectra are often found among colonial birds, such as weavers and some of the estrildid finches. According to Morton, the songs of birds living in the lower levels of tropical forests are not only low pitched; they are generally also pure in tone, in contrast to the wide spectra and modulated sounds of grasslands birds. A similar situation seems to apply to primates. The vocalizations of rhesus monkeys, baboons, gorillas, and chimpanzees, which form relatively compact social groups, tend to contain wide spectra. On the other hand, pure tonelike sounds are produced by some cercopithecan and colobus monkeys, which live in forests and move in rather loose groups.

The design features of signals may be influenced not only by the signal environment but also by other ecological relationships. Certain forms of signals may be most advantageous for species recognition, but they may also alert predators. Various methods are used to resolve the conflict between the need for species recognition and vulnerability to predation.

Many songbirds, some gallinaceous birds, and mammals such as ground squirrels produce a long, pure tonelike sound toward a bird of prey flying overhead. The same animals utter a brief impulsive sound in the presence of a terrestrial enemy. Although these calls are not species-specific, this example demonstrates how sounds used by prey are designed with consideration of the strategies of different predators. Using human psychophysical evidence, P. Marler reasoned that the physical structure of the aerial alarm call makes it difficult to localize, whereas the anti-terrestrial enemy call contains cues suitable for localization. This was the first attempt to discover adaptive significance in the structure of animal signals. It is to

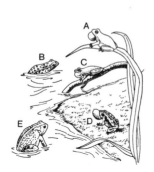

Fig. 18-4. The courtship calls produced by male frogs in the evening and during the night are unique to each species. Hence, females can find conspecific males even when several species live in one pond. A. Squirrel tree frog; B. Leopard frog; C. Tree frog; D. "Eichen Kröte" (toad); E. American toad.

the advantage of prey to know where a terrestial enemy is located before deciding to flee, whereas the best defense against a swift aerial predator is to take cover without delay.

The discoverer of the enemy must use different strategies in the two different situations. It can tell other members of the species the location of a slow-moving terrestrial enemy without being caught by the predator. For this purpose, it must use a signal which is easy for other members to localize. The discoverer of a swift aerial enemy must protect its own life while informing others of the danger. This can be accomplished by using a signal which is hard for the predator to localize. The nonlocalizable property of the signal does not bother other members, since for them taking cover at once is more important than knowing where the enemy is.

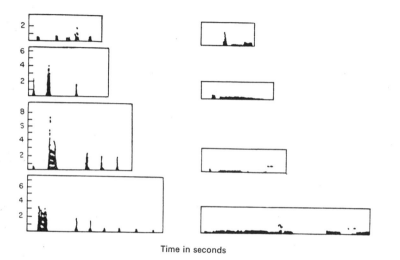

Time in seconds

Fig. 18-5. Many prey animals give specific warning calls when specific predators are detected. Here, sound spectrograms of alarm calls given by males of the domestic chicken are represented. The ordinate represents the frequency of the calls in 1000 cps, the abscissa represents time in seconds. Calls depicted in the left side of the figure are made, when a terrestrial predator approaches. Calls depicted on the right are made if predator comes from above, for example, a hawk. Both sets of calls are louder whenever the releasing stimulus is stronger (nearer). Hence, other animals not only obtain information about the kind of predator, but also how close it is.

There is a general trend that signals for species recognition are conspicuous. They can make their carriers vulnerable to predation. The simple fact that conspicuous signals evolved despite predation must be explained. There are several possible explanations which are not mutually exclusive. Conspicuous signals of prey may not be conspicuous for predators, especially at the distance where predators first defect prey. When prey and predators belong to widely different groups, prey often use sensory cues which are completely undetectable to their predators. As recent work by T. Eisner and his co-workers suggests, butterflies may use ultraviolet markings on their wings to recognize species. They are invisible to vertebrate predators. Also, the activity rhythm of prey may be such that predators do not encounter prey in the act of exchanging conspicuous signals.

One of the interesting solutions of the conflict is sexual dimorphism in which the male has conspicuous characters, while the female is cryptic. Incubating, brooding, and nursing females are more vulnerable to predation than nonparticipating males. Also, lost eggs are harder to replace than sperm. Therefore, natural selection favors female crypticity which guards against predation, and male conspicuousness which facilitates species recognition. In this connection, mimetic butterflies adopt ingenious solutions; mimicry is restricted to the female among butterflies in which the visual characteristics of the male are used in species recognition, whereas mimicry occurs in both sexes among those species which use olfactory cues in recognizing the mate.

In summary, the design features of animal signals are determined by a number of factors such as the nature of their evolutionary sources, functional properties of sense organs, the need for reproductive isolation, individual recognition, physical laws regulating the transmission of signal energy, physical properties of the habitat in which communication occurs, and prey-predator relationships.

19 Electrical Stimulation of the Brain— A Method in Neuro-Ethology

The workings of the brain, and especially its function in controlling behavior, has long fascinated scientists. Early information on these subjects was obtained with a very crude method, the destruction of entire brain portions by means of ablation. If we surgically remove a part of the brain and then find definite disturbances in behavior, we may well conclude that the ablated portion is at least partly responsible for controlling this behavior. For example, G. Schneider removed the superior colliculi from the tectum (the dorsal surface of the midbrain) of hamsters, and found that the animals could still perceive shapes but were unable to localize these figures visually in relation to other objects. This means that the superior colliculi of the midbrain serve to create a map of the visual field, which is necessary to create a map of the visual field, which is necessary for visually relating objects and forms to each other in space. But the actual discrimination of perceived forms depends on the visual cortex, located at the posterior end of the forebrain. When this region is destroyed, the animal can no longer perform discriminative responses.

The ablation method provides us with very limited information about how such brain or nervous structures actually work. A better means of studying these functions is to record the activity of single nerve units, or neurons. Nerve cells transmit and evaluate information by means of electrochemical processes. This makes it technically possible to record the electrical potential (Fig. 3-5) of active neurons with the aid of very thin, fine-tipped wires (electrodes), and to amplify this activity to the point where we can detect and measure it. We may then examine the function and "competence" of certain brain cells. For instance, experimenters were able to show that certain neurons in the diencephalon of cats respond to stimuli normally triggering attack behavior—such as meeting a rival tomcat or being pursued by a dog—with an increase in the rate of impulses, and that this rise in activity occurs regularly before the animal actually attacks. These nerve cells obviously respond to stimuli indicating danger, and at the same time probably release the entire sequence of muscle actions that we observe as attack behavior.

Electrical stimulation of the brain—a method in neuroethology, by J. D. Delius

Fig. 19-1. This rhesus monkey was required to choose between objects of various shapes, and was rewarded for taking the rectangular one. Monkeys with their visual cortex removed could no longer perform these tests.

Fig. 19-2. Recording of electron action potentials (lower trace), taken with a very thin electrode from a nerve cell in the basal forebrain of a rhesus monkey. This monkey was trained to pull and push a lever in order to get food. (The upper trace shows movement of the lever.) Note that the nerve cell fires before the animal actually moves its arm; this cell could therefore be a source for the neural command signals controlling the movement.

The overwhelming number of nerve cells found in the brain of vertebrates makes it impossible to study single-cell activity in these animals

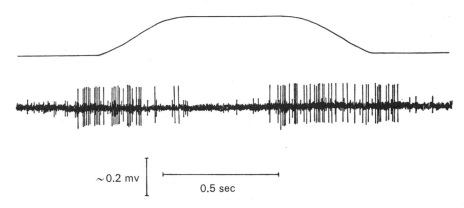

~0.2 mv

0.5 sec

with the recording method described above. Instead, we are usually forced to record the simultaneous activity of many such brain cells by using large electrodes, such as with electroencephalographs. By attaching electrodes to the surface of the head, we can record the electric currents produced by thousands of neurons in the cortex of the forebrain, detecting these currents through the skull. Though imprecise, this method is popular in clinical practice, which demonstrates its usefulness in helping to discover the physiological processes underlying behavior. For example, there are changes in the electroencephalogram (EEG) during dreaming, where brain waves are similar in appearance to those recorded while the subject is awake. This indicates that at least the cortex of the forebrain works in a similar way while the subject is awake and while he is dreaming.

In addition to measuring brain-cell activity normally displayed by animals, we are now able to influence this activity artifically with electrical and chemical stimulation, that is, to increase or inhibit the production of impulses in the brain. Ethologists have little to gain from studies in which single cells in vertebrate brains are stimulated but the stimulation of many neurons can induce an animal to perform more or less complete behavior and movement patterns.

From early in the 19th Century, scientists had known that nerve tissues respond to electrical stimulation; between 1870 and 1900 this method came into use for the purpose of artificially evoking behavior. The initial attempts—at first on strictly restrained animals, and later on subjects which were allowed to move about relatively freely—were indiscriminate and without systematic procedure. A real breakthrough did not come until around 1920 with the work of W. R. Hess in Switzerland. Hess used brain stimulation techniques on cats in their waking state who had almost no restrictions in movement. His research was honored in 1945 with the Nobel Prize for medicine, and his methods have been used in many laboratories.

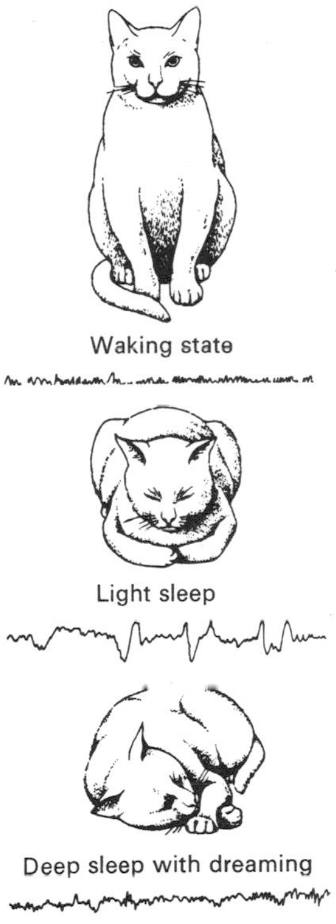

Waking state

Light sleep

Deep sleep with dreaming

Fig. 19-3. Electroencephalograms (EEG) of a cat. The EEG of deep sleep resembles that of the waking state (paradoxical sleep). We know from similar experiments with humans that dreams occur during paradoxical sleep.

We can now take a look at how electrical-stimulation experiments are done. The best subjects are animals whose behavior has already been studied and described in detail, so that behavior patterns elicited by brain stimulation can be compared with those occurring in the organism's natural environment, and their meaning better interpreted. Furthermore, it is important to know a great deal about the subject's nervous system. The best source for this information is a stereotaxic atlas, which is a precise map of individual areas in the brain. Such an atlas contains carefully

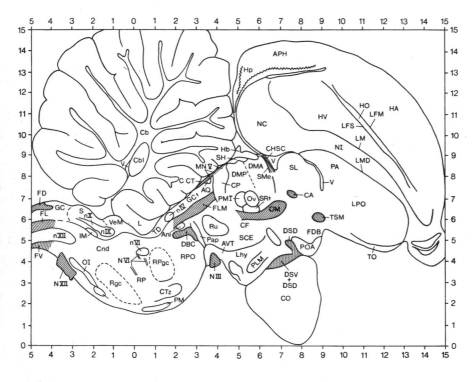

Fig. 19-4. Drawing from a stereotaxic atlas, representing a section of the brain.

labeled photographs or drawings of stained sections of the brain, localized within a three-dimensional set of coordinates whose axes and points of origin lie in relation to certain fixed measuring points on the skull. Electrodes consist of thin, straight wires that are electrically insulated with enamel except for a tiny area at the tip. By using an instrument that enables us to aim the tip of an electrode at precisely defined areas of the brain, we can insert the electrode through a small hole drilled into the skull. The degree of error in reaching our intended structure in the brain is small; it derives from individual differences in shape and size of the skull. These electrodes are then attached to the skull surface and combined with a tiny electrical plug that is also glued down. In addition, we attach a large surface electrode to the skull underneath the scalp.

Once the animal has recovered from the operation, we connect it with thin wires to an electric impulse generator, an instrument for producing electrical stimuli. This allows the subject a relatively large degree of mo-

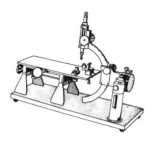

Fig. 19-5. Stereotaxic machine for inserting electrodes into the brain. The head of the anesthesized animal is held in position by the ear bars.

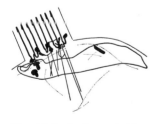

Fig. 19-6. A pigeon with electrodes attached to the skull. The conducting wires are connected to a miniature socket tightly cemented to the skull. Below: Drawing of an X-ray photograph (exograph) shows the seven implanted brain electrodes.

bility. The latest technique is to replace this machine with a miniature stimulation device attached to the animal's head and controlled with a wireless apparatus, allowing complete freedom of movement. In an experiment, we conduct an alternating current lasting from about five seconds to about two minutes between two depth electrodes whose tips are lying close together. Voltage is increased gradually until we observe a behavior elicited by the stimulation. Only rarely do we need to administer more than 0.2–0.3 milliamperes, and lower values are usually enough. The common practice is to test each electrode for many trials over a fairly long period. Animals with implanted electrodes live as long as intact animals and—except when they are being electrically stimulated—show completely normal behavior. They do not suffer any pain because the brain itself has no pain receptors and the surgical wounds in the skull are small and heal quickly.

Stimulation by means of electrodes almost always evokes some kind of behavior. Usually the behavior is stable; that is, a certain electrode more or less triggers the same behavior in repeated trials. For instance, when experimenters provided electrical stimulation with an electrode whose tip lay in a certain area of the motor cortex of a cat, the animal raised its leg, and over more than a thousand trials this was a very reliable response. If, on the other hand, we stimulate another area called the medial thalamus (a section of the diencephalon), the same cat will become less aroused, begin to doze, look around for a sleeping place, lie down, curl up, and fall asleep. Stimulation of another point in the midbrain causes only a contraction of the pupils. We also seem to encounter areas in the brain where stimulation does not lead to any observable behavior. But a closer examination has shown that many of these "silent" electrodes do have some effect, if only to raise the heart rate or to increase the production of saliva.

We know from brain-stimulation experiments that had to be performed on people undergoing neurosurgery that subjective experiences can also be elicited. One patient, whose oculomotor nucleus in the diencephalon was stimulated, told the observer that he saw a thick red stripe pattern on a green background directly in front of him; but he did not show any obvious changes in behavior. In other cases, however, such "hidden" reactions can indeed be associated with observable behavior. Another patient who was stimulated at a certain point in the forebrain made violent movements with his right arm; when questioned, he said he was trying to catch a butterfly that was fluttering about in front of him.

Our next question is: How far can we consider behavior patterns induced by artificial brain stimulation to be normal or natural? This is difficult to answer, but we can obtain some indication by testing whether conspecifics, and expecially members of a subject's group, respond appropriately to this kind of behavior. In order to examine the issue, D. Ploog and his associates carried out some experiments with squirrel monkeys. Using a wireless method to stimulate the hypothalamus (located in the diencephalon), for example, they elicited a certain movement sequence that they thought might function as a threat gesture. The animal that was artificially stimulated, and which had full freedom of movement, was finally avoided or occasionally threatened by others in the group. If the stimulus was repeated often enough in a low-ranking squirrel monkey, even the highest-ranking group member began to stay out of its way. It seems that the subject's social status was increased through the stimulation experiment.

In order to release a certain behavior by way of an electrode, we must always apply a certain minimum voltage. This is called the threshold stimulus current, and it varies according to the type of electrode and what kind of current is used. But even when the same stimulus current is used throughout an entire series of tests, the stimulus threshold changes from one trial to another. Frequently the minimum voltage required merely to release the behavior will rise with repeated stimulation, but then fall again after cessation. This phenomenon resembles the habituation effects (see Chapter 17) occurring when normal external stimuli are repeatedly presented, and in many cases these two effects are probably based on the same nervous processes.

The threshold stimulus voltage

Other changes in stimulus threshold are even more interesting. It can often be demonstrated that natural stimuli which release a completely different behavior will nonetheless increase the threshold current if they are presented along with the artificial brain stimulation. As an example, I found that a herring gull which had been frightened by my approaching hand required a threshold current of 80 microamperes through an electrode in order to show its characteristic threat posture, while an unmolested animal needed only 40 microamperes. This supports our assumption that nerve cell networks responsible for mutually exclusive behavior patterns will inhibit each other, ensuring that only the most strongly activated behavior is allowed to discharge. Surprisingly enough, researchers have found a similar system at work in the spinal cord, which prevents the simultaneous tension of antagonistic limb muscles.

Changes in stimulus threshold

When the stimulus current is raised above the threshold, we can usually observe an increase in whatever behavior is released. A herring gull's threat posture expressed only weakly by ruffling the feathers on its back when we administered the threshold stimulation, showed a marked change when the current strength was increased to 100 microamperes: The bird lowered its wings, pointed its bill down, and even began to attack an "invisible" rival. Ethologists have made many more observations like this one. Our example lends support to a theory which states that the organizational structure—the "intenal" program which regulates the performance of complex behavior patterns—consists of a system of thresholds where every threshold is responsible for a certain element in the behavior sequence. A central motivational mechanism has to pass each of these thresholds individually for every element, so that the unit can be integrated into the behavior pattern as a whole.

Increasing the stimulus voltage

Experiments of this kind, where the intensity of stimulus currents is gradually increased, do have their limitations because the thresholds of the nerve cells and the pathways lying a little further from the electrode tip are probably also surpassed. This may result in very different, often oppos-

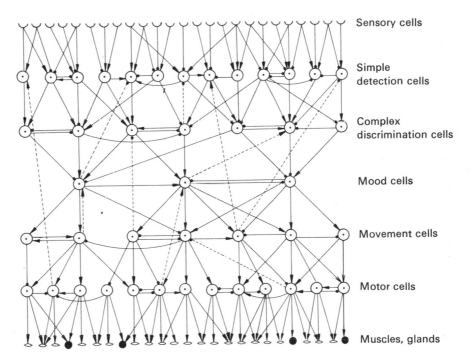

Sensory cells

Simple detection cells

Complex discrimination cells

Mood cells

Movement cells

Motor cells

Muscles, glands

Fig. 19-7. Schematic representation of a hypothetical model indicating the hierarchical structure of the nervous system. Arrows represent drive effects, the dots, inhibiting effects. The broken lines of connection interrupt the hierarchy. We must not think in terms of individual cells, however, but of a large number of cells with similar functions.

ing behavior patterns being elicited. We can illustrate such effects by using electrodes whose location deep in the brain can be shifted without having to anesthetize the animal. The fact is that stimulus points releasing the most diverse movement patterns often lie no farther apart than 0.1 mm. When I stimulated a herring gull with an electrode tip lying in the anterior part of the forebrain, the bird showed its species-specific food-searching behavior: trampling with the feet, an activity normally used to bring small shells, crabs, and worms to the surface of shallow tidal pools. But if we fed the gull before this test, it would not display food-searching behavior even with a very high stimulus strength; on the contrary, our electrical stimulation only caused it to make very fast turning motions on the spot. Feeding the gull prior to stimulation probably increased the threshold value, and this increment could not be compensated for by stronger stimulation without eliciting another, different reaction from neighboring brain tissues. By the way, it is interesting to note that very high stimulus currents induce fits that do not occur in normal animals, but which bear a high resemblance to pathological states such as epilepsy.

There is also another way in which brain stimulation at the same locus can lead to different results: E. von Holst and U. von Saint Paul found that a stimulus current of constant intensity had to be administered for a period of up to one minute before a food-searching hen could be made to sit down by means of a sleep-inducing electrode. But if this hen was already sleepy, the very same stimulus current had the same effect in a much shorter period. A similar effect is obtained when drowsiness is induced artificially through prior stimulation with the same electrode. Ethologists refer to this time lapse between stimulus and response as latency.

Fig. 19-8. By stimulation with an electrode placed into the brain stem of a cock, E. von Holst and U. von Saint Paul were able to elicit sleeping behavior. It is interesting that this sleeping reaction begins only some time after stimulation (latency). Behavior patterns assigned with a dot were triggered individually by stimulating other loci in the brain.

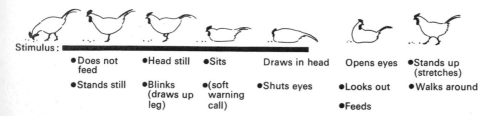

Stimulus:

●Does not feed	●Head still	●Sits	Draws in head	Opens eyes	●Stands up (stretches)
●Stands still	●Blinks (draws up leg)	●(soft warning call)	●Shuts eyes	●Looks out	●Walks around
				●Feeds	

If we know the stimulus field in the brain for a certain behavior pattern, we can find the shortest possible latency period by testing different strengths of current. This measure is particularly interesting in cases where similar behavior patterns can be elicited in very diverse areas if the brain. For stimulation in the anterior diencephalon of a pigeon, which elicited cooing associated with courtship behaviors, the shortest latency was five to ten seconds. By contrast, the same calls but without courtship behavior were released in the lateral midbrain (mesencephalon) with latencies of less than one-quarter of a second. This indicates that there is a nervous pathway leading from the diencephalon to the mesencephalon, taking a route from those areas responsible for integrating the cooing sounds into courtship behavior as a whole, to the centers inducing the animal to make the sounds, and finally to the muscles actually producing them. This notion is confirmed by experiments in which the relevant section of the midbrain was ablated: Pigeons that have undergone this operation remain silent.

We can frequently observe that a behavior pattern elicited through brain stimulation will continue to discharge even if the stimulus current is interrupted. For example, pigeons continue the cooing ceremony mentioned above, often for several minutes after stimulation has ceased. These so-called after-effects suggest that artificial stimulation of the brain acts in a way similar to natural, external releasing stimuli, and that the evoked behavior discharges as a fixed sequence.

On the other hand, a natural stimulus may have effects other than directly triggering a behavior; it can also help to influence an animal's disposition or "mood" (see Chapter 17). D. Harwood and D. Vowles examined this kind of after-effect in ring doves. Upon stimulation of various centers in their forebrain, these birds displayed either threat or fear behavior. Once the experimenters had succeeded in evoking such a behavior, they used repeated stimulation at two-minute intervals to determine the minimum voltage (threshold stimulation) needed to re-elicit the

Fig. 19-9. Diagram of the transverse section of a herring gull's brain. The dots indicate the position of electrode tips eliciting threat postures during stimulation. Loci indicated by crosses did not give rise to threat behavior.

movements. In this way they discovered that, with some loci, the threshold stimulus value steadily declined until after about ten minutes it reached its lowest point. After that, it rose again and reached its initial value about twenty minutes after the first stimulation. At other stimulus fields, however, the threshold values first increased and then decreased again, although with the same temporal progression. Harwood and Vowles were also able to show that threshold changes of this kind occurred even when the initial behavior was triggered by natural, not electrical, stimulation, which in this case was the appearance of a spider. It seems that stimulation released either fear or aggression in ring doves, and that this mood persisted for some time.

Several experiments with lesser black-backed gulls and herring gulls point to certain relationships that may explain these after-effect phenomena. I hit upon a number of stimulus points where threat behavior could be elicited in gulls; these movements consisted of bowed posture with ruffled feathers and lowered wings. After-effects lasted for up to twenty minutes, but here they were not so much a matter of threshold changes as of inhibiting the fleeing responses usually displayed by our gulls in the experimental cage. When we examined sections of the brain under a microscope in order to find where these stimulus points were located, we discovered them in a very limited part of the forebrain, close to the ventricle which is filled with a slowly-circulating cerebrospinal fluid. Interestingly, the ventricular wall is thicker at this location than anywhere else in the brain, because here it contains glandlike cells that probably secrete their substances into the fluid. It is therefore quite possible that the after-effects of stimulation, creating what seems to be a particular mood, are the result of secretions into the inner cerebral ventricle. These substances are specific hormones called liquormones. At other points they are absorbed by receptor nerve cells into the ventricular wall, where they cause a change in receptor activity and so facilitate whatever behavior is controlled by these cells.

This hypothesis is well supported by other experimental results. For example, cerebrospinal fluid was taken from sleeping goats and injected into the brain ventricle of goats in their waking state, who then became drowsy—no doubt as a result of a liquormone inducing sleep.

Regular hormones that circulate in the blood stream can also influence activity in the brain, affecting behavior in a certain way (see Chapter 21). These effects are also accessible to experimentation, where specific hormones are injected into certain parts of the brain with small canules. When in heat, spayed cats no longer invite a tomcat to mate in their species-typical way. But R. P. Mitchell was able to reproduce this soliciting behavior in spayed cats by injecting them with minute amounts of estrogen (the female hormone normally secreted by the ovaries) into the anterior hypothalamus, a part of the diencephalon. The same amounts of estrogen had no effect in other areas of the brain. Michael even demonstrated that radioactive estrogen injected into the blood stream rather than directly into the brain was absorbed only by this reactive area in the diencephalon. This shows that some hormones can activate certain nerve cells in such a way as to release the behavior controlled by these cells.

The stimulation of some portions of the brain triggers complex behavior patterns that consist of an ordered series of simple movement elements. Other brain areas produce only parts of sequences, while still others release no more than single elements normally belonging to a pattern. B. Åkerman studied the influence of electrical brain stimulation on aggressive behavior in pigeons. When certain loci in the brain were stimulated, a pigeon took on a motionless and alert posture; then it ruffled the feathers of its neck and body, lowered its tail, crouched, beat its outstretched wings at a real or "imagined" enemy, and emitted a short, deep sound ("wao" call) that sounded like suppressed coughing. But when he stimulated other points, he could only elicit the first part of this behavior sequence—up to the crouching with lowered tail. In other sections, only single elements of this pattern can be released, such as lowering the tail or beating with the wings. Amazingly enough, some of these movements become embedded into other, nonaggressive behavior sequences, for example, ruffling the feathers and lowering the tail, both of which also belong to the pigeon's courtship behavior.

Very early in the history of ethology, scientists recognized that the neurophysiological mechanisms controlling behavior are organized in a hierarchical system. At the highest level are the motivational centers, which determine whole groups of functionally related activities (behavior systems), including sexual behavior, hunger, and so on. Lower down in the hierarchy are systems responsible for smaller and more restricted behavior elements, such as locomotion, body posture, facial expression, heart beat, etc. On the next level down are centers controlling even smaller units of behavior, such as leg movements, extending the neck, piloerection (hair-raising), etc. Subordinate to all the other levels are mechanisms which only control individual muscles. Different leg movements sometimes use the same individual muscles; the same leg movements in turn may be involved in different forms of locomotion, and the same locomotory activity can be associated with different states of motivation (for example, certain congruencies between courtship and fighting actions). For that reason we can understand how the control of lower centers by higher ones sometimes leads to a variety of overlaps.

It seems that physiological studies, such as those carried out by F. de Molina and R. Hunsperger, have also provided evidence of this kind of hierarchical structure. These two researchers have made a map of all the points in the brain where electrode stimulation releases threat and fleeing responses in cats. These points are located in a number of brain structures that anatomists know are connected in a descending hierarchy. They are the amygdala in the forebrain, the medial hypothalamus in the dience-

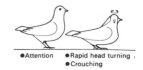
●Attention   ●Rapid head turning
             ●Crouching

●Erection of the   ●Ruffling the feathers
 head and body   ・Walking
●Attention

●Erection of the   ●Ruffling the feathers
 head and body   ●Lowering the tail
●Attention

Fig. 19-10. Three behavior sequences (above, fleeing; middle, courtship; below, threat) that B. Akerman was able to release by stimulating different fields in a pigeon's brain. Some behavior elements (dots)

Stimulation experiments with cats

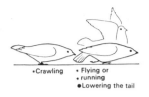

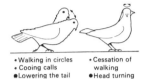

• Crawling   • Flying or
            • running
            ● Lowering the tail

• Walking in circles   • Cessation of
• Cooing calls   walking
● Lowering the tail   ● Head turning

● Crouching   Wing-beating
• Wing erection   "Coughing"
• "Growling call"

occur in two or more sequences. Other elements (asterisks) can be triggered by stimulating various loci in the brain; these elements will then occur singly or in association with other behavior patterns.

Simultaneous stimulation of two areas in the brain

phalon and the central grey in the mesencephalon. The amygdalal influences the diencephalon by way of a nerve pathway known as the Stria terminalis. In turn, the central grey is influenced by a complexity of nerve pathways originating in the diencephalon. When we interrupt this descending system at various points by some physical means, any electrical stimulation of structures lying above the point of interference will no longer be effective; that is, it will not elicit threat or fleeing behavior. By contrast, stimulation of structures lying below that point will continue to have the same effect as before.

Studies by W. Roberts and R. Carey have shown that electrical stimulation of the brain can have effects other than just triggering movement sequences. Using cats as subjects, they discovered that stimulation of the field in the diencephalon which controls prey-catching behavior would not only elicit the movement pattern for killing rats, but would also influence a number of other behaviors. For instance, these cats learned to respond to stimulation by running through a maze just to have the opportunity of attacking a rat at the other end. Without stimulation, they showed no interest in the rats and did not learn to run the maze. However, this result was not based on any feeling of hunger. The animals had plenty of cat food, but as soon as they were stimulated they stopped eating and pounced on the rats. Most of the time they did not even eat them. This indicates that stimulation triggers a "predation drive" manifested in the way these cats perceive rats: From the cats' point of view, the rats change from neutral objects (i.e., they are ignored) to a reinforing object with definite reward characteristics.

But the results of brain-stimulation experiments are not always in agreement with the hypothesis of a hierarchical organization. For example, J. Flynn and his co-workers stimulated a certain locus in the cat's diencephalon, releasing a pre-catching behavior where the cat lies in wait for a rat, seizes it and then kills it. One element in this chain is biting at the rat. As it happens, however, just this element fails to be released when a lower movement system is activated. Instead, Flynn was able to show that brain stimulation only increased sensitivity in the skin around the cat's muzzle, and that the biting reflex was not elicited until the mouth made direct contact with its prey. In this case, then, stimulation influences the degree of tactile sensitivity, resulting only indirectly in a behavior element.

In other experiments, it was demonstrated that other functional relationships could also disrupt the hierarchical organization of the nervous system. For example, when two separate areas of the brain responsible for different behavior patterns are stimulated at the same time, we can observe a whole series of different behavioral phenomana. In the simplest case, the elicited pattern appears to be a compromise between these two behaviors. Then the two are either merely superimposed, as when food-pecking and head-turning in chickens are triggered simultaneously, or they blend with each other, as with a simultaneous activation of random looking around and watchful staring. These observations were made by von Holst and von Saint Paul.

More frequently, however, one response was found to depress the other, preventing it from occurring. Von Holst and von Saint Paul described this phenomenon for the activities incubating-fleeing and fleeing-eating, where simultaneous stimulation leads to either one or the other pattern, but never both at the same time. As we have mentioned, observations like these indicate that neural pathways responsible for producing these behavior patterns are cross-connected by inhibiting pathways, so that only the system most strongly activated can manifest its behavior.

Occasionally it happens that, when two stimulus fields responsible for different behaviors are activated simultaneously, the animal displays a completely new pattern. For example, J. Brown and his associates described two loci in a cat's brain that always responded to individual stimulation with threat behavior, inducing the animal to growl and flee. Yet simultaneous stimulation of these two loci produced only attack behavior, with hissing and hitting with the paws but no threat or fleeing movements. This suggests that threat and fleeing behavior mutually inhibit each other, but that both together facilitate those neural mechanisms responsible for attack behavior with hissing. In this way we can actually produce behavior in brain-stimulation experiments that closely resembles the conflict behavior occurring in other contexts (see Chapter 25). The studies we have described generally show that these cross-connections represent an important part in the hierarchical organization of behavior.

Naturally, there are methodological problems we may have to contend with when studying these phenomena. For example, in stimulating certain areas in the rat's diencephalon with electrodes, we can elicit both eating and drinking, though not at the same time. On the one hand, we could speculate that the two behavior patterns are controlled by the same network of neural connections, but at different times; on the other hand, it was also possible that the electrically stimulated points actually belong to two independent neural circuits that just happen to lie adjacent to each other, so that our stimulus current activates both in the same way. But then S. Grossman discovered a method for stimulating these two systems separately. When he used a canule to inject noradrenaline, a transmitter substance, into the appropriate section of the hypothalamus, his animal subject engaged only in eating. Conversely, the introduction of acetylcholine (another transmitter substance) through the same canule led only to drinking behavior.

In order to learn, the nervous system has to be flexible (adaptable). We would therefore expect to find such plasticity in the results of brain-stimulation experimented as well. E. Valenstein and his co-workers were in fact able to offer some evidence for this. With electrodes implanted in the diencephalon of rats, they evoked a number of different behavior patterns, with each electrode releasing only one. Admittedly, some of these

Fig. 19-11. By stimulating a certain locus in the diencephalon of a cat, J. Flynn elicited prey-catching behavior.

Conflict behavior

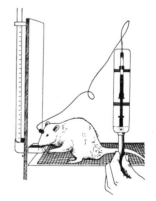

Fig. 19-12. Chemical stimulation of the brain. A minute amount of acetylcholine (one of the neural transmitter substances) was injected into a certain area of the rat's hypothalamus by means of a canule. As a result, the animal started to drink.

reactions were largely determined by certain objects present in the cage (for example, water when drinking behavior was released, wooden blocks for gnawing, etc.) Furthermore, when rats were repeatedly induced to perform gnawing movements without giving them something to gnaw on, they went through hundreds of trials and then switched to another behavior—one they could do in the cage, such as drinking when water was available.

Surprisingly, once a rat had changed to another response, it maintained this new behavior even when presented with objects (blocks) appropriate to the initially evoked gnawing reaction. This suggests that here the electrically stimulated neural structures had "learned" to perform another task in lieu of the original function. This would mean, though, that they must have somehow changed their connections to the lower nervous centers.

Drawing inferences for the capacity to learn

These experiments make us wonder whether the same can happen with behavior patterns released by natural stimuli. If so, our attempts to associate a certain activity with a particular nervous structure would of course be meaningless. There are in fact a number of cases where electrode tips lay at virtually the same points in different individuals, yet released totally different behavior patterns.

Experiments in electrical brain stimulation can provide us with even more direct data on learning. As far back as the 1930s scientists demonstrated that the stimulation of certain areas in the brain could act as a conditioned stimulus in Pavlovian terms (see Chapters 1 and 22), perhaps replacing the sound of a bell to which a dog secretes saliva because the animal had learned to associate this sound with the presentation of food. It is not surprising that electrical stimulation can produce such effects, if we take as an example a certain activation of the auditory cortex in the diencephalon which probably produces sensory hallucinations, causing the animal to respond as to normal acoustic stimuli.

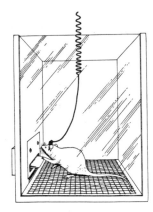

Fig. 19-13. A rat in a self-stimulation experiment. Every time the animal pressed the lever, it received a brief electric shock through the implanted electrode. This kind of reward may at times be so effective that some rats (and other animals) operate the lever as fast as possible, driving themselves to complete exhaustion.

In 1950, J. Olds and P. Milner discovered that brain stimulation could also be used as a "reward," or reinforcement, for teaching animals a certain behavior such as pressing a lever, where they would previously have received food. The two researchers found that animals learned many tasks, including pressing a lever or running through a maze, even when their reward was no more than a brief stimulus current through a particular electrode. Later studies showed that only certain stimulus points in narrowly defined areas of the brain could produce this effect, but that some of these—such as the cerebral cortex, the diencephalon, and the tectum—showed the amazingly strong results. To obtain an electrical "reward" through electrodes implanted in these areas, rats have been known to operate a lever continuously for many hours with a rate of up to 5000 lever-presses per hour, sometimes driving themselves to total exhaustion.

At first glance we are very much surprised that electrical stimulation alone, without the motivation of some ostensible drive, can actually serve as a reward. In a normal experimental situation a monkey (or some other animal) will learn to press a lever in order to obtain peanuts—but only if he is at least moderately hungry. Yet if he is rewarded by having these brain areas stimulated, the monkey will perform the same task even when his drives are fully satisfied. All kinds of theories were created to explain these results; only recently have we come closer to solving this apparent contradiction, through the discovery that (contrary to previous opinions) a number of normal extrinsic stimuli could act as rewards without the animal showing any apparent drive. For instance, a monkey learns to press a lever even when his reward is solely the opportunity to catch a glimpse of a running electric model train.

We also know of certain brain areas in humans where stimulation is experienced as rewarding. During clinical tests like those preceding neuro-surgical operations, C. Sem-Jacobson found that patients sometimes experienced an electric current running through an electrode as pleasant, and—when permitted to do so—continuously operated the stimulus switch. Some of these patients reported sexual arousal or other pleasant feelings, or frequently just a general feeling of well-being which resulted from electrical stimulation. Unfortunately, our hopes of treating depression by administering this kind of brain stimulation remain futile, because certain undesirable side effects (habituation and addiction) cannot be eliminated.

Brain stimulation applied to humans

In summary, we believe that notions of some day using electrical brain stimulation to control man's behavior for good or bad are utopian. This method is far too crude for actually manipulating the infinitely complex and highly developed computer that is the human brain.

20   Neuro-Ethological Research in Invertebrates

Neuro-ethological
research in inverte-
brates, by
M. Dambach

The field of neuro-ethology studies the neural basis of behavior patterns. In multicellular animals, behavior patterns are organized within the nervous system. A neuro-ethological analysis therefore begins with the behavior of an intact animal, then relates such behavior to the elements of the nervous system—the nerve cells and their interrelationships. The nervous systems of many invertebrate animals have a relatively simple structure, and are thus more accessible to experimental scrutiny. The history of neurophysiology also demonstrates that many vitally important discoveries were made on specimens (preparations) of invertebrates: The ionic theory of resting and action potentials, for example, was tested and expanded by means of studies on the giant nerve fibers of cuttlefish (Cephalopoda; squids of the genus *Loligo*) (see Chapter 3). Important discoveries concerning the function of sensory cells arose from studies on extensor receptors in crabs (crustaceans), and on the eye of the horseshoe crab (*Limulus*), to name only a few examples.

Comparison with vertebrates

As more studies were made in the course of time, it became more obvious that the sensory and nerve cells of invertebrates had very similar characteristics to those found in vertebrate animals, although certain differences do appear in their fine structure, their dimensions, and their biochemical features. The process of transforming sensory stimuli into nervous excitation, whereby messages are encoded and conducted in the form of nervous impulse sequences, and the transmission of this excitation to other nerve cells (neurons) and to muscle cells is, however, basically similar. We now consider it a fact that the behavior patterns of multicellular animals and of man are ultimately determined by the activities of nerve cells. Transmission, processing, and storing of information (memory) are inextricably linked to the functions of nerve cells and their connections (synapses). The influence of hormones on the transfer of information—primarily important in the control of long-term phenomena—will be discussed more extensively in the following chapter.

If, in principle, the nerve cells of all the various animals and those of

Fig. 20-1. Neural network of the hydra.

man function according to the same elementary laws, we may well ask what causes the differences in diversity and development of behavior patterns between simple invertebrates, on the one hand, and humans with their capacity to think and speak, on the other? The nervous systems of snails, crabs, and insects contain approximately 10,000 to 1 million neurons, that of the human brain, about 10 billion. It is these enormously large numbers of neurons and their resulting, hardly imaginable possibilities for interconnection which provide the basis for the higher nervous skills. On the whole, lower animals will have the same problems as vertebrates in their fight for survival. They, too, must be able to perceive external stimuli, obtain food, escape from danger, and reproduce. In this way their achievements may be compared to those of vertebrates, though carried out with the greatest possible "economy": No more than eighty motor neurons help to propel the flight of the desert locust, and only four auditory sensory cells enable the moth to flee from a hunting bat (see Chapter 8). We can see, then, that invertebrate animals, with their simple nervous systems and in part rather large neurons, are particularly well suited for the study of neuronal mechanisms involved in the processing and storing of information at the level of single cells.

Instead of a general survey, we shall describe individual examples from four animal groups which have been studied particularly closely (coelenterates, annelids, insects, and mollusks) to illustrate the neural analysis of behavior patterns. In this way we shall try to convey an understanding of the field of neuro-ethology.

After the sponges (Porifera), which do not have either sensory or nerve cells, the coelenterates, (Coelenterata) represent the next highest level of organization in the animal kingdom. There are two kinds, the octopus and the jellyfish. Coelenterates respond to stimuli in a variety of ways. They overwhelm their prey and move about using different methods: by creeping, "turning somersaults," or—as in the jellyfish—by rhythmic contractions of the bell. In order to comprehend the outstanding features of such behavior patterns, we must know a few things about the structure of the nervous system. The first discoveries were made in the 19th Century by E. A. Schäfer, the brothers O. and R. Hertwig, and other scientists. The detection of particular nerve cells was accomplished by a method still used in some cases: various staining techniques found to apply specifically to nervous tissues in higher animals. The nerve cells lie at the base of the outer layer of epidermal cells (ectoderm), and are interconnected by a wide-meshed neural network. In the freshwater hydra, this network of nerves is evenly arranged over the entire body, while in the medusa it is a little more dense along the margin of the bell, creating a band where the nerve cells are mainly oriented in a longitudinal direction. Similarly, such specific bands proceed from the marginal sensory bodies in a radial direction.

Assuming that nerve cells function in the transfer of information, we

▷
Aside from inducing the growth of the "courtship dress," hormones also control aggressive and sexual behavior patterns. Above: A cock marks his territory by crowing. Below: Turkey performing a typical courtship movement.

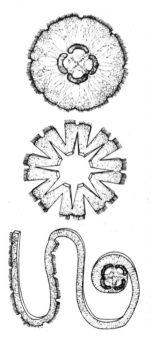

Fig. 20-2. Dissection experiment on the bell of a jellyfish (*Aurelia*), according to J. Romanes. A wave of contractions, released by stimulation, travels via nerve fibers along the sectioned bands. This is possible only when the nervous system has a network structure.

◁

The male stickleback's nest-building behavior and zig-zag dance depend for the most part on the influence of hormones. Above: The male brings nesting material and tamps it down with his nose. Center and bottom left: The zig-zag dance, which is the male's courtship behavior, is sometimes oriented toward the female and at other times toward the nest. Bottom right: Three females and a male at the nest.

may predict that in a neural network, stimulation can be conducted from every point in the body to every other point. This prediction was substantiated by the English scientist J. J. Romanes in 1874 in the course of a classical experiment. Out of the bell of a jellyfish (*Aurelia aurita*), using a very careful sectioning technique, he created two band-shaped structures: One was long and slender with two ends, the other zig-zag shaped and closed within itself. If we stimulate these bands at one point, all the muscles contract in a wave that runs from one end to the other or along the zig-zag band. As we have seen in other experiments, the excitation does spread over the nerve cells, and not over the muscle cells as in a relay race of single contractions. A necessary condition for universal stimulus diffusion in the neural network is that the points of contact (synapses) between nerve cells allow impulses to pass in both directions (Fig. 20-2).

Although the structure of a nervous system may indicate a great deal about its capacity, certain phenomena have served to demonstrate that nerve cells and nervous systems do not function unvaryingly. Instead, they display a plastic array of actions and reactions. Intense contact will cause a sea anemone (Actinaria) to contract, a process which can be recorded with simple instruments on a drum. A more precise analysis becomes possible, however, if we stimulate the water lily electrically rather than mechanically, for better control of stimulus intensity and duration. Experiments like this were carried out on hermit crab anemones (*Calliactis*) and other anemones of the genus *Metridium*. It was found that the strength of a stimulus—provided that the stimulus lay above a certain minimum value (threshold intensity)—proved far less important for the release of a contraction than was the number of stimuli and the time intervals between them. More precisely: A sea anemone does not respond to a single stimulus, regardless of intensity; it will contract only when a second one is applied. The reaction then becomes stronger the more quickly the second stimulus follows the first. We could say that the first stimulus "prepares the way" for impulses released by a second, or by other stimuli which may follow. This process is referred to as facilitation.

If we extend the time interval between the facilitating stimulus and the next one, the sea anemone's contractions will grow weaker and then stop. Thus, after a certain time the nervous system will have "forgotten" the facilitating stimulus, and the second stimulus once again meets the sea anemone "unprepared." In the case of natural, mechanical stimulation, a single contact may release a contraction, probably because such a stimulus releases not one nerve impulse, as in the case of electrical stimulation, but whole volleys of discharges. Out of the numerous impulses in a volley, the first ones will facilitate and the following ones will then induce a contraction.

New ideas about the functional organization of the nervous system have emerged from studies on polyp colonies (polyparies). There are colonies in which individual polyps form a neural association with one another, for example in the case of the genera *Cordylophora* and *Pennaria*. If a polyp stalk is stimulated mechanically or electrically, the first to contract are those polyps which lie closest to the area of stimulation, followed by ever-more-distant ones. The wave of contractions spreads out over the colony with a speed of 1-3.5 cm per second. But the notable factor is that reactions of the polyps become weaker with increasing distance from the point of stimulation. The most distant ones remain unaffected. This decrease in neural excitation with increasing distance has been referred to as decrement. R. K. Josephson succeeded in shunting off electrical impulses released by electrical and mechanical stimulation in the ramification (stolon) of an athecate (*Cordylophora*). With an increase in the distance of the recording electrode from the point of stimulation came a decrement in the number of impulses within a volley. There was also a corresponding decrease in polyp contraction. Only one explanation is possible: The polyps receive these impulses as signals to contract, and the polyps closest to the stimulation respond with greater intensity because they receive more impulses. But what happens to the "lost" impulses? They are "used up" at the synapses to facilitate passage for the ones to follow.

At the sight of a swimming jellyfish, most people are impressed by the almost uncanny appearance of uniformity which accompanies the pulsation of its bell. Muscle cells which lie in a ring-shaped arrangement in the bell contract rhythmically. Water ejected from the bell cavity causes a recoil by which the jellyfish moves about (locomotion). Neurophysiological investigations have shown that the rhythmical pulsations of the jellyfishes's bell arise according to the principle of a nervous pacemaker: A neural center acts autonomously to produce spontaneous rhythmical excitations. By means of elimination experiments, we can show that the pacemakers lie in the eight marginal organs attached to the bell, which also contain sensory organs of gravity and ganglion-like agglomerations of nerve cells. A jellyfish can be divided by means of very deep incisions into eight equal sections, each containing a marginal sensory body. All eight of these sections will then contract independently of one another in their own rhythm. However, since the intact jellyfish beats with great regularity without "stuttering"—the eight pacemakers must be coordinated among one another in such a way that they function as one. The previous assumption, that the fastest determines the beat and the others remain the background as "reserve" pacemakers, has been proven incorrect. All eight of them function simultaneously and will readjust one another back to zero after each contraction. The first pacemaker to reach the stage of discharge then releases the next pulsation. The jellyfish is the most primitive organism with a neural pacemaker.

However, spontaneous behavior in coelenterates is by no means limited to simple rhythmical contractions. Sea anemones perform slow movements that cannot be explained as mere reactions to external stimuli. Simple stimuli may release whole sequences of coordinated movements.

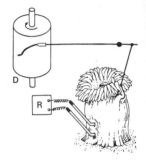

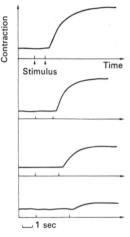

Fig. 20-3. Electrical stimulation of a sea anemone, and illustration of the contraction to demonstrate the pathway. R. Instrument for applying stimuli; D. Recording drum. The more closely the second stimulus follows the first, the stronger is the contraction of the sea anemone.

Pulsations of the jellyfish's bell

Fig. 20-4. A *Pennaria* polyp colony. In response to an electrical stimulus (R), the polyps closest to the stimulus contract first, followed by the more distant ones. The intensity of contraction decreases with distance from the stimulus.

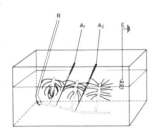

Fig. 20-5. Experimental procedure in recording impulses from the stolon of a polyp colony (*Cordylophora*). R, stimulus electrode; A_1 and A_2, recording electrodes. The response triggered by the stimulus (lowering of tentacles) decreases with distance from the stimulation. Correspondingly, there is a decrement in the number of impulses within the volley triggered by the stimulus.

The sea anemone *Stomphia coccinea*, for example, will detach itself from a background in response to the approach of enemies (starfishes, sea slugs) or to electrical stimulation, and will swim a short distance through the water by means of several brisk sideways strokes. All these observations show that a nervous system in its most elementary form is already capable of making its owner more than a mere reflex machine, and to give it a spontaneous and complex set of behaviors.

The nerve cell processes (axons) responsible for the conduction of stimuli in coelenterates are very thin. As a result, conduction is very slow because the speed of transmission depends on the thickness of the fiber. In addition, time is lost during the transfer of excitation at the synapses, and so it is understandable that these animals are somewhat slow to respond. But for many invertebrates it is a matter of survival to respond "quick as a flash" when danger threatens, either by disappearing into a tunnel or by swimming or jumping out of reach.

This problem is solved by means of giant fibers, which are nerve cells with particularly high-caliber axons specialized for rapid conduction. There is no way of speeding up the processes occuring in the receptor cells, the transfer of excitation to the muscles, or the process of contraction, and so the only possibility of saving time lies in the intermediate section. Giant fibers are found in animals living in tunnels, such as sea worms (Polychaeta) and comb jellies (Phoronida), but also with insects; one example of the function of giant fibers among insects is the escape activity of cockroaches when something touches their anal rings (cerci). In cephalopods the giant fibers allow for a quick and simultaneous contraction of the pallium muscles during recoil swimming. Three such fibers in the upper part of the abdominal nervous cord mediate the jerk reflex in earth worms, the sudden contraction of the body when the anterior or posterior tip is touched. Conduction speeds were determined by tracing the giant fiber impulses at different points. The middle fiber conducts with a speed of 15–45 m/sec, and in the intact animal from front to back, whereas the lateral fibers conduct at 7–15 m/sec from back to front (range of temperature 18–24° C). The earthworm uses its system of giant fibers only in the case of "emergency,"—during escape—and never in its normal crawling activity.

The movements of animals or their external organs (legs, fins, wings, tails, ears, antennae, etc.) take place both in space and time. Behavior may therefore be divided into a spatial and a temporal pattern. In terms of neurophysiology, inquiring into the way in which such movement patterns are released and combined to fulfill a particular function is a way to gain an understanding of the interplay of receptor (sensory) cells, nerve cells, and muscle cells. As we have indicated, there are behavior patterns which are released as well as more or less rigidly controlled in their sequence by sensory stimuli, and in such a case we normally speak of reflex behavior. In recent years it has become more obvious that many behavior sequences are based on ready-made "programs" in the nervous system of an animal. These programs either begin to run spontaneously or continue to run according to autonomous laws after being triggered by (sometimes necessary) signals from sensory organs (see Chapter 3).

For the purpose of analyzing simple movement patterns at the level of nerve and muscle cells, scientists have found crabs and insects to be suitable subjects. These organisms display many behavior patterns that are constant in form and therefore easy to quantify, and that continue to discharge even when the animal's freedom is restricted by experimental manipulation. In addition, the exoskeleton of these animals lends itself to an easy attachment of electric wires leading to electrodes, which in turn may be implanted into nerves or muscles for the recording of action potentials. Several neuro-ethologists have begun their investigations by focusing on simple, rhythmical movement patterns such as running, swimming, respiratory movements, stridulation, and flight.

D. M. Wilson has carried out a detailed analysis of the neural bases of flight in the desert locust (*Schistocerca gregaria*). Flying desert locusts move both pairs of wings with a frequency of 17–30 beats per second. We can let a grasshopper "fly" in the laboratory if we attach the insect by its thorax to a small rod, then suspend it in a wind tunnel. The grasshopper's flight apparatus begins to discharge as soon as the animal loses contact with the ground and at the same time feels an air current by means of sensory cilia attached to its head. In nature the animal may be exposed to a similar releasing stimulus situation when it jumps through the air. There are two groups of muscles which provide power for the movement of the wings; those which raise the wings (levator muscles) and those which lower them. Both kinds of muscles are attached directly to the base of the wings. Other insects such as beetles, Hymenoptera, and flies move their wings in a different way, by means of so-called indirect flight muscles. We have been able to film the wing movements of grasshoppers.

But how can we determine at what points in time the individual flight muscles contract? A very clever method has been developed to investigate this problem. Through previously drilled openings in the thoracic exoskeleton, thin electrodes (steel or copper wires 0.03–0.05 mm thick) are inserted into the muscles being studied. With the help of electrical instruments (pre-amplifiers, cathode-ray oscillograph) we may record the action potential of the muscles. A muscle will receive the command signal to contract in the form of nerve impulses, transmitted by means of motor nerves from the thoracic ganglion. A nerve impulse triggers a muscle impulse, and this in turn leads to muscle contraction. More precisely, the muscle of a grasshopper consists of several sub-units, each of which is supplied by its own motor nerve fiber. It is important to note, however, that a recording of the muscle potentials also provides a simultaneous record of the activity of the motor nerve cells, since the correspondence between nerve impulse and muscle impulse is 1:1.

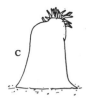

Fig. 20-6. Swimming movements of the sea anemone *Stomphia coccinea*, triggered by contact with a starfish. A. Normal posture; B. Pulling in the tentacles as a first response

to the stimulus; C.
Elongating the body;
D–G. Detaching itself and
swimming away by means
of sideways strokes.
H. End of the swimming
sequence.

Our next question concerns the nature of the "program" which sends command signals from the control center to the musculature. Recordings of impulses from the wing-lowering muscles and the wing-levator muscles, combined with the curve produced by movement of the wings, will show the following: The lowering muscle will start working shortly before the down-beat, the levator muscle, shortly before the up-beat. Thus, both muscles operate in strict alternation with each other, and only for a short time. By contrast, the paired wings of a segment will operate synchronically, whereby the anterior and posterior wings of one side are slightly displaced in time in relation to each other. Anatomical examinations have revealed sensory cells (extensor receptors) in the thoraxex of grasshoppers. These receptors, located near the points of wing attachment, are arranged in such a way that every movement of the wings will provide a stimulus and will release a volley of nerve impulses. These impulses are then conducted to the thoracic ganglia. Since the extensor receptors provide feedback to the central nervous system after each beat of the wings, we may require whether such feedback is necessary to facilitate the next wing movement.

In order to test this assumption, we need only eliminate the extensor receptors by separating them from their nerves. The results are astonishing: If we eliminate only one of the four extensor receptors (there is one at each wing base), hardly any changes occur. Upon eliminating several receptors, we find that the "flight engine" continues to run rhythmically but with a lower frequency. If all four extensor receptors are put out of operation, the wings will beat about ten times per second, only half their usual frequency. We conclude, then, that the rhythmical basic pattern for flight is created by a "generator" in the thoracic ganglion, and that this generator will continue to function even when the extensor receptors stop providing feedback about the movement of each individual wing. The signals generated by the receptors merely influence the frequency of beats.

In nature, of course, the flight of the grasshopper will not be as uniform as under the constant conditions of the wind tunnel. Conceivably, animals flying in a swarm will occasionally touch one another and beat their neighbours' wings. We may assume that such external influences on wing movement, too, will be registered and signaled by the extensor receptors. Such messages probably serve to trigger the operation of muscles which in combination with other receptor organs, restablize the movements of flight.

Rhythmical wing movements of insects are not confined to flight. Crickets (Grylloidea) and locusts (Tettigonioidea), for example, use their wings for stridulation (which produces the typical chirping sound), and grasshoppers (Acridoidea) will employ their legs in the same way (see Chapter 18). During stridulation, male crickets will move their diagonally upward-oriented forewings against each other in a particular rhythm. When the wings move inward, the stridulating crest of the right wing

Fig. 20-7. Flying desert
locust.

with its row of teeth will grate against the stridulating edge of the left wing, causing both wings to vibrate. As a result, special wing structures begin to resonate and amplify the radiation of sound waves. Every inwards movement produces a sound impulse (syllable). A single "cree" contained in the song has two to five of these subelements. Depending on the biological situation, a male is able to stridulate three different and easily distinguishable songs: to attract females, to warn rivals, and to court.

As with the flying grasshopper, we can also record electrical impulses from the wing muscles of the singing cricket. Simultaneous recordings of the song and of muscle impulses will enable us to examine the temporal relationship between single events. Muscles for opening and closing the wings are turned alternatively on and off, but a sound (syllable) is produced only by the activity of the closing muscle. The tone intensity, indicated by the amplitude of the recorded sound phenomenon, may be increased by using a greater number of muscle units and thus increasing the expenditure of force. The command signal for singing orginates in the brain. By means of electrical stimulation in certain areas of the brain, we can artificially release the species-specific song. The song for attracting females also depends on a signal indicating the presence of a spermatophore in the genital region.

Numerous experiments carried out by F. Huber and his associates have shown that the thoracic ganglia has a high measure of independence in the organization of songs. In this area, the function of the brain does not seem to go beyond signalling to the organism when and what it should sing. As is the case with the flight of grasshoppers, there is a group of nerve cells in the thoracic ganglia with the capacity to produce autonomous rhythmic excitations; these impulses are timed and transmitted to the wing muscles in such a way that they ultimately effect a species-specific song.

In crickets, both sexes have auditory organs located on the shins (tibia) of the forelegs. Does a male cricket have to hear its own song in order to sing properly? In other words: Are the details of the song, or the general readiness to sing, under the control of auditory feedback? We may examine this question by experimenting with the auditory system and sound-producing apparatus. If we amputate the male cricket's forelegs, which carry the auditory organs, the animal still sings in the correct manner. If we interfere with the singing apparatus by fixing, weighting down, or amputating the wings, we may disturb or even prevent the production of sound, yet—and this is the main point—the pattern of muscle impulses remains the same. In other words, the neural program will run its course regardless of whether the sound apparatus is working.

Nature itself provides us with an interesting experiment. In cricket stocks we sometimes come across males with dwarfed anterior wings. But although their stubby wings cannot touch, these animals will often "sing" mutely for hours, without producing a sound. We know that the wings move in an exact song rhythm because we can record the pattern of

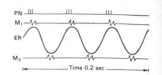

Fig. 20-8. Diagram illustrating the temporal sequence of events during a grasshopper's flight. PN, up-and-down movements of a wing; ER, impulses from the extensor receptor, and muscle impulses from the wing-lowering (M_1) and wing-levator muscles (M_2).

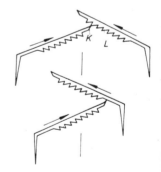

Fig. 20-9. Male field cricket (*Gryllus campestris*) in a singing posture. A toothed stridulating crest on the underside of right forewing grates across the inner edge of the left one.

Fig. 20-10. Greatly simplified diagram of the process of stridation. The cricket moves both forewings inward against the center of its body (vertical lines). The individual teeth of the right stridulating crest (L) touch the stridulating edge (K) of the left wing.

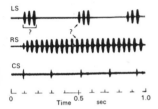

Fig. 20-11. Oscillogram of the luring song (LS), the song for warning rivals (RS), and the courting song (CS) displayed by the field cricket (*Gryllus campestris*).

Fig. 20-12. Activity of two singing muscles during sound production by the field cricket. The impulses of the muscle moving the wing outward (a) and of the one moving the wing inward (b) alternate with each other. Sound is produced only by the inward movement, as seen by the simultaneous recording of the song's oscillogram (c).

Fig. 20-13. A short-winged form of the field cricket often found in laboratory stocks. Although their wings cannot touch, these animals "sing" mutely, with the proper rhythm.

muscle impulses. We might add that the field cricket, too, has extensor receptors which "fire" during singing with every movement of the wings. In contrast to a grasshopper's flight, however, no changes occur in the song of the cricket when the extensor receptors are eliminated. Cricket songs function as signals and are species-specific. It is therefore important that these songs retain as constant a form as possible so as not to be misunderstood. The rigid "central programing" which has now been demonstrated seems to represent one way of ensuring this constancy. The field cricket (*Gryllus campestris*) is unable to fly, and only the house cricket (*Acheta domesticus*) has the capacity to fly for a short time. Yet it is interesting that even the field cricket, when released into an air current, will adopt a typical flying posture; the anterior wings are seen to vibrate and to be positioned differently than when the animal is singing.

By recording impulses from the relevant muscles, we can show that the beat frequency is 30 Hz. Therefore, the time intervals between wing beats correspond precisely to the intervals between syllables of the song, which themselves are created through individual movements of the wings. This provides us with an important argument in favor of the assumption that stridulation, as a behavior, emerged with the wings from the movements of flying. At the same time, we have here an example of the parsimonious functioning of the nervous system: Various rhythmic activities such as flying and singing, and also breathing and running, may be controlled by means of a very small number of "basic connections."

Other well-documented examples of elementary behavior patterns displayed by arthropods and whose sequences are subject primarily to some kind of central control are the movements of the abdominal legs (pleopoda) of crabs while swimming, the sound production of cicadas, and the mating behavior of the praying mantis. Experimenters have also demonstrated that the running movements of insects are also based on endogenous, rhythmic excitations. But in cases like this, movement coordination as well as changes in stride following amputation are additionally influenced by messages from receptor cells which register the position of the legs (proprioceptors; see Chapter 6).

The organization of cricket songs is an example of creating sound symbols by means of a relatively rigid program of muscle movements. From a biological point of view, the songs of crickets and grasshoppers represent —as is actually the case with most of the sounds produced by animals— auditory signals with a specific meaning for the receiver (see Chapter 8). With crickets, the males are the senders or transmitters, and females or other males the receivers. We may therefore distinguish between a system within the animal that creates sound symbols and generates the species-specific songs, on the one hand, and a system that evaluates these symbols, registering the signals and recognizing them as species-specific. Since the process of evaluating or assessing auditory signals in the auditory organs and nervous system of invertebrates has been examined most thoroughly in Orthoptera, we shall turn to this animal group for examples to illustrate the phenomenon. In most species of locusts and crickets, only the male will produce songs; some exceptions include the male cricket (*Gryllotalpa*) and the katydids (*Ephippiger*). In grasshoppers, too, the females are able to make sounds and will often enter into a real duet with the males. Locust and cricket males will often spend several hours singing the songs that attract females. J. Regen had shown as early as 1913 that the biological meaning of these songs is the attraction or luring of females. Females that are ready to mate will even run toward a telephone receiver which emits the luting song of a male conspecific.

But how does the female recognize the species-specific song, and how does it locate its source? We may examine these questions with the use of behavior-physiological and electrophysiological methods. In a behavior experiment, we can use "auditory decoys"—artificial sounds whose parameters can be modified electronically—and note their effectiveness. In various kinds of crickets exposed to such decoys, we have found that the significant feature for song recognition is the intervals between syllables (the rate of syllables). The same applies to locusts. The songs of three species of the genus *Homorocoryphus* living in the same area differ only in their sequence of syllables, and experiments have shown that females will respond only to auditory decoys that contain their species-specific sequence. With grasshoppers, we have not as yet discovered what parameters constitute the recognition traits. Nor has it been possible in behavior experiments on crickets to define whether the tone frequency of their songs carries informational value.

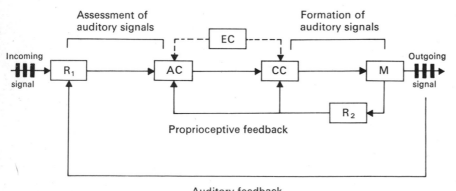

Assessment of auditory signals   Formation of auditory signals

Incoming signal → R_1 → AC → CC → M → Outgoing signal

EC

R_2

Proprioceptive feedback

Auditory feedback

> Top left: lion cubs playing with their mother. The lioness, in turn, playfully puts on an expression of aggressive biting. Top right: rough-and-tumble play in young baboons. Bottom left: young foxes playing. Below right: Although these fox cubs are a long way from sexual maturity, the animal on the right playfully tries to "mount" his partner, the first step in male mating behavior.

Experiments with auditory decoys

Fig. 20-14. The functional relationship involved in the cricket's auditory communication. R_1, auditory organ; AC, assessment center of the central nervous system for the evaluation of incoming signals; CC, command center of the central nervous system where signals are formed; M, musical organ; R_2, proprioceptive sensory organ; EC, endogenous control system which determines general response readiness (e.g., hormones).

We have said that a syllable is produced by a single stroke across the stridulating crest. As a result, the individual "teeth" generate vibrations that, as tone or carrier frequency, construct the syllables, giving the individual "cree" its characteristic pitch. A frequency analysis has shown that the luring song contains mainly tone frequencies of 4000 Hz, while the courting song has frequencies of 16,000 Hz. Electrophysiological tests on

Frequency differences between the luring song and the courting song

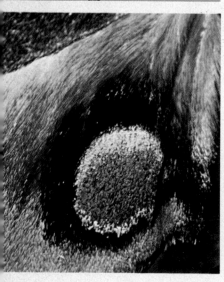

◁

Left: Protective coloration and "fright posture" in the eyed hawk-moth. This insect blends with the background in its resting posture (above), but when frightened it will spread its wings to display eyelike patches (center). The sight of this pattern is enough to frighten small birds away. Below, photographic enlargement of such an eye pattern. Top right: Warning coloration in spotted salamanders. These animals are not edible for any predators and therefore have no need to camouflage themselves; on the contrary, their conspicuous appearance is far more useful, reminding the enemy to stay away after his first unpleasant experience. Bottom right: When frightened, the fire-bellied toad turns over on its back and displays its belly with its warning coloration. The skin on the toad's back, which is normally in view, has a camouflage or protective coloration.

Perceiving shock vibrations

the sensitivity of the tympanic membrane organ, made with electrodes placed on the auditory nerve to record impulses, have demonstrated the existence of primarily two groups of receptor cells: those which selectively respond to a frequency of 4000 Hz, and those which are geared to 16,000 Hz. The biological significance of this amazing correspondence between the auditory system and the species-specific songs may lie in the distinction between the luring song and the courting song. The receptor cells naturally register all other parameters of the song (length of syllables, intervals between syllables, length of verses and intervals). The pathway of messages from the auditory organ to the brain leads over certain nerve cells (interneurons) in the abdominal nervous cord. At convenient loci, for example in the neck, neurons such as these may be "tapped" with electrodes and their impulse patterns traced. At this level, there has already emerged a specialization for the transmission of certain messages, so that some units of information convey the syllabic rhythm of cricket songs and others only the rhythm of verses.

The process by which a female seeks out the source of a sound has been closely examined in the cricket *Scapsipedus marginatus*, from eastern Africa. This species's song consists of five-syllable verses produced at intervals of 1.6 seconds. Notably, the female does not approach with regular movements but intermittently: Movement occurs only during the verse intervals, after the animal has reoriented itself in the direction of the source. If one auditory organ is impaired or destroyed, the female will run in a circle in the direction of the intact organ, which means that normal orientation requires both organs. In locusts, special "relay switches" of auditory interneurons have been found in the abdominal nervous cord. These connections lead to a sharpening of contrast between the auditory impression on the left and on the right side, thus facilitating the determination of direction.

Orthoptera hear and stridulate in order to communicate with conspecifics. This does not mean, of course, that their ears are unable to receive other sound phenomena such as those produced by enemies, as long as their frequencies fall within the range of tympanic membrane organ sensitivity. It is interesting that, in addition to their auditory organs, Orthoptera and other insect groups possess special receptor organs for the perception of ground-shock vibrations. These so-called subgenual organs lie in the tibia of all six legs. By making electrophysiological recordings at the corresponding leg nerves of crickets, grasshoppers, and cockroaches, it was discovered that these organs are extremely sensitive: With shock frequencies of around 1000 Hz, they will respond to ground vibrations as small as a few angstrom, even less in cockroaches (1 angstrom = one ten-millionth of a millimeter). In the abdominal nervous cord of crickets, thick and therefore rapidly conducting interneurons (giant fibers) have been found which are connected to the subgenual organ and transmit messages about ground vibrations to the brain. In turn, the brain sends a command signal to the thorax, causing the animal to stop making sounds or to scampter to the safety of a cave.

Chapter 8 contains a detailed report on the special auditory warning system of moths (Heterocera), for example that of night moths, which is precisely tuned to the direction-finding sound emissions of their enemies, bats. The advantages for neuro-ethological analysis offered by the ear of a night moth are, first, the simplicity of its anatomical construction, and second, the organ's association with a biologically highly significant behavior pattern. We now have the opportunity to probe into behavior using sensory processes as a starting point.

The analysis of a grasshopper's flying movements, on the other hand, was began with a study of motor processes. In between the sensory input —the messages coming from the receptor organs (afference)—and the motor output which involves command signals to the musculature (efference), lies the central nervous system. According to recent estimations, the brain of a bee contains 100,000 neurons. It would be therefore impossible, from a technical point of view alone, to record simultaneously the activities of all neurons taking part in a more complex kind of behavior, let alone interpret the meaning of these impulses. This is why attempts are being made, using as subjects vertebrates as well as invertebrates with highly developed brains such as cuttlefish, to investigate the influence of certain areas of the brain on behavior. These areas are seen as relay centers, and their function is examined by means of elimination or electrical stimulation.

However, if we want to trace the basic processes such as perception, motor coordination, learning, and memory to their most elementary features, we arrive automatically at the level of single cels. The most suitable subject for experiments of this kind would be an animal whose behavior patterns could be clearly defined and whose nervous system is constructed of relatively few and large neurons. To some extent, these requirements are found in certain sea snails. The sea snail *Aplysia*, which belongs to the Opisthobranchia, has now become a classical subject for electrophysiological experiments. This animal has a simple nervous system made up of several cephalic ganglia and an abdominal ganglion, all of which are connected with one another. In addition, nerve bundles with sensory and motor axons proceed from the ganglia. Since the various nerve cell bodies have natural colorings, it has been possible to classify them and to investigate their function.

Out of a large number of detailed studies on *Aplysia*, we shall describe the gill-retraction reflex. The gill, or branchia, which enables the animal to breathe in the water, is partially covered by a mantle fold which also encloses remants of a shell. When the edge of the mantle or its extension, the anal siphon, is touched very lightly, the branchia will retract underneath the mantle. This reflex, designed to protect the branchia from injury, may be quantitatively recorded in a simple way by placing a photo-cell

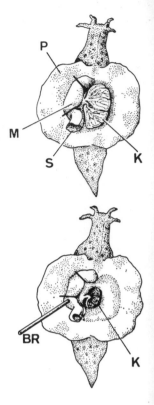

Fig. 20-15. Gill-retraction reflex in the sea snail *Aplysia*, triggered by mechanical stimulation of the edge of the mantle. Note the size of the gill before (above) and after (below) stimulation. P. Patapod; M. Mantle S. Siphon; K. Gill; BR. Mechanical stimulation.

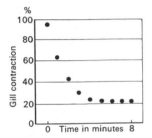

Fig. 20-16. Habituation in the gill-retraction reflex of *Aplysia*. The intensity of gill contraction decreases with repeated stimulation. The first stimulus produces the strongest response (95% of the greatest possible contraction). In the intervals between stimuli, the gill spreads out again.

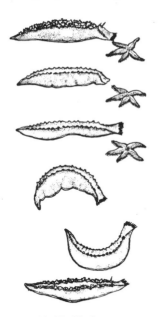

Fig. 20-17. Various stages in the escape response of the sea snail *Tritonia* following physical contact with a starfish. Top to bottom: Retraction of body appendages; flattening and broadening the head and foot like a paddle; free swimming by cruving the body alternately down and up; settling down; and moving along.

under the branchia and applying a defined jet of water as a mechanical stimulus. This experiment will have the following results: If we apply equally strong stimuli within certain time intervals, the gill retraction reflex will become weaker each time and eventually cease altogether. After a certain period without stimulation, the system will recover, and the reaction starts over again with full intensity.

Such cases of a behavioral response decreasing with the repetition of equally intense stimuli is referred to as habituation (see Chapter 17). The phenomenon of habituation is found at all levels of organization in the animal kingdom, including man. It may be seen as the most elementary form of learning. By preparing an *Aplysia* specimen, we may examine where on the body habituation takes place. We have now discovered that the gill-retraction reflex is controlled only by the abdominal ganglion, and that three kinds of cells are involved: receptor cells in the edge of the mantle and in the siphon, interneurons (relay cells or intercalary cells), and motor neurons (cells that send command signals to the musculature) which activate the gill-retraction muscles. If we apply a direct electrical stimulation to the motor neurons in a rhythm that normally leads to habituation, the result is always a complete gill-retraction reflex. This means that the habituation effect cannot be based on muscle fatigue. Furthermore, we may demonstrate by mechanically stimulating the skin, and recording impulses from the axons of receptor cells, that habituation is not based on receptor cell fatigue (sensory adaptation), either. These and other experiments indicate that short-term changes occur at the synapses between receptor cells and interneurons, or between interneurons and motor neurons. Recognizing this fact has given us a starting point for relating short-term changes in elementary behavior patterns to processes at the neural level. The gill-retraction reflex of *Aplysia* is an example of plastic behavior.

As a final example of neuro-ethological analysis, we shall describe an instance of fixed, programed behavior. This is the flight or escape behavior of *Tritonia*, another sea snail belonging to the Ophistobranchia. This animal grows to a length of about 30 cm, carries two rows of plumed or tufted gills on its back (dorsal side), and has two structures on its head resembling tentacles, so-called rhinophores. One of its enemies is a certain species of starfish. Contact between these two animals triggers a flight reaction in *Tritonia*: It pulls in the gill plumes and rhinophores, flattens the anterior end and ventral side like a paddle, performs swimming movements by alternately arching its ventral and dorsal side, and then sinks down passively to resume its original posture. Unlike *Aplysia*, this animal has three pairs of ganglia which lie so close to one another that we may conceivably speak of a brain. The nerve bundles proceeding from this brain contain efferent motor and afferent sensory fibers running from the receptor cells at the body surface.

In his studies on the isolated brian of *Tritonia*, A. O. D. Willows discovered which of the nerve stumps relays sensory messages to a particular cell, and, in turn, which one relays motor messages from the cell to the musculature. Willows had to go to considerable lengths in his experiments to accomplish these results. The basic procedure was to electrically stimulate the various nerve stumps, at the same time using microelectrodes to record resulting excitations in the cell bodies of the brain; stimulate the cell bodies and record impulses from the nerve stumps; and finally, stimulate the cell bodies and record excitations from neighboring ones. The results enabled Willows to draw up a diagram of connections, a "relay chart" of the brain showing that in all individuals of *Tritonia* examined, the nerve cells are arranged and connected with one another in an almost unvarying schema.

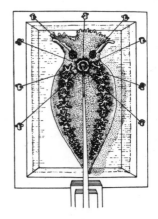

Fig. 20-18. The sea snail *Ttitonia*, fixed to a shallow basin with threads in such a way that its body can move in all directions. Note the two gill plumes on the animal's anterior end, and the two rhinophores between the head and the trunk.

To discover the role played by individual brain cells in the release and control of behavior patterns, a *Tritonia* snail was suspended in a flat aquarium in such a way that it could move on the spot, but its brain itself remained fixed and accessible for the insertion of electrodes. Electrical stimulation of single cells then released very specific behavior patterns, for instance contraction of the gill plumes at the left side when a certain cell in the left brain hemisphere was stimulated. Correspondingly, the symmetrically positioned cell at the right side would release a contraction of the right gill plume. Other cells would trigger a unilateral curving of the body, and still others a symmetrical reaction. But the most surprising effect occurred while stimulating a group of small cells in the area of the so-called pleural ganglion (a group of nerve cells at the sides of the body). No more than a brief period of electrical stimulation, lasting one-half of a second, resulted in a discharge of the entire sequence of flight behavior over a period of about 30 seconds. The only possible explanation was that the stimulated cells would activate, according to a fixed program over time, those individual motor neurons responsible for certain movements. Furthermore, the experimenter was able to show that this group of trigger cells became active when *Tritonia* came into contact with a starfish. By means of simultaneous recordings from individual brain cells, and by making films of the animal, it is possible to relate the discharge patterns of particular cells directly to the animal's corresponding movements.

It was recently found that by stimulating the trigger cells even in a totally isolated brain, the entire neural program of flight behavior could

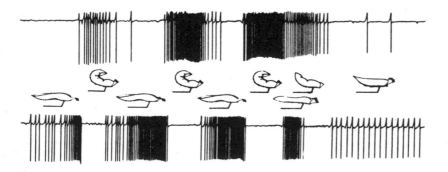

Fig. 20-19. Simultaneous recordings from two neurons in the brain of *Tritonia*, which stimulate the bending of the back (upper track) and of the belly (lower track) while swimming. The illustrations in the center indicate the specific body postures that correspond to the recordings in the snail which is attached to a rod.

be released—even though the brain was cut off from the corresponding effector organs (organs that respond to a received and transmitted stimulus) and therefore also from any possible sensory feedback. We may certainly conceive of the idea that such feedback occurs under normal circumstances, to be utilized for the control of finer details in the flight response. The basic form of this behavior, however, is probably determined by central programing.

21   Hormones and Behavior

The preceding chapters have illustrated that, in addition to external stimuli, internal factors play a vital part in the manifestation of behaviors. These factors primarily include hormones and their effects. When speaking of hormones, we generally refer to substances that are formed in special cells and secreted internally into the blood stream. In this way they are transported to various organs, where they perform their characteristic controlling functions. Similar to the nervous system, hormones serve to coordinate the large number of individual tissues and organs which together make an organism. But while the effects of nervous control are quick and, as a rule, short-lived, those of hormones have a slower onset and will last for a longer period of time.

Hormones are usually described in conjuction with metabolic processes: They will regulate, for example the level of blood sugar or the balance of water. But there are also clear indications that some hormones have a strong influence on certain behavior patterns. One example is the relationship evident between the size of the germ glands (gonads)—organs which produce sexual hormones—and behavior. In many vertebrate groups, the gonads do not develop fully until the spring; this is often the time when individuals of various animal species who have lived together peacefully in herds or swarms become aggressive, leave the group, and form their own territories. When the gonads are fully matured, the males of many species begin to court the females, which leads to a wide array of courtship behaviors and, finally, mating. During the rearing of young, however, the gonads become involuted and all these behavior patterns subside. In late summer these animals gather once again into peaceful societies.

By removing the gonads themselves, we can demonstrate that it is really the changes in gonadal development which are responsible for those of behavior. The behavior of castrated animals during the reproduction period is never the same as that of intact conspecifics. A capon (castrated cock) is no longer able to crow (see Color plate, p. 297) and it neither

Hormones and behavior, by R. Sossinka

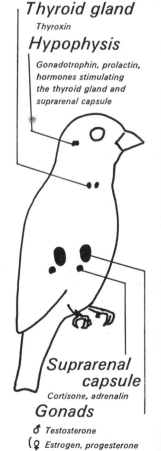

Thyroid gland
Thyroxin
Hypophysis
Gonadotrophin, prolactin, hormones stimulating the thyroid gland and suprarenal capsule

Suprarenal capsule
Cortisone, adrenalin
Gonads
♂ Testosterone
(♀ Estrogen, progesterone

lowers its wing nor shows any other courtship behavior. Ducks whose ovary has been removed cease to perform the typical movement patterns inviting the partner to mate. But all these behaviors which have been lost will reappear if castrated animals are given the missing sex hormones by artificial means. Castrated mallard drakes which are subsequently injected with testosterone (the hormone produced by the male gonads) once again display their typical courtship behaviors; capons treated with testosterone will begin to crow and mount again; and, by the same token, spayed hens injected with estrogen (a follicle hormone) once again display their species-specific posture inviting the male to mount. Thus, the absence of hormones after castration will prevent the occurrence of certain courtship behaviors, but this process is reversible: After a sufficient hormone treatment, these behaviors will reoccur. This observation leads us to conclude that behavior sequences may be influenced by the action of hormones. Our next questions are: Are hormones able to regulate all areas of behavior, and how strong may their influences be?

In the case of sexual behavior, a large amount of control does lie with the momentary level of sex hormones in the animal, as many examples will show: The females of one grasshopper species will allow copulation only when under the influence of a hormone belonging to the Corpora allata (Fig. 21-2); without this hormone they will ward off all males. In male cockroaches, copulation movements are blocked through the brain, but a hormone belonging to the Corpora cardiaca (Fig. 21-2) may remove this inhibition and allow the behavior to occur. A male stickleback will perform its typical movements of territorial defense, nest-building, and the zig-zag dance (see Color plate, p. 298) every time its gonads produce a sufficient amount of male hormones. In the same way, the sex hormone level of male parakeets determines their loud singing and their courtship behavior. Male rats copulate successfully only when they have produced enough testosterone. Here, we have also found that the individual elements of the sex act (mounting, intromission, ejaculation) seem to have different threshold values. Testosterone may induce frequent mounting even in female rats, a show of male sexual behavior. In fact, quite a number of vertebrate females will perform male sexual behavior if treated with androgens (a collective term for male sex hormones). Thus, hens which have received a sufficient amount of testosterone will perform male courtship movements and even crow. Chicks of both sexes, too, will crow and court if injected with androgen (Fig. 21-3).

On the other hand, we cannot as a rule induce female sexual behavior in male animals by the injection of estrogen. Occasionally, a treatment with estrogen may even produce the opposite effect: Castrated pigs begin to show elements of male behavior that were absent before the injection. In some cases, control of female behavior is relegated to several hormones rather than one. Thus, female rats go fully into heat only when, in addition to the follicle hormone (estrogen), they are also injected with the luteniz-

Corpus allatum
Juvenile hormone

Corp. cardiacum
neuro-hormone

Prothorax gland
molting hormone

Fig. 21-1. (left) and Fig. 21-2 (above). Location of several important glands and the hormones they produce.

ing hormone (progesterone). So while male sexual behavior can almost always be induced by testosterone, the control of corresponding female behaviors in many animal groups does not seem to be the exclusive domain of one single hormone (estrogen).

Similar conditions apply to brood behavior, which has been especially well documented in fishes, birds, and rodents. Nest-building behavior in female rabbits, for example, may be triggered by a certain ratio of estrogen to progesteron (1:4000). If this ratio is changed (1:400), the females will cease to build brood nests. In female mice, too, the lutenizing hormone progesterone plays a deciding role in brood-nest building. But in male sticklebacks, nest-building behavior is controlled by gonadotrophins (hormones produced by the pituitary gland or hypophysis) which also stimulate the testes to produce testosterone, as well as by androgens. Paradise fish will build their typical bubble nests only when the pituitary hormone prolactin is present, in addition to gonadotrophin and androgen. In discus fish, as well as in sticklebacks, prolactin releases the characteristic fanning movement which serves to provide the fry with currents of fresh water. In birds, too, prolactin facilitates brood-care behavior, though perhaps in part only by suppressing the further production of sex hormones and their associated sexual behavior patterns.

Fig. 21-3. A chick's response to a proffered hand: A normal chick watches (above), while a chick treated with testosterone tries to copulate (below).

Many brood-care behaviors seem to occur quite independently of the hormone indicator, because those external stimuli which release the behaviors are very strong. Thus, in the presence of newborn (neonates), young female rats as well as castrated and intact males showed the typical movements of brood-care behavior such as licking, warming, and carrying into the nest. The young may be nursed, of course, only by adult females, whose mammary glands must first be activated under the influence of prolactin. Furthermore, the pituitary hormone prolactin is essential for the production of crop milk in pigeons and the layer of mucus in discus cichlids, both of which provide food for the young.

There are still other areas of social behavior more or less under the influence of hormone action. We can observe this influence particularly with aggressive behavior patterns. The mere fact that the number of intra-specific duels markedly increases during the gonadal growth period—for many species, in the spring—indicates the presence of hormonal influence. Since in most animal species the males are far more aggressive than the females, we may well assume that testosterone plays a significant part in the level of the aggressive drive. And in fact, castration—for example in male domestic animals—does lead to a considerable reduction in aggressive behavior; for a long time people have taken advantage of this to transform "wild" stallions and bulls into "tame" geldings and oxen. Conversely, androgen injections lead to an increase in fighting readiness: Male fish (sword tails), amphibians (salamanders), reptiles (lizards), birds (pigeons, gulls, turkeys), and mammals (rats, mice, monkeys), when subjected to corresponding experiments, tend to react in the same manner. Differences occur only in females: While those of some species do respond to testo-sterone injections with increased aggressiveness, female mice did not respond at all.

There is also considerable variation in the effect of female sex hormones on aggressive behavior displayed by both sexes of the above-mentioned species. A large number of mammalian females are particularly peaceable at the time of estrus (ovulation); they will also display a very low level of aggressiveness when brought in heat artificially by means of hormone in-jections (estrogen and progesterone). Similarly, domestic chicks of both sexes will avoid duels if treated with estrogen. In other groups of birds, however, the female sex hormone did not seem to have any influence: Neither canaries, night herons, nor quail showed any change in their readiness to fight.

Strongly coupled with aggressive behavior is the dominance hierarchy that arises in social groups of animals as a result of single fights. Since testosterone influences the fighting instinct, we may speculate that this hormone will also have a marked effect on the dominance hierarchy of a group. Many experiments have supported this. For example, when a low-ranking pigeon was treated with a male sex hormone, it soon rose to a higher status; with sufficient doses it even rose to the highest rank in the order. Furthermore, its status was maintained throughout the treatment. Similar results were obtained with chickens, mice, and even chimpanzees.

The influence of hormones, however, is not restricted to specific be-havior patterns such as those of sex, brood-care, and aggression. In some cases, hormones may regulate the general activity level, the frequency of movements, and even the intake of food. In many articulate animals (arthropods), for instance, we have been able to observe a clear corre-spondence between the periodical production of brain hormones and the animal's activity cycle. In the bristle worm *Platynereis*, a brain hormone induces the animals to leave their dwelling tubes upon sexual maturity, and to swarm into the open water (see Chapter 11).

Vertebrates, too, respond to variations in their hormone balance with a change in general activity. Castrated drakes, for example, remain largely passive and swim about far less than intact animals. Their behavior returns to its normal level only after treatment with testosterone. In labyrinth fishes, this hormone and the thyroid hormone thyroxin lead to a general restlessness and an urge to migrate. On the whole, in fact, thyroxin seems to have the function not only of facilitating metabolic processes, but also of inducing a general restlessness in many vertebrates. According to some researchers, the phenomenon of flight unrest—an increase in the general level of activity which ultimately leads to the spring and fall migrations of birds of passage—is dependent on this thyroid hormone. Other scien-tists believe that flight unrest is caused by the sex hormones. Their theory holds that only a certain level of testosterone functions as a releasing agent; below and above this level, migration will not take place.

In a similar way, hormones control the exaggerated urge to eat which leads to an accumulation of surplus fat for migration. Generally, in addition to nervous feedback, the intake of food and water is regulated by various messenger substances. It should be noted, however, that although these substances—for example, secretion—have played a part in the rise of the hormone concept, they are no longer defined as hormones in the strictest sense.

In addition to these general behavior patterns, hormones influence those special qualities and skills frequently referred to as "mental," that is, controlled by the central nervous system. In higher vertebrates, even their learning ability is dependent on certain hormones. Although experiments on the influence of testosterone on learning have produced contradicting results, other studies have clearly indicated that learning in rats is facilitated by protein hormones of the pituitary gland. When this gland was removed animals subjected to conditioning experiments (training) showed a decrement in performance. But when they were treated with a certain section of the protein hormone ACTH—a long chain of molecules, stimulating the adrenal cortex—performance of these subjects rose back to the level of intact control animals, and memory skills were restored.

Effect on learning ability

In summary, we may conclude that a great many areas of behavior—perhaps even all—are subject to influence from specific hormones. The degree of this influence, however, varies not only among animal species and between the sexes, but also within behavioral categories.

Our next question is, how can we explain why a certain hormone causes some behavior patterns to appear more frequently than others? What, in fact, is our general understanding of the mechanism by which hormones operate on behavior? To begin with, a brief description of morphological and physiological relationships may help clarify the way in which hormones come into contact with the nervous system—that agent directly responsible for an animal's behavior. Some hormones are produced internally within endocrine glands and secreted into the blood stream, whereby they are carried to various parts of the central nervous system. A number of other hormones, however, are produced within special sections of the central nervous system itself. In this way, some nerve cells have become specialized, transformed, and are now—as so-called neurosecretory cells—able to manufacture their own hormones. In vertebrates, cells like these may be found predominantly in the farther reaches of the hypothalamus (an important relay center in the diencephalon; see Fig. 21-4), and their secretions, called neurohormones, are stored in part in the pituitary gland. The situation is complicated by the fact that a large part of these pituitary hormones have control over the hormone production of other endocrine glands. For instance, the gonadotrophins stimulate the ovaries to produce estrogen, and the testes to manufacture testosterone. These hormones, produced in the genital organs, in turn give negative feedback to the hypophysis; in other words, larger amounts of estrogen or testosterone inhibit the further secretion of gonadotrophins.

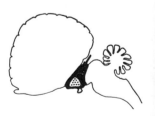

Fig. 21-4. Diagram of a mammalian brain. Black: Diencephalon with attached hypophysis (pituitary gland). Dotted: Main concentration of neuro-secretory cells.

These facts alone indicate that the various hormone systems influence one another and are interwoven. Added to this are the interactions between hormones and the nervous system, taking place at different levels and in many different ways. We can suggest six basic ways in which hormones influence behavior:

1. Structuralizing effect:
   (a) New releasing structures are formed;
   (b) relay connections and pathways are finalized in the brain.
2. Stimulating effect:
   (a) Hormone action is specifically aimed at stimulating centers performing specific functions;
   (b) general response readiness is enhanced;
   (c) messenger substances originating from neurosecretory cells are put into operation;
   (d) external organs are sensitized.

The first method is very indirect, with its roots extending more into the field of developmental physiology than that of ethology. A researcher may encourage or induce the development of very specific forms out of embryonic tissues and organ potential. Again, the role played by sex hormones has come under particularly close scrutiny. In a mammalian or avian embryo, whose future sex cannot yet be determined, the presence of testosterone will induce the growth of primary and secondary sex characteristics. Even adult animals may develop secondary sex characteristics under the influence of the male hormone; many birds, for example, grow brood feathers which then become lost after reproduction. In the absence of testosterone, female traits will develop even without the influence of estrogen. It is only when the presence or absence of certain hormones has given rise to these releasing characteristics that sexual partners will respond. This process becomes very clear with hormone-controlled releasers appearing for only a short time, such as the typical swellings of female chimpanzees when in heat, the pheromones ("social" hormones, those that function in social communication) on the antennae of female cockroaches, or the "red belly" displayed by male sticklebacks (see Color plate, p. 298).

Testosterone will have the same effect on the central nervous system. Again, it is the hormonal influence which determines during an early phase —often prior to birth—whether the animal will later develop predominantly male or female characteristics. And again, the absence of hormones will lead to feminine development, the presence of testosterone, to masculine typing. If pregnant guinea pigs or rats are treated with testosterone until their young are born, the sexual centers in the brain of all newborn— even females—will bear the masculine stamp, due to their contact with the artificial androgens transported in the mother's bloodstream. Both the cyclical nature of gonadotrophin influence typical of females—responsible for the regular periods of ovulation—and the feminine sexual behavior patterns are strongly inhibited or even absent. If the females born in this experiment are treated with testosterone—they cannot produce this hormone themselves, since the females have only ovaries—they will behave like males. Conversely, newborn male rats that are castrated will no longer be under the differentiating influence of testosterone; in later life, these animals will be subject to the cyclical control of gonadotrophins just like females, and they will show female behavior patterns toward other males.

But there are other areas of behavior, too, where certain hormones may wield a permanent influence established during an early phase. Following a hormone-induced sex reversal, changes have been observed in the play behavior of rhesus monkeys and the marking behavior of mice. A male dog who was physiologically "castrated" in the womb—that is, treated with a prepared substance to completely eliminate the influence of testosterone—and observed in adulthood did not urinate by raising his leg, but by squatting like a female, even though he had testicles.

Fig. 21-5. Diagram of possible loci where hormones may influence the discharge of behavior patterns (open arrows, organizing effect; plain arrows, regulating effect): 1. Change in the transmittor; 2. Organizing effect on the CNS; 3. Direct and specific influence on the CNS; 4. Direct but nonspecific influence on the CNS; 5. Indirect influence on the CNS; 6. Influence on peripheral (receptor) organs. (Other connection in the CNS are only indicated.)

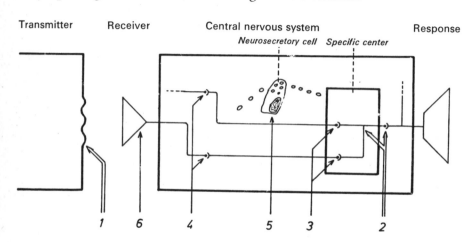

Aside from this formational (determining) influence on special centers in the brain, hormones may also have temporary effects, in this way facilitating reflexes or behavior sequences whose nervous pathways are already complete. We know that very specific behavior patterns may be released in the brain by means of electrical impulses, providing that our aim is accurate and we hit the appropriate center (see Chapter 19). Correspondingly, we can also stimulate these brain centers by means of certain hormones. Instead of electrodes, tiny hormone crystals are inserted into the brain; they dissolve slowly and penetrate into adjoining tissues. If one of these hormones reaches a center that is sensitive to the substance, and the situation is right, the frequency of corresponding behavior patterns will increase. We can facilitate the isolated discharge of different elements of courtship behavior in pigeons, depending on the area hit by the testosterone particles. If we implant testosterone crystals laterally into the diencephalon of male rats, aggressive behavior increases. If the crystals are placed into a closely adjoining area, the pre-optic region, the number of copulatory movements will increase. Finally, if the experimental testosterone reaches both areas in the brain simultaneously, an appropriate situation will release the entire repertoire of courtship behavior—strong support for the proposition that courtship is a combination of single aggressive as well as sexual elements.

Aside from this direct, specific influence on the central nervous system, we know of equally direct but nonspecific effects attributed to certain hormones. In general, these may increase the excitability of large areas in the brain. The lutenizing hormone progesterone, for instance, will have such an effect on female rats, as recordings of electrical activity in the brain (EEG) have shown. Animals treated with this hormone respond more quickly, not only to sexual stimuli (progesterone is included among the sex hormones) but also to practically all other stimuli.

There are still other hormones, however, that exert an indirect influence on the central nervous system. In a number of cases, the long latency period between injection of the substance and the first effects would in itself indicate that some intermediate reaction processes are taking place. After cats were injected with radioactively marked estrogen, it was observed that after a few hours the presence of this hormone increased in some areas, then gradually disappeared. But a definite activation of the sexual drive did not occur until four to six days later. Here, we are assuming that the estrogen induced the formation of a messenger substance, and that a change in behavior could only occur when a sufficient amount of this substance had accumulated. By means of detailed examinations on individual cells, we can ascertain that estrogen does facilitate the neurosecretory activity of some of these, while inhibiting others such as those producing gonadotrophin (negative feedback). Scientists have discovered quite a few of these messenger substances, most of which in turn release a pituitary hormone.

There is one other way in which hormones may influence behavior. Instead of acting directly on the brain, they can operate on external organs, effecting changes. This also leads to different forms of sensory feedback. Testosterone, for example, induces changes in the skin surface of a rat's penis, which then becomes more sensitive to touch and transmits stronger messages to the brain. As a result, the brain responds to this "exaggeration" with a corresponding sexualization of behavior. In a similar way, hormones share in the control of a pigeon's behavior in feeding its young.

There are a number of ways, then, in which hormones can exercise control over behavior. Under natural conditions, in fact, any one hormone may operate simultaneously in a variety of ways. For instance, it may sensitize an external organ, increase excitability within a brain center, or modify releasing structures. Conversely, an organism is normally subject to the simultaneous influence of many different hormones, so that it becomes very difficult to determine which substance is responsible for what effect. F. A. Beach, one of the leading scientists in this field, made the intricate nature of hormone action very clear when he stated that as yet

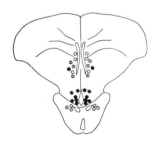

o No response
o Following behavior
o Invitation to nest-building
● Following behavior
  and invitation to
  nest-building

Fig. 21-6. Location of crystals of testosterone implanted in a pigeon's brain, and the courtship behaviors they release (cross section).

no behavior was found which depends on only one hormone, and no single hormone is solely responsible for any one effect.

Furthermore, the degree of hormonal influence varies not only among animal species and even subspecies, but also among individuals of the same subspecies —depending on the animal's previous experiences. If we castrate tomcats that have had no sexual experience, all sexual behavior disappears. On the other hand, castrated tomcats who had copulated several times prior to the operation will show a decrease in sexual activity, nonetheless remaining at a higher level than inexperienced subjects. We may conclude,

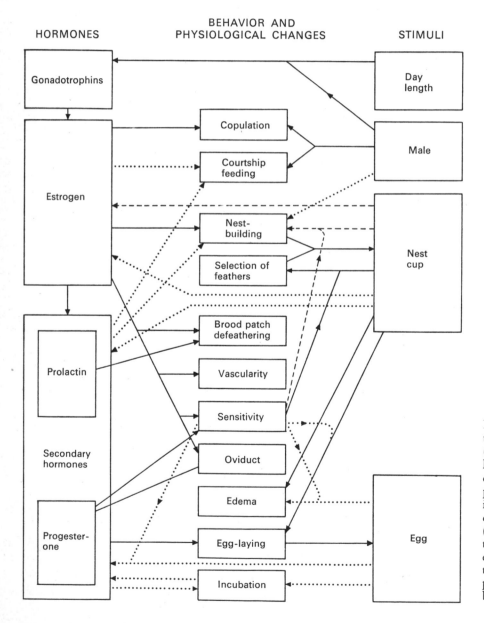

Fig. 21-7. Relations between external stimuli (right column), internal hormonal conditions (left column), and various physiological and behavioral responses of a female canary during reproduction (after Hinde, 1965). Continuous lines: positive effect. Broken lines: negative effect. Dotted lines: probable relationships. beside drawing.

then, that experience can modify the degree of hormonal influence. This may apply more to higher-developed mammals than to lower species, just as in general the direct influence of hormones is increasingly replaced by cerebral control, the higher the species under study lies on the evolutionary scale. For this reason we suggest great caution in generalizing from our examples.

Until now we have talked only of the influence that hormones exert on behavior patterns. Conversely, however, the performance of behavior patterns can affect hormone levels. This becomes possible through the capacity of some nerve cells to respond in two ways. Most nerve cells operate neurohumorally; that is, they manufacture substances that enable the transfer of excitation from one nerve process to another (synapse), conducting electrical impulses. Some nerve cells, however, can operate neurohormonally, producing messenger substances (mostly polypeptides) that, as neurosecretions or "releasing factors," in turn either release or block the release of hormones. Thus, these cells are able to transform nervous impulses into hormonal messages. In this way, the song of a male parakeet can stimulate the formation of gonadotrophins in the female to such an extent that she produces mature eggs. In canaries, too, females will produce increased amounts of a follicle-stimulating hormone (another gonadotrophin) in response to male courtship behavior. Similarly, seeing and feeling eggs during incubation stimulate birds to formate prolactin. In some species of bird, nest-building activity stimulates the production of a pituitary hormone (LH). The phenomenon known as "social stress," also basically results from behavior-induced hormone fluctuations.

Evidently, then, aside from the interplay of different behavior patterns and single hormone systems, there are ways in which hormones affect behavior as well as instances where the performance of certain behavior patterns influences hormone production. While experiments are aimed to isolate relationships and to provide for controls in only one direction, the processes we find in nature involve a simultaneous and meaningful interaction of all these influences.

We may use canaries to illustrate this point. In females, the formation of gonadotrophins is stimulated first by external factors such as length of days, but later mainly by the courtship behavior of the males, especially their song. This leads to gonadal growth and the production of estrogen, which in turn causes a sexualization of behavior: The female will respond to courting and allow copulation; she also begins to build a nest. This nest-building activity and the sight of the nest indirectly stimulate the production of progesterone and later also prolactin. The influence of progesterone enables the eggs to mature and leave the body to pass into the nest. On the other hand, prolactin functions mainly in the formation of the brood patch: Feathers fall out from the breast area, and the skin becomes spongy —due to vascularity and—sensitive to the touch. The female is therefore able to warm the eggs better, and also better able to feel them. This tactile

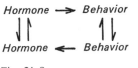

Fig. 21-8.

stimulation leads to stronger brood-care behavior and increases the production of prolactin. When prolactin has reached a sufficient level, it inhibits the secretion of estrogen, eliminating sexual behavior to make way for the behavior patterns of feeding the young.

This greatly simplified version of interactive phenomena was designed to illustrate the fundamental process: A behavior pattern—or some other stimulus—leads to the production of a hormone that (at least quantitatively) modifies behavior; this in turn stimulates the production of another hormone which again wields its influence over other behavior patterns.

22   Learning and Play

Learning and play,
by B. Hassenstein

A boy and his friend are riding horseback through the woods, approaching an area where lumberjacks are at work. Suddenly a tree crashes to the ground close by. Both horses naturally shy away and start running—an innate reaction. But the boys stay in control and ride on. A little later, one of the horses shies again, but this time its rider cannot see why. The other boy, however, has been here often and gives an explanation: This is precisely the place where, some time ago, the animal had stumbled and been frightened. Now the horse shies every time it passes that spot, even though the obstacle has long disappeared. Thus, we can assume that the animal's central nervous system had stored some part of that previous experience. The memory trace—the engram—was now expressed in behavior. The horse is frightened, we might say, by a danger that is in the past but which has been stored in its memory.

As we know, animals, including man, do not by any means remember everything they experience. If we want to teach a dog certain rules of behavior, or if we want to try to learn an important mathematical formula, many repetitions are needed to create the desired engram. But other experiences seem to "stick" right away, without repetition, like the horse's startling experience. In humans isolated memories from childhood may still be alive decades later. Herein lies the basic problem of learning theory, which may be stated as a variation of the way K. Lorenz had formulated the question: Out of the uncountable number of impressions impinging upon the organism, how does the learning mechanism manage to select those for storage that are conducive to survival, while ignoring others whose memory would interfere with the individual or the species?

Initially we may answer this basic question easily enough. There are two kinds of information that the nervous system could store: the messages received from all sensory perceptions, and the internal command signals triggering all behavior patterns. If the nervous system is to "forget" worthless material but store valuable information, each message must first be evaluated. Furthermore, the assessment "valuable" must automatically be followed by the command: "Store it!" With this basic procedure in mind, we can look at a number of ways in which learning takes place.

In the early years of this century, there were lively discussions about whether fishes could hear. K. von Frisch, who had previously devised a way to study the sensory functions of bees by means of conditioning procedures, settled the controversy with a clever method utilizing the animal's ability to learn. He kept a dwarf sheatfish that lived in a small tube at the bottom of its aquarium. Von Frisch would feed the animal by holding the food close to its mouth on a rod. One day he began to accompany the feeding with a whistle. The fish had never responded to whistling before. But five days after he had begun the experiment, von Frisch whistled before feeding the dwarf sheatfish. The animal quickly left its shelter and swam around in the open water, making searching movements. Von Frisch gave the article a title in which he described these and other observations with a notable flair: "A dwarf sheatfish that comes when I whistle."

There are other, similar examples we may cite before attempting to clarify the "evaluation" principle in these phenomena. Over a period of time, jays were given food in one particular cage. Here, experimenters deposited freshly killed loop moth caterpillars which resembled twigs (protective mimicry), together with small pieces of wood from the caterpillar's food plant. Then the hungry jays were let into the cage. During preliminary experiments, these birds immediately noticed stationary grasshoppers or moving caterpillars, but now they took an average of fifteen minutes before making chance discoveries—for example, by stepping on them—the stationary caterpillars, which they would then eat. As soon as this happened, however, the jays immediately changed their behavior: They began to search specifically for objects in the shape and form of these caterpillars, and they quickly found all of them. In the course of their search, the birds would also hit upon twigs, but these were quickly abandoned—a definite sign that their search was oriented to physical appearance. Behavior change in these jays—the sudden attention paid to twigs—was brought about by a favorable experience. The animals had learned something, and this changed their "normative response."

Furthermore, there are certain things that must be learned by all individuals of a species, at least in the wild, if they are to survive. In this way, even a learned behavior may be considered typical of a species, or an even-more-general group: For example, foals, young gazelles, and other ungulates (hoofed animals) start to look for their mother's nipples shortly after birth. The young have an innate tendency to search at "something vertical" and "something horizontal." Both added together (rule of heterogenous summation) "in a corner," i.e., in the axilla of the forelegs and the flexure of the hind legs; but they find milk only in the latter area. After a fair number of attempts, the young learn the proper orientation and

Fig. 22-1. a. Nilgai antelope: 31 minutes after birth, the young searches its mother to find the source of food. These behavior patterns are inborn, but the newborn animal does not yet know where to look.

b. 62 minutes after birth, the young has found the right place to suckle. From now on, it will know where to look without making mistakes.

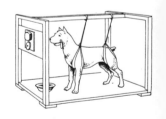

Fig. 22-2. Apparatus used in studying classical conditioning according to Pavlov's method: The experimental animal is tied to the frame; a device for measuring the amount of saliva secreted during the experiment has been surgically attached to the subject's mouth. Note the position of the bell and the food bowl.

cease making errors, provided that the female does not push the young in the right direction—as experienced females do.

In this context, we may also cite the "classical" experiments of I. P. Pavlov (1844–1936), who demonstrated the conditioned reflex: When a hungry dog sees or smells food, its salivary and gastric glands secrete saliva or gastric juice. If a bell is sounded while the dog is eating, something changes in the connections of the nervous system. This process is not immediately or outwardly noticeable. But we can infer this change because, after repeatedly combining the bell with the food, the sound of the bell alone will trigger salivary and gastric secretions. The repeated temporal coupling of the primary ("unconditioned") and secondary ("conditioned") stimuli had led to the formation of a "conditioned association" between the sound of the bell and the secretion of gastric juice. But there is more to this phenomenon: When the dog is released from its harness, it runs to the source of the stimulus—be it bell, metronome, or some other object—and begs at it for food, barking and waving its tail. The result of the learning experience becomes manifest in the dog's species-specific appetitive behavior, which in this case is social food-begging.

We can derive a general principle from these four examples of animals that have undergone learning (dwarf sheatfish, jay, newborn gazelle, and Pavlov's dog): In all cases, the animals had learned as a result of some favorable experience, namely, the satisfaction of hunger. During or shortly before this pleasant experience, there was always a sensory impression: The dwarf sheatfish heard a whistle, the jays saw the twiglike caterpillars, the young gazelle found the precise location of its mother's nipples, and the dog heard the sound of a bell. In every case, the result of learning was that the animal began its appetitive behavior, in the course of which it would orient itself as much as possible to the perceived stimulus.

We may now formulate one of the first principles of learning: If an organism perceives an initially neutral stimulus situation before or during the satisfaction of a drive, a learning process may take place so that this stimulus situation thereafter gives rise to or becomes the actual goal of the associated appetitive behavior. In this way, the appetitive behavior is coupled with a new releasing and directing stimulus, based on experience; it becomes a "conditioned appetence" (short for "appetitive behavior based on experience, or learning"). Now we can also state the assessment principle whereby a command signal is produced, meaning: "This information must be stored!" It says that all those units of information are worth storing that the organism receives shortly before or during the satisfaction of a drive; this information may, after all, be a useful guiding signal in the future.

In all the cases described, hunger was the deciding force, and the decisive reward was its satisfaction. But other drives can be involved. Some fishes and songbirds can be trained to enter a certain place where they are

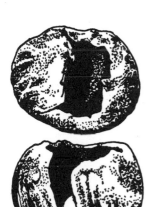

Fig. 22-3. Hazelnut shells cracked open by squirrels. The fourth nut ever opened by one young squirrel is shown from the top (above) and base (center). This shell is covered with many irregular grooves. Below, by contrast: The work of an experienced animal that gnawed a hole into the tip of the nut and then split the shell by lever pressure.

allowed to fight with a rival (or with their own mirror image, an imagined rival). Hamsters respond best when their reward is an opportunity to climb around wooden blocks of different sizes, giving them an opportunity to follow their drive for orientation and play. Ants learn best in association with carrying pupae into their nest; that is, with an opportunity to "fulfill their function within society." One chimpanzee who had received sixty bananas during a learning experiment did not eat them, but gave them back to the experimenter; his learning had been motivated by an "interest in the problem *per se*," and not by any readiness to learn that might have resulted from some other drive condition.

In addition to the phenomenon of conditioned appetence, whereby animals learn to associate certain stimuli with their appetitive behavior, another kind of learning may take place: An initially neutral behavior or behavioral element is made to serve the appetitive behavior. When K. von Frisch was a student, he kept a Brazilian parrot. He would let the bird fly around for a while only when he saw that it had just defecated in its cage. In this way he made sure that his pet did not leave a "calling card" in the room. The bird soon learned to obtain its "reward" by producing tiny amounts of feces at will. It was immensely comical to watch these attempts. The action of pressing took on reward value, and occasionally the bird would "beg" in this original way even outside the cage when it saw an attractive morsel or had some other urgent wish. By means of a learning process, the behavior element of defecation was made to serve an entirely different drive which the parrot wanted to satisfy.

In a similar way, we can train animals with learning capacities to perform a variety of behavior patterns, though naturally within certain limits. Dogs, for example, "shake hand"; dolphins "stand" in the water, projecting their trunk far out of the water and beating rapidly with their tail. All we have to do is reward the animal right away, as soon as it performs the desired behavior for any reason whatever. Another example is that of a small monkey in a zoo who was always pushed aside by the bigger animals. In his desperation, he started to jump up and down on one spot. The zoo visitors noticed him and began to throw him food over the heads of the other monkeys. As a result, this animal associated the drive to obtain food with his "successful" behavior: The more hungry he was, the more frequent became his up-and-down jumping behavior—a rather unique way of begging for food. In one experiment, a dog was given food only when—as a response to mild electric shock—he raised his leg. His learning manifested itself through his subsequent behavior of raising that leg whenever he was hungry.

The most prominent scientist researching this kind of learning is B. F. Skinner. One of Skinner's expriments, based on a description by K. Grossmann, is the following: The experimental subjects were pigeons, deprived of food and placed individually in a row of cages. Every fifteen seconds, a feeding machine dispensed small amounts of seed, independent

Fig. 22-4. The trainer has conditioned a dolphin to remain straight up out of the water and to offer its keeper a fin. The animal "stands up" by beating its tail rapidly back and forth. The conditioning of such a complex task requires the trainer to reward each new step as it is learned.

of the animal's behavior. After a while, the pigeons began to act strangely: One kept turning about its own axis, another spread its left wing and left leg, a third stretched its neck up as high as possible, a fourth made peculiar bowing movements in a corner, and still another jumped around on one spot. Initially the presentation of food coincided with some spontaneous behavior pattern, so that this behavior was apparently "rewarded." It therefore increased because, of course, the reward was inevitably repeated. Thus, every pigeon learned something different, and each subject behaved as though its eccentric actions were a direct cause for the appearance of food. Skinner compared this phenomenon to superstitious behavior in people.

The learning principle involved in the behavior of the "pressing" parrot, the jumping monkey, and the "superstitious" pigeons is different from that of conditioned appetence. A reward is not given in association with a new stimulus, but for the performance of an initially neutral behavior element. By means of this learning process, the new behavior element can be made to serve the goal of the appetitive behavior, for example, the satisfaction of hunger. In such cases, the hungry animals do not look for a particular stimulus, but rather they perform a certain behavior—jumping on the spot, lifting a leg, or going through various body movements. Thus, if an initially neutral behavior element is repeatedly rewarded, it will be associated with and utilized for the satisfaction of the hunger drive. Henceforth, the new behavior element is displayed more frequently the stronger the activation of the drive.

Skinner designated this kind of learning "operant conditioning." The German term *bedingte Aktion* (conditioned action) indicates that this learning takes place in the efferent (motor) section of the nervous system. But the evaluation criterion is the same as with conditioned appetence: "It is worthwhile to retain the memory of that behavior which occurred before the satisfaction of a drive, because this behavior could have been the cause of satisfaction." Again, evaluation or assessment is based on drive satisfaction.

The learning of new stimuli and the learning of new behavior elements are based on two different principles; yet both methods help the appetitive behavior to attain its end, and they may therefore be combined; that is, they may occur together. In such cases they can be so tightly interwoven that the individual elements within the whole complex of behavior are no longer noticeable. This applies, for example, to the way squirrels learn to crack hazel nuts. According to I. Eibl-Eibesfeldt, these animals have an innate tendency to take nuts and similar round objects between their forepaws, turning them over and over and gnawing at them. The squirrel must learn, however, to gnaw the nut open at the most convenient and efficient spot. Although these animals are able to perform the movements of gnawing and cracking from the start, they must learn how to use this "fixed action pattern" (*Erbkoordination*, innately coordinated movement pattern) to open nuts in the most efficient way. Experienced squirrels do it with a minimum of effort: They gnaw a furrow in the broad side of the nut starting at the top, then use their lower incisors like a lever to crack the nut in two. By contrast, inexperienced animals will gnaw many superfluous grooves until the nut breaks open at any side. But even at the beginning they will attempt to crack the nut, and they will repeatedly attempt to use their front teeth as a lever. This movement pattern, however, leads to success only when the furrows are properly placed. The first sign of learning occurs when the furrows are made parallel to the grain, or texture, of the nut shell, and concentrated on the broad side. In learning this technique, the squirrel goes the path of least resistance, because it is harder to gnaw against the grain than with it, and the incisors will slide off the more strongly curved side of the nut. In this way, the animal's activity is directed or oriented by the structure of the nut into a certain channel. In addition, the squirrel keeps trying to crack this object, and it will remember any successful movements. In this way, most squirrels ultimately arrive at the same efficient technique for cracking nuts. But there are individual variations. For example, some squirrels learned to crack a hole in the nut with a few furrows made on top of one another. One animal achieved its goal quickly by gnawing at the base of the nut. It kept doing this for a while, but eventually learned to gnaw a hole by making a few superimposed grooves. In the end, the squirrel used the latter technique to gnaw at the tip, where the nut is thinner (Fig. 22-3).

Almost fifty years ago, J. B. Wolfe designed and started an extensive series of experiments aimed at producing high levels of learning and performance in chimpanzees, comparable to human behavior. These anthropoid apes were given round chips about 3 cm in diameter, to be placed into a food-dispensing slot machine. The chimps quickly learned this task, a first step for obtaining food. Next, they learned to distinguish between different kinds of tokens: One kind could be exchanged for food, another for water, a third for an opportunity to play with their keeper, a fourth to open a door, and so on. In a typical situation, a chimp would be busy collecting tokens, when he would suddenly hear a familiar conspecific calling in the next cage, or he would be disturbed in his own cage. Having learned that blue chips can be used to open the door to the next cage, the subject immediately picks several blue tokens out of his pile, goes to the door, and opens it.

A third step in the experiment is to make the chimps "work" for their food tokens. For instance, they may be required to manipulate a lever device to obtain chips, but they have to wait a while before exchanging them for something else. The subjects accumulated a supply of tokens and treated the chips like food, thus making a transfer of function from the primary link in the behavior sequence to a secondary one. For the chimps, these tokens had acquired a certain "worth," a "food valence," a symbolic value for desired social contact.

Fig. 22-5. A chimpanzee places his food token into a slot machine. His reward will appear in the lower tray.

Conditioned aversion

We have already cited one example of learning based on a bad experience: the horse that shied away from a spot where it had previously been frightened. It did not want to go near a place where it felt uncomfortable; in fact, it would have preferred to run away. Another example of learned avoidance behavior (conditioned aversion) is the case of three young eider ducks that used to swim together, then come ashore to groom themselves. One day, one of them was caught by its keeper and placed alone in a cage. There it became restless, running back and forth and "crying": It emitted the so-called "cries of abandonment." After this happened eleven times in two days, the young duck followed the other two only to the shore, then stayed in the water. When the keeper did manage to lure it to the land, he repeated his act and again isolated the duck from its siblings. On the thirteenth occasion, the animal stayed far out on the surface of the water and refused to come ashore. Thus, it "anticipated" the danger and was able to avoid it. The duckling transferred this negative "tone" or valence associated with a particular situation to the accompanying circumstances, and therefore managed to avoid a dangerous situation before it actually happened.

According to reports by K. Immelmann, the same learning mechanism seems to operate in zebra finches and perhaps also in other birds, where males commonly restrict their courtship to females. If juvenile males are separated from their parents shortly after becoming independent (fortieth to forty-fifth day of life), they will later court males and females alike. But if they remain with their parents or other adults after that, their father will drive them away from the vicinity of the nest. This is the time when the young birds imprint into their memory the characteristics of the male's "dress" (feather patterns and coloration); they will combine this impression with another one: The owners of such dress are aggressive and must be avoided. Because of this experience, maturing zebra finches become "goal-inhibited" toward males in their courtship, and so will court only females.

Indirectly, the phenomenon of conditioned aversion has played an important role in the evolution of certain animal species. Songbirds can recognize the animals they prey on quite well by appearance; fly-snappers, for instance, learned to distinguish the stingless drones from the sting-bearing worker bees, and to catch only the former. Here, it is evidently worthwhile for aversive (armipotent) or bad-tasting insects to develop some typical and conspicuous appearance. Predators will then learn to watch out after the first bad experience, providing they can, in fact, learn from experience. Aversion to their enemies in many animals, for example wasps, is accomplished by bright colors and conspicuous markings. The animal will behave so as to make their coloration most effective. An inexperienced predator, having once caught such an animal, will retain a visual memory and associate it with the bad experience: It will henceforth recognize this odd-looking creature and avoid it altogether. This provides the wearer of such a "warning dress" with additional protection. Frequently, it even happens that several armipotent animal species display the same warning dress. But we may also find that nonarmipotent animals wear the same coat, which is then called a "pseudo-warning dress." Behavior shaped by the predator's experience seems to be stable enough to keep them away from animals wearing the pseudo-warning dress, as well.

All these examples can be reduced to the following common denominator: Stimuli or stimulus configurations that have occurred in conjunction with a frightening, painful, or otherwise unpleasant experience will henceforth be avoided by the animal. We may consider this behavior as conditioned appetence with a negative sign—conditioned aversion. If the now-aversive stimulus previously represented the goal for an appetitive behavior, that behavior will have become goal-inhibited: Inhibition now applies, not only to stimuli preceding a good experience, but to those preceding bad experiences as well. This inhibition is then coupled with a behavioral command: Watch out and get out!

Aside from goal inhibition, there are also conditioned inhibitions (those based on learning or experience) of behavior elements: If a behavior pattern keeps having unfavorable consequences, it may become associated with an inhibition. One way to teach a dog not to pull at its leash is to give him a collar that tightens automatically and causes pain whenever the dog begins to pull. The dog then develops an inhibition that becomes coupled with the urge to pull, and henceforth that impulse meets with internal inhibition.

If we compare learning from bad experiences with learning from good ones, we can see the following similarity: In both cases there is an association or coupling with initially neutral stimuli or initially neutral behavior elements. We can therefore organize the four kinds of learning described above in the following chart for cross-classification:

	Learned:	
Basis for learning:	*The releasing stimulus situation*	*The behavior element*
Good experience	*Conditioned appetence (approach)*	*Conditioned behavior (action)*
Bad experience	*Conditioned aversion (avoidance withdrawal)*	*Conditioned inhibition (inaction)*

We may elaborate by describing an experiment in which several animals in the same situation showed different kinds of learning. A brief glance at the table (above) will show what happened:

In observational cage, eight yellowhammers were given eyed hawk-moths as prey. These insects have a particular way of responding when frightened by enemies: Accompanied by a hissing sound, they quickly

▷ By using positive reinforcement (reward), we can train a dolphin to perform some amazing physical tasks. Certain natural movement patterns must be rewarded step by step, each in turn, until the animal has learned to complete this high jump of several meters over a horizontal staff.

Conditioned inhibition

Yellowhammers prey on eyed hawk-moths

◁
Above: A pig-tailed
macaque has been trained
on Sumatra to pick coco-
nuts. The animal is wearing
a long chain that reaches
to the ground. Below: a
work elephant in northern
Thailand.

open their wings and display large, colored eye patterns or patches—a
sight that will frighten songbirds and make them fly off. All the yellowhammers were exposed to the moths several times. After a short time, six
of these birds came to ignore the insects' fright reaction and they hesitated
less each time before eating them. But the other two birds became more
frightened with increased experience; finally they backed away from the
moths as soon as they saw them, even without the latter's fright reaction
(see also Fig. 17-5).

The general rule is: When a situation carries positive as well as negative
traits for an organism, repeated experience may strengthen either one or
the other, depending on circumstances. This will then guide the animal's
behavior in the corresponding direction. The organism may become
"habituated" (accustomed) or "sensitized" (disaccustomed) to a situation.
This applies to all higher organisms with a sufficiently differentiated capacity for perception and learning. In the example given above, we would
refer to the behavior change in the six "brave" birds as conditioned appetence (approach) and that of the fearful birds as conditioned aversion
(avoidance, withdrawal).

"Punishment" and
"reward"

Readiness to learn and the animal's performance generally increase
with the strength of either the reward or the punishment. In the case of
punishment, however, learning and performance increase only to the
point where the situation as a whole becomes very frightening for the
animal. When this "overriding tension" takes over, learning diminishes
rapidly. Animals and probably humans with a chronic, overwhelming
anxiety will then become seriously deficient in their readiness to learn,
resulting in apathy. For instance, young dogs or monkeys that were raised
in isolation, without the care of a mother animal, later become overly
fearful (hyperanxious), aggressive, and socially unadaptive (asocial).
Furthermore, they have great difficulty in assessing any kind of experience.
For example, it will take them much longer than other animals to avoid
touching a painful object which is electrically charged or too hot.

Comparing lower animals and higher animals with respect to how
easily their learning readiness can be activated, we can state this general
rule: The higher the organization of an animal species, the stronger the
effects of reward and the weaker the effects of punishment on learning.
Animals at the highest level of the phyletic scale also possess additional
behavior patterns or complexes with a corresponding capacity for learning: exploratory behavior, curiosity, and play. When these behaviors are
activated, learning and performance will be at their best.

On the other hand, it is crucial that the value difference between the
relevant traits—i.e., the positive or negative valence that these traits may
hold for the animal—is large enough to activate the highest degree of
learning readiness. In one case, dogs were first trained to distinguish between forms or shapes by means of patterns which the experimenters
drew on their subject's food bowls. nsCertain patter indicated food in the

bowl, and others were meant to show that the bowl was empty. Attempts to teach the dogs to open one kind of bowl and not the other met with little success. But the experimenters later realized that this was not due to poor form perception but rather to the fact that the dogs simply opened all the bowls, knowing that they were sure to find the food that way. The subjects were much better able to differentiate when forced to jump over a ditch against hanging doors bearing these patterns. Doors with the "right" patterns would open, but those with the "wrong" patterns would not, so that here the dogs fell into the shallow ditch. With this approach, the "value differences" among the various traits were greater, so that the dogs were induced to perform their best by drawing very careful distinctions among the different patterns.

Experimenters can also make the opposite error: If the animal's drive to obtain its expected reward is particularly strong, it may override the entire situation, and performance goes back to zero. If a dog or bird is shown a very desirable piece of food through the bars of a cage, the door of which is open, the animal will try over and over again to reach the food through the bars, although it knows the detour that would lead to its goal. Only when the experimenter holding the food moves away from the bars a little, will the animal suddenly turn and take the way it has perceived in order to reach its reward. The tension that had caused the animal to try reaching the instinctive goal directly had subsided, and our subject was now able to use what it had learned, in this case recognition of place.

Oddly enough, it seems easier for phylogenetically higher animals to combine and remember numerous details of a situation simultaneously than to form a single, isolated association. For instance, a dog being trained to obey the command "sit!" will initially learn all the details of the entire situation as he first encounters it: place, time of day, tone of the trainer's voice, and so on. Subsequently, the trainer must extinguish in the dog all these superfluous situational signs.

In all the previous examples, temporal associations were of primary importance: The initially neutral stimuli or behavior elements had to occur shortly before or during the good or bad experience; only then did the animal learn. But learning may also take place as a result of concurrence or contiguity of temporal and spatial events, that is, without either good or bad experience. However, there has been considerable disagreement over this. The problem is, how can we prove that a situation judged as neutral did not actually have a positive or negative valence for the animal? In all likelihood, this controversy will be resolved only when we have discovered the essential processes taking place in the brain. And so we are left with only the following criterion for learning, when it is based on nothing more than temporal associations: The conditions for the learning processes in question must not include any changing states of readiness, any drive satisfaction or, any changes in threshold; in other words, the only variable affecting the animal's behavior must be the temporal relation

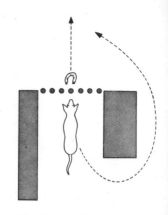

Fig. 22-6. Schematized representation of the detour experiment, viewed from above: The sausage has been placed so close to the cage that the dog is virtually "riveted" to the bars. But when we pull the sausage back in the direction of the arrow, our subject becomes free of this overwhelming force and takes the detour he has learned to obtain the food (curved arrow).

Conditioned reflex

Fig. 22-7. Trained elephant. Above: headstand. Below: sitting up.

Fig. 22-8. Two keepers force a young elephant to take up the position they want him to learn and then actively perform on his own. Standing on a pedestal prevents the animal from moving off the spot.

Fig. 22-9. An older animal that has already been trained is brought in to help train a younger one.

between events. For that reason, the learning processes belonging to this category are really very elementary ones.

When we touch a frog's foot, it retracts it. This behavior is a protective reflex, an example of an "extrinsic motor response" (muscle reflex controlled, not by the muscle's own sense organs, but by receptors lying in other parts of the body). If we take a fine brush and touch a certain part of the frog's body skin every time we touch its foot, a conditioned reflex will develop. This means that, henceforth, merely touching the body skin is sufficient to cause a retraction of the leg. Here, only the spinal cord needs to be functioning, since even a "spinal" frog, whose spinal cord has been surgically cut off from the brain, will learn this conditioned reflex.

A jackdaw had a nesting cage whose entrance was hard to find. At first it always had to look very carefully to find it. During this period the act of searching and looking around was imprinted into the bird's nervous system. When it finally learned the way and really came to know it, the bird still continued to move its head around searchingly in front of the cage entrance, as in a magic ritual. In such cases we speak of motor learning. Acts following one another in time are connected so that henceforth the animal discharges them in fixed sequence. This coupling takes place at that neurophysiological level where the organisms respond to the eliciting stimuli.

We can make an animal perform the same action sequence over and over again, perhaps by a repetitive series of external stimuli or by some external coercion. As a result, the single acts will be coupled with one another, so that in the end they are performed in the same sequence even when the initial controlling factors are removed. For example, if a starfish is turned on its back and allowed to use only two of its arms for turning back over, it will continue to restrict itself to these two "practiced" arms for a while even after we let it free. After then conditioning trials daily for a period of two weeks, this restraining factor stayed in effect for another week.

Within the boundaries of their territory, animals often move about far more quickly and skilfully than they do in a strange area, where they have to reorient themselves with every turn. A mouse that has been alarmed while out in the open and then makes a straight dash to its nest acts "blindly," following a "well-engraved" action sequence. When placed into unknown terrain, however, the mouse is forced to move about slowly and cautiously. It learns the escape routes between the entrance to its nest and the various territorial boundaries by constantly scurrying back and forth.

Horses are trained to perform gaits that they would never use in nature, or only in special situations. This is accomplished through motor learning, just like people train themselves to perform deliberate movement patterns like driving a car or swimming with a crawl. These movement sequences ultimately become so ingrained with practice that they discharge "instinctively." The same applies to many trained circus acts: By means of direct physical force and guidance (in elephants, for example, with the aid of trained conspecifics), by baiting and luring, or by alternately triggering flight and aggression (alternately passing into and moving out of the critical distance), the animal is induced to perform certain movement patterns over and over again until it has learned them by rote. Subsequently, these "aids" are reduced, until all that remains are minimal intention movements by the trainer, to which the animals—having "learned from experience"—respond correctly as signals for the actual behavior (which the trainer no longer carries out).

There is a particularly odd learning process, based neither on reward nor on punishment, that plays a part in the brood-care behavior of the sand wasp *Ammophila*. The female digs a nest in the ground, drags a paralyzed caterpillar into the hole, and lays an egg on top of the victim. About the second day, when the larva has hatched, the mother wasp brings another caterpillar. She will then add three to seven more caterpillars, bringing one at a time, and finally closing off the nest hole. The female starts a new nest even while she has one or two older ones to look after, and so she usually has two or three nests with larvae at different stages of development. Every morning she visits each nest, initially without carrying prey. The number of caterpillars she sees at each location will not necessarily correspond to the age of the nest, since the experimenter could have added or removed some. But these caterpillars do determine the female's behavior for the rest of the day: After her initial "scouting" excursion, no change in circumstances will make the mother wasp behave differently. For example, if three caterpillars are still missing at her first visit, she will bring three in the course of the day and then close off the nest—regardless of whether in the meantime, the experimenter has added or removed any caterpillars. The wasp's learning process is always rigidly confined to her daily "reconnaissance" visits.

Mother lionesses bear their young at some hiding place far from the pride. The cubs stay there for about ten weeks, keeping very still while the mother is out hunting. Upon her return, she suckles the cubs. We know that when they are full neither hunger nor any other drives such as those of self-preservation will be activated. Yet the cubs remain very active, running around in the vicinity of the lair and exploring everything they meet: One of them may suddenly discover a stick, a small bush, or a tuft of grass, then reach for it with its paws and roll over on its back. Often, another cub will want to join in. Two closely observed lion cubs played for two hours with an abandoned ostrich egg; others splashed around with their paws in a brook, and tried to run on the bank along with the current. A young lion cub may also show behavior simply indicating a readiness to play. Running toward another cub, he will pounce on his playmate; work him over and rough him up with his paws; seize him by the cheek, the ears, or the back of the neck; lick him; and roll around with him on the

Fig. 22-10. A sand wasp is seen curving its abdomen forward, stinging a caterpillar it has caught, to paralyze the victim's nervous system.

Curiosity behavior and play

Fig. 22-11. Two lion cubs play with their mother's tail.

Fig. 22-12. Gorilla mothers will also play with their young. This is Achilla with her baby Jambo, in the zoo at Basel, in 1961.

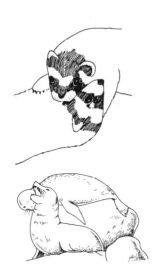

Fig. 22-13. Young polecats as well as seals aim directly at their playmate's throat when biting. But when hunting, adult seals never do this to their prey. On the other hand, their ancestors would have shown this behavior during predation, since they were terrestial carnivores.

ground. The mother may invite one of her cubs to play with her tail by moving it back and forth. Or she may stub a little one with her nose or paw; once the cub struggles and kicks around with all its legs, the mother starts licking it.

Older lion cubs will enact all the phases of their inborn hunting behavior during play: stalking, preparatory posture for a quick attack, pouncing, chasing the partner in flight, or fighting. Blows of the paw during this kind of play—always with retracted claws—are soft and all in fun, yet they are well aimed. Playful bites are obviously aimed at the playmate's throat or nape, as if he served as a "model" for prey. But a lion cub will never bite hard enough to injure its partner.

Erratic locomotory behavior not only raises the probability that an animal meets with some strived-for situation; it also allows for other kinds of experience that may become useful for survival. To this end, many higher animals have developed a specific exploratory behavior that is not triggered by any particular instinctive action readiness. When exploring, the animal can be completely satiated; if it does find a piece of food, it may test the morsel for edibility. But then, rather than eating it the animal will simply remember the spot, hide the food, or carry it to the den. When any mammal is placed into strange surroundings, it will at first do nothing but explore. Only when it knows the area does the animal turn to eating or playing. Many animals mark their new surroundings while exploring, mostly using glandular secretions.

When an animal encounters a new object or situation within already familiar surroundings, this novelty —providing that it is nothing frightening—can release specifically directed exploratory behavior. Corresponding to certain human urges, we call this curiosity behavior. In many cases organisms explore novel objects in certain species-typical ways: Squirrels gnaw at the object, while dogs sniff at it to learn more about it. Curiosity behavior points to the most highly developed behavior patterns, which we may call a drive for making experiences that will serve only indirectly to satisfy vital needs.

The category of play can include almost all behaviors contained in the organism's repertoire. That is why there is a particularly broad range of knowledge to be gained during play. In exploring, the animal becomes familiar with new surroundings, while it uses curiosity behavior to learn about novel things in general. In this interplay with its environment, the organism can develop an individual response and behavior repertoire. Compared to other kinds of behavior more prominent in the wild, zoo animals use much more of their time for play—an indication that play helps to occupy time not—or, in young animals, not yet—spent in satisfying other needs. Play behavior occurs in a "relaxed field," when no other kinds of action readiness are activated.

Only animals at a very high level of evolutionary development, and primarily mammals, are actually capable of playing; this applies especially to the young. Juvenile play behavior consists mainly of instinctive actions or parts thereof. But when they are performed in play, there are several ways, in which they differ from those carried out "in earnest": They occur more frequently, often continuing without signs of fatigue and without their readiness decreasing; and the behavior does not stop when the consummatory goal is reached. On the other hand, the animal switches randomly and spontaneously from one bit of behavior to another, performs pattern fragments in sequences other than those required for "serious" action, or modifies even the behavior parts. All this indicates that the instinctive behaviors we see in play are not instigated by their usual readiness (e.g., by hunger in hunting play and defense readiness in fighting play), but rather they are activated by a specific readiness for play. There is also another reason for assuming that an animal can be motivated by an actual "play readiness": Just like young animals try to get their parents or human caretakers to feed them, they have special means for clearly inviting play. This shows that animals play simply to play, and not for the goals normally associated with any of the instinctive behaviors that may occur at this time.

Fig. 22-14. A young marten is curiously peeking into a closet. There are no limits to what may serve as an object for curiosity; novelty is the only criterion.

During play, the animal may focus predominantly on behaviors belonging to one large category of action. In hunting and fighting games, animals often choose objects that—if only very slightly serve in the role of partner, fleeing prey, or opponent. Thus we see young carnivores playing with balls, and squirrels tearing branches from trees. When an animal accepts a person as a playmate, the human may be sure that he is protected by the animal's "social biting inhibition." While playfully chasing each other, young squirrels act as though in earnest, and the "fleeing" animal always tries to escape from its partner's view. Squirrels in captivity tend to create movement games frequently consisting of difficult runs and jumps, such as somersaults, requiring a great deal of practice. All individual movements are joined into a recurring chain of actions which may then be repeated hundreds of times.

Fig. 22-15. Animals will even accept a familiar person as a playmate.

Ethologists have observed how some mammals invent their own games: Pine martens, otters, and sea lions "slid" on smooth surfaces, a badger turned somersaults, and seals—even those in the wild—would throw pebbles in the air and then catch them. Chimpanzees do everything possible with a new and accessible object, from examining it with their teeth to rubbing it against their skin or the ground. In this way, chimps even discovered how to "draw." When given crayons or colored pencils, these animals quickly saw that rubbing would make the color come off. One of them discovered a completely new variation of this "drawing game": He colored his entire surroundings and even tried to color himself black.

An animal is always encouraged to play when the object or partner reacts in some way, for example when a special sound is produced, when the play object is seen in a certain way that stimulates the animal, or when

Getting a response

a partner does something particularly odd. Such a "response" stimulates the playing animal to repeat its previous action. And because in the context of play it responds every occurrence following the animal's own activity, and since it repeats that recent behavior, the organism learns the lawful consequences of its own actions. This innate operational principle can be regarded as a precursor of one methodological dictum in scientific experimentation: The mere repetition of experiments enables us to distinguish a chance or random concurrence from a lawful relationship between events.

Imitating behavior

Play is also based on the drive to imitate the behavior of parents or playmates. In Basel, for instance, J. Hess observed how a young gorilla followed its mother with its eyes and then performed the same actions she had. For example, the mother repeatedly dipped her hand into a puddle, then scraped off the liquid with the other hand and licked it; the youngster sat by and watched. First its eyes followed the mother's movements, then they wandered ahead of the mother's action to the next step. In this way the young gorilla learned how things should proceed "in theory" by rote or memory. The playful tendency of an animal to imitate the behavior of conspecifics, thereby translating perceived events into its own behavior and thus learning what it observes, supplements the genetic transmission of information with traditionally passed-on information gained by individual experience.

It would be difficult to prove conclusively that in nature play behavior has selective value. From a theoretical point of view, such proof may even be impossible. However, the functional characteristics of play behavior in a broader sense (exploration, curiosity, play, and imitation) do seem to enable organisms to gain the greatest amount of information without danger—experiences that would be of future value in cases of biological need, in situations that are serious rather than playful. At the same time, the animal learns to improve and perfect its general skills during play, and stays "in practice," so to speak. From this point of view, too, the reader may well understand why we cannot irrefutably demonstrate a "direct" biological function for play behavior: Under these conditions, every behavior element in play holds no more than potential applications for the future. From an "immediate" point of view, all play behavior must therefore represent its own end. As for the biological function of play, it is best served when the behavior involved is geared to openness. After all, the best way to gain useful information is to pick up all kinds of information with as little restriction as possible, maximizing the probability that something useful will be included. On the other hand, this openness must have some limitation to avoid a possible overloading of the memory-storage system. It is also very interesting to study the nature of the evaluation system, based primarily on three principles that we have already described: First, in play, animals prefer novel objects and situations; second, animals prefer that which responds to their own actions; and third, animals prefer that which may be imitated. Clearly, these criteria are meaningful enough.

Fig. 22-16. The person who plays with a young member of a carnivorous species may generally rely on the infant's biting inhibition. This is a lion cub playing with its keeper.

There are still many unanswered questions in the realm of learning theory. We still do not know what happens within the nervous system during learning, nor have we located precisely the substances carrying whatever information is contained in the engram. Furthermore, we do not know whether memories are ever stored in a single area of the brain (localized); a number of findings indicate that they are "diffused" throughout the brain matter. But even a person who knows something about learning automatons will have difficulty understanding such a diffuse system, because machines will show a definite loss of certain pieces of information after part of their storage system is removed. By contrast, brain operations on animals have had the effect of a general performance decrement rather than the loss of individual engrams. Detailed microscopic examinations of parts of the brain that are responsible for memory as a whole have yielded little so far. Vertebrates, arthropods and cephalopods display a similar but rather nondescript structure: many thin fibers connected in all directions with all the adjoining elements—indeed anything but revealing architectonics.

And, finally, we do not even know if there is such a thing as forgetting —that is, the erasure of engrams, such as with tape recorders—or whether, by contrast, everything once stored remains with the animal and is merely filed into an "archive" where it remains under normal circumstances. Supporting this theory is the fact that very old people sometimes remember childhood scenes that may have been dormant for half a century. In the coming decades, we can anticipate some very interesting discoveries from research on learning processes in animals and man.

Memory, the brain, and the nervous system

▷
Physical contact with the mother is an important precondition for the normal development of behavior in young primates. Here, a baboon mother with its child.

▷▷
Imprinting and sensitive phase. Top left: Choice experiment on imprinting of the following response in young procosial animals. In a circular runway, two different models hanging from a rod are moved around the apparatus. The duckling knows one of these (here the black one) from an earlier experiment, while the other is new. As we can see, the subject follows the familiar one. Top right: Even objects bearing no resemblance to a natural mother duck can serve to imprint a duckling, providing that the model, such as this ball, is moving. A loudspeaker says, "Come, come." For initially releasing the following response, only motion and rhythmic sounds are necessary; shape, color, and size of the model are unimportant. Middle left: Model experiment for sexual imprinting. The society finch (center) raised by zebra finches ignores the conspecific model, instead courting the zebra finch model and attempting to copulate with it (right photo). Below: A mallard drake raised by large white domestic ducks is sexually (Continued on p. 345)

23   Nonverbal Thinking

Nonverbal thinking, by O. Koehler

Man is descended from apelike ancestors. Among all animals living today, the anthropoid apes bear the closest resemblance to humans in anatomical structure, functioning of organs and behavior patterns. But man also has a great deal in common with organisms at a lower level of evolutionary development. Medical drugs are often tested on rats or rabbits; discoveries made on frogs in the physiology of movement, and on fruit flies (*Drosophila*) or even bacteriophages in the realm of genetic research, may be generalized to man.

The only factor that clearly distinguishes us from other animals is our capacity for speech, and everything it makes possible: self-responsibility, and tradition, religion and creed, art and science. And yet, our ability to speak, and therefore our mental powers, are based on the inventions of our animal ancestors, above all on their "nonverbal thinking." Nonverbal thinking includes all "inner life"—pleasure, pain, memories, wishes, plans, likes and dislikes, hopes, and disappointments—all unnamed qualities still directly experienced in a nonverbal manner not only by higher animals and human infants, but also by human adults.

Psychology, ethology, and physiology

We have a long way to go before we will understand the physiological processes that lie at the base of subjective experience. Since each person can experience only his own self, human and animal psychologists can arrive—epistemologically speaking—only at analogical conclusions. Yet these scientists have found that there are indeed congruencies and transitional stages. It is fruitless to quarrel about the exact beginnings of mental qualities (inner life) within the evolutionary scale. Some philosophers avoid this dilemma by speaking of a "universal animation," or "universal soul." It is more expedient to describe behavior in physiological terms, so far as this is possible at our present state of knowledge. Ethological terms such as appetence, IRM (innate releasing mechanism), hunger, drive, readiness, and so on are working notions or concepts that have proven useful even though the physiological mechanism underlying hunger in the fly is different from that in the sheep. At the same time, we may choose at

◁
(*Continued from p. 342*) imprinted to the white plumage and tries to tread a domestic duck, even though it looks very different from mallard females in both size and color.

which level of biological organization we begin to assign to a physiologically explained drive a subjective affect or emotion, as derived from our own experience. Even our affects are experienced initially without words.

Some people contend that words alone make possible the use of concepts. But this notion would apply only to someone already speaking, and even the concepts are meaningless without some perceptual awareness. In order to learn how to talk, a child has to undergo the reverse process: it must begin with familiar notions such as "my mother," "bottle," or "doll," then associate these ideas with certain sounds appropriate to the situation. Furthermore, these sounds or names may then be used as symbols outside of the original situation. By overcoming the differences (abstraction) and by focusing on the mutual points and (generalization) in several apprehensions, a child forms concepts and learns to apply words to them. Also, it is obvious that a child makes value judgments long before it has learned to say "yes" or "no" and "sweet" and "sour." A little boy who could not talk could understand his grandmother when she said: "Get the teddybear!" To the adult's surprise, he then proceeded to fetch—when asked—everything whose name he already knew. By contrast, dogs require length, step-by-step training to teach them to obey eight verbal commands in random order. The child was able to generalize the verbal concept "fetch" without difficulty, but the dogs needed special learning for this. On the other hand, both immediately generalized the notions of "no" and "shame on you."

Organisms do not pass on their traits simply as fixed qualities. Instead, these qualities unfold within hereditary limits in conjunction with external influences: Both hereditary and environment determine the ultimate form of almost any trait. This applies not only to physical shape and coloration, but also to behavior patterns (see Chapter 15). Nonetheless, for the sake of brevity we customarily say, for instance, that the species-specific locomotory patterns which develop quite uniformly are innate.

Heredity and environment

By "innate releasing mechanisms," (IRMs) we mean the potential range of variation, representing the way an organism may perceive certain external situations and respond with an instinctive behavior. IRMs are described by their effective sign stimuli. In robins, the red breast feathers of a strange conspecific may release fighting behavior in a bird whose territory has been violated. Pecking behavior in young herring gulls can be released by the red patch on an adult's beak (see Chapter 17).

Sign stimuli

The higher the evolutionary development of a species, the more its members can alter and refine many of their IRMs by learning. The IRM provides animals with an innate idea of the significance of certain external stimuli. The animal begins by learning all the relevant details of a situation, for example, the characteristics of a territory it may have claimed, or the personal traits of its mate, parents, or young. Next, these nonverbal conceptions give rise to the nonverbal idea, "my territory," "my mate," or

Nonverbal idea

"my young." Soon the bird makes the nonverbal judgment: "Only my mate or young belong in the vicinity of my nest. Away with all other conspecifics!"

The common capacity to think nonverbally closes the "unbridgable gap" between man and animals. Only organisms already utilizing concepts to which names might be assigned could have been capable of developing language. The fact that, amazingly enough, all the necessary qualities just happened to combine in man alone should not give us cause for arrogance.

Ethologists by no means deny or ignore the differences between human and animal behavior. But it is not their task to draw artificial boundaries. Their aim is rather to examine all comparable features between animal methods of communication and human language, in order to trace the evolutionary path of this one special quality of Homo sapiens. Here, as in many other cases, we have a great deal to gain by looking closely at the child's ontogenetic development, in this case his acquisition of species-specific forms of communication.

Bird songs

Songbirds with their good vocal capacities are extremely well suited for comparative studies. People who object that birds are, after all, only a sideline on our evolutionary family tree have failed to consider that convergences (Chapter 2) are often as revealing as homologies (Chapter 38). Ethologists use all possible means to study phylogenetic processes, and even parallel developments form an important part of the total picture. Insects did not inherit their chitin from fungi, nor Tunicata their cellulose from plants, nevertheless, the substances are found in these animals.

Isolation experiments

F. Sauer raised young white throats from the time of hatching in isolated conditions (deprivation experiment). The males were kept in sound-proof chambers so that each could hear only its own voice. All subjects started to "say" *tseep* at around the ninth day when hungry, and three days later all made the sound *eedat*. By the time they could fly off, twenty-one other sounds were added, in the same sequence and associated with the same moods for all subjects. We therefore assume that these calls are innate. After eating, the birds will even sing the first two calls together in alternating order. Other sounds are gradually added, and so the males compose their juvenile song by constantly interchanging the sounds and trying them out. Once the bird has reached sexual maturity, however, this multiplicity changes, becoming structured into rigid tonal motifs that signal to all others the owner of the territory. Most people know only this territorial song, in no way reminiscent of the bird's rich juvenile compositions.

Unaltered juvenile song in deaf birds

In basically similar ways, the blackbirds (*Turdus merula*) raised in auditory isolation by E. and I. Messmer composed their own juvenile song. Even a young bird that had been deafened very early would utter the same sounds in the same situations as other blackbirds of its age, without ever having heard itself. On the other hand, hearing is necessary if the elements of the juvenile song are to be used for the formation of the territorial song. and, of course, there can be no imitiation of species-specific (nor alien) sounds without the ability to hear (auditory intactness). The nestlings learn their parent's feeding sounds at a very early age. This enabled the Messmers to condition young birds to respond to certain whistling melodies: "When we let hungry nestlings listen to radio music and whistled along at the same time, they slept on peacefully; but as soon as we switched to the training melody, even with radio still playing, all five heads shot up with their mouths open wide."

Chaffinches raised by W. H. Thorpe as Kaspar Hauser subjects developed an undifferentiated beat, measured a little too short. By constrast, one-year-old chaffinches in the wild conform in their territorial song to that of the father and neighbors. This process gives rise to local dialects. One such dialect, perhaps originally an imitation of the call, of the pied flycatcher, was passed on unchanged in the Teutoburg Woods for at least twenty years. As early as the middle of the 16th Century, one Baron von Pernau, living near Coburg, had kept young chaffinches who for the first year of their life were exposed only to the songs of the tree pipit. He then let them free to settle in a part of the woods not inhabited by any other chaffinches. Henceforth, generations of chaffinches continued to sing only like tree pipits, because the young birds in their one sensitive phase had never heard how their conspecifics "really" sing.

Mockingbirds, on the other hand, may choose any model at all; sometimes they repeat imitated sounds within the proper context of a situation. So it is with our grey parrot, which had learned to imitate its previous owner who answered the telephone by saying "hello," and has saluted his departing friends with "good-by then." In the same way, our bird now said "hello" when the telephone rang, and repeated the phase "good-bye —then" to our guests until they really left. The parrot also combined two words it had learned separately, *cooks* and *cuckuck*, to form a new word, *cooduck*. Every time it wanted to sleep it would then screech out this word until we covered its cage and turned off the light. This command, invented by the bird and uttered in the most penetrating manner, had an absolutely reliable effect. In a similar way, a small child learns to understand the meaning of his first words within the context of a situation, sometimes inventing new words himself.

In a circus, a lady performer asks her grey parrot up to thirty-five questions in random order, and the bird "answered" each correctly. For example, to the question: "What is your name?" it would say "Lora Eston." Or: "How do you greet our audience?" it answered: "Good evening." It could also answer a few questions in another language. We could not discover any hidden aids. This excellent feat of learning corresponds to the duet songs of birds in the wild—a method of communication that, according to W. H. Thorpe, has been found in roughly 120 species and is particularly well known in African butcherbirds. Both partners

Formation of dialects

Imitating sounds within their proper context

Duet singing

learn the entire verse, and then each sings half of it either at the same time the other bird is singing or alternately. E. Gwinner reported that one male raven frequently barked like a dog, while his mate gobbled like a turkey. Whenever they were separated from each other, just like the deaf blackbird raised by the Messmers, she would start to bark and he would gobble, as if calling each other by name.

P. Marler observed that, white-crowned sparrows (*Zonotrichia leucophrys*) develop all their species-typical sounds without a model and even without hearing themselves, combining them to compose their juvenile song. The only feedback they have is through their "muscle sense." White-crowned sparrows that can hear will learn additional sounds, as well, and they never forget them even if they are deprived of hearing later on. M. Konishi discovered that these birds, having heard a model at some time in the past, need only to repeat this verse and to actually hear themselves singing it once after their change of voice. People who become deaf have the experience with language—a correspondence between different species giving some impressive evidence for the value of phylogenetic convergences.

The thirty-five sounds used by a monkey species cannot be compared with the 40,000 words in a human language, though the mimicry or play of expressions is comparable in monkeys and man. But the fact that even anthropoid apes cannot imitate sounds is probably based on differences in brain structures rather than in vocal apparatus. As noted recently, the larynx (voice box) of mammals and newborn humans is situated at such a high position that the epiglottis (laryngeal cover) touches the soft palate (velum). This keeps the respiratory and food tracts separate from each other, so the organism cannot swallow the wrong way. It is only in man that the larynx descends, especially in the first year of life, so that for a short part of the way both tracts become fused into a new chamber, absent in other mammals. In dummy experiments on the vocal apparatus of rhesus monkeys, P. Liebermann, D. H. Klatt, and W. H. Wilson (1969) claim to have shown that the resonance of this space is essential for clear vocalization. Other statements that only man has muscular fascicles leading to the vocal cords have not been substantiated. Furthermore, the snout or nose of the growing mammal changes its shape much more than our mouth cavity, which becomes larger but keeps its shape. In Steinheim man, the hard palate was arched to the same extent as in modern man, whereas that of Neanderthal man was much flatter. Thus, according to F. A. Kipp, Steinheim man must have had fairly well-developed speech. while Neanderthal man would hardly have talked at all. Perhaps this was the reason why Neanderthal could not compete with Cro-Magnon man in the late ice age. But when my dentist flattened my palate like that of Neanderthal, and even filled it in horizontally like that of rhesus monkeys, I found that although I had some difficulty speaking, I remained perfectly intelligible.

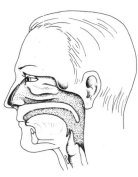

Fig. 23-1. In orangutans, the epiglottis remains connected with the soft palate. In this way the respiratory tract is separate from the food tract, and the organism cannot swallow the wrong way. In humans, the epiglottis descends and an opening is created that supposedly enables us to talk. Respiratory and food tracts lead through a common chamber.

In order to compare pre-language sounds made by small children with the development of song in birds raised without models, U. Grabowski and M. Dieseger-Knoll recorded the babbling of their first children on a tape recorder. In addition to hearing themselves, the infants were exposed only to adult speech and radio music; no one was allowed to babble to them. Whenever possible, every new sound was immediately recorded, noting the date, possible cause or occasion, and accompanying behavior. Many of these sounds could be easily identified as vowels, diphthongs, and consonants. The infants developed all the sounds necessary for speech without help but—in contrast to the birds described earlier—in different individual sequences. At about five months of age, they began to make repeating sounds, such as *da-da-da* or *dvay-dvay-dvay*, as if practicing.

As early as 1942, R. Jakobson, following on the findings of A. Grégoire, wrote: "At the height of its babbling speech, the child is able to produce all imaginable sounds." But, to our general surprise, this wealth of sounds becomes lost to a great extent when children start imitating words, as with birds when they develop their territorial song. According to Jakobson, a child imitates the things he hears only with considerable exertion, often requiring a long time to learn to produce all the palatal, hissing, and flowing sounds. Some children start doing this early close to the end of their babbling stage; others take months or even years even though they may have learned to understand language much earlier. Similar to the way that newborn individuals and those born prematurely can paddle in the water without help, while older children have to take swimming lessons, those learning to speak late, having lost everything they could previously babble, must learn these sounds all over again. Jakobson claimed that when a child has problems with his speech, the first sounds he loses are those he learned to imitate last; the sounds learned earlier are retained the longest. Conversely, recuperating children relearn these sounds precisely in their ontogenetic sequence.

If infants in all societies really do babble the same way—babies born deaf babble in the "normal" manner, too—then the entire phonetic repertoire of man should be innate, just like the juvenile song compositions of birds. In order to learn how to talk, a child need only learn the words of his language, which acquire meaning when the infant hears them spoken at the same time that he experiences the matching nonverbal idea. After that, he has no trouble learning grammar.

Animals communicate by means of expressive movements, touching and feeling, making sounds, and transmitting odors. This enables them to say: "Come here, go away, here I am, help me, feed me, there is danger, let's fly away," and many other things that conspecifics have to know about each other. In addition, the external circumstances help to convey meaning. Humans, too, when placed in surroundings where everyone else speaks a foreign language, have to resort entirely to the innate means of communication that are shared by all people.

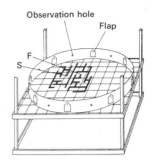

Fig. 23-2. In this elevated maze, mice were required to learn the shortest path from start (S) to finish (F).

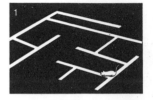

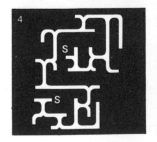

Fig. 23-3. Four variations of the same maze. Without additional training, the mice ran to the finish point almost without errors.

Forming new concepts nonverbally

The only animal that we presently know is capable of communicating information outside of the situational context is the honeybee (Chapter 33). By means of its waggle dance, the leader of the dance "tells" her co-dancers in the colony what she has just discovered in the outside world: the nature, value, distance, and direction of some promising food supply or a new home. This information is conveyed with such precision that the others can find the exact route to their goal. These insects are born with the ability to express meaning in this way, and to understand it; but they cannot learn new "words." Humans however, can create and assign names to as many new ideas and concepts as they wish. But man is not completely alone in this, for we have seen earlier that higher animals can also acquire new concepts. We have described IRMs as innate notions. In addition, higher animals experience (as we do) the constancy phenomena, subjective spatial relations (front-back, right-left, above-below), and all perceptible forms of comparison (lighter-heavier, bigger-smaller, etc.). But do they also form new nonverbal concepts?

Waggle dance of the honeybee

One familiar example is a dog learning to open a door purely by self-training: The animal becomes impatient and jumps against the door from the inside accidentally pushing down the handle with his forepaw and thus opening it. It is remarkable enough if, after this incidental success, he should henceforth open the same door in the same way. But if the animal then starts opening any door from the inside, or even from the outside by moving backward, we know that he has definitely made the proper mental connection, or transformation.

The learning capacity of free-living animals may best be observed on so-called civilization followers—species that become used to the proximity of humans and take advantage of their facilities. The songbirds in our garden know exactly where we put their seeds, expect to be fed at a certain time, and return to the same place in the fall in anticipation even though they were not given seeds all through the summer. Young birds may be completely spoiled in this way, because they do not need to find out the natural sources of food when they learn to become independent. Occasionally, a pair of titmice will move into a mailbox and many a great spotted woodpecker learns to open these artificial nests at the very spot where it can find and eat the young titmice. Gulls, sharks, and dolphins follow fishing boats and steamships, waiting for morsels of food.

Animals can do amazing things in their orienting behavior (see Chapter 11), starting with crabs and spiders. These organisms know precisely that part of the riverbank or seashore where they live, can find it equally well from water and from land, and have their own particular shelters. Each year, migratory birds return to their familiar brooding places (see Chapter 13). In the fall, ringed plovers leave their beach, and the area becomes crowded with people; the next spring, everything has been laid waste, with the surf rolling over their former nesting places—yet they remain. Young salmon migrate from their fresh water streams into the, sea then return after several years to spawn. We now know that they are guided by scent, which leads them through currents and rivers back to the very brook where they were raised. They must learn where specific scents belong, and then remember this "scent melody" for many years to play it back in reverse on their way home (see Chapter 7).

Maze experiments are another way of finding out what figurative conceptions animals have of familiar paths. In an elevated maze made of aluminum strips, mice had to learn to avoid all the blind alleys and, without turning back, to reach the other end by the shortest way. There they were picked up with a large spoon and brought back to their cages; this was the only reward. The shortest way was 402 cm, and the total length of all the strips was 725 cm. When the mice had mastered this task, the maze was distorted and modified in four ways (Fig. 23-3): 1. Doubling the linear dimensions of all the pathways but keeping angles the same; 2. changing the angles to alternately 45° and 135°; 3. reconstructing the maze as a mirror image; 4. rounding off all the right angles. In every case where a mouse was presented with one of these modified maze forms, it hesitated at the first point that looked different, then ran back to the starting point, making a few other similar errors. But then suddenly it understood: Without further training each mouse would run the changed path to the end almost without mistakes. In the two distorted mazes they cut across all the sharp angles; while running the "doubled maze" for the first time, they stopped in the middle of the longest pathway. When the mice were presented with each of the new forms once and then in random order, their errors were almost nil.

Blind mice learned as quickly as sighted ones. In several blind animals who could run the initial training maze without errors, the experimenters then destroyed the primary visual center in the brain. The mice recovered quickly and showed no loss of learning, but they were no longer able to adapt to the modified mazes. These results lead us to conclude that a conditioned mouse has something which W. Köhler terms a "figurative image of the path," and is able to transpose it the same way as we do with our writing. Even when an animal has learned something without the use of sight (peripherally blind), it seems to need its visual center to create such an image or conception.

From 1926 to 1959 we carried on experiments on how well animals could learn to "count" non-verbally, with the careful exclusion of irrelevantly helpful clues that the experimenter may give unwittingly. When presented with very different kinds of tasks, which these animals usually learned without negative reinforcement (punishment), all the species we investigated reached the same upper limit for the two capacities—simultaneously discriminating different quantities, in short, "seeing numbers"; and operating with quantities, presented in succession. This upper limit was five in the pigeon, six in the parakeet and jackdaw, and seven in the Amazon parrot, the magpie, the squirrel, the raven, and man (without

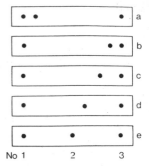

Fig. 23-4. A pigeon is trained to choose a set of two objects, and always picks the two whether left (a) or right (b), even when the food grain on the inside is moved toward the center (c, d). But once the inside grain is placed directly in the center, so that the sets are eliminated, the pigeon chooses grain 1 and 2 as often as grain 2 and 3 and grain 1 and 3; in other words, the animal picks up two successively and leaves the third.

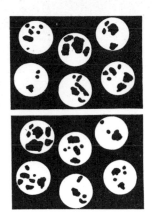

Fig. 23-5. Example of a simultaneous-simultaneous task. There are 2 of 100 slides shown to ravens and 100 human subjects (presentation was of equal duration). Top center of each slide has the standard pattern, here with 4 and 6 patches. Out of the 5 surrounding patterns, the subject has to choose the one containing as many patches as the standard. Ravens and humans did equally well on this task.

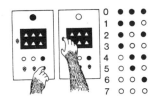

Fig. 23-6. A chimp sees 6 triangles and "writes" the binary number 6 by erasing the right light of the 3 (black). The light above the triangles then lights up (white) to indicate that the choice is correct. After the chimp presses the button to the left of the triangles, his reward drops out through a lid. Right: Numbers 1–7 in binary notation.

using his ability to count verbally). One grey parrot was able to reach seven in the simultaneous trails but eight in the successive ones; this same limit has also been reached by H. Zeier's best pigeon, trained with the Skinner method (see Chapter 22). But Zeier's pigeon was given the advantage of rhythmical stimulus presentation, whereas our animals were forced to deal with completely random (arhythmical) sequences in their successive trials, making the task considerably more difficult.

Any normal child will have no trouble understanding that three sheep seen together, three successive bolts of lightning, or three strokes of a church bell one after another all have in common the number three. The child does not need any help in making the connection between the simultaneous and sequential meaning of this number. If he turns his eyes from the cookie jar to his mother, and she raises two fingers, the youngster will reach for two cookies—he operates with visually perceived numbers. Except for one pigeon, which had been "counting" simultaneously and then managed to make the simplest possible transition spontaneously to successive counting, our animals had a great deal of trouble learning to make this connection even though they had already mastered both the simultaneous and the corresponding successive tasks. A variety of associations are possible, and we did succeed in teaching them to the subjects:

1. Simultaneous-simultaneous tasks require the animal to look at a group of different quantities and pick out the only one with the same number of patches as the standard pattern. Jackdaws learned to do this up to five, ravens, magpies, squirrels, and our grey parrot up to seven. The opposite task, namely, to pick out the only different quantity from a group of otherwise similar ones, was mastered by the squirrel up to six.

2. Simultaneous-successive: Operating with visually perceived quantities, for example, picking up as many baits as indicated on the patterned lid. Jackdaws learned to do this with two and four, and the Amazon parrot, with two and three. In dealing with a series of tones (when a recorder, tuning pipe, organ, or piano was played back singly from a tape recorder), the grey parrot compared one tone, while with two-part music it compared two. Musically inclined people, by comparison, were able to recognize trios 39 percent of the time, quartets 20 percent, and quintets 12 percent.

3. Successive-simultaneous: Visually perceiving operated quantities, for instance, picking up bait and then choosing a group of spots with the same number. Our Amazon parrot accomplished this task with one, two, three, five, and seven, the squirrel with two, three, and five, and the magpie with three and seven.

4. Successive-successive: For example, operating with a series of visually perceived objects. The grey parrot learned to respond to two successive light flashes by looking for two pieces of cookie in a row of bowls with varying contents; three flashes caused him to look for three pieces of cookie. After that, he learned just as easily the double tasks of three flashes

to four cookie pieces, 4:5; 5:6, and even 6:7. After the double task involving 2:3, the grey parrot responded just as well to tones of the recorder or pipe as to the light flashes, without having to learn the new connection first: this transfer was therefore spontaneous. With a rhythmical stimulus presentation and a different pitch, the animal frequently learned the association better than before. Since the gray parrot does not hesitate to equate things it sees with things it hears, we may safely assume that it has the nonverbal concepts of two and three.

P. Lögler had a very difficult time convincing others that this hetero-modal transposition (the transfer of visual percepts to the auditory modality, and vice versa) really took place, although he was not the first to demonstrate this phenomenon. Back in the 1920s, N. Kohts, working in the Suchumi monkey station on the Black Sea, taught her chimpanzee Joni to make choices from visual samples which it then transferred spontaneously to the tactile modality: Without looking, the chimp was required to pick out of a bag containing variously shaped stones the one matching a sample form, such as a triangle. This learning task gave the animal no trouble at all. By contrast, G. Ettlinger's experiments with rhesus monkeys in 1960 were unsuccessful. Better results were obtained with white-fronted capuchin monkeys (*Cebus albifrons*) tested by B. P. Blakeslee and R. Gunter in 1965. One orangutan and two chimpanzees trained in two-way choices by R. K. Davenport and C. M. Roger in 1970 learned to choose, by touch, like-shaped forms matching up to forty different sample objects or their black-and-white images. The apes accomplished this in 500 trials, while our best animal subjects needed fewer than 100. But even human subjects, when exposed to a similar experimental design not involving verbal explanations, often took quite a long time finding and remembering the solution.

If tasks like these are integrated into a natural situation, the animal frequently learns to perform them right away. For example, by pointing his finger, H. Croze (1970) drew the attention of his tame carrion crows to three white cockles lying close together, under which he had hidden some meat. As a result, these birds started to look for white shells and stones dispersed over a large area out in the open. The same thing happened when Croze showed the birds yellow or red shells. He needed to present each new pattern only once for the crows to learn what they had to look for. Wild crows lured with bait behaved in the same way. One bird raised by hand even learned to discriminate easily the finest of details, which the wild crows would overlook. Such behavior of hunting for the "search image" (see Chapter 15) belongs to the natural repertoire of other birds as well, which have periods where they concentrate on momentarily abundant sources of food. Choosing from samples, hunting for a search image, and set perception (spotting and isolating certain parts from complicated patterns) require the capacity to form nonverbal and in part transferable concepts, or to acquire such concepts when presented with a

Transposing visual stimuli to auditory and tactile ones

Learning faster in the wild

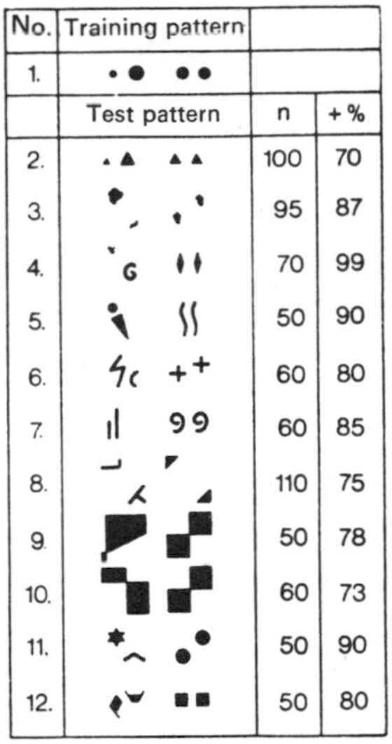

No.	Training pattern	n	+%
1.	• • ••		
	Test pattern	n	+%
2.	• ▲ ▲ ▲	100	70
3.	• , , •	95	87
4.	' G ∤ ∤	70	99
5.	❮ ⌇⌇	50	90
6.	4c + +	60	80
7.	ıl 9 9	60	85
8.	⌐ ⌐	110	75
9.	▪▪ ▪	50	78
10.	▪ ▪▪	60	73
11.	★ •	50	90
12.	♪ ▪▪	50	80

Fig. 23-7. A small civet cat was trained to discriminate two identical from two different dots (No. 1). In the first twelve transpositions after training, the cat spontaneously chose the dissimilar patterns. n = no. of tests. +% = percentage of "correct" choices.

Fig. 23-8. The "theme" derived by E. Tretzel's crested larks from the whistling of a shepherd. Notation is two octaves lower.

sample pattern so that corresponding objects are recognized in different surroundings.

In 1965, C. B. Ferster required two chimpanzees to choose from two sets of three lights each, the one indicating the number represented by a sample pattern. Here, light-off meant zero, and light-on meant one. If at least one light was burning, seven different hook-up combinations were possible; in binary notation these represented the numbers one through seven. Chimpanzees would certainly not know this, nor would most human subjects. According to Ferster, the chimps mastered this problem after "hundreds of thousands of trials"; in the end they even managed to "write" the "number" they saw, using the same notation. This amazing feat of learning, as Ferster himself admits, is of course not really counting objects in the arithmetic sense. Although humans are able to assign numerical terms to these different patterns, the names themselves have nothing to do with a clear concept of numbers.

According to A. Rossmann (1959), fishes can only discriminate one from more than one. N. Pastore reported (1961/62) that two canaries had learned to pick out the smaller of two visually perceived quantities, and to choose the second or third in a row of identical objects even when other objects were interspersed with these in several ways. Without interference from other objects, the birds mastered this successive task up to five. One canary even learned to step into the ninth of a row of equally baited chambers.

J. C. Lilly (1967) claimed that one dolphin, hearing a person recite ten nonsense syllables to him in rapid succession, answered with an equal number of his own calls. A rhesus monkey belonging to E. Kühn (1953) was able to discriminate 13 from 12 dots of equal size, a task that human children, according to E. Becker, can master up to 17 from 16. But these authors did not control for possible unintentional aids or clues as well as we did in our experiments.

In his book, *Behavior Mechanisms in Monkeys* (1933; deserving far more notice), H. Klüver wrote of crab-eating macaques, capuchin monkeys, spider monkeys, squirrel monkeys, and ring-tailed lemurs (a prosimian species from Madagascar) that learned to choose the heavier object, the rougher buzzing sound, and the visually larger area—all the way down to a ratio of merely 1.06:1, regardless of shape which may be square, circular, polygonal, dotted circular, or even cubic. A small civet-cat (*Viverricula malaccensis*; family Viverridae) was reported by B. Rensch and G. Dücker to have transposed the concept "crooked versus straight," which it had learned from a sample, to sixteen pairs of figures, and the notion "even versus odd" to thirty-three new pairs—all without difficulty. S. Knecht's parakeets were able to distinguish fourths from fifths in a series of tones, regardless of pitch—that is, these birds could pick out intervals on a relative basis. Animals therefore have the nonverbal concept of the comparative (degree of comparison) just as we do.

Melody is defined by musicians as a certain sequence of intervals, in most cases rhythmically arranged, taking shape in time and thus transposable in its absolute pitch (keys), in tempo, and in timbre or tonality; its theme may be varied in all its dimensions in a virtually unlimited number of ways. Could a person who has memorized all thirty-two of Bach's Goldberg variations correctly guess the theme in its original form? Yet crested larks, according to E. Tretzel, managed to create and remember a pure and constant theme out of the many variations in the whistling of an obviously tone-deaf shepherd. Even his sheep dog immediately recognized the theme of the birds' song.

As with our every-day speech (not to mention poetry and song), the songs of birds have rhythm, emphasis, and melody. In transposition experiments run by J. Reinert, his Indian elephant learned faster and more than his parakeet; a jackdaw gave the best performance. All three animals have what we call absolute pitch, yet they recognized their sample melody even when it was played in a different key, with a new beat, and in a different tonality. The elephant even put up with changes in intervals, as long as his "positive" three-tone signal went up-down, and his "negative" one down-up; nor did it matter when the quarter-notes were split into triplets or eight notes or combined into chords. The jackdaw distinguished four-four time from three-four time, both hummed and played, and in any key; it grasped the most subtle features of rhythm. W. H. Thorpe compiled a particularly long list of impressive examples illustrating auditory concept formation and musical facility in birds.

In 1921, W. Köhler published his pioneering work on chimpanzees in Tenerife. He did not doubt that his chimps showed true insight when they solved an unfamiliar problem, not by accident in the course of trial-and-error manipulation, but by seeing a connection and having the "right idea" immediately when faced with the problem. For instance, the chimpanzee Sultan fitted two sticks together to reach at a fruit, while others piled boxes on top of each other to climb up and reach for the fruit. Even in ourselves we can recognize all the stages of problem-solving, from trying out solutions by direct trial-and-error manipulation, to seeking various alternatives by thinking about them, to hitting upon just the right idea in a flash of insight.

Until recently, only man was considered capable of using tools. Yet the nest-building weaver ants have a technique for rolling up tobacco leaves: While some hold the edges of a leaf together, members of a second group will each hold a thread-spinning larva in their mastigatory ridges and pass it back and forth like a weaver's shuttle. Spider wasps (*Ammophila*) seal the entrance to their egg chamber by beating it shut with a stone held in the mandibles. It may be assumed that both these behavior patterns are purely innate. On the other hand, the Galapagos woodpecker finch, which uses a cactus spine to poke around in a tree trunk for insects, may have to practice until it knows how to choose a suitable thorn, how to

Fig. 23-9. One-year-old chimpanzees who had been raised in a cage and are now adapting to life in an open enclosure. Here, they are learning to climb a tall fir tree by means of an existing track and a branch they fetched for themselves.

shorten it when necessary, and to store it in an appropriate spot. I. Eibl-Eibesfeldt observed how a satiated woodpecker finch in his cage stuffed boiled or live meal worms into cracks and furrows, then found a tool and poked them out again. Similar behavior was seen in mangrove finch by E. Curio and P. Kramer. J. van Lawick-Goodall reported how individual Egyptian vultures living in the wild took a stone in their beak, lifted it up as far as possible, and hurled it repeatedly onto an ostrich egg with a downward jerk of the head, until the shell broke. Other vultures looked on and joined in the feeding. Some birds take ants in their beak and wipe them through their feathers. Hand-reared warblers belonging to F. Sauer played with stones by letting them fall on glass; and N. Tinbergen wrote about herring gulls that would collect pieces of iron from the yard of a screw factory, letting them drop on well windows, breaking the glass. In both cases, the birds probably had no other motives but to hear the pleasant sound of clinking glass. Bower birds collect all sorts of colorful things to decorate their nest. While floating on their back, sea otters place a stone (retrieved from the sea bottom for that purpose) on their stomach, then proceed to crack open any food with a hard shell by smashing it against the stone.

Baboons crack open crustaceous animals and hard kernels by using stones; they also throw stones in self-defense, and patch up wounds with grass. Capuchin monkeys crack nuts between two stones. B. Rensch observed how one of these monkeys, who liked to play with various objects by placing them in different ways, danced for joy as he managed to get a potato masher to stand on one end on its small plane surface. A. Nolte and G. Dücker watched an infant capuchin monkey invent various games with tools by trial and error, during which it may have used some behavior imitated from its mother. A. Kortlandt compared throwing and hitting behavior using sticks in gorillas, chimpanzees, gibbons, five other Old World monkeys, and capuchin monkeys, which are natives of South America. A chimp was observed to mash up a piece of apple with a croquet ball, hammer a nail into the ground, use a branch as a lever to lift a board, use the same branch for vaulting and climbing, and to use a knife as a file or for peeling a branch. According to M. Meyer-Holzapfel, these were all games which the chimp played for their own sake, and not for any other purpose. The free-living chimpanzees that J. van Lawick-Goodall has studied for many years at the Gombe Stream Reserve would scratch open the entrances of termite nests, insert a thin stick, and wait until the termites had attached themselves to it by biting, then pull out the stick and lick off the insects. The chimps would start breaking off their sticks on the way to the termite hill, measuring them off and occasionally using them over again. One of them used sticks of up to 1 m long to work on a freshly opened ground nest belonging to the ferocious *Durania* ant. To get water out of a tree hole, the chimps sometimes crush leaves into sponges, dip them in the hole, and suck out the water. H. Hofer has listed ten other methods used by chimps to obtain drinking water.

Tool-using in monkeys

N. Kohts presented a chimp with a metal tube and a variety of objects that he might use to get at bait hidden in the tube. The animal had greater success with sticks suitably modified than with bent wires. Frequently he would shove several of these objects one after another into the tube, always the longest object first, until he got the bait. But in 111 trials, only twice did this subject join objects together, in the way that W. Köhler had described with his chimpanzee Sultan. Again, the activity itself was often more important than the reward.

J. Höhl's female chimp Julia learned to open fourteen different containers in a certain sequence, and each with a special key. Here, one box contained the key to the next one, and only the final box offered a reward. Next, the experimenter varied the sequence and positions of the boxes, and also offered the chimp a choice of keys which enabled her to shorten the sequence. Without further practice, Julia made nonverbal associations and drew the right conclusions. Out of fourteen additional locks with a far more ingenious construction, Julia opened seven right away, and the rest soon after, using trial-and-error techniques, displaying an amazing degree of manual skill. "We are justified in speaking of intentional behavior using forethought," writes Döhl. In the course of even more difficult trials, Julia was able to handle detours with fifteen different subgoals, even when their position was changed. In time, Julia learned to cope with progressively more intricate mazes involving two winding sets of passages, of which only one led to the exit. The chimp's performance equaled that of students. In this task, she would always look for the exit first, using it as a starting point for visually pursuing the complicated passageway back to the start.

At the Delta Primate Center (Louisiana), E. Menzel and H. Hofer observed the process by which one-year-old chimpanzees, which up to that time had been kept in cages, became accustomed to an open-air enclosure. These animals approached a 3-m-long ladder only with great hesitation. When the chimpanzee Rock, who was already familiar with the enclosure, climbed up a tall pine tree, the others stared and screamed with fear. The trees were protected with an electrically charged wire fence against having their bark peeled by the animals, but Rock showed the younger chimps a way to get around the wire: He would produce a straight branch about 4 m long, hold it at one end while climbing up one of the 3-m-high iron tracks running through the area, drag up the branch, lean it from the track diagonally against the tree, and climb up on it. The "pupils" soon joined in this venture (Fig. 23-9).

In this way, "traditions" arise: One individual invents something and the others imitate him. The higher the animal stands on the phylogenetic scale, the less restricted is its behavior. Monkeys quench their thirst, for example, in many ways: They may lap up water with their tongue; dip a hand into water and lick it off; lick water off of leaves, small sticks, or the fur of a conspecific; dip their own tail tuft, a sponge made of leaves, or a

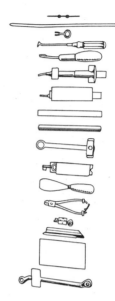

Fig. 23-10. The female chimp Julia learned to open various containers with these tools. Each box contained the tool for opening the next.

rag into water and then suck on it; suck water through a straw or from the hollow of their hand; and finally they may learn from humans how to drink from cups, bottles, and spoons.

Similarly diverse behaviors were observed among forty-two orangutans in zoos, studied by F. Jantschke. As an example he cites eighteen ways in which these animals used sheets and rags. The male orang Katjong (Nuremberg) was particularly imaginative in his play, always tying up ropes and sheets with different knots and undoing them again—for purposes very similar to what we would be doing, like making hammocks or swings.

All attempts to teach chimpanzees human words have failed. Even gibbons, whose vocal ability is very good, do not seem to imitate strange sounds. The Gardners therefore embarked upon a very fruitful venture when they decided to teach their female chimp Washoe the American Sign Language (ASL) used by deaf-mutes, whose manual signs represent words rather than letters. Washoe lived freely in her own trailer and yard, spending her days almost exclusively in the company of familiar people who used only ASL in her presence. She learned these signs without any negative reinforcement, and was rewarded by increasingly good communication, the joy of success, and the pleasure of playing and being tickled (which she liked most of all). Soon she understood and gave intelligent responses to the questions: "What is this?" and "What do you want?" After twenty-two months of continuous teaching, Washoe knew thirty-four signs. Just as small children go through stages of calling every animal a "doggie" and every machine a car, Washoe sometimes generalized her concepts. For instance, she labeled tobacco, tobacco box, and the smell of roast alike as "flower;" until she learned the gesture for "odor" or "smell;" After having learned her eighth or tenth sign, Washoe began to form and correctly use her own sentences of two or more words, such as: "Come tickle right away, quickly, more!" Deaf-mute visitors were able to communicate with her from the start. To entice her down from a tall tree, the only way was to play "sad"—whereupon she immediately came down to comfort the person.

Today, Washoe lives together with thirty-two other chimps at the research station run by W. B. Lemmon in Oklahoma, where the ASL experiments, supervised by R. Fouts, are carried on more extensively. Several other animals have already learned these signs. Fouts, taught Washoe while she was living with the Gardners. There, at the age of five, Washoe had already mastered several hundred ASL signs, used them to create her own multiword sentences in proper situational contexts, always making intelligible statements, expressing wishes, or asking questions. The Gardners have reproduced the full sequence of 132 of these signs, giving their meaning as well as extensions and limitations of meaning. These include seventy-three nouns, among them fourteen names belonging to Washoe's caretakers, twelve food labels, twelve terms for toiletry or grooming ob-

Fig. 23-11. In the Nuremberg Zoo, the orangutan Katjong developed an amazing variety of ways to tie and knot ropes and sheets.

jects, and five for animals; there are also twenty-two verbs, nineteen adjectives (including four colors), five adverbs, and four pronouns. Washoe can use the negation "no!" and the inquisitive pronouns "who" and "where that," and will point, for instance, to the part of her body where she wants to be tickled by this or that person. She will also, as another example, make the sign for "drinking" when seeing the picture of a drinking person. For herself, she uses the sign "Washoe" the twenty-sixth sign she learned or "me" (her thirty-second sign). All this leads us to assume that Washoe uses her signs in the same way that very small children use their words. In the future, other chimps may learn these signs faster, since even learning takes practice.

D. Premack taught his female chimpanzee Sara to place variously shaped tokens representing certain words onto a magnetic board. She was much more restricted than Washoe, and could not invent new signs of her own. When the experimenter placed words one under the other to make a sentence, Sara would answer to the right of the board and would obey written commands such as: "Sara, put banana pail apple plate!" She learned the procedure of choice from sample, and transferred this correctly to the tokens, performing less well, however, in the course of repeated trials. The way her vocabulary was put together was similar to that of Washoe, Sara, too, was able to deal with conjunctions, for example: "If Sara gives Jim apple, Mary gives Sara cherry." But we start to wonder when reading that, in response to the question: "What is the relationship between the token apple and the object banana?" Sara was reported to have answered: "Not its name." It simply does not sound like a chimpanzee. Washoe is more convincing because she always acted like a chimp. Furthermore, we can be certain that Washoe was not given any unintentional aids or clues, while with Sara this possibility has not been eliminated.

Could we say, then, that these chimpanzees have learned to talk, write, and read? After two conventions held specifically to discuss this problem, no consensus on the definition of language was reached. I would hold that, in her constant interaction with humans, Washoe learned to express her nonverbal thought processes in words and to form simple but situationally correct sentences consisting of several words, very much like small children do. This contradicts my previous earlier statement that animals, such as mockingbirds which use human words in their proper context, could never pass beyond one-word sentences, and that only a small child could do better than that. On the other hand, chimpanzees are different from children in that these animals do not seem to strive for a larger vocabulary.

The case of Helen Keller has become world-famous as an example of how a person could communicate through sign language. At age 18 months, illness left her blind, deaf, and mute. By means of manual signs that she had invented herself, and to some degree perhaps had also been taught, Helen Keller was able to communicate particularly with her mother in very limited fashion. When the child was six years old, the

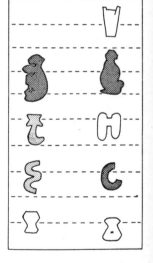

Fig. 23-12. D. Premack taught the female chimp Sara to attach different figures representing certain words to a magnetic board. Here are two sentences. Left: "Sara jam bread take." Right: "Not Sara honey biscuit take."

teacher Anne Sullivan began to instruct her, and immediately started to "write" letters of the blind alphabet into Helen's hand. For example, she would spell out the word "doll" when the child happened to be holding a doll in her arm. The word stage was therefore passed over entirely. After three weeks, as Helen was given the sign sequence "water" in one hand while with the other hand she felt water pouring from a pump, she understood in a flash that for every nonverbal concept there is a letter sequence to name it. She then embarked upon a veritable barrage of questions, just as happens frequently with other children, and thereafter her vocabulary increased at an amazing rate.

There are small children who learn to spell at a very early age, and may even start right off with a typewriter. Others have a difficult time at school because they cannot adjust to the look-see method (sentence, word, letters). The important role that sign languages can play is illustrated by the ancient Egyptians and the Chinese, who created great cultures with the help of their language symbols written without letters.

24 Imprinting as a Form of Early Childhood Learning

The preceding chapter has illustrated that animal species may gather experience in many different ways, and that individual learning processes in turn are characterized by certain distined properties. Animals may learn through their own experiences through "trial and error," or simply through observation and imitation. They learn new movements or combinations of movements of certain vitally important environmental conditions. Finally, the various habits of an animal species may have given rise to dispositions for learning certain tasks with greater or lesser ease.

Learning processes also vary in that there are phases of an animal's life when they are likely to be most effective, or to which their occurrence may even be entirely limited: A certain experience can have a very different effect on an individual, particularly in the nature of permanence, depending on the stage of life in which the experience occurs. We are not actually concerned here with the well-known phenomenon that learning capacity itself decreases when the animal gets fairly old. With the exception of humans, this development seems to play a significant role only with a very small number of animal species whose life-span is unusually long. What has really captured our interest, however, is the fact that many animals and, quite evidently, even humans pass through time periods in their individual growth when certain learning experiences are far more significant than at other times. This means that if the individual encounters the same kind of experience outside of this period, it will not be influenced as much, and perhaps it won't learn from the experience at all. Such a specified period of time is referred to as a "critical phase" or "sensitive period."

Many of these learning phases occur at a remarkably early age, often in the very first stages of life. This means that they arise much earlier than ethologists had assumed in the previous decades. Today we know of some cases where learning has taken place even before birth.

The most impressive demonstration of this kind of learning phase being important in early childhood is the phenomenon of imprinting Konrad Lorenz introduced the term "imprinting" in 1935 to denote certain learn-

Imprinting as a form of early childhood learning, by K. Immelmann and C. Meves

Critical phase or sensitive period

▷
Most animals threaten by displaying their weapons to the opponent or by making themselves more impressive in some other way: This threatening stump-tailed macaque (*Macaca arctoides*), for example, bares his teeth. At the same time, blood rushes to his head, so that the naked skin sections around his eyes glow and become more noticeable.

ing processes which belong to the realm of social behavior. The idea of imprinting is based on the following observation: Members of many animal species are to a great extent already born with a capacity to perform the movement patterns of their "social" behavior that is, sexual behavior or mother-child interactions. Since these behavior patterns are innate, they are typical of a species and, for the most part, unmodifiable. Yet the individual does not innately recognize the "object"—such as the sexual partner or the parents—at whom these behaviors are to be directed. He must learn his social objects, and this learning may take place at a very early age. The effects of early learning in the realm of social behavior, in turn, can be remarkably persistent.

There are two main behavior systems where imprinting processes are well known: the "following responses" displayed by young precocial animals, and the sexual responses of a number of animal groups. In bird species defined as "precocial," such as ducks, geese, gallinaceous birds, rails, and wading birds, the young have more-or-less fully developed sense organs when they hatch. They are able to follow the mother or both parents actively after a very short time. In contrast to "altricial" animals, which are born blind and which remain in their nest for an extended period, these precocial young must recognize their parents quickly and accurately. Yet often they are born without or with very limited recognition of adult conspecifics. As a result, the first thing they will do is follow any moving object. The characteristics of this object are imprinted rather quickly, and thereafter the young animals follow only this object.

Normally, of course, the very first moving object that the newly hatched bird sees is its own mother. But we may alter this situation experimentally by presenting the young with a human being, another species of bird, or an artificially moving object as the first one it is allowed to see. The bird will then direct its following initial response toward the experimental substitute, which therefore becomes fixated as the "mother object." The fact that even artificially moving objects can release a following response enables us to study the qualitative and quantitative aspects of this kind of imprinting more closely. Most prominently, E. Hess has been doing this kind of experiment. He induced mallard ducklings to follow a dummy whose steady motion led the subjects around in a circular runway. In this way, he was able to record the age at which his birds could best be imprinted, and how long they would have to follow the object for the imprinting to become permanent. We can also test the effects of obstructions, negative stimuli, drugs, and other artificial influences on imprinting by the experimental use of dummies. Again, this imprinting of objects, which always occurs rapidly and at a very early age, has been well documented in a large number of species including many kinds of ducks and geese, turkeys, domestic hens, rails, and wading birds.

Learning processes similar to imprinting seem to occur in humans, too. Evidence for this may be found in the realm of psychopathology, through

◁
Different forms of threat behavior: Above: Flamingoes and other birds direct their beak at the rival when threatening, but pull it back when showing submission. Middle left: The frilled lizard props up its anterior body and spreads the ruff, a huge collar which usually lies folded against the skin. Middle right: In a highly intense threat posture, the western African land crab (*Gecarcinus*) displays its pincers to the rival. The effect of this pose is further enhanced by the red color. Below: Two male elephant seals, whose scars bear witness to many previous fights, raise their bodies at the territorial boundary and emphasize this visual display with loud roaring sounds.

experience with children who were adopted or placed in foster homes after spending their infancy in an institution. This is not at all unusual in our part of the world. A large proportion of these children develop serious problems, which may arise as early as public school but come to the fore mainly at puberty. They are based on a general lack of ability to participate in responsible interpersonal relationships and to show consideration for others. Often, too, these problems arise from not being able to observe thoroughly and accurately. In most cases, then, social development is severely impeded in young people who share this kind of background.

A foster family may have raised other children who, in their infancy, were under the care of primarily one nurse or parent figure. This would apply to the parent's own children, or to those adopted after a period of individual rather than institutional care. In such cases we might gain an even better understanding of the developmental phenomenon we have just described by comparing this "control group" with children adopted into the same family but subjected to a variety of caretakers in their first months of life. Again and again it was found that socialization is far less successful with youngsters who in their infancy were nursed by an ever-changing series of caretakers.

On the basis of these observations, we may conclude that the infant human also passes through a sensitive phase during which he becomes attached to whatever person cares for him. This phase can be "by-passed" entirely if the infant has too many caretakers or if he is simply neglected. The result is maladapted behavior characteristic of social deprivation. Disorders will stubbornly persist into later life and may be alleviated only in part by long-term, well-directed therapeutic efforts. In such cases the main problems lie in developing a capacity to form attachments, to "follow," to make social adjustments, and also to make accurate observations and to show interest and follow through on any one issue. The imprinting of following behavior in man and animals begins when the motor, sensory, and neural functions necessary for this learning process have matured (see Chapter 4). In humans it is the sensory receptors that allow the infant to perceive and, beyond that, to focus on another person about six weeks after birth. We assume that the "sensitive" phase in humans begins with the "looking stage" (as K. Bühler called it) and ends six months after birth. A duckling learns to recognize the objects to be followed shortly after hatching, during a period of only a few hours. By contrast, the human infant must undergo this process of recognition learning by repeated contact with the "object" during an "imprinting period" of several weeks. Consequently the imprinting process in humans is less rigid than in other animals, and not immediately irreversible. On the other hand, susceptibility to many kinds of interference is considerably increased. There is no doubt that imprinting is a far more differentiated process in man than in other creatures, yet we can still detect certain features parallel to corresponding early learning experiences in the higher vertebrates.

Studies with "control groups"

Fig. 24-1. Examples of altricial and precocial animals:

The wryneck (above) is altricial, while the lapwing is precocial.

Newborn mammals can also be divided into these two types. Above: rabbit (altricial); below: hare (precocial).

Fig. 24-2. In the imprinting apparatus developed by E. Hess, a model of a duck is attached to an arm and moved in a circle by mechanical means, using a clock mechanism. A loudspeaker is built into the model. The ducklings, kept in a dark room from the time of hatching until the experiment, are exposed to the model at different ages and for different lengths of time.

In addition to the following responses displayed by young of many species of birds, we have observed similar imprinting phenomena in the area of sexual behavior, that is, where the recognition of the sexual partner is concerned. Experimenters who placed young birds of one species into the care of another have found that at a later age their subjects preferred to mate with individuals belonging to the species of their foster parents. Quite often they will completely reject their own conspecifics. Birds raised by hand will court their human caretaker, demonstrating a sexual fixation on humans.

F. Schutz examined this issue by experimenting with various species of ducks. Ducklings were hatched in an incubator, then placed with a mother belonging either to the subject's own species or to a different one, or they were raised with the young of another bird species. Schutz then released his ducks onto a large pond and allowed them to choose their mates among numerous other kinds of water fowl. In the majority of cases, experimental birds would mate with a member of their foster species. In this way, ducklings were imprinted not only onto other species of ducks, but also to geese and—with less consistent results—to gallinaceous birds. Conversely, geese, domestic chickens, jungle fowl, and coots could be imprinted to members of other species.

Imprinting of the following response occurs only with precocial animals; but even altricial birds may be imprinted sexually to another species. Strong evidence for this phenomenon was found in exchange experiments with various species of (Estrildidae), a group of small, seed-eating song birds inhabiting the tropics of the Old World. Two species are especially suited for this kind of experiment, the Australian zebra finch and the society finch, a domesticated bird from eastern Asia. Between these two species we may exchange eggs or newly hatched nestlings, so that the young are raised by foster parents. The result would be that our experimental birds, upon reaching maturity, choose to mate only with members of the foster species. Conspecifics would always be ignored or even attacked.

These experiments serve to demonstrate that certain kinds of experience will play a decisive role in the way ducks as well as weaverfinches eventually choose their mates. Furthermore, since in each case the young animals had been separated from their foster parents as soon as independence was reached, we know that the salient learning processes take place in very early childhood.

Imprinting of both the following object and the sexual partner is common especially within the class of birds, and it is here that ethologists have made their most intensive studies. Yet we know of very similar phenomena in mammals, and isolated cases have also been found in other groups of animals.

At this point we may ask: How are imprinting processes different from other kinds of learning? There are really only two criteria, both of which

Fig. 24-3. This male emu raised by hand in the Zurich Zoo, follows his caretaker around for courting; he treats the person like a conspecific female.

Konrad Lorenz described as early as 1935. Most important is the fact that imprinting always occurs within a specified period of time, which for the most part, is narrowly-defined and arises early in the organism's development. We are dealing with a most striking example of what we have called sensitive periods. The existence of such periods leads us to the second main criterion for imprinting, that experiences made outside of these phases do not have the same influence. Accordingly, knowledge acquired during a sensitive phase cannot be altered at a later date. We refer to this latter criterion as irreversibility.

Both of these essential properties defining imprinting—the sensitive phase and irreversibility—have been thoroughly examined and analyzed in numerous experiments. Ethologists have discovered that the sensitive phases for imprinting both following responses and sexual objects vary from one species to another and even within the same species. For instance, the sensitive period in mallard ducks for the imprinting of a following object lies between their eighth and twentieth hour of life, while the critical phase for sexual imprinting begins at an age of 5 to 19 days. Two more examples: the following response of young tufted duck may be imprinted at an age of 26 to 38 hours, the sexual responses of zebra finches, at an approximate age of 15 to 40 days.

After this sensitive phase has ended the animals cannot "relearn" by imprinting another image of what the mother or the future mate should be like. Thus, five-year-old turkey cocks raised by hand still preferred to court humans, even though as adults they had been placed with females of their own species. Some of the mallard ducks raised by parents of another species remained attached to partners of their foster species even after nine years, although they had been given opportunities to mate with their own kind. Male zebra finches raised by society finches were kept for years thereafter only with conspecifics, yet they retained their sexual attachment to the foster species: As soon as they were allowed to choose again, they immediately proceeded to court the society finches, ignoring their fellow zebra finches and even attacking them. Their behavior was the same as before the period of enforced intraspecific contact. To date, the oldest male in this set of experiments was fixated upon the species of its foster parents as rigidly at the age of six years as it was immediately after reaching sexual maturity. Yet this individual, too, had been placed with conspecifics for several years, while in the subsequent choice experiments he was allowed to be with a female society finch for only a few hours. Since the life span of these animals must be shorter in the wild than in captivity, and the stability of learning processes would not seem to decline when the animal gets older, we assume that object fixation in zebra finches becomes permanently irreversible once the sensitive phase is over.

One note of caution should be added: The fact that an animal is imprinted to a certain species does not itself imply that the individual restricts its relevant behavior (following responses or courtship behavior)

▷
Examples of territorial and competitive situations: Top right: Two kittiwakes fighting over a food morsel. Top left: The regular intervals between dwelling holes dug by individuals of many crab species shows that these animals defend not only their holes but also the surrounding territory against conspecifics. Center: Before the females arrive, male sea lions divide up the coast among themselves by fighting. Only those who have secured a territory can form a harem. Below: The tiny nesting territories of flamingos, consisting only of self-made mudhills with an egg laid on top of it, are tightly packed together.

exclusively to members of this imprinted species. We can only say that it prefers that species if given a choice. If the animal has no opportunity for such an attachment, it may well follow its own species or an entirely different one, and the same goes for courtship. Such behaviors directed toward a "substitute object," however, are known to occur in other situations as well; hence they are not peculiar to the imprinting process.

There is an additional interesting feature which applies to sexual imprinting, and again we see a possibility for comparing the developmental processes of animals with those of man. Sexual imprinting is distinguished by the fact that there is a temporal discrepancy between object fixation and performance of the appropriate behavior. In all animal species examined, the sensitive phase is completed before sexual maturity. Zebra finches, for example, attain this maturity at the age of 70 to 90 days, yet they become imprinted to their mating object as early as 15–40 days. Mallard ducks undergo sexual imprinting between 8 to 10 weeks of age, but perform their courtship and copulatory behaviors no earlier than 4–5 months, sometimes even as late as 6 months or more.

Inevitably, this interval between object fixation and the onset of actual performance—without doubt one of the most essential and important criteria of sexual-imprinting processes—raises the question of an accompanying "reward": Since the animal is not yet sexually mature at the time of imprinting, that is, since it cannot perform the appropriate behaviors no "sexual reward," such as successful copulation, is forthcoming. Instead, the necessary "reinforcement" which ultimately leads to the development of an object preference must be "borrowed," so to speak, from another functional group. Reinforcement, then, may be totally unrelated to sexual performance per se, and consist of such experiences as being fed by the parents or foster parents, having contact with these social objects, and so forth. This problem is fascinating, especially in the light of a variety of learning theories prominent today. The phenomenon does, however, require a more thorough investigation. The special characteristics that make imprinting different from other kinds of learning processes naturally pose a great many questions, mostly in the area of physiology. Primarily, the sensitive phase poses a problem for the researcher. It is difficult to understand why a particular learning capacity should arise only within such a brief and rigidly defined period of time. Theoretically, a wide variety of influencing factors could be at play, based on the motor sensory, or neural development of the organism. With regard to motor development, it has been speculated that the sensitive phase for following behavior in young precocial birds begins at that moment when the individual's motor capacities have matured enough for it to follow its mother.

However, such a connection cannot likely be made for sexual imprinting in altricial animals, since in this case the sensitive phase begins at a time when the young are still in the nest, and demands on their mobility are at a minimum. It is far more likely that there are associations between recep-

Sexual imprinting

◁

Examples of territorial identification: Top left: By means of bright colors and rhythmic up-and-down movements with the anterior body, the male common agame communicates to other males that this area belongs to him. Middle left: Male ghost crab building a "signaling pyramid." It demarcates the copulation hole, and "courts" and "threatens" in the animal's stead. Meanwhile, the male can withdraw into the hole. Right: With a gland located at the base of its tail, the cheetah produces a secretion which it rubs on trees and bushes at various points within its territory. Below: The roaring of stags is a familiar example of acoustic territorial identification; just like chemical marking, this behavior is best suited for areas difficult to survey (forests, grassy country, etc.), where optic signals would be useless.

tiveness to imprinting, on the one hand, and sensory or neural development, on the other. For example, the sensitive phase for sexual imprinting in zebra finches begins a few days after their eyes have opened. We do know that an animal does not necessarily attain its full capacity for visual perception at the time its eyes first open, since in many cases the successive neural connections mature a little later. Thus it is possible that imprinting receptivity simply begins as soon as the animals are first able to receive and process the visual stimuli coming from their parents. On the other hand, this explanation would not hold for the beginning of sexual imprinting receptivity in precocial young, since their imprinting to a following object —a feat requiring a similar learning capacity—precedes the onset of sexual imprinting.

We can find the same inconsistency in speculations concerning the end of the sensitive phase. In many cases it was suggested that the following response terminates with the maturation of antagonistic drives, especially the instinct to escape. At a certain age this drive would prevent the animal from approaching strangers, and thus the necessary conditions for imprinting would no longer exist. Finally, it is possible that irreversible engrams may be formed only while new connections (synapses) can arise between individual nerve cells (see Chapter 3). Since in many animals this latter process in turn seems to be restricted to the early phases of childhood, we could imagine that the stage of irreversibility, and thus the closing of the sensitive phase, is determined by the conclusion of this kind of neural growth. These issues point the way to a wide area of research in which studies have just begun.

Similarly, we still lack adequate information about sexual imprinting processes in humans. There has been no systematic research, and since, we can only resort to incidental experiences and the description of as many cases as possible to bring us closer to an answer.

Nevertheless, the phenomenon of sexual imprinting in a variety of animal species may help us to increase our understanding of what Freud has termed the Oedipus complex, observed by a host of psychoanalysts. The Oedipal phase is defined by psychoanalysts as a period in the child's development, occurring around the age of five, when the reference persons —usually the parents—imprint the individual with the attitudes he will show toward his partners later in life. If the child has certain frightening or repulsive experiences at this age, heterosexual relationships in adulthood may become blocked or perverted. Beyond that, the adult may inappropriately generalize his behavior toward other persons by applying the same attitudes that he learned from his parents at the critical age. This phenomenon is called transference. Psychoanalysts work primarily to help patients overcome the rigid transference schema of their Oedipal stage by means of insight and "reconditioning." Yet even Freud spoke of the libido's tenacity at clinging to the object it has once chosen.

According to these observations, can we not say that even human

Fig. 24-4. On the studies of F. Schutz:

From a few hours after hatching, a mallard duckling and a chick are put raised together. However, this would not be necessary for sexual-social imprinting; that is, the two would not have to be together under imprinting conditions in the first two weeks. Furthermore, there is a difference between chickens and ducks, because chickens are imprinted far more easily.

Despite their attachment, the two animals differ in their behavior toward water. Whereas the duckling immediately feels at home in the water, the chick trys to avoid it. These behavior patterns are innate.

The cock is imprinted in his sexual-social behavior on ducks, and chooses the shore of the pond as its habitat because the ducks spend most of their time here. It keeps wading into the water, sometimes up to its belly, in order to be close to the ducks, which it regards as conspecifics.

children undergo learning processes that are difficult to reverse and have a bearing on the later choice of sexual object or partner, hence resembling imprinting? In any case, practical experience tells us that after the fourth year of life, the parential models—especially in the child's attraction or aversion to the same-sexed parent—have a preforming influence on that person's sex-specific behavior in later life. The learning and recognition of a generalized object therefore seems to occur—probably the same as with ducks—in a different functional area, namely, on the parents. If the models are poor, these patterns may be shifted or distorted, so that after puberty a person will suppress sexual drives directed at members of the opposite sex. If the basis for a heterosexual object choice in childhood is too narrow or totally lacking, the person may even make accidental associations between genital sensations and inappropriate objects. This is how fetishism, pedophilia, homosexuality, and sodomy can arise. Such "perversions" are as persistent as the misguided cases of imprinting on "unnatural objects"—mentioned earlier—such as same-sexed companions or individuals of other species, which we have observed in animal experiments.

Yet we must warn against assuming exactly the same kinds of developmental processes or faulty imprinting for humans as for animals. The much longer duration of sensitive periods in man, and the fact that resistance to therapy does occasionally break down with long-term treatment in cases of neurotic inability for commitment and learning as well as of neurotic perversion, both indicate that we cannot simply transfer the concept of imprinting to man without any further reservations. For example, recognizing and discriminating the mother from other people, then experiencing affection and therefore a desire to imitate is a learning process determined very much by the human child's positive emotional involvement. We know this because a child can suffer emotional harm not only through the lack of a caretaking mother but also through a cold and unloving one, where the infant loses all motivation to freely express its own desire for affection. Though such an oft-repeated "bad experience" in early childhood can "stick like glue," it may still basically be corrected by better experiences in later life.

Differences in development between animals and humans

Psychophylogeny

It seems, then, that the ability to gather experiences as a basis for future behavior during sensitive phases emerged in the course of higher evolution. B. Rensch speaks of psychophylogeny. This aspect of evolutionary development allows us to compare and to evaluate our findings productively: Ethologists can point to simple structures in animals suitable for helping us understand conditions of human development, for the fact that early childhood conditions are particularly important in the development of human neurotic behavior shows: 1. the great importance of drives in the first years of a person's life, and 2. the dependence on certain biologically determined, "natural" conditions of growth; without these experiences, problems in behavioral development are likely.

Aside from the "classic" imprinting phenomena, the fixation of objects for following and future sexual response already described by K. Lorenz, similar occurrences have now been observed in a number of other functional areas as well, including some outside the "social" sphere. Fixations similar to imprinting may be found particularly in the ecological realm. For example, in various animal groups individuals reared on a certain kind of food may show a later preference that persists even after an interim feeding with other kinds (food imprinting). Furthermore, the choice of a certain habitat is often based on experiences in early childhood —a phenomenon variously referred to as home, biotope, environmental, or geographical imprinting.

Impressive examples for the latter may be drawn from the class of birds, where a large number of species show a typical "site tenacity," or loyalty to a particular place: Whenever possible, they will return to their place of birth or childhood or very close to it for breeding. Long-distance migrants spending the winter in central or southern Africa also prefer to rest in flight or spend their winter period in biotopes looking much like their home area, sometimes maintaining these for many years. As H. Löhrl demonstrated in his exchange experiments with collared fly catchers, site tenacity again does not derive from innate factors, but is acquired during early childhood.

Site tenacity in birds

The tremendous significance of experiences in childhood and youth can be taken even further: Recent investigations have shown that, in addition to recognizing conspecifics and showing a preference for certain ecological conditions, other far more basic behavioral traits in animals may depend on experience. This applies, for example, to motivational factors, in other words, to the strength of "drives." This association is particularly strong with the aggression drive (see Chapter 2). The Australian zebra finch is one example. Within this species there are aggressive as well as "peaceful" males, the characteristic being transmitted to the offspring. Exchange experiments have shown that this transmission process is not purely genetic. Even the young of peaceful zebra finches may develop a very high level of aggressiveness if they are raised with bellicose foster parents. Since the young are driven from the vicinity of the brood nest immediately after reaching puberty—the intensity of this act being dependent on the male's aggressiveness—the level of the young's aggressive drive in adulthood is obviously determined by the degree of their own aggressive experience with the father (or stepfather). A similar example was described by G. P. Sackett in rhesus monkeys: Young animals who in their first two to three months of life were raised by very aggressive mothers later showed a markedly high aggression drive, displaying more than twice as many aggressive behavior patterns as conspecifics raised by peaceful mothers. Such a strong influence of early childhood experiences in the area of fighting behavior was also found in other species of mammals and birds.

Aggressive tendency in zebra finches

In recent years, human psychologists in particular have offered a va-

Aggressive behavior in humans

riety of detailed evidence to suggest that aggressive behaviors may also arise in humans as a response to early childhood frustrations. It has been demonstrated that children reared with violent methods tend to show rough behavior patterns and criminal violence (S. and E. Glueck). On the other hand, these observations give only one side of the story. Increased aggressiveness is also found very much in people reared under conditions where the mother and father never act aggressively toward their children, and never oppose them. Even the wild infant chimpanzee Flint, in the Gombe National Park, whose aging mother could not raise this late-born offspring properly and who was unable to wean him—according to Jane van Lawick-Goodall's description—developed an "aggression inhibition" with aggressive outbursts, just like we see in "spoiled" children. The fact that "aggressiveness" may be learned in a way similar to imprinting still does not prove that it is basically a matter of upbringing. The assertiveness of "making room for oneself," necessary for self-preservation, will logically grow into a generalized and exaggerated character trait if it has been suppressed too much and too violently in childhood; but it may also burst forth in a sudden fit of temper, as if "in vacuo," when alternate avenues for activity have been blocked through too much permissiveness.

Need for companionship vs. social inhibition

In a very similar way, an animal's need for social companions and its ability to be part of a group are to a large degree determined by early experiences. Again, we can use the Australian zebra finch as an example: Here, individuals that, after becoming independent at about five weeks, were separated from their parents or step parents and totally isolated until sexual maturity at three to four months were later found to be far less social than animals always kept in groups, as is natural for that species. Even in group aviaries, the previously isolated birds remained separate, never or very rarely taking part in any social acts. This characteristic was most prominent in hand-reared animals that were given no opportunity for social contact even before reaching independence. They showed such a high degree of desocialization that some of them, when placed into a group aviary, died within a few days although they displayed no overt signs of illness and initially had a normal intake of food. Evidently they could not cope with the social stress (see Chapter 36), that is, the physiological burden associated with meeting conspecifics.

A wealth of basically similar findings attests to the experimental determinants of sociability, giving us an impressive demonstration of how much all vertebrate groups resemble each other in this respect. In domestic dogs, for example, the degree of sociality in adult life depends on experiences occurring between the third and fourteenth week of life, this phase reaching its high point in the seventh week. Domestic chicks reared individually by hand up to their tenth day of life either could not or could only with difficulty be induced at a later stage to follow a hen; they stayed separate even when they were with groups of others of the same age. Finally, fish fry of the cichlid *Tilapia* must form a bond with their parents during an early sensitive phase. After this period, lasting until about the twelfth day of life, no bond can arise.

Humans, too, may show later socialization problems if they are too greatly isolated after infancy. As we have mentioned, a person's desire for social contact will tolerate the most from an inadequate mother-child relationship during the first five to ten months; but—for other reasons —human children may, in a way similar to the zebra finches, develop into anxious, shy loners if between two and six years they are not part of some social group (siblings, nursery school) and thus learn to assert themselves in an appropriate way and to adapt. Accordingly, these children also have difficulties when starting school. Often they develop anxieties about school, showing psychosomatic problems, stomach pains, fainting spells, refusal to eat and vomiting. These are all manifestations of fear, or anxiety, that indicate a lack of sufficient preparation for the social interactions and adaptations necessary in a school environment.

Similarly, many other developmental processes within the realm of social relationships are, in a very general sense, found to be determined by experience. Thus, in some animal species the development of normal sexual behavior, brood-care parental responses, or the escape drive may be influenced significantly by impressions obtained in childhood and youth. Even the level of general activity can depend to some extent on experiential factors. Finally, evidence has been found for various animal groups that a lack of sensory stimulation in the early stages will also decrease the future capacity to learn, so that animals from a stimulus-deprived and very monotonous environment learn more slowly and do not retain what they have learned as well as do individuals raised in varied surroundings. These examples clearly show the extent to which environmental influences may affect the development of greatly diverse behavioral characteristics that are directly or indirectly concerned with social interactions. Moreover, the highly stable nature of early childhood experiences that was first observed for imprinted following responses and sexual patterns is obviously a feature of other traits, as well.

In humans, too, the willingness and ability to learn and the level of activity is determined by appropriate stimulation in the first few years of life, although positive results occur only when other important factors have been considered as well: Too many demands at an early age and overstimulation can lead to symptoms like unconscious resistance or protests, stereotypes, or apathy. If a child has no bond with a person it can talk with, language development will be retarded and vocabulary limited. With an excess of mechanical substitutes (technicalized toys, television), the activity of children decreases so much that their very capacity becomes "atrophied," and their general ability to be active is stunted. Even in humans, activity itself is extremely difficult to learn at a later stage, and an early deficit may never be compensated fully if the person has not had enough practice in his first six years of life. The child who constantly watches television

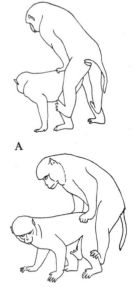

A

Fig. 24-5. H. F. Harlow carried out many studies on rhesus monkeys, especially for investigating the so-called deprivation syndrome. Sequence A shows the copulatory behavior of a normally raised male, while sequence

Biological meaning of early childhood learning processes

and is otherwise "spoiled" by our affluent society is a typical example.

Our next logical question concerns the biological meaning of early fixation processes. What are the advantages that may have caused these kinds of learning processes to develop in the course of evolution, even though the rigidly determined sensitive phase and the relatively irreversible nature of these influences in later life do in some ways seem pretty inflexible? We can answer this question quickly enough by referring to the way members of such animal species grow up: Evidently these are all species with definite patterns of brood care, or parental rearing of the young, where the offspring remain in a family unit during childhood and experience close contact with conspecifics (parents, siblings, and other members of the family or social group). In this way the young accumulate skills and experiences for later life more easily than after the family has dissolved. It is therefore advantageous for the individual to be especially receptive to such influences at precisely this early stage, and to retain its experiences later. For this reason, increased sensitivity during a certain period, and subsequent stability—precisely those properties characterizing many of the early fixation processes—would certainly be highly adaptive.

In addition, however, there is a possible disadvantage, which will practically never arise under natural conditions but may become all too obvious in an experiment: As a necessary consequence of the more-or-less rigid sensitive phases and the stability of early learning, it is far less possible to correct such childhood fixations, as it is to modify the effects of other kinds of learning. Hence the animal may suffer serious and permanent developmental problems if by chance it is not exposed to the normal environmental influences of early childhood.

Behavioral problems through abnormal conditions in early childhood

Such behavior problems may come about in two ways: Either the animal is not allowed to have the necessary experiences at the right time, or it can not make them with the right object. For the first problem we can find very striking examples with the imprinting of following responses. For instance, if young ducklings are deprived of contact with a "mother object" by being isolated during their sensitive phase, they lose their opportunity for ever establishing a real mother-child bond: In later experiments, these animals show fear and escape responses toward a model that during their sensitive phase would have released following behavior and established an imprinted response.

Many people also seem to react permanently with fear and flight and an aversion to contact—perhaps culminating in complete withdrawal— if during their sensitive mother-recognition phase they are separated from the mother and instead presented with other "faces." This developmental problem is often seen in children who during the imprinting phase are suddenly removed from their mothers over many weeks, perhaps through a stay in the hospital. Without an intense treatment for recovery, problems such as delays in speech development, reduction of the ability to observe and respond, and suspicious attitudes based on fear of others will arise.

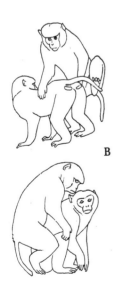

B shows how a male raised in isolation tries to mate.

Phases of imprinting may also be "missed" in the realm of sexual behavior. Two male zebra finches, for example, which were hand raised in near darkness and therefore could not gain social experiences during their sexual imprinting phase, later courted everything flying past their cage—from bumblebees to birds the size of pigeons—and could not be induced to restrict their courtship to conspecific females even after living with others of their kind for many years thereafter. Thus, if an individual does not learn to recognize its future sexual partner very precisely during the early stages, it cannot make up this knowledge later even with long-term experiences.

A very similar indifference can arise in people if suitable objects are lacking in the imprinting phase for sexual objects. This may occur, for example, in girls growing up without a father, older brothers, grandfathers, or male friends, being reared only by a single mother. Some of these girls have Lesbian tendencies—this can also happen if the father is present but was seen as repulsive by the little girl—and most of them become afraid of men when they reach puberty. If, because of this fear, heterosexual impulses are repressed, problems arise such as psychosomatic or so-called "hysterical" symptoms.

Similar problems in development occur when animals had the opportunity for social experiences at the right time, but had made them on a "wrong" object. We have mentioned cases of mistaken imprinting on members of other species, on humans, and even on inanimate models. Just like the missed imprinting periods, these cases of faulty object imprinting are "irreparable": imprinting on a wrong object cannot be reversed even after many years of living with conspecifics.

On the whole we can see that, because of their early sensitive phases and their subsequent stability, imprinting processes do have considerable advantages under natural circumstances, but that they are also subject to disturbance and thus may quickly lead to grave developmental problems when environmental conditions are changed. This fact has much to bear on human development when we compare our findings with influences in people's early childhood: Increasingly, our environment —not the least our social environment—is becoming subject to artificial change.

In this context, some very interesting results were obtained—interesting because we can make direct comparisons—by a number of different investigators, especially the school of H. F. Harlow on rhesus monkeys. If young monkeys are raised away from their mothers, they quickly show serious disturbances. Together these are called the "deprivation syndrome," expressed in movement stereotypies and general restlessness, aggressive reactions, extensive apathy and lack of exploratory behavior, various compulsive habits, tearing out their own hair, and many other abnormal traits. As adults, most of these animals were unable to copulate because they attacked their partner. Many females never became pregnant although they were kept up to seven years with experienced and very

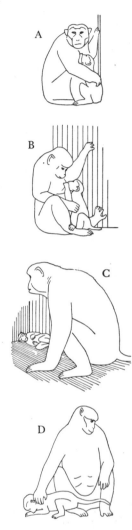

Fig. 24-6. Mothers raised under normal conditions care intensively for their babies (A and B). Females raised in isolation neglect their infants and may even attack them (C and D).

"patient" males. Those females who did bear young were very poor mothers, mistreating their children and suckling them rarely and with great resistance, so that these infants had to be fed with a bottle as well.

Numerous investigations have revealed similar "deprivation symptoms" on children living in institutions from the time they were born. These youngsters, too, try to relieve the tension caused by lack of a "natural situation" through self-pacification. Frequently they develop a *jactatio capitis* (throwing their head from side to side, often also including the trunk), a behavior occurring mainly when they are put to bed; but with severe tension this may also happen during the day, rocking their upper torso back and forth while sitting or rocking from one leg to the other while standing. Some children may also wrap curls in their hair, pull out hair and even eat it, chew their fingernails, lick themselves, scratch their body, or manipulate their genitals. But we must note that today these kinds of deprivation symptoms also appear more and more, increasing at an alarming rate, in children from families—the degree of disturbance rising with the amount of time the mothers spend away from their infants, do not breast-feed them, and impede their motor development, their exploratory drive, and their creative needs for independence with substitute objects and general overstimulation.

But even Harlow's observation that socially deprived monkeys later failed as mothers can be applied to human psychology: Mothers raised in institutions are far less likely to make the necessary sacrifices for their children in the first few years of life; these mothers may neglect them or even place them in institutions, so that a "generational neurosis" may result.

The developmental deficiencies we have described could to some extent be avoided in rhesus monkeys if the animals were isolated during the first six months of their life for no more than three months, in the second half year for no more than six. With longer periods of isolation, resulting damages could not be corrected even if the animals were kept with conspecifics for many years thereafter.

This indicates that normal mother-child interactions are necessary for the growth of all systems of social behavior in both rhesus monkeys and humans. Accordingly, every disturbance in these interactions will have far-reaching effects on later social, sexual, and parental behavior, because a "missed" socialization phase in early childhood can never be corrected in monkeys, and in humans only with the greatest deliberate effort.

Even here, then, we can clearly recognize a "missed imprinting period." In this case the findings on young rhesus monkeys fit directly into studies on imprinting in other groups.

Fig. 24-7. A young rhesus monkey is separated from its mother and is raised only with models. The animal prefers to be with the cloth-covered model (left), maintaining contact with it even while drinking from the wire one (right).

25 Forms, Causes, and Biological Meaning of Intraspecific Aggression

When we say that someone is behaving aggressively, most people think mainly of acts of violence and physical fights. Others tend to judge behavior less by its appearance than by the motivation behind it. They would say that a person is also expressing some kind of aggression, for example, when he "locks his teeth into" an activity, or "attacks" a problem and then "masters" it. Still others associate this concept with feelings of anger, rage, aversion, and hate, with a hostile and aggressive attitude, or with a destructive desire. But ethologists cannot judge behavior according to emotions that may accompany it, because feelings in animals are difficult if not impossible to investigate.

It would also help to clarify the issue if we separate the other two criteria, as well. When we mean the form of a behavior (that is, what the organism is actually doing), we shall speak of aggressive acts or behavior patterns; in contrast, we shall use the terms aggressive motivation, aggression drive, or aggressiveness to describe all the internal physiological states that together with the external releasing stimuli lead to a particular behavior pattern. In this context, the terms motivation and drive are used synonymously.

When looking at an animal's entire repertoire, how do we define which behavior patterns are aggressive? Several examples may help to answer this question, and we shall refer to them again in the chapter.

Fiddler crabs (*Uca*) live in sandy and muddy tidal zones of all warm ocean waters, forming dense colonies. Both males and females dig vertical tunnels down to the ground-water level, where they seek refuge during high tide and when threatened by predators. If one crab comes too close to the holes of another member of the colony, the owner rushes at it, grabs its pincers, and pushes the opponent back or even tries to twist off the other's pincers should the intruder resist.

Marine iguanas (*Amblyrhynchus cristatus*) live in various subspecies on the Galapagos Islands. During their egg-laying period, females fight for the sandy spaces between the rocks, where they bury their eggs. After a

Forms, causes, and biological meaning of intraspecific aggression, by H.-U. Reyer

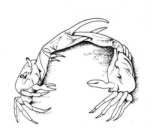

Fig. 25-1. Male fiddler crabs hook into each other's pincers when fighting. Should a pincer break off, the animal grows a new, small one with the next molt, and the other pincer grows larger. This is why in some males the left pincer is larger, in others, the right.

Fig. 25-2. Marine iguanas fight by ramming and pushing each other's heads.

short contest where strength is measured by the opponents pushing each other's head, one animal usually gives way. Sometimes, however, these encounters escalate into more vicious fights, where bites are exchanged.

The Northern American red-tailed hawks (*Buteo jamaicensis*) hunts by waiting on high trees until it can spot prey animals—mainly rodents—on the ground below, and then swooping down on them. Should another hawk use the tree perch of a pair of these hawks, its owners quickly fly to the scene, bear down on the intruder, and push it off their tree if it does not immediately fly away.

Here is our last example: During the breeding season, sea lions (*Otaria byronia*) gather on South American coasts. The males arrive first, and they engage in biting fights until the strongest bulls have divided the coastal area among themselves. This arrangement then allows them to acquire a harem, since the females have to land in the bulls' territories (see Color plate, p. 369).

The motivation behind these behavior patterns is hard to determine. The forms of behavior (grabbing pincers, head-pushing, driving another away from a tree, or biting) vary as much from one species to another as do the situations in which they occur. Yet these patterns have similar effects and functions. This is why we classify an animal's behavior as aggressive if the other animal is injured, thrown into submission, or driven off. We also include behavior patterns closely associated with the others in time, where external circumstances remain constant (see Chapter 2).

The term "agonistic behavior" has become popular with ethologists. It comprises everything that leads to a resolution of conflict between individuals; this means aggression as well as escape behavior and all patterns resulting from a superposition of these two kinds of behavior (threat and submission movements).

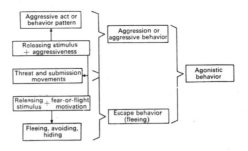

Fig. 25-3. Schematic representation of the elements that go into making agonistic behavior.

Aggressive behavior in situations of intraspecific competition

Among animals, many conflicts arise when two or more individuals strive for the same objects. These situations, occurring mostly between members of the same species, are described as competitive, and the animals taking part are then competitors or rivals. Aggressive behaviors may also appear during play (see Chapter 22), and they can be evoked experimentally, for example, by administering pain.

As our examples have shown, many different things can be objects of dispute: dwelling holes, egg-laying sites, tree perches, females, and so on. In most cases, animals defend not only these objects but also a certain area around them. Hawks usually start to attack when a stranger approaches up to within a certain distance from their tree. Such an area, where rivals are excluded by the mere presence of one or more owners, is called a territory. A dispute over territories is the most frequenct cause of aggressive encounters.

A territory may be occupied by a single animal (fiddler crabs, marine iguanas), a pair (red-tailed hawks), or a harem group (sea lions). It may be occupied by a family group, as with the hunting territory of North American wolves, or by even larger units, as with the sleeping site of a baboon troop (*Papio ursinus*). In some species a territory is subdivided into smaller areas. For example, the territory of a male belonging to the cichlid species *Lamprologus congolensis* contains several smaller territories occupied by females. While the male is allowed to swim through the "property" of any of his females, none of these is allowed into the territories of her neighbors.

Aside from this intraspecific territorial behavior, which is very common, we know of less frequent examples where members of different species mark off territories against each other. This applies to animals that use the same kind of food or other environmental features. These are interspecific territories, and we know they occur in a number of chat species (*Oenanthe*) and with the colonies of many species of ants. Usually, however, different species do not compete, and so their territories overlap.

The boundaries of territories can often be recognized by natural objects such as the edge of a forest, isolated trees, hedges, paths, dells, streams, stones, and similar landmarks. Animals overlooking their entire territory —perhaps from high up in the water (fishes), from treetops (birds), or in open country (sea lions)—will defend all parts of it, though intensity decreases from the center to the periphery. Thomson's gazelles, however, fight most along the periphery of their territory. On the other hand, mammals living on the ground and in densely vegetated biotopes not only have a more limited view but are also confined to a few passable trails where they can move relatively fast. These animals primarily defend this network of paths and the places they connect, which are sleeping and feeding sites, places for urinating and defecating, objects they can rub against, places for sunning or courting, water holes for drinking and bathing, and so on. Figure 25-5 can easily be modified to apply to the territories of foxes, badgers, bears, and other carnivores, and also to some ungulates.

An animal will visit these places according to a more or less rigid "schedule" at different times of the day. On its regular prowling or foraging trips it will drive away any rivals. But at other times, when the animal is occupied in another section of its territory, the same rival can use the same path or site without interference and will now drive other conspecifics away. In other words, these territories are divided in time as

Territories

Fig. 25-4. The territory of a male demoiselle is delineated by landmarks. Dotted lines show the spawning site.

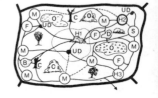

Fig. 25-5. Territory of a mammal. H1, H2, and H3, home site of the first, second, and third order; O, obstacle; UD, site for urination and defecation; C, comfort site; B, bathing site; F, feeding place; S, storage place; D, drinking place; M, marking spot. Heavy line: territorial boundary. Dotted line: area of tolerance. Continuous line: paths or runways. Broken line: subsidiary paths.

well as in space. Areas may overlap, as with the hunting territories of domestic cats, but rivals can never occupy them at the same time. We do not know for sure how these animals avoid sudden encounters and fights. P. Leyhausen suspects that the urine marks which cats deposit along their trail inform any conspecific following along the path that someone has already been there, and how close he is. A fresh mark could mean "section closed"; an older one, "you can proceed with caution"; and a very old one, "passage free"—much like block signals in a system of railway lines.

Territorial behavior in related species

Related species—for example, members of the same phyletic family— often vary in their territorial behavior. Thus the kittiwake (*Rissa tridactyla*) uses its territory for pair-bonding, courtship, copulation, incubation, and raising the young until they can fly, while the laughing gull (*Larus ridibundus*) selects different sites for several of these activities.

Territorial behavior: differences within a species

There are differences within a species, too, which depend on the individual, on age, and on sex. Animals that are stronger and more pugnacious generally acquire larger territories. Young ones do not yet take part in altercations, and old ones no longer participate. Where males and females come together only for courtship and mating—as in ruffs (*Philomachus pugnax*), birds of paradise, and other phyletic groups—it is only the males that stake out and defend territories, seeking to lure one or more females into their boundaries. If the partners stay together during the raising of young, in many songbirds and substrate-brooding cichlids the male does most of the defending of territory and will fight mainly at the boundaries, while the female takes care of her brood and gets involved in defense only when an intruder comes too close to the young. In rare cases, the roles are reversed, as with the cichlid *Tilapia macrocephala* and the Andalusian hemipode (Turnicidae). There are other bird species where both sexes defend their common territory, but each does so mainly against rivals of its own sex.

Territory and environment

We can see by these examples that there is no one kind of territorial behavior, nor can we speak of "the" territory as such. Even variations in environmental factors can lead to different systems. Any aquarium owner knows that, with a multitude of fish species in his tank, he can increase the number of territories by adding more plants. Nature itself obviously provides for such variations in the distribution and number of "visual screens." These "partitions" influence the shape and size of territories and therefore also the density of population in a particular area. In turn, population density—also determined by other factors such as food and predators— affects the size of territories. This is why we find a few large territories in scarcely populated areas, but numerous smaller territories in densely populated ones. But there are limits to how much territories can shrink. In three-spined sticklebacks (*Gasterosteus aculeatus*) the number of territories in a limited area can only increase as long as each of these covers at least 45 cm^2.

In order to understand this phenomenon, we have to describe another observation made with a great number of animal species: When territory owner "a" somehow enters the territory of his neighbor "b," b attacks and drives the intruder, a, back over the boundary. In his eagerness to pursue the rival, however, owner b may easily end up in a's territory, and now a turns around and chases b back over the line, again shooting over the mark and ending up in b's territory. This whole procedure is repeated several times. But the contestants venture less and less into the other's territory, slowly come to a halt, finally facing each other and threatening but not attacking. This line, characterized by an equilibrium of forces, is the territorial boundary. N. Tinbergen demonstrated this relationship with a simple experiment (Fig. 25-6). He showed that fighting force and readiness (the aggressive value, as coined by J. v.d. Assem) decreases from the center outward. Figure 25-7 helps to illustrate the theory: If we know where the border is, we can compare the aggressive values and predict who will win at what place (e.g., a in A, b in B). We can also determine ahead of time where a newcomer has the best chances of becoming established— right on the border (C) between two territory owners. For one thing, the aggressive tendencies of the two neighbors is equally low at this point, and second, what aggression they do show is partly directed against each other. In order to conquer this particular site, the newcomer has to show an aggressive force at least as strong as that of a and b combined at this point. But the closer the territories lie to each other, the higher the aggressive value a newcomer has to possess in order to win territory. When a certain density is reached, this task becomes impossible, because the fighting force of established owners is now too strong at the boundary and their defense is too hard.

Naturally, this model for explaining why territories cannot grow smaller than a certain size is a simplification, since it does not consider other factors such as suitability of site, opportunities for hiding and shelter, and other environmental factors. Nonetheless, we do observe that new territories are usually started on the periphery of other ones, and that, due to the increase in fighting force which accompanies ownership, the action radius gradually increases.

Many animals, even if they are not territorial, will attack any conspecific approaching nearer than a certain distance—and this occurs regardless of where the animals may be at that moment. We then speak of individual distance. According to P. Marler, in chaffinches (*Fringilla coelebs*) this "personal space" extends from 18 to 25 cm between males and from 7 to 12 cm between females and between the sexes (Fig. 25-8). Higher-ranking females maintain a greater individual distance than lower-ranking ones, certain animals, regardless of social rank, are tolerated at a closer range than others, and "unassuming"individuals are allowed closer than those whose approach behavior is threatening. Aside from these variations associated with sex, individual characteristics, and social behavior,

Territory and fighting force

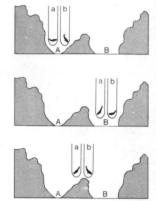

Fig. 25-6. When two sticklebacks with neighboring territories are placed in test tubes and presented to each other, the one that attacks will be the owner of the territory where the test tubes are held (above and center). At the border, the two opponents threaten each other (below).

▷
Individual distance in dragonflies sunning themselves on a stalk.

▷▷
Many behavior patterns of fighting have evolved convergently, that is, in very similar forms in different species, independently and unrelatedly. Above: the stag beetle owes its name to the similarity of its "antlers" and fighting form to those of deer. Below: Although we must count the fighting of wapiti stags among the ritualized types of combat, occasional fatalities occur when two rivals cannot disengage their antlers.

Individual distances

◁

Two phases in the fighting of giraffe bulls. The animal to the right gathers force and throws its head with bonelike protuberances against the neck of its rival. While the horns of younger animals are still covered with fur, those of older ones have often become exposed at the tips, and injuries can occur. This style of fighting is used only against conspecifics. Giraffes fight off predators with their hooves; this is one of many indications that the head weapons of ungulates did not develop for defense against predators.

Dominance hierarchies

◁◁

Above: The way kissing gouramis rush at each other with their mouths has the appearance of kissing, but is really an element of combat. Below: Hippos also push their wide-open mouths against each other when fighting. They often follow this up by biting their opponent in the neck —an example of a damaging fight.

the limits of tolerance may fluctuate periodically. For example, this border line comes increasingly closer to the mate until copulation, while conspecifics of the same sex are attacked at an ever greater range.

These minimum distances may then result in the "military" formations of certain animal groups, as anyone knows who has ever watched starlings flock on telephone wires or dragonflies sunning themselves at the edge of a pond (see Color plate p. 385).

Some people compare individual distance with a small territory that the animal always carries around; certain authors, including H. Hediger and P. Leyhausen, view the possession of a personal space, which no conspecific may violate, as a phylogenetic stage preliminary to territorial defence. According to this notion, animals became territorial by committing themselves to a certain place and thereby transferring their flexible individual distance to geographically defined boundaries. For instance, some observers have noted that certain fishes and birds shift from territoriality to merely maintaining an individual distance when the population density increases.

Territoriality is characterized by the fact that whether or not one animal is superior to another depends on place and sometimes also on the time of an encounter. Leyhausen calls this a relative social hierarchy. On the other hand, when an animal is always dominant regardless of place and time, he speaks of an absolute social hierarchy, generally known as rank order or dominance hierarchy. Aside from the defense of territories and individual distances, the formation of dominance hierarchies is yet another source of aggressive interactions. Under natural conditions we find this kind of organization almost exclusively among members of social species; but in forced communities, as in captivity, even normally solitary animals will create dominant and submissive ranks. If we put several iguanas together in a cage smaller than their minimum size of individual territory, one male will soon dominate over the others.

More frequent than this two-way division into ruler and ruled is a truly graded hierarchy of ranks and dominance. Such a rank order is usually defined by assigning animals letters of the Greek alphabet. The highest-ranking group member is the alpha animal, and next in line is the beta animal which dominates all others except alpha. Next is the gamma animal, which is dominated only by alpha and beta, and so on down the line. When this order can be clearly discerned right down to the lowest-ranking member, the omega animal, we describe it as linear. In many cases, however, this hierarchy is interrupted by three-way relationships or by loops. A further complication may also arise when—as with baboons or chimpanzees—two or three males form a group of their own and thereby dominate together over an animal that would subjugate each of them on an individual basis (Fig. 25-9).

In some animal societies, males and females create separate rank orders. If they form a joint hierarchy, the males usually dominate, though there are exceptions. In groups of dwarf mongoose (*Helogale undulata rufula*) the oldest female takes the lead, followed by the oldest male.

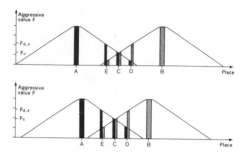

Fig. 25-7. The aggressive values F (= fighting force) of many animals (black for a, dotted for b) are highest in the center of their territories and decrease toward the periphery. The border (B) lies wherever $F_a = F_b$ In the top diagram the territories are larger and their centers lie farther apart.

Formation of dominance hierarchies

The first to study how dominance hierarchies are formed was T. Schjelderup-Ebbe, in 1922, using domestic chickens. If hens that do not know each other are brought together in a group, all-out fighting ensues. Each will then remember against whom she lost, avoiding those individuals in future. The alpha hen is the one that won against all the others, beta lost only to alpha, and so on. Often enough, physical strength is not the only factor that determines position in the hierarchy. K. Lorenz reports how low-ranking female jackdaws (*Corvus monedula*) rise in status when pairing off with high-ranking males. Female baboons attain a higher position when in heat or when they have young; and juveniles may take up the status position of their mother by way of tradition. The dominant males in a baboon troop do have to earn their rank mainly by physical force and endurance, but after that they keep it by virtue of intelligence and experience even when they grow old and weak.

Effects

Once the dominance hierarchy is established, life in the group may become so harmonious that we would hardly notice any social ranking at all. However, if we know that a hierarchy exists and what it looks like, we can detect the signs even during "peaceful" periods. The leader of a pack of wolves can be recognized by the way he holds his head, ears, and tail, and by the submissive manner with which subordinate pack members approach him. In other cases—as with the female chaffinches mentioned above—dominant animals maintain a greater individual distance than subordinate ones. Furthermore, high-ranking animals always enjoy certain

Privileges and duties of high-ranking animals

privileges. They are the first to approach a source of food, they obtain the best sleeping or nesting sites, and they are usually the only ones to mate with females in heat. Quite possibly, however, they also carry out certain duties that lower-ranking animals do not have. In the groups of Japanese macaques (*Macaca fuscata*), the dominant members lead and supervise the others. They settle fights, protect females and young from being accosted by younger males, care for the young when their mothers are pregnant again, and confront particularly strong and vicious predators; less dangerous attackers are warded off by a group of subordinate males that also have the role of scout and leader during the group's travels, just as with baboons (Fig. 31-6).

Morphological features of high-ranking animals

Another way many species show status position is by morphological and physiological traits. A high-ranking gambusia fish (*Gambusia hurtadoi*) can be recognized by strong yellow color on its dorsal and caudal fins, and the dominant rabbit buck can be distinguished by its enlarged chin gland which it uses to mark its territory. This scent marking is carried out mainly by high-ranking animals in other species as well. Furthermore, it seems that in some species individuals recognize the status position of others in their group by certain characteristics. V. Geist found that wild sheep assess their conspecifics according to horn size, and that newcomers can take up their position in a group of sheep without any fighting. This contradicts the common view that hierarchies could only appear among animals that know each other individually, because each must remember who its superiors are. It would suffice that an animal simply compares certain traits of any other individual with its own. We have not determined whether this always needs to be learned, like in deer, or whether some species can manage it without prior experience.

Biological meaning of aggressive behavior and of territories

What is the function of intraspecific aggression? Charles Darwin had already suggested an answer: The stronger and healthier individuals are selected for reproduction in their competitive fights for females, territories, and social dominance, providing the best guarantee for offspring capable of surviving. This kind of selection through intraspecific conflict has led to the natural breeding of particularly strong animals, especially males well able to defend themselves. Fights may serve to obtain certain objects (especially territories), to maintain individual distance, and to gain a position of dominance. In the following paragraphs we shall discuss how these functions are biologically significant.

The answer to the question, "when do individuals acquire territories?" may indicate their function. Many animals do not occupy a territory until they are sexually mature. This observation suggests a strong association between territorial behavior and reproduction, supported by countless instances among animal species where adults do not show territorial behavior in the intervals between breeding, and where individuals without a territory do not reproduce. Another way to study the functions of territory is to examine the activities taking place in such a defended area. Ruffs and birds of paradise court and mate in their territory, while herring gulls mate in one territory but brood in another. In these cases we can associate territory not only with reproduction as such, but more precisely with courtship, brooding activities, etc. These and other observations have led ethologists to ascribe to territories the following functions, although not all are exclusively associated with functions.

(a) Familiarity with a certain area is an advantage in foraging for food and escaping from predators; it also increases fighting force against rivals.

(b) When partners leave the area for short periods (e.g., while looking

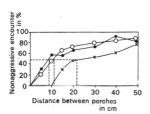

Fig. 25-8. Determining individual distance by measuring the distances where fifty percent of approaches trigger an attack. Crosses: distance between males; dots: between females; circles: between males and females.

for food) or until the next breeding season, it is easier for the mates to find each other again, if they have a common "meeting place." We know, for example, that storks form permanent pair-bond relationships because both the male and the female are "attached" to the same nest. Thus a territory facilitates pair formation and maintenance of the pair bond, which may be an important factor in reproductive success.

(c) Effective defense of territorial boundaries prevents or reduces interference in courtship, copulation, nest-building, and rearing activities.

Yet another indication of territorial function is the kinds of objects defended. In small areas they are easy to determine—as in the territories of colony breeders like gulls and flamingoes, where boundary lines demarcate little more than the nest (see Color plate, p. 369). But how do we recognize the most important object in an extended territory, which contains all the things an animal needs? The answer lies in the frequency with which the owners occupy certain parts of their territory, or in the places where fighting intensity is highest. In many species, the central object of the territory is the breeding site. In particular, this applies to animals with special needs and activities, burying their eggs in the sand (marine iguanas), incubating eggs on the confined spaces of rocks (gulls), or raising their young in hollow spaces (jackdaws, starlings, and numerous mammals). In other territories, the focal point is a place to seek refuge (fiddler crabs) or the main source of food (red-tailed hawk, some species of hummingbirds and butcherbirds). A great many discussions have involved the food-providing function of territories. For years, ethologists have tried to investigate these relationships with precise statistical methods. In a comprehensive study of seventy species of birds, T. W. Schoener came up with the following results:

—Need for food consumption is positively correlated with body weight, and body weight is positively correlated with size of territory, i.e., larger species as a rule have to consume more food and defend larger territories (Fig. 25-11).

—Birds of prey usually occupy larger territories than herbivores or omnivores with the same body weight.

—Birds of prey living in areas with fewer prey animals occupy more extensive territories than others of their species living in more abundant areas.

In addition to these correlation studies, we can also perform experiments. If we increase the yield of heather by fertilizing a certain area, Scottish willow grouse (*Lagopus lagopus scoticus*)—which feed off these plants—cover smaller regions than before.

However, these results do not imply that all territories serve the function of food provision. For example, some birds decrease their territorial range just at the time their young hatch and need to be fed; others look for food in neutral areas; and still others, such as robins, often violate boundary lines without opposition just when foraging for food.

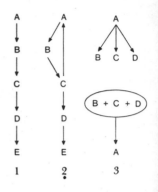

Fig. 25-9. 1: Linear dominance hierarchy. 2: Linear dominance hierarchy with a triangular relationship. 3: A is superior to B, C, and D as long as everyone is by himself (above). But A becomes subordinate when B, C, and D join together as a group. Arrow always points from dominant to subordinate.

Meaning of territories with regard to food foraging

Fig. 25-10. Bighorn sheep recognize the status position of their conspecifics by the thickness of the horns, which increases with age (order of increasing age and rank from right to left).

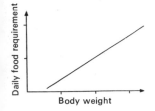

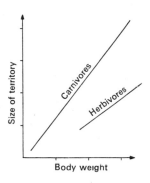

Fig. 25-11. Correlations between body weight and the need for food consumption (above), and between body weight and the size of the territory (below).

Fig. 25-12. Researchers try to determine which stimuli release aggressive behaviors by performing dummy experiments. When we place both a stuffed juvenile with brown breast feathers and a bundle of red feathers into the territory of a robin, the owner attacks and threatens the red bundle more frequently and strongly than the brown-breasted model.

Furthermore, we must not assume that every advantage gained by territorial behavior is necessarily the direct result of evolutionary selection —although the term "biological function" can easily encourage this way of thinking. The fact that an animal may find certain objects in its territory or may remain unmolested by others can be seen as a byproduct, a feature that means nothing if these objects are in good supply or if individuals would leave each other in peace even without territories.

On the other hand, very few quantitative studies have been made on the correlation between territory and reproductive success. One of the few exceptions is J. v.d. Assem's work on three-spined sticklebacks, where males with larger territories are in fact more successful in nest-building, courtship, and breeding activities. When, as in this case, individuals with proper territories raise more young than those without territories or with smaller ones, then the latter animals will eventually be driven off. In this way, territorial behavior may have developed by favoring the superior individuals, a consequence of natural selection. However, some biologists, particularly V. C. Wynne-Edwards, believe that this kind of evolutionary development could only be explained by group selection, in other words, by selection of the best-adapted populations. Their assumption rests on the fact that territorial behavior provides advantages not only for the individual but also for the population as a whole:

(a) By delineating territories, conspecifics are kept at a distance, causing the species to spread over a larger area. In order to illustrate one of the advantages of this spacing-out, K. Lorenz gives an example of occupational distribution among people: "If a fairly large number of doctors, merchants, or bicycle mechanics want to attract sufficient business within a certain region out in the country, the members of each of these occupations would do well to establish themselves as far as possible from others of their kind." Just like business people have only limited markets, the vital resources in the habitat of many animals (e.g., food or dwelling holes) are limited within a particular area. These resources would be over-exploited and eventually exhausted if all the members of a species were to concentrate in one of these areas. When conspecifics repel each other, new habitations are found and other resources come into use.

(b) Since in territorial species only members that do obtain and defend some property will usually reproduce, territorial behavior is an important mechanism for controlling number of offspring. By way of the number of territories held in a population, the number of young is determined by the supply of objects necessary for survival; and the fact that territorial behavior can be adapted to existing conditions—for example, when food is plentiful there are more and smaller territories, while under conditions of scarcity fewer and larger territories appear—ensures that resources are utilized to best advantage.

(c) In addition, individuals without territories form a reserve to compensate for losses when territory owners fall prey to predators, accidents, or disease. This phenomenon was clearly demonstrated in an experiment where all the birds that fed on a particular kind of worm in a certain forest area were removed. Each territorial pair was eliminated as soon as it appeared, yet the territories had new occupants every single day.

(d) Studies made by Tinbergen and his co-workers point to another advantage of spacing out: Camouflaged eggs and the birds's young are better protected from predators. The researchers distributed gull eggs at various distances from each other in open country, and found that carrion crows destroyed more eggs the closer they lay. Like other predators, these crows will look for more objects near where they first found something. But they soon abandon the search if the prey objects are sparsely distributed.

V. C. Wynne-Edwards and several others regard these regulation mechanisms as the reason for why territorial behavior is selective; they consider any advantages for the individual as incidental. But proponents of the natural-selection hypothesis, who are far more numerous, hold that any consequences for the population are incidental to the selective advantages for individuals. In addition to these opposing views, there are considerable differences in opinion regarding the relative importance of the territorial functions mentioned earlier for individuals of a species. Disagreements arise partly because of differences among the species studied by various researchers, and partly because our knowledge of the interrelations between environment and behavior is still incomplete.

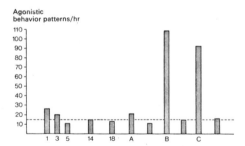

Agonistic
behavior patterns/hr

Fig. 25-13. The number of agonistic behavior patterns per hour in a group of captive rhesus monkeys. After the animals are put together, this frequency decreases within five weeks to an average of fifteen. But it increases again every time the stability of the group is disturbed: A, cage made smaller; at B and C, two adult females were added each time.

According to P. Marler, the primary significance of individual distance also lies in providing the organism with certain vital resources, mainly food. He observed that aggressive encounters resulting from violation of personal space occur mostly in the winter, when animals gather at the few places where they can still find food.

Meaning of individual distances

Many scientists hold that dominance hierarchies serve to prevent constant fights within a group. Although this stabilizing effect does seem to occur, it is probably not the only function of rank ordering. The privileges of high-ranking animals with regard to food and sleeping places, for example, ensure that at least those animals remain well fed and rested that are required to fulfill special functions within the group and in the group's

Meaning of dominance hierarchies

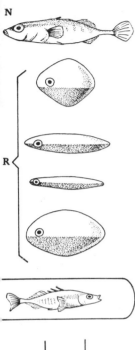

Fig. 25-14. Red-bellied stickleback models (R) are attacked more frequently and strongly than a realistic-looking model without red (N). Models presented vertically have a greater effect than those in a horizontal position.

Negative aspects of aggressive behavior

defense. According to A. Rasa, the hierarchy in a group of dwarf mongoose corresponds to the different demands made upon its members: The alpha female is the only one bearing young; the beta male often confronts intruders by himself; and the young that are next in rank require more food than the older animals who are at the bottom of the line, since the younger ones are still growing. When sufficient food is available, even the omega animal can eat. But when food becomes scarce, members are excluded in the reverse order of their usefulness to the group. This may seem unjust and cruel by human standards, where there should not be any distinctions between important and unimportant individuals. But from a biological point of view this mechanism has a definite function: Whenever there is a conflict between "everyone too little or some enough," that is, between the interests of the individual or those of the species as a whole, the decision must go in favor of species survival.

In cases where the traits of high-ranking animals—for example, physical strength, health, etc.—are determined genetically, it is certainly an advantage for these individuals to have first rights to females and preferred breeding sites. As mentioned before, the species as a whole benefits if those members reproduce that are best able to defend themselves and their young. In more highly developed organisms, where intelligence and experience also determine status (or at least its retention), it is probably useful for even strong and healthy individuals to be excluded from reproductive activities until they have made experiences compatible with those of older members, thus attaining their "social maturity" in addition to their sexual one. Under these conditions, the group should as a rule benefit from the subordination of most members to the leadership of the higher-ranking (more experienced) ones. It is a well-known observation that chimpanzees and monkeys imitate only higher-ranking individuals or those with the same status, and that jackdaws pay little attention to the fright reactions of young birds, but flee immediately if the alarm is transmitted by an older bird. On the other hand, we know just as little about the more precise meaning of this kind of authority for a social group as we do about another—probably quite positive—consequence of social hierarchy: the division of tasks.

Many of the positive effects resulting from aggressive actions can be attained by other means as well, especially when we talk of the regulation of population density. Figure 25-15 schematizes some of the alternatives and shows that, when applied to the entire animal kingdom, aggressive behavior patterns are only one set of regulating factors among many—albeit one that became more and more important in the course of evolution, until now the highly developed vertebrates have come to show the effects of crowding sooner on a social and physiological level than on a physical one.

According to what we have described so far, one might well get the impression that an animal should be more successful, the greater its pugnacity. But too much aggression also has negative effects. One of these is the sheer waste of time. An individual constantly engaged in conflicts, or continually controlling the boundaries of a huge territory, is obviously all the more limited in its search for food, its courtship and nest-building activities, feeding of the young, and so on. Studies on a number of bird species and on sticklebacks have shown that this "aggressive neglect" may endanger reproduction. On the other hand, if as in some insect states (e.g., termites) there is a distinct division of labor, so that certain members act as "professional soldiers" and would not take part in foraging for food or reproducing anyway, then the time that these individuals spend fighting with neighboring groups is by no means wasted.

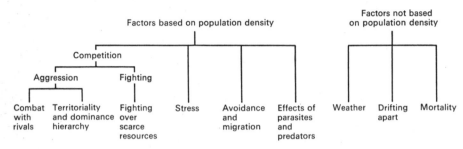

Fig. 25-15.

Another disadvantage of too much aggression is the danger that besets an animal that fights frequently. Since it must concentrate all of its energy on the fight itself, the animal becomes more vulnerable to predatory attack. In any case, it runs the risk of being injured or killed, especially if it loses the contest. The species as a whole would then lose its reserves needed to compensate for losses.

A further danger in being overly aggressive is that such an individual may never be able to mate and reproduce because a potential partner would always be running away, or that it might endanger its own offspring by attacking them and thus lowering its reproductive success. But if very aggressive individuals raise fewer young than others do, the percentage of their hereditary traits within the gene pool of all the offspring population will decrease in the same way as that of individuals with too little aggressiveness. As a result, what will predominate are the genetic traits of animals whose degree of pugnacity allows for the greatest advantages with the least disadvantages. We shall describe this kind of "balanced aggression" within the next few paragraphs. Before going on to discuss this point, we must explain why of all the members of its species (young, females, males, etc.) an animal regards only a few as rivals, as objects of aggression.

As with other behavior patterns, aggressive acts are released by certain stimuli (see Chapter 17). When an ant approaches the entrance to a nest, she is frequently and repeatedly examined by guards with their antennae and mouth implements. If her scent is not the colony's, she will be attacked. In addition to odorous substances, animals mainly use visually perceived

Releasing stimuli

shapes, colorations, positions and movements, and auditory signals as sign stimuli and releasers: Male sticklebacks respond to a red belly and the head-down position (Fig. 25-14), robins to the color of the breast feathers (Fig. 25-13) and to their species-specific song.

But the triggering of aggressive behavior does not depend merely on the rival's stimulus characters. The same bird that is attacked at 50 m or more when he crosses into another's territory is actually able to come as close as a few meters to the rival himself, provided he does not cross the boundary line. This demonstrates—in the same way as Tinbergen's test tube experiments and the existence of an individual distance—that the spatial relationship between different stimuli is a decisive factor in releasing behavior. Furthermore, the timing of their presentation is important as well. The same red-bellied model that triggers intensive attacking behavior in male sticklebacks during the spring will have no effect at all under winter conditions. We shall describe the various motivational states for aggression a little later. Finally, other stimuli presented at the same time will have an influence on whether an aggressive behavior occurs.

Animals living in families or in larger closed social groups must be able to fight or flee, because they have to engage in conflicts with neighboring groups; but this behavior has to be buffered against members of the animals' own group in order to maintain their bond. This requires individuals to discriminate between friend and foe. Members of anonymous closed associations recognize each other by supraindividual characteristics such as clan, nest, or colony odor; in individualized groups, everyone knows the others personally. A great many experiments have shown that aggression can be decreased by a common group characteristic as well as by individual recognition.

If an ant from colony A is given the scent of colony, she will be allowed into nest B. If she is then brought back to nest A sporting her new chemical "membership badge," her former comrades will attack her.

J. Lamprecht separated cichlid parents (*Tilapia mariae*) with a partition that they could not see through, leaving each partner with about half of the young. When he waited five hours and then removed the obstacle, only a few weak attacking movements would occur between the partners. But if he separated them and took out the male, for example, replaced it with a strange male, waited about five hours until the newcomer had "adopted" the fry, and then removed the partition, the number of attacks were more than three times higher and also much more intense. On the other hand, if the experimenter allowed the strange individuals to see each other through a glass partition before putting them together, the number of attacks directed against the other through the glass steadily decreased, until two days later the male and the female could be safely placed together. Using this method, Lamprecht even managed to "pair off" two males with each other.

These experiments demonstrate not only that animals familiar with

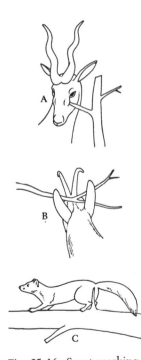

Fig. 25-16. Scent marking with glandular secretions. The glands lie in front of the eyes in the Indian black buck (A), behind the horns in the chamois (B), and underneath the base of the tail in tree martens (C).

each other show less aggression toward the partner, but also that this tolerance is attained only through habituation (see Chapter 17). Similar behavior is found among territory owners, where as a rule each soon gets to know its neighbors and hardly reacts any more when they meet at the common boundary.

In this context, it is again relevant to see how territories are delineated. In addition to natural boundary marks, many species demarcate their borders by other means. For example, the sand pyramids constructed by the crab *Ocypode saratan* serve to keep conspecifics away (see Color plate, p. 370). Mammals in particular deposit scent signals at certain points in their territory. For this purpose, many use special gland secretions that are wiped onto grass, bushes, trees, and other objects. Others mark with urine and/or feces, often excreted in a ritualized manner. The male pygmy hippopotamus (*Choeropsis liberiensis*) distributes its dung by vibrating its bristly tail from side to side, mixing the feces with a stream of urine sprayed back at the same time.

Many authors also tend to speak of territorial marking when an owner proclaims his presence by means of special positions or movements, by displaying certain color patterns, or by making special sounds. It seems more to the point, however, to define behavior patterns like these, where the effect ceases when the animal leaves, as threatening behavior. We then follow R. Schenkel's example and speak of marking only when the animal actually deposits a more permanent mark which conspecifics can still detect later. Most important, however, we must not assume that in every instance where an animal does give off a mark the purpose is to ward off rivals. For one thing, urine and dung are deposited in very diverse kinds of situations, and second, it has not been demonstrated beyond any doubt in even a single mammalian species that certain odorous substances have a repellent effect on conspecifics. Nonetheless, it does seem that lions and domestic cats at least become more cautious and alert after smelling a strange mark.

Perhaps the marks that are deposited at different points in the territory are designed to help its owner become more familiar with the terrain, and to orient himself. In cases where members of the family or group are also marked with urine (wild rabbits, guinea pigs) or gland secretions (dwarf mongoose), this marking behavior may help—as with the nest odor of ants and the clan odor of rats—to distinguish friends from enemies.

Despite these limitations, there is no doubt that some animals avoid certain "boundary posts" (especially natural landmarks) because they may have learned that they will be attacked if these marks are crossed.

Dominance hierarchies work according to a similar principle, namely, to reduce the number of aggressive interactions. The difference is that individuals do not avoid certain marks symbolizing the prerogative of another, but rather yield to the dominant conspecific itself. On the other hand, dominance hierarchies are certainly not essential nor absolutely

Marking of territories

Fig. 25-17. The flightless Galapagos cormorant brings "gifts" to its mate, in this way diverting the partner's aggression away from itself and toward the object.

Fig. 25-18. Appeasement behavior derived from parent-offspring interaction: licking the mouth of the partner in tree shrews.

Fig. 25-19. The appeasement posture of baboons is derived from the female's invitation to mate.

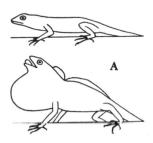

Fig. 25-20. Threatening by enlarging the contours of the body: The false anole, a South American species of lizard, props itself up and inflates its throat.

Fig. 25-21. Male squids threaten with the broad side of the outer arms while taking on a striped pattern.

effective in keeping aggression at a low level. We know of animal species such as the African wild dog (*Lyacon pictus*) that do not show any obvious rank order, yet live together more peacefully than, for example, baboons whose social structure is strongly hierarchical but who are still engaged in constant altercations.

Another way that partners can avoid fighting with each other is to redirect any released aggressions onto other objects. The flightless cormorant (*Nannopterum harrisi*), a native of the Galapagos Islands, always returns from trips to the ocean carrying a tuft of seaweed, a small piece of wood, or a starfish. If we take these objects out of a bird's bill before he reaches the nest, his mate will receive him with hard thrusts and snapping movements that would otherwise have been directed at the "gifts." Eibl-Eibesfeldt suggests that these objects serve as a kind of "lightning rod." In many cichlids and anatids, aggressive behavior triggered by stimuli coming from a mate are instead directed to outsiders. Lorenz proposes that this reorientation is one of the phylogenetic beginnings for the ritualization of greeting gestures that serve to bond members of a group. Joint aggression against outsiders helped to reduce aggressive behavior within the group, providing a starting point for individualized relationships.

Innumerable other mechanisms serve to reduce the frequency of aggressive acts on a short-term basis (e.g., during courtship) or consistently (in permanent groups). The common feature is that, in addition to presenting a stimulus for aggression, the partners always show stimuli that also activate other behavior tendencies. The differences between mechanisms lie in the kinds of stimuli used and the behavior patterns released, which we classify as appeasement postures. Many of these movements and postures are derived from parent-offspring interactions—especially from feeding and body-care behavior, as anyone can see from observing the way rhesus monkeys in a zoo relax when grooming each other. Figure 25-18 shows another example. Such "childlike" behavior patterns (infantilisms) have a pacifying effect because young are not attacked. Sexual stimuli coming from a mate also reduce attacks on the partner, at least during courtship, and so elements of sexual behavior provide a further means of bonding between members of a group (Fig. 25-19).

In addition to the methods we have described for reducing the frequency of aggressive encounters, there are many "inventions" designed to utilize the advantages of aggressive tendencies while at the same time limiting the negative effects. Most prominent are the behaviors of threat, ritualized combat, submission, and escape.

Behavior patterns of threat or display are those designed—like fighting itself—to challenge, intimidate, or drive away an opponent, but without physical contact. A threatening animal will generally make itself larger and more impressive. Movements and postures that we can observe with particular frequency on the borders of territories will often be accompanied by a display of certain body characters. These physical features may be the actual weapons, as in many birds that direct their beak to the opponent, in cervids that present their antlers, or in hippos, carnivores, and monkeys that display their teeth (see Color plate, p. 363). Special color patterns may also be presented, for example the red belly that male sticklebacks show their rival in a vertical position, or the conspicuous display coloration of many fishes, reptiles, and birds (see Color plate, p. 370). Among invertebrates, an example is the male squid (*Sepia officinalis*), which threatens with the broad side of its outer arms and at the same time takes on a purple-and-white-striped pattern; another example is the male fiddler crab that displays by rhythmically moving its enlarged and brightly colored pincer up and down (see also land crabs, Color plate, p. 364).

In many cases particular sounds are emitted to enhance the threatening effect, forming perhaps the only signal in terrain with poor visibility. Familiar examples are growling dogs, roaring stags, male gorillas beating their chest, and birds singing. D. Lack told of how a newly arrived robin engaged in a song duel with a territory owner right in the middle of dense brush, finally fleeing from an opponent it had never seen. Not all bird songs have this repellent effect, of course, and we are often as much in doubt about these as about the threatening nature of most chemical marks.

We cited the vertical position and red belly of sticklebacks as well as the song robins as examples of stimuli both releasing aggression and intimidating the rival. This two-way effect has been demonstrated experimentally for numerous threat signals. Concomitant with this phenomenon is a double motivation for threat behavior, which becomes evident when we look at the form of the behavior patterns. The most simple occurrence is a superposition of attack and escape movements. In many bony fishes (teleosts) the tendencies to swim toward the rival and at the same time turn away from him lead to a corresponding compromise position, ritualized in the course of evolution to lateral display. The ominous thickening of a laughing gull's neck results from simultaneously tensing muscles used in striking with the beak and others which usually pull the beak back. But laughing gulls and other animals have a variety of threatening movements, corresponding to various proportions of aggressiveness and fear. In other cases the drive conflict discharges by means of redirecting aggressive acts or by displacement activities (see Chapter 15). Sticklebacks, for example, express conflict by biting the substrate.

Under natural circumstances threat precedes fighting, and often enough contests are decided only by such displays. Primarily, this occurs within a dominance hierarchy and between owners of neighboring territories, where animals had already tested their relative strength in some earlier fight and now only needed an occasional reminder of who is stronger and has more privileges. Threat duels also occur between individuals whose strength is obviously very different. Threat displays have a challenging effect when two roughly equal rivals face each other, and

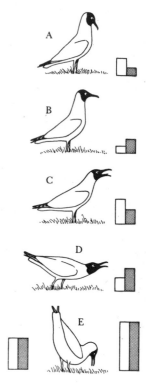

Fig. 25-22. Different display postures of a laughing gull. Columns represent the proportion of attack drive (shaded) and escape drive (light).

Motivation of threat

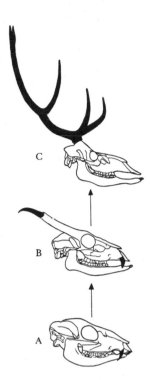

Fig. 25-23. Some deer living today give us an idea of how, in the course of evolution, antlers may gradually have replaced the canine teeth as weapons. A. Musk deer; B. Muntjac; C. Red deer.

Fig. 25-24. Just like its ancestors, the Manchurian sika threatens with its canines, even though these have degenerated into tiny structures.

fighting ensues. After displaying to each other for a while, cichlid fish direct pressure waves at the opponent's flank by beating their tail from side to side. After that, the rivals swim around each other in ever smaller and faster circles; finally, some *Tilapia* perform mouth-pushing movements and then push each other back and forth, while other cichlids engage each other by the jaws and tug away in a contest of strength (see Color plate, p. 387). This kind of fighting, where the stronger of the two is determined according to fixed rules and without either opponent being injured, is termed as stylized or ritualized.

Not every bloodless contest, however, is a ritualized fight. Some encounters are harmless if only because the contestants have no dangerous weapons or are well protected from whatever weapons they do have. The force of pincers belonging to stag beetles, for example, is far too weak to cause their chitin-armored body any harm (see Color plate, p. 386).

For that reason we propose to limit the term ritualized fight to species that bear weapons that are potentially harmful and even lethal. The fence lizard (*Lacerta agilis*), for example, has pointed teeth but never injures its opponent. The two rivals alternately present their well-armored neck, and the other seizes it by his teeth. Strength is measured by the force of this bite, which is never aimed at any other spot. Additional examples are shown in Figures 25-33 and 25-34.

We suspect that the ancestors of present-day ruminants used their long, daggerlike canine teeth for threatening and fighting, just as we can still observe in cherrotains (*Hyemoschus*), which have no antlers, in musk deer (*Moschus*), and in water deer (*Hydropotes*), all of which also show a number of other primitive traits. We can only guess how this might have led to the ritualized frontal fighting technique (pushing, butting, and ramming with foreheads) of present-day horn and antler-bearing ungulates. F. Walther suggests that perhaps the first small forehead weapons might have evolved in connection with predator defense; he emphasizes, however, that the use of horns exclusively in intraspecific aggression can have influenced their further evolution. These head weapons then replaced canine teeth in intraspecific fighting as well, becoming more and more predominant in the course of evolution. Even today there are species that could serve as a prototype for this phylogenetic development. Muntjacs (*Muntiacus*) have small antlers, but they still fight with their canines and also present them while threatening. Dybowski's deer and red deer perform the same kind of threat movement, although they fight with their antlers and despite the fact that their canine teeth have degenerated into tiny vestiges. Male mountain goats (*Oreamos americanus*) fight by standing in a reverse-parallel position to each other, gathering force with their head and throwing it with full weight and horns forward against the rival's flanks. Although these animals have a few safety features, such as thick hair and skin, serious wounds and even death may result. Walther holds that mountain goats stand right on the border line, beyond which longer horns used in this manner would become increasingly maladaptive. Thus, for the long-horned species to survive, contestants had to direct their attack to the least vulnerable part of the opponent's body, namely, the horns themselves.

This represented a first step from damaging combat to ritualized fight. According to V. Geist, this primitive frontal fighting technique gave rise to two different methods: The pushing contest typical for antelopes, and the ramming commonly found in buffalo, sheep, and ibex (steinbock). In each case, the forehead structures (antlers and horns) have been adapted to the fighting method of the species (see Color plate, p. 386).

Such weapons may also grow to excessive dimensions, as can be seen in deer and kudus (*Tragelaphus strepsiceros*). In exceptional cases during a contest, the branched and twisted antlers can become so intertwined with each other that the rivals are no longer able to disengage, leading to death for both combatants. It is also assumed that the giant deer (*Megaloceros giganteus*) became extinct at least partly because of its extremely large set of antlers. For one thing, their yearly shedding and renewed growth required a huge amount of metabolic energy, and second, a head ornament with a span of 4 m could only be "carried off" in open ice country. Inevitably, they became maladaptive as soon as the animal's habitat changed into forests. This is one example of how in the course of evolution compromises have to be made in response to opposing selection pressures, not only in behavior patterns but also in the weapons used.

Even among the numerous species that pass from preliminary threats into a full-fledged damaging fight and wound each other by biting, scratching, stabbing, and so on, the combatants usually do not kill each other. Admittedly, the fights between hippopotamus bulls, with their powerful canine teeth, give the appearance of being extremely dangerous (see Color plate, p. 387), and bloodshed is not unusual. But their wounds heal quickly, and the thick skin and the fat layer protect against more serious injuries. Similarly, pinnipeds and mountain goats have this adaptive protection, and male lions use their mane to catch the bites and blows of their rivals.

Once an animal recognizes the superior strength of his rival—either through the other's threat signals or by fighting—he either flees or takes on a submission posture, which usually looks like the opposite of the threat pose. Species that threaten by emphasizing their body size will then make themselves as small as possible; animals like squids, cichlids, and some reptiles, which can change color, lose their body coloration; animals that threaten with a raised head will submit by lowering it, and vice versa; and individuals that point their weapons toward the opponent when threatening will hide them during submission. The effect of submission postures, like that of fleeing behavior, is based on the removal of aggressive stimuli. The victor will pursue his fleeing rival for a short distance, perhaps up to the boundary line of his territory. In cases where the rival

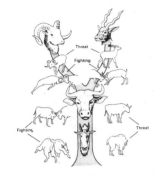

Fig. 25-25. The evolution of horns in the family of Bovidae: bottom, mountain goats; next, buffalo; then a division into sheep (left) and antelopes (right).

Fig. 25-26. Members of species that threaten by emphasizing their body size, like the gnu, reduce their apparent size as much as possible when taking up an appeasement posture.

Fig. 25-27. The kittiwake threatens with its beak open and pointing down (animal at right); it submits by closing the beak and pointing it up.

Fig. 25-28. Threatening animals frequently show their weapons to the opponent, while submitting individuals hide them. After a fight between two hippos, the loser closes its mouth and lowers its head, while the winner threatens with its mouth wide open and its head raised until the other has retreated.

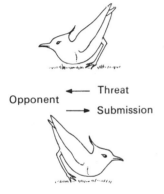

Opponent   ← Threat

→ Submission

Fig. 25-29. In plovers, the threat posture and submission pose differ only in their orientation with regard to the opponent.

Fig. 25-30. The submission pose of a defeated turkey resembles the female's invitation to mate. Because of this, the winner sometimes tries to copulate with the loser.

submits, the winner of a fight will stop attacking, although he often maintains a threatening posture until the loser has withdrawn.

When one animal merely "turns off" its aggressive stimuli, however, the aggressive mood of the other partner wanes only very slowly. For that reason, as Lorenz suggests, it would be "suicidal for an animal to suddenly offer its opponent a highly vulnerable part of the body, since the attacker is still very much in a fighting mood." In order to really inhibit aggression, according to Lorenz, a mere submission posture is not enough. Such a pose must also have an appeasement effect, eliciting other behavior tendencies which derive primarily from the functional groups of courtship and parent-offspring interaction. We can often observe how male fence lizards and turkeys try to mate with a submitting rival, because the loser's appeasement posture is the same as a female's invitation to mate, or how a jackdaw that turns its beak away in submission is scratched on the back of its neck by the victor, because it simultaneously displays its long, silken, and light-gray feathers which seem to act as an additional signal and thus appease the opponent.

Much as so many threatening movements can be interpreted as conflict between attack and flight, numerous submission postures also seem to derive from a superposition of these two kinds of motivation. However, we know less of how the submission movements might come about than we do of the threatening movements.

It is somewhat difficult to understand how such inhibitions to attack a yielding partner have evolved. After all, we might say, does not a contestant who is first to submit or who inhibits his urge to bite run the risk of being worsted by uninhibited rivals? How can the genes responsible for this behavior survive in the population? According to W. Wickler, however, these difficulties exist only because "we are in the habit of seeing the 'egoism' shown in a fight as more basic, simply because we have known about it longer." We have already pointed out that excessive aggression is harmful even to the individual showing it. Animals have always had to be tolerant at least toward their own mates and young, since otherwise the species would not have survived. A more likely explanation, as Wickler suggests, is that the behavior patterns of submission are not "secondary inventions to guard against aggressive tendencies that would endanger the species as a whole," but rather the converse, "remnants of intraspecific tolerance that have always been present." With increasing aggressiveness, which animals could then afford to develop, these mechanisms of tolerance may have evolved into inhibitory ones. During this process it was possible that "old mechanisms were replaced by new ones, just like in evolutionary history organs and bone structures of different origins replaced each other in different parts of the body."

The fact that elements of sexual behavior as well as those of parent-offspring interaction are commonly found in appeasement postures lends support to Wickler's hypothesis, as does our observation that aggressive behavior occurs most frequently among organisms at higher phyletic levels.

As a rule, animals behaving aggressively aim at driving their rival away rather than killing him. Indications of this are found in the common occurrence of threat behavior, in the numerous duels that are harmless and ritualized or even damaging but not fatal, and in the fact that winners in a fight will spare their rival if he submits or flees. Certainly there is the occasional death, as when two musk oxen collide so strongly with their heads that one of them breaks his skull—but these cases are more accidental than intentional, just like the wounds that mountain goats inflict upon each other. On the other hand, we must not ignore the instances where opponents often do kill each other because less harmful methods are of no avail. This happens particularly in captivity, where the loser is often prevented from retreating far enough and leaving his rival's territory, or where low-ranking animals cannot stay far enough out of their dominant rival's way (Fig. 25-31). Every zoo director knows that he cannot keep two stags in the same enclosure during the rutting season without running the risk of their stabbing each other to death.

Adaptation of fighting behavior

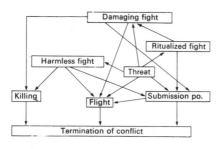

Fig. 25-31. There are many possible transitions between the single types of combat (direction of arrow). If a conflict cannot be terminated by means of submission postures and flight, killing the rival is the only alternative.

This is why the common practice of dividing species into those with ritualized combat and those engaging in damaging fights is actually misleading. Even animals that usually test their strength in a stylized duel without injury are capable of inflicting fatal wounds. Here, the various forms of aggressive acts are only points on a continuous scale where behavior patterns differ in their stimulus threshold (Fig. 25-32). The adaptive value lies in the animal's ability to utilize these forms in their appropriate situations. It is not threat or ritualized combat in themselves that are favorable to the species, but rather threat or ritualized fighting under certain circumstances. In other situations—for example, where a conspecific becomes pathologically uninhibited, or where populations are so dense that stress results—it may aid the survival of a species more if some are "sacrificed" for the sake of the population.

There are also circumstances where—especially with lack of space—conflicts between conspecifics without dangerous weapons can nevertheless be fatal. The opponents simply go on fighting until one of them dies of exhaustion. In this case, as Wickler writes, "neither one destroys the

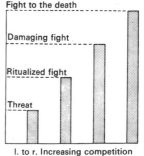

Fig. 25-32. Schematic repersentation of the relationship between the intensity of conflict and types of combat.

other, but each destroys only himself." This is why he stresses that the capacity to give up and withdraw—a feature that people have deliberately bred out of, for example, fighting fish and fighting cocks—is as important to the species's survival as the inhibition against killing a rival. Tinbergen points out that the chances of getting a territory and reproducing will increase whenever an animal yields in the face of stubborn resistance and looks somewhere else.

Our discussion has made it clear that there is not only one kind of aggressive act. Differences appear not only in form, but also in the context of a behavior, the stimuli that elicit the act, and the function it serves. On the other hand, is there one kind of aggression, one motivation underlying all kinds of conflict? This is a fundamental question that has led to lively disputes and discussion. Today we can differentiate four major viewpoints, according to which aggressive behavior patterns are based on: 1. a primary drive; 2. a motivating force controlled by external stimuli ("secondary drive"); 3. frustration of a primary drive; 4. learning processes.

Primary drive
—innate
—endogenous-
    spontaneous
—can build up
—appetence
—vacuum activity

A primary drive would result from the fact that some operator in the central nervous system continuously produces action-specific energy for the performance of certain behavior patterns, and on an innate basis. Since this energy is produced even in the absence of exteroceptive stimuli, it is called endogenous-spontaneous. As a result, the specific action readiness increases until suitable stimuli appear to release the final or consummatory act, which depletes this energy (drive satisfaction). The longer the animal must wait to perform the consummatory action, the more energy is stowed, or "dammed up" (storing of drive energy). Consequently the animal will first show a searching or appetitive behavior (see Chapter 15) aimed at satisfying its drive, and then will respond to increasingly less specific stimuli (lowering of stimulus threshold) until finally the behavior is performed in vacuo (vacuum activity). The validity of this model has been substantiated a number of times for feeding and reproduction. K. Lorenz, I. Eibl-Eibesfeldt, and other scientists hold that it also applies to aggression; P. Lehyausen claims that the model also applies to fleeing behavior.

"Secondary drive"
—innate
—reactive
—can build up

In contrast to this, P. Marler, W. J. Hamilton, and others propose that fluctuations in aggressive motivation (= energy) do not result from endogenous-spontaneous activity but occur in response to external stimuli. The animal's aggressiveness is supposedly stimulated by the sight of a rival or of a spot where fighting once took place. Without appropriate stimuli, no appetitive behavior and no vacuum activities occur. These interrelationships are sometimes described as "secondary drive"—a term also used in human psychology, but for indicating acquired drives that motivate human behavior.

Frustration-aggression
hypothesis

In 1939, J. Dollard and his associates formulated the so-called frustration-aggression hypothesis. As a rule, ethologists only speak of frustration (involuntarily foregoing the satisfaction of a drive) when a behavior sequence already set in motion is obstructed by some external event, preventing the animal from reaching its goal. Psychologists often use this term for every kind of obstruction that prevents access to an imagined or tangible goal. Both goal and obstruction may be conscious or unconscious. In its presently valid form, this hypothesis states that: 1. frustration increases the likelihood of aggression, and 2. aggression is one of several possible reactions to frustration.

 The hypothesis makes the assumption that part of the energy associated with an innate primary drive (sex, hunger) is transformed into aggressive energy when the satisfaction of this primary drive is frustrated. The resulting aggressive act can serve to eliminate any obstruction and thus allow for the satisfaction of primary drives. The degree of frustration that an organism will tolerate before showing aggression will depend on earlier social experiences.

 A fourth group of scientists holds that no aggressive drive energy exists. Aggressive acts in the sense of unprovoked attacks are primarily the result of learning. J. P. Scott and others suggest that young animals either imitate adults or are subjected to painful bites and pummeling from others in their competition for food, in playful interactions and so on, which they try to ward off by defensive movements. In both cases, the success of these actions is reinforcing (see Chapter 24) and, as a result, aggressive behaviors become habitual.

 People often try to play these viewpoints off against each other. This leads to fruitless discussion, since in reality each of the theories described has a kernel of truth. We should therefore take a closer look at the differences, and ask:

1. Is aggressiveness innate or learned?
2. Can aggressive energy be stored? If so,
 (a) does the build-up occur independent of exteroceptive stimuli or does it result from the summation of external influences?
 (b) Can aggression occur in vacuo?
 (c) Can organisms have an appetence for fighting?
3. Is aggression autonomous or is it fed with energy from other drives?

 Before discussing these questions, we should point out that aggression cannot be measured directly but only by the evaluation of aggressive acts: Aggressive motivation must increase with a lowering of threshold for the releasing stimuli, with an increase in appetitive behavior (where applicable), a decrease in the time needed for a fight to begin under standard test conditions, an increase in frequency, duration, and intensity of fighting, a decrease in the effectiveness of inhibitory mechanisms, and a decrease in submissive postures and flight.

 1. The first question, whether aggressiveness is inborn or acquired, has to be answered with "yes and no." Breeding experiments have shown that it is at least partially determined by hereditary factors: If an aggressive mouse or rat population (A) and another population of nonaggressive

—both innate and learned
—reactive
—can build up
—derives energy from another drive

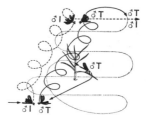

Fig. 25-33. The aerial combats of male dragonflies fighting over territories is an example of ritual fights. Without ever touching, they circle each other continuously and spiral higher and higher until one of them flees. T, owner of territory; I, intruder.

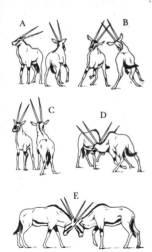

Fig. 25-34. The long sharp horns of gemsboks are rendered less harmful by the way these oryx antelopes fight. After preliminary threatening (A), the rivals "fence" with their horns (B–D), cross them, and try to push each other back (E)—a typical example of a ritualized fight.

Fig. 25-35. Effect of winning and of losing on aggressiveness. Dots: animals that were forced to fight first against more aggressive partners. Circles: Animals fighting first against less aggressive partners; conditions were reversed after eleven days.

animals (N) are bred among themselves, even the second generation of descendants of A-animals are clearly more aggressive than the N-strain. Here, maternal care is only of secondary importance: When N-young are raised by A-females and vice versa, there are hardly any differences in behavior to distinguish the young from their natural parents. On the other hand, this does not mean that social experiences and treatment in childhood as well as learning processes are insignificant. Scott was able to produce pacific male mice if he raised them with females and also lifted them up every day by their tail and stroked them. Conversely, he could increase their aggressiveness by creating for them success in fighting. If he put an A-mouse and an N-mouse together, the aggressive one immediately rushed at the nonaggressive one. But as soon as the latter started to defend itself, the attacker was taken away, giving the N-mouse the impression that it had won the fight. After five days of training, the previously peaceable male attacked even females and young. Negative stimuli—in experiments mainly pain through electric shocks—can also raise aggressive motivation. But in this case it depends very much on the degree of negative stimulation. Some dosages have an attenuating effect, just like repeated failures suffered under natural conditions.

2(a) With regard to whether aggressive energy can be stored, we can cite two species of cichlid fish whose behavior has become quite famous, *Etroplus maculatus* and *Geophagus brasiliensis*. Males without brooding experience live peacefully with their females as long as they can attack other conspecifics in the aquarium—if only through a glass partition. If we prevent a pair from seeing any others—that is, if we isolate it visually—the male will attack his mate and kill her. Lorenz interprets this phenomenon as an endogenous build-up of aggressive energy. If this energy cannot be constantly discharged on outsiders, it increases and finally breaks through the barrier of inhibition that normally prevents the fish from attacking his mate.

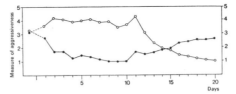

These and other findings do suggest an increase in the tendency to fight, but this increment does not necessarily come about through endogenous-spontaneous processes. W. Heiligenberg and his co-workers were able to demonstrate with two other cichlid species, *Pelmatochromis subocellatus* and *Haplochromis burtoni*, that adult males isolated with younger fish for several weeks became less aggressive. When he then presented the males with adult conspecifics or with models resembling them, aggressiveness increased once again (see Chapter 17). In *Haplochromis*, this externally influenced build-up can be traced primarily to the orientation of the rival's black eyestripe, which is missing in young. Heiligenberg's results also suggest another way of explaining why *Etroplus* and *Geophagus* males attack their mates when isolated. Males and females of these and other pair-forming cichlid species look very much alike. The partner therefore displays signals like those of a real opponent that possibly increase the male's tendency to attack. This is why the isolation of pairs, often used as supposed illustrations of endogenous-spontaneous energy build-up, is not really a good experimental procedure for studying the issue of spontaneity, although the fact remains that the male does not attack his female as often when other males are present.

A better method is to isolate individuals, as E. Courchesne and G. W. Barlow have done with hermit crabs (*Pagurus samuelis*). After different periods of isolation, the subjects were given the chance to fight against conspecifics that had been kept in groups. The experimenters found that, for one thing, the previously isolated animals showed aggressive behaviors like extending the pincers (E), holding tight (H), and pulling at the shell opening (P) more often than control animals, and that the frequency of aggressive acts increased with the length of the isolation period. The greatest increment occurred first with E (after 3 days), then with H (after 8 days), and finally with P (after 30 days; Fig. 25-37). Since we know that the stimulus threshold is lower for E than for H, and the threshold for H is lower than that for P, these results could support the idea of a continuous build-up of endogenous-spontaneous aggressive energy. On the other hand, the isolated subjects seemed to react more sensitively to other releasing stimuli as well (for example, their escape behavior was different), so that this experience may have brought about a more general sensitization. This reservation also applies to the multitude of experiments (especially with mice) that have shown that young animals raised in isolation become unusually aggressive.

Such individuals that had never been provoked or frustrated by members of their own or other species, and that never had the chance to learn either by imitation or through success, fought much more intensively than normally raised animals on their first meeting with an opponent. But in these cases it seemed to be a matter of general developmental disturbance rather than a specific influence that the isolation experience would have on aggressiveness. The subjects are often more active and generally more irritable and excitable; they show increased muscle tension, have stronger reflexes, tremble frequently, and in many cases have enlarged adrenal glands as well as different levels of various chemical body substances. We shall discuss these conditions, all of which are symptoms of stress, later. The temporary isolation of adult mice leads to similar effects. For that reason, many of the isolation studies that only measure the influence on aggression and do not examine other effects as well are also poorly suited

Isolation experiments

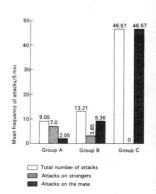

Fig. 25-36. Mean number of attacks made by male cichlids during the reproductive period. In Group A there were several conspecifics in the tank; in B, they were separated by a glass partition; and in C, the pair was totally isolated.

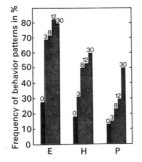

Fig. 25-37. Frequency of behavior patterns "extending pincers" (E), "holding tight" (H), and "pulling at the shell opening" (P) in conflicts between hermit crabs. Black: animals held in groups. Shaded: animals isolated for 3, 8, 12, and 30 days.

Testosterone

Corticosterone

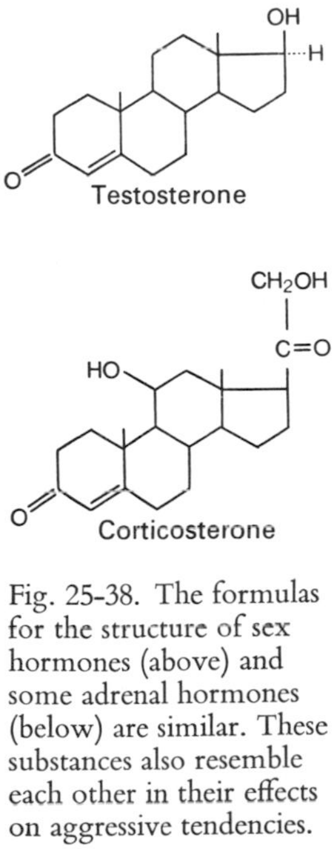

Fig. 25-38. The formulas for the structure of sex hormones (above) and some adrenal hormones (below) are similar. These substances also resemble each other in their effects on aggressive tendencies.

to help explain whether aggressiveness is stored in an endogenous-spontaneous manner.

2(b) We have not found any evidence for "vacuum" aggression. There are several chemical substances that lead to an increase in the level of aggressiveness, but not one that would release aggressive behavior without the corresponding external stimuli.

2(c) Are there any indications that animals look for rivals or for opportunities to fight (appetence)? The experiments of E. v. Holst and U. v. Saint Paul may support this assumption. They stimulated various loci in the brain of domestic chickens with fine electrode tips, causing the animals to walk around restlessly and look for very specific objects. When offered food, water, mates, or even rivals, the subjects responded to only one of these each time, depending on the stimulus field. Another method of testing appetence is based on the fact that animals can only be conditioned by using situations or activities that lead to drive satisfaction. It turns out that very different species of animals do perform a certain behavior if they are rewarded with the opportunity to see a rival or to engage in a fight. Conditioning is often as successful as when food is used for reinforcement. Fighting fish (Betta splendens) can be taught to swim through a ring if they are subsequently presented with their own image in a mirror or with a threatening model. Hamsters and mice learn to enter the correct runway in a T-maze if a rival is placed at the goal. Experiments using situations that release escape responses as a reward have not been successful, pointing against the notion of an appetence for fleeing.

The validity of these experiments depends very much on the use of proper methods, which tends to be extremely difficult here. For one thing, it must be established that the animal's appetence is really for a rival or an opportunity to fight, and not for something else. Second, we must decide whether the conditioning effect is based on the sight of a presented stimulus or—as Lorenz believes—on the actual performance of behavior released by this stimulus. It is, after all, possible that these isolated subjects strive to see anything at all, and then react aggressively if it turns out to be a rival.

Although none of the studies have eliminated all sources of error, there are strong indications that an aggressive appetence can exist. But this is not the same as an endogenous spontaneity of aggressive motivation. It is more likely that the first few random encounters raise the level of aggressiveness—for example, that the entrance to a maze is learned as a part of the rival's stimuli during these first meetings—and that this constantly visible (conditioned) stimulus repeatedly arouses the animal's aggression. For instance, mice chose the correct runway (with an opponent at the end) only fifty percent of the time during their first daily trials even after lengthy training. But the rest of the day their ratio of "hits" reached ninety percent and more. It seems that their aggressive motivation has to undergo a "warm-up effect" by extrasystemic means. After that it becomes a drive to look for an appropriate situation and continue the behavior. Animals that were separated after a short period of fighting even crossed over electrically charged grids in order to reach their opponent again. The amount of electric charge they are willing to suffer then becomes a direct measure of their level of aggressive motivation.

3. The notion that frustration of a primary drive could facilitate aggression has been substantiated in many experiments. For example, a mouse that has learned to find food at the end of a maze, will immediately attack a conspecific if it is prevented from reaching its goal by some obstacle. However, this can be stated only for the frustration of a behavior sequence already in motion, and even then aggression does not always result. A low-ranking animal frustrated by a high-ranking one usually responds with submission; and should it ever attack a conspecific in this situation, it certainly will not be the dominant rival (who is the cause of its frustration), but rather an even-more-subordinate animal. In addition to this kind of social influence, the effect of a frustrating incident will also depend on individual differences. The A-mice, bred for pugnacity, fail to show any rise in aggression after being subjected to a frustrating situation that does increase aggressiveness in the N-animals. In other words, frustration is only one of many possible causes for aggression, and aggressive motivation that is already high apparently will not increase further in the face of a frustrating experience.

Now, among all these factors, is there one that would justify the assumption of a drive energy specific to aggression? We know that hunger and food appetence are triggered in the hypothalamus, a part of the diencephalon, when the blood-sugar level sinks below a certain point. An increase in body salt leads to thirst and a search for food. In all these cases we can pinpoint a physiological correlate for action-specific energy. This has not been possible with aggression. We do know of certain parts of the brain, of organs, and of chemical substances that are associated with aggressive behavior, but the way they work together is still largely a mystery. We shall briefly describe a few of these, with a note of caution that some of the results contradict each other.

When male mice are castrated, so that they cannot produce any more male sex hormones (androgens), their aggressive tendencies decrease (Fig. 25-40). They increase again if the animals are injected with testosterone and other androgens. These results have also been obtained with other animal species, and support the notion that male sex hormones have a direct influence on aggressive behavior. Further evidence derives from the observation that primarily males show aggressive behavior, under natural conditions as well as in response to isolation or electric shock. But even if females are injected with these hormones, their own aggression usually will not increase—unless they had received a dosage on the day they were born. Similarly, males that are castrated immediately after birth do not respond to androgens as adults unless the experimenters inject them with

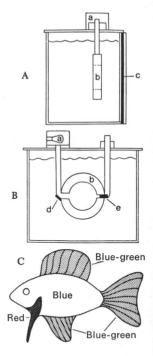

Fig. 25-39. Experimental apparatus for demonstrating appetence for a rival in fighting fish. A: lateral view. B: frontal view. As soon as the fish swims through the ring (b), it interrupts a beam of light coming from the lamp (a) and directed by means of a mirror (d) to the photo cell (e), whereupon a rival or a threatening model becomes visible on the back wall.

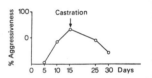

Fig. 25-40. Mice made aggressive by fifteen days of isolation become more pacific again after being castrated.

these hormones immediately after castration. This implies that androgens have some indirect as well as direct influences on behavior. We are fairly certain that these indirect effects consist of raising the later sensitivity of certain neural elements in the brain while the animal is still young. In other words, these elements become sensitized in childhood, and the effect is a long-term one. We do not know precisely how this happens.

Another important organ associated with aggressiveness is the adrenal gland. When this is removed—again in our familiar experimental subject, the mouse—aggressiveness decreases. Conversely, high-ranking individuals as well as those made aggressive through isolation have heavier adrenal glands than low-ranking animals or those raised in groups. Furthermore, both the weight of adrenal glands and the level of pugnacity increase with the length of isolation. Hormones produced by the adrenal cortex (corticosteroids), which are chemically similar to the sex hormones (Fig. 25-38), and the catecholamines (noradrenalin and dopamine) produced by the medulla of this gland serve to increase aggressiveness, while adrenalin is apparently associated with fear.

Fig. 25-41. Schematized longitudinal section of a mammalian brain (left); cross section through the cerebrum, or forebrain (right). Dotted: cerebral cortex; shaded: areas where stimulation can elicit aggressive behavior patterns. 1. Hypophysis; 2. Hypothalamus; 3. Thalamus; 4. Roof of the mesencephalus; 5. Formatio reticularis; 6. Limbic system with amygdala (7).

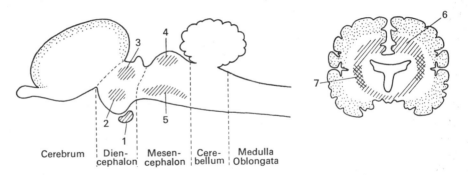

Conversely, again, it was found that increased aggressiveness—whether caused by genetic factors, isolation, or a preceding fight—raises the concentration of these substances in the body. This is why we really do not know which is cause and which is effect, or even whether both effects are not merely parallel consequences of other, as yet undiscovered influences. What is certain, however, is that an increased production of adrenal hormones can also sensitize certain areas of the brain, temporarily or on a long-term basis, especially since we know that adrenalin and noradrenalin lower the excitation threshold at the synapses (connections of individual nerve cells). A threshold-raising effect, with a concomitant inhibition of aggression, seems to result from the action of serotonin, a hormone secreted by a large number of tissues.

These changes in the excitation threshold brought about by sex hormones, adrenal hormones, and other substances do not just influence aggressive behavior, however, (see Chapter 21). And conversely, the concentration of these substances (and with it sensitization toward releasing stimuli) are not determined solely by aggression but also by the general activity level, by whatever activities are taking place at that time, by all kinds of stress factors, and by seasonal periodicity (see Chapter 12) and other endogenous fluctuations (e.g., hunger, thirst, and sex drive). The latest practice is to mark these chemical substances radioactively in order to follow their path into the various parts of the brain, and to determine more precisely where sensitization takes place and what brain sections are associated with aggression. This method is still in its beginning stages; but another technique, that of electrical brain stimulation (see Chapter 19), has provided us with some insights. Threatening movements, attack, and fighting can be elicited in cats and rhesus monkeys if, among other areas, the stimulus electrodes are placed into the following sections:

(a) The hypothalamus, a subsection of the diencephalon that constitutes an important center for the control of vegetative functions such as hunger, thirst, breathing, sleeping, etc.

(b) The thalamus (the largest neural center in the brain stem) and the extended spinal cord (Formatio reticularis tegmenti), which form relay centers for the transmission of incoming sensory signals to various parts in the brain—in particular to the cerebral cortex, which contains the fields for cognitive processes, memory, and voluntary actions.

(c) The limbic system (especially the amygdala), which surrounds the brain stem. By way of a large number of nerve fiber pathways, this group of structures and regions is interconnected with the hypophysis, hypothalamus, thalamus, and cerebral cortex, and also with the sense organs by way of the spinal cord.

The overall results of research into aggression do not allow us to make a final statement, but certain preliminary answers may be given to the three questions we asked earlier:

1. Aggressiveness is determined by both genetic and social factors.
2. Aggressiveness can fluctuate.
   (a) Such variations may occur both in response to exogenous factors and on the basis of endogenous-spontaneous fluctuations (hormone changes, seasonal periodicity, and so on). There has been no evidence that this spontaneity is specific to aggression.
   (b) There is no conclusive evidence that aggressive behavior can appear in vacuo.
   (c) A number of experiments have demonstrated that animals can have an appetence for a rival or the opportunity to fight. As a rule, however, this appetence seems to be dependent on a prior aggression increment brought about by exteroceptive stimuli (warm-up effect).
3. Unlike for hunger, thirst, and sleep, there does not seem to be any control center in the brain to produce endogenous-spontaneous aggression-specific energy. Instead, the numerous nerve networks associated with aggressiveness overlap, and are maintained by other regions in the brain and by nervous and hormonal influences from the

▷
Examples of asexual reproduction: Top left: Paramecium in the process of asexual (vegetative) division. Bottom left: Precious coral as an example of colony formation by means of vegetative reproduction (budding). Top right: *Volvox globator*: the daughter globules in the back part of the cell colony develop into reproductive cells (gametes). Bottom right: Salp (Salpida) as an example of a tunicate with alternation of generations (metagenesis).

▷▷
Two examples of indirect transfer of spermatophores: Top row: Mating in the whip scorpion *Mastigoproctus giganteus*. Left to right: The male (left animal) takes hold of the female and deposits a spermatophore. The female (right animal) takes up the spermatophore with its genetal orifice; the stalk is just barely visible underneath the female's posterior end. The male assists in the taking in of the spermatophore; the process of supporting or assisting, as seen from behind and below. Bottom row: Indirect transfer of spermatophore in *Heterosminthurus bilineatus*. Left to right: The male is hooked under the female. The male deposits the spermatophore. The female drinks the sperm, while the male has dipped its antenna into it.

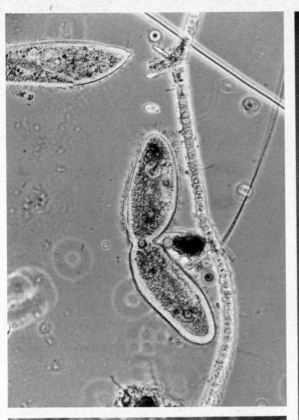

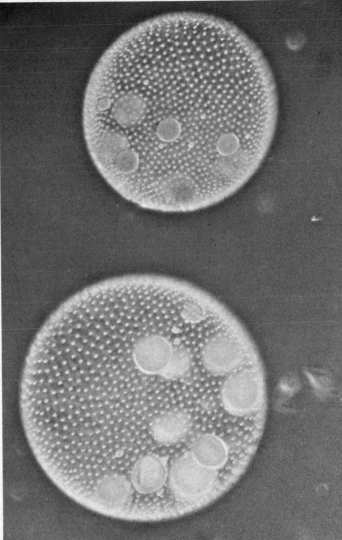

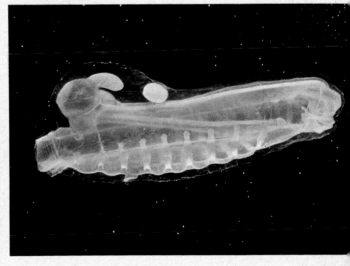

body. These networks in turn wield their influence on a multitude of structures and behavior patterns.

By means of all these connections joined together mainly by the limbic system, aggressive tendencies can be influenced by internal and external stimuli, conscious and unconscious processes, hunger, thirst, drowsiness, pain, other behavior tendencies, physiological and social circumstances, genetic factors, learning processes, memories, frustrating experiences, and much more. Each of these factors has some influence on the stimulus threshold, but it is the interaction between them all that determines whether and how an aggressive act is performed. This sort of network of connections, which orients behavior according to whatever situation prevails as a whole, should have a far greater adaptive value than a relatively fixed aggression drive, where energy builds up through endogenous processes and leads to periodic searching behavior. "We can easily understand," writes W. Wickler, "that such appetitive behavior is an important part of those drives where deficits are corrected, be it food in the individual or individuals in a species." It may also be useful for an animal whose aggressiveness has already been aroused by a rival to be alert and seek a final outcome to the encounter. But why should an organism that has not met up with any rivals go out periodically and look for them, especially since here the disadvantages in the absence of competitors would outweigh the positive aspects? "We might assume that the drive for aggression was initially directed against predators, and an animal searching for enemies that it had not seen for a while would simply display a very necessary vigilance against camouflaged predators; perhaps fights between conspecifics keep animals fit for emergencies. But were aggression against predators and aggression against conspecifics originally the same?

There are several arguments for the notion that intraspecific aggression developed from defense reactions against predators. In any case, this is more likely than the view that it derived from the predatory behavior of the animal itself; if the latter were the case, herbivorous animals should not be aggressive. But in the first place, these interrelationships have not been examined nearly enough, and second, the above-mentioned theory would not imply that aggressive responses against predators and those against members of the animal's own species would necessarily be the same today. The multitude of physiological and social factors associated with either kind of behavior must certainly have stood under very different selection pressures, ultimately following separate evolutionary paths. We have already gathered brain-physiological evidence that even intraspecific aggression is controlled in different areas (territorial defense, competition for females, defense of young, response to pain, and so on) by different chemical substances and nerve networks. In other words, not only are there variations in situational context, form, and releasing stimuli, but even in the nature of aggressiveness itself. But if different types of aggression appear even within a species, we should assume, in Wickler's words, that

◁

Hermaphroditism and sexual dimorphism as examples of the ways in which males and females can differ. Top left: Simultaneous hermaphrodite (edible snail). Top right: Successive hermaphrodite (rainbow cleaner fish). Center left: The differences in size and color between male (above) and female (below) of the arachnid species *Eresus niger* shows the high degree of sexual dimorphism often seen in spiders; body length of the female is 20 mm, that of the male, 9 mm. Center right: Female of the worm *Bonellia viridis*; the males of this species are microscopically small. Below: Winter moth; left, male; right, female, which in this species is unable to fly.

"for other reasons as well, it is likely that aggression as a mechanism developed independently several times, that is, convergently, somewhat like the wings of insects and birds. . . ." In different and perhaps even closely related species, aggression may be constructed differently, "just like insect wings and bird wings both enable the animal to fly, but are constructed in very different ways."

This is why we cannot simply argue from one species directly to another. Nevertheless, comparative studies of many kinds of animals—and on people in different cultures—may serve to provide insight into overlapping features, congruencies, and general principles.

26   The Biological Significance of Sexuality

The biological signi-
ficance of sexuality,
by F. Schaller

One of the main features of organisms is their ability to reproduce and multiply. This ability is linked to the presence of so-called nucleotide substances, the basic building blocks of viruses, bacteria, and the nuclei of all plants and animals. As nucleic acids, these remarkable substances had already been discovered and described by J. F. Miescher in 1869, but their nature and meaning were not fully recognized until the second half of this century, through the work of biochemists and geneticists. In 1953, J. Watson and F. Crick received the Nobel Prize for uncovering the structure of DNA (deoxyribonucleic acid) and finally breaking the code. They found that they were dealing with thread-shaped macromolecules, twisted around each other in pairs to form what they called a double helix. Each thread with its nucleotide building blocks represents the exact negative of its partner. Under certain metabolic conditions provided by the cell environment, these threadlike structures reproduce their own mirror images, and in this way become the carriers and agents of the most basic process of life, the identical reproduction of the self. Hence, reproduction and hereditary transmission are no more than the capacity of living systems constantly to duplicate themselves and propagate their own kind.

The basic biological phenomena of the double-helix structure of genetic substances and the doubling that chromosomes accomplish during cell division do not in themselves have anything to do with sexuality; yet there are already certain common principles, the paired or bipolar nature of these duplication processes, and the sexual reproductive phenomena among organisms. On the most basic level, sexuality always involves the bipolar combination of germ substances, and this is its real purpose even when obscured—as is so often the case—by all kinds of preparation, accompanying behaviors, and patterns of performing the conjugal act itself.

We tend to take the bipolar nature of sexuality—the general phenomenon where two sexes are involved in reproduction—so much for granted that we hardly ever question it. We do find forms of sexuality in animals, including humans, that would nowadays perhaps be labeled as

Fig. 26-1. The nucleotide substances in viruses and bacteria correspond to cell nuclei in plants and animals. With a chemical procedure we can make them darker to contrast with the rest of the cell, as with *Bacillus cereus*, above.

"group sex," but if we analyze this further we always find that any such groupings basically consist of interactions between two partners. A few decades ago M. Hartmann, working in comparative research on sexuality, proposed three major laws: 1. the law of universal bipolarity of the sexes; 2. the law of universal potential for bisexuality; 3. the law of relative strength of male and female determining factors.

With his third major law, M. Hartmann defined the key set in the phenomenon of sexuality: the principle of "male" and "female." What do these opposing concepts mean, and how did organisms evolve the male-female dichotomy?

For a start, let us return to the basic issue of reproduction and propagation before turning all our attention to sexuality. As we know, reproduction and propagation come about only through division, more precisely, cell division. This can be seen most clearly in the single-celled animals, which can multiply again and again through binary or multiple fission without even a trace of any sexual processes. In some forms that are more complex, for example, ciliates which are attached to shells, we may find partial divisions of cell bodies (budding); but even here the result is an exact distribution of the chromosomal matter. On the other hand, even multicellular animals may reproduce through asexual "vegetative" cell division. In plants this method is practically universal; but even in a large number of multicellular animals—though these are usually at a fairly low level of organization—we can find budding and vegetative reproduction. These organisms are frequently sessile forms, and often ones that grow in colonies, such as sponges, corals, moss animals, and tunicates, but also parasites like cysticercus of the dog tapeworm. Here, certain body cells always retain their capacity to divide and then differentiate, so that at all times or at least for a certain period "young" individuals can emerge from the body cells of old ones.

But even among bacteria, with their characteristic and almost proverbial ability to reproduce and propagate, we sometimes find processes that seem to have rather the opposite effect. Two bacteria (for example, *Escherichia coli*) will come together and form a connection with each other by way of a plasmodesma, or cytoplasmic bridge. One of the partners (the donor cell) then transfers part of its nuclear hereditary substances to the other (the recipient cell). In this way the recipient cell gains additional traits, both for itself and for its descendants, which had previously belonged only to the donor cell.

Hence this process, known as bacterial conjugation, leads to at least a partial recombination of genetic traits in one of the two partners; in this way it resembles sexual reproduction, because the sexual processes of all cellular organisms always consist of some recombination of genetic material between two partner cells. But in sexual organisms the genes of the two partners combine fully: Two reproductive cells (gametes) or nuclei are totally fused into a synkaryon, or fertilized egg cell (zygote). In the

The principle of "male" and "female"

▷
Examples of courtship and mating behavior in invertebrates: Above: Mating circle in a pair of damselflies (family Agrionidae). The male holds the female by the neck with his abdominal pincers. The female then bends her abdomen inward until her genital opening reaches the male's spermatophores. Below: Courtship and mating in lycaenid butterflies (*Lycaena*).

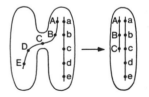

Fig. 26-2. Schema of a bacterial conjugation process. Two bacteria come together and form a plasmodesma (cytoplasmic bridge) through which genetic material (here the genetic traits A, B, and C) moves from the donor cell to the receiver. The formation of this link becomes visible through electron-microscopic photographs. We know that a transfer of genetic substances has taken place because the receiver cell now has traits it did not have before, and which the donor cell did have.

most simple case (in algae) the partners may look exactly like their conspecific asexual (vegetative) counterparts. This is called hologamy, and in such cases we do not refer to male or female sex cells but rather to + (plus) or — (minus) partners in a copulation. With increasing level of organization, however, there is also an increasing tendency to develop gametes very different from their vegetative cellular counterparts, and these become increasingly differentiated into two distinct types: egg and sperm cells. Fertilization at this stage is called merogamy. This differentiation between the two types of gametes begins with an external resemblance (isogamy), leads into a condition where the gametes are dissimilar (anisogamy), and culminates in the formation of eggs and sperm (oogamy). In all multicellular animals, sex cells are always clearly male or female. Carriers and producers of these gametes may then show sexual differences of their own, but this is not always the case.

Before we start to describe the tremendous variation and abundance of sexual phenomena, however, we must delve a little into the question of how gametes are formed and differentiated. As we have mentioned, the "purpose" of these sex cells is to fuse genetic material into new combinations. The joint product of two gametes is a zygote. In multicellular animals, this fusion is called fertilization, and the zygote is the fertilized egg cell (ovum). When two cells combine into one, their number of chromosomes and genes automatically doubles. Since the number of chromosomes will then continue to multiply in a geometric progression, this number has to be reduced accordingly when the reproductive cells are formed. What in fact happens is that the gamete mother cells have one stage during their so-called maturation division when they always refrain from doubling their chromosomes (and thus the DNA substances), with the result that their daughters and further descendants receive only "half" the number of chromosomes, or a single set. Thus, multicellular animals, which always develop from fertilized ova, usually carry double sets of chromosomes in all their body cells (these are diploid), but only a single set within the nuclei of their mature germ cells (these are haploid).

In single-celled animals and lower plants, this reduction of chromosomes (meiosis) often occurs immediately after copulation between the gametes. In these organisms all the individual cells growing from the zygotic division are haploid, and only their zygotes are diploid for a short period of time. In speaking of "individuals at the cellular level," we are using only a biological interpretation. An individual is a unique, clearly recognized, and well-defined representative of his species. This applies to man just like to any earthworm. After all, more than anything else it is the process of sexual reproduction, with its constant recombination of genetic material, which ensures the individuality and uniqueness of organisms. Strictly speaking, however, this does not apply only to the adult specimen of a species but goes right back to the fertilized ovum, where the recombined nucleus (synkaryon) already contains all the individual genetic

Above: Greeting ceremony of the Cape gannet (*Sula capensis*). Left, the joint bill-raising typical of many gannets. Middle left: Courtship behavior of the male bowerbird (*Chlamydera cerviniventris*), from New Guinea. By showing off the "bower" ornamented with berries, colored stones, and other colorful objects, the male lures females. Middle right: Preliminary mating behavior of Sandwich terns (*Sterna sandvicensis*). The male (right) circles the female in an erect posture. Below: "Communal courtship" in flamingos, consisting of stretching and preening movements with fixed form and sequence. This group behavior reinforces social cohesion, and actually has nothing to do with mating.

traits of the future organism. Seen in this way, there are several levels of biological individuality: 1. the individuality of gamete and zygote cells; 2. the individuality of single cells and cell generations that result from a division of gamete and zygote cells (in single-celled organisms and simple cell colonies); 3. the "higher" individuality of multicellular organisms, whose bodies consist of a large number of cells with the same genetic material but differing in structure and function; 4. the secondary individualities of animal species that form colonies and states, where asexual reproduction creates single individuals that combine into new and more complex individuals by proximity and division of functions (for example, coral colonies), or where sexual reproduction leads to generations of single animals so interdependent that they appear to survive and reproduce only as a social unit (for example, bee colonies).

At these different levels of biological individuality, the nature of sexuality is expressed by the fact that the male-female dichotomy is no longer limited to the gamete generations: Increasingly, sexual differences start to show in the external appearance, body structure, and behavior of individuals at higher phyletic levels.

At the very first, we find sexual differences in the areas where gametes are formed and stored, as with algae, for example, and among the lower animals with sponges and coelenterates. "Germinal layers" for male and female gametes (antheridia and archegonia) appear. Next, the "sex organs" become more complex and better suited for their purpose; special appendicular glands, and facilities for collecting and passing out the gametes, show increasing differentiation into male and female. As a result, this sexual bipolarity becomes more noticeable externally as well, so that in insects and most mammals we can immediately distinguish males from females. In other words, structure and behavior become increasingly "sexualized." There are two basic kinds or levels of sex characteristics: 1. what we call primary sex characteristics, that is, the special features of the actual genital organs; 2. the secondary sex characteristics, which involve other parts of the body and usually appear, through the indirect effects of hormones (for example, lacteal glands in female mammals, the structure of the larynx and the resulting voice of howler monkeys, the claws of the male stag beetle, and so on). When the sexual partners are clearly different we speak of sexual dimorphism. In certain rare cases we may find amazing differences in size, with the males almost always smaller than the females. In the marine worm *Bonellia*, for example, the females may grow as long as 1 m, including their proboscis, while the "dwarf males" reach only 1–3 mm (Fig. 26-8 and Color plate, p. 416); in anglers of the genus *Edriolynchus*, the dwarfed males are permanently attached to the females' belly.

In other ways, however, males of a species are often more "privileged" than the females: In many cases they have more highly developed and more capable sense organs and locomotor systems, which enable them to

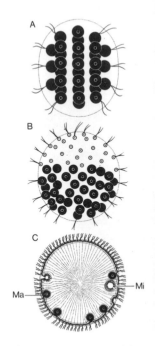

Fig. 26-3. Increasing differentiation of sex and somatic cells, as well as of "male" and "female." Sex cells are particularly prominent in the colony-forming flagellates of the family Volvocidae. A, *Eudorina elegans*: All thirty-two cells in this colony are "responsible" for metabolic processes, acting as somatic cells, as well as for reproduction, acting as sex cells. B, *Pleodorina californica*: Half of the 128 cells have lost their capacity to divide (somatic cells, light). The other half now takes care of reproduction (sex cells, dark). C, *Volvox globator*: The sex cells—surrounded by thousands of somatic cells with flagella—are differentiated into "male" microgametes (Mi) and "female" macrogametes (Ma) (see also Color plate, p. 413).

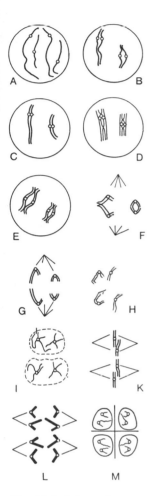

Fig. 26-4. Schematic representation of meiosis (reduction division), a process where the diploid set of chromosomes in the somatic cells is reduced to a haploid set carried by the gametes. Chromosome strands (A) pair off in parallel fashion (B, C). Then each chromosome splits longitudinally. The sister strands of the neighboring chromosomes may cross each other (D). As the germ membrane disintegrates, the chromosomes drift apart (E–H). The resulting two diploid sets of chromosomes (I) divide into 4 haploid sex cells (K–M).

play an active role in finding the females, holding on to them, and transferring the sperm successfully. We need only think of the huge eyes found on drones or male horseflies, and of the fully developed flying apparatus of male winter moths, whose females have only stubby little wings. On the other hand, females are usually built more "elaborately" in connection with brood care functions, a range of activities often left entirely to this sex. Examples are the formation of brood chambers in various crabs, and the pouches found in marsupials.

If we look at the entire animal kingdom, we find that the tendency to display male and female characteristics in separate individuals is not really universal. In many animal groups, particularly those at a relatively low level of organization, we see the expression of Hartmann's second major law dealing with the universal potential for bisexuality. Both sexes are fully developed in one and the same individual. These hermaphrodites are called transitory or consecutive when the developmental and maturational phases of both sexes occur one after the other, and simultaneous when the male and female sexual apparatus functions at the same time. More common is the transitory type of hermaphrodism, evidently to prevent self-fertilization. In most cases the male phase of maturation precedes the female one (proterandry), as in tape worms, earthworms, and land snails. Occasionally the reverse is true, as in true salps and some fishes (proterogyny).

The cases of hermaphrodism among lower vertebrates are very interesting. The common hapfish (Myxine), belonging to the cyclostomates, has a permanent tapelike hermaphroditic gonad. In young animals—28 to 32 cm—the posterior, male section of this sex gland is the first to develop fully and function; then it degenerates, and when the animal reaches a length of 35 cm the anterior section of the gonad becomes enlarged, producing large, oval eggs. In marine bony fishes there are many cases of hermaphrodism, and all three types occur: simultaneous hermaphrodism, proterandry, and proterogyny. The simultaneous hermaphrodites found among certain sea basses will alternate in their spawning behavior between male and female. They can reverse their sexual role within a few seconds, even changing skin color in a flash.

Among bisexual fishes, the most frequent forms of hermaphrodism are those in which the female maturational phase occurs first. Here we shall focus mainly on the lipp fishes (Labridae). In some of these fishes we find two types of functional males (diandry): One is male from the start and always stays that way (primary males); the other emerges from a female individual (secondary males). The testes of these two types are clearly different in structure, and there are also obvious differences in external appearance. Thus, for example, the magnificent secondary males of the rainbow wrasse (Coris julis) are so distinct from the primary males that we would hardly recognize them as belonging to the same species. The transformation processes of gonads and skin color are closely associated in time, as well; they occur only when the animal has reached a certain body size.

This is why, too, the secondary males are always larger than the primary ones. In addition to the differences in development and body structure, social and sexual behavior may also vary. The inconspicuous primary males are "group spawners," while the magnificent secondary males mate only with one female. We have been able to show that this sex reversal of amphisexual fishes is determined by hormones. In frogs and toads we can also find strong tendencies toward hermaphrodism and sex reversal, usually of a kind of proterogyny.

At the lower end of the zoological system and in the lower plants—that is, in single-celled organisms—we naturally find many cases of mixed sexuality. In some groups hermaphrodism and sexual activity are very much hidden, as with heliozoons. One such organism, *Actinophrys*, builds a temporary capsule within which it divides into two daughter cells; these, however, recombine into the same individual as before (pedogamy). The pedogamous heliozoon may also break down into several pairs, but once again these recombine into further pairs to form zygotes (as with, for example, *Actinosphaerium*). In principle this may be compared to autogamy, where an animal does nothing but undergo nuclear fission and refusion. One example for this latter process is the flagellate *Trichomastix lacertae*, a parasite of lizards.

One step higher is the hermaphrodism of ciliates. These have two kinds of nuclei: a large nucleus, which is actually a collection of many nuclei, the metabolic center of the cell roughly corresponding to the body cells of multicellular organisms; and a small nucleus with only reproductive functions.

When two ciliates mate, they dissolve their large nuclei and divide their small nuclei in such a way that each partner creates a so-called stationary nucleus and a migrating nucleus. Next, the conjugal partners establish a cytoplasmic bridge, exchange their migrating nuclei, and separate again. In this way, each remains the same "individual" as before, but with a newly arranged small nucleus that carries some of the genetic material of the ex-partner. In itself, this hermaphroditic behavior of ciliates shows an absolute "equality," but in some cases this has also given rise to a secondary diclinism (separation of the sexes). In campanularia (*Vorticella* and others), which are attached to stalks, certain dwarfed individuals detach themselves, swim to the sessile larger ones, transfer their small nucleus to the larger partners, and then die. Accordingly, the passive recipients no longer create a migrating nucleus but only a stationary one. Among higher single-celled organisms we can already find a secondary diclinism that probably arose from a preliminary stage of hermaphrodism.

The natural occurrence of hermaphrodism and amphisexual development and activity must not be confused with abberrant or incomplete dioecious (separately sexed) differentiation, whose products are called intersexes. These are sexually functioning individuals simultaneously displaying both male and female characteristics, either because they remained

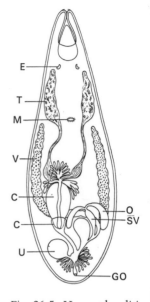

Fig. 26-5. Hermaphroditic sexual apparatus of the turbellarian *Acrorhynchus sophiae*. Each animal carries testes (T) and ovaries (O). The sperm cells move through canals into the male copulatory organs (C) and to the genital opening (GO). During copulation, the two partners exchange sperm cells. Each animal fertilizes its own eggs from the seminal vesicle (SV), surrounded by nutritive cells arising in the vittelaria (V). Uterus, U; mouth, M; eye, E.

Fig. 26-6. Slipper shell as an example of transitory hermaphroditism. The youngest animals are males, later turning into hermaphrodites and then into females.

fixed as hermaphrodites on the way to full bisexuality or because they belong to an otherwise dioecious species but suffered some kind of disturbance (usually hormonal) that prevented a full expression of their one sex. Such "accidents" of intersexuality and incomplete sexual development lead us to the question of sex determination: How does one or the other sex actually develop in the individual?

The fact that intersexes occur demonstrates that organisms carry a general potential of bisexuality, and that the fixation of one sex or the other is not a matter of black and white. This complies with Hartmann's law of the relative strength of male and female determination. But since cases of intersexuality are still relatively rare, there must be mechanisms to help determine and ensure whatever species-specific sexual form (hermaphrodite, male, or female) the individual will ultimately grow into.

We cannot understand the mechanisms of sex determination, however, without knowing something about genetics. We know that the structural and functional features of organisms arise through the interaction of genes and environment. The same also goes for the various ways that sex can be determined. We distinguish three basic methods: 1. the phenotypical method (modificatory), 2. the genotypical-polyfactorial one, and 3. the genotypical-monofactorial one.

With the first method, modificatory sex determination, the genetic basis of all individuals provides them with a roughly equal bisexual potential. Whether any one animal grows into a male or a female is not determined by its genes but by external influences. A classic example for this relatively rare situation is the marine worm *Bonellia*, which we briefly described earlier. Held separately, all larvae at a certain stage develop into large females. But if during this time any of them has direct contact with females or even encounters females in its immediate vicinity, the larva develops into a tiny, dwarfed male. Their fixation as males is caused by hormonelike substances (pheromones) that the females secrete primarily at their proboscis. Thus, we have found that proboscis extracts do have a masculinizing effect on *Bonellia* larvae; by varying the duration of this hormone action, we can create any number of different intersex forms (Fig. 26-8 and Color plate, p. 416).

The second method, genotypical-polyfactorial sex determination, is based on the fact that the genes already carry a program for greater tendencies to develop either into a male or a female. Depending on the species, however, a different number of genetic factors is responsible for sex determination, and therefore the number of individuals on either side is very arbitrary; the ratio of males to females may fluctuate to almost any degree. We know that this is the case with amphipods, isopods, and fishes. In the most extreme situation, one clutch may produce all males (arrhenogeny) or all females (thelygeny).

The most common method of sex determination is the genotypical-monofactorial one. This is typical for most higher organisms (i.e., flower-

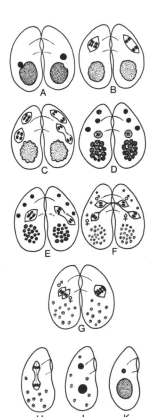

Fig. 26-7. Hermaphroditism in paramecia. A: Beginning of conjugation. B and C: First and second division of the micronuclei. D: The macronuclei degenerate. E and F: In each animal three of the four resulting micronuclei degenerate; the fourth divides again, resulting in male migratory nuclei and female stationary (resting) nuclei. G: The migrating nuclei are exchanged and fuse with the stationary nuclei. H, I, K: Reformation of macronuclei and micronuclei in the now-separate animals.

ing plants, insects, and vertebrates). Their cell nuclei contain independent sex chromosomes, which are distributed in such a way during the formation of gametes and zygotes that there is always a 1:1 ratio between males and females. The most familiar case is the X–Y mechanism of mammals, including humans. Here, the female sex is determined at the time of fertilization by the combination of two X chromosomes, while males arise from the combination of X and Y chromosomes. Among insects, on the other hand, we often find the X–O mechanism, where females carry two X chromosomes in their cell nuclei, and males carry only one. There are other cases, however, where males are characterized by two sex chromosomes that are the same, while the females have two different chromosomes (X–Y) or only one (X–O); this is typical of butterflies, for example.

In all these cases of genotypical sex determination, then, the individual's sex is decided by genetic factors at the time of fertilization. The gametes (eggs and sperm cells) contribute either an X or a Y chromosome (or none, as with the X–O mechanism) to the copulatory process. This random combination of chromosomes then results in the normal statistical 1:1 ratio of the sexes.

A very good piece of evidence for the occurrence of genotypical-monofactorial (heterochromosomal) sex determination is offered by the strange pseudohermaphrodites (gynander) sometimes found among insects. These are individuals where either one side of the body is male and the other female, or the body is clearly composed of male and female parts which form a mosaic pattern. This would naturally result from some "accident" during the individual's development, but the actual cause is different from the phenomenon of intersexuality. In gynanders, the single body halves or parts are different not only in appearance but also in the genetic content of their cells—female cells having XX chromosomes and male cells XO chromosomes.

However, it seems that even in animal species where sex is genetically determined, the sexual state is quite labile. It is a well-known fact that especially in mammals the genetically fixed sex characteristics (primarily the secondary ones) can largely be "reversed" through disturbances in their embryonal development, and even at a later stage through castration or through treatment with sex hormones. With this kind of flexibility, we have a strong indication of just how deeply the universal bisexual potential is rooted even in animals where later evolutionary developments have given rise to mechanisms of sex determination that supposedly ensure a clear-cut unisexuality in each individual.

By contrast, we have not found any problems of sex determination in hermaphroditic animal species, since here the bisexual potential naturally finds its full expression. But even here there are often tendencies toward separate sexes: Some individuals become more masculine, others, more feminine. We notice this particularly where hermaphrodites form perma-

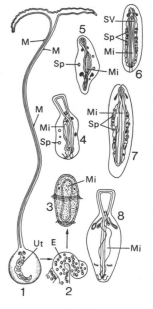

Fig. 26-8. The classic example for modificatory sex determination is the marine worm *Bonellia.* The various states: 1. Adult female, with three dwarfed males (M) attached to her long proboscis. 2. Eggs in a slime strand. 3. Undifferentiated larva. 4. Strongly feminine intersex with clear tendency for proboscis formation 5. Intermediate intersex. 6. Strongly masculine intersex. 7. Normal male. 8. Young autonomous female. Middle intestine, MI; seminal vesicle, SV; stage of spermatogenesis, S; uterus, U: The lower chamber contains the eggs (E), the upper the dwarfed males.

Fig. 26-9. By way of "accidents" in the first embryonal cell division of the fertilized egg—e.g., in *Drosophila* through loss of an X chromosome—certain individuals of many insect species become peudo-hermaphrodites (gynander). In each animal below, the left half is female and the right half is male:

Dovetail (*Papilio dardanus*).

Parthenogenesis

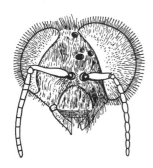

Head of a honeybee.

Sexual differentiation and reproduction is typical for higher levels of organization

nently mated pairs, such as blood flukes, blood parasites, which also penetrate humans and cause, among other conditions, the serious tropical disease called bilharzia. We are certain that the ancestors of these organisms were hermaphroditic; indications of this are also found among all their relatives, including the liver flukes which are definitely hermaphrodites. When two larvae of these trematods form a pair, one changes into a male and carries the thread-shaped female in his ventral groove. What we do not know, however, is the mechanism that decides which of the two partners becomes male and which becomes female. But we assume that here, too, external factors are at play.

Finally, we want to mention a special case of sex determination, one we have known about for some time because it applies to the only "domesticated animal" among insects: the bee. Any bee state usually harbors only one reproductively functional female, called a queen. When she lays her eggs into the brood cells, she uses her forelegs as a measuring device to decide which cells are destined for future females (workers and queens) and which for males (drones). In the first kind of cells she then lays fertilized eggs, in the second, unfertilized ones. With her sac filled with sperm from the nuptial flight, the queen can add sperm to any eggs ready to be laid. The fertilized eggs always produce females; the unfertilized ones produce males.

In itself, the fact that even unfertilized bee's eggs can produce healthy individuals is not unusual. It is not at all uncommon among animals for unfertilized eggs to have at least a potential for growth. This phenomenon is known as "virgin birth" or parthenogenesis, and is often found in species that are parasites or seasonal animals with the need for very quick and intensive reproductive activity. This is the case with, for example, plant lice and water fleas, which spend the summer reproducing purely through parthenogenesis. Even the eggs of frogs and other animals can be stimulated artificially to develop without fertilization, by administering ultraviolet rays or simply by pricking them with a needle.

All these examples show that egg cells, with their single set of chromosomes, do basically carry the complete genetic program for an individual's species-typical development. This principle also applies to the nucleus of spermatozoa; scientists were able to remove the nucleus from egg cells, then fertilize them with sperm, and stimulate this combination of egg plasma and sperm nucleus to develop into a basically normal individual. This reproductive phenomenon is called androgenesis.

Even though certain groups of lower organisms seem to get along without sexuality, and although a number of multicellular animals can actually reproduce asexually for many generations through fission and budding, there is no doubt that sexual differentiation and reproduction is typical for all higher stages of evolution. This observation leads us to three basic questions:

1. Do all the forms and manifestations of sexuality known throughout

the entire world of organisms, or at least within the animal kingdom, have the same phylogenetic origin? That is, are they homologous?

2. Are the known universal rules of sexuality—especially bipolarity and bisexual potential—the only viable ones? Or would there be other forms of sexuality that are plausible and biologically meaningful?

3. What is the purpose—that is, the biological meaning and advantage —of sexuality? In other words, what is its selective value?

We can safely answer the first question affirmatively. The chemical and structural nature of all living systems is so uniform, their reproductive mechanism so much the same, that the chromosomal reduction and re-combination processes—basic to all sexual patterns—certainly share the same origin and developmental history, as well.

We are less sure about the second question. It is quite plausible that at least in the beginning of sexual evolution other forms of reproduction were "tried out." The example of partial and unilateral gene combination resulting from what we have called bacterial conjugation demonstrates that an exchange of genes would in itself already have a selective advant-age, without necessarily ending in a narrowly determined bipolar combi-nation. Furthermore, we could imagine that in the past, evolutionary "experiments" occurred where not only two kinds of germ cells but possibly four or more were produced, thus giving rise to four or more "sexes" Even today there are bacteria among which, under certain con-ditions, several individuals get together in a star-shaped formation. In all cases we know, however, the partners in these star formations exchange or recombine genetic material only in pairs. Yet we could imagine certain bacteria that exchange genes multilaterally.

An example for this is the smut fungi (Ustilago). These organisms do not produce free gametes; instead they pair off with each other by fusing whole threads (mycelia) together and merging their nuclei two at a time. Crossing experiments have shown that these fungi contain not only two, but four types of mycelia, although copulation takes place only between 1 and 2 or 3 and 4. This looks suspiciously like four-way (tetrapolar) sexuality.

In addition, there are cases where it seems that gametes with incomplete bipolar or multipolar development may copulate with one another in groups of more than two. In Trichonympha, a flagellate existing as a parasite in the intestine of termites, we can observe four germ cells in a chain for-mation. It looks as if the two middle partners are "male" in the front and "female" in the back. But if in such cases all four nuclei were to combine, we would have tetrapolar sexuality in the narrow sense of the term. As a consequence, however, the organisms would also have to reduce their number of chromosomes accordingly during gamete formation. It is diffi-cult to imagine how this would be accomplished. Thus, it seems logical that the bipolar mode of sex differentiation represents the only economical solution to the problem of recombination.

Fruit fly (Drosophila).

Bacterial conjugation

▷
Two scenes from the synchronized courtship dance of two male great birds of paradise (Paradisaea apoda apoda). Above: Both males waiting in the court-ship position; the females are expected to come from the right. Below: The colorful plumage is pre-sented to the female, who is now sitting on the same branch. The males continue to display their plumage until the female moves very close, nudges a male with her beak, and plucks at his feathers. The male then jerks around and uses his beak and wing-beating movements to prepare her for copulation.

◁
Invitation to mate, and
copulation in lions. Above:
Female soliciting a male.
Center: Copulation (right:
neck bite). Below: Final
copulation phase; left:
male hissing; right: both
animals hissing.

Fig. 26-10. Castrated
animals give a clear
demonstration of the
labile sexual state inherent
even in species where sex
is genetically determined.
The castrated drake
(bottom left) differs only
slightly from the normal
one (top left). By contrast,
the spayed female bears a
strong resemblance to the
male. This is because the
female sex hormones
suppress the "species
plumage" typical of these
ducks, namely, the male
display feathers. The plain
appearance of the females
(top right) is of course an
advantage during brood
care.

Natural selection
through sexual
preferences

In considering the biological meaning of sexuality, the first question
we might ask is, what is the point of all these interesting and elaborate
forms of sexuality found in most higher organisms? It would take pretty
well a lifetime to describe even roughly the variety of structural, functional
and behavioral phenomena associated with sexuality. The higher the level
of organization, the more time and energy is devoted to sexual reproduction. Even if we assume that this is not simply the result of selection pressure, but in many cases rather of a surplus of energy so that it really means
a kind of "luxury" for the species, the implications of sexuality, at least,
are still hard to ascertain and understand.

The basic biological significance is clear enough: The production of
gametes with its separation and recombination of paternal and maternal
traits, and the formation of zygotes (fertilization) with is new combination
of grandparental genes were and still are the essential requirements for the
modifiability and therefore adaptability of organisms. But the process
could have stopped there. Why did organisms have to undergo a further
sexualization, with all the complications that have led to such curious expressions of form, color, and behavior? We can name some of the selective
factors: Sexuality also ensures that what has been accomplished does not
"decay." In other words, sexuality helps not only to modify but also to
preserve some of the organism's genetic potential. It allows the genetic
code typical of a species to be transmitted from one generation to the next,
protects the genes from being mixed with those of another species, and
leads to a further selection and fixation of ever more "species-specific"
forms.

All higher forms of sexual differentiation also act as barriers between
species. We need only think of the lock-and-key type of copulatory structures found in insects or spiders, and of the synchronized as well as parallel
forms of behavior displayed by sexual partners in fishes or birds. The high
degree of specialization shown by these partner systems could only have
emerged through intraspecific selection (selective breeding or natural selection); we may assume that it was also designed to prevent hybridization.

Sexuality plays an essential part in making a species ever more "typical" of itself. Ethologists know from observations and experiments that
many secondary sex characteristics act as sexual releasers, and that these
signals can even be presented as "supernormal" models which have a
stronger effect on individuals than the natural objects or signs. Here we can
start to understand the mechanisms of sexual selection: If, for example,
female oyster catchers prefer rolling larger eggs into their nest rather than
smaller ones, it means that over many generations there will be a selection
for larger eggs and animals; if male baboons with long manes have a
greater effect on females and command more respect from other males
than do rivals with shorter manes, then selection will be toward longer
manes. This is the only way, for instance, that hamadryas baboons could
have evolved to be so different from other baboon types. Of course, this

kind of "luxurizing" development could only occur where the simple external selection pressures for survival have been "screened out" by the natural selection of highly adapted and well-functioning types, to the point where a species can "afford" certain intraspecific complications.

On the other hand, there are also more direct reasons for certain special forms of sexuality to emerge. Flatworms, which have an otherwise very simple structure, show remarkably well-developed and differentiated hermaphroditic sex organs—especially if they are parasites (liver flukes, tapeworms)—that often take up the greatest part of their body mass. Quite obviously, this "sexual hypertrophy" has to do with their parasitic way of life. Because they are otherwise so rudimentary, these organisms have produced hardly any types capable of finding and entering a host on their own power. In addition, they must change their host at least once or even several times during their life cycle in order to attain reproductive maturity. Each time it is only a matter of accident whether such a worm can start its existence and continue the life cycle. These parasites would probably have been extinct by now if they had not enlarged their sexual apparatus to suit their needs, enabling them to produce a far greater number of eggs and descendants.

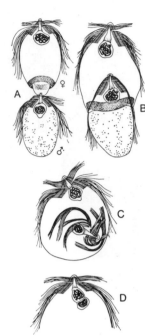

Relationships between sexual reproduction and ecology

We find such economical relationships between sexual reproduction and ecology in all parts of the animal kingdom. All species whose individuals care for their brood produce far fewer eggs and young than those that leave their offspring to fend for themselves; we need only compare the titmouse with the carp. Everywhere we find animals laying and incubating their eggs or bearing their young during periods when external, seasonal conditions and supply of food are most favorable—(e.g., insectivores such as warblers and bats reproducing during spring and summer, while seed eaters such as crossbills may not reproduce until winter. Under extreme conditions the sex biology of a species may also show unusual features. For example, seed-eating birds living in areas with unusually erratic precipitation will have continuously functioning gonads, in contrast to their relatives in regions with regular rainy periods. In this way, the former species can start a breeding season immediately after a rainfall which then provides a final "push"; this is the case, for example, in Australian zebra finches. On the other hand, anthropoid apes, humans, and domestic animals have grown so independent of their environment that they scarcely have any more "natural" sexual activity periods or reproductive cycles.

Humans, as the "luxury creatures" of evolution, show in their own behavior where sexuality may lead once it has been "liberated" from the selection pressures of mere survival. Only if a person consciously wants it that way does sex serve the purpose of reproduction; otherwise the individual uses it directly and indirectly for enriching his or her life and for reinforcing social bonds and status positions. The individual is also able to separate pleasure from biological function. As a result, human sexuality

Sexuality "liberated" from selection pressures

◁

Fig. 26-11. Schematic illustration of copulation in the flagellate *Trichonympha*. A, The male gamete (below) attaches itself to the "reception mound" of the female gamete (above). B, Next, the mound is inverted and the male gamete penetrates to the inside. C, Both gametes fuse with each other, but the cell nuclei are still separate. D, The nuclei are starting to fuse. This process, illustrated with two individuals, has also been observed during Stage A between four individuals linked to each other in a row.

Fig. 26-12. Broad schematization of the germ-cell development of a multicellular organism in the course of one generation. In the process of fertilization, the haploid egg and sperm cells give rise to a diploid zygote. This undergoes numerous cell divisions, in turn giving rise to the somatic cells and original germ cells of the embryo, all of which are again diploid. With increasing specialization—ultimately forming the complete organism—the somatic cells lose their capacity to divide, then age and die. Their genetic material is lost. From the original germ cells, the male spermatogonia and the female oogonia differentiate; these in turn undergo meiosis to produce haploid sperm and egg cells. This route, known as the germ tract, ensures that genetic traits are passed on from one generation to the next.

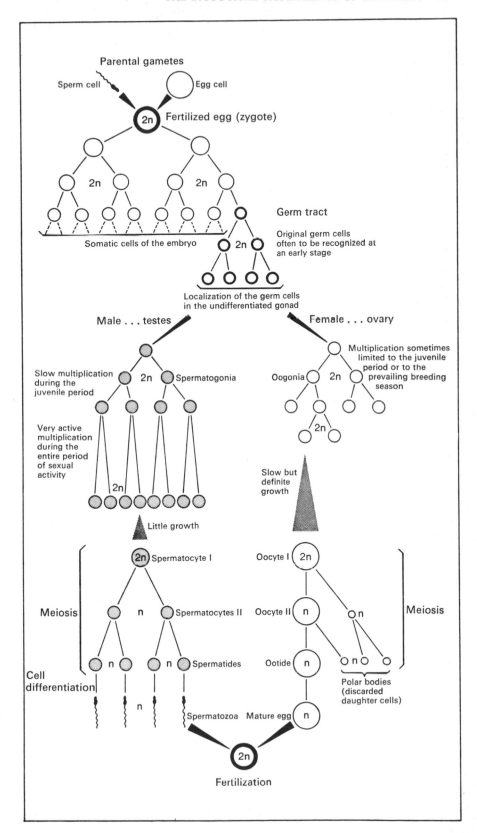

may reach a dangerous "drive" level. Not only that: people may deliberately exploit this overly strong "drive" in others by putting it to commercial use; and businessmen in the "sex industry" profit from the fact that instinctive reactions to stimuli hardly ever decline, much less become satiated. Particularly sexual releasers and responses, in contrast to other drive satisfactions, are always enjoyed anew. This applies not only to humans, by the way, but also to their domestic animals, whose sex life is often just as excessive and seemingly erratic due to the elimination of the natural reproductive cycles and the natural bonds between mates and family members.

We can derive one more function of sexuality from the concept of a "germ tract" (Fig. 26-12). What we mean is that germ cells are potentially "immortal," just like single-celled organisms, which are theoretically able to reproduce forever. Once a new organism arises from a fertilized egg through cell division, it also means that sooner or later it must die: The cells making up its body become specialized as muscle, skin, and nervous tissue, for example, thus increasingly losing their capacity to divide and regenerate. The multicellular organism ages and dies. But from the very start it carries within itself—clearly separate and unaffected by the aging process—the original germ cells and their descendants, the egg and sperm cells. We might say that these embody the neverending thread of life, leading through all the mortal individuals. We could carry this concept to its extreme and say that all individual creatures have only one basic purpose, to carry the germ tract. And in fact, the "business of reproduction"—that is, everything concerned with the continuation of the germ tract—is one of the most compelling expressions of life that we know of, both in animals and in ourselves. It was therefore inevitable that sexuality, as an agent for polarizing all these structural and functional reproductive devices, became a dominant element of life. Without it, existence on earth would be incomparably poorer. This applies also to humans and our psychological or spiritual world; for we, too, stand under this strange spell and act according to its force, again and again, without even knowing it.

A germ tract and its carriers

27 Indirect Transfer of Spermatophores in the Animal Kingdom

F. Schaller

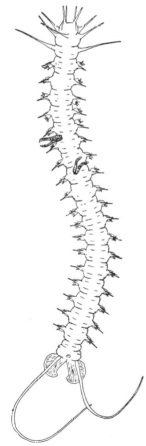

Fig. 27-1. Female bristle worm *Hesionides arenaria* (2 mm long) with two paired spermatophores on the back.

Among animals, an indirect transfer of spermatophores occurs when the male deposits its sperm in droplets, packets, or stalked receptacles on the ground so that the female can take it up and fertilize her eggs. This form of sexual behavior is possible only in water or relatively moist air, because spermatozoa (seminal or sperm cells) cannot live long in dry surroundings. After all, the most primitive kind of egg insemination occurs in those aquatic animals where both eggs and spermatozoa are simply discharged into the open (for example, sea worms, echinoderms, and fishes). No direct physical contact between the sexes is necessary, but the partners stimulate each other by discharging their genital products so as to achieve the biologically necessary synchronization.

In addition, sperm receptacles or carriers (spermatophores) are found in animals that transfer the sperm directly to or into the female genital orifice, as is the case with crickets and grasshoppers as well as crabs and cephalopods. The males of these species have developed special organs for the transfer of sperm, either at their genital orifice (penis) or by modifying various appendages (into so-called gonopodia). Yet they do not simply discharge the sperm or semen as fluid, but rather transfer it to or into the female orifice in more-or-less complicated packets.

In the transition from life in water to life on land, the formation of spermatophores became necessary especially in animals that did not evolve their own copulative organs. Often, the process of sperm transfer in these organisms is very troublesome, and drawn out to such an extent that the highly vulnerable spermatozoa would be in danger of drying out if they were not enclosed and protected within special receptacles.

Still at the preliminary stage of transition from water to land, some annelids such as small, sand-dwelling bristle worms, secondly, organisms related to earthworms, and thirdly, leeches have evolved spermatophores and corresponding mechanisms for transfer. For example, the males of the bristle worm *Hesionides arenaria* (L 2 m), with a genital orifice at their head, produce double spermatophores in the form of 60-175-mm-long tubes which they attach to their females at any point on the skin. Within a day the sperms penetrate the skin and travel through the female's body cavity to her eggs. One important aspect of this transfer process is that the female resists the male, who must therefore act very quickly. Evidently, the males do not recognize females as such, since they also attach spermatophores to males of their species.

Biologists have long been familiar with the reciprocal mating behavior of the hermaphroditic earthworms and leeches. In earthworms, very simple sperm packets are transferred to the partner by way of temporary furrows in the skin. Many jawed leeches produce spermatophores which they attach to the partner's skin in combination with a tissue-dissolving secretion. The double tubes containing spermatozoa then contract and squirt their contents through the skin into the body cavity, where the sperms reach the eggs by means of chemical stimuli. In this process, however, many of the sperm cells are consumed by phagocytes; this may be why these leeches undergo several mating sessions, during which they will attach up to ten spermatophores to each other.

Even in parapoda (phylum Onychophora, family Peripatidae and related species), which live only on land, the method of spermatophore transfer is the same. The males glue their packets, often many of them, on some point on the female's skin. After about seven days, amebocytes have dissolved the skin at that point, allowing the sperms to penetrate and fertilize the eggs in the ovaries. Again, the females remain completely passive, and the males are not selective in the "mating," since they will also attach spermatophores to males and juveniles.

But in all the cases we have been describing, we cannot really speak of spermatophore transfer in the narrow sense of the term, because the sperm receptacles are not deposited on the ground. On the other hand, a "classical" example is the behavior of salamanders. The males of the pond, alpine, or great crested newts acquire a splendid "nuptial gown" in the spring, then begin their restless search for a female. When they find one, they will stand directly across the female's way, using the tail to fan odorous substances (produced in their cloaca) in the partner's direction. A female that has been "approached" in this way starts to follow the male around. Such a mating procession may go on for a long while, and if during this time the female's snout touches the base of the male's tail, the male will stop briefly, push its cloaca to the ground, and deposit a cup-shaped receptacle with a sperm packet at the top. Both partners then continue walking in close contact with each other, the male leading the female right over the spermatophore in such a way that the female inevitably touches the packet with its swollen cloaca. This contact, probably combined with the signal effect of special "luring substances" or attractants (pheromones), triggers a searching and "swallowing" behavior in the rims of the female's cloaca. This movement generally ensures that the sperm packet is taken in.

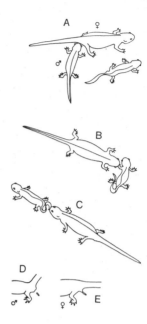

Fig. 27-2. Mating behavior of salamanders: A, the male smells at the other to see if it is a female, then positions itself in front of the female; B, the male fans attractant substances with its tail in the female's direction; C, the mating march; D, the male deposits a spermatophore; E, the female takes up the spermatophore.

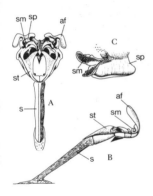

Fig. 27-3. Spermatophore of the Amblypygids *Tarantula marginemaculata*. View from the top (A) and from the side (B); the female comes from the left to take over the sperm. C, Sperm plate with preparation of a spermatophore specimen. af, armlike processes; s, stalk; sm, sperm packet with sperm; sp, sperm plate; st, spermatophore.

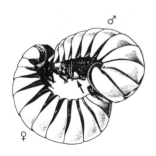

Fig. 27-4. The African millipede *Sphaerotherium dorsale* uses its walking legs to transfer the sperm (arrow).

Fig. 27-5. Male glomerid (left) with a mud ball.

In these aquatic salamanders, then, we can define two essential traits characterizing a typical indirect transfer of spermatophores: 1. The male depositing a sperm packet on the ground; 2. the female actively receiving the spermatophore.

Most mechanisms and methods of sperm transfer obviously evolved in the lower terrestrial arthropods (articulate animals). In many species that have become totally separate both from water and from relatively humid air environments, we find complete copulative organs allowing for internal insemination of eggs in the womb. Other species transfer their sperm indirectly by means of secondary sexual apparatus (gonopodia), for example, spiders. Finally, there are more primitive forms, still bound for the most part to moist or wet environments, whose methods of transfer are indirect and complicated and where the sperms often remain in the open for some time. Evidently, these indirect mechanisms for transfer of spermatophores have developed independently both in millepedes and primitive insects and in arachnids.

Among the millepedes (Myriapoda), many diplopods transfer their sperm directly by means of gonopodia. The African millipede (*Sphaerotherium dorsale*, which has no gonopodia, uses its cursorial (walking) legs to transport the sperm. During this process, the male holds onto the female's genital orifice by means of forcipated telapodia (terminal legs formed like tongues). In this inverted position, the male and female genital organs lie almost 2 cm apart. The male then discharges a sperm droplet about 1 mm in diameter from the frontally positioned, paired genital orifices, and transfers the droplet with its tarsal joints from one pair of legs to another, down the line until it reaches the female's genital orifices. This is probably the most primitive known case of sperm transport carried out with the help of normal—i.e., as yet unspecialized—limbs in terrestrial arthropods.

Just as noteworthy is the sexual behavior of the related millipedes (Glomeridae). The males use their telopodia to attach themselves to the genital orifices of the female, then curl up temporarily to polish the telopodia with their jaws. Next, the males pick up a soil particle (generally a ball of mud or feces) with their frontal cursorial legs, rolling it along the ground and gnawing at it to round it off. They then hold this particle to their own genital orifices at the second pair of legs, for ten to twenty seconds, depositing a sperm droplet directly onto the ball. Now the particle, and with it the sperm droplet, is transported over all the legs to the telopodia, where it is held in such a way that the droplet lies directly in front of the female's genital orifice. Finally, the male inserts the sperms into the female by means of simultaneous and rhythmical movements of the limbs. The soil particle which has been functioning as sperm carrier falls away. We may define this intricate procedure as tool utilization for the purpose of transferring sperm.

Sperm transfer in *Polyxenus lagurus* (order Pselaphognatha) is completely indirect. These diplopods generally inhabit the bark of trees. The males use their two "pennile appendages" so spin a double zig-zag filament across a small indentation where they suspend two bare sperm droplets. Next they turn about and spin with their glands, lying at the seventh body segment, two thick parallel filaments, after which they move away. These latter threads evidently contain signal substances to attract sexually mature conspecifics, and the mature females crawl along here while searching for the sperm droplets at the end. Upon finding the sperm, the females take up the droplets into their genital openings. The males and females of *Polyxenus lagurus* have no physical contact at all with each other, and their "mating" behavior is the most indirect method possible for the transfer of spermatophores. Sexually mature males are also stimulated by the signal filaments to direct their search into the zig-zag threads. Here, they will consume the sperm droplets of their predecessors, renew or repeat the web at the same spot, and deposit fresh droplets. This behavior is biologically significant because its guarantees a constant supply of fresh sperm for the female.

Many centipedes (subclass Chilopoda) have a typical indirect method of sperm transfer. The males of true centipedes (scolopender) spin a web following preliminary mating rituals, and deposit a bean-shaped spermatophore into it. The female remains in close contact with the male throughout this time, and then takes up the sperm. In this case, the web is not only a sperm carrier but also a signaling device directed toward the receptive female.

In the central European *Lithobius forficatus*, the preliminary mating ritual and the transfer of spermatophore has the same basic appearance as with the true centipedes. But they have much closer contact between the partners: Having completed the web and deposited the spermatophore, the male turns around with the anterior part of its body and, using its feelers, gives the female a contact sign to proceed forward. In addition, the web of the male *Lithobius* always contains some rows of filament arranged to be conspicuously broad and evenly diagonal to the animal's longitudinal axis. These special filaments undoubtedly have signal value of the female.

We find the sex biology of symphilids (subclass Symphyla) significant mainly for phylogenetic reasons: These small, pale, and blind ground-dwelling animals among the myriapods are considered more likely to be related to the wingless primitive insects (Apterygota) than any other species. Even when no females are present, the males deposit simple sperm droplets on stalks about 1.5 mm long. The animals then move on, performing strange movements and twists with their posterior area. Mature females of pauropods normally consume the "heads" of these spermatophores. However, the sperms are not swallowed, but instead are accumulated in special mouth pockets. In this way, the females may collect up to eighteen sperm droplets in a day and store them in their mouth pockets. The eggs are inseminated only when they leave the body. At this time,

▷
Reinforcement and maintenance of the pair bond by mutual preening. (top: a member of the parrot family, the chattering Lori *Domicella garrula*, male on left) and grooming (below: Hamadryas baboons).

Fig. 27-6. *Polyxenus lapurus* with two penis appendages.

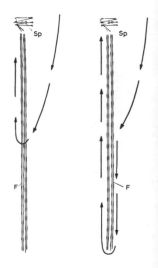

Fig. 27-7. Using its penis appendages, the male of *Polyxenus lapurus* spins a double zig-zag thread and hangs two sperm droplets (Sp) from it. Next, it spins two threads (F), on which the female will crawl in search of the droplets.

◁

In monogamous species both males and females take part in caring for their young.
Top left: The pillbugs (*Sysiphus schaefferi*) form a ball from dung and bury it. Below the surface, the female makes it into a pear-shaped structure which includes a small brood chamber into which she deposits an egg. The soon hatching larvae feeds on the dung which is protected against drying out.
Top right: In the dwarf cichlid both partners guard the spawn. They fan the eggs, remove infertile eggs, and later defend their young. This bond between mates is frequently established and maintained by special greeting ceremonies.
Center: A male European bee-eater feeds his female.
Bottom: During each exchange while brooding their chick, Adelaide penguins greet one another according to a strict ritual. They stretch their heads upward, beaks open, and utter short, choppy, trumpet-like calls.

the female uses its oral apparatus to pull each egg separately out of the genital orifice; then she deposits it at a convenient spot and goes to work on the egg with her oral implements. During this time, the surface of the egg becomes lubricated with a few small drops containing sperm. Thus the eggs of pauropods are fertilized outside of the body of both male and female, with a particularly indirect method characterized by the activity of the female.

We have discovered a very simple kind of spermatophore among the primitive insects, namely, in the Poduridae (subclass Collembola). Here, the males deposit bare sperm droplets on stalks, without the presence of females. The females must find the spermatophores, wipe them off the stalk, and take them in. This procedure is made easier by the fact that each male may produce more than 100 spermatophores during its reproduction period. In addition, the males provide for a constant supply of fresh sperm by consuming any spermatophores, their own or others, that have become stale, immediately replacing them with fresh ones. The droplets are examined with the animals' antennae.

A few species of Poduridae have developed behavior patterns that result in at least a one-sided approach between the sexes, from the male to the female. One example is *Podura aquatica*, which lives on the surface of water. Here, the male simply looks for a female and examines her for mature eggs. If the female is carrying ripe eggs, the male deposits a fence of spermatophores beside its partner, then drives the female in that direction so she will wipe up one of the droplets. The female, on the other hand, remains completely passive even when approached by a male.

Such approach behavior which aids in the transfer of spermatophores is found to a great extent in the Dicyrtomina which are related to the Poduridae. The males of *Dicyrtomina minuta*, for example, seek out a female, examine it by smell, then, like the Poduridae, deposit a fence of spermatophores around the partner, so that the female must ultimately come into contact with a sperm droplet. Here, too, the females show no interest whatever in their "partners," who will nontheless go as far as defending "their" female against other males by "jostling."

The *Sminthurides aquaticus* manages to establish very close contact and engages in real pair-formation. It lives on the surface of water and is characterized by the greatly diverging appearance of the sexes (sexual dimorphism). The males are much smaller than the females, and they have specialized clasping antennae, which they use to attach themselves to the feelers of the females, often "riding" their partners for several days. But a male may also take the lead, testing the female in a long movement ritual for its receptivity. When conditions are right, the male will deposit a spermatophore on the water and carefully draw the female across it, moving backward, to allow the partner to take up the sperm into its genital orifice. But a male may also lead its partner to the spermatophore in a semi-circular turn. Thus, we find in the Poduridae every transitional stage of indirect spermatophore transfer, from a complete absence of contact between the partners to a tight pair-formation with close physical contact (see Color plates, pp. 414–415).

Some species of Thysanura have external genital appendages, and so it was once generally assumed that they engage in real copulation. But in fact these organisms have not reached that stage in their mating behavior; rather, the males use their appendages only for spinning. One example is the sugar mite (*Lepisma saccharina*). Following a complicated ritual of running and touching, the male of this species spins a few threads transversely to the female's direction of movement, attaching one end to the ground and the other to some higher point. As a result, when the female moves around, it cannot avoid getting its raised tail bristles caught in these filaments. As soon as it feels the threads, the female jerks to a half and begins to search around carefully, for it is right here, under the cohesive filaments, that the male has deposited a sperm packet on the ground. Now the female will have no trouble finding the spermatophore and taking it into its genital orifice.

The sexual behavior patterns displayed by the various groups of arachnids are very different, corresponding to the marked polymorphism of this animal group. Scorpions are famous for their long mating walks (*promenade à deux*), during which the male holds the female tightly by the pincers. While performing this preliminary ritual, the male stabs its partner several times in the wrist. Toward the end of the mating walk, a sperm packet is formed, deposited by the male, and received by the female. The male presses its genital orifice against the ground, attaches the stalk of the spermatophore, pulls out the packet by lifting its body, and finally moves backward to lead the partner over to the exposed spermatophore. The female, in turn, feels for the spermatophore with its combs and positions itself so that the paired sperm receptacles lie directly in front of its genital orifice. Finally, the female opens the covering of the sperm packet with a sudden lever pressure and squirts the semen into its own genital cavity. At this point, the male immediately releases the female, which in turn consumes the sperm packet. The spermatophores of male scorpions are formed in special gland pockets out of two "mirror-image" halves which are glued together when discharged.

The mating behavior of whip scorpions or Pedipalpi (order Uropygi) is more diverse than in scorpions. Detailed investigations have been made on the mating patterns of a small tropical species, *Trithyreus sturmi*. Here, the male runs after the female and feels it all over (palpation) while making trilling sounds. The female responds by turning around, whereupon the male follows suit so that now the female sits behind it. In this position, the female uses its mouth claws (cheliceria) to hook onto a button-shaped appendage located at the male's posterior end, then lets the male pull it along. Next, the male deposits a spermatophore containing, like that of scorpions, two balls of sperm. With repeated jerks (palpitations) of the posterior end,

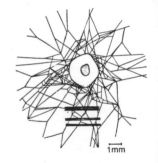

Fig. 27-8. After the preliminary mating ritual, the male *Lithobius forficatus* constructs a net, where it deposits a spermatophore which is then taken up by the female.

Fig. 27-9. Sperm transfer in symphilids.

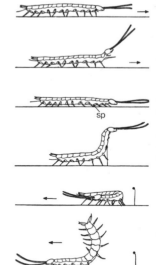

The male deposits the spermatophore (sp).

the male finally pulls its attached partner across the spermatophore, and the female receives the sperm.

In the large whip scorpions of the genus *Telyphonus*, which live in southern Asia, the male feels out the female while making trilling sounds, then uses its pincers to grasp the female by her long and crossed first pair of legs. In this way, the male often drags its partner back and forth for days. After repeatedly feeling the female's genital orifice, the male turns about the female's genital orifice, the male turns about slowly, stepping over the anterior part of the female's body. Now the male is standing with its posterior end in front of the female. The male deposits a spermatophore and moves forward with its partner until the female can receive the sperm. At the end, the male strokes the female with its tail appendage for about twenty minutes; then the pair separates.

The mating ritual of the South American whip scorpion of the genus *Mastigoproctus* lasts for many hours, with the discharge of spermatophores alone taking three hours. Here, the male no longer leaves it entirely up to the female to take in the sperm, but stuffs the sperm packets into the female's genital orifice with both palpi (see Color plates, pp. 414–415).

The sexual biology of a number of species of the whip scorpions belonging to the order Amblypygi has been examined. The sexual partners of the Malayan species *Sarax sarawakensis* perform lengthy palpation rituals before a transfer of spermatophores takes place. During this time, however, there is no physical contact between the partners. At the end of these rituals, the male turns around and deposits its spermatophores, maintaining contact with the female by means of its backward-oriented first pair of legs. Next, the male turns about a second time and signals the female with quivering pedipalpi (palpation legs). The female responds by moving forward, after which the male yields backward slowly to lead its partner directly to the spermatophore.

The same basic procedure is followed in the pair-formation and indirect transfer of spermatophores among other species of whip scorpions. By contrast, the males of *Tarantula marginemaculata* are more active in the transfer of sperm. First, a male will perform strong movement rituals in front of the female, as if making false attacks. Then the male turns about and deposits its sperm receptacle. Next, it steps across the spermatophore and rubs its anterior body against the receptacle. The female now moves repeatedly toward the spermatophore, performing swinging and circling movements over it with its raised legs. Finally, the male grasps the female with its feelers (palpi) and pulls her across the sperm receptacle. With the clawlike appendages of its genital orifice, the female quickly tears out the sperm packets and moves away. In addition, the female will normally consume the sperm carrier (Fig. 27-3).

The active part played by males of the South American genus *Admetus* in the actual transfer process is even more extensive. After depositing the spermatophore, a male will turn back to face it, enclose it with both pal-

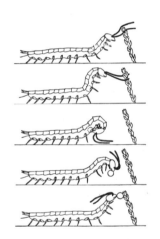

The female lays an egg, fertilizes it, and deposits it at some suitable spot.

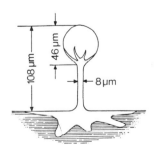

Fig. 27-10. Spermatophore of a Podurida.
1 m = 1/1000 m.

pating hands, break open the sperm receptacles with the claws of its mouth. In this group of animals, the spermatophores have such a complicated structure that the male's preliminary work seems to be necessary to allow the female easy removal and reception of the sperm packets.

Male spiders do not produce spermatophores. Instead, they deposit a sperm droplet onto a web, fill the semen into a specially formed appendage attached to their oral palpi, and transfer the fluid into the female's genital orifice.

Scientists have known about the mating rituals and methods of indirect spermatophore transfer of the false scorpions (order Pseudoscorpiones) for longer than they have known of any others. The males of several species deposit their stalked spermatophores even without the presence of females, leaving the females to carry out the reception process by themselves. By contrast, house scorpions and their relatives (*Chelifer* and *Chernes*) perform mating dances before the transfer of a spermatophore takes place (Fig. 27-14). A male house scorpion will dance in front of a more-or-less passive female and deposit its spermatophore. At the same time, the male will invert two long glandular tubes or ducts from its genital orifice, apparently to lure and stimulate the female by chemical means. Not until a short time before reception of the sperm droplet will the male grab its partner and make it come into contact with the droplet through strong shaking movements.

In the genus *Dactylochelifer*, too, mated pairs will "dance" and display other behaviors similar to the house scorpion. But before depositing its spermatophore, the male releases its partner, which in turn starts to approach the sperm carrier by itself. When the female has nearly reached the spermatophore, the male rushes forward, hooks into the female's genital orifice with the very long claws of its front legs, and virtually pulls her along to the spermatophore.

The males of the genus *Dendrochernes* will grasp the female's hand and hold onto it during the entire mating dance as well as the subsequent transfer of sperm. The partners may repeat this mating ceremony for several hours. Mating behavior is similar in *Dinocheirus*, from Florida. A male belonging to this species will hold its partner with one hand and stroke her back and pincers with the other. During this time, the male deposits a spermatophore and pulls the female in that direction. When the female has received the spermatophore, the male will shake her intensively with his free hand. Another variation is to be found in *Withius*, where the male grabs his partner first by her palpi, then by the anterior body, and pulls her close several times while quivering with his third pair of legs. Next, the male deposits a spermatophore, pulls the female across it, and presses her tightly onto the carrier. The males of this species produce a particularly complicated sperm carrier which includes a leverlike opening mechanism.

In mites (order Acari) we have found various methods of indirect

Fig. 27-11. During their long mating march, the male scorpion stabs its partner several times in the wrist.

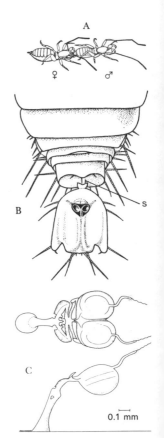

Fig. 27-12. A. Mating march of the whip scorpion *Trithyreus sturmi*: B. Posterior end of the male with a "button" for the female to hook into (S = spermatophore); C. Spermatophore, top and side view.

spermatophore transfer which have obviously evolved independently of one another. Frequently, limbs are used to shove spermatophores or simple sperm packets to or into the female's genital orifice. For example, many Gamasida transfer sperm by using their chelicera (feet ending in pincers or claws) which are constructed to suit this purpose. Many water mites use their third pair of legs to transfer the sperm. Others will produce stalked spermatophores and transfer them while maintaining close contact with the female. Among moss mites (Oribatida), too, there is a great variety of mating behaviors.

Some water mites produce stalked spermatophore droplets. The most detailed studies have been made on the species *Arrenurus globator*. The males, which look completely different from the females, adopt the so-called preparatory posture when making contact with a female: they will extend only their posterior body appendage toward the partner. If the female is ready to mate, the male moves in her direction and immediately becomes glued to the female by his posterior body appendage with a sticky secretion. After an extended mating ritual, the male deposits a spermatophore, then performs a finely coordinated forward movement to bring the genital orifice of its attached partner directly to the sperm packet. The male then facilitates sperm reception in the female by careful rubbing movements.

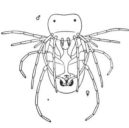

Fig. 27-13. Water mites (*Arrenurus globator*) during mating.

When we look at the wealth of behavior patterns involved in the indirect transfer of spermatophores described in this chapter, several facts stand out: Almost all animals which employ such a method inhabit particularly moist environments. The higher the population density of a species, the less contact there will be between sexual partners. Most animal groups which employ the methods of spermatophore formation and indirect transfer of sperm will occupy a low position on the phylogenetic scale; at least in the area of sex biology, these animals have not yet completely emerged from the water. The manifold peculiarities evident in the behavior and body structure of such organisms will not likely be based on a common evolutionary history (homology), but will rather have resulted from similar environmental pressures which have given rise to certain similarities (analogy).

In summary, we may derive three basic rules of sex biology from the examples described above: 1. The fertilization of eggs within the womb has emerged predominantly during the evolutionary transitional stage from life in water to life on land. 2. The mating ritual of animals without real copulative organs will cover a particularly broad range of peculiarities. 3. There are definite relationships between environmental conditions and the sexual forms of physical structure as well as of behavior.

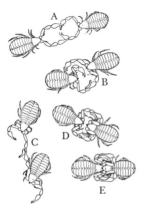

Fig. 27-14. Mating dance of the false scorpions *Chernes cimicoides*.

It would not be an exaggeration to claim that the study of indirect methods of spermatophore transfer leads us into an area of the comparative biology of sex that surely counts among the most fascinating and imaginative in the whole field of biology.

28 Forms and Functions of Courtship

Courting birds and fishes have always fascinated people with the apparent grotesqueness of their movements or the beauty of adorning body parts and display coloration. Scientists use the term "display" for behavior patterns which often emphasize special structures and colors. German ethologists have coined the word *Ausdrucksbewegungen* (expressive movements) because they express moods and tendencies that communicate between individuals. In this chapter, courtship refers to all expressive movements involved in pair-formation and copulation. However, courtship movements in this sense are often closely tied to expressive movements with other social functions, as we shall see in the following paragraphs.

Everyone has probably encountered the phenomenon of courtship at some time and has been fascinated by it, perhaps through direct observation of animals, perhaps through books or films. In order to examine courtship sequences scientifically, we must know about as many different patterns as possible. One very prominent example is the courtship of peacocks, easily observed in any zoo.

The peacock courts on a clearing, shielded by trees in the background. He unfolds and lifts the extremely long tectrices of his tail to form a train in which the "eyes" contrast with the rest of the feathers, in turn shimmering green to red-brown depending on light. Each of these eyes has a core of dark blue, surrounded by an area of turquoise and one of brown, and the whole pattern is bordered by violet and green-gold. In a zoo it is not unusual to see several cocks opening and raising their trains at the same time. But they always stand clearly apart, never directly beside each other.

What is the meaning of this movement? Is it perhaps a signal to attract hens? But the train is opened only after a hen has already approached. In order to answer this question, we have to take a very close look at how the cock and hen behave toward each other.

The hen moves around in the vicinity of the cock, obviously very intent in her search for food. She seems to ignore the courting cock completely. But he does not simply stand motionless; he approaches the hen

Forms and functions of courtship, by R. A. Stamm

Fig. 28-1. The stage of "ecstasy" in the courtship ceremony of a peacock.

Fig. 28-2. External fertilization in the bitterling: The female deposits the eggs with her ovipositor into the gill cavity of a pond mussel. They are fertilized there by the male's spermatozoa, which the mussel draws in with its respiratory activity.

Fig. 28-3. Copulation in the gray heron: The partners come into physical contact with each other, enabling the male to deposit his spermatozoa into the female's genital organ.

in a peculiar way, walking backward with small and "nervous" steps, at the same time leaning his train over the front and fanning rhythmically with outspread feathers.

If the hen does not want to mate, she will truly ignore the cock and soon depart. He will then fold up his tail. But if the hen is receptive, she will respond to his backside approach by walking toward him and around to face the front part of his train. She performs this movement quite casually while continuing to peck for food. The cock has been watching, and as soon as the hen appears laterally at the edge of his train he will turn toward her with a jerk, tilt the train as far over as possible, and make it quiver in periodic "ecstasy." But still the hen does not look up. If she keeps on pecking the ground, he will be induced to continuously quiver his tail; if she stops, the cock once again turns his back and starts the whole game anew. The cock's turning away, the hen's approach to his front, the turning of the train toward the hen, and ecstatic tail-quivering is repeated several times until finally she invites him to copulate by offering her lowered back end. The cock then rushes toward her and treads her.

This impressive sequence of courtship behavior in peacocks is a ritual. Like the rituals performed in human cults, it is a well-formed series of movements that symbolizes a "scene from everyday life." And like many human rituals, it is exaggerated so much that the core elements representing this scene can no longer be easily discerned. But we can interpret this pattern correctly by observing close relatives of the peacock, other pheasants.

In our example we notice that the hen pecks for food, sometimes with the greatest intensity when she is standing directly in front of the cock and in the "focal point" of his train. We observe in ring-necked pheasants and domestic chickens that females frequently approach the cocks when the cocks are pecking on the ground for food. The hens then also peck, at the same spot where the cock had "led the way." This is the very same behavior performed by chicks when following their mother who is searching for food. In the three species of peacock pheasants which belong to the argus pheasants (*Poplylectron*), we can see that the cock even holds a food grain toward the hen for a short period of time.

In the interaction between a hen and her chicks, the hen's pointing out of the food and the young's imitation pecking are biologically meaningful because here the chicks are encouraged to take in food and perhaps also to learn to recognize certain kinds of food particularly fast. Courtship behavior obviously makes use of this situation: The cock symbolically takes the role of the mother hen while the hen plays the part of a chick. The complex courtship antics of the peacock are really an immense elaboration of behaviors normally involved in the transfer of food. The symbolism of this ritual has gone so far that food is no longer either offered or shown; the cock only "pretends" that there is food in the focal point of his arched train. The hen, too, often pecks only symbolically.

But why do peacocks go to all this effort in their courtship rituals? Why has the food-offering scene become so terribly complicated? Can the goal of courtship not be attained just as efficiently if the cock simply offers the female a morsel of food, as do ring-neck pheasants? These birds, after all, reproduce just as much as peacocks; in other words, their simple behavior patterns lead to basically the same ends as the peacock's more complicated ones.

At this point we can only guess at the answer. Males of all pheasants tend to grow colorful plumage. Here, the differences between species— just like their differences in courtship behavior—probably serve mainly to let females recognize members of their own species, so that the hens do not try to mate with cocks of another species. It is also quite possible that the females exert some kind of selective pressure on the species as a whole by preferring males with more conspicuous plumage and more intensive courtship behavior, responding to these more readily. Another suggestion is that males with greater coloration and more intensive courtship movements have an intimidating effect on other males, who are then inclined to stay away and not disturb the mating process, or who are even prevented from mating. And in fact, peacocks may court successfully only when their train feathers are fully developed, at the age of three years.

Other conditions are also necessary for successful mating in peacocks. Although free-living peacocks in India court virtually all year round (except in the very dry areas), the hens do not allow copulation until the monsoon rains have started in May. This is important because the young can be safely reared only during the rainy season when plants grow, blossom, and bear fruit. Uninterrupted courtship requires the cock to defend a territory and to drive away all rivals. This territory remains quite stable, and the cock will maintain it even when he cannot see any hens and therefore cannot court. The males leave their territorial guard positions only at night to gather peacefully in trees, where they sleep. More than anything, this difference in behavior between day and night demonstrates clearly that territorial conflicts serve a practical purpose, in this case preventing others from interfering with courtship and copulation activities.

By looking at the example of courtship in peacocks, we have seen how ethologists study this kind of behavior: 1. The sequences are described in detail, especially the interaction between partners; 2. the conditions under which this behavior occurs are investigated, as well as those facilitating interaction between the partners; 3. the functions of this behavior are examined; 4. attempts are made to trace the phylogenetic development of the present behavior from its origins in the history of the species.

A courtship situation does not have to be very noticeable, nor do we necessarily perceive it as either beautiful or bizarre. Even certain single-celled animals display a complex preconjugation behavior, such as *Paramecium* and shellfish (*Stylonychia*), which both belong to the ciliates. "Whereas paramecia normally do not react when coming into random

Fig. 28-4. In sea horses, the female inserts her genital papilla into the male's pouch, filling it with orange-red eggs in less than 11 seconds.

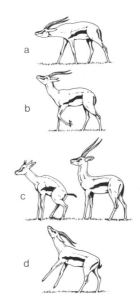

Fig. 28-5. Mating behavior in the Thomson's gazelle: A territorial buck approaches the individual females in the herd (a). He raises his head in closer approach (b). The female micturates, and the buck smells the urine and flehms in order to test her estrus

contact with conspecifics," writes W. Rose, "they now show a behavior I would describe as preliminary conjugation: These paramecia swim around each other for up to half an hour, roughly on the same spot by constantly changing direction. They can lose each other, then find each other again or continue this ritual with a new partner. Three or four individuals may also do this together . . . . Once two small and thin animals have met (only these really conjugate), the preliminary ritual enters a new phase. The partners no longer swim around each other but remain side by side, separated by a distance equal to the length of their cilia, their anterior ends pointing in the same direction, swimming back and forth on the same spot for about a quarter of an hour. During the final minutes they come closer and closer, then suddenly jerk and cling together." Undoubtedly we can call this a true preliminary mating behavior.

It is very common for individuals of a species to be divided into males and females (single-celled organisms may in addition have even more complicated sexual relations), which have to mate so that genetic information can be combined in the offspring (see Chapter 26). Sexuality is characterized by the union of sex cells. This may happen in different ways:

Type 1: Egg and sperm cells are simultaneously ejected into the water, where they unite. Whole groups of diverse aquatic animals show this type of sexuality, such as ctenohpores, echinoderms, chaetopods (bristle worms), mussels, many fishes, frogs and so on. But there is also a large number of animals living in oceans and fresh water that do not belong to this type, including aquatic insects; these are descended from terrestrial animals.

Type 2: The sperm cells are carried to the egg cells which remain inside the female's body. This can happen in two ways: In type 2a there is copulation, physical contact between the male and the female (Fig. 28-3 and Color plate, p. 432); in type 2b the sperm cells are deposited in packets (spermatophores) and actively taken up by the female (see Chapter 27).

Internal fertilization of egg cells is essential particularly where the eggs must have hard shells to protect them from dehydrating—i.e., in animals living on land. But impermeable shells can only develop after the eggs are fertilized. Internal fertilization is also found in many aquatic animals, however; this is usually tied in with special brood care such as the depositing of fairly large yolk reserves.

If we remember these types of interrelationships, it will not be difficult to discover even simple precoital patterns. We can look for any kind of behavior that leads to the fertilization of egg cells. Here at least two functions must be served: 1. The sexual partners have to find each other; 2. they have to coordinate their activities so that egg and sperm cells are combined.

Finding each other is not a problem where the species' way of life ensures that conspecifics of both sexes automatically live together within a narrowly defined area, as with sea urchins in the coastal spray water zone of oceans. In all other cases the partners must show a behavior that is ac-

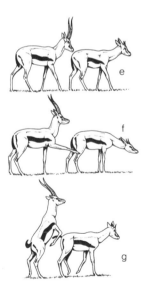

stage (c). He stamps his forelegs (d). The two partners walk one behind the other (e); during this "mating march" he kicks with his foreleg into the direction of the female's hindlegs or between them (f). Mounting (g) and copulation follow.

tively oriented toward the other. Either males and females must stay together for a long time or permanently in pairs or swarms, or they must seek each other out for the period of fertilization, recognize each other as suitable partners, and synchronize their movements.

Encounters may be controlled by external events causing all individuals of a species to gather "automatically" in a precise spot at a precise time, as do Palolo worms (see Chapter 12). But even in Syllidae, which are close relatives of the Palolo worm, the partners respond to each other directly and attract each other with light signals. This demonstrates that there is probably no case among animals that is clearly the most rudimentary or even the most original behavior. If we examine the biology of so-called primitive species in detail, we discover that even here the interplay of physiological factors and frequently also the behavior can be highly differentiated.

Fig. 28-6. At some time before egg-laying, the male toad attaches himself to the female's back and in this way stimulates her to release the eggs.

The process of simply depositing sex cells into the water is in itself of some interest to ethologists, and should certainly not be viewed as primitive. Even where sex cells are produced in very large numbers, as with salmon or mackerels, fertilization would be virtually impossible if these cells were merely shed randomly at some time or in any place whatever. Here, the male and the female must still coordinate their activities: After the partners find each other, or ensure that they stay close together, the egg and sperm cells must be released simultaneously. This may be induced by external factors: In the bitterling, for example, the presence of a live pond mussel (into which the eggs are laid) is a precondition for the growth of the male's spawning dress and the female's ovipositor, both indicating readiness to spawn (Fig. 28-2). In other cases the sexes perform some behavior on each other, the female on the male or vice versa.

Fig. 28-7. The male midwife toad wraps the string of eggs around his hind legs and carries it around for quite a long time.

We shall give two examples to show how the difference between internal and external fertilization—as far as the complexity of courtship is concerned—may be completely blurred. The first concerns the mouth-breeding cichlid fish of the genus *Haplochromis*. Here, the female takes the eggs into her mouth cavity immediately after spawning, when they are not yet fertilized. Next, the male swims past the female's mouth, with his body leaning to the side and his anal fin spread out pointing toward her. This fin has a row of yellow spots that greatly resemble this species's eggs in size and color. The female stubs her nose against these "dummy eggs," as though she wants to pick up some that she has "forgotten." At that moment, the male ejects a stream of sperm fluid which the female inevitably takes in. In this way the eggs in her mouth are fertilized.

In our second example the animals' behavior bears a startling resemblance to real copulation, despite the fact that fertilization is external. Sometime before the eggs are released, male frogs and toads climb on their females's back and cling to it. By making certain movements or sounds, the two partners synchronize their behavior in such a way that the eggs are fertilized immediately after shedding. In Anura, mating behavior is also

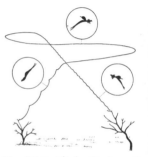

Fig. 28-8. Flight display of the paradise whydah.

Fig. 28-9. The male Fisher's whydah has shorter wings than the whydah bird and does not fly as well. Shaking and whirring, he courts his female in a sitting position.

Fig. 28-10. Courtship of *Ecsenius bicolor*. The male leaves his dwelling tube and swims with vigorous jumping and seesaw movements toward a female. If the courtship takes place in a horizontal position, the jumps are broad; if the courtship is vertical, they are short.

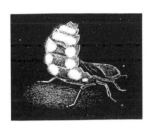

Fig. 28-11. Female firefly *Phausis splendidula*, which produces light signals with its lower abdomen.

complicated by the fact that the male and female may each take roles in caring for the eggs. Males of the midwife toad, for example, wrap the string of eggs around their hind legs; in the pouched frog, males place the eggs on the females's back where they are covered by a special fold of skin and mature within these "pouches."

The phenomena we have been describing represent the basic patterns of courtship, we might say. These can be detected even where partners use impressive rituals that go far beyond merely finding each other and working in synchrony for the purpose of fertilization. A walk through the world of animals will tell us much about the various ways that what we have basically termed precoital activities can be exaggerated and extended.

We shall try to organize these facts by describing a few characteristic forms of courtship, their functions, and the way they come about. A courting animal can draw attention to itself with a whole variety of sounds, movements, odors, body appendages, color patterns, and other signals. Of these we will mention only a few:

Crickets and grasshoppers chirp; male mosquitoes find their females by means of a flying sound (see Chapter 8). Some fishes also make species-specific sounds, like gobies (Gobidae) and the croakers (Sciaenidae) that make conspicuous sounds by rapidly vibrating special muscles. Most people are familiar with the croaking of frogs and toads. Sounds or calls are important with nocturnal owls. Another well-known example is the overwhelming abundance of bird songs.

In wolf spiders, males wave to the females with both palps. Male jumping spiders perform real dances during their courtship. Fiddler crabs derive their name from the conspicuous claw-waving movements that remind us of someone playing a fiddle. Other expressive movements include the zigzag dance of sticklebacks, the threat ritual of cichlids, and the strange courtships of female sea cucumbers, which wrap spirally around their males. We can also observe a variety of courtship dances and flying movements in ducks, plovers, gulls, birds of paradise, and many other avian creatures. Some elements of precopulation rituals in ungulates can also be included.

Very often, sounds and movements during courtship are combined with a display of colorful patterns and bizarre appendages. We may note, for example, the play of colors in cuttlefish, the colorful patterns of coral fish and other groups of fishes, the elongated colored fins of fighting fish, the colorful feathers of male ducks and wild chickens, and the magnificent plumage of male birds of paradise (see Color plate, p. 431).

Smells are important signals for many animal species, which often manufacture their odorous substances in special glands and release them through special organs. These odors can serve to attract the sexual partner from a distance, like the pheromone released by female mulberry silk moths or silkworm (see Chapter 6); others are effective only for short distances. In the giant snakes (boa constrictor, python, and anaconda), the male recognizes his partner by the scent from her anal glands. A very large proportion of mammals are macrosmatic. For example, the tenrec secretes a milky fluid from its orbital glands, the marmot releases an odorous substance from its anal glands, and male dogs follow the scent trail of a bitch in heat.

In a number of species belonging to very different phyletic groups, members of the opposite sex signal each other with light pulses, like the Syllidae we mentioned earlier, among insects, the fireflies, various deepsea fishes called lantern fish, the salmon fishes Myctophoidei, and probably also the Anomalpidae. Depending on the species, the light-producing organs are usually grouped in certain patterns on the body. In some cases light can be turned on and off by the individual himself, following a time pattern that is again typical of the species. This allows members of different species to distinguish between outsiders and conspecifics, a process we have examined particularly in fireflies.

Courtship behavior very often contains elements belonging to the realm of brood care. For instance, the courting male stickleback lures his female to the nest and induces her to deposit her eggs. We must remember that in almost all multicellular animals, sexual behavior is closely connected with the production of offspring: Mating immediately leads to fertilization, in turn leading directly to the start of embryonal development. It should therefore be functional to combine behavior patterns like nest-building, taking over of the brood by the male, or provision of food for the future young with mating itself.

Our brief survey is enough to give us a varied and, at first glance, rather confusing picture. But we can order these many different phenomena by examining the functions served by the various courtship rituals. Naturally, the first task is to bring the sexual partners together. This is accomplished with conspicuous courtship ceremonies of the kind already mentioned—songs and other vocal expressions, the courtship flights of many birds and insects, pheromones, and many other signals. In addition, however, there are expressive behaviors to ensure that any attempt at mating with an unsuitable partner is frustrated.

Successful mating requires that three conditions be met: 1. Both partners must belong to the same species; 2. with the exception of hermaphroditic species such as land snails, the partners must be of opposite sex; 3. both must be willing to mate. In many animals it is also necessary that, wherever possible, proper conditions for rearing the offspring are established before mating begins.

Since courtship ceremonies are always very conspicuous, they not only attract the sexual partner but may also draw the attention of predators. This alone is a good reason why courtship has to make a direct impact on the proper partner, so that the sexes find each other as quickly as possible. In some cases predators actually exploit the courtship signals of their prey. Predaceous fireflies, for example, imitate the light signals of other species

▷
In the flamingo family, both parents take care of the young.

▷▷
A birth taking place in a sea lion family. Top left: Emergence of the newborn from the womb. Top right: Making contact with the mother. Bottom left: Suckling the young; gulls will eat the afterbirth. Bottom right: Father, mother, and child; notice the size difference between male and female, an expression of sexual dimorphism in sea lions.

in order to mislead, attract, and then eat them. Another kind of "deception" is found in the orchid *Ophrys*: This flower not only imitates the appearance but also the smell of receptive female digger wasps or bumblebees; when the respective male insects land on them in order to mate with what looks like a partner, they instead pollinate the flower.

To ensure that the right partners get together, all courtship expressions in their various sense modalities are species-specific, usually sex-specific, and strictly limited to the mating season. Conspecifics respond only to the "proper"—that is, species-typical—combination of signals, and only when they themselves are ready to mate. The following examples give a more detailed picture.

Male grayling butterflies (subfamily Satyrinae) initially pursue every other butterfly passing by, the sign stimuli being appearance (as dark as possible), size (as large as possible), and locomotory path (not straight). If this object turns out to be a female grayling, it will soon land. Now the male positions himself in front of the female and performs various movements: rhythmic movements of the wings, fanning the wings in the female's direction, and circling his antennae. Probably the wing movements themselves carry the male's special scent to the female, an odor released by his scent scales. In addition, the male flaps his wings together so that the female's antennae are caught between them and her olfactory organs come into close contact with the male's scent. If the female is unreceptive, she will either move away or flap her own wings several times; the male then stops courting immediately.

In the fritillary butterfly (*Argynnis* and related genera) it was found that the male reacts not only to the visual appearance of his partner but also primarily to her odor. Like the grayling, a male fritillary will pursue "wrong" objects, but he continues to court only if he catches the proper scent.

Several islands in the North Pacific are inhabited by two closely related species of albatross: the Laysan albatross (*Diomedea immutabilis*) and the black-footed albatross (*D. nigripes*). Wherever possible, these two species live in different parts of the islands, but at times they do move close together and will then brood in mixed colonies. When this happens, other factors come into play to prevent mixed breeding. Although the species have rather different colorations (the black-footed albatross is dark-brown to black, the Laysan albatross, white with dark wings and tail), occasionally one will try to court the other. But these attempts are cut short by a number of behavior patterns. First of all, the sounds and calls are different; presumably the chicks are imprinted to their species-typical calls by hearing their parents while still in the egg. Second, the courtship ceremonies are highly divergent. When two Laysan albatrosses meet, they both extend their bills vertically into the air. Pair-formation then comes about only if the male is taller than the female; if not, the female is driven off. However, both male and female black-footed albatrosses are bigger than any Laysan

Examples of brood-care behavior. Top left: Female false scorpion (*Mastigoproctus giganteus*) with larvae, carrying them on the hind part of the body for protection against predators. Below left: When changing her location, the leopard female carries her young along by holding them with her teeth by the nape of the neck. Right, from top to bottom: Mouth brood care in the cichlid *Hemihaplochromis multicolor*: a) Female carries young in her mouth; b) She spits out the young; c) By positioning her body diagonally, she induces the young to return to her mouth cavity when night falls or danger threatens.

Examples of brood-care behavior. Left from top to bottom: A South African sand wasp (*Ammophila*) bringing a paralyzed caterpillar into the brood chamber: The wasp opens the entrance to the brood chamber, which she has built and kept closed (above), drags the caterpillar into the tunnel (center), and seals the entrance with a large pebble (below). Top right: Female of the spider species *Stegodyphus sarasinorum* feeding two young ones mouth-to-mouth. Below right: Brood care in the discus fish. At the beginning, the young feed from the body slime of their parent (a kind of "nursing" in fishes).

male. As a result, no Laysan male would ever choose a black-footed female as his partner.

Once they meet, the partners make contact with their bills. In Laysan albatrosses this involves a gentle interlocking of the bills, while black-footed albatrosses beat their bills together from left to right, a movement that would immediately put any Laysan albatross to flight. These two species also differ in other behavior patterns. In the black-footed variety, both partners present both of their wings simultaneously, while in the Laysan species the partners alternate and each present one wing and then the other. In fact, the whole "temperament" is different: Black-footed albatrosses make slow and regular movements. Finally, the partners have to be "engaged" for one or two years before copulating, and therefore two unsuitable mates would separate at some point during this time, should the initial barriers ever be crossed.

The isolating effect of behavior differences has been demonstrated in other animals as well, such as the fruitflies (*Drosophila* and the viviparous tooth carps *Poecilia* (the guppy and its relatives). The idea that differences in appearance and behavior between closely related species really does serve the purpose of sexual isolation can also be derived from the fact that these differences are especially prominent where a number of such species live in the same area. Conversely, they can almost disappear when a species lives in an area by itself. For example, the mallard duck in Hawaii has lost its colorful plumage; its species-specific courtship behavior however, has not disappeared.

For an unmated animal it seems more productive to keep looking for a partner rather than to interfere with an already mated pair. In a number of species, however, several males usually mate with one female, as with pike fish and the common grass snake. In other species, one male mates with several females, as with many weaver and hummingbirds. But in general, successful mating requires a precise interplay between the partners and an exact temporal synchronization of certain behavior patterns. The presence of other conspecifics would disrupt this process and jeopardize the entire mating procedure. In order to avoid such interference, many courtship movements appear as a threat to rivals and may even be associated with a real attack. This function is so important that in some cases courtship becomes an elaborate affair primarily for that reason.

Male cuttlefish (*Sepia*) threaten and display to each other, for at short time acquiring a more prominent display coloration while the female, in her usual subdued colors, waits nearby. Iguanas and agames threaten with rapid head-bobbing movements that have a species-specific rhythm and also serve to identify the males to their females. Fights between males precede pair-formation in the grey heron and copulation in the ruff (which does not form pairs). On the other hand, the peculiar fighting rituals of black-throated divers (*Gavia*) are perhaps no longer associated directly with choosing a mate, since they also occur outside of the mating season in fall.

Fig. 28-12. Courtship of the grayling butterfly *Eumenis semele*:

The male positions himself in front of the female and by various movements induces her to take up the copulatory posture; he fans his odor, produced by the scent scales on his forewings, in her direction (left: scent scale; right: two ordinary scales).

By bowing down, the male clasps the female's antennae —which include olfactory organs—between his forewings.

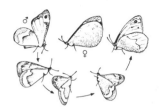

After this bowing movement the partners lock into each other for copulation: The male moves laterally and quickly around the female, bending his abdomen in her direction. The female then hooks her mating organs onto the male's copulatory appendages, which project from the tip of his abdomen.

Fig. 28-13. Two Laysan albatrosses raising their bill. The male (left) is clearly larger than the female.

Fig. 28-14. In the Laysan albatross, the partners make contact by gently interlocking their bills.

Fig. 28-15. The female tree frog arches her back inward, indicating to the male that she is about to release her eggs.

In some crabs, in dragonflies, and in a large number of vertebrates, the necessity of keeping rivals away has led to a system of territories. In this way the males of a population are evenly dispersed and wait for a suitable sexual partner (see Chapter 25). In species that live in groups, the same purpose may be served by a dominance hierarchy, which allows some animals better opportunities for mating and prevents lower-ranking individuals from interfering with the mating activities of higher-ranking ones.

As a general rule, the final phase of sexual activity—releasing the sex calls outside of the body or performing copulation—cannot be enacted without considerable timing and preparation. The sexual partners have to increase their own as well as each other's mating readiness. The initial stages of preparation occur on a long-term basis, with the advent of individual sexual maturity and with the frequently periodic receptive states, occurring in annual or monthly cycles. These processes depend largely on the stages of growth and maturation, but also on certain extrasystemic factors such as lunar phases and climatic conditions (springtime, rainy seasons, etc.). In addition, however, the behavior of future partners can also be important.

When analyzing behavior, we are particularly interested in the short-term interactions between sexual partners from the first encounter to actual copulation. Often we can clearly distinguish two separate elements: 1. Preparation for the release of sexual behavior patterns in a narrower sense—in other words, those leading to fertilization or the release of eggs; 2. overcoming defense and escape tendencies resulting from the physical closeness that mating partners must maintain, but that individuals normally try to avoid.

The first problem always has to be solved—regardless of whether fertilization is internal or external. For example, one partner must induce the other to go to the spawning place or to take up a mating posture. This may be done in different ways, by emitting chemical, tactual, acoustic, or optic signals. In sea urchins and oysters an arousing messenger substance (pheromone) is released together with the sperm, stimulating females and other males to release their own sexual products. In tree frogs the female has a "signaling posture"—curving her back inward—that tells the male she is about to deposit her eggs. The male then slides back until his cloaca touches the eggs (probably an additional safety measure), and he fertilizes them. In copulating birds and mammals the female's mating posture indicates to the male that he can approach and perform the act.

The second problem applies particularly to those animals normally keeping a distance between themselves and conspecifics—that is, where individuals live either alone or together in a group or territorially organized society but which usually do not come into physical contact. This otherwise typical interindividual distance must be overcome if mating is to take place. One way is through aggression, where one partner subordinates the other. But frequently we can also observe appeasement gestures that act as signals for a peaceful approach. Some animals, including spiders, have to take extra precautions to avoid being eaten as prey by their partner.

For many vertebrates, particularly for fishes, birds, and mammals, the development of a more stable pair-bond relationship also provides courtship behavior with additional functions. Behavior patterns that we designate as courtship, performed between individuals of opposite sex, would now go beyond the period immediately before copulation. At this point we have to distinguish between pair-formation, behavior patterns characterizing the relationship between partners (pair-maintenance), and rituals associated with copulation (mating rituals in the real sense).

We shall mention only briefly the functions of a more-or-less permanent pair bond: preparing for copulation, care of young, and mutual protection for the partners. By means of these pair-bond relationships, individuals of both sexes are more evenly distributed within a population, and sexual stimulation occurs only when a suitable mate is found; thus, copulation becomes separated from the process of finding each other. If the partners live in an environment where offspring can be raised only during a short period of time, the familiarity they have gained with each other in the course of pair-formation enables them to be aroused fairly quickly. This may be observed especially in animals where the sexes live together but do not engage jointly in brood care, for example, in pike, cobras, and mallards. Conversely there are animals living in families but without pair-bonding, for example, mother families like those of gallinaceous birds and cichlid fish, and father families like those of sticklebacks and emus.

We may note that conjugal pair-formation without a "family life" is found not only in vertebrates but also in invertebrate animals. One example is the water flea (*Rivulogammarus pulex*). We can often see a pair of these water fleas where the male—usually somewhat larger than the female—straddles his mate's back. He stays in this position for weeks, but there is no copulation: The male has simply taken possession of an immature female for future reproduction.

Far more frequently, however—at least in vertebrates—we find that pair-bonding is associated with raising a family. This means that the offspring can be taken care of by the joint activities of both partners, who often perform different tasks. Since conditions must be right for the rearing of young, the courtship ceremony itself has to include signals indicating that proper measures have been taken for brood care. A different solution to this problem is found among some invertebrates that do not form stable pairs: After copulation, the female stores the spermatozoa up to several months in a special sperm pouch, fertilizing the eggs only when conditions for her offspring are suitable. Examples are the bumblebee and the land snail.

In nest-building animals, the nest may be completed before pair-for-

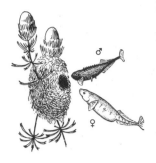

Fig. 28-16. The male nine-spined stickleback shows a female the entrance to his nest.

▷
Examples of brood parasitism. Top left: Cuckoos do not rear their own young, but instead deposit the eggs into a nest belonging to "host parents." Here, a wagtail (left) is feeding a young cuckoo. Bottom left: A pair of nightingales attacking a cuckoo model. Under natural conditions, this reaction serves to prevent the cuckoo from laying eggs in their nest. Top right: Ichneumon fly laying its eggs into the pupa of a butterfly, which is then eaten by the larvae from the inside. Center and bottom right: Female false scorpion *Neobisium muscorum* in the nest with her larvae before and after feeding. The larvae suck the female dry.

Fig. 28-17. Male bower-birds meticulously prepare a spot on the ground for courtship. With gifts and dances they lure females into this "bower," where copulation takes place. Subsequently the female builds her nest in a tree by herself, lays the eggs, and raises her young.

Fig. 28-18. Courtship feeding in terns.

◁

Top left: Feeding in hole nesters (here, nuthatches). Top right: The wren avoids making its nest dirty by taking the young's feces (enclosed in a thin membrane) directly from their anal opening, dropping it at some distance from the nest. Below: Some mammals suckle their young while lying down (left, wild pig), while others do it in a standing position (right, plains zebra).

mation or at the latest before copulation. If the male is the one to build the nest, his courtship may consist of attracting the female to his site and showing her the nest structure; he may even present her with several nests to choose from. Correspondingly, the female would then take part in the ritual by viewing and accepting or rejecting the nest. Courtship behavior focusing on the nest is found, for example, in sticklebacks and many weavers. But if the partners jointly build their nest in the period between pair-formation and copulation, these behavior elements are lacking in courtship; at the most, we may find that the male indicates his readiness for further nest-building activities by symbolically fetching and showing nesting material.

An even closer connection between mating preparations and brood care can be seen in courtship feeding, a very common practice among birds of various kinds. The male almost always offers the food. As demonstrated in a number of species, this feeding behavior enables the female to build up the tremendous reserves needed for growing her eggs. Since she can remain in the nest before the eggs are laid, she is spared the additional exertion which would endanger the formation of eggs. In addition, the female has more time to protect her nest, either by driving away inquisitive conspecifics (in colony breeders) or those still looking for their own nesting site, or even by warding off predators.

In the various bird species, we find that courtship feeding has undergone a greater or lesser degree of symbolization. While black-headed gulls still engage in real feeding, males of the parrot *Agapornis* merely balance a food kernel on the tip of their tongue before copulating, holding the bill wide open and making themselves appear larger by ruffling the feathers. This position does not allow an actual transfer of food, but feeding may occur in another way: It is independent of this symbolic food offering and takes place only after copulation. We have already mentioned the various degrees of symbolization in pheasants. The peacock no longer feeds his hen nor even shows her food; he only "pretends" to offer it.

It is clear that in many animals living together with a future sexual partner is a form of protection. As a pair they are better able to defend against predators, provide against climatic conditions, and find food. These advantages apply to all forms of communal life in the animal kingdom. In addition, many species combine social facility with sexual activities and the functions of brood care, by simply having large numbers of individuals belonging to both sexes live together from the start.

The process of integrating pair-bonding and family life into a larger, superordinate social system again leads to more complex and varied courtship behaviors. The task is to form as many pairs as possible, and to prevent mutual interference with copulation. Furthermore, not only the partners but also group members in general must recognize each other and synchronize their moods. For instance, black-headed gulls all brooding at the same time will raise more young than those breeding in temporal succession. The process of raising offspring places certain demands on the large social system—for example, when individual pairs defend territories against one another. These territories have to be large enough to provide one family with sufficient food; but they also serve to protect the brood, because predators have a harder time finding widely dispersed nests.

However, there is one problem with a territorial system: On the one hand, a potential mate has to have access to the territory of an unmated animal, while on the other hand all other conspecifics are driven off. This is why often enough a territory owner will initially greet a female with aggression. But instead of being intimidated or scared off, the female must now appease the male. As a result we commonly find this solution to the problem of pair-formation: Threatening behavior that discourages rivals actually serves to attract sexual partners; in other words, it lures the mate while maintaining its original signal function toward others. Typical examples are the threatening courtship patterns of woodpeckers, the thematic song patterns of songbirds, and among mammals the ritualized fighting of pigs and horned animals.

Other elements of courtship behavior are perceived as decidedly amicable. These kinds of signals are found especially in species where pairs are formed within existing social aggregations. Male laughing gulls do make threatening movements when a female laughing gull approaches, but these are no longer oriented in her direction; the frontal threat posture is now displayed to the side of the female. The fact that the male does not intend to attack is also demonstrated by his averting the beak—a weapon —combined with showing the back of the head (Fig. 28-21). The aggressive tendency is discharged instead against other conspecifics passing by, which are suddenly attacked and pursued. But this attack is incomplete; the animal counters its aggressive swoop by sailing upward, perhaps as a display toward the female. The same occurs in ducks, except that the roles are reversed. While the drake expresses a friendly mood by swimming in front of the female with his head emphatically averted, she threatens in a direction away from her partner and toward other males ("inciting"). In many pair-formation rituals the partners appease each other and induce a friendly mood by mutual grooming.

In many cases the act of copulation arouses aggression in partners who normally lead a peaceful coexistence, because they are not used to such close physical contact. For that reason the stimulation to mate by means of courtship behaviors may also involve overcoming aversion or escape tendencies in the partner. Depending on the species, this occurs either through aggression, by forcing the partner to submit, or through appeasement and the display of amicable intentions during the approach.

Similar functions may be seen in certain postcopulatory behaviors, like the feeding of the female in *Agapornis* and the "bill-averting" movement of laughing gulls after copulation, a behavior increasing with the success of the mating act.

Fig. 28-19. A male herring gull (left) regurgitates food for his female.

Fig. 28-20. The mating ceremony of the greater kudu contains ritualized fighting behavior interspersed with gestures of threat and display. This male is standing behind the female; he displays by holding his head high and turning it to the side.

Fig. 28-21. The partners in a pair of laughing gulls induce a friendly mood by turning their bill, or weapon, away and at the same time displaying the back of the head.

Fig. 28-22. "Inciting" in sheldrakes: The female (front) threatens away from the partner and toward another pair.

Fig. 28-23. In the post-copulatory behavior of gray herons, the partners repeatedly cross their necks and the male offers his female twigs for building the nest.

The postcopulatory behavior of ducks contains elements used to demonstrate social cohesion even within the flock, such as "head-up, tail up" and "nod-swimming." Immediately after copulation the drake pulls his head and neck up onto his back and then makes strong nodding movements. In general we can say that postcoital rituals serve to recreate the calm and familiar relationship characterizing the pair bond, so to speak, which was temporarily disrupted by the excitement of copulation.

In forming associations and maintaining group cohesion, it is always essential that conspecifics find each other, recognize each other, and synchronize their movements and activities. This is why certain courtship movements also serve a function within the group as a whole. In other words, they not only determine the nature of a sexual partnership but also influence the larger social unit, which may comprise several or even a great number of mated pairs and where the boundaries are not always clearly distinguishable. The prominent, complex, and very much species-specific "social game" of ducks is only vaguely associated with the processes of pair-formation. This also applies to the group combats of black-throated divers, the group ceremonies of flamingos, and the communal dances of chorus manakins which belong to the South American manakins; in the chorus manakins, several males court in front of one female (although her role may be taken symbolically by a young male).

In addition to stimulating its own partner, a courting animal may also arouse its neighbors. In gulls, for example, the time when eggs are laid is far better synchronized in larger colonies than in smaller ones, probably because the courting animals stimulate each other. It has been demonstrated that this synchronization of breeding activities is essential for ensuring the greatest reproductive success in these birds. Pairs breeding too early or too late raise far fewer young, because they lack the communal protection of neighboring pairs and are therefore more vulnerable to predators; this also makes them too anxious and leads to neglecting the brood.

Now that we have discussed the various functions of courtship, the diverse ways these elements are brought together, and how they originate, we shall take a brief look at the regularities that may appear in the way courtship behaviors are combined during the course of evolution. We can state three basic principles: clarification, association, and simplification.

Courting animals draw attention to themselves, they "put themselves on display," because they must attract a partner and show that they are conspecifics. Courtship songs, flying displays, and dance rituals, all these terms indicate the display character of courtship. Movements are performed more intensively than usual, individual phases are emphasized, and the sequences are often repeated over and over again. The effects of such movements may be enhanced by colorful patterns, adorning body appendages, and various sounds or calls.

Behavior patterns of different origin may be combined, or associated, to serve the various functions of courtship. In this way, patterns of appeasement and brood care may appear together, or the display of recognition traits may be combined with attempts at mounting, the indication that an animal is ready to mate.

And finally, behavior patterns with an originally different goal can be simplified to the point of mere symbolism. Examples are the showing of food rather than actual feeding, and mock attacks without an object rather than actually driving a rival away.

In this chapter we have discussed some of the most diverse functions of courtship. Especially since the 1960s, this functional or ecological viewpoint has led ethologists to perform more and more studies in the wild, with more precise methods of observation. After all, any interpretations of behavior, even the most plausible ones, must be tested on real-life organisms in their natural habitat. For example, when we claim that courtship behavior leads to mutual stimulation and to a synchronization of breeding activities within a group, then we must validate this assumption with exact observations and experiments.

Quite apart from this steady progress in scientific testing, however, the natural observer will always be amazed at the manifold forms of courtship. He will marvel not only at the beauty and diversity of courtship movements but also at the complex relationships between functions of "daily life" and the way these movements are expressed. Far from taking the magic out of natural events, modern forms of behavioral analysis reveal their beauty and complexity to an even greater extent, giving us constant warning that we must deal with nature far more carefully—we might even say humanely—in the future.

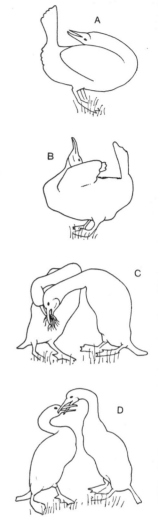

Fig. 28-24. Courtship behavior of Brandt's cormorant: A. The female pulls her neck to the back and raises her tail feathers. B. The male courts the female in a similar posture, showing the bright blue patch on his neck. C. After copulation the male brings the female nesting material, which they both carry to the nesting site. D. These birds do not feed each other, but instead they make contact with their bills.

29   The Nature and Function of Pair-Bonds

The Nature and
Function of Pair-
Bonds, by
Carl J. Erickson

In his *History of Animals,* Aristotle observed that the female turtle dove remains faithful to her mate throughout her lifetime. The ecclesiastical scribes of the Middle Ages, apparently inspired by such exemplary fidelity, embellished this observation in their bestiaries. These ornate compendia of animal lore seldom failed to include the story of the dove's remarkable constancy. For example, one 12th Century version tells us that "It is truly believed that when a turtle dove is widowed by the loss of her spouse, . . . she grows bitter about love itself—which produces more sorrow out of death than sweetness out of loving. So she refuses to repeat the experience, nor does she break the bonds of chastity or forget the rights of her wedded husband. She keeps her love for him alone, for him she guards the name of wife." The monks then added: "Mere women are seldom able to come up to the standard of doves."

As centuries passed, the fact and fiction of animal social life became increasingly entangled. Anecdotes describing animal love (both fulfilled and unrequited), jealousy, and grief became legendary. Yet in spite of strong public curiosity, scientific analysis of these phenomena has been peculiarly slow to develop. For example, it is not yet clearly known whether Aristotle's claim was entirely correct. Certainly we know that not all female turtle doves take but one mate in their lifetime. But is it not possible that the claim is true in some circumstances? Scientific investigation of many problems concerning animal social behavior has only recently begun, and the answer to this question remains uncertain. Nonetheless, we are in the formative stages of a research area that promises to be exciting and provocative.

The problem of the
pair-bond

Turtle doves are not the only animals that establish extended heterosexual relationships. Such associations, or "pair-bonds," are found throughout the animal kingdom. Because these associations vary in duration and are not always confined to a single mate, ecologists have attempted to discern the particular biological advantages of the various mating systems that are found in different environmental contexts. On the other hand, ethologists and psychologists are often sensitive to the fact that the processes controlling pair-bond formation can differ among species, and they have attempted to discern the mechanisms of pair-bonding in terms of the developmental, physiological, and stimulus conditions affecting the relationships. Because of the diversity in their principal interests, the ecologist, ethologist, and psychologist have emphasized different aspects of the pair-bond and, as a result, have created some confusion through their slightly varied use of the term. It may be appropriate to say that the ecologist has focussed on the "pair" while the psychologist and, to some extent, the ethologist have attended to the "bond." The ecologist is inclined to apply the term "pair-bond" whenever a male and female remain together beyond some minimum length of time. His primary considerations are usually the duration of the association and the number of individuals paired. The psychologist and ethologist, however, are often less interested in the temporal aspects of pairing than they are curious about the processes binding one individual to another, and consequently they may be reluctant to conclude that a pair-bond has been formed unless there is clear evidence that the animals recognize one another as individuals and have formed a genuine social attachment. As a result, the ecologist may refer to an existing pair-bond when the ethologist or psychologist would not, and occasionally the opposite will apply. Although the discrepancy exists, it should not seriously affect our understanding of the major issues. First, let us consider the variety of pair-bonds from the ecological viewpoint.

Due to the efforts of countless amateur ornithologists, our knowledge of breeding behavior in bird species is relatively complete. In contrast, we know relatively little about a great many other vertebrates and invertebrates. For this reason our discussion of pair-bonds will emphasize the mating relationships of the better-studied groups.

The variety of mating systems

In many species no pair-bonds are formed. In fact, in some cases the sexes never come into contact for mating. Among many aquatic animals, for example, the male and female release their gametes into the water, and no social interaction is required for reproduction. Among some fishes such as the cichlids, however, extensive pairing relationships may develop.

In birds we find a wide array of mating arrangements. Although a few species are entirely promiscuous, the great majority develop heterosexual associations of some duration. This does not necessarily imply, however, that these are monogamous relationships. In many instances a single male may breed with several females in the same season, an arrangement known as harem polygyny. In very rare circumstances polyandrous relationships, in which a single female may pair with several males, can occur.

Monogamous relationships are commonly defined as those in which during a single breeding cycle, mating takes place with only one individual. According to this definition, the overwhelming majority of bird species are monogamous. David Lack, who has compiled perhaps the most complete census of avian pair-bonds, estimates that among all bird species, nine-tenths have such relationships. Among the passerines the incidence may be even higher. In contrast, monogamous associations are quite unusual in mammals. An extensive survey by John Eisenberg notes a few monogamous species among the wolves, foxes, and beavers. There is also some evidence that a few primates such as the marmosets and gibbons form monogamous relationships.

The significance of pair-bonds in animal life

Mating relationships, like other characteristics of animal behavior, provide animals with advantages for survival and for the production of offspring. In monogamous systems it is common for the male to participate in the care and protection of young. Thus the newborn have the benefit of two parents rather than one. When the pair-bond is extended beyond a single breeding cycle, there may be additional, although less obvious, advantages. For example, it is not uncommon for breeding cycles to be shortened and the production of offspring increased in cases where the male and female have bred together previously. John Coulson spent many years studying a small gull, the kittiwake *Rissa tridactyla*, on the Northumberland coast of England. He found that the number of eggs laid, the frequency of successful hatching, and the overall efficiency of breeding were directly related to the number of occasions upon which a male and female had previously bred together. Similar effects have appeared in laboratory studies. In an experiment performed by Carl Erickson and Robert Morris at Duke University, male and female ring doves *Streptopelia risoria* were allowed to breed and raise young in laboratory cages. After the successful raising of two broods all birds were placed in isolation cages for an interval of two weeks. At the end of this period half of the pairs were simply reintroduced to one another in breeding cages and allowed to breed for a third time. The remaining males and females were also allowed to breed but not with their former mates. This was accomplished by merely exchanging partners. As in Coulson's study, those birds breeding together for a third time hatched more of their eggs successfully than did those birds of similar age and breeding experience that were paired with unfamiliar animals.

This study produced another interesting result. It was observed that when reintroduced to former mates, males exhibited some displays much less vigorously than when placed with strange females. The reason for this difference in performance is not entirely clear, but from this an additional advantage of retaining a mate can be suggested. Many animals have displays that identify them with a particular sex and species. These displays serve the very important function of preventing matings between members of different species. When individuals already recognize one another as a result of having bred together before, these displays become superfluous. They can even be disadvantageous, since they prolong the breeding process and may possibly attract the attention of predators. This then may account for the reduction of display in the initial phases of breeding in established pairs. Whether such changes occur in other species in their natural environment requires further documentation.

The advantages of monogamy seem to be so great that at first it may be puzzling that not all species enlist the male in the service of parenthood. But as in the case of other biological characteristics, any aspect of social behavior must be evaluated with the entire fabric of circumstances in which the animal lives. A proper understanding of pairing relationships emerges only through an examination of the ecological context as well as the evolutionary and cultural history of the animal under consideration. By taking account of the whole constellation of such factors when comparing the reproductive behavior of one species with another, we begin to grasp the significance of particular mating styles.

In recent years several studies have illustrated the fruitfulness of the comparative approach to social behavior. One of the most elegant of these was conducted by John Crook, of Bristol University. Crook spent several years in Africa and India comparing the habits and habitats of several species of weaver birds in the subfamily *Ploceinae*. These birds are known as weavers because of the elaborate, intricately woven nests they construct. Crook studied these birds because, although closely related taxonomically, the various species exhibit several mating styles. Moreover, they live in a variety of habitat such as forests, swamps, and grasslands, as well as dry and humid savannahs. Thus the weavers offered an unusual opportunity for discerning correspondences between ecology and social organization.

When he examined the forest-dwelling species, Crook found that a high proportion of them were monogamous. In this habitat the food supply is often dispersed and rather limited in supply throughout the year. In most cases these weavers were insectivorous. Because of the food limitation, male aid is an important factor in rearing young. In the humid savannahs, however, polygyny is the rule. The weavers of this area are typically seed eaters, and their breeding coincides with the super-abundance of food that occurs at particular times of the year. Crook suggests that with such a surfeit of food available, females are able to raise young successfully without help from the male. Consequently, males are free to mate with more than one female.

Crook also found that polygynous birds were frequently colonial. Many of the species build elaborate hanging nests, and some build as many as several thousand nests in a single tree. In a few species, such as the buffalo weaver (*Bubalornis albirostris*), adjacent nests may share common walls and appear as a single avian "apartment house." Crook observed that of fifteen colonially nesting species, thirteen were polygamous. In contrast, of nineteen solitary nesting species, at least sixteen were monogamous.

Crook's analysis was more extensive and complex than can be described here, and any brief description suffers from some oversimplification.

▷
Aggregations of animals where food sources are particularly favorable. Above: During low neap tides, crabs of the species *Dotilla sulcata* gather in large numbers to forage for food in the tidal zone of the Red Sea: left, several groups; right, close-up of one of the groups. Below: Vultures aggregating to feed on a gnu carcass in eastern Africa. During the day, these birds localize dead animals from high above with their very sharp vision. Within minutes, vultures from a circumference of several kilometers gather at the very spot as soon as they notice that one of their number has darted toward a find.

Nevertheless, some generalizations apply. He found that monogamy occurred most often in species that lived in the forest, had dispersed cryptic nests, and lived on insects that were of rather constant but limited supply. On the other hand, polygyny was frequently associated with colonial nesting in open country. Breeding in the polygynyous species usually coincided with the appearance of abundant seeds but sometimes insects as well, such as occur along shorelines and riverbanks. In sum, food supply seems to be an important factor in determining whether a species develops a monogamous or polygynous mating system. In fact, it may be the case that members of a single species will be polygynous or monogamous depending upon available resources. Edward Armstrong found in the winter wren (*Troglodytes troglodytes*), typically a polygynous species, that on the islands off the British coast the races are monogamous and the male participates in the care of young. Armstrong attributed the monogamy of island wrens to the marginal food supply.

Food availability is not the only determinant of mating relationships. Crook and many other investigators have noted the importance of such factors as the precocity of young at birth and the pressures of predation as contributing determinants. Many species of fowl, grouse, and pheasant are promiscuous. Males and females meet briefly for copulation and have no further association with one another. The female builds a nest and incubates her eggs without aid. After hatching, the young require relatively little care since they are mobile, feathered, and capable of feeding themselves with little help.

In recent years there has been considerable discussion among biologists concerning the origins of mating systems involving multiple pairs such as those that occur in polygynous arrangements. At one time it was argued that polygyny was due to a disproportionate birth ratio of females to males. It now seems clear that mating systems do not necessarily reflect this sex ratio. Moreover, in many polygynous species there are about equal numbers of males and females produced. In those cases where the number of adult females seems to exceed the number of adult males, the imbalance is more often a result than a cause of polygyny. It seems that the males of these species often do not breed until their second or third year and thus are absent from the mating groups. Presumably this allows them to become stronger and more capable of gaining territories and female mates. In these species young males may thus ultimately benefit in terms of their own survival and the production of offspring if they postpone breeding for a year or two.

Polygyny seems to offer an obvious advantage to a male, as the more female mates he can acquire, the greater is his contribution to the gene pool of his species. Jared Verner and Mary Willson collaborated in an attempt to distinguish those conditions in which females may also benefit from a polygynous relationship. They have noted that only fourteen of the 291 species of North American passerines are regularly polygynous

◁
Aggregations that serve to protect the individual. Above: Schools of young sea catfish (*Plotosus anguillaris*), left, draw closer together when danger threatens (right), making it appear to the predator as though they are one large body. Center: Brood colony of emperor penguins (adults, black and white, young animals, brown). The crowding together of brooding animals helps to keep them warm and also assures that young are never isolated. Bottom left: On very hot days, butterflies often gather to swarm about on the wet sand near water in order to drink, as in this photograph taken on the Amazon River. Bottom right: Males of the burrowing bees (*Halictus quadricinctus*) often aggregate in sleeping groups, the biological significance of which is not fully known.

or promiscuous. Thirteen of these fourteen species breed in marshes, prairies, or savannahlike habitats; that is, these breeding areas are typically in open country. They suggest that in these areas the territories of the males vary widely in the quality of nesting cover and other resources. Thus it may be more advantageous for females to form polygynous relationships with males in favorable territories than monogamous associations in less favorable areas. Verner and Willson argue that abundant food is not sufficient reason for monogamous systems to be abandoned. They feel that unless there is competition for limited resources such as protected nest sites, monogamy will still be the rule.

It is probably safe to assume that no single evolutionary explanation can accommodate the diversity of mating systems among animal species. George Bartholomew, of the University of California, has spent many years investigating the seals and sea lions of the Pacific coast of North America. Nearly all of these pinnipeds exhibit harem polygyny. Bartholomew has tentatively but persuasively suggested that harem polygyny in these animals is determined to a large extent by their unique characteristic of breeding and feeding in two very distinct habitats, for these animals copulate and give birth on land but feed in the sea. This combination of features, when linked to more general mammalian characteristics, provides some insight into the mating organization. For example, marine mammals are usually large, with extensive insulating fat stores which reduce the heat loss when feeding at sea. Large size and fat stores can also be advantageous in a polygynous mating system because the males can then fast for long periods while defending their territories and engaging in breeding. Those males that can remain on their territories the longest and continue breeding effectively will produce the most offspring. Thus, large size will be favored in the selection process, which accounts perhaps for the fact that most male pinnipeds are far larger than the females. Bartholomew has shown that although the combination of offshore marine feeding and terrestrial birth is a primary determinant of polygyny in the pinnipeds, a full understanding of the evolution of this mating system must include an appreciation of many other factors such as the animal's extreme gregariousness, seasonal mating patterns, and ontogeny.

Thus far our discussion has focused upon the adaptive significance of animal pair-bonds and the various mating systems in which they are found. In so doing we have postponed several important questions. For example: How do we know when a pair-bond has been formed? What factors aid or hinder pair-bond formation and maintenance? Do all pair-bonds manifest themselves in the same way, and do they all depend upon the same mechanisms? Scientific investigation of these problems is in its infancy, but some important insights are emerging from current research.

Most of the investigators mentioned previously have not been primarily concerned with the physiological and behavioral processes controlling breeding behavior. This is because they are mainly interested in

▷ Arthropods living in social groups. Top left: Communal structures consisting of many individual webs of the spider species *Cyrtophora citricola* represent a transitional stage on the way to real social behavior. Top right: Communal web of several spiders of the species *Stegodyphus mimosarum*. Middle left: Wasps of the genus *Belonogaster* build a common nest from wood chewed up into paper and attached by a thin stem to better defend it against ants. The larvae mature inside the cells, which are open at the lower end. Middle right: Weaver ants (*Oecophylla*) make nests out of leaves which they join together with silk-spinning threads. Suppliers of this "yarn" are the colony's larvae, which the workers use as silk-producing shuttles. Bottom left: Swarms of the honeybee (*Apis mellifera*) occasionally construct combs in the open when a suitable shelter cannot be found; but they abandon this exposed structure as soon as a better site is available. Bottom right: Earth-mound nest of a western African savannah termite, (*Cubitermes*). The "mushroom" tops help the rain run off the nests. In some termites, this structure serves as an "air-conditioning" system that regulates temperature, ventilation, and humidity in the inner nest.

Mechanisms of pair-bond formation and maintenance

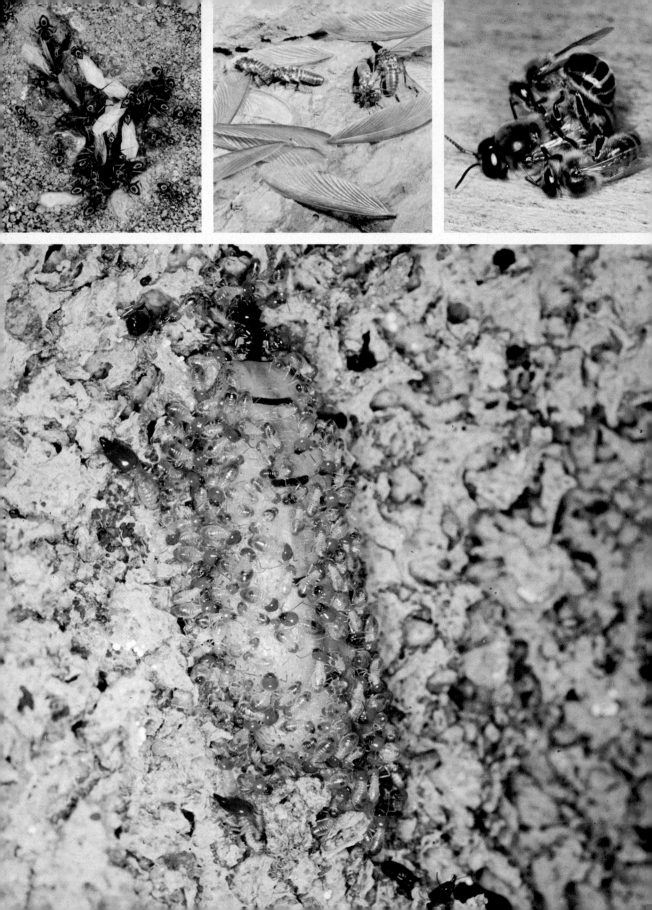

◁

Highly developed social behavior in insects. Top left and center: The winged reproductive females of ant (left) and termite (center) societies discard their wings after the mating flight, just before starting a new colony underground. Top right: Honeybee colonies reproduce drones only during a certain season.

What are the characteristics of a pair-bond?

These males have a single task: to fertilize the "queens" (reproductive females) during their "wedding flight." Once they have served this purpose, the drones become useless "hangers-on" and are driven out of the colony by the workers: many are stung to death. This photograph shows a drone—easily recognized by its large eyes—being evicted by two worker bees. Below: Division of labor in termites, as in many other social insects, is associated with differences in the morphology of the various "castes." Here, a termite queen with her greatly swollen abdomen full of eggs (the black head and thorax are pointing upward) is seen amid numerous small workers and a few "soldiers" (with strong mandibular pincers on their brown-red head portion).

determining why a particular mating relationship provides special advantages in one environment but not in another. From this viewpoint it is relatively unimportant, for instance, whether two species are monogamous as a function of similar or different biological machinery; in either case the consequences in terms of adaptive value may be very much the same. In fact, because they are interested in discerning the survival value of the various breeding systems, ecologists have often been more concerned with polygynous and promiscuous relationships than they have with the less puzzling monogamous types. In contrast, the complex behavioral processes underlying the pair-bonds of monogamous and durable polygamous attachments often arouse greater curiosity among ethologists than do the promiscuous modes of mating.

A pair-bond may be defined as a social interaction that is maintained between an adult male and female for some variable duration coincident with breeding. Thus, the companionships that arise in the nonbreeding seasons are not commonly classified as pair-bonds unless they develop into breeding relationships during the reproductive seasons. Similarly, the attachments that bind parent and young are not usually considered pair-bonds. For many investigators, however, the term "pair-bond" connotes more than this loose definition suggests. For them, pair-bond formation reflects the development of attachments in which the mate is clearly recognized as an individual and becomes preferred to others as a mate. In simple anthropomorphic terms the male and female appear to be "in love" with one another. Such assumptions must be made with caution. Many animals exhibit intricate, adaptive social interactions that upon initial inspection seem to have the quality of psychological attachment, but further examination may warrant reinterpretation. The behavior of some decapod crustaceans—a group including shrimps, lobsters, and crabs —may be taken as an extreme example. According to Von Buddenbrock, when the male and female are ready to mate, they crawl into a hole in a coral or sponge and mate monogamously for the remainder of their lives. Their fidelity may have little connection with the social attachments of other animals, however, for once inside the cavity both male and female grow too large to escape. The deepsea angler is a monogamous fish, but the attachment of male and female is more than figurative. The male attaches himself to the female's body and becomes an obligatory parasite upon her. Once attached, their bodies virtually fuse together, and the male becomes entirely dependent upon the female for nourishment.

Mechanical constraints as extreme as these are relatively rare and are cited merely to emphasize the point that the processes maintaining heterosexual pair relationships may vary from one species to another. More commonly the processes are subtle and complex. Many invertebrates, for example, develop intricate co-operative interactions for reproduction which are free of the external constraints such as those mentioned above. Yet these associations also retain an impersonal, inflexible character suggesting controlling processes different from those existing in many verte-brates. Wolfgang Wickler and Uta Seibt have used the apt term "anony-mous monogamy" to describe relationships of this type. Anonymous pair associations may be more typical of vertebrate species than is com-monly believed, however. Among many birds the same mate is retained from year to year, and it is generally assumed that a firm social bond sustains the relationship. But many of these same species return to the same nest site, and it is entirely possible that the continuing relationship is due as much to individual attachment to this nest site as it is attributable to an attachment of one animal to the other.

Some years ago Theodore Schneirla, of the American Museum of Natural History, referred to such impersonal and relatively inflexible social relationships as "biosocial." On the other hand, for those interactions involving learned adjustments, he used the term "psychosocial." These terms are a bit unfortunate in the sense that they imply that one process is organic and the other is not. Moreover, the dichotomy suggests two types of social processes when if fact there may be many. Schneirla in-tended neither of these inferences, however, and his insight was important in that it emphasized the multiple processes that could bind animals in co-operative endeavor. As previously discussed, breeding relationships involving pair-associations are widespread and likely have been independ-ently evolved many times. There is no reason at present to assume that similar mechanisms apply in the bonding processes of different species.

Nonetheless, many animals develop complex modes of interaction suggesting a basic attachment to a single individual. For example, a male may become aggressive toward all females other than his mate. Conversely, his mate alone may be the object of his courtship displays, his caresses, his copulatory attempts, and his protection. Her disappearance or death may produce symptoms of distress. These features suggest that something more than simple, adaptive neuromuscular adjustments char-acterize these interactions. One is compelled to acknowledge the participa-tion of rather complex perceptual and motivational processes in the pair-bond.

Consider, for example, a simple experiment performed at Duke University. A large number of female ring doves (perhaps closely related taxonomically to the turtle doves described by Aristotle) were assigned as mates to an equal number of males. Each pair was placed in a separate laboratory cage and allowed to mate and raise young twice in succession. At the completion of the second breeding cycle all birds were individually isolated for up to eight months. At the end of this period they were tested for mate recognition and preference. This was accomplished by observing the animals in small groups. For example, three males and three females were taken from isolation and transported to large outdoor cages in a nearby forest. All birds in each group had previously mated with one of the animals of opposite sex in the group. The birds were then observed to determine which of the previously mated doves located themselves in close proximity to one another, and which of them copulated, built nests together, and incubated the same eggs. This procedure continued until all of the original pairs had been examined. The results left no doubt that the ring doves could identify their former mates, for without exception all birds paired with the same individuals with whom they had mated in the laboratory. It should be noted that mates were not selected originally by the birds themselves but were assigned by the investigators. Nevertheless, it was these assigned mates that were chosen as partners in the outdoor tests. This indicates that as a consequence of mating in the laboratory, the doves had *acquired* a preference for these particular mates. Moreover, the possible influence of a common attachment to a nest site was eliminated by testing the animals in an unfamiliar environment. This study demonstrates that ring doves learn to recognize their own mates as individuals and acquire a preference for them as partners in subsequent breeding periods. It seems that it is this development of learned preferences for particular individuals that often differentiates the pair-bonds of birds and mammals from those of invertebrates.

The determinants of pair formation

Establishment of a pair-bond is often attributed to a wide variety of factors. Courtship displays in particular are most frequently cited as promoting social attachments. This emphasis seems to be because pair-bonds are often established in a brief period shortly after the meeting of male and female, a period in which these animals usually display most vigorously. At this time courtship displays may be accompanied by several other activities which may also promote pair formation. For example, the male and female may spend much of their time in preening or grooming one another. Many animals also engage in feeding one another during the early mating period. This is especially common among bird species. One of the birds will pass food from his beak to that of his mate. In some cases no food is actually passed, and the animals merely touch or rub their beaks together. Copulation, in addition to providing for the fertilization of the ovum, may enhance the attachment. Wolfgang Wickler, in his book *The Sexual Code*, suggests the importance of copulation in this regard. He cites Hans Kummer's extensive observations on the hamadryas baboon as evidence. In this species the male must copulate several times before ejaculating and impregnating the female. Only the highest ranking male ever completes the series of copulations; nevertheless, young males often mount and copulate with females, short of ejaculation. Wickler suggests that copulatory attempts by the young males enhance the social attachments within the group. He emphasizes that copulation may be very important in maintaining the bond between male and female.

Although most ethologists have focused upon the behavioral interactions of the sexes as being of primary importance in pair-bond formation, it should be noted that the internal states of the animals may also be of considerable significance. At certain stages of maturation or when particular hormones are circulating, the animals may be especially disposed to form bonds. It may be for these reasons that not all monogamous animals form bonds coincident with the commencement of the first breeding cycle. For example, Jürgen Nicolai has shown that some tropical finches, such as the violet-eared waxbill (*Uraeginthus granatinus*), form monogamous pairs long before breeding begins. In fact, pair-bond formation takes place before they are thirty-five days old, a time when they are still being fed by their parents. In contrast, adelie penguins (*Pygoscelis adeliae*) do not seem to settle down to firm monogamous relationships until they have bred and raised young several times.

Several investigators have suggested that once it is established, a pair-bond is sustained by the familiarity of the animals with one another. Such an explanation has appealing simplicity. Familiar individuals are often more predictable and less threatening than unfamiliar animals. Nonetheless, although familiarity may be an important contributant to pair-bond maintenance, it seems clear that other factors are also essential. Le Resche and Sladen have found that young adelie penguins establish companionships with "keeping-company" partners; but when they are ready to breed, mates are seldom drawn from among these animals. Similarly, Nicolai has shown that the young bullfinch (*Pyrrhula pyrrhula*) forms a bond with a nestmate when very young but later selects a breeding partner from another group of animals.

It is now becoming clear that pair-bond maintenance may be contingent upon a number of factors, familiarity being but one of these. A fascinating study concerning this issue was performed by John Coulson. He found that although the majority of kittiwakes return to the same mate in successive breeding seasons, about one-fourth of the animals select new mates even though their former mates (presumably familiar) are present in the colony. An examination of their breeding histories revealed that a high proportion of those animals choosing new mates had reproduced unsuccessfully in the previous season. In most cases the bird's eggs had failed to hatch. Although it is possible that both the failure to remate and the failure to breed successfully were due to a deficiency in the initial pair-bond formation, it seems not unlikely that retention of a mate depends somewhat on successful breeding. It would appear to be biologically adaptive for animals to dissolve the bond and select new mates when young are not produced successfully. It also appears possible that other contingencies may operate at various stages of the relationship. For example, anecdotal observations of pigeons and doves indicate that pair-bonds dissolve more readily in the interval between breeding cycles or on occasions when the male has been defeated in a fight.

Many of the behavioral interactions mentioned in connection with pair-bond formation have also been implicated in pair-bond maintenance, and presumably without the continued exhibition of these interactions the pair-bond would be jeopardized. Julian Huxley was one of the first

Pair-bond maintenance

▷
Food exchange and brood care in social Hymenoptera. Above and below: The members of an insect state (above, wasps; below, red ants) engage in a constant exchange of information and food. The complex rituals of touching and the sharing of chemical signals ensure that all members of the colony remain linked with the flow of information, and are constantly informed about conditions in the nest. Part of the information making the rounds may also concern the presence or absence of a queen. Middle left: Various stages in the ontogenetic development of wasps, from the young and pale pupal phase to that of the fully colored and mature organism ready to hatch. Middle right: Ants (here ants of the genus *Camponotus*) always carry their pupae to the areas in the nest that provide the most favorable conditions for their growth (temperature, humidity).

to recognize this possibility. In his classic report on the great crested grebe (*Podiceps cristatus*) he stated, "What is the good of all these divings and posturings, these actions of courtship, the 'expressions of emotion?' To what end are colors and structures developed solely to be used in them, and what return is got for the time and energy spent in carrying them out? They are common to both sexes, and so have nothing to do with any form of true sexual selection; they are self-exhausting processes, not leading up to or connected with coition, and so cannot be sexual excitants in the ordinary sense of the term.

"It must be, however, that they fulfill some function; and I believe I know what this function is. I believe that the courtship ceremonies serve to keep the two birds together, and to keep them constant to each other."

Just as courtship, mutual grooming, courtship feeding, and copulation have been implicated in the initial formation of the pair, the maintenance of the relationship has also been attributed to these behavior patterns. In recent years there has been a growing interest in the role of vocalization in pair-bond maintenance. Many animals have vocal greetings which are given when they meet or are reunited. Paired animals may develop elaborate vocalizations as a means of recognizing one another and maintaining contact with one another. Paul Mundinger, of Rockefeller University, has shown that as two adult pine siskins (*Carduelis* [*Spinus*] *pinus*) spend more and more time together as a pair, their calls become increasingly similar to one another. Such mutual imitation occurs even when an American pine siskin male is mated with a European siskin (*Carduelis spinus*) female.

Other species may use other vocal means of recognizing one another and maintaining contact. For example, some of the African shrikes (*Laniarius*) develop elaborate antiphonal duets that are remarkable for their musicality and flutelike timbre. The male's phrase is answered within milliseconds by the female's, and the phrases may alternate many times. Each pair of birds develops its own repertoire, which may contain up to fifteen or more songs. Thorpe and North note that occasionally when one of a pair loses his mate, he will sing not only his own song but also that of his mate.

It is possible that many other behavior patterns contribute to pair-bond maintenance. The various aspects of reproductive activity such as nest building, incubation of eggs, and care of young, when shared by both the male and the female, may strengthen the bond between them. However, it is important to inject a word of caution at this point. There is, in fact, little precise or quantitative evidence to support the view that pair-bonds are directly affected by the activities listed. Most of the behavior patterns described have other, more obvious functions. Courtship displays may be used in declaring and maintaining a territory, establishing a nest site, or stimulating the female to lay eggs or engage in copulation. Mutual grooming and preening are often associated with the removal of

◁
Nesting communes in weaver birds. Above: The great communal nest structures of the sociable weaver (*Philetairus socius*) which inhabit the dry areas of southern and southwestern Africa, may be as long as 4 m and nearly 3 m wide. These nests are often used for decades at a time. The structure is held together by a strong and thick roof made of thorny branches; it provides excellent protection. Bottom left: On the underside of the communal nest we find entrances to the various chambers of individual birds, serving as brood nests and sleeping places. Up to 500 of these highly social birds literally nest "wall-to-wall." Young birds from the first brood help their parents feed the younger siblings. Bottom right: Another form of brooding in colonies is shown by the African steppe weavers (*Textor*), also extremely social birds. Their single nests, nearly as round as a ball, hang only a few centimeters apart from each other.

parasites. Courtship feeding may actually provide nourishment, and copulation has the basic purpose of insemination and the production of offspring. The evaluation of any additional contribution of such activities to pair-bond strength is difficult. The problem is further complicated by the fact that estimation of pair-bond strength in any quantitative sense is a demanding task, and one that has seldom been attempted. Yet the challenge may well be worth the effort. Ultimately our understanding of societies—both animal and human—must rest upon an analysis of the processes which bring individuals together and bind them in co-operative endeavor.

30   Parental Behavior

Behavior in relation
to raising of young,
by R. Sossinka

One of the most important tasks of all organisms is to provide for a sufficient number of offspring so that natural losses are continually replaced, and the survival of the species is guaranteed. Not counting asexual methods of reproduction (see Chapter 27), there are two basically different means by which a species may produce enough young. The first method is based on the probability that if the number of eggs produced is sufficiently high, many young will hatch, and of these at least some will reach sexual maturity even in the absence of parental care. Especially in the environment of sea water, which is extremely favorable to life, a great variety of animal species simply eject their eggs into their surroundings. Instead of caring for their brood, these organisms have greatly increased the number of their eggs. The mussel *Crassostrea*, for example, deposits more than 14 million eggs, and the ling fish *Molva* lays as many as 28 million and more.

Of course, this method is very uneconomical because the losses are exceedingly high. Even though the eggs are generally minute, there is still a considerable waste of valuable material. Because of this, many animal groups have—quite independently of one another—paved the way for a different kind of development that aims at reducing the number of eggs and at the same time improving the means for their protection. The maturing of young is no longer left to chance. Instead, the parents attempt to reduce certain kinds of "danger": They may protect their brood from cold and dehydration, ward off enemies, prevent the young from starving, and so on. The means and effectiveness of such protective measures vary from one species to another, always being highly adapted to the animal's total way of life. In essence, although the general term brood care is most common, we may distinguish between two stages of brood provision: a simple care of young, and the raising of young.

When parents do no more than care for their brood at the most elementary level, they have no contact with the young at all. The parents —or in most cases the mother—merely find or provide for a proper spot in which to lay the eggs, and if necessary will ensure that the young, once hatched, have some kind of food supply. This is the limit of parental duty, and the young must fend for themselves once they leave the egg.

By contrast, animals that raise their young not only care for the eggs but also tend to their offspring during the early stages of life. Parents now have direct contact with the young and will undertake a number of different tasks, such as feeding, protecting, cleaning, and "instructing." These behaviors in relation to raising of young may be carried out by the mother, father, both parents, or whole groups of adults.

In this context, some animal groups have come up with an additional "invention": Instead of laying eggs in the open and then going to great pains to protect and provide for them, the developing young are kept directly at or even inside the body. The eggs are then well protected from harmful environmental influences and predators by being contained in brood chambers and pouches, perhaps between the widened leg appendages of isopods, in the small paired sacs or pouches at the posterior end of copepods, or in the abdominal pouch of male sea horses. In some species, the eggs may also remain in the genital organs, perhaps maturing within the oviduct as in *Peripatus*, sharks, and mammals. In all these cases, food may be provided for the ovum or egg in the form of yolk supplies, or within the oviduct from mother to young by means of a placenta. These adaptations, by which young are tended to without behavioral activity on the part of the mother, are called passive brood care. In the most ideal cases, the entire development of the embryos takes place inside the womb (uterus), so that the mother gives birth to small but nevertheless independent offspring (e.g., geckos). These young do not need any further care but will immediately take to the world at large by running, crawling, or swimming away; such individuals are referred to as extreme precocial (also antophagous or nidifugous) animals.

Fig. 30-1. *Cyclops* (above) with paired egg pouches, and a male sea horse (*Hippocampus*) with a full abdominal brood pouch.

There are many intermediate stages between these two alternatives of laying the egg in the open and then actively tending to it, on the one hand, and keeping the egg in the body (passive brood care), on the other. In many cases, parturition will occur long before the young is fully developed, so that the creature that is born still needs intensive care and raising; this is a so-called altricial (also insessorial or midiculous) animal.

Our next question is: What behavior patterns serve the raising of young; or What do animals do to ensure the survival of their offspring? We may first examine the behaviors involved in the simple care of young. The most elementary form of brood care occurs when the mother doesn't just drop her eggs anywhere, as for example, fleas do, but rather she selects that place or spot in which the eggs may best develop. For instance, many dragonflies immerse their eggs in the water, because this is the environment needed by the larvae to grow; some species of Hydrophylidae build small floats of slime, in which the eggs are embedded to float on the surface of the water. Frogs and salamanders, which spend the greater part of the year on land, also go into water for spawning. Conversely, sea turtles come out of the water just to bury their eggs in the warm sand, because reptile eggs cannot develop in water. There is even one bird group displaying a corresponding form of brood care. The cocks of Megapodiidae build a large mound of damp leaves, which generates warmth by means of the fermentation processes. The hen buries her eggs there leaving it to the cock to regulate temperature within the brood mound by ventilating it or stamping down the leaves. When the young has finished developing inside this "incubator," it hatches from the shell, laboriously fights its way out of the pile of leaves—often many meters thick—and runs off into the jungle without any further contact with the parents. In this way, the parents provide for proper conditions of temperature and humidity, as well as for protection and camouflage by laying the eggs in some suitable location.

In many cases, a parent will even make provisions to supply the young with food at their later time of hatching. Butterfly caterpillars attach their eggs to the plants that later become a source of food for the new caterpillars. One kind of beetle—the blossom weevil—gnaws passages inside apple blossoms and lays its eggs in them, so that the larvae can feed on the growing fruit when they hatch. Other beetles even provide "food packages" for their unhatched young. Dung beetles, for example, build underground passages, deposit an egg at the end, and fill the tunnel with horse dung or similar matter so that the larvae can feed on it. The dung beetle even packs the dung into balls, buries them, and lays its eggs on the packets (see Color plate, p. 442). Many members of the Hymenoptera provide their future young with food supplies. Ichneumon flies, for instance, have a long, flexible sting with which they drill into their victims, such as caterpillars and pupae. The eggs are then laid through the sting into the living host animal; when the young hatch, they eat it up from the inside. Once matured, the young leave their host, who by that time is usually dead (see Color plate, p. 463). (Since a number of ichneumon fly species reduce certain harmful insects this way, their activity is important for the biological fight against pests.) Sand wasps have an even more complicated way of providing for their future young. First, they dig underground passages or brood chambers, then look for a particular prey animal, such as a caterpillar, which they will paralyze with their poisonous sting aimed at the vicinity of the victim's nervous system. Next, the wasp carries its prey into the chamber, lays an egg on it, and covers up the passage. The hatching larva, fully hidden and protected, will feed on the prey which remains paralyzed and fresh (see Color plate, p. 457). "Solitary bees," that is, species that do not form societies, also deposit food supplies into brood chambers, in this case consisting mostly of nectar and pollen.

Such delectable piles of nectar and pollen, on the other hand, will attract quite a few predators and scavengers. For that reason there has been an evolutionary trend away from the habit of leaving one big food

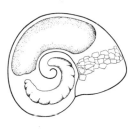

Fig. 30-2. Brood chambers of solitary bees (*Osmia*) containing food supplies. Above: Ground hole in which the bee has laid out leaves of poppy blossoms; she has deposited pollen packets and her larva in these. Below: Larva and pollen supply sealed within a snail shell.

supply, and to start providing the young with several small portions over a period of time. This trend constituted a transition from elementary care of the brood to actually raising the young (see Chapter 31). The final result of this development is the direct feeding of young, in which the parent ceases to create food supplies altogether and instead delivers the food bite by bite directly to its offspring.

Such behaviors in relation to raising young which serve the purpose of feeding occur frequently in the animal kingdom. There are spider mothers, for instance, which regurgitate a small food droplet to be passed on to the young (see Color plate, p. 457). Burying beetles (*Necrophorus*) do this, too, while some cockroaches feed their young with substances out of the hind-gut, mainly symbiotic organisms (microscopic unicellular animals that live in the intestines and prepare hard-to-digest matter for digestion). In many cases, young will simply partake of the food that their parents are eating. In some scorpion species, the offspring eat virtually no more than the "crumbs" falling from their parent's food. Most of the time, however, the parents transport food to the young, with birds being a familiar example. Many songbirds carry flies, caterpillars, or earthworms in their beak and shove them down their nestlings's throats (see Chapter 31), while other avian species, such as herons, first swallow the morsels or prey, then fly back to their nest and regurgitate the food for the young. In this way, food can be partially predigested and made more digestible to the young. Still other animal species have gone one step further: They actually produce their own food to be given to the offspring. In bees, for example, the larvae are fed during the first few days by secretions from a special gland formed in worker bees. The discus fish, too, feeds the young with substances from its own body: Glandular cells in the skin secrete a mucilage consumed by the young (see Color plate, p. 457). Pigeon nestlings are fed with a milklike substance. Their parents produce this "crop milk" with glands lying in the crop, regurgitating it to feed the young. And finally, we all know of that large animal group which has been named according to its method of feeding offspring: the mammals, whose mammary glands (converted from original sweat glands) produce milk for the suckling young (in German, mammals are called *Säugetiere*—suckling animals) (see Color plate, p. 464). In this way, many forms of relationships between parents and their growing offspring have developed throughout the animal kingdom by means of the feeding methods.

There is another area of brood-care functioning that has given rise to behaviors in relation to raising young during evolution of species. This is the function of guarding the eggs. Many spiders, insects, and other arthropods (articulate animals) remain near their eggs to ward off predators. In some species, the female stays to protect the young, even after they have hatched. For example, certain orb web spiders wrap their eggs into a carefully layered cocoon and then continue to guard them. When

Fig. 30-3. Female cuttlefish guarding her eggs, which she has attached in clusters to the roof of a cave.

the tiny offspring hatch, their mother provides aid by carefully opening the cocoon and letting them free. Sometimes the mother will even carry the young around on her back, as do also some scorpions and whip scorpions (see Color plate, p. 458). Sticklebacks also guard their nest. Once the young fish have hatched, the male will pursue any that may wander too far from the nest, take them in his mouth, and spit them back into the nest. Similarly, mammals often carry their little ones back into the nest or transport them to a safe place when danger threatens. Many species such as rodents, dogs, and cats have developed a special method for carrying their young: The mother uses her teeth to pick up the young animal by the skin of the nape, without hurting it (see Color plate, p. 458). The best protection for offspring is provided by some mouth-breeding fish. As indicated by the description, these fish carry their eggs within the mouth until hatching, and even then the offspring are taken back into the mouth between their exploratory swims, or return on their own into the dark opening of the parent's mouth when in danger (see Color plate, p. 458).

Parental protection serves not only to guard against enemies, but also against adverse environmental influences like unfavorable climatic conditions. Female earwigs watch out that their eggs, deposited in cracks in the earth, do not dry out, and if necessary will carry each of them individually to a damper spot. If necessary, male sticklebacks provide the eggs in their nest with fresh, oxygen-rich water by positioning themselves over the nest and fanning the water with their pectoral fins. Warm-blooded animals (birds and mammals) must also ensure that their eggs or young are kept warm. They accomplish this with carefully constructed nests and with their own body heat. The giant penguins of Antarctica take particular care to protect their eggs or young from snow and ice at all times by keeping them warm in a special abdominal fold. Should a young bird be separated from its parents at any time, several adults without chicks of their own will immediately come to the rescue.

Many parents even protect their offspring from contamination and therefore from the danger of becoming sick. This can already be observed in some arthropods, where the mother frequently rearranges her egg pile and discards unfertilized eggs or those covered with fungus, by eating them. We may observe similar behaviors in a number of brood-caring fish. With many birds we even speak of nest hygiene. The parents of these species remove the feces from their nest. The nestlings contribute to this "house cleaning" by conspicuously sticking their lower abdomen in the air and spreading their feathers in this area. The clump of feces which is then voided is covered with a thin membrane, so that the parents can pick it off and carry it away (see Color plate, p. 464). Almost all mammals clean their young by licking, and newborn are freed of any embryonic membranes. The parents will also stimulate their young to urinate and defecate by licking the offspring's anal region.

Fig. 30-4. Giant penguin with a chick in its abdominal fold.

And so there is a variety of behavior patterns by which numerous animal groups provide for the physical well-being of their young. But beyond that, the actions of raising young can be extended to include the "mental development" of offspring, wherein the parents convey information and pass on experiences. This occurs rarely outside of the vertebrate group, with some species of butterfly among the few exceptions. In these cases, the caterpillars grow up eating only the leaves of those plants in which the mother has deposited her eggs. If we attach the unhatched eggs to a different suitable plant, the hatched caterpillars will feed only on these "false" leaves and refuse to eat the original ones. In many avian and mammalian species, parents convey information about which food is suitable and beneficial, albeit much more "directly." Throughout their lives, rats and also northern ground squirrels, for example, will continue to prefer those kinds of food which they received from their parents. Cats learn from their parents how to catch prey by going hunting with them and imitating the movements of seizing and biting prey in the back of the neck. And finally, anthropoid apes raise their children in a way highly reminiscent of humans. Not only do young apes learn a great deal by imitation, they are also encouraged to practice their climbing skills and discouraged from doing things that are dangerous. Lastly, there is learning through imprinting processes, which are also tied to parent-young relationships. Examples of this kind of learning are song acquisition in zebra finches, and the recognition of the mother or the future mate in greylag geese and wax bills (see also Chapter 24).

We could list many more examples. There is almost no animal group without at least some species that perform special brood-care behaviors. There are even groups—mostly highly developed ones like the Hymenoptera, birds, and mammals—whose members raise their young without exception. We can hardly comprehend the variety of behavior patterns occurring in this context, from actions designed to divert enemies away from the brood, to the artful construction of complicated nests. These amazing skills are not based on anticipation or deliberation on the part of individuals, but almost always represent innate, instinctive action patterns. They are discharged in response to certain releasing stimuli, usually in combination with some corresponding motivation. Long action sequences may in this way be linked together as a series of instinct patterns enacted by the animal without conscious deliberation or insight. This may be demonstrated quite well in the sand wasp. When a female sand wasp carries her paralyzed prey to the underground passage, she deposits the victim beside the entrance, then crawls into the hole and examines it. Next, the wasp comes back out to drag her prey into the passage and, finally, she covers the hole. If during this procedure we move the caterpillar a little to the side, the wasp looks for it and once again deposits the victim at the entrance. Now the whole action chain is re-enacted from the beginning, and the wasp examines her nest all over again. This procedure is repeated to exhaustion every time the prey is moved. In animals at a higher level of evolutionary development, these sequences are often less rigid, although they still require certain releasers to set their corresponding instinctive movements in motion. One example of this process is the signals transmitted by songbird nestlings in the form of conspicuous throat markings. Another instance is that of newly hatched jays which respond to a shaking of their nest by raising their head straight up and gaping, pulling their beak wide open. The shining reddish color inside of the young bird's beaks induces their parents to give them food.

This signal is so compelling that cuckoos make use of it for their brood parasitism. Many cuckoo parents cannot perform real brood-care behaviors, and so the females deposit their eggs into a suitable nest belonging to another species. Here, the young cuckoos usually hatch before the offspring of their host, and they throw the host's young out of the nest. Then they proceed to open their throat up wide, sending a very effective signal that induces the "foster parents" to feed them at once (see Color plate, p. 463).

The young, as well, must be able to perform certain instinctive behaviors, and to recognize from the start some of the releasing stimuli coming from their parents. Gull chicks, for example, beg for food by picking at a red spot on the tip of their parent's beaks. Experiments with models have shown that the only releasing factors in this parental signal are the red color and a clear contrast with the background: the color of the rest of the beak, or even of the model's head, is unimportant.

As the offspring grow older, they learn to recognize their parents by other characteristics as well, just like the parents learn a great deal about the appearance of their children through day-to-day experience. These learning processes usually result in a relationship of personal recognition, based in birds on color or vocal signals, and in mammals, often on individual odors. A mother hen will not notice a strange chick placed with her brood during the first three days after hatching. After that period, however, she has learned to distinguish the down color of her own chicks well enough to notice and chase away any others. By contrast, a mother sea lion recognizes her young by its smell, and drives off strange ones attempting to suckle with her. Generally, there is an overlap of responses to innate releasers and of acquired trait recognition. This may lead to strange contradictions in behavior. For instance, one mother duck rushed to the rescue of a duckling calling in distress and drove off the enemy; then she saw that the very chick she had just saved was not her own, and she pecked at it.

While the release of certain behaviors is necessary, other action patterns and responses must be inhibited when parents are raising young. Members of predatory species run the danger of mistaking their young for prey. For that reason, some brood-caring spiders do not eat at all while guarding their eggs or young. Birds of prey are inhibited in their prey-catching activities only in the vicinity of their nest, enabling small songbirds to build their own nests right under the "inviolable precinct" or refuge of the predator's domicile. Other behaviors, too, are subject to inhibition. Pheasant hens, for example, do not take to the trees at night as long as their chicks are unable to fly.

In many of these cases, not only releasing stimuli are important; there must also be a certain kind of motivation. Especially in the context of raising young, motivational variables are predominantly under the influence of hormones. Thus, even the behaviors of nest building are often influenced exclusively by hormonal variables, as demonstrated in experiments with mice. (Further examples are given in Chapter 21.)

In the course of evolution, then, the pressure of natural selection has always given rise to that system suited to each species for ensuring the production and survival of a sufficient number of offspring. In accordance with their individual demands, members of different species display a great variety of behaviors in relation to raising young. Differences are also found in the division of labor between the mother and the father. In some cichlids, both parents share equal time in caring for their brood, and both display the same behaviors in guiding the young; this is a so-called parent family. Birds of prey, on the other hand, already show the begging of a division of labor: The male usually catches prey and brings it to the nest, while the female tears off small pieces from the booty and gives them to the young; this is called a mother-father family. Many predatory animals have taken another step: The male still defends the family's territory, but he no longer maintains direct contact with his offspring; this is a male-mother family. The ultimate stage of this development is the mother family, where the male takes part only in the mating process and leaves all brood-care behaviors to the female (for example, ruffs), and conversely, the father family, where care of young is left exclusively to the male—who will not let the female come near, as with sticklebacks and emus.

In addition to the many direct advantages that the development of brood-care and brood-raising behaviors will hold for an animal species, there is the indirect advantage that such parent-child relationships can lead to the formation of real societies.

31 The Evolution of Social Behavior in Animals

The evolution of social behavior in animals, by H. Markl

Organisms without any social tendencies are very rare. We know of few animals that avoid all contact with others of their species, or drive them away whenever they approach. Even these species, however, are forced to seek or tolerate such contact at some time in order to breed. By contrast, a large number of species belonging to the most diverse evolutionary lines and stages are to be found in groups or even densely clustered droves—an oyster bank, swarms of grasshoppers, schools of herrings, flocks of starlings, the brood colonies of gulls, or herds of ante-lopes, to give only a few examples. When we try to discover how such aggregations form and then stay together, we find that certain environmental conditions are usually responsible, allowing many individuals to do certain things at a certain place where highly favorable conditions prevail. Perhaps the food sources are very good, the young are well protected, the gathering of many gives each more security against attacks by predators, or the sexes are brought together for reproduction while responding (innately or through experience) to specific signs in that particular environment.

Evolutionary stages of group behavior

We can say, then, that such aggregations basically arise because each animal responds to environmental stimuli in the same way, and because conditions allow the survival of many in the same area. Naturally, group cohesion may be reinforced secondarily by the way members react to each other, so that, for example, the group as a whole takes action against a predator. In most cases, however, the size of an aggregation is determined primarily by the availability of space and food. As long as there is sufficient food, new members may join the group without friction, and in principle all members can be replaced by other conspecifics. Thus, individuals within an aggregation do not recognize each other as belonging to the same community; there is no "membership card," even though certain individuals may know each other personally. On the whole, members stay together because they are all attracted to the same place or to certain external conditions, forming an anonymous association just like people going to a theater or shopping in the same store.

This form of group behavior is very common among animals. In some cases, however, it gave rise to a qualitatively different kind of social life, on a higher level of social interaction. This chapter will describe the basic features of such "personal" associations, whose organization is found in groups where two or more adult animals and their offspring live together on a consistent basis and tend to be "exclusive." Members of these groups recognize each other and distinguish between fellow members and outsiders, who are more or less rigorously excluded from taking part in group activities. Such higher societies therefore display a relatively closed structure, making it difficult, if not impossible, for new members to join. In almost all cases, such groups also involve a certain task differentiation, which may be described as rank relations (dominance hierarchies) or, in other groups, as a division of labor. These systems of differentiation may best be seen structurally as a distribution of roles. Looking at this kind of behavior from an evolutionary point of view, the most important aspect is the division of roles for breeding purposes. For example, certain higher-ranking individuals may have better chances or even exclusive rights for reproductive mating. Furthermore, at least those animals on a very high level of social development would not, under natural conditions, even be able to survive; this applies to individuals as well as to the species as a whole. The interactions within such groups then form a vital basis, not only for the survival of its members, but also for the normal and healthy development of young within the context of their social environment.

Within this kind of highly developed animal society, we find the origins of human beings and our own social system. In fact, scientists are firmly convinced that the very traits we call human could develop in their basic form only with the conditions of life in such closed societies. This means that any research into the evolution of higher forms of social behavior takes on a special significance for man's own attempts to understand himself. This applies even when we study animal societies further removed from us on the phylogenetic scale, because the solutions applied to various problems of social life have been basically similar across species and are to be found everywhere. In the same way that very different species have evolved organs for similar functions to meet the demands of similar environmental conditions—the use of eye structures is an example —the forms of organization in social life also represent adaptations to the requirements of a species's survival. Ethologists may critically compare these organizational principles across species, a useful exercise even if it should only sharpen our understanding of the way different animal groups sometimes solve the same problems in rather different ways.

This "higher" social life is found among insects, especially in many Hymenoptera and in termites, and also among vertebrates in certain birds and mammals. In addition to the social groups of many rodents, lagomorphs, carnivores, and ungulates, the primates deserve our special attention, since our ancestors evolved within this order. On the other hand, we shall restrict our descriptions mainly to the state-forming insects, because Chapters 34 and 35 provide a detailed discussion of basic features in the social life of vertebrates. This provides us with the challenging opportunity to show how—contrary to popular opinion—even the study of social evolution in lower animals can yield valuable information for an understanding of social behavior in higher animals and man.

First of all, let us take a look at the nonsocial behaviors that may have provided roots for the growth of a higher social life. As in any other evolutionary process, we can assume that new developments did not take place by sudden leaps and bounds, but rather through small changes and stages of adaptation. Fundamentally, we may distinguish three major kinds of relationships that determine behavior between conspecifics: aggression, sexual attraction, and parent-offspring interactions.

Conspecifics will frequently avoid each other, or attempt to drive each other away. When this behavior is combined with defense of a certain area, we speak of a territory (see Chapter 25). The motivation underlying such repulsion of others is called aggressiveness. All behavior designed to inhibit conspecifics, that is, to restrict their development and especially reproductive activity, can be described as aggressive. Within this context, we would also consider mutual avoidance or any self-imposed restrictions as the result of internalized aggressive influences. There is good reason for explicating the concept of aggression by using general and neutral terms like rejection and inhibition, because we want to divert the emphasis away from the "brutal" and damaging kind of combat aggression—a relatively infrequent occurrence among animals—to the overall function of aggression: preventing a rival from gaining access to a food source, a mate, or a place of shelter. In other words, aggressive behavior does not necessarily lead to injury beyond this basic inhibitory effect. Aggression plays an important role even in the behavior of social animals (see Chapter 25); but, since it is a negative force leading to repulsion and avoidance, this tendency cannot have been a factor in facilitating life in groups.

Even in animals that behave aggressively toward their conspecifics, this tendency cannot always dominate: During breeding it is essential that the partners are attracted to each other. Sex can therefore be a cohesive force between members of a species. On the other hand, it is quite evident that higher forms of social behavior cannot have evolved primarily from extensions of sexual behavior. Sexual attraction may well have provided an additional bond by inhibiting aggression; yet in a peculiar way it is tied closely to aggressiveness against other conspecifics, since at least during copulation there is no room at all for third parties. Any intruders are driven off, frequently by both partners. Competition for opportunities to mate and reproduce therefore has a disruptive effect on larger societies, so that sexual relations between conspecifics must also be excluded as a possible starting point for higher forms of social behavior.

We are therefore looking for a source of attraction between conspecifics that is based on the relationship of more than two individuals, yet does not extend anonymously to all others of the species but rather discriminates clearly between those that belong and those that don't. This kind of relationship is found between parents and their offspring, as well as between the young growing up together. We can, in fact, demonstrate that throughout the entire animal kingdom, higher forms of social behavior have always developed from the brood-care relationships between members of a family. It should therefore be no accident that particularly mammals, where brood care (or the raising of young) is so basic that the entire class was named with reference to this behavior (note also the German word for mammal, *Säugetiere*—animals that suckle their young), and Hymenoptera, who also spend a great deal of time providing for their offspring, show such a strong "propensity" for evolving higher social orders.

But what is the process by which this bond between members of a family, leading to mutual support and cooperation, comes to be applied to larger groups as well? Could we not argue that each individual, according to the rules governing evolutionary processes, should be concerned primarily with its own reproduction? Any animal working together with conspecifics in a social group would also be benefiting the offspring of these others in everything it did; the individual living in society behaves "altruistically" toward its fellows, without regard for its own best interest. What are the benefits of such altruism in social animals? In other words, what is the selective advantage for such traits to evolve?

In order to understand this particular issue, we must delve a little into the language and subject matter of population genetics. This area is concerned with the combined genetic potential of a group of interbreeding individuals, called a population, and the changes resulting from mutation, selection, and other factors. Since changes in the genetic potential of a population are defined as evolution, this particular field of study provides the exact basis for research into evolutionary processes. Within the context of population genetics, we speak of altruism when an animal facilitates the reproductive chances of a conspecific at the expense of its own. This occurs whenever members of a group help each other and, by doing so, disregard their own welfare. This altruism may involve anything from sharing food when food is scarce to defending others at the risk of the animal's own life. It may also mean that the animal relinquishes its own chances for reproduction—quite possibly without choice—in favor of those of others in the group. We can easily see how this tendency benefits all the other members. On the other hand, how can genetic traits favoring altruistic behavior survive and become dominant in a population, when these very traits discourage the reproduction of their bearers? W. D. Hamilton demonstrated, with the help of mathematical models, that such altruistic tendencies can be preserved, in the course of evolution, only within the family unit. This is because in order for these genetic traits to

Population-genetic factors in the evolution of "altruistic" behavior

▷
Chemical signals in the feeding behavior of insects. Above: A wasp is licking a sweet substance (a drop of honey or sugar water). Middle: The trilling of an ant induces aphids to secrete a drop of honeydew. Below: Adult staphylinids (here *Lomechusa strumosa*) imitate the food-begging signal of worker ants. In response, one of the worker ants feeds the "guest" from the front, while the symbiont secretes a "pacifying" substance from the back. This substance is lapped up by another worker ant.

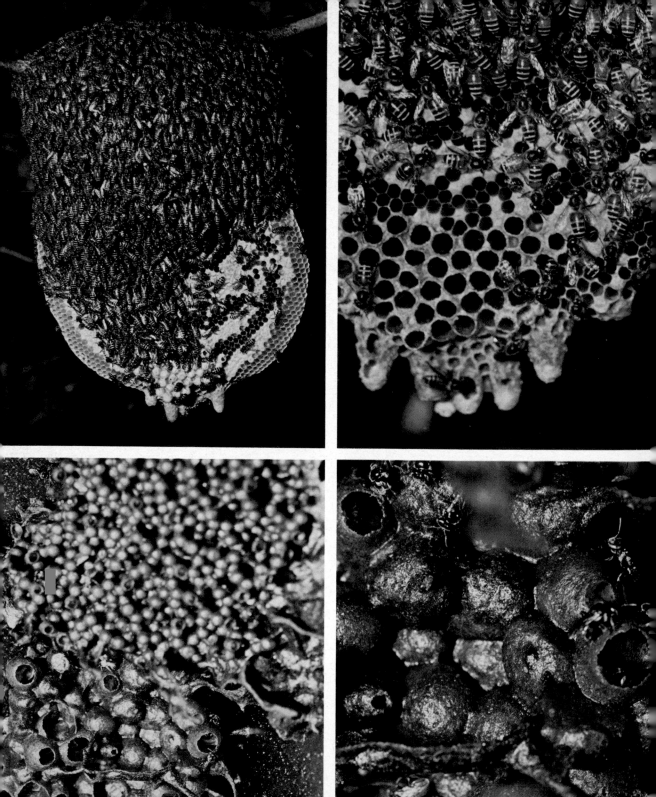

Top left: A colony of dwarf honey bees (*Apis florea*). These build only one comb, attached on a branch in the open. At the bottom the experimenters have brushed the bees aside a little, to expose a section of the comb. Top right: Enlarged view of the same comb. Note the small cells, containing broods of worker bees, underneath the larger drone cells, and several queen bee cells at the edge of the comb. Bottom left: Nest of stingless bees, here *Trigona iridipennis*. Above, the clump-shaped structure of passageways; below, the storage "bins" containing pollen and nectar. Bottom right: Enlarged section with some of the workers.

Drawing conclusions from the population-genetic theory of higher social development

reproduce and multiply, the "sacrifice" of an animal's own reproductive opportunities in favor of another's must be more than compensated by the effect that the very same kind of genes will now be reproduced in a greater proportion. The vehicle for this effect would be the increased reproductive chances of the beneficiary. However, the probability that these same genetic traits are present in another member of the species is sufficiently high only with relatives. Quite obviously, this fact also offers an explanation for the phylogenetic development of parental readiness to act altruistically on behalf of their offspring.

This state of affairs can indeed be quantified in exact numerical relations. Every individual shares a mean characteristic portion of genetic material with each of its relatives, depending on degree of relatedness. We call this the inheritence coefficient or relatedness coefficient (r). For example, a father shares with his children on the average one-half of his genetic traits (r = 1/2), since the other half is passed on to the children through their mother. Between grandparents and grandchildren, or between cousins, the coefficient is 1/4. From the standpoint of population genetics, when an individual completely sacrifices its own chances for reproduction in favor of relatives, it would do so by working for these others and feeding them, for example, or by losing its life in defending them. The genetic traits causing such altruistic behavior would therefore survive in the population only if the reproductive chances of beneficiaries were to be more than doubled, in the case of siblings, or more than quadrupled in the case of cousins. The more distant the genetic relationship, the less "profitable" the sacrifice. An animal's own reduction in reproductive opportunities by ten percent would have to be compensated by an increase of more than twenty percent in the chances of its siblings, if these are the beneficiaries. This intriguingly simple theory can be regarded as the population-genetic basis for the fact that higher, altruistic social behavior can arise only within the context of family units. The theory holds very important implications for the way such social groups are organized.

The first implication is that, in societies having evolved according to these principles, each individual should display a decreasing readiness for supporting and defending another at the expense of its own interests when the degree of relatedness itself decreases. This is clearly the case, for example, in state-forming Hymenoptera. Some evidence to this effect has also been offered for mammals, although investigations of these kinds of relations are very difficult to perform in the wild. In human societies, this decreasing willingness to sacrifice one's own interests for others depending on kinship is frequently taken for granted, as may be evidenced by some of our legal codes. So-called "primitive" societies in particular tend to place a heavy emphasis on kinship relations, which are often spelled out in the most minute details and which play an important role in the regulation of social life.

If higher forms of social behavior arise from brood-care relationships, we may derive a second consequence: Whenever only the females of a species take care of their young—and this is the rule—we find a problem of how males fit into the evolution of social life, since if males play no direct or indirect part in caring for the young or guarding them, they are not only dispensable—with the exception of the mating act—but even a burden to the household of any population, where they are simply useless hangers-on. We will return to this point and discuss how it affects the distribution of roles as well as that of the sexes.

If the degree of relatedness between individuals is to have any consequence for certain kinds of behavior readiness, members of a family group must recognize each other as such. This familiarity is an essential determinant of communication in a social group. We can suggest five main principles that enable group members to know each other:

a) Being tied to a stable, common area (place loyalty), which members recognize by certain traits possibly created by their own activity (e.g., scent markings), and where the animals can leave their dependent and less mobile young, may provide a basis for group cohesion and mutual recognition. Wasps (Polistinae), for example, recognize others from the nest by the ease and certainty with which these conspecifics fly toward the nesting place.

b) A frequent device for mutual recognition—found in ant and bee colonies and up to mammalian groups—is the use of scent signals, or chemical messengers, that act as "membership cards" (see Chapter 32).

c) Sound signals are very specific and easy to modify. At present we have only a few indications that such signals are used as group characteristics in animals; among humans, however, the many dialects and languages are an absolutely fundamental feature in recognizing members of particular social units.

d) Similarly, visual signals—like decoration and dress—will serve as distinct features in characterizing human groups.

e) Superimposed upon all these specific traits may be the factor of personal and mutual recognition, where individuals simply know each other—again by olfactory, visual, auditory, or other traits. We know this to be the case in higher bird and mammal groups. Individual recognition may also be a precondition for recognizing the degree of relatedness beyond simply knowing the group members, with corresponding differences in behavior.

Group members have to recognize each other as such, then, in order to share in the benefits of altruistic social behavior. On the other hand, altruistic behavior must not be extended to outsiders if the respective genes are to survive. Thus, foreigners to the group are to be noted—usually by their lack of membership traits—and excluded from the benefits of altruistic support. This leads us back to intraspecific aggression: In order for higher social behavior to develop within a group in the first place, rejection between groups must continue, complemented by attraction and a willingness for mutual support among group members. In this way, discriminating against outsiders lies at the very root of altruistic behavior. Repulsion between groups may be manifested in brutal fighting—that is, destroying an outside intruder—such as we observe in many social insects. But this does not necessarily happen, nor do such drastic measures have to occur in territorial groups: Animals can accomplish the same purpose by engaging in fully ritualized aggression, where signals are exchanged from a distance and the groups then avoid each other. We can find this form of aggressive behavior in many primates living in the wild, for example, where more than one group is able to live in the same area. Groups like these almost never attack each other; as a rule, they avoid each other and will show their readiness for aggression only by means of "threatening" displays. We find nothing in the evolution of social behavior to force groups into damaging each other by fighting. On the contrary, we can see a definite selective advantage in the development of behavioral mechanisms that allow animals to decide on rights and privileges without brutal combat. In this way, a number of groups may coexist even in very tight spaces—like hamadryas baboons on their sleeping cliffs (see Color plate, p. 531)—providing that altruistic behavior and in particular brood care is more or less limited to closely related members of the family, as is the case in hamadryas males who each provide only for their own females and offspring.

The meaning of intergroup aggression for the evolution of higher social behavior would certainly throw light on the biological basis of many a human reaction pattern. But in every case we must critically examine to what extent results obtained with studies on animals can also be applied to humans. Should anyone wish to use these findings for explaining or even justifying war, for example, it would be a grotesque, thoroughly unscientific, and even dangerous abuse of otherwise genuine research efforts. What we are aiming for, on the contrary, is an objective study of whatever influence biological dispositions may still have on human behavior.

On close examination, we find two aspects to the discrimination against outsiders: Members of a group reject strangers, but they also act aggressively toward one of their number who does not behave according to the group's norms or who no longer shares the group characteristics, such as appearance or scent, because these traits were eliminated through illness. One of many such examples is the repulsion of albinos. Again, this aggressive behavior of social animals toward nonconforming members of the group, observed in a countless number of cases, actually lies at the root of social evolutionary development. Propaganda efforts aimed at strengthening group ties easily make use of this phenomenon, by encouraging hostility against "outsiders." The issue is a highly significant one for human psychology and sociology. Again, however, we must emphasize that a certain behavior readiness—especially in the primates, who possess an exceptional ability to learn, and in particular in humans—does not imply that the animal necessarily behaves in a certain way. We are never forced to act by our dispositions, but recognizing biological tendencies opens the way to rational control.

Aggressive behavior can also play a fundamental part within the social group itself: Competition between members may result in a distribution of roles. In this way, the capacity of an individual for performing certain tasks will be used to maximum efficiency, providing the best organization for distributing tasks and making life within a social group far more advantageous than solitary existence. Intragroup aggression can take a variety of forms. We know of insects among which a few members in the group turn others into slaves by some chemical means (see Chapter 32). On the other hand, in many primates we find that playful altercations between growing youngsters influence role division in a decisive way, for example, the dominance hierarchy and the corresponding evaluation that an animal will make of its own abilities and those of other group members (see Chapter 35). This form of social organization constitutes part of the beginnings of communication mechanisms within a society, because there are many processes of information exchange that accompany the determination, recognition, and acceptance of roles.

Part of the communications repertoire of a social unit therefore arises through ritualized expressions of aggression and subordination. The creation of a rank order provides the higher-status animals with better chances for reproduction, but it also implies that the subordinate elements in the group "sacrifice" themselves from a population-genetic point of view by contributing to the more favorable development of young sired by the higher-ranking members. We should note, however, that this "altruism" is in no way voluntary; the subordinate members of the group simply have no choice.

Finally, according to our theory, a third source for mechanisms of communication must lie in the parent-child interactions during the rearing period, assuming that all truly social behavior does represent an extension of family relationships. This last conclusion, too, has been repeatedly confirmed through observation of social animals. Insects most extensively use the communication processes associated with feeding—that is, offering food, indicating a prey, begging for food, and so on—for their social interactions. We can observe similar behavior in vertebrates; for example, mouth-to-mouth feeding, which probably gave rise to the act of kissing. In fur-bearing mammals, an important part is also played by forms of communication arising from social grooming (mutual grooming and "delousing") and from the exchange of signals during courtship and mating. But we also know of insects among which signals associated with sexual behavior have acquired an extended meaning: The queen bee, for example, makes use of similar odorous substances, or secretions, when going on her

Fig. 31-1. Various mechanisms of social communication that may be derived from brood-care behavior:

a) A termite worker feeds a large soldier.

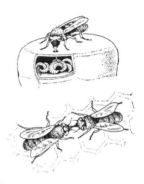

b) Above: A bumblebee feeding larvae. Below: Two worker bees feeding each other.

c) Mouth-to-mouth feeding and greeting kiss in chimpanzees.

d) Male savannah baboon showing brood-care behavior by grooming an infant.

e) Mutual "delousing" between two adult female baboons.

f) Papua mother feeding her child mouth-to-mouth.

mating flight and when interacting with the worker bees. All these mechanisms that facilitate contact between members of a group are extremely well suited for reinforcing bonds within the extended family group, the social unit. Such ties may also be strengthened by the deliberate display of signals characterizing the group, as with animals using scent markings and, similarly, with humans singing together as a group or performing dance rituals.

We would assume from this population-genetic model that individuals sharing genetic traits support each other altruistically, without reservations. In extreme cases, we see this happening in animal colonies where all the individuals, or elements, come into being through asexual reproduction from one founding animal.

In order to test whether our assumptions about the evolution of social behavior are accurate, we would have to describe what we know about social life and the social evolution of a great variety of highly developed animal groups. Higher forms of social behavior have evolved many times and independently of one another: among insects they developed in termites and in at least twelve or thirteen kinds of Hymenoptera, and among vertebrates in birds and in mammals, where we can see a particular variety of social forms. We shall give some examples to show how this higher kind of social behavior developed without exception from the brood-care (that is, rearing) behavior of nonsocial ancestors.

State-forming Hymenoptera are especially useful for explaining evolutionary processes in social behavior, not only because of their diversity and developmental level, but also because among them we find ample support for the population-genetic theory of social evolution.

All Hymenoptera—even the solitary species—provide for their brood. In the most primitive case, the parent will simply deposit its eggs where the larvae would be most likely to find suitable food, while more complicated methods involve the building of special mounds, underground structures, and so on where paralyzed prey animals or a pollen-nectar mixture will be deposited specifically for the larvae. Another method is to spare one's energy, and instead of collecting food reserves and building structures, smuggle the brood as parasites into another species. In various branches of brood-care development, we also find a new principle emerging for the protection of young: Many females care for their common brood or for the offspring of the "queen." In the latter case the female "servants" or laborers are what, among insects, we classify as workers. How did this development come about? In wasps and bees we can find so many intermediate stages, from sexually mature females to workers no longer active in reproduction, that we are able to reconstruct the process quite accurately.

It starts with one female continuously decreasing the number of its young and providing better and better care for the fewer offspring. One key point in this development is probably the degree of protection against brood parasites, mentioned above. The selection pressure exerted by the presence of these enemies has caused the sand wasplike hymenoptera, the ancestors of social wasps and bees, to develop ever-more-extravagant protective measures, with the result that a lone female could only care for relatively few offspring. On the other hand, this process also meant that a larva would not get its entire food reserves packed into its cell from the start, but that now it would be fed at different times during ontogeny. This step is extremely significant, because it means that a female will now maintain contact with its growing young. It enables the female to care for several larvae at the same time, especially if their cells are constructed close to each other, and also to protect them better. Even more protection is afforded if the female deposits an egg and does not feed the larva until they hatch, as, for example, in members of *Stictia* or *Microbembex*. And finally, the meat-providing ancestors of the social Vespidae changed from giving larvae the entire prey animal to offering morsels chewed up to bite size. This meant that food would no longer be lying around to attract brood parasites. The development of brood care in bees followed a similar path, except that here the larvae were made to adopt the same food as adults (pollen and nectar).

As the mother started to care continuously for more than one offspring at a time, it became inevitable that new adults would hatch while the mother was still expanding the nest and servicing other larvae. We know that many solitary bees habitually build their own nest in the vicinity of the old one—a sign of place loyalty in these animals, whose way of life requires that they always return to the same small area after their foraging trips. In the burrowing bees *Lasioglossum inconspicuum*, the nest entrances of different females have underground connections, although each female provides for her own brood. When several females tolerate each other—because they are related in the closest way possible—and jointly use one system of nests, the safety of the brood is greatly increased, since the females alternately stay behind to guard. In nests of *Pseudagapostemon divaricatus* there is only one entrance, guarded alternately by several fertile females of equal status. If it now happens that some females concentrate more on foraging and others more on guarding and laying eggs, we have a transition to "altruistic" reproductive behavior within a group, and therefore to a division of labor. The way roles are distributed results primarily from the fact that the first-born daughters of one mother are less fertile, due to inadequate food reserves. They are "spayed," so to speak, by malnutrition. This represents a special kind of aggression, albeit totally unintended.

The next stage may well be represented in *Augochloropsis sparsilis*: A few of the workers never mate and are therefore eliminated from reproducing female offspring. The ultimate "spaying" occurs in the worker females of the burrowing bee *Evylaeus malachurus* and bumblebees (*Bombini*); in *Evylaeus marginatus*, colonies survive for over five years because of the queen's long life-span. Colonies of the stingless bees, honeybees, and some wasps of the subfamily Polybiinae virtually go on forever, because they rear fertile members that return to the nest after mating; the colonies multiply by means of dividing the workers into new groups among various queens.

Some researchers have suggested that the primitive stages of social behavior in bees would also have emerged through unrelated females moving together into a brood colony, since advantages would be the same. Quite probably, however, any animals nesting together in such harmony would in fact be related to each other. Furthermore, when looking at the ontogenetic growth of all higher Hymenoptera states, we find that they all developed from exclusive family associations.

What we have said about bees also applies in a similar way to Vespidae, where even in the higher species each colony owes its existence to one founding female. The burrowing Polistinae wasps are extremely interesting. Several females often work together to start a nest. But soon this cooperative attitude changes to serious rivalry: One female, which may have had more fertile ovaries from the beginning, becomes dominant and aggressively prevents the others from reproducing, for example, by eating their eggs. The first daughters to hatch are undernourished and therefore workers. Now the "alpha" female will probably drive all her original helpmates out of the nest. It is quite possible, by the way, that these seemingly helpful colleagues were really a kind of brood parasites whose original aim was not to help each other but to obtain a nest they were incapable of founding on their own. Their "altruism" would probably be worthwhile from a population-genetic point of view, since these females are likely to be sisters that come together for a new nest near the one where they were born.

We know fairly little about the development of social behavior in ants, but even here we can say that all colonies of primitive ant species are founded by a single female. This makes the colony an exclusive "family business," so to speak.

All Hymenoptera states are rigidly closed units. Members often recognize each other by the colony odor, and they are merciless in their aggression against members of other groups. We can see in the burrowing wasps (*Polistes*) that even within a social unit, aggressive behavior may be effective in a direct way: It leads to a division of labor between the alpha female—often the only one to lay eggs and the one who also receives the best food—and the foraging "lower charges." We find the most ingenious solution to this problem among the highest-evolved Hymenoptera: The queen of a honeybee colony produces a chemical substance (see Chapter 32) serving a number of purposes—sterilizing the worker bees (an extremely aggressive act, from a population-genetic point of view), preventing them from constructing queen cells that would produce rivals to the queen, and giving the queen a sexual attractant to lure the males on her mating flight. This drug is eagerly accepted by the worker females; it makes the rounds of the entire colony, and even inhibits aggression among the workers themselves (see Chapter 32). This transfer of chemical information from one animal to another, often combined with giving food (trophallaxis), is a fundamental form of communication generally found in insect states.

W. D. Hamilton has suggested an explanation of why the Hymenoptera, among all others, so frequently followed a social development: The mechanism for determining sex in these animals is producing males from unfertilized eggs and females from fertilized ones. As a result, the daughters of a pair are more similar to each other (inheritance coefficient or relatedness coefficient, $r = 3/4$) than are the siblings of parents both possessing a double complement of chromosomes. If we are correct in assuming that altruistic behavior among animals is all the more likely the closer they are genetically related, then the tendency of Hymenoptera to engage in social life is easier to understand: A female who becomes a worker will promote her own genetic potential more effectively by caring for her sisters ($r = 3/4$) than providing for her own descendants ($r = 1/2$). This requires, of course, that alternate sibling generations can learn to know each other in the first place and stay together in their nest. Hymenoptera make this possible through their extended brood care. In other arthropods where sex is determined in a similar way, but which do not provide this context of elaborately drawnout brood-care behavior which enables siblings of different generations to recognize each other, no social organization has developed.

The fact that males take virtually no part in caring for the brood of a colony may be seen as a logical inference of this theory, since the genetic traits they share with their descendants are in a mean ratio of only 1/4 to 1/2. Furthermore, the inheritance or relatedness coefficient between aunts and nieces is only 3/8, less than between sisters (3/4) or between mother and daughter (1/2). This corresponds to our observation that the typical Hymenoptera society has only one queen, and that queens act with extreme aggression to suppress the chances of daughters being hatched from other females, even from sisters. The fact that some groups do support more than one queen probably results from adaptation to special circumstances. In these cases we usually do not know whether all the queens have equal opportunities for reproduction, or whether the colony in fact consists of several "tribes" or subunits.

It would also follow from this hypothesis that workers foster their own sons ($r = 1/2$) over their brothers ($r = 1/4$); this fits the observation that in many Hymenoptera states, many or all the males hatch from eggs produced by the female workers. And finally, the females are less altruistic toward their brothers ($r = 1/4$) than toward their sisters ($r = 3/4$). We can observe that females in a bee colony, for example, restrict the occurrence, life span, and number of males to the utmost minimum. Since the males

▷
Above: Gnu herd in the Serengeti. Bottom left: Reindeer migrating at the North Cape. Bottom right: Blesboks in South Africa. The rump patch probably serves the purpose of intraspecific communication.

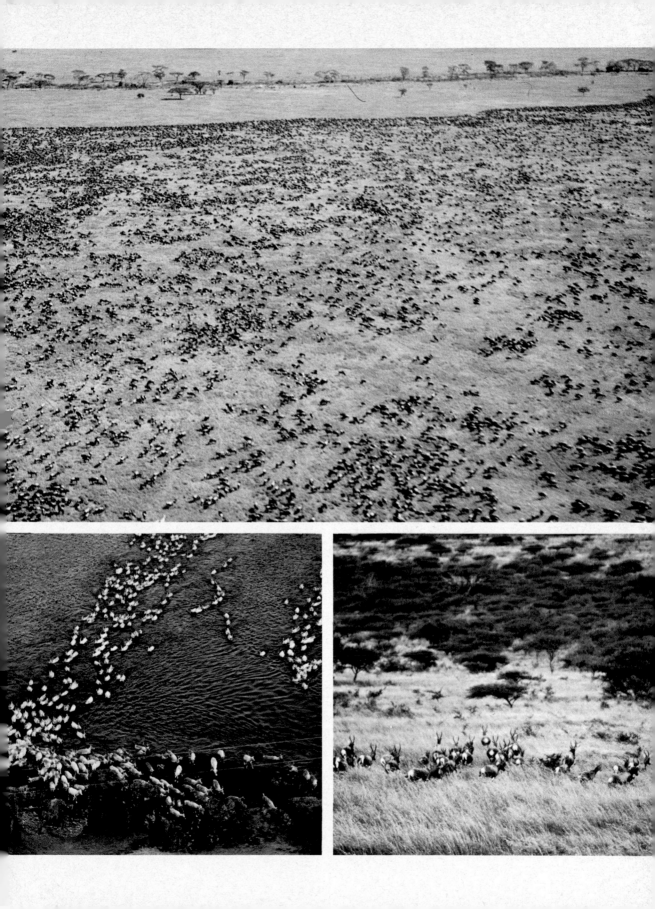

take no part in either brood care or protection of the colony—the females can do this much better with their poisonous sting—the only function left for them is mating; they are necessary for the reproduction of females. This forces them into a "sexual slavery," and once they have fulfilled their role they are driven from the nest, as in honeybees.

The evolutionary progression of social life in termites is much harder to unravel, because all species living today are social. Where colonies do not multiply by fission, as they do in honeybees, they are always established by one pair, a male and a female. The sex ratio of descendants is also fairly equal. The typical feature of a termite society is the fact that workers are not adults but less mature animals. These workers remain at their undeveloped stage due to the chemical influence of the adult reproductive members. Only very few may reach sexual maturity and reproduce at certain levels of juvenile development. The termite state, again, is one large family, but of a different kind: Here it is the young that are enslaved. Since both male and female larvae occupy the various castes (worker, pseudoworker, and soldier), termite societies have no "male problem."

How did this strange form of organization develop? Here, selection pressure was provided by the termites's mechanism for food consumption. All lower termites feed on plant material and digest it with the help of flagellates living symbiotically in an intestinal cavity. This chamber is emptied every time the termite sheds its skin. (Higher termites let fungus do the work of "predigesting" plant material.) The newly molted animals must immediately replace the flagellates, and do this by eating the dung of other colony members, which always contains healthy symbiotic animals. This explains the dependence of young on their parents, and the odd way that brood care is performed, "through the anus." Our next question is: Why do the worker classes consist of "inhibited" larvae? An indication may be found in the semisocial brown hooded cockroach (*Cryptocercus punctulatus*). This insect also digests its woody plant food by means of flagellates living in the intestine. But their symbionts reach a stage where they can be transferred to another cockroach only when the host animal is molting. This means that only larvae in their molting stages are able to provide the newly hatching brood with these vital symbiotic creatures. The younger larvae therefore need the older ones and not the adults. When these cockroaches found a new colony, they always have to carry with them larvae capable of molting; otherwise the colony would not survive. Should this process also have occurred in ancestors of termites, who were probably closely related to cockroaches, we would understand how larvae came to be used as a worker caste.

Among birds we find examples enough to show how population structure and social relations may depend on living conditions, especially on the availability of food, predator pressure, and the peculiarities of a species' reproductive habits. Many birds tend to brood in colonies situated at some suitable spot. The mere fact of living in such aggregations will

◁
Top left: Somali wild ass in the Danakil desert in Ethiopia. A mare with two foals. Top right: Grant's zebras in the Ngorongoro Crater. Family group, with a stallion at right. Middle left: Pride of lions in the Serengeti. Middle right: Herd of impalas, with a territorial buck to the right. Bottom left: Elephant mother family. Bottom right: Outside of the rutting season, red deer stags form bachelor groups.

provide the individual with certain advantages, like the watchfulness of many pairs of eyes. In most cases, however, these colonies are anonymous and open systems, though the nearest brooding neighbors will recognize each other˙ personally. Other colonies, such as those of jackdaws, are structured according to a dominance hierarchy where animals all know each other personally; and the family units of geese are truly social groups with a closed structure. Nesting communes are another kind of avian society, where several birds rear all their young together and sometimes even use a common nest. Such nesting groups may behave antagonistically toward other groups of the same species. We find the most interesting social relationships among birds like the ani (*Crotophaga ani*, a species of cuckoo). Here, fifteen to twenty-five birds, probably a genetically related clan, communally tend one nest and defend a certain area harboring the growing young of several females. Those birds that are first to hatch participate as adults in the care of younger siblings. Another remarkable feature among anis in a group is that they vigorously preen each other.

In birds, then, we find numerous tendencies to engage in a highly developed kind of social life. On the other hand, these animals did not give rise to any evolutionary line showing greater phylogenetic development; that is, there was no evolutionary "breakthrough." Yet their truly social tendencies can all be traced to brood-care relationships and family structure.

Mammals may form temporary or permanent anonymous associations —for example, a variety of pinnipeds or ungulates, whose herds often contain family groups—but, in addition, we can find all evolutionary stages from solitary life to the highest social organization. The social life of primates is an excellent example, of course, but our attention is also drawn to the social behavior of some rodents, lagomorphs, carnivores, and ungulates. Unfortunately, we know very little of the social life of whales, which does seem to show a high degree of complexity. In whales, high-ranking males often mark other members˙of their group, as well as paths, lurking places, or territorial borders, by using odorous secretions and/or urine.

Many species of rodents live alone or in mated pairs for variable lengths of time during the mating period. But some show a prominent and well-developed family structure. Beavers, for example, form families with the two parents and their numerous juvenile offspring. The clan organization of brown (or Norway) rats includes several generations. The tolerance and need for contact shown in these rats between members of one clan is starkly contrasted with their frequently strong aggression against outsiders. We can observe rank order relations within the group, but these are expressed by ritualized displays rather than by direct combat. The result is a fairly stable system of differential privileges applying to food, running paths, resting places, and mates. These clans are easily compared with the social groups of higher primates, not only in structure,

Fig. 31-2. Cross section of a termite hill. The king cell is located in the center. The fungi chambers are connected through narrow passageways, while the wider passages are built for ventilation. The narrow passageways leading out from the ground, right and left, are entrances and exits.

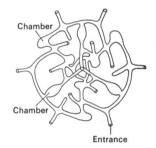

Fig. 31-3. Schematic drawing of a rat's burrow.

but also in their ecological adaptability. Our own competition with rats will bear witness to the hardy survival value of their social organization.

The highest level of social structure is to be found in the "towns" of prairie dogs (*Cynomys ludovicianus*), sometimes reaching a population of over 1000 animals. Probably the main reason why prairie dogs are less scattered over the globe than rats is that prairie dogs are tied to stable burrows. Wherever they build and concentrate, prairie dogs alter the vegetation so that it becomes more favorable to their food consumption. The social organization of these rodents was probably even more encouraged by the advantages of a communal alarm system, affording protection against their numerous predators. Spatial organization in a town seems almost human: The basic unit is the "coterie," harboring from two to thirty-five animals. Members usually include one alpha male, perhaps a subordinate male (probably a son of the dominant animal), two to three females, and immature young. In the spring, coteries with more than one male often split up in the wake of many fights, and the younger males usually move away to establish new coteries at the periphery of the town. On the other hand, the parents may move out during the latter part of the summer if the young have become too bothersome. The adults will then start a new burrow and leave the old established home to their growing offspring. In the closely related marmots, conditions are much the same, except that "towns" are not as common as families living in "isolated coteries."

Among lagomorphs, the most detailed studies have been made on social life in common rabbits (*Oryctolagus cuniculus*) imported to Australia, and in the American cottontail (*Sylvilagus floridanus*). At the start of their breeding season, three to five female rabbits—not more, because otherwise they would fight—form a breeding group and establish a burrow. They defend a territory around this common burrow against other females. A few males attach themselves to these groups, establishing a dominance hierarchy maintained through aggressive interactions, which serves mainly to regulate access to the females. Unfortunately we do not know very much about how females in a breeding group might be genetically related. Since there are always only a few mutually tolerant animals that form groups, these might well have learned to recognize each other from being together earlier—perhaps having grown up in the same warren? We would then find a social structure much like the lower burrowing bees. Cottontails, which do not dig burrows, are less territorial than the European common rabbits.

The social behavior of some carnivores shows an amazingly level of complexity. A good example are the canine species (Canidae). The simplest type of social organization here consists of one mated pair and their juvenile offspring, as in the fox; at the other end of the scale, wolves and the African wild dogs *Lycaon* live in large clans that probably always arose through actual family groups. With wolves, we can see a clear division

Fig. 31-4. Schematic drawing of a prairie dog "town."

Fig. 31-5. The food-begging behavior of young African wild dogs (*Lycaon*) (above) gave rise to the way adults greet each other by nudging the partner with their nose (below).

of labor in their co-operative hunting and in guarding the cubs. On the other hand, wild dogs do not have these role differentiations. Around their burrows they are strongly territorial, but various clans have overlapping hunting ranges sometimes as large as several hundred square kilometers. When different packs meet in these areas, they usually retreat from each other without fighting. Observations of the various interactions between these wild dogs provide additional support for the hypothesis that social behavior developed from relations in the brood-caring family: As in social insects, members of a wild dog pack exchange food all around; this includes returning hunters, cubs, and guards. This constant food exchange after a hunt serves to reinforce the social bonds in a group, and food-soliciting behavior (nudging with the nose and licking at the corners of the other's mouth) is clearly seen to have developed from the begging behavior of young into actual greeting rituals. Quite possibly, members of carnivore families were encouraged to stay together on a long-term basis because of the increased success afforded by cooperative hunting.

We refer the reader to Chapter 35 for a detailed account of social behavior in ungulates.

The behavior and social life of primates—the most interesting of mammals for humans—are described in Chapters 24 and 35, where the authors also discuss their significance in exploring the antecedents of human behavior. In support of the population-genetic theory, all research results uniformly point to the fact that primate societies evolved on the basis of broadened and extended relationships between parents and young. In the context of our own discussion, however, there is one point we would like to consider in more detail: The role of males in primate societies. Do these animal groups also have a "male problem"?

▷ Forms of ecological adaptation to the animal's diet. The bills and beaks of many birds are adapted to their specialized feeding habits. Left, above and below: With its long, flat bill, the spoonbill plows through shallow water areas to catch insects, small crabs, and other aquatic animals. Right, above and below: James' flamingo uses its tongue to pump water through its bill, which acts as a filter. Weirlike lamellae keep the food (algae, small crabs, and single-celled creatures) inside the bill when the water is pumped back out.

Fig. 31-6. In circumstances where a primate group needs protection from big cats and other dangerous predators, the presence of many males is an advantage, since they will actively defend the group against enemies. This drawing represents a baboon troop being threatened by a predatory cat. rM, high-ranking male; M, male; F, female; Y, young.

One of the first things to note is that the females and infants of higher primates living in trees, and especially juveniles with their long growing period, are constantly threatened by predators. As a result, the males not only guard their groups against mingling with others, but also acquire the special task of protecting and guarding group members against alien enemies. In most primates this role differentiation has led to a conspicuous-

ly larger physical development in males. However, the number of males that will find this kind of "employment" within any primate society very much differs according to ecological conditions.

Where the environment offers plenty of food, and a large number of adult males provide protection against predators such as big cats—this applies, for example, to steppe or savannah baboons which live in relatively open terrain, and to rhesus and Japanese macaques—many males can indeed be integrated into a group, and solitary animals are rare. Much the same circumstances prevail among chimpanzees and gorillas. Where food is scarce—for example, in the dry mountain habitats of hamadryas baboons or geladas—the proportion of females—which always live in groups—may be four to nine times that of associated males. The same applies to species like the patas (red guenons), where males do not actively defend the group but rather protect it against predators by their vigilance. When danger threatens, patas males quickly induce their families to flee; the rest of the time their main role is to keep family groups separate from each other.

In other words, group structure and the proportion of males that can be integrated into a unit will depend on environmental conditions as well as other factors. Yet we can say generally that, in most higher primates, the number of males who find a definite and stable role within a social group is always less than that of females, on the average about half. Two points may be relevant in accounting for this observation:

1. In the comparative study of human societies, it is commonly noted that young men—in whatever culture area—frequently have a hard time finding a place in their social organization. There have always been attempts to group them into young men's associations or clubs, or keep them separate from the rest of society, and to assign them with tasks where they cannot compete with the roles and activities of older men. Such duties often turned out to be the most dangerous. We should seriously consider if these findings do not show some traces of the same problem found in primate societies where there is not enough room for all the young males, so that a certain percentage of them have to be excluded.

2. A second reason why males with leading roles in a primate social group number less than adult females is the difference in growth: In all primates, the females gain their full social status as adults as soon as they reach sexual maturity. Males, on the other hand, are not fully grown at puberty, but undergo a longer process of physical development and maturation where they also gain more experience. In other words, they must complete a "training period" before they can obtain a leading role within society. However, the long period between sexual and social maturity does not condemn most primate males to abstinence, since females will allow the "juveniles" to copulate with them during the ascending and descending phases of their estrus cycle. It is during estrus, when

◁
Forms of ecological adaptation to avoid competition for food. Animals living in the same habitat concentrate on different kinds of food. The giraffe (top left) feeds primarily off branches at the very top of this acacia, while the gerenuk (top right) reaches only the lower branches. Elands feed off bushes (middle left), higher grasses are consumed by zebras (bottom left), and, finally, the short stubs remain to be grazed by Grant's gazelle (bottom right).

the females are fertile, that mating is restricted to the fully adult males. In this way the lower-ranking juvenile males may engage in sexual behavior and gain experience without actually reproducing.

In primates as in other animals whose social organization we have studied, the higher forms of social life evolved from patterns of family relationships between parents and offspring. Yet a wide spectrum of social behavior can be observed, from the monogamous pair-bond of gibbons to the one-male "harem" groups of hamadryas baboons, and finally to the sexual encounters between various mating partners in savannah baboons and macaques. These variations in social organization depend heavily on the ecological conditions to which a species has adapted. All forms of social behavior are in some way responses to the demands of a particular habitat. Seen in this context, we cannot expect to understand the origins of human social structure without examining the ecological —and in humans this also means economical—adaptations of a society to its environmental challenges.

32   Chemical Communication in Social Insects

Chemical communication in social insects, by B. Hölldobler

In many social insects, communication by chemical means plays an important role. Chemical communication takes place with special glandular secretions called pheromones. Some of these are received as odorous or taste substances by the animal's chemoreceptors (see Chapter 6) and release a certain behavior. Others affect metabolic processes and trigger a physiological reaction; these are responsible, for example, for certain forms of development such as the various castes in state-forming insects.

The most complex animal societies are probably found among insects. In addition to termites, we find the highest form of state organization primarily among bees, wasps, and ants, all of which belong to the Hymenoptera. Here we distinguish not only reproductive and worker animals, but a further division of labor among workers as well. The organization of an insect state will function smoothly only when each individual is completely integrated into the social structure. State-forming insects achieve this integration with chemical signals that control the activities of getting food, raising offspring, distributing food, defending against enemies, warning of danger, looking for new nesting sites, and other vital functions.

As we know, in solitary insects such as butterflies males and females are brought together by means of highly potent attractant substances (see Chapter 6). But the sexes must find each other for reproduction even in the state-forming insects. Again, chemical communication plays an important part in this process.

At certain times of the year, the young and usually winged reproductives of ants will swarm out for their nuptial flights. Copulation takes place. The females are fertilized and found new colonies or enter already existing ones; the males die. Because of the short life span of males, it is essential that the nuptial flights of both sexes are exactly synchronized in time. For instance, the female reproductives of carpenter ants (*Camponotus herculeanus*) make their flight precisely when most of the males start out. This mating exodus occurs during a relatively short period, increasing

Fig. 32-1. Reproductives (female front, male behind) of the red ant *Formica polyctena* during copulation.

the chances for males to encounter the females in the air. The explanation for this amazing synchronization lies in a strongly odorous secretion released by the males's mandibular glands when starting their flight: This substance stimulates the females and induces them to swarm out as well.

In most cases, however, we still do not know just how these ants finally meet for copulation in the air. Only very recently, scientists were able to localize a true sex pheromone in some myrmicine species (subfamily Myrmicinae), released by the receptive females and serving not only to attract the males but also to stimulate them for mating. Surprisingly enough, this sex attractant is secreted by one of the sting glands. In one species of carpenter ant living in Florida (U.S.A.), if males ready to swarm are offered a small wooden stick with this secretion, they are attracted in large numbers and will even try to copulate with the stick. When ready to mate, the usually wingless females of a central European carpenter ant show a characteristic behavior (*Locksterzelverhalten*): Sitting calmly at some conspicuous place, they raise their abdomen and simultaneously protrude the sting to release the sex pheromone; this attracts males. Hence, it seems in several species of carpenter ants a certain gland originally associated with the sting and serving a defensive function had in part changed into a sex-pheromone gland.

Male bumblebees also release an odorous secretion from their mandibular glands during the nuptial flight. By selecting conspicuous leaves and branches and marking them with this pheromone, they create odor trails along which they will fly over and over again. The trails and the substance vary from one species to another. The females are attracted by their species-typical scent marks, follow the trails, and thus copulate with the males. In turn, the females also appear to release a pheromone that makes them attractive to the males. In one species, male bumblebees tried to copulate far more frequently with a model carrying the appropriate female scent than with one lacking this odor.

The most complete information has probably been obtained concerning the sexual behavior of honeybees. From their greatly developed mandibular glands, the queens release a whole series of substances. One of these, the unsaturated fatty acid 9-oxo-*trans*-2-decenoic acid, is a sex pheromone which the young queen bees use to attract drones on their nuptial flight. In experiments where young queens, extracts from queens, or synthetic 9-oxodecenoic acid were placed on poles or helium baloons 6 to 20 m above the ground near so-called drone gathering places, it was found that the drones were attracted to the substitutes as much as to the real queens. We should point out, however, that drones respond to these odor signals only outside of the colony hive. When embarking on their nuptial flights, the males leave the hive and gather at certain places, which are used over many years by each new generation of drones from the neighboring colonies. In finding these spots, the males orient not only to certain landmarks on the horizon but also to a special "assembly odor"

Fig. 32-2. Pheromone glands of an ant (subfamily Dolichoderinae): 1, mandibular gland; 2, metathoracal gland; 3, anal gland; 4, hind gut; 5, sting poison gland; 6, Dufour's gland; 7, Pavan's gland (trail gland of Dolichoderinae).

Fig. 32-3. Receptive female of the myrmicine ant *Harpagoxenus sublaevis* raising her abdomen to attract males.

Fig. 32-4. Exocrine glands of the honeybee: 1, mandibular gland; 2, Nasanov's gland; 3, sting glands; 4, wax glands.

secreted by other drones. The young queens approaching such assembly places release the sex pheromone 9-oxodecenoic acid from their mandibular glands. This arouses the drones, which immediately swarm out to follow the queens and thus ensure fertilization.

Swarming in honeybees

But before a young queen may start out on her nuptial flight, the old queen must have left the colony; two queens will not tolerate each other in the same hive. Thus, shortly before the young queen hatches, the old one emigrates with a large number of workers to find a new site. This is the "swarming" event referred to by beekeepers. When swarming, it is important that the workers do not lose their queen; they manage to stay together because errant worker bees are also attracted by the queen's 9-oxodecenoic acid. Cohesion within the cluster, however, is maintained primarily by a second major element of the mandibular gland secretion, namely, 9-hydroxydecenoic acid, which is less volatile than the attractant.

Tandem running in termites

The social organization of a termite state is completely different from that of a hymenopteran state. Thus, sexual behavior also varies. The winged reproductive animals fly out of their nest, but they soon descend back to the ground and break their wings off. This is when mating starts. Pair-formation has been studied with particular care in the genus *Reticulitermes*: The females first show "attracting behavior," raising their abdomen about 90° but otherwise remaining still. As soon as a male appears and touches her, however, she lowers the abdomen and starts running. The male follows, touching the female with his antennae. While performing this "tandem running" behavior, the pair looks for a suitable place to build the founding chamber for a new colony. Only when this task is completed does the "founding pair" seal itself into the chamber for copulation. According to previous studies, a certain chemical signal released by the female apparently determines both the "luring" behavior and the tandem running. We do not know, however, which gland produces this pheromone.

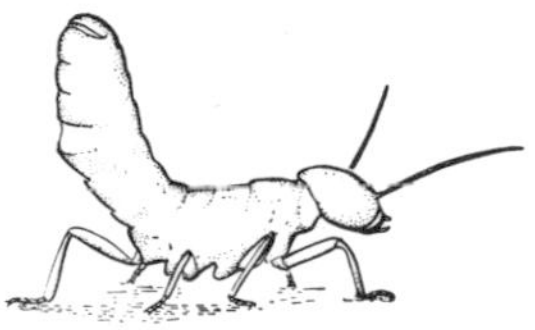

Fig. 32-5. Attracting behavior in the termite *Reticulitermes flavipes*.

The queen substance

In addition to controlling sexual behavior, pheromones are also the primary agents for the strict division of labor found in insect societies. Again, the most familiar example is the honeybee. By means of the mandibular gland secretion known as the "queen substance" and mentioned above as a sex pheromone, members of a colony are constantly informed about the presence of their queen. When a state loses its queen, the entire colony becomes restless; the workers soon begin building special cells. These are called queen cells or royal cells, where young larvae are raised as replacement queens. At the same time, some of the workers develop their own ovaries and lay eggs. But this does not help to rear new queens, since only drones will hatch from these worker eggs. Normally, when a queen is present in the hive, the queen substance is constantly licked off of her body surface by the workers and distributed over the entire colony. This chemical inhibits development of ovaries in the workers and prevents the raising of "rival" queens.

Hence, it is the queen substance which controls the basic division of labor between the fertile queen and sterile workers in a honeybee colony. State organization in ants, bumblebees, and wasps involves a similar system of castes, and there is a great deal of evidence to show that this differentiation is also based on pheromones.

Determination of caste membership is far more complicated in termites. We shall limit our discussion to the best-known species, the primitive dry-wood yellow-necked termite (*Kalotermes flavicollis*), and will focus only on its reproductive caste. In contrast to hymenopteran states which are based on females, termite societies have both a king and a queen, as well as workers and soldiers of both sexes. We shall examine only the question of why termite states usually contain no more than one royal pair, and how the appearance of two or more pairs is prevented.

Long and careful studies have revealed that a complex functional system of pheromones is primarily responsible for controlling the reproductive castes in *K. flavicollis*. Figure 32-6 gives a more detailed representation: At the top is the royal pair, below them the so-called pseudergates, larval stages that perform worker functions until they develop further. The arrows represent pheromones secreted by the king or queen; the direction of the arrows indicates the path of the substances. Pheromones 1 and 2 are taken in orally by the pseudergates and passed on to others through the anus. In this way the chemicals are continually distributed throughout the entire colony; they prevent pseudergates from becoming reproductives. Another king pheromone stimulates female pseudergates into becoming female reproductives, but this chemical is effective only if the inhibitory pheromone 2, released by the queen, is lacking. In other words: If the royal female ceases to be present, there is no more female-inhibitory pheromone; the male-stimulative pheromone now comes into effect, ensuring that female pseudergates will quickly produce a replacement queen.

In the rare event that too many reproductives should arise, they will fight each other intensively—males against males and females against females. They recognize each other by means of a sex-specific "recognition pheromone" (4 and 5). It was also found that king and queen stimulate each other to release the inhibitory pheromone (1 and 2).

A strong division of labor also exists among worker animals in an insect state. Various means of communication are necessary for such an organization to function smoothly. Whenever rapid action is required, as when danger threatens or new sources of food have been discovered, individuals use special signals to arouse their nestmates into certain kinds of action. In many cases these signals, too, are of a chemical nature.

For example, if we molest a worker bee at the entrance to her hive by holding her with a pair of tweezers, other bees nearby will become aroused, running frantically back and forth and touching the irritated worker. Others rush into the hive, and very soon more than 100 bees try

Termite castes

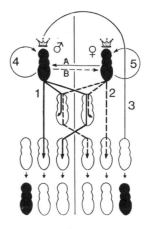

Fig. 32-6. Pheromones determine the reproductive castes in the primitive yellow-necked termite (*Kalotermes flavicollis*).

Fig. 32-7. Honeybee raising its abdomen (*Giffsterzeln*) to release poison. The sting is extruded and the alarm pheromone is released; the whirring of the wings speeds up dispersion of the chemical.

Fig. 32-8. Alarm and defensive behavior of the red ant *Formica polyctena*. The abdomen is pushed forward, and at the same time the ant ejects its poison and alarm secretion.

Fig. 32-9. Soldier of the nasute termite *Ryhncho-termes peramatus*.

to burst out of their nest. The same effect can be obtained with the crushed sting apparatus of a freshly killed bee. Within seconds the bees rush out of their entrance, drum their antennae on the sting apparatus, and in some cases take up a guard position. This demonstrates that the sting apparatus releases an odorous substance that alarms the nestmates.

When in danger, the harvester ant *Pogonomyrmex badius* uses its greatly developed mandibular glands to secrete a substance whose main element is 4-methyl-3-heptanone. This chemical triggers alarm reactions in nest-mates. In order to prevent small and purely local disturbances—where only a few workers release their alarm pheromone—from throwing the entire colony into a state of alert, it is important that this chemical's effect fades quickly and that its range is fairly small. In harvester ants the alarm substance has a maximum range of 6 cm and disperses over this area within 13 seconds; the signal fades completely after 35 seconds. With greater danger leading to wider disturbances, more animals are directly affected and therefore more of the alarm substance is released. As a result, the chemical has a greater range and its alarm effect is prolonged.

Only highly evolved colonies of hymenopterans have a chemical system of alarm. This mechanism is lacking in the more primitive states of bumblebees and burrowing wasps. Within the bumblebees, the alarm substance is produced in different glands, depending on the species. Some even manufacture it in the poison gland, where the alarm substance is partly mixed with the poison but not identical with it. However, in some ants, such as *Formica* and *Tapinoma*, the poison itself may indeed serve as an alarm substance. When releasing the chemical, these animals often show a certain behavior resembling defense or attack.

Termites, which usually live underground, report danger with mech-anical vibration signals; they seem to place less emphasis on chemical alarming. On the other hand, some researchers claim that the defensive substance secreted by nasute termites (*Nasutitermes*) with the noselike structures attached to their head also functions as an alarm pheromone. Other termite species create so-called alarm trails. For example, when worker larvae of *Zootermopsis* have discovered a caved-in wall on their nest, or some other source of danger, they mark the path from this spot back to the nest with a chemical trail composed of secretions from their sternal gland. In this way, workers and soldiers are alarmed and will follow the trail to eliminate the danger.

Insect colonies are in great competition over the available sources of food. Thus it must be an advantage for nestmates to be informed as quickly as possible about new food discoveries and prey objects so that the booty can be retrieved quickly by the communal efforts of colony members. Again, chemical communication plays an important role in achieving this aim.

Thanks to the pioneering studies of K. von Frisch and his school of researchers, we know that honeybees employ a highly complex dance language in communicating information about food (see Chapters 10 and 33). This dance is the most important means of communication; yet even here a chemical signal provides additional information to help newcomers orient themselves toward a food source. Bees that have found a rich source will extrude a scent organ, called Nasanov's gland, from their abdomen, from which they release a chemical attractant whose effective elements are citral and geraniol. The same attracting signal is also used at the hive entrance to lure young bees back home from their first orienting flights. Finally, it also helps the orientation of scouting bees when they discover a new nesting site.

For communicating information about food, the chemical method plays a far greater role in the stingless bees (genus group Meliponini). We shall describe the system used by *Trigona ruficurus*. When a forager discovers a new source of food, she will first of all fly back and forth several times between the food site and the colony. After that, she will fly back to the nest, stopping off in intervals of 2–4 m at some blade of grass, a pebble, or similar object to deposit a scent marker with her mandibular glands. Once back at the nest, she joins others who have now become alarmed by her behavior, and she leads a large group of nestmates back along the odor trail to her goal.

Chemical communication about food is particularly prevalent among ants. We can distinguish three types: In "tandem running," a foraging worker solicits a nestmate, who then maintains close contact using her antennae and follows the leader to the food source. In group foraging, which occurs, for example, in the North American carpenter ant *Camponotus socius*, a forager leads a whole group of nestmates along a chemical trail to the food source. There are also cases where the trail secretion alone is enough to induce others to follow the path.

The tandem-running technique has been studied with particular care in the tropical ants *Bothroponera tesserinoda* and *Camponotus sericeus*. When a successful forager returns to the nest, she will solicit others individually to follow her. One of the nestmates will then attach herself to the forager by constantly touching the latter's abdomen and hind legs with her antennae. If we disrupt this connection between the leading and following ants, the leader will immediately stop or make looped searching movements. Once the ants have found each other again they will continue on their tandem run. Following behavior is triggered by a surface pheromone secreted by the leader, as well as through a contact stimulus. Thus the leader ant may easily be replaced by a substitute, for example, the abdomen of a dead ant, or a glass ball treated with the surface pheromone. Follower ants will pursue such models for many meters.

It gets more complicated when we look at the communication behavior of group foraging. When a foraging worker of the subtropical ant *Camponotus socius* discovers a new food source, it will lay a chemical trail on its way back to the nest. This secretion is produced in the hind gut.

Fig. 32-10. Trail behavior of a termite belonging to the genus *Zootermopsis*. Above: Normal position of the abdomen. Below: During trail-marking the abdomen drags on the ground, while a trail substance is secreted from the sternal gland.

Fig. 32-11. Honeybee raising its abdomen, with its Nasanov's gland extruded.

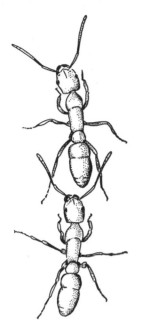

Fig. 32-12. Tandem running in ants of the species *Bothroponera tesserinoda*.

Fig. 32-13. The fire ant (*Solenopsis saevissima*) laying a trail. The trail substance is secreted with Dufour's gland through the extruded sting.

Ant guests (myrmecophiles)

But the trail substance alone does not arouse the others in the nest. The forager must perform a certain waggling movement, which stimulates the nestmates to follow the scout out of the nest, then along the orienting trail to the food source.

Odor trails are probably the most common means of communication. In the fire ant (*Solenopsis saevissima*), foragers that have discovered a food source will mark their homeward path with a secretion deposited through their extruded sting and originating in the accessory gland of the sting apparatus. If we take this substance and draw an artificial trail across a glass plate, hungry ants will follow it looking for food. The number of ants laying a trail will increase with the amount and value of the food, and this in turn will increase the concentration of odorous substance so that even more foragers are stimulated. Since the trail pheromone fades relatively quickly, its concentration rapidly falls below the threshold value after the food source is exhausted, so that finally no more foragers start out on the trail.

Biologically, it is advantageous to the fire ants that their trail substance fades relatively fast, because their food places change rapidly. By contrast, the jet ant (*Lasius fuliginosus*) is better off if its trail substance has a very long fade-out time, because here the trails lead to constant sources of honeydew. Ants of this species deposit their trail substance from the hind gut. In fact, the origin of trail substances is as varied as that of alarm pheromones. The most diverse glands may be brought into play here, sometimes even several glands at the same time. Leaf-cutting ants (*Atta*), for example, manufacture the trail pheromone in their poison gland. This chemical is so potent that only 0.33 mg would suffice to lay a trail completely around the world.

In termites, too, trail-following behavior is very common. Trail pheromones are produced, for example, in the sternal gland of *Nasutitermes*, *Zootermopsis*, *Reticulitermes*, and other genera. The trail substance of *Zootermopsis* not only triggers alarm, but also leads newcomers to the food places. Since the trail secretion of termites has a relatively long fade-out time, its primary purpose is obviously orientational. Body vibrations or other added mechanical signals seem to play an important role at least in some species for inducing nestmates to follow a trail.

Despite the variations in chemical means of communication used by insect states, there are a whole series of "boarders," such as certain insects, isopods, spiders, and mites that manage to live right inside a colony. Ants are particularly beset with many such guests, called symbionts (more specifically, ant guests are called myrmecophiles). We have only recently been able to explain how this puzzling form of symbiosis comes about: Scientists have discovered that in the course of evolution these guests obviously acquired the communication mechanisms of their hosts, so that they do not only understand the ants's signals but also imitate them with a high degree of accuracy.

One example is the nitidulid beetle *Amphotis marginata*, which intercepts food-carrying ants on their trails and begs for food by imitating their own signal. With its antennae and mouth parts, the beetle drums on the head and especially the mouth region of its ant host, inducing it to regurgitate a food droplet. Ants beg each other for food in a very similar way: The recipient palpates the head of the donor with its antennae and at the same time stimulates mainly the food-carrier's mouth region with its forelegs. Laboratory experiments have shown that *Amphotis marginata* recognizes its host by odor, and finds the food trails by means of their specific trail pheromone.

Guests living right in the nest chambers of ants are even better adapted to their hosts. Among these are the staphylinid beetles *Atemeles* and *Lomechusa*. Their larvae reside in the brood chambers of red ants (*Formica*); they not only eat their hosts's brood, but they are even fed by ant nurses. They imitate the begging behavior of ant larvae, but more than that, it seems that they also secrete a pheromone releasing brood-care behavior in the nurse workers. As a result, the beetle larvae are nursed just as well as the ants's own brood, sometimes even better.

Fig. 32-14. Food-begging behavior in the myrmecophile larvae of *Atemeles* toward its host ant *Formica*.

The adult staphylinids live directly in their ant hosts's nests, as well. These now imitate the food-begging signals of ant workers, and in this way they receive an abundance of crop food. In contrast to *Amphotis marginata*, which has a strong back armor or carapace to protect it from the ants's attacks, the staphylinids avoid attacks by using chemical means: First, they eliminate the host's aggressiveness by secreting a "pacifying" substance from the tip of their abdomen; this secretion contains protein, and the ants like licking it up. Second, the guests use glands at the edges of their carapace to produce a "pseudo-" or "dummy" pheromone that induces the ant hosts to adopt and care for the beetles.

State-forming insects play host to a great variety of such symbionts, which are often closely integrated into their social system. These "social parasites" have accomplished the feat primarily by understanding and imitating the communication system of their hosts, so that they are mistaken for members of the hosts's own species.

Fig. 32-15. Food-begging behavior of the adult myrmecophile *Atemeles* toward its host ant, *Formica*.

33 The Language of Bees

The language of bees, by M. Lindauer

The previous selections have shown that animals may communicate with others of their species in a great variety of ways. Many attempts have been made to compare these forms of communication with the human language. To clarify the differences, however, we will begin by setting forth the basic characteristics of language:

Characteristics of human language

1. Language serves to transmit information—for example, "something is burning."

2. This transmission of information occurs independent of time and place. I may tell my neighbor in Toronto: "Two weeks ago a house burned down in Buffalo."

3. Language makes use of symbols for concrete statements: The word "house" may symbolize a farmhouse as much as a cottage, a bungalow, or a skyscraper. Thus, concepts are formed on the basis of generalizations or abstractions.

4. These concepts and symbols (semantemes) can be interchanged in many ways, put together in different combinations, and modified: The concept of "house" may stand for a residence, a place of work, a building where a convention is being held, or it may refer to a person's place of birth or death.

5. Language usually does not consist of a monologue, but requires a transmitter (sender) and receiver. We may ask our partner a question, declare our intentions, express mood, or ask him or her to do something.

What does the bee's dance have in common with real language?

The dance of bees is a form of communication which has certain elements in common with real (or spoken) language. These give the language of bees a quality surpassing other forms of animal communication:

a) The dancer (female) conveys information: She tells the other workers in the hive that she has found a worthwhile source of food. Any worker bees not busy at the moment are then stimulated to fly off and look for this food source.

b) The bee's transmission of information is independent of the time and place of the event's occurrence: For example, if the foraging bee has been gathering nectar at a spot 2.5 km from the hive, she will "tell" about her find only after returning to the colony.

c) The dancer uses symbols to indicate the position and value of the crop: Rhythmical waggle movements vary in their temporal sequence to correspond precisely to the distance of the find from the hive. The quality of the yield, on the other hand, is indicated by the speed of the dance.

d) The sign elements represent pieces of information that may be interchanged—for the gathering of nectar, pollen, or water, or even for the discovery of a new and suitable nesting site.

e) The dancer not only tells the others about her find; she also induces them to help her harvest the crop.

The essential difference between the language of bees and that of humans, then, is that bees use a rigid coding system which does not allow them to depict more complicated events through the possible exchange or modification of symbols. Furthermore, the constituent elements of the bee's language are innate and cannot be newly arranged, learned, or passed on by tradition. Lastly, the bee—like all other animals—is incapable of asking her partners any questions, nor would a partner give a direct answer. Yet it is just the qualities of mutual interchange and the variations in conversation which are the basis of a real language.

K. von Frisch was the scientist who discovered the language of bees in their strange dancing movements and who thoroughly examined the content of this language in what has become a classic study.

On sunny days there are always a few foraging bees flying off to find new sources of food. These foragers are older bees, experienced in orientational flying due to their previous gathering activities. Out of 100 flying bees, only three or four become selected for this difficult and dangerous activity. The majority stay at home so that they can provide their services promptly in harvesting a new source of food.

Once a patrolling bee has found a flowering crop, she gathers nectar or pollen and returns to the hive, where she calls all the idle workers together and starts her dance. With frenzied and tripping steps, the dancer darts about in a kind of figure-eight, running a semicircle first to the left and then to the right, repeating this pattern for up to three minutes (Fig. 33-10). This "round dance"—the simplest one—already contains four important pieces of information:

1. The dancer has found a worthwhile source of food and is now inviting her partners to help with the harvesting.

2. She tells them what kind of flowers she has visited. This information is conveyed in the scent of the flowers carried in her body hairs. The other dancers touch her with their antennae and take in the scent.

3. The forager conveys through the speed or tempo of her dance what the food consists of and whether it is worth harvesting. In fact, she alarms the others only when there is plenty of nectar and pollen and when the nectar contains a lot of sugar. This means that in the spring, when the

Fig. 33-1. View of a normal bee colony, with combs arranged side by side (from a photograph of *Apis indica*, native to Sri Lanka [formerly Ceylon]).

Fig. 33-2. An observational hive, with glass panes allowing a view of all the bees and combs.

Fig. 33-3. Stand for marking material. A fine brush is dipped into a color solution and used for marking individual bees.

Fig. 33-4. While the harvesting bees gather at the food dish, the experimenter marks them with his color solutions.

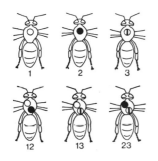

Fig. 33-5. By using different combinations of white, red, blue, yellow, and green, the experimenter can give each bee an individual number.

major crop is in bloom, bees will dance only for a nectar source with high sugar content, about sixty or eighty percent of the possible yields. But in the fall, when harvest crops become very scarce, the foragers will dance with the same speed even when an experimenter has diluted the nectar to one-tenth of its concentration. For the dancing to be released, it is also important to the bee whether the food is hard or easy to gather. If she is forced to exert considerable effort for even a grain of nectar, visiting dozens of flowers to fill up her honey crop, her motivation to dance will decrease correspondingly.

4. Finally, this dance conveys the location of the find. In contrast to the waggle dance, which we shall describe later, the round dance tells only that the food source is very close to the hive.

It seems incredible that a bee is able to transmit so much detailed information using only simple dance movements, and in the gloom of the hive. K. von Frisch first provided evidence in 1920, when he began to publish his experiments and findings on the dance language of bees. Since then, new experimental results have accumulated, but little has changed in the methods which had proved highly successful from the start. We shall describe them briefly.

First, let us take a look into the hive, making sure that we can recognize and observe the bees that frequent our experimental food tables. Normally the honeycombs are arranged side by side, making it difficult to see into the alleyways (Fig. 33-1). Von Frisch therefore constructed his own "observation hive" in which the combs lie one above the other instead of side by side, and with glass panes on both sides to allow free observation of the comb surfaces (Fig. 33-2). Now we proceed to mark each bee by putting tiny spots of colored paint on their backs and abdomens while they are taking in food from the dish. With different colors, we can provide each bee with a personal "number." Five colors allow combinations representing the numbers 1 to 599. Instead of natural flowers we offer the bees an experimental food dish containing a strong sugar solution. In the vicinity of the dish we scatter a few drops with a pleasant flower scent, such as lavender. This ensures that our particular group of bees always finds its way back to the proper place.

In order to attract a host of gatherers to our food table, we approach the flight hole of the hive with a few drops of honey and lure away the bees that are momentarily "unemployed." Next, we start to move the table away very gradually, first centimeter by centimeter, then step by step until we reach our designated location. Even if only one bee has followed us all the way, it means success: Within a few hours an entire host of helping workers swarms about the food dish. By means of our marking technique we ascertain that they all come from the same hive. Thus, the first bee has managed to alarm (or inform) the others in her colony.

Now we can construct experiments for testing the first four statements that we made concerning the information conveyed by the round dance, and its effect on the other workers in the hive:

Experiment number 1: The dance signals the kind of food and whether it is worth harvesting. During one of the seasons when the yield is poor, such as in September, we induce a host of ten marked bees to swarm about our table, where we feed them a very concentrated sugar solution. To ascertain the alarm effect of ten foraging bees within a period of one hour, with the food located 10 m from the hive, we simply catch all unmarked newcomers during this time and lock them in a cage so they cannot alarm any others. During several experiments, our ten foragers brought an average of 75 to 100 newcomers to the table.

The next day we change this experiment by presenting the bees with sugar water diluted 32 times. At this time of year, such a dilution is just barely acceptable for feeding, but is not sufficient for releasing the dance. Thus, there are no newcomers to the table. We can also make the harvesting more difficult by placing six layers of filter paper over the dish, so that the bees have to exert a great deal of effort to obtain the sugar solution out of the paper. Again, there are no newcomers, and by observing the hive we confirm the assumption that the foragers do not dance.

Experiment number 2: The dance contains information about the kind of flowers visited. We place our food dish on top of some filter paper impregnated with a few drops of rosemary oil. We also add a tiny amount of rosemary oil to the sugar solution. We now want to test whether the newcomers have been informed about the rosemary flower scent. Round about the hive at a distance of 10 m, we place a variety of scented plates with filter paper impregnated with either rosemary, lavender, jasmine, peppermint, or some other scent. We observe that the newcomers search only at the plates smelling of rosemary.

Experiment number 3: The round dance conveys the message that the food source is located very close to the hive. Again we impregnate the food with rosemary, but this time the control dishes (smelling only of rosemary) are placed at varying distances from the hive, from 5 m to 1 km. As a result, the newcomers that had been alarmed by the foragers's round dance search for the source only in the vicinity of the hive, a perimeter of about 0 to 50 m.

Our harvesting bees, however, do not gather food only in the vicinity of the hive; their activity extends to a perimeter of 2–3 km. Occasionally, bees will carry nectar and pollen to the hive from a distance of 6, perhaps even 10 km. To indicate a source located at such great distances, the foragers use another kind of dance, the "waggle dance," which includes a precise description of location (range and bearing) in addition to all the other kinds of information. We shift our food dish, step by step, from a distance of 10 m to 30, 50, 100, 200, and 500 m. At the same time, we observe the dancing activity in the hive and note that the form of the dance changes markedly during the transition from 80 to 100 m. Between

▷
Group behavior in baboons. Top left: During the dry season, hamadryas baboons dig for water in the sand of river beds. Their conspicuous drinking posture serves to inform others that the search was successful. Top right: One-male group of an Ethiopian hamadryas baboon sitting on the troop's sleeping cliff. The brown females and young always stay close to the gray-maned male who is twice their size. Bottom left: Part of a herd of gelada baboons in the high plateau of Ethiopia. These monkeys are digging for roots in the alpine meadow. Bottom right: Gelada one-male group resting at a social gathering spot on the edge of the high plateau. The male can be seen sitting to the right and grooming (delousing) one of his females. Immediately behind them are the cliffs where they spend the night.

◁
Group behavior in baboons. Above and center: Guarding behavior in hamadryas baboons: A male looks for a missing female (top left), stares at her threateningly (top right), attacks (middle left), and bites her in the back of the neck (middle right). Bottom left: Concentrated feeding has triggered a fight over females. This altercation has spread throughout the entire troop. Two subtroops are fighting each other. Bottom right: Threat in conjunction with a defensive behavior in free-living hamadryas baboons. The weaker female in the middle threatens her stronger rival (left), but at the same time presents her anal area to the male to preventing him from attacking (right).

Fig. 33-6. A swarm of bees has gathered under a rock precipice, remaining there until those preparing a new hive give the signal to move (*Apis dorsata*, a giant honeybee from Sri Lanka [formerly Ceylon]).

the two loops of the figure-eight, the bee inserts a so-called waggle run. Here, she runs straight ahead for a moment and waggles or shakes excitedly from side to side with her whole body. Then she makes a loop to the left, another waggle run, then a loop to the right. Again, the waggle dance may go on for several minutes (Fig. 33-9). If the food we offer is particularly rich, the dancer will accompany her waggle movements with buzzing sounds. The most important information conveyed by the waggle run is the location of the food source, with its exact range and bearing.

The dancer indicates the distance from the goal by varying the rhythm of her waggle dance. By gradually shifting our food table from 100 to 10,000 m away from the hive and comparing the waggle dances performed by the harvesting bees, we obtain a relation of distance to dancing rhythm as indicated by Figure 33-7. A brief glance at the graph tells us that this rhythm is much faster when the food source is close to the hive. Thus, for example, the bees will make eight complete runs for a distance of 200 m, three runs for 2000 m, and only one and a half runs when the source is 10,000 m away.

In order to be that accurate in their indication of distance, of course, the dancers themselves must first make a precise measurement. For a long time, scientists assumed that the bees do this by measuring the time spent in flying from the food source to the hive. Experiments have shown, instead, that the bees make this computation by means of the energy they exert in flying the distance, rather than the flight time. A host of bees was required to fly from their hive up a hill to obtain food, while another group had to fly down the hill. The uphill foragers indicated a greater range than the downhill ones, although both food tables were equally far away. In the same way, the dancers indicated a greater distance when they had to fly into a headwind than when they flew with a tailwind. We have not yet been able to explain how the bees can measure their energy consumption so accurately.

Now, the information about range would be of little use to the other bees if the dancers did not also convey the direction, or bearing, of the food source. To examine how this is done, we lure our host of gathering bees to a food table 600 m to the south, and observe their waggle dances for a whole day. The result is that the direction of the waggle runs changes from morning to evening in a counterclockwise fashion, correlating precisely with the sun's rate of change in azimuthal direction (the azimuth is the vertical position of the sun relative to the earth's horizon). In the morning, the waggle run points 90° to the right, by noon has shifted to a perpendicular bearing, and at six p.m. it ends up with a straight waggle run 90° to the left. Thus, a dancer "translates" the solar bearing from a horizontal to a vertical one, giving each dance a vertical straight run either to the left or to the right depending on whether the direction of flight is either to the left or the right of the sun. An upward waggle run then means that the food source lies in the direction of the sun, a downward run that it lies in the opposite direction. A waggle run 80° left of the perpendicular indicates that the food source lies 80° left of the position of the sun, and a run that leads 90° right of the perpendicular indicates a source lying at an angle of 90° to the right of the sun. In other words, the dancer angles her waggle run to coincide with the angle between the sun and the food source, demonstrating the fundamental role of the sun as a compass in the activities of bees (see also Chapter 10).

Not only does this solar function aid the bee's orientation, but it also facilitates communication with the other bees. Since a bee must transpose the solar angle into the dimension of gravity, we have here an amazing example of a high degree of sensory as well as central nervous functioning: The bee must, after all, translate the angle of solar bearing accurately from one sense modality to another. Since these discoveries were made, researchers have found that many other groups of animals are capable of this impressive feat: If ants, dung beetles, or certain small crabs are blocked from the sun, but given the opportunity to use gravity for their orientation, they will act as bees do and translate their previously visually oriented movements into the gravitational mode.

In Chapter 10 we discussed how bees orient their movements on a cloudy day according to the direction of oscillation of polarized sunlight. We shall mention only briefly that the foragers are able to compute a straight path to their food goal even during sidewind drift and when forced to make detours. With all our intelligence, we humans may accomplish such a feat only with the help of expensive instruments.

In his classical stage and fanning experiments, von Frisch was able to determine how far the fellow dancers pick up and make use of the dancing forager's instructions regarding range and bearing of the food source. To his surprise, he discovered that ensuing flights outside the hive followed a more accurate path than that indicated by the dance. We can only conclude that these other dancers, which after all do not venture forth until they have followed several dances in succession, actually "compute" the mean of all these pieces of information, averaging them out to obtain more precise instructions.

Communication through the round and waggle dances serves not only the function of food gathering, but also that of finding new places to build a hive. When a swarm of bees migrates from the mother colony, the members first gather in a cluster near the original hive (see Color plate, p. 500). If the beekeeper does not watch out at this point, two to three dozen foraging bees will soon venture into the surrounding area to scout for suitable nesting sites. They will start by fanning out in all directions, until some of them find an old hollow tree or similar holes in ruins or an empty fox hole. These scouting bees fly back to the cluster and perform a waggle dance, telling the others about the range and bearing of their find in the same way that food foragers do. The bees in the swarm that are alarmed by this dance fly off and examine the potential nesting site.

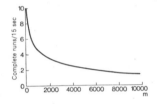

Fig. 33-7. Distance curve, reproduced by counting the waggle runs per unit of time. Abcissa: Distance from food source to hive. Ordinate: Number of waggle runs per 15 seconds.

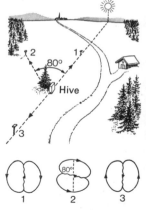

Fig. 33-8. The "code" for communicating direction: Pattern 1: The food source lies in the direction of the sun; the waggle run on the vertical comb is therefore oriented vertically, toward the top. Pattern 2: The food source lies 80° left of the position of the sun at that time; the waggle run is oriented to 80° left of the perpendicular. Pattern 3: The foragers fly from the hive to the food source away from the sun; the waggle run is directed toward the bottom of the comb.

Fig. 33-9. Waggle dance, with a conspicuous straight waggle run inserted between two semicircles, or loops.

Fig. 33-10. The simplest form of dance used by bees to communicate with each other is the round dance. It indicates a source of food close to the hive. Note how the other dancers touch the lead dancer with their antennae.

Fig. 33-11. Except for *Apis mellifica carnica*, bees perform a "sickle dance" when switching from a round to a waggle dance. This intermediate phase is the beginning of a directional communication, though still very rough.

If they "agree" that the spot is suitable, they will try to enlist others to examine it as well. To study this procedure, I (Lindauer) marked the dancing scouts with dabs of colored paint and drew up the indicated nesting sites on a surveying board; in several cases we actually found the site before the swarm had taken it over. We could hardly offer more convincing proof that we can understand and interpret the language of bees.

During these experiments, however, we were surprised to find that the nest-seekers indicated not just one but several potential sites in their dances. In one case we counted twenty-four different reports within the cluster. How then do the 20–30,000 bees in a swarm reach a consensus on which of the sites they will accept? The swarm itself cannot split up; it must remain as a unit and move along with its queen to build a new hive. The bees eventually reach a general agreement. The dances become more and more uniform, until finally they indicate only one nesting site. Now the cluster breaks up and the members of the swarm move together into their new home. It can also happen that the dancers end up concentrating on two potential sites at the same time, but again they are forced to reach an agreement, and so one of the two "factions" gives in to the other.

How do the bees achieve their consensus? We have discovered that the scouts always make the decision. And, as it happens, their final choice is always a good one: They will invariably decide in favor of the best site reported. But in order to communicate to one another where this best spot would be, each dancing scout indicates not only the position but also the exact value of its find. If the nesting place is of top quality, the scout who has discovered it will dance with a great deal of tempo and endurance, going on for hours and even entire nights. By contrast, reports about an inferior lodging would be communicated only by a hesitating, brief, and feeble kind of dance lasting only a few minutes. As a result, more newcomers are recruited for the really good spot right from the start. The bees that received these messages fly out and look over the different sites, so that on their return to the cluster they will try to enlist other newcomers for the best place. The really important factor, however, is that even those scouting bees that started out dancing for mediocre nesting spots are receptive to the reports of their fellow scouts, and they will then examine these other sites for themselves. After comparing these spots, they will end up dancing for the best one, along with the scout that originally discovered it. In other words, the scouting bees in the swarm have to be

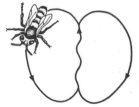

flexible and willing to "change their opinion." In this way they are fundamentally different from the food-collecting bees, which insist on remaining with their foraging group as long as there is any nectar left in that yield, and which show no interest at all in the dances of other groups.

All subspecies of honeybee and all species that belong to the genus *Apus* communicate by means of round or waggle dances. But we can observe certain variations in these dances depending on the subspecies or species. Heredity determines the basic pattern of the dance, and the correspondence between tempo and the distance of the find.

A significant preliminary stage to the directional signaling of our honeybee may be found in the southeast Asian dwarf honeybee (*Apis florea*). Unlike the European bees, this species cannot translate the angle between their flight path and the sun into the field of gravity. No doubt this has something to do with their primitive way of life. *Apis florea* nest in the open, attaching their single honeycomb to a branch. The foraging bees perform their dances only on the horizontal platform on the ridge of the comb, which is enlarged to suit this purpose. Since the "dance floor" is horizontal, the dancers can point directly to their goal: They simply do their waggle run so that it lies at the same angle to the visible sun as the proposed flight path.

If we remove the comb from its branch and tip the "dance floor" to a vertical angle, the foragers stop dancing immediately and run to the top of their comb as it now stands. They will resume their dancing there only after stamping out a new horizontal platform. Furthermore, these dances lose directionality as soon as we either block the bees's view of the sky or cover every horizontal plane on the comb. We may therefore conclude that such dances performed on a horizontal plane with a view of the sky represent a stage preliminary to those acted out in a gloomy hive, which necessitate a transfer of information from memory to the field of gravity.

The group of stingless bees (Meliponini) living in Brazil have developed a unique means of alarming their conspecifics and indicating the direction of a find. It has always been found that the foragers can guide newcomers to any particular food crop they have discovered, but with varying success. Once a forager has made a find, it will give an alarm in the hive by making excited and undirected zigzag runs, frequently also buzzing. The alarmed newcomers then either search randomly according to the odor of the find that the dancer imparted to them, or follow a trail of scent markings to the goal. In the latter case, those foragers making a find will deposit scent markings on rocks, clumps of dirt, or blades of grass, using a secretion from their mandibular glands. This trail of scent markings is constantly renewed as long as the yield is profitable. If we prevent the foragers from marking—for example, by forcing them to fly over a pond—none of the other collecting bees will come to our food table on the other side, even if the dancers alarm them for several days.

▷
Male tree shrew, *Tupaia belangeri*, with his tail hairs smooth (top right) and erected (bottom right). Left: Kidneys of male tree shrews under control conditions (above) and after two days of confrontation with a dominant conspecific (below). Under stress conditions (below) the capillaries are distended and contain few red blood corpuscles or none at all. The renal tubules are also distended. Microscopic sections are 4/1000 mm thick; stain: hematoxylin-eosin.

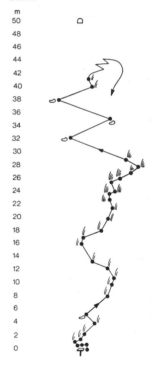

ⱡ Scent marks on a grass stalk
◠ Scent marks on a stone

Fig. 33-12. A successful harvesting bee deposits scent marks on its way back to the hive. It then lures the bees it has alarmed away from the hive along this scent trail to the food table.

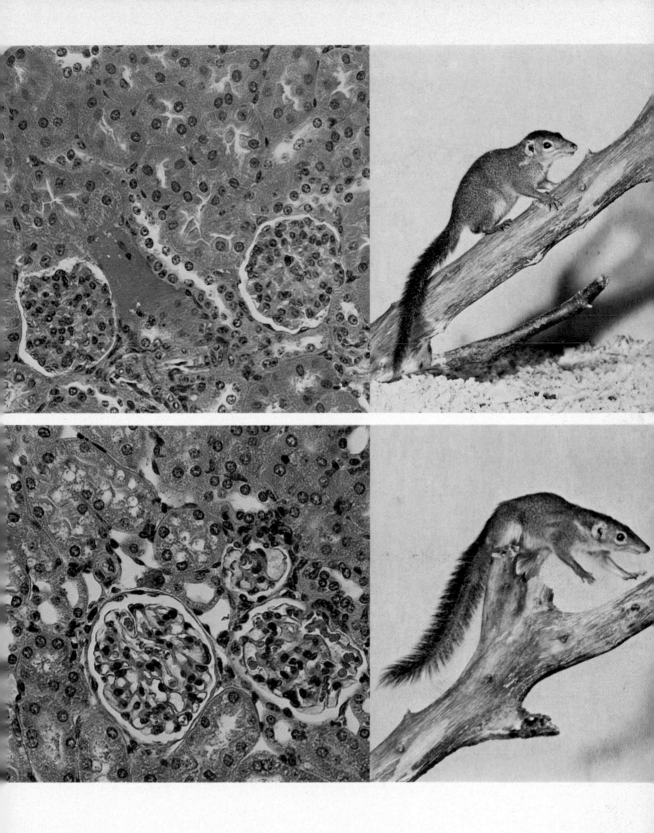

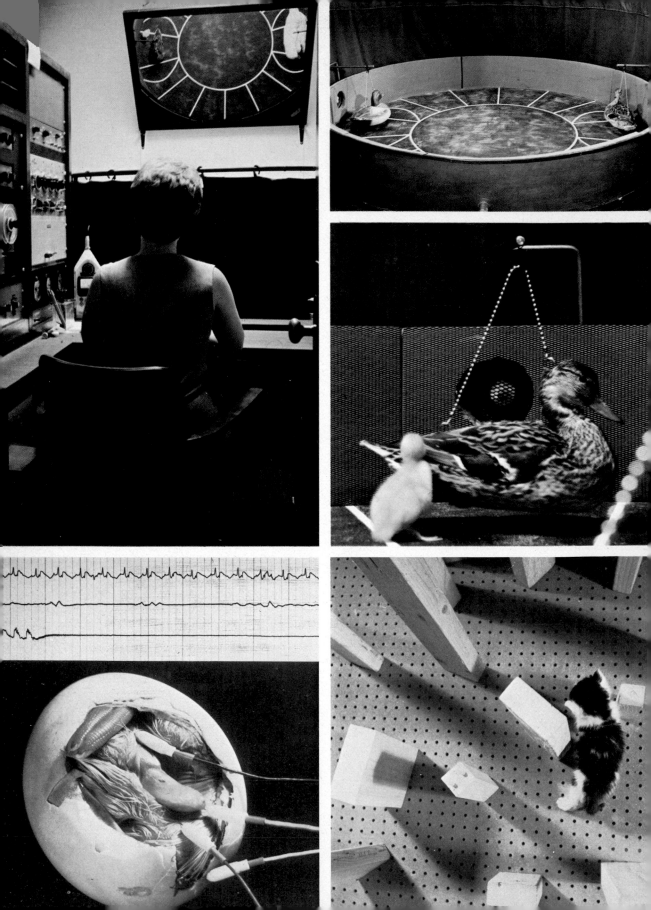

But if we stretch a rope with some brush across the water, scores of new-comers will find their way to the goal (see Color plate, p. 500). This is a far more primitive means of communicating the location of a food source: A dancer can point the way only in a direct manner. But there is another very important component that we find in the language of sting-less bees: the zigzag runs used for alarming other collecting bees. This type of activity often precedes the actual dance in the honeybee, and it is quite possible that zigzag runs formed the basis for specifically oriented dance patterns in later stages.

◁
In a laboratory experiment, mallard ducklings that did not experience their mother or any mother substitute were allowed to choose between a female mallard duck and a stuffed white Peking duck, with both models made to sound the mallard duck contact call by means of a tape recording. The duck-lings preferred the colored models to the white ones, but they did not show a preference for any color in particular. Top left: Observations are made with a mirror fixed above the experimental table. Top right: Mallard drake (left) and duck. The duck-lings choose neither of the models when both are made to sound the mother's contact call. Middle right: In this case the duckling chose the female mallard duck. Bottom left: A Peking duck embryo is prepared for the test one day before hatching: Electrodes are attached to bill and upper torso—the recording curves are shown above the egg— to study behavior (bill-clacking, middle curve) and physiological processes (heart beat, upper curve) as well as the embryo's sounds (lower curve). Bottom right: Young cats must first have explored the experimental room visually and through autonomous movements before they are able to pass between obstacles and not bump into them.

34   Group Formation in Ungulates

The social behavior of mammals can be classified as follows: 1. Solitary animals. Males and females meet only during the mating periods. Example: Malayan tapir (*Tapirus indicus*). 2. Pairs. Males and females stay together for certain periods of time even outside of the mating season. Example: red fox (*Vulpes vulpes*). 3. Social groups. Males, females, or members of both sexes form smaller or larger associations.

Within the category of social groups there are three kinds of associations: In open anonymous ones, members of a species form aggregations, but know each other only as conspecifics and not as specific group members. Group numbers can increase or decrease by the movement of individuals, where newcomers may readily join while other members may leave. Example: brindled gnu (or wildebeest; *Connochaetes taurinus*). In closed anonymous associations, group members recognize each other by a common smell. Strange conspecifics are rejected, driven away, or even killed. When the size of such a group passes a certain limit, it may split up. Example: black house rat (*Rattus rattus*). In individualized societies, members know each other personally by smell, appearance, or sound. Outsiders are allowed to join the group only under certain conditions. Example: plains zebra (*Equus quagga*).

However, the individual forms of social organization are in no way limited to any particular animal groups. Different forms may be found within groups of related species, and conversely, very similar forms may appear in rather remote phyletic groups. This derives from the fact that the social organization of a species is adapted to its environment. Hence, it is advantageous for all species living in a certain habitat to adopt a particular kind of social life. We shall begin our description of the various interrelationships between environment and social organization by using the African antelopes as an example. These animals are particularly well suited for this discussion because they have penetrated all habitats from rain forest to desert, they include a large number of species, and they have been studied by ethologists in fairly great detail.

Group formation in ungulates, by H. Klingel

Forms of social organization

The most important ecological factors in the social organization of animals are the variety and amount of food, its spatial and temporal distribution, and the ease with which an animal can survey its habitat.

In horned animals (family Bovidae), which include the antelopes, we can find three major forms of social life:

Family territories

1. Family territories: Individuals of a species live either alone (solitary) or in more-or-less stable pairs. They occupy well-defined and permanent territories from which other conspecifics are rigorously excluded. Under natural conditions we would find at the most one pair of adults with young. The offspring are later driven off by their parents and have to find other accommodations. The owners mark their territories with glandular secretions especially of the preorbital glands, and also by special defecation sites. The territory itself serves two functions: It provides its owners with food throughout the year, and it offers a sanctuary where they can reproduce without too much interference from conspecifics. Since the boundaries are stable and outsiders are at a disadvantage within another's territory, aggressive interactions take place only at the borders.

Mating territories

2. Mating territories: Here, only the males are consistently territorial. The females travel about in smaller or larger, but always variable groups, remaining in a male's territory only for short periods of time and forming loose, temporary pair-bonds. We cannot really speak of "harems," because mating ties are not permanent. Not all adult males occupy a territory; the surplus males, primarily younger and weaker ones, form bachelor herds which constitute reserve pools for new territory owners. With this form of social organization, territories are virtually no more than places for mating; they are marked with glandular secretions and dung.

Herds

3. Herds: The third form of social life is mixed herds. Adult males, adult females, and young all live together, although they may not necessarily know each other personally. In some species old males leave the herd. These herds are not tied to any particular place, instead moving around within sometimes huge home ranges that are not defended against others. The groups may split or merge with other herds, but these changes in size and individual membership do not affect the basic structure of herd organization. Again, it is the strongest males that reproduce, but this time they dominate directly and not by the "roundabout" way of territorial ownership: During the mating season, only winners in the fights for dominance—usually highly ritualized duels—are allowed to mate with females, while the losers are forced to abstain.

In Africa, the most important habitats for antelopes are: 1. Rain, highland, and gallery forest (also called fringing or riverine forest); 2. dry forest and bushland; 3. grassland; and 4. semidesert and desert. In the forests, visibility is usually only a few meters, providing both small and large animals with excellent hiding places. Visibility is far better in dry forests and bushland; even smaller animals have only limited opportunities to hide in this kind of country. There is almost nowhere at all to hide in grassland, semideserts, and deserts.

In the rain forests and tropical high forests, food is available to plant-eating (herbivorous) animals quite evenly throughout the year. But antelopes can make only limited use of this because they are restricted to herbs, to the accessible leaves and buds of bushes and lower trees, and to fallen leaves and fruits.

The forest areas of Africa contain a large number of antelope species, but their populations never reach a high density. In these regions, it is mainly forest duikers (*Cephalophus*) and other small antelopes that occupy stable territories, either alone or in pairs. Larger antelopes also appear, like the bushbuck (*Tragelaphys scriptus*) and the bongo (*Taurotragus euryceros*). These feed on leaves, buds, and flowers blossoms, but they also eat fruits and bark and occasionally animals, such as birds. Their territories are so large that the owners can find food throughout the year; thus they do not have to migrate. These forest dwellers move through the underbrush with their head held low, hiding when danger threatens. Their body structure is adapted to the way they locomote, with the hind part of the body built higher than the front. Males and females are roughly equal in size and have a very similar appearance.

Dry forests and bushlands

The dry forests and bushlands, which are adjacent to rain forest regions, have very few species of plants. Usually they contain only two or three kinds of trees or bushes. Herbs are common, and the ground is covered with a layer of grass. These habitats typically receive one or two periods of heavy rain throughout the year. Since the variety of plant species is so very limited, food supply is poor and fluctuates greatly due to climatic changes, since almost nothing grows in the dry seasons.

A number of medium-sized antelope species live in these habitats, feeding more or less off leaves and buds as well as grass. Examples are sable antelopes (*Hippotragus niger*), roan antelope (*H. equinus*), impala (*Aepyceros melampus*), nyala (*Tragelaphus angan*), gerenuk (*Litocranius walleri*), and several species of waterbuck (*Kobus*); the greater kudu (*Tragelaphus strepsiceros*) can probably be included here too. Only the bucks are territorial in these animals; larger female herds have been sighted only for impalas. These associations are not stable, but may shift even within a few hours by members joining or leaving; thus they are open anonymous associations. The home ranges of female herds are much larger than the territories of bucks. During the mating season the hinds (females) linger with a number of males, one after the other, and mate with them.

As in the rain forests, population density is small in dry forest and bushland. The animals are spaced quite evenly over their territories throughout the year. Shifts in the distribution of population may occur only in the dry periods when individuals flock to the water holes. When danger threatens, however, these antelope species respond very differently from forest dwellers. They "freeze" at the first approach of a predator, and so blend in visually with their surroundings. Only when the predator comes closer do they run away quickly, but only for short stretches; then they "freeze" again.

In the grasslands it is truly the grass vegetation that determines over-all appearance. Herbs are of little significance, and trees and shrubs play no part at all in the diet of antelopes that inhabit these areas. The food fluctuates in a regular annual pattern: There is an overabundance in the rainy seasons, and scarce resources during the dry periods. The social organization of antelopes native to this habitat differs little from the organization of dry forest and bushland dwellers: The bucks are territorial, while the females travel in anonymous herds. However, these herds are much larger than those of the other antelope groups, often numbering into the hundreds and sometimes even the thousands. The major species are gnus (or wildebeest; *Connochaetes*) springbok (*Antidorcas marsupials*), topi (*Damaliscus lunataus topi*), Buffon's kob (*Adenota kob*), Thomson's gazelle (*Gazella thomsoni*), and Grant's gazelle (*G. granti*). Because of the regular fluctuations in food supply, these animals are forced to migrate seasonally within the greater part of their distribution areas, frequently combining whole populations into gigantic herds. During the rainy season they keep to the steppes, where they seem to find particularly fresh and tasty grasses; but in the dry periods these antelopes migrate to the neighboring bushland and dry forest areas, feeding off dry grasses within reach of permanent water holes. They reproduce during the rainy seasons, and it is generally only during the rut that bucks form territories.

Example: Gnus

The degree to which the social organization of one species may be adapted to changing circumstances was demonstrated by the studies of R. D. Estes and R. M. Watson on gnus (wildebeests) in the Ngorongoro-Serengeti region of Tanzania. Inside the Ngorongoro Crater the seasonal variations in food supply are not nearly as great as on the Serengeti Plain, where there is also a severe lack of water during the dry period. Gnu bulls living in the Ngorongoro Crater occupy stable territories where some of them may remain for several years, even though the rutting season lasts only a few weeks.

This is very different from the behavior of gnu bulls belonging to the large migrating population of the Serengeti. Whenever the herd stops on its way to the feeding places—during the rut—these males form their territories within minutes and are then ready for breeding. As soon as the herd moves on, the bulls leave their territories and join the other animals. During the next stop this whole procedure is repeated, until the mating season comes to an end. Here, the territories are usually held for only a few hours, or at the most for a few days. Nonetheless, this form of reproductive behavior is quite successful, as evidenced by the population of more than a million wildebeests in the Serengeti. Permanent territories would fulfill no biological purpose and would even be disadvantageous under these conditions, since this social system would keep the strongest bulls apart from the cows.

Most antelopes inhabiting grassland have no way of hiding from their enemies, since the country is open everywhere. Instead, they watch any approaching predator carefully, try to avoid them, and if necessary flee over longer distances in a tightly knit group. Females defending their young may also attack the enemy, depending on the relative sizes.

The hardest living conditions are found in semidesert and desert. Vegetation consists of no more than a few bushes, plants, and grasses, with a complete absence of plant life over stretches of various sizes in between. In the semideserts there is still some fairly regular rainfall, although it certainly does not rain in all areas every year. Only very occasional precipitation occurs in the desert regions, and some areas receive no rain at all for years.

Semideserts and deserts

Among the few antelopes that live in deserts, three species are particularly well adapted to drought conditions: addax (*Addax nasomaculatus*), oryx (*Oryx gazella*), and eland (*Taurotragus oryx*). These belong to the third kind of social organization we described: they do not stay in one place, but form mixed herds that migrate to wherever rain has fallen most recently, providing a temporary abundance of food. Oryx and eland are not exclusively desert animals, they also inhabit more humid grasslands and even bushlands and dry forests.

In addition to these antelopes, a number of gazelle species inhabit the semidesert and desert regions. The following are closely related to Thomson's gazelle: Dorcas (*Gazella dorcas*), slender-horned (*G. leptoceros*), Spekes (*G. spekei*), Pelzeln's (*G. pelzini*), Heuglin's (*G. tilonura*), and red-fronted gazelle (*G. ruficollis*). Relatives of Grant's gazelle are the Soemmering's (*G. soemmeringi*) and dama gazelle (*G. dama*). We know virtually nothing of their social organization. Most likely the bucks—as with Thomson's and Grant's gazelle—secure territories only for short periods of time during the mating season, while the adult females live in small groups. As soon as living conditions get worse, the bucks leave their territories and migrate with the rest of the population to better pastures. Defensive behavior of semidesert and desert dwellers is much like that of grassland dwellers.

Adaptive value of social organization

In judging the adaptive value of any of these social organizations, we must delve a little into the phylogenetic history of horned animals. The ancestors of the various groups of antelope and gazelle were species of the genus *Eotragus*, living at the time of the Miocene (Middle Tertiary). They were small, resembling the forest duikers in appearance and way of life. It is therefore likely that they also had a similar social organization to the forest duikers, and that all other forms of social life evolved from this original one.

For inhabitants of rain forests and similar habitats, it is best to remain in one particular area where sufficient food is available throughout the year. In this way, an animal gains a precise picture of its home range, which provides a decided advantage in avoiding predators. The individual maintains its food supply by defending the territory against conspecifics. Furthermore, this type of behavior leads to an even distribution of the species. Since emigrating juveniles colonize unoccupied areas, overpopulation and therefore shortages of food are avoided. The strongest individuals secure the best territories, while the weakest are driven off. This is favorable to the species's survival.

By contrast, consistent territoriality would be a disadvantage for any animal living in semidesert or desert regions, since it has to migrate in response to irregular food supplies. Thus oryx, addax, and eland are ideally suited to these living conditions because they are no longer territorial, instead living in real herds. On the other hand, desert gazelles as well as antelopes and gazelles living in grasslands have made a "compromise" between their hereditary and, for mating purposes, essential territorial behavior and the migratory behavior necessary for maintaining their food supply: The strongest individuals form short-term mating territories, but outside of the rutting season they migrate along with the rest. This is the only way for grassland dwellers to utilize the food resources within their home range, which vary with the rainy seasons.

In habitats where animals cannot hide from predators, it is favorable to form groups. Only one member of a group is enough to alert all others with its behavior when danger threatens, and this eliminates most surprise attacks. Furthermore, a predaceous animal has fewer chances of making a kill from a whole group, because the sheer number of prey animals acts as a distraction and confuses the hunter. Single animals are much easier to catch.

Animals inhabiting dry forests and bushlands

Inhabitants of dry forests and bushlands have adopted an intermediate form of social organization. The territoriality of the males facilitates spacing and ensures that only the strongest reproduce in the more favorable sites. The herd formation of females offers protection against predators.

In summary, then, we can say the following: From rain forest to desert, food supplies become less abundant and varied; temporal fluctuations become more pronounced, and visibility of the countryside increases. Antelopes and gazelles have adapted to these conditions. Their social organization ranges from territories occupied by single individuals or pairs, to male territories and female herds, and all the way to a loss of territoriality with true herd formation. These rules also apply in a general way to other plant-eating mammals, and are certainly not limited to African species.

Bovine animals

This relationship between ecological conditions and social behavior holds true even within a species. The African buffalo (*Syncerus caffer*) divides into three subspecies, the forest-dwelling dwarf or forest buffalo (*S. c. nanus*) and two forms inhabiting the steppes, the grass or Sudan buffalo (*S. c. brachyceros*) and the Cape buffalo (*S. c. caffer*). In the forest-living subspecies, individuals stay in one place and live alone or in small groups, while in the steppe-dwelling forms they join into large mixed herds. On the other hand, as far as we know, even the forest inhabitants do not have territories. This indicates that bovine animals, which inhabit mainly open country, have lost their territorial tendencies. The European bison (*Bison bonasus*) lives in the forest and, like the forest subspecies of African buffalo, also forms small groups. By contrast, his American relative, the bison (*Bison bison*), is a steppe-dwelling animal that forms large herds. Small groups are also found in the Asiatic wild oxen (*Bos* spp.) which generally inhabit terrains with poor visibility. In its final retreat to the primeval forests of eastern Europe, the extinct aurochs (*Bos primigenius*) even seemed to have become solitary. By contrast, domestic cattle (*B. p. taurus*), descendants of the aurochs, are very much herd animals.

In deer, as with antelopes, the most primitive species such as musk deer (*Moschus moschiferus*) and muntjac (*Muntiacus muntjak*) are solitary and territorial forest dwellers. In other deer that inhabit less dense forest areas, the males are evidently territorial at least part of the time, while the females travel in groups, for example, fallow deer (*Dama dama*), barasingha (*Cervus duvauceli*), and sambar (subgenus *Rusa*). On the other hand, the more advanced forms such as red deer (*Cervus elaphus*) and reindeer (*Rangifer terandus*) are not territorial, instead winning possession over females by fighting among themselves and creating a dominance hierarchy. Moose (*Alces alces*) inhabit swampy forests, sometimes live solitarily, and are probably territorial.

Two species of deer living on the steppe show that there are exceptions. In roe deer (*Capreolus capreolus*), both sexes are territorial for certain periods, while the South American pampas deer (*Odocoileus bezoarticus*) live in pairs or small family groups. In their social organization, both forms seem better adapted to forest areas with good visibility than to their actual habitat. Only recently did the roe deer become a steppe dweller; its Siberian subspecies (*C. c. pygarus*) still inhabits the forest.

All the examples we have discussed refer to horned animals and deer, both belonging to the even-toed ungulates, or artiodactyls. Among the other artiodactyls we find the hippopotamuses of special interest, because they include a solitary form living in the forest, the pigmy hippo (*Choeropsis liberiensis*) and a herd-forming species inhabiting the steppe, the hippopotamus (*Hippopotamus amphibius*). The same can be found in giraffes: The okapi (*Okapia johnstoni*) lives alone in the forest, while the giraffe (*Giraffa camelopardalis*) lives in groups on steppes or wooded savannas.

Among the odd-toed ungulates, or perissodactyls, we also find solitary species inhabiting the forests and highland terrain, such as tapirs (*Tapirus*) and the Javan rhinoceros (*Rhinoceros sondaicus*); perhaps we can also include the Sumatran rhinoceros (*Dicerorhinus sumatrensis*), although we know nothing about its social behavior. By contrast, the other forms of rhinocerus, which inhabit open country, live in small groups. Of all the perissodactyls, only horses (*Equus przewalskii caballus*) form larger associations. Aside from the square-lipped rhinoceros (*Ceratotherium simum*), they are the most prominent grassland inhabitants of this mammalian order, ranging into semideserts and even desert regions. Two very different kinds of social organization can be found among equids.

Plains zebra (*Equus quagga*) and mountain zebra (*E. zebra*) as well as the horse live in permanent, nonterritorial family groups consisting of a stallion, one to six mares, and foals. Surplus stallions form bachelor herds. The adult members of such a family group know each other personally and stay together for many years—the mares until they die, the stallions until they become weak with age. These groups are therefore individualized closed associations, true "harems" in this case. Offspring leave their families according to a fixed pattern. At puberty the young mares are "kidnapped" by another stallion against the will of the family "head" and forcibly integrated into another family; they may also start a new family with one of the "bachelor" stallions. By contrast, young stallions leave their families voluntarily between the ages of one and three years and join one of the bachelor groups. Families as well as bachelor groups often merge into large anonymous herds; but even here they maintain their individualized bonds, because each member of a family knows the others.

A different social organization is found in Grevy's zebra (*E. grevyi*), in the African wild ass (*E. africanus*), and probably also in the onager (*E. hemionus onager*), where some of the stallions are consistently territorial. Their territories are relatively very large. For Grevy's zebra we measured territories ranging from barely 3 to more than 10 km², and for wild asses our estimate is many times this size. These are the largest single territories known to exist among ungulates. Most likely they are marked primarily by the presence of the owner himself, who may be distinguished from his conspecifics by particular behavior patterns, calls, and a conspicuous posture with the head held high. For orientation, the owner uses mainly large dung heaps deposited inside the territory and at the borders, where he always defecates.

Unlike all other territorial mammals, stallions belonging to these last three equid species—with some exceptions—defend their territories only when mares are ready to mate are making their stop inside a border area. Intense fighting then ensues within a boundary stretch about 50 m wide, but only between neighboring stallions. Between times each tries to drive the mare into the center of his territory. There are no real winners or losers in these fights, because the contest ends as soon as the mare leaves the border area or wanders into one of the territories.

Despite their territoriality, these stallions tolerate all conspecifics including other males. Within his territory the owner has the highest rank. Should another stallion approach an estrous mare, he will be driven off, but he need not leave the territory itself. We can observe this behavior even when no mares are present, in conflicts obviously serving to reinforce the dominance order. Serious conflicts never occur inside a territory, because strange stallions accept the priority of its owner and do not fight him. As with antelopes inhabiting dry forests, bush areas, and grasslands, the territories of these three equid species serve mating purposes.

Since the stallions are strongly attached to their territories, the population splits up into many smaller units as soon as living conditions get worse. Mares, foals, and weaker stallions migrate to better pastures, while the territorial stallions—the only ones that could reproduce—stay behind until all the food resources are finally exhausted and even the last water holes have gone dry. Only then do they relinquish their territories and migrate, but not necessarily to where the mares are. This is not unfavorable in zones with fairly regular seasonal cycles, because the breeding season coincides with the rainy periods. In Grevy's zebra, however, some populations live in semidesert and desert regions, just like all onagers and African wild asses. Thus they inhabit regions with irregular or very scattered precipitation. The fact that these animal species are successful anyway probably reflects less on their social organization than on their physiological adaptation to desert life, especially the ability to feed on dry grasses and to get along with very little water, containing salt or soda.

As with antelopes, the ancestor of equid species was a small duikerlike animal that inhabited dense primeval forests with poor visibility. This primitive horse (*Hyracotherium*) probably also resembled the duiker antelopes in its social organization. When the equids changed their diet from leaves to grass, during the Middle Tertiary, they evidently adapted their social organization to the new habitat: Territories grew larger but came to serve only the purpose of mating; the females formed nonterritorial groups. Grevy's zebra, the wild ass, and the onager remained at this level, while Burchell's zebra, the mountain zebra, and the horse gave up all forms of territoriality and developed a new kind of social life, the stable family group.

The ancestral horse

In our discussion we have placed special emphasis on those environmental conditions that affect the social organization of a species. In many cases, particular social systems are extremely well adapted to the specific animal's environment, while in other cases such adaptations are less obvious or even entirely lacking. In these latter forms, then, the social organization is determined mostly by genetic factors rather than ecological ones. For a species to survive, of course, adaptations of group behavior to a habitat are not the only important ones; morphological and physiological traits can compensate for inadequate sociological adaptations.

Animal sociology is a relatively recent branch of ethology. Although some mammalian groups have indeed been studied in greater detail (ungulates and especially primates, elephants, carnivores, and rodents), our knowledge of this area as a whole is still very much in the beginning stages. The best-known species are mainly larger animals living in open country, which are easily observed in the wild. Smaller species and those inhabiting forests and oceans are far more difficult—if not impossible—to observe directly, so that we know much less about their social behavior. In recent years scientists have developed a method that opens the way to hitherto undreamed-of opportunities for investigation: telemetry (see Color plate, p. 19), in which radio transmitters are placed on the animal. This technique should provide us with a whole range of new and significant information about the social organization of mammals and other animals.

35 The Organization and Function of Primate Groups

Viewed from the outside, most primate groups are relatively closed associations of conspecifics inhabiting a limited territory. The members of this group know the sources of food and water, and the areas of danger within their territory, planning their route each day so that they can take in a sufficient amount of food while at the same time avoiding danger as much as possible. Their daily march always ends at some elevated sleeping site, on cliffs or trees that nocturnal predators would find hard to climb. The band knows its neighboring groups of conspecifics, but it usually avoids them or threatens them if they meet at the territorial peripheries.

The organization and function of primate groups, by H. Kummer

The group's internal organization is a complicated network of rank and role relations, of alliances and rivalries, of close bonds between near relatives and looser ties between animals living in the center of the group as well as those moving about on its periphery. Unlike most ungulates, which regroup annually for the purpose of reproduction (Chapter 34), the typical monkey band maintains the same structure year after year. Mating and child-bearing take place throughout the year, and the young maintain close contact with males and females of all ages until, several years later, they have matured and take up their ecological and social roles as adults.

Special features of social life in monkeys

Monkeys have a very high capacity for learning and a long period of juvenile dependency as compared with other mammals of their size. Because of this, the members of a monkey group are able to occupy various roles quickly and at any time when a vacancy arises. Every group has its special network of relations, shaped by the peculiarities of its members and thus undergoing change when these individuals age, leave the group, or die. The uniqueness of whatever social condition prevails at any time cannot be met simply with rigid or fixed and innate behavior patterns. Rather, each individual must get to know his or her other fellows in the group "personally" through observation and direct contact or communication. Each individual has to learn who belongs to what particular role, whose rank is going up or down, and who would support whom in what situations. Scientists who have closely observed groups and individuals have found that each monkey combines all these experiences and is able to predict quite accurately the behavior of the other group members in any current situation.

Primates are not equipped with any complete or fixed innate behavior programs like those of many fishes, birds, and even insects. It is likely that their simple movement patterns and calls used in communicating with one another (for example, behaviors of threat, submission, flight, and approach) are inborn. But they do not, for example, have stereotypical behavior sequences to be used in courtship. Monkeys need close contact with conspecifics, where they are nurtured and given many opportunities to learn. In isolation they would not, for example, acquire the sexual behavior patterns normal and typical of their species.

Furthermore, even in dealing with their nonsocial environment, monkeys and all other primates possess very few complete and innate techniques. The infant seems to learn what is edible by observing its mother and older siblings. Monkeys do not store food like some rodents do. They must gather their food each day and eat it right away, and only chimpanzees occasionally carry food around for more than a kilometer. Lower primates build neither nests nor shelters, in contrast to many other animals. The large anthropoid apes do build nests, but they construct them far less artfully than, for example, weaver birds build theirs. Male baboons have the strength and the powerful jaws to catch and eat young antelopes. But they practically never hunt together with other males, nor do they share their catch with females or young as many social predators do [Some baboons in Kenya have been observed hunting as a group and to some extent sharing the nest.] Thus, the hunting behavior of baboons is of little use to the troop as a whole.

Fig. 35-1. The overlapping ranges of olive (or anubis) baboon troops in Nairobi Park, Keyna. The double letters stand for individual groups in their particular area, ranging in size from twelve to eighty-seven animals. Each core area is used exclusively by the resident group. This is the area visited most frequently and it also contains the regular sleeping trees. When two neighboring troops meet at the overlapping peripheries, they do not fight, but instead avoid each other. Baboons are not territorial animals.

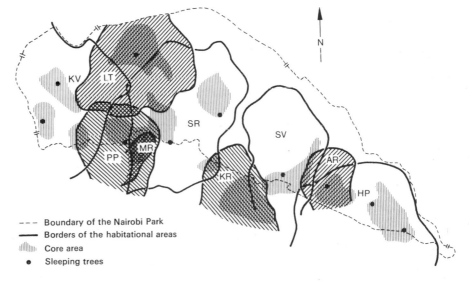

--- Boundary of the Nairobi Park
— Borders of the habitational areas
||||| Core area
• Sleeping trees

On the other hand, Jane van Lawick-Goodall observed several times how male adult chimpanzees of the Gombe Stream Reserve encircled and caught colobus monkeys or young baboons with apparent deliberation, and then gave some of the prey to others begging for it. But their hunting activity does not follow any set plan and is unlikely to be innate, any more than the various kinds of tool use in chimps which vary from one place to another (see Chapter 23). And yet it was a primate—man—who created the most highly evolved technology on earth. The majority of primate species, the so-called lower monkeys who will be the focus of our discussion, have had no part in this.

Research on primates in the field began with describing the social systems of whatever species was being studied. Around 1965, it became evident that these species-specific systems were subject to regional variation, and researchers started to ask about the possible adaptive value of a particular social system functioning within a certain environment. At the same time, Japanese primatologists induced Japanese macaques (*Macaca fuscata*) to develop group-specific behaviors that these monkeys would then actually pass on through tradition.

At the beginning of this chapter when we described how a band of monkeys would march around within its territory, we wrote as if this group were a closed unit with a common nervous system or a single leader. But these impressions are misleading. It is highly unusual for one particular animal to lead a group of monkeys. The members choose their common route by communicating with one another through sounds and gestures, or, perhaps more frequently, by watching the behavior of others in the group. Decisive movement by a high-ranking animal—either forging ahead or hesitating—or the falling back of a young monkey can cause the group to change its direction.

We may say that there are two levels of activity in a monkey group. The individual animal gathers food and finds its resting places within a

The individual and the group

Fig. 35-2. Daily marching routes of baboons, green vervets (or grass monkeys), and patas monkeys (or red guenons) near Chobi in Murchison Falls Park, Uganda. Every night, the baboon and vervet groups return to their high and safe sleeping trees on the banks of the Nile. Entire groups cluster on a few trees to spend the night. By contrast, the patas monkeys are true savannah dwellers and sleep on low-standing trees in open grassland (see rectangles). Their groups are smaller to begin with, and members easily disperse, so that each little tree harbors only one or two animals. This sleeping arrangement reduces the danger of predator attack.

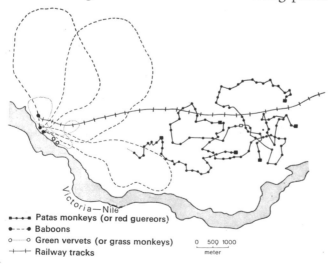

Victoria—Nile

●━━●━━● Patas monkeys (or red guereors)
●━ ─ ─● Baboons
○┈┈┈┈○ Green vervets (or grass monkeys)
┼━┼━┼ Railway tracks

0   500 1000
      meter

perimeter of about 20 m, and avoids dangerous spots. Correspondingly, the group as a whole extends its activities to a perimeter of several hundred meters or even several kilometers. Each adult monkey eats, drinks, flees, and climbs up to its sleeping place without any help from others. Thus, no one passes food to another adult animal, builds a shelter for anyone else, or gives physical support at a time of danger. Only infants are nursed and fed by their mothers, carried about, or rescued from danger by adults.

But before the individual members of a group can exploit any particular area, the group members must explore it. And so the primary task of a group is to exchange information. When a baboon comes upon a small water hole, it conveys this information to the others by its conspicuous drinking posture, with the hind part and tail sticking up. Similarly, the nearest neighbors are attracted by the excited hand movements with which a baboon digs up a delectable tuber. Beyond that, the discoveries made by one group member can also serve the others on a more long-range basis: During a drought, an old male can lead his group to a distant pond which he remembers from a previous visit during a similar dry period (see Color plate, p. 531).

Depending on the species and locale, a monkey group may contain two to several hundred members. The smallest groups are found among gibbons (*Hylobates*), the small South American titi monkeys (*Callicebus*), and the marmosets (tamarins and marmosets; family Callitrichidae), which also belong to the New World superfamily. Although gibbons and titi monkeys are only very distantly related, both species form monogamous bonds, and the group then consists of the father, mother, and young. The situation is similar with marmosets, although here the older juveniles usually take part in raising their younger siblings, so that small groups are formed, usually with six to eight animals. All of these cases apply to territorial, forest-dwelling species, and both parents defend their group territory against conspecific outsiders. In gibbons, neighboring groups meet regularly at their common borders and loudly chase each other back and forth over the boundary line. The average territory is about 2 km in diameter for gibbons, but only 70 m for titi monkeys.

More commonly we find forest-dwelling species that form somewhat larger groups of a few adult males and several females with their young. The size of these groups usually ranges from ten to twenty members. Again, these species generally show territorial behavior or at least emit vocal signals indicating the position of each group. Examples are the black-and-white colobus monkeys or guerezas (*Colobus polykomos*) and the howler monkeys (*Alouatta*) which are known for their habit of "fighting" only on a vocal level. The territories of these species are also small, and groups travel in trees at a daily distance of only a few hundred meters.

Larger groups of around seventy animals typically inhabit the edge of forests, open woodland, and savannahs. Examples are the green vervets or grass monkeys (*Cercopithecus aethiops*), baboons (*Papio*) with the ex-

Group structure and composition

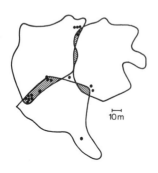

Fig. 35-3. Territories of three groups of orabussu titi monkeys (*Callicebus moloch*). In contrast to the range or habitation areas of baboons, these territories overlap hardly at all. To our knowledge, the residents or territory holders defend their boundaries. In any case, the daily "fights" in which groups threaten and chase each other back and forth (black dots) do take place along the overlapping peripheries (striated regions).

ception of the hamadryas baboon (*P. hamadryas*), and the rhesus monkeys (*Macaca mulatta*). These species travel through fairly large and greatly overlapping ranges or habitation areas of roughly 5–10 km², but they do not defend them as territories, preferring to avoid neighboring groups traveling within the overlapping border areas. We assume that there are two reasons for this: First, the inhabitants of each region are always less familiar with the far-reaching border areas, and this leads to caution rather than aggressiveness. Second, savannah habitats do not offer the abundance of food which for the forest-dwelling and territorial gibbons often leads to violent disputes at the boundaries.

In more scanty areas we find a tendency for multilayered societies to evolve. The smallest unit is always the unimale group, consisting of one adult male with several females and young. In the patas or red guenon monkeys (*Erythrocebus patas*) of the open savannahs, the one-male groups live separately. In geladas (*Theropithecus gelada*), which inhabit the high, treeless Ethiopean grasslands, the smaller groups gather during the bountiful season to form herds of several hundred animals. The hamadryas baboons (*Papio hamadryas*) of the thornless steppes and semideserts by the Red Sea even show a three-level organization: Several unimale groups combine into a permanent band, and several bands gather in herds on the face of sleeping cliffs; there we find a gathering of up to 700 animals. The habitation areas of these species may be as large as 40 km². They are never defended as territories. The hamadryas baboons living in the southern Danakil-Plain travel an average of 13 km in a day (see Color plate, p. 531).

Since the size of a group varies to such a large extent from one species to the next, we must assume that, among other things, group composition is adapted to life in a certain environment. After all, the social organization of a species must be "economical," that is, suitable to meet the demands of prevailing ecological conditions. Those primates living in open habitats the majority of them terrestrial, usually have a smaller population density than those living in the forest. The former also exploit much larger areas per group and travel over a greater distance per day than the latter. This is understandable because there is less available food in open areas. On the other hand, it is difficult to see why the groups of open-space primates, despite a smaller population density, are larger than those of forest-dwelling monkeys. We must conclude that the optimum group size depends not only on the amount of food per square kilometer; it is also important whether this food supply is evenly or unevenly distributed, and whether its abundance and distribution varies with the seasons. Of similar importance is the distribution of sleeping sites. Other factors determining group size depend on the way in which a particular species manages to avoid running into predators.

The most important factor, however, is still the food supply. With a certain food density, a large group is forced to travel farther to satisfy all

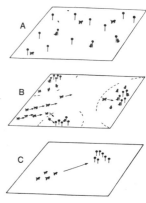

↑ Food plant
-- Limits of information exchange

Fig. 35-4. Schematic illustration of various ways that size of a group may be related to the distribution of food. A. Food is available in small portions, so that only one monkey can feed at any one spot. These portions are evenly distributed, and there are enough portions to satisfy all members of the band. Here, groups can be either small or large. B. The food portions are larger, so that a number of animals may feed from them together. Even larger groups can make good use of this arrangement if the members temporarily disperse and communicate their finds to only a few others in the vicinity. C. The distance between food portions is greater than the maximum distance for communication between monkeys. In this case, an independent small group is the most adaptive unit; its size should be adapted to the average size of the food portions.

its members than is a small group moving in the same formation. Thus, when food density is low, only small groups will be able to survive. Furthermore, communicating a food source to conspecifics is a good thing only when this food is concentrated in certain spots and there is enough to satisfy others as well as the finder. If the food is clustered in small quantities and evenly distributed, such as blades of grass in a meadow, there is nothing to discover or communicate. It is just as pointless, and even detrimental, when the finder of a pod-carrying bush attracts twenty other group members through his behavior, when the yield is just enough for one animal. Information regarding a find should reach just enough conspecifics to ensure a maximum exploitation of the food source.

It seems that there are three ways for primates to avoid the clustering of too many animals at one source of food. First of all, except for chimpanzees, primates usually make no sounds when they find something to eat. The finder draws only the attention of his or her nearest neighbors who can see him or her at that moment. Second, when feeding, monkeys do not tolerate the presence of lower-ranking group members within a certain distance. These more subordinate animals are forced to look for another feeding place when the first is already occupied. However, if the food sources lie very far apart, this mechanism is no longer useful, because the lower-ranking animals would end up as hungry spectators at every feeding place, or they would have to leave the group. Under these conditions, there is only the third solution: adapting the size of the group to the size of individual food sources. The optimum group size is one in which all members can satisfy their hunger simultaneously. The same applies to the size and distribution of sleeping sites.

How is group size dependent on the need to avoid predators? When some group members do run into an enemy, it is best that they inform as many conspecifics as possible, because the predator could simply disappear for a short time and then approach again. And so monkeys seeing danger will convey this information with loud calls that carry over long distances. Furthermore, vervets use different alarm calls for snakes, for aerial enemies, and for terrestrial ones. Therefore, large groups are best suited for spreading information, as well as for intimidating predators. We would expect, then, that in areas with a large number of predators, the monkey groups in turn are as large as the food supply and sleeping sites will permit.

This does seem to be the case with primates inhabiting open spaces. Vervets and baboons living in savannas—areas that contain relatively ample food supplies but also many predators—do travel in large groups. Baboon troops are well guarded by their very aggressive and armipotent males; they often venture far into the treeless grassland areas. These baboons usually try to flee when sighting a large predatory cat, but they cannot run for very long. Observers have watched adult males in Nairobi Park (Kenya) approach a cheetah and drive it away. In Amboseli Park (Kenya) baboons fought with a leopard, and some were injured; but the

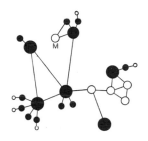

● Adult male
● Sexually mature female
○ Female less than 2 years old
○ Juvenile or subadult male

Fig. 35-5. Schematic network of relations within a band of hamadryas baboons. All the sexually mature females are the exclusive "property" of some adult male forming a one-male group (black shapes). Social grooming and sexual activity between adults take place only within the one-male group. Relations among these "families" are determined solely by the males, and consist primarily of "signaling": The male presents his buttocks to the other and in this way inhibits aggression. The subaltern male (M) is a subsidiary member of the unit; his leading male allows him to groom the females but not to engage in sexual activity with them.

leopard ran off. The huge males of the South African chackma baboons (*Papio ursinus*) occasionally maim or kill dogs that come close to the herd.

Because of the normal ratio of males to females in the one-male units, a number of adult males are left without females. Thus in some species we find the formation of more-or-less extensive all-male groups, similar to those formed among some ungulates (see Chapter 34). It is after all sufficient for one male to reproduce with several females. Furthermore, separating surplus males from male-female groups serves to protect females and young from too much competition for food, since the males are much stronger.

However, in looking at the way that group structure helps to fulfill certain functions, we still cannot explain the social behavior patterns with which these animals create large or small multi-male or one-male groups. They certainly do not consciously adapt their social structures to a particular habitat, through insight into ecological conditions and relationships. A population creates its groups in the way that it "has to" on the basis of innate social responses and the influence of social tradition. We can see this in the social systems where small groups recurrently join into larger ones and then split up again, a procedure well suited to conditions where the supply and distribution of food sources vary from day to day and throughout the year.

It seems that chimpanzees have accomplished the optimum adaptation to whatever habitat they may live in. These primates may form groups of just about any size and composition at a moment's notice: large mixed groups, all-male units which travel over particularly large areas, less mobile mother groups with young, or a female in heat staying with one or several males. We also come upon solitary individuals of either sex. Apart from very strong and enduring ties such as occur between a mother and her offspring of all ages, any combination is possible.

We would expect to find such an expedient organization in lower monkeys as well. The fact that these animals are never as flexible is probably due to their limited social capacity for adaptation. Thus, for example, groups of hamadryas baboons or geladas can split up only along pre-established "seams." One reason is that the males claim their females —and in geldas vice versa, too—as exclusive property and never allow them to join other groups. Large anthropoid apes are more tolerant of this.

Patas monkeys, geladas, and hamadryas baboons use different behavioral means to build up their one-male groups. The patas monkeys seem to have the simplest methods, since they do not form herds and thus avoid the constant danger that females of one unit will be taken away by rivals. All that happens is that the leading male of each group drives away all the other males, including his own sexually mature sons. Evidently, males are intolerant of each other mainly in the presence of females, since patas males without females often join together in all-male groups. We can observe

How group composition comes to be

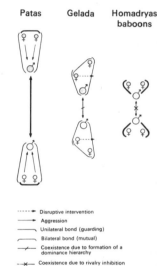

| Patas | Gelada | Homadryas baboons |

----→ Disruptive intervention

——→ Aggression

——⌐ Unilateral bond (guarding)

—— Bilateral bond (mutual)

—— Coexistence due to formation of a dominance hierarchy

—×— Coexistence due to rivalry inhibition

Fig. 35-6. Different ways of forming one-male groups. In the patas monkeys (or red guenons), the females stay together, forming the core of the group. The male stands apart and keeps other

the disrupting effect that females have on male compatibility in other species as well, for example in entellus langurs and hamadryas baboons.

If the males of geladas and hamadryas baboons were as intolerant of each other as the patas monkeys are, they would create similar one-male groups, but then they could not form herds. Experiments in enclosures have shown that here the males engage in a fight for dominance until the weaker one submits; after that they accept each other. Patas males, by contrast, have no submissive behavior or postures, and the rivals fight until one of them is finished. The formation of herds, so advantageous from an ecological stand point, is therefore made possible through the behavior mechanisms of submission and the fact that males create dominance hierarchies among themselves. Here we are not dealing directly with either ecological pressures or intelligent behavior, but rather with an innate capacity that developed in the course of evolution.

In the societies of geladas and hamadryas baboons, where we can observe this so-called "fission" effect, the one-male groups remain closed and stable even within the herd, so that fission always occurs along clear-cut lines. As far as the ecological requirements are concerned, it does not matter how this is accomplished. And in fact, the two species again use different mechanisms. In both societies each female is the closely guarded and exclusive property of one male; but in hamadryas baboons, the male alone has the task of watching and defending his females. When his group is on the move, he will look back frequently. Should one of his females stay too far behind or come too close to another male, her leader will stare at her, beat the ground with his hands, or rush up and nip her in the back of the neck with his canine teeth. The female reacts by screaming and following her male closely back to the group. In this way, the one-male unit always stays together within the band or herd and may split away any time the male chooses (see Color plate, p. 532).

On the other hand, gelada females, which also belong to one-male groups, are allowed to disperse throughout the herd. From time to time they will gather around their male, and in the meantime they will frequently look around for him. When he leaves the herd, his females follow with their young. The male in turn intermittently surveys the herd. Should he notice that one of his females has moved too far off, he rushes toward her but he does not bite. Instead, she turns toward him, crouches, and screams. He then sits down, and the female starts delousing or "grooming" him while he produces an enormous "yawn," displaying his long canine teeth.

We can see that the ecological mechanism of the one-male group operates both in geladas and in hamadryas baboons: The males claim possession of a number of females, but control is more unilateral in hamadryas baboons than in geladas. This in turn creates a certain danger for the hamadryas baboons: When two unfamiliar bands meet at a sleeping cliff, or when we use food to lure a band into an unnaturally tight space,

males away from his group. In geladas, the male guards the highest-ranking female, who in turn guards the next-highest-ranking female, and so on down the line. Each lower-ranking female is strongly inhibited from making contact with the male because of intervention from higher-ranking females. The males establish and maintain clear-cut ranking relations among one another, and therefore get along within the herd. In the one-male groups of hamadryas baboons, all females are the passive property of some male who guards them closely. Conflict inhibition between rivals prevents one male from taking another's females.

Rivalry and inhibition

the males end up fighting for the females (see Color plate, p. 532). There is always the danger that an entire herd will get involved in fights; but hamadryas baboons have developed a precaution against this. Once a male has realized that a strange female belongs to another member of the band, there is a social inhibition that prevents him from taking possession of the female even if he is stronger than his rival. Thus he will leave the female to her partner, turn his back, and show all signs of a behavior conflict: scratching himself profusely, looking at the sky, or shuffling pebbles on the ground with his fingertips. This inhibition against competing with a rival serves to protect the existing pair-bond and will fail only where the owner of the female is very much weaker than the strange male. Occasionally such transfers are necessary, because otherwise an "upper stratum" of old males would end up owning all the females, while the younger and stronger males would remain without mates.

A group of a given size and composition has one means for dealing with the environment that is not available to higher animals living as solitary individuals: The group can alter its shape and spatial arrangement. The bands of some terrestrial monkeys mass together in a tight unit when crossing unfamiliar or dangerous country. Olive (or anubis) baboons (*Papio anubis*) in Nairobi Park form protective male fronts that act like a shield when encountering enemies. Gelada herds, which often move along the upper edge of cliff faces, are frequently flanked on the more dangerous inland side by a troop of unmated males. The females and young keep closer to the precipice of the cliff; this is where the herd flees when threatened. Often two one-male units of hamadryas baboons will travel together, with the females arranged in a straight line between the two males. But should the males start to fight, the females immediately form rows to the outside. In general, the females always choose a safe position at any particular moment. On their daily march, a baboon male often lets the weaker members of the herd pass and then forms the rear guard himself.

The less abundant and more widely dispersed the essential food sources, water holes, and sleeping sites in a habitat, the more important is the quality of leadership in a group. If a particular day's route is poorly plotted, more energy could be expended than taken in. As far as we know, primates choose their routes by constantly exploring during their daily march, and also on the basis of memory.

From their studies on yellow baboons (*Papio cynocephalus*) in Amboseli Park, J. and S. Altmann suggest that at least some of the herd members know the geographic layout of their habitat, its food sources and areas of danger, and the status of these conditions at any given time. If this is the case, the route can be planned to good advantage—at least in general outline—even before the herd starts to move. Observations on hamadryas baboons support this assumption. Every morning, while still at the sleeping cliff, the herd sends runners off in two or three directions. At the tip of

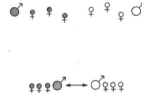

Fig. 35-7. Two ways that two one-male groups in hamadryas baboons may be arranged: above, when conflict arises between the the two leading males; below, while foraging for food.

Fig. 35-8. Rivalry inhibition in a hamadryas baboon. This animal has turned away from his rival's group (here a male-female pair, not shown), lowered his gaze, and begins pushing around some stones lying on the ground.

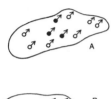

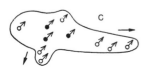

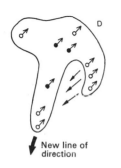

▼ New line of direction

🗢 Outline of the troop
♂ Initiative-taking males
♂ Decision-making males

Fig. 35-9. Changes in troop arrangement of hamadryas baboons just before starting on a march. A. Resting phase during the early morning. The high-ranking males (decision-makers) sit near the center, while the lower-ranking ones (initiative-takers) stay closer to the periphery of the troop. B. Some of the initiating males are forming an offshoot heading east. C. A second group of initiating males form another offshoot heading south. D. One of the decision-making males starts moving south, inducing the entire herd to begin marching in that direction.

these "offshoots" there are several males, mostly young adults with an obvious urge to push on in their particular direction. These scouts may enlist other males for their route by approaching them, making a fast turn, and then moving off a few meters in their intended direction. Once such a runner loses too many followers, even these lead males will retreat back to the herd. Their behavior serves only as a suggestion; the decision lies with older males who watch the various offshoots from the center of the herd. When one of these old males rises and moves off in the direction of a particular one, it takes only seconds for the whole herd to stream in his direction and to start the march.

We do not as yet know how the monkeys, just setting out on the march keep up with what is happening within their habitat. The Altmanns' work does show that the daily plan can be modified because of previous experiences at certain points. One cluster of sleeping trees was abandoned after two members of the troop were killed by a leopard. While these trees had been used on thirteen of the fifty-seven preceding nights, they remained empty during all of the sixty-eight control nights after the fatal incident.

Within the group, division of labor is not always an advantage. In food gathering and flight, for example, no individual in the group can be replaced by others. On the contrary, certain activities require that all group members do the same thing at the same time. When a group visits a water hole in open country during the dry period, each member has to drink the water even if he is not yet thirsty. The next opportunity, after all, may not come until the following day. Thus, if an animal does not make use of the water hole, it runs the risk of dying of thirst, or it is forced to go alone on a perilous trip to another water hole. Such risks are avoided because of the wide-spread tendency of social animals to do the same thing at the same time. In ethology, we call this phenomenon "transfer of mood," while in everyday language we would speak of contagion. It is this transfer of mood that causes birds to fly off together, and even individuals in human groups often conform to the behavior of the majority.

There are other functions, however, which require one or a few animals to act in the place of all the others. It would be pointless, for example, if all members of a group sat on a bare termite hill at the same time and watched out for leopards. We assume that this does not happen because of a social inhibition, preventing an animal from performing a certain activity if it sees others doing it.

The most powerful form of social inhibition is that produced by rank, or status. Whenever two animals have a simultaneous intention to do something, the more subordinate animal will yield to the dominant one if the latter claims this activity or its object for himself or herself alone. This applies particularly to desirable mates and preferred sleeping spots. In extreme cases, rank orders or dominance hierarchies are established through individual fights, usually remaining unchallenged and unaltered for several months thereafter. In this way, the number of conflicts is reduced because all individuals concerned are aware of the priority rights which remain stable for a certain period.

Not all such rights, however, are assigned within the same dominance hierarchy, or subject to any rank ordering at all. A nearly adult male is practically always superior to a half-grown one in the food-gathering hierarchy. On the other hand, the younger male has more sexual freedom because the adult males are more tolerant with him than with the almost mature rival. In a group of olive baboons in Nairobi Park, three adult males had formed a coalition which included one old animal with worn-out canine teeth. Thanks to the support of his allies, the old male usually beat a fourth male in competing for females, even though this outsider was superior to the old monkey in individual strength. A young rhesus monkey is able to beat his peers and even larger and stronger juveniles in rough and tumble play if his mother has a higher status than the mothers of his playmates. He attains his own rank by the fact that his mother will come to his aid as soon as he screams. Furthermore, it is not always the highest-ranking and most aggressive males who may perform the greatest number of mating acts: In one group of chackma baboons, the highest success with females went to an older, nonaggressive, and only third-ranking male. Generally, high-ranking males terminate conflicts in the group by threatening one of the opponents. On the other hand, one group of bonnet macaques (*Macaca radiata*) studied by P. E. Simonds had a relatively low-ranking old male who was extremely effective in inhibiting intragroup conflict, even though he threatened very little.

The taking up of roles will depend not only on status or rank, but also on age, sex, and personality. For example, age determines which male has the decision-making role in leading a band of hamadryas baboons. Old males determine the marching route of the troop even when they have lost their females to younger males and no longer possess their full physical capacities.

In contrast to rank positions, which remain stable for several months, some roles may shift from one hour to the next. Therefore, a role as such is not innate; but dispositions for certain behavior patterns and roles are indeed genetically determined. These may differ between males and females, and may also change with age. In general, male primates tend more than females toward rough-and-tumble play and contests, to explore, to leave the group, to show territorial behavior, and to display defensive behaviors. Females fight less, but they perform more care of young and more social grooming even before reaching sexual maturity. Within the group, females function to maintain social bonds between members.

Experiments with pig-tailed macaques (*Macaca nemestrina*) in enclosures showed how much the exploratory and defensive dispositions of males could depend on age and rank. When an empty cage was placed in the

Rank ordering

Distribution of roles

enclosure, it was predominantly the juvenile males who approached. Such curiosity behavior toward a strange but nonthreatening object or conspecific may also be observed in baboons among the sexually immature males. Where the cages contained an object of a possibly threatening nature, for example a snake, the lower-ranking adult males were the ones who approached. Only when there was an open threat to the group, such as an attacking person, did the highest-ranking male thrust himself forward to counterattack.

In the field, few primatologists have made such experiments to study roles. E. W. Menzel presented Japanese macaques with buckets containing alternately food, a mildly aversive object (such as a doll), or a combination of both. The monkeys could see what was inside only when very close to the bucket. The adults left it to the young to investigate, watching them for their reactions. On the basis of such observations, the adults would then either come closer or stay away.

Shaping social behavior

To a large extent, the social and ecological skills of an adult are the product of many years of socialization within the group and of experience with his or her habitat. Female rhesus monkeys raised without mothers grow into adults with such severe behavior disturbances that they push their own newborn away (Chapter 24). E. A. Mason observed that one-year-old rhesus monkeys born in the laboratory are socially less active than experimental animals of the same age who up to that time had lived in their natural environment. The lab animals display less grooming, more frequent aggression, develop less stable social bonds, and often show abnormal sexual behavior. In all hamadryas baboon herds studied in the field, the males guard their females; but males born in zoos do not show this watchdog behavior, although they live together with older males who could serve as models. We may compare the ecological and social skills of free-living primates with a profession, where individuals must learn to combine single, innate behaviors and to use them in their proper situational context through many years of observation and experience.

An infant monkey learns mainly through observation. For example, it may examine a certain spot on the ground where its mother has just dug up some food, then smell at this food while the mother is chewing it. Learning also takes place in adult Japanese macaques who watch their juveniles looking into the bucket. Male hamadryas baboons start their career as leaders of the herd during their juvenile period while traveling with a one-male group at the side or back, without helping to decide on direction. As young adults, these males travel increasingly at the head of the group, gaining experience about what the older male considers to be the best path by watching his behavior (either following or hesitating), and then acting accordingly. Only after years of experience, when they in turn become older males, do these baboons make decisions for the group and finally for the entire herd.

Individuals in a group must learn all the rank orders and alliances of which they form a part, and keep relearning them when changes take place. It seems quite clear, however, that some early social experiences also determine future behavior in a young animal. D. S. Sade, for example, described the following developments in a group of free-living rhesus monkeys on Cayo Santiago, an island near Puerto Rico: "In the second half of their first year of life, the young monkeys start to wrestle with each other. They can beat others of the same age whose mothers occupy a lower rank than their own mother, and conversely they are beaten by those whose mothers have a higher status. . . . As adults, the females arrange themselves into the hierarchy precisely one step below their mother."

Maturing males, too, frequently occupy a rank close to that of their mother. Beginning with puberty, however, they may rise or fall in rank independently. Rank assignment based on the status of the mother was also found in Japanese macaques. In this case, the oldest daughter has an adult rank immediately below the mother, and her sisters—really half-sisters—are ordered down the line according to age. Strangely enough, the situation is just the opposite in rhesus macaques on Cayo Santiago: The youngest daughter "passes" her older sister before the second year of life, so that the youngest of the adult half-sisters always ranks highest, next to the mother. In both species, however, we know that the daughters do not inherit their status from the mother, but rather attain it through their mother's social influence.

The extent to which the behavior of adult primates is determined by social experiences in childhood is so great that real "traditions" may arise within a group. One case—now famous—is the "potato washing" behavior of Japanese macaques on the small island of Koshima. In 1935 the female Imo, then barely two years old, developed a technique for removing sand from the sweet potatoes that people on the project regularly threw on the beach. Instead of doing a scanty job, rubbing off the sand with her hands, she carried the potato to a brook, dipped it into the water with one hand, and washed it with the other. In the years following, this new technique slowly spread throughout the group, and in addition, the animals came to prefer more and more the salty ocean water to that of the brook. It was the juveniles of Imo's age who learned most quickly, and they passed the method on to their mothers and older siblings. This habit, then, spread around by way of "kinship" lines, that is, through particularly close and personal relationships. This was probably one reason why none of the adult males adopted the washing habit. Certain other families learned to wash potatoes and to use additional techniques more easily than others did—evidence for the fact that hereditary propensities help to determine the individual's ability to learn.

The infants of mothers already in the habit of washing potatoes learned not only this technique but also to swim and dive, since now their mothers carried them to the water right from the start. Later on, some of these

Fig. 35-10. A Japanese macaque washing potatoes. This habit is a local tradition created by the group of macaques on the island of Koshima.

youngsters could even swim to a neighboring island. The potatoes they ate had nothing but a salty flavor, and some began to treat their potatoes not only by washing them but also by dipping them in the sea water after each bite to enhance the taste. With these modifications, the new tradition began to show one characteristic of human culture and civilization: A new habit easily creates still further opportunities for learning, and these again lead to other changes, or innovations, in behavior. When this ability to learn and to gain insight into certain cause-and-effect relationships is great enough, it may lead a species to ever-accelerating modifications in behavior, perhaps culminating in changes of habitat.

In man, the basis for such modifications must have been there for thousands of years. Using a conspecific as a social tool antedated the use of technical tools by several million years and probably paved the way for this later development. But the skill of using technical tools in turn opened up habitats to man greatly exceeding the range of distribution possible for any other primates. The new ecological niches then stimulated further local developments, ultimately resulting in a unique multiplicity of ecological and social forms of life. As man's material standard of living improved, and cultural values became more and more important, he also continued separating his social systems from the direct environmental pressures of natural selection. But the social systems in turn developed internal selection pressures. In the extreme condition of modern human societies, the process of sociocultural development finally assumed an "inner life" of its own, forgetting that all social systems and forms of social evolution remain ultimately dependent on the ecological system of our planet.

36   Social Stress in Mammals: Tree Shrews

Each animal species has its own social structure, established and maintained through continuous contact among neighboring individuals. In mammals, where this social environment is most important, every animal forms a part of its neighbors' environment and continually affects their behavior and physiological state. Thus, any change within a society—for example, through an increase in population—may influence the physiological condition of all its members. This applies both to "social" animal species—those living in groups or herds—and to "solitary" animals, among which each may occupy a well-defined territory separate from all the others.

In many cases the normally aggressive behavior patterns whereby a social structure is formed and maintained are observed easily enough, and in some mammals they have been studied in great detail. But as yet we know very little about the factors of physiological change, since intraspecific stimuli have different effects on individual animals.

In this way, stimuli or behaviors acting as stimuli may have a contagious effect, or cause changes in mood. We see this especially in social animals, where all group members often show the same behavior even if some individuals do not have the corresponding physiological need at that particular time. Thus, for example, the nutritional standard of social animals held in groups is markedly higher than that of isolated ones, because the group members stimulate each other to feed.

Conspecifics, or stimuli associated with them, may also act on the central nervous system and the hypophysis to influence certain physiological processes, expecially the functioning of the gonads, and thus affect fertility. This includes the way many animal species control their rutting cycles through visual or olfactory stimuli produced by the males (Whitten effect). Growth rates and the advent of sexual maturity itself may also be facilitated in both males and females by the presence or odor of conspecifics. Conversely, in some species of mouse the embedding of a fertilized egg within the uterus may be prevented by the odor of strange males (Bruce effect).

Social stress in mammals, by D. v. Holst

"Alarming" through
stress

These phenomena result only from intraspecific stimuli, but in addition, the organism may be generally activated (alarmed) through many harmful factors such as cold, lack of oxygen, contagious diseases, and even fights and high population density. In 1950, H. Selye labeled these alarming effects as stressors; the corresponding body state would then be called stress.

Regardless of the cause of alarm, the body always has a more-or-less consistent pattern of response, called the "general adaptation syndrome" (syndrome = diagnostic category arising from a combination of various characteristic symptoms). The sympathetic nervous system is activated, and the adrenals secrete hormones from their medulla (adrenalin, noradrenalin) and cortex (corticoids). This means that within a period of from seconds to a few minutes the heart begins to beat faster and stronger and the blood pressure rises; blood circulation through the skeletal muscles increases, and at the same time less blood reaches the skin, the gastric-intestinal tract, and the kidneys. In response to the heightened need for energy, the liver releases more blood sugar; muscles and fat are turned into sugar and stored as glycogen (a substance corresponding to plant starch) in the liver as well as, to some extent, in the muscles. At the same time, the synthesis of body substances and the activity of the gonads are inhibited.

When subject to acute stress, then, the body undergoes a change in priorities: The organism is prepared in the best possible way for expected or present action (for example, fight or flight), while any activities not directly related to combatting or adapting to this stress are inhibited.

If situations of this kind occur only rarely or briefly, the animal is not affected in any significant way. But if the body is under constant or long-lasting stress (for example, through continuous fighting or cold temperature), it will adapt with a change in its physiological state. The activity of the adrenal medulla and the cortex rises according to degree of stress, while the functioning of other organs—especially the gonads—decreases in a corresponding manner. Excessive stress may then lead to a total collapse in adaptive capacity, ending in death.

Stress and the
constancy of individual
numbers

For some time now, ecologists and ethologists have also had a special interest in the causes and effects of stress, because this process should help us understand the control of natural populations. As yet we know very little about the mechanisms of regulation. It is puzzling that although each animal species is capable of producing far more offspring than necessary for the next generation to be as large as the previous one, the number of individuals remains constant over generations (except in man). Until recently the causes were thought to be climatic influences, lack of food, predators, and epidemics. No doubt these conditions can lead to great losses and even to the disappearance of entire populations; but they do not give an adequate explanation for the constancy of individual numbers, since both in nature and in the laboratory, populations do not simply increase indefinitely even when all conditions are favorable. Furthermore, this population constancy is rarely or never maintained by eliminating surplus animals but rather by limiting the rate of reproduction. This is why a number of ecologists, foremost among them, V. C. Wynne-Edwards, came to the conclusion that, in the course of evolution, animal species must have developed selective mechanisms whereby populations gauge themselves to a particular density. This would represent the number of individuals that could live in a certain habitat over long periods of time without exhausting its food reserves.

In 1950, J. Christian proposed that this self-regulation occurs through stress. According to this theory, an increase in population density leads to changes in social interactions, in turn creating a "social stress" for the animals. Members of the population then adapt to this social stress with changes in hormone levels, with the result that fertility and the capacity to survive decrease with a rise in population density. Christian even claimed that exceedingly high levels of social stress would lead to a breakdown of orgasmic functions, a phenomenon particularly noticeable in the mass deaths of certain vole species.

In the meantime, evidence has been gained for various rodents which strongly suggests that populations regulate their own density by social stress mechanisms. Due to the lack of documentation, this claim has not yet been substantiated for other species. Nevertheless, scientists have demonstrated that in all species studied—including man—hormonal stress reactions or associated physiological changes would result from social interactions, whether in a laboratory or in the wild. Unfortunately, we still do not really know just how conspecifics create stress for the individual, since these effects generally come to light only when the animal has been killed and subjected to detailed physiological studies. Any conclusions that we draw about the relevant stimuli must therefore be treated with caution, since in every social situation there are always several influences acting at the same time. We might consider, for example, the sheer physical effort involved in a fight, the number of contacts to which an individual is subjected, and the meaning of these stimuli to the animal. Furthermore, even simple handling procedures, such as catching an animal for blood samples to check its hormone levels, may produce such a stress that the experimenter can no longer determine which changes are significant.

I would now like to discuss some of the effects of social stress, and how they come about. As an example, I shall use mainly the laboratory studies made on tree shrews, or tupaias (*Tupaia*). Tree shrews are diurnal mammals about the size of squirrels, with long and very furry tails. Their taxonomic status is ambiguous. Some zoologists classify them as prosimians; but they also show a large number of traits possibly linking them to insectivores or lagomorphs. Tree shrews are widely distributed in the forests of southeastern Asia, living singly or in pairs within territories that they defend vigorously against intruding conspecifics. They are particularly well suited for studies on social stress, because they provide us with an external indicator of organismic responses to arousing (alarming) social and other kinds of stress. We can therefore use simple observation to determine the social interactions responsible for stress responses in tree shrews, and to measure their duration.

Under normal conditions, tree shrews have their tail hairs flat against the skin; but as soon as they are aroused in any way, perhaps through strange conspecifics, unfamiliar noises or scents, or extreme cold, their tail hairs immediately become erect and bushy (see Color plate, p. 537). This bristling of the tail hairs results from an excitation of the sympathetic nervous system and the simultaneous secretion of hormones from the adrenal medulla. If we take a twelve-hour day as our observation period and count the length of times that an animal raises its tail hairs, we get a "tail-bristling value" that tells us the percentage of time during which our subject's sympathetic nervous system was excited on that day. In a habituated tree shrew the tail-bristling value (abbreviated to TBR) remains more or less the same over many months, as long as there are no changes in the animal's environment. But this TBR value reaches a new level if the animal is placed in a different and unfamiliar cage, if members are removed from a group, and so on; the TBR value may then lie anywhere from 1% to 100%. Furthermore, there is a close interconnection between the amount or duration of daily excitation and the various changes in behavior as well as changes in the animals's physiological state, probably based on corresponding hormonal changes. In this case it makes no difference whether arousal was caused by conspecifics or by other factors, such as experimental influence or street noises.

If we move a pair of tree shrews into a larger cage separated from neighboring ones by a wire mesh, both animals will at first be uneasy in their new surroundings (TBR value up to 50%). But after one or two months they will have become used to their enclosure, as well as to their neighbors and other disturbances. At this point, none of the behaviors observed on a daily basis will release any significant arousal or tail-bristling, not even prey-catching or sexual activity, including copulation. Fights or conflicts in general do cause the animals to bristle their tails, but such incidents do not occur between partners if the pair is well matched. Even between neighbors, aggressive interactions decrease so that after a while there is hardly any conflict at all. Both partners in a pair therefore end up with low TBR values (consistently less than 5%). Under these conditions the female will bear one to four offspring every forty-five days. The young are fed immediately after birth; they will consume up to fifty percent of their body weight in milk at this time. Then the mother cleans her babies, eats the afterbirths, and leaves the nest, returning only every forty-eight hours to feed them for about five minutes.

If for any reason the female's TBR value rises to more than 20%, she

Fig. 36-1. Birth of three baby tree shrews. Two of the infants are seen suckling at the mother, their stomachs already bulging; one of them is still attached to the umbilical cord. The third infant is already "greeting" the mother with typical licking movements at her mouth.

will still bear young every forty-five days; but within a few hours after birth, either she or a conspecific will enter the nest and eat the young. This cannibalism evidently results from the lack of a secretion released from a gland on the mother's sternum, with which she normally marks the young after birth and which protects them from others in the group. If we place marked young from a strange female into the nest, they will not be harmed.

Fig. 36-2. A tree shrew mother eats her own baby, only hours after she had given birth, fed, and groomed it.

Observations on females with a TBR value of about 20–30% showed that, in addition to their normal mating behavior, they regularly displayed abnormal male copulatory behavior. These females get very excited and drive conspecifics, lick their genitals, mark their head with the sternal gland, mount, and perform copulatory movements—all in exactly the way males normally do. While the males usually do not tolerate such behavior, the females do not resist, allowing themselves to be "copulated" with as by males.

Once the TBR value has risen to more than 60%, the females stop giving birth altogether. This results from two processes related to each other: For one thing, the females are less willing to mate, strongly rejecting any male attempts at copulation; for an other, they become infertile because eggs no longer mature in their ovaries.

The reproductive capacity of the males also decreases with a rise in TBR values. We can already recognize this externally from the position of the testes: With TBR values of less than 50% the testes lie outside of the peritoneal cavity in the dark-colored scrotum, and cannot be retracted into the peritoneum. There is a large production of sperm which fill the epididymis. With TBR values of approximately 50–70% the testes still lie in the scrotum, but any disturbance—and resulting tension of the abdominal wall—will now cause them to recede into the peritoneal cavity. As soon as the alarm is over, the testes return into the scrotum within a few seconds. Under these conditions they are clearly lighter and smaller than with lower TBR values, and the production of sperm is decreased; but the males are still able to reproduce.

Stress affects fertility

When the TBR value is higher than 70%, the testes recede into the peritoneal cavity within a few days, remaining there for as long as the high TBR values persist. At the same time the testes lose a great deal of weight—about 80% within ten days and they no longer produce sperm; the animals then become sterile. Organs dependent on the male sex hormones, such as the seminal vesicle, prostate gland, and sternal gland, soon atrophy completely. The scrotum also degenerates, and after fourteen days can no longer be recognized; the animals are now no different from immature males.

With increasing TBR values, young animals are delayed in their sexual maturity. For example, while an undisturbed male tree shrew reaches puberty at approximately sixty days, an animal with a TBR value of 50% does not enter this stage until about 100 days. With still higher TBR

Stress delaying the onset of puberty

values, puberty does not occur until the stressors responsible for this delay cease to operate. The growth rate of juveniles also decreases with increasing TBR values. The body weight of adult animals is to a large extent determined by environmental factors and it, too, is therefore related to the TBR level. All these symptoms, however, may disappear again once the disturbance has vanished and TBR values have fallen back to normal levels.

If a certain condition leads to a daily TBR value of more than 90%, the animal's weight will steadily decrease at a rate equal to or greater than what it would be if the animal had nothing to eat, although it takes in as much food as usual. After from two to sixteen days it has muscle cramps and paralytic seizures, then falls into a coma and dies. The cause of this death is poisoning from an excess of urea (uremia), which results from kidney failure due to lack of blood supply to the kidneys; this condition may be so severe that the kidneys end up without any blood at all in their vessels. The stricken animals then die after two days, just like subjects whose kidneys have been surgically removed (see Color plate, p. 537).

When a tree shrew pair lives in a fairly large enclosure, it will produce offspring—as already mentioned—every forty-five days. These leave the nest at about thirty days of age and form a family group with their parents and, sometimes, other siblings. At night all the animals sleep together in one nest, and they will often lie together during the day as well. In larger families we may observe up to ten tree shrews lying in several "layers" on top of one another. When they meet, members of a family group often "greet" each other by one stubbing the other on the corner of its mouth and then licking up the saliva.

The relationship between parents and offspring does not change greatly even after puberty. Only the licking behavior when greeting each other now occurs less frequently, and becomes restricted to family members of the opposite sex—i.e., it occurs between father and mother, between father and daughters, and between mother and sons (Fig. 36-7). Aggressive encounters or any kind of dominance relationship may not occur if we observe the animals for several months. This is particularly remarkable, because the young are attacked by strange conspecifics as soon as they reach puberty.

Social interactions between parents and their young seldom or never lead to arousal. Hence the TBR values of the parents do not change, even though the number of animals living in the enclosure increases each time a set of young leaves their nest. But once the offspring reach sexual maturity, the parents's TBR level rises markedly—in the mother whenever a young female reaches puberty, and in the father when a young male becomes sexually mature. If it so happens that the offspring of one female are all males, their puberty has no effect on the mother and she continues to bear young. Only the puberty of a female offspring will cause the

Fig. 36-3. Six tree shrews resting together in a family group.

Fig. 36-4. Adult tree shrews in a family group greeting each other by licking at each other's mouth.

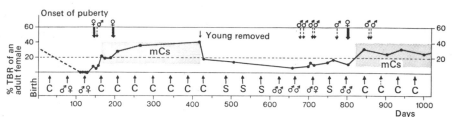

mother to reach a TBR value of more than 20%, preventing any further young to be reared because they will be killed and eaten after birth.

This rise in TBR values of parents after their young reach puberty indicates that sex hormones are involved. We can demonstrate this with experiments: If young males are castrated before their puberty, the father's TBR value does not increase; castration of young after puberty results in the father's TBR value sinking back to its original level. If we than inject the castrated animals with the sex hormone testosterone, the father's TBR value rises again (Fig. 36-7). Before sexual maturity, juveniles generally do not engage in scent marking, but after puberty their marking activity increases by a large amount and then remains at this high level. In castrated young males the frequency of marking decreases again to almost zero, but it may be boosted once more by injections of the male sex hormone, depending on the size of the dose.

As with females, the males do their marking with a glandular field in the sternal region, producing a substance that they spread around the enclosure in a very conspicuous way. This is how the animals mark their personal area, rubbing the secretion on branches and trunks, sleeping boxes, food dishes, and so on. In males, the development of the sternal gland as well as the performance of sternal marking is dependent on the male sex hormones. The increase in parental TBR values is based on scent marks distributed by the sexually mature young all around the enclosure, and these odors remain effective for days even when the young are removed. The scent marks of other conspecifics can also lead to higher TBR values —for example, when the cage was marked for several days by a strange tree shrew pair without any direct contact taking place between these animals and the regular occupants of the cage. Up to a certain point, the scent marks of animals of both sexes can have a summated effect: For example, while the puberty of one daughter in the enclosure raises the mother's TBR value by about 20%, this value will increase to approximately 40% when a second daughter reaches sexual maturity (Fig. 36-5). On the other hand, any additional maturing animals of the same sex—or rather, the addition of their scent marks—will raise the TBR value only very slightly.

Female tree shrews only rarely mark objects in their cage with the sternal glands; the secretion produced by these glands obviously serves to protect the young. By contrast, sexually mature males mark their area many hundred times a day. Their secretion is deposited almost exclusively

Fig. 36-5. Relationship between the sexual status of conspecifics in a cage and the tail-bristling (TBR) value of a female, as well as her reproductive and mating behavior. Each arrow at the bottom indicates a birth. Where the young were reared, we have indicated their sex. All the other young did not survive long enough; they were either eaten (C) or they starved (S). Births 14 to 16 would have starved just like birth 17 if the young had not been given additional food by the experimenter. As shown, the female's TBR value rises only when a young female in the cage reaches puberty (upper arrows); the sexual status of males has no effect on the female's TBR value.

Sternal marking

Fig. 36-6. Male tree shrew engaged in sternal marking

Fig. 36-7. Relationship between the TBR values of a tree shrew pair and the sexual status of both its male young. As shown, only the father's TBR value rises as the young males reach puberty. At the same time, father and sons no longer lick each other in greeting, while the father's marking behavior does not basically change in frequency.

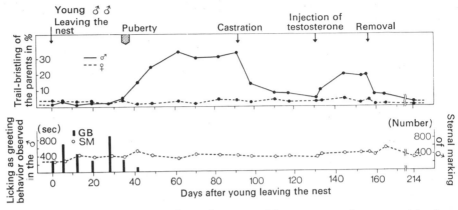

The marking value

at the borders separating them from neighboring tree shrews, with the intensity of marking activity increasing, the higher the aggressive tensions between the animals concerned. But in general the degree of marking performed by a paired male on a daily basis will remain stable for many months. Neither does the frequency of marking change very much after the male offspring have reached puberty; only the distribution of the father's scent marks changes: He will now engage more frequently in marking objects inside the cage, as well. If we castrate the young males, the obviously arousing effect of their own scent marks will cease. In such a case, the father once again restricts his marking activity more or less to the boundary lines.

Once again, we can demonstrate the influence of other scent marks on marking activity: If we place a male into a small, clean cage for a certain period, he will mark it a few times. This marking value differs from one animal to another, but remains fairly stable for individuals over several trials: Depending on the animal, we can observe one to twenty-five marking acts in ten minutes. If this cage has already been marked by an immature male or a castrated adult male for ten minutes prior to the test, using urine, feces, and secretions from the various skin glands, the marking value of the experimental animal will change little or not at all. Neither do the scent marks of female tree shrews or of alien species have any significant influence on marking. But if the cage was previously marked by a sexually mature and functioning male, our experimental animal will now mark far more frequently—depending on the animal, fifty to 300% more. As shown by experiments with castrated animals that have been injected with different amounts of male sex hormone, the marking frequency of the experimental animal increases with the level of testosterone contained in the prior scent markings (Fig. 36-9).

Male and female tree shrews viciously attack strange conspecifics of their own sex, both in the laboratory and in the wild. But even in family groups, sooner or later conflicts arise between individuals of the same sex: between father and sons, and mother and daughters. These fights are brief and fierce, but rarely bloody. As a rule it takes only a few minutes to decide who is dominant. The younger opponents are usually sub-ordinate, but occasionally the loser will be an animal weakened by age. The loser then flees to some sheltered place, leaving its refuge only for feeding. During the fight both animals are extremely aroused, as shown by the bristling of their tail. Within ten minutes after the fight, the victor usually gives no more indication of arousal and ignores the loser complete-ly; but the defeated animal lies practically all day in his hiding place, following the winner's movements with his head and continuously raising the hairs on his tail. Even on following days, if the enclosure is fairly large, the two animals very rarely or never fight. And yet, the presence of the victor creates a stress for the loser—and only for him—because he bristles his tail each time he catches sight of his antagonist.

If the cage is large enough and offers many places to hide, the defeated animal can go on living even if the winner is still there. But males as well as females do lose weight in accordance with their TBR values, and grow sterile within a few days. Juveniles grow more slowly or not at all. However, these harmful effects can be eliminated: An animal stressed in this way will recover completely if separated from the victor. The cage in which this tree shrew is placed can be quite a bit smaller than the space in which he could move about even with the antagonist present. Hence it is not lack of space that creates these ill effects, but rather the interactions with his conspecific. In fact, even in larger cages the loser of a fight—especially if male—will be constantly aroused by the presence of the winner (TBR value above 90%) and will die of uremia within a few days.

The fact that the constant arousal of the subordinate animal, and there-fore its death, is caused by the continuing presence of the dominant one and not by physical exhaustion or injuries from the fight is also shown by the following results: If an adult male is placed into the cage of a strange conspecific, the "intruder" is subordinated by the occupant within less than three minutes. If we then separate the two animals with an opaque screen, the loser will recover from the fight as quickly as the winner. Even when a subordinate animal is subjected to fighting several times every day for two weeks, it will not die nor even lose much weight if separated from the victor after each fight. On the other hand, if the two animals are separated by a mere wire mesh, so that the loser cannot be attacked but still sees the winner constantly, the subordinate subject will keep its tail hairs erect all the time, and will die shortly thereafter. In contrast to the density effect described earlier, the determining cues here are optic ones; it is not merely the smell of the victor that leads to the loser's death. In this kind of situation, then, death is not directly the result of social interactions and their physiological effects, but rather of central nervous processes based on experience and learning. In other words: The loser dies of constant anxiety.

Thus we can clearly distinguish two processes whereby conspecifics may create a stress situation for individual tree shrews, and which play a

Fig. 36-8. Tree shrews viciously attack strange conspecifics of their own sex. Fights are short and intense, but usually with-out bloodshed. Here, a male is fixating his rival on the wire mesh; then he leaps up in a violent attack. Both animals are extremely aroused, as we can see by their tail-bristling.

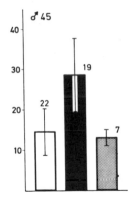

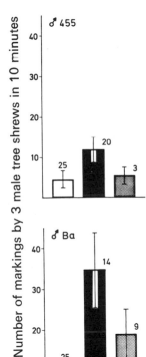

Number of markings by 3 male tree shrews in 10 minutes

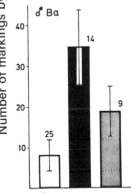

Fig. 36-9. Sternal marking behavior of three male tree shrews and the way it is influenced by scent markings from conspecifics. White column: Cage is unmarked. Black column: Cage has been marked by a sexually mature male. Grey column: Cage has been marked by an adult castrated male.

possible part in the control of population density: 1. density effect, and 2. dominance effect. The effect of density is nonspecific; that is, it is the same for all tree shrews of one sex in the same situation, so that all of these individuals will have roughly the same TBR values. In many cases these two processes overlap; a dominance relationship, for example, presupposes the presence of a same-sexed conspecific and therefore makes a slight density effect inevitable.

We know that varying degrees of density have similar effects on all other mammals in which this phenomenon was investigated. Again, however, there are many other species in which the pure density effects are usually overlapped by those of personal dominance relationships and aggressive encounters. Information about the population density at any time may be gained from single perceptual stimuli or from a combination of various stimulus effects, depending on the species. But the basic effect is always the same: An increase in the number of animals within a given space, which inevitably leads to a greater frequency of contacts, results in a higher level of social stress. Although even with other animal species the effects of density alone are not as marked as those of aggressive interactions, they are enough to inhibit or even prevent any further population increase. The females then produce less milk and neglect their young; group members may become cannibalistic and both sexes show abnormal sexual behavior. These patterns are especially noticeable in species where individuals live more-or-less peaceably in groups or herds, yet do not simply multiply in unlimited numbers, as with many mouse species, deer, and elephants.

Dominance relationships and their effects have also been observed among vastly different animal species. In tree shrews, which in nature normally live singly or in pairs and occupy territories, the continuing presence of a dominant conspecific creates such a stress that the subordinate animal soon dies. Similar effects are known for other territorial mammals, such as rats, mice, and voles. But even in species whose members normally live peaceably in groups with a fixed dominance hierarchy, we find the same basic effects among the more subordinate animals, but to a lesser degree: The lower the rank of an individual (the more subordinate he is), the higher is his adrenal activity level and the lower his body weight and reproductive capacity. Such an animal is generally more susceptible to disease, and so on. In the most extreme cases, very low-ranking individuals —from mice to monkeys—are completely infertile, seem listless and physically stunted, and usually die prematurely.

Changes in the hierarchy lead to changes in the animals's physiological condition. For instance, high-ranking male rhesus monkeys have a higher testosterone level than lower-ranking ones, and their adrenal glands show weaker responses to stimulation. If we take previously high-ranking individuals and put them in a group where they are forced to take a lower status, their testosterone level declines. Conversely, change to a higher status is correlated with a corresponding rise in testosterone.

Until now we have discussed only the effects of alarming stimuli. Animals and man can adapt to these within a few minutes to several hours. Even organs such as the adrenal glands can respond to an increased production and secretion of their hormones by nearly doubling their size and weight within forty-eight hours. But although we have been gaining considerable knowledge about the effects of acute stress, we still know very little about how an organism responds to otherwise harmless stress conditions that prevail for months or even years. Our ignorance is partly due to the fact that the animals that experimenters like to use generally have a short life span. Furthermore, long-term studies are so costly in space and technical materials that they are usually not feasible. Yet it is really the more lasting and relatively harmless kinds of stress that are of special interest to man, and I would therefore like to describe a few results worth mentioning in this context.

As we have said, the presence of a sexually mature daughter creates a mild but measurable stress effect for a female tree shrew. The mother does continue to bear young, but she eats them after birth. If we eliminate the stress by, for example, removing the daughter, the mother will not harm her next litters. However, should this prior stress situation created by the daughter's presence be extended for a period of more than four to six months, the mother will no longer rear her young even when the daughter is removed, because she produces too little milk and feeds the young only at very irregular intervals; the babies than starve to death within a few days after their birth. Not until twelve months after this relatively short and harmless stress situation has been eliminated does the mother again produce enough milk and feed the young on a normal enough schedule to rear them properly.

Many studies on other animal species have also shown that if the mother is stressed before the birth of her young (for example, through overpopulation, daily fighting, electroshocks, or chemicals), the growth and fertility of these offspring is adversely affected, should the young survive. J. Christian kept laboratory mice of both sexes in a fairly large cage. These animals multiplied very quickly and reached a stable population density after approximately five months. In the following two months there were thirty more litter births, but none of the babies survived. When pregnant females were removed from this overcrowded group and kept individually under favorable conditions, forty percent of them still did not rear their young. Once again we can see that this phenomenon does not result from direct disturbances through too much contact with conspecifics, but rather from some kind of physiological changes due to overpopulation, which affects the females in some unknown way.

Christian and C. D. Lemunyan made an even more surprising finding in 1958: They kept male and female mice for forty days at a very high density and then placed them in pairs into smaller cages, The females then had young, which developed more poorly than normal, as expected. But

Effects of prolonged stress

even the young of these original offspring showed inferior growth and development, despite the fact that both they and their parents were held under the best conditions possible. This was a case, then, where a brief period of stress experienced by females affected even their grandchildren.

While in these cases the offspring were affected through environmental influences bearing on the mother, stress factors, such as fighting between other conspecifics in the same cage, or dominance fights, may influence their further development even after these young have reached independence. These damaging effects—less body weight, late onset of puberty, etc.—cannot be corrected later, or at least not to any significant extent, even when the animals are kept under the best possible conditions. We then get adult tree shrews, for example, with body weights ranging from 150 g to more than 300 g depending on their degree of stress during childhood, and who also show marked differences in gonadal and adrenal activities as well as in their entire metabolism as compared with conspecifics or even twin brothers raised under normal conditions.

Effects on general responsiveness

Stress affects not only the body, but also curiosity and learning behavior and the responsiveness of the animal in general. This occurs most prominently when the stressors arise through the mother during fetal development, or come about in earliest childhood. Lighter kinds of stress, such as mild noise, some overcrowding, or isolated fights, generally seem to improve curiosity and learning and to decrease excitability; in this way animals learn to cope better with their environment. But when stressors become stronger, as with very high population density or constant subordination in aggressive encounters, the reverse is true: The animal's level of alertness and exploratory activity decreases; individuals often become listless, they no longer respond to conspecifics, and they are less ready to learn, if they are ready at all. This is why—as we have found with rats, mice, voles, rabbits, tree shrews, and monkeys—such animals cannot escape the source of their stress, such as a dominant conspecific, either through migration or learning.

In 1962, J. Calhoun used brown (or Norway) rats (*Rattus norwegicus*) to carry out the most extensive investigations to date on the effects of excessive social contacts. Brown rats live more-or-less peaceably in large colonies. The individual members try to isolate themselves as much as possible from the others only when feeding. If all the members of a colony are fed at only one spot, each animal will take a fairly large morsel, carry it to a protected corner, and eat it there. But we can prevent them from carrying their food away from the feeding place by covering it with a wire mesh so that each animal may bite off only tiny pieces. Thus a hungry rat must always crowd around the food site together with its conspecifics in order to get a bite. No fighting occurs under these conditions. The satisfaction of hunger—a positive experience—soon becomes associated with the initially negative factor of close contact with conspecifics. As a result, contact with the others now takes on a positive tone as well: "Food can only be enjoyed in a crowd." Thereafter, even when many feeding places are distributed around the room, all the animals mass around one spot and ignore the other places.

After a few months, however, the negative effects of these excessive social contacts start to show. Although the animals have enough space to live comfortably in large numbers, we now start to observe a complete breakdown of all the behavior patterns necessary for survival: The males show abnormal courtship and mating behavior, some indiscriminately pursue juveniles, males, or females and try to copulate with them, others withdraw completely and are totally apathetic even toward females in heat. Still other males suddenly attack young and adults alike, of both sexes. Females no longer build nests, they feed and groom their young erratically or not at all, and many eat their babies. As a result, infant mortality is more than ninety percent. In many cases the embryos die during pregnancy, leading to poisoning; roughly half of these females are killed by this. The small number of young that do reach sexual maturity have very few or no offspring themselves, and these few offspring never survive more than thirty days. These effects persist in the young even when almost all of the conspecifics have died—that is, when the social contacts are no longer excessive. Inevitably, the population dies out—and this only as a result of the forced learning of certain social interrelationships! These experiments have now been repeated several times, on mice as well as rats, and always with the same devastating results.

In addition to these kinds of stress reactions, mammals will show all the damaging effects of civilization known in humans, such as stomach ulcers, arteriosclerosis, high blood pressure, and heart and kidney ailments merely by being subjected to too much contact with conspecifics. For example, emotional stress and conflict situations can within a few days induce stomach ulcers in rats and monkeys that are so severe that some animals bleed to death. All other symptoms appear only with relatively long-term kinds of stress.

This was impressively demonstrated by, among other things, the findings of Soviet scientists who have been studying monkeys—especially baboons—for several years. Chapter 35 offers a detailed account of social behavior and dominance relationships in baboons. If we separate the highest-ranking male from his group by means of a cage partition, he will at first seem highly aroused and anxious. After a while the external signs of arousal abate, but an inner tension remains. For example, there is a markedly higher secretion of hormones from the adrenal cortex. If the females and young are fed so that he can see them, he will make heated attacks to try to prevent them from eating. An even stronger reaction is released if we place a strange adult male with his former group. Now our subject chases around the cage, makes continuous threatening sounds, and attacks his rival viciously through the bars. After a few months, males kept under these conditions develop neurotic behavior, high blood pres-

Breakdown of social behavior patterns

sure, and vascular and cardiac ailments; some die as a result of these symptoms.

High blood pressure and other chronic disturbances are found regularly in many animal species whose populations are overcrowded, both in captivity and in nature. For example, the blood pressure of members in a freely multiplying population of voles increases with the population density. After a while, when the population has reached a more or less stable maximum density, the average blood pressure of these animals is roughly twenty percent higher than the original level. J. P. Henry and his associates studied the relationship between social stress and high blood pressure in more detail: If, after weaning, laboratory mice are housed singly or in pairs, the blood pressure of all these subjects remains at the same low level until death. But if they are kept in a more dense population, or exposed each day to a different number of arousing social interactions, their blood pressure rises according to the amount of stress. If these stressful conditions are eliminated after a few months by putting the animals in solitary cages, their blood pressure returns to a normal level. If, however, the animals are separated only after five to nine months, this high blood pressure will not fall. Among other things, their heart and kidneys are damaged and the animals develop arteriosclerosis.

Implications for man

We have tried to examine the origin of social stress and its effects on mammals. The physiological condition as well as the behavior of an individual are—within the limitations of his genetic potential—always determined by environmental stimuli. Most important among these are usually conspecifics and their signals, though undoubtedly there are also nonsocial conditions that act as stressors, such as constant harrassment by predators, loud noise levels, or climatic factors. Every change in an animal's environment—be it alterations in social structure resulting from increases in population, individual decline in social status, or the like—will lead to consequences harmful to the animal concerned, but at times favoring the survival of the population as a whole.

Humans react to stress in basically the same way as do other mammals. Everyday stimuli or aggravations such as tension on the job, frequent contacts with other people both at work and during leisure, noise, air pollution, and various annoyances or agitations which everyone is exposed to lead to activation of the organism. It has been shown that excessive and long-term emotional stress in people can lead to sterility, kidney diseases, and even death. Hence the physiological state of human individuals is probably also the result of all the externally impinging stimuli and of people's emotionally determined arousal levels.

Indeed, our environment has been changing at an ever-accelerating pace in the last few decades, primarily through industrialization and increasing urbanization. Do these changes have adverse effects on us, too? Unlike with animal groups, where we already know a great deal about the relationship between population density and social environment, and the resulting pathological effects, we can do little more than speculate when it comes to humans. Studies of recent decades have certainly provided us with clear-cut statistical correlations between population density, housing conditions, and social status, on the one hand, and certain diseases, life expectancy, psychological damage, neglect of children, and frequency of criminal behavior, on the other; but these statistics alone do not prove that there is a direct causal relationship.

Furthermore, most people are able to cope with even extreme conditions of overcrowding and industrialization without showing direct signs of damage, at least as far as we know. But this does not necessarily mean that our present state is harmless. Many environmental influences have up to now been considered insignificant, but what will their effects be if they persist over many decades—as they undoubtedly will? The general increase in diseases related to civilization—ailments such as heart, kidney, and circulatory malfunction for which no medical explanation has been found—are known in mammals as consistently resulting from excessive and long-term stress. Biological findings point to these dangers, and raise issues that we must pursue jointly within all the various fields of science and technology, taking man as our target.

37　The Development of Behavior

The development of
behavior,
by G. Gottlieb

Aims and assumptions

One of the major aims of the study of the development of behavior is to gain an understanding of how adult behavior realizes itself. This aim carries with it the assumption that certain, if not all, adult behaviors are determined or influenced by events which occur earlier in ontogenesis (= history of the individual). According to this viewpoint, early experiences or events explain many important aspects of adult behavior. The familiar saying, "The child is the father of the man," is a succinct statement of this point of view. Sigmund Freud's theory of psychoanalysis is a particularly striking example of the supposed continuity of early and later behavior; experiences of unusual gratification or excessive frustration during the early oral, anal, or genital periods of infancy and childhood are said to lead to the manifestation of particular kinds of neuroses in adulthood, as well as influencing adult character structure in non-neurotic individuals. While this aspect of Freudian theory is no longer as completely accepted as it once was, it does serve as a dramatic example of the view that continuity is an important aspect of the development of behavior. Whereas continuity in human personality development is usually inferred from observing the adult behavior of human beings and then reconstructing their past history through personal interviews, as we shall see later, there is some very clear experimental proof for the continuity point of view in the development of behavior in animals, where direct and highly controlled experimentation is both possible and practical, as well as ethical.

Most developmental theorists and investigators realize that the development of behavior involves the emergence of new functions or abilities which are not necessarily derived from previous stages. Therefore, the other major aim in the study of the development of behavior has to do with the understanding and analysis of transitions where discontinuity is a much more prominent and important feature than is continuity. As a matter of experimental strategy, the search for continuities in development has been emphasized somewhat at the expense of discontinuities. The reason for this is that it is much easier, and much more direct, to demonstrate experimentally that a current behavior has been derived or influenced by previous events in the history of the animal than it is to obtain convincing evidence that a current behavior is entirely free of influence from past ontogenetic events. Since the experimental study of behavioral development is itself in its infancy, it is presently much more practical and useful to attempt to establish firm continuities than it is to prove the "null hypothesis," i.e., that functions arise entirely *de novo*, without any direct dependence on any of the previous activities in the life of the organism.

Thus, at the present time in the short history of experimental behavioral science, the "child is the father of the man" theory of behavioral development (strict developmental continuity) is more heavily emphasized than are the developmental discontinuities. Fortunately, discontinuity-theory is not without its supporters; understanding of behavior development will almost certainly include discontinuities and continuities in its wake, but we are nowhere near that goal now, nor will we be so in the foreseeable future.

The most prevalent misconceptions concerning the development of behavior have to do with the influence of phylogeny on ontogeny and the related problem of the role of genes in behavior development.

Common misconceptions

"*Ontogeny recapitulates Phylogeny.*" This generalization, sometimes called the biogenetic law, means that the succession of stages through which an individual passes in his ontogenetic development repeats the sequence of adult stages through which his remote ancestors passed in their evolutionary development. E. Haeckel originally invoked the law to explain the particular sequence in which certain structures make their appearance during embryonic or larval development. The law has sometimes also been used to explain behavioral development (e.g., by G. Stanley Hall).

That individuals of a given species usually pass through a number of behavior stages in a regular succession or order is not to be doubted; that these regularities are to be explained by phylogeny is a misconception. These regularities can only be understood by a direct experimental analysis of the conditions under which they occur; by an analysis of ontogenesis itself. The belief that phylogeny explains ontogeny proved a great stumbling block to the introduction of the causal-analytic (experimental) method of study in embryology (J. Oppenheimer; G. deBeer), so we can benefit from that historical example. The older point of view was that, if phylogeny explains ontogeny, an experimental analysis of ontogenesis is superfluous. We now know that this idea was mistaken: It is only through experimental analysis that we can come to understand and fully appreciate the phenomena of development. Phylogeny does not illuminate the *mechanics* of behavioral development at all. This point may become clearer as we analyze the misconceptions about the role of genes in behavioral development.

"Innate or learned"

Nature vs. nurture, or instinct vs. learning. This misconception assumes that the developmental process is controlled by two completely different factors, one of which is phylogenetically derived (genes) and the other of which is ontogenetically derived (environment).

It is common to hear or to read that some behaviors are genetically determined whereas some other behaviors are environmentally conditioned during the lifetime of the individual. The misconception is that the latter can be no less genetically determined than the former. There can be no behavior without an organism and there can be no organism without genes. No one knows of any organism, part of which was constructed by genes and the other part of which was constructed by the environment.

A much more subtle misconception implied by the idea of the genetic determination of behavioral development is the rather horribly over-simplified view it imparts of the so-called gene behavior pathway. When one uses the phrase "genetically determined" it is as if there were a direct or close relationship between the expression of the genes and the emission of behavior during development. Quite to the contrary, it is an absolutely humbling experience when one merely considers all the different levels of analysis (molecule, cell, tissue, organ, organ system, organism) which must be traversed to get from the genes (DNA) to the overt behavior of an organism. Add to this the temporal dimension of change which is of such importance in developmental study, and one becomes dubious about the precise meaning of the phrase "genetically determined."

At the present stage of science we do know (1) the genetical complement must set limits—currently unknown—on behavioral development and (2) the species-specific genetic complement substantially increases the likelihood that the capability for certain forms of behavior will occur later in development. We simply do not yet know the developmental mechanisms whereby (1) and (2) are realized—these are as yet unanswered investigative questions in the domain of developmental genetics. Thus, we must respect the "distance" between the genome and behavior and also try to avoid thinking of part of an organism's behavioral development as stemming from genetic determination and some other part from environmental determination. The capacity for the organism to react to the environment is as genetically determined as its inability to react to certain features of the environment. It does not seem that the genes could be any more responsible for one than the other.

"Maturation or experience"

This fallacious dichotomy is related to the previous one on nature vs. nurture. The misconception in this case is that genes control anatomical maturation while the environment controls experience. Since there is evidence (to be mentioned later) that experience affects the maturation of the brain and other structures during development, the proper way to visualize the relationship in developmental terms is maturation *and* experience, rather than maturation *vs.* experience. As Zing-Yang Kuo, T. C.

Schneirla, and Daniel S. Lehrman have tried to make clear, maturation always takes place in some "experiental" context, where experience is defined broadly to include the various stimulative aspects to which organisms are subject during prenatal and postnatal life. One of the chief aims of developmental analysis is the specification of how early conditions or stimulative events interact with organic maturation to exert particular effects on later behavior.

There are a few fairly well-founded principles of development. Chief among these are the following. **General principles**

The behavioral capabilities of newborn animals are typically so very well adapted to their usual life circumstances that they have a preadapted quality. Indeed, investigations of embryonic and fetal development show that the behavioral capabilities of the neonate do arise during the prenatal phase. In fact, many of the behaviors which the newborn must manifest if it is to survive actually become functional (or capable of function) well before the time of birth or hatching—for example, the sucking reflex in mammals. **Forward reference**

The phenomenon of precocious functional maturation is so general that several writers have broached the idea that many prenatal and early postnatal phenomena are "preparatory" or "anticipatory," that is, that early neurosensory and neuromuscular maturation has a "forward reference." While this descriptive principle is certainly valid as far as it goes, it does not have any explanatory value, of course. That is, it is still necessary to undertake an experimental examination of early development in order to determine the factors which affect or control the development of any behavior, regardless of how adaptive it may be. The principle of forward reference incorporates the well-known view that early neural and behavioral development affects later behavior, specifically that the adaptive behavior of the neonate emerges from a background of prenatal development.

Generally speaking, as development proceeds, the organism's behavior becomes more versatile and more highly differentiated. This simply means that as the repertoire of behavior increases with age, so does the capacity for the elaboration of fine details in the perceptual sphere and the performance of finer muscular movements in the motor sphere. (The means or mechanisms whereby developmental changes in versatility and and differentiation come about are among the most significant questions in the field.) **Versatility and differentiation**

There are periods or stages in development when the organism is maximally susceptible to certain kinds of stimulation, or when ease of mastering certain behavioral tasks is much higher than at other times in the life cycle. These stages are sometimes referred to as critical periods of development. It is not correct to think of these behavioral stages as occurring at a highly fixed time in the life of each member of the species, nor as being of an invariant length. The best example of the various **Optimum stages** features of an optimum stage is the learning of spoken languages(s) by children. Although this occurs most readily and with the greatest facility when the child is very young, the capacity continues into adulthood, albeit with diminished ease. (The biological and ontogenetic behavioral variables which control optimum stages of development is another investigative question of the first order.)

Individual differences

There are large and significant individual differences in all quantitative aspects of behavioral development. In many instances, these differences become even more marked as development proceeds.

Despite significant individual differences in the quantitative aspects of development, the *sequence* in which behavioral stages follow each other in any given species is remarkably constant when typical developmental conditions prevail. (The systematic experimental alteration of species-typical developmental conditions can shed great light on the various factors which participate in the sequencing of behavioral stages.)

Function

The role of function (use, exercise, practice, stimulation, experience) has always been a key question in the development of early behavior, especially where species-specific, instinctive, or innate behavior is involved. Some theorists have held that function is necessary merely to *maintain* innate behavior once it has developed fully, whereas others have held that function also enhances or even channels behavior *during* the developmental process itself. Certainly function is necessary for maintenance. The extent to which it also plays a role in the maturation process itself is, however, still an active investigative question. Several recent experiments (reviewed below) now suggest that function may play a constructive as well as a maintenance role in the complete differentiation of behavior.

Primary methods of study

The strategy for conducting research on behavior development involves, first, a description of the more prominent behaviors at each age they occur under natural or "real life" conditions, and, second, the controlled experimental manipulation of certain developmental factors which the investigator has reason to believe will influence a given behavior at a certain age. In a successful developmental experiment, some aspect of a behavior will be shown to covary directly with some feature of the animal's previous developmental history that the investigator has manipulated in a systematic way. Every time that happens, our knowledge of development is increased in a very definite way.

The main experimental manipulations in developmental studies involve the subtraction, addition, replacement, and displacement of prior events, activities, or experiences which the investigator believes to be influencing the development of the behavior in question.

Subtraction method

The subtraction method may be used in many different ways. For example, in order to test the hypothesis that accuracy of pecking in chickens is achieved gradually via practice during the first few days after hatching, an investigator deprived newly hatched chicks of practice by keeping them in the dark and hand-feeding them for several days so they had no opportunity for pecking: then he tested their pecking accuracy. (He found that practice does make some contribution, but other factors are also involved.)

It is important to note that the subtraction method is not purely that —it is not solely taking away or reducing a particular event—but it also involves *replacing* the activity in question with some other one (hand-feeding and rearing in the dark in the above example). We need to remain alert to the possibility that it may be the particular replacement rather than the subtracted event per se which is the responsible factor.

Although it has many other uses, the addition method—which is also a variant of the replacement method—is most often used in developmental studies of perception and learning in rodents, where one group is experientially deprived by placing them in an impoverished environment (e.g., an empty home cage with no companions or objects) and the experience of another group is enhanced by rearing them socially in a large cage with numerous brightly colored, mobile objects. (The latter group usually outperforms the former group in later problem-solving or learning tasks, indicating the contribution of the social and physical rearing environments to the development of intelligent behavior.) Addition experiments are sometimes criticized on the ground that they really show the deleterious psychological effect of deprivation (that is, subtraction) rather than an increase in normal ability through enrichment (addition). A clear interpretation of this question can only be had when the normal or usual developmental levels of experience are known for the particular species under consideration.

Addition method

As will be obvious from the foregoing, this method involves the substitution of some other event or activity for the usual one in the course of ontogenesis in order to determine the influence of the usual event (or in some cases, to determine the "plasticity" of the organism). A familiar example of this method occurs in imprinting experiments, where newly hatched birds are presented with a model rather than their own parent, or in other cases, with foster parents from another species rather than parents from their own species. In one of the more dramatic uses of the replacement method, individual kittens were reared with a rat to determine the influence of this experience on their later rat-killing behavior. Some of the kittens later killed rats and some didn't—none of the kittens killed the particular rat they were reared with.

Replacement method

For a careful and precise quantitative delineation of optimum stages, it is necessary to move typical developmental events forward or backward in time: this is the displacement method. In order to determine a particularly critical or sensitive period for imprinting, for example, groups of young animals are exposed to the parent object for a short time at only one age period after birth or hatching and later tested for their attachment to the object as a function of their age at the initial exposure. That age period which is associated with the highest measurable attachment in

Displacement method

the later test is called the critical or sensitive period. Sometimes the displacement methods will reveal an age period before and after which the exposure will not result in later attachment to the parent object; most often the displacement method merely reveals an age period which is relatively more effective than the other ages. The nervous system is thought to be particularly "ripe" for the task at hand during optimum stages of development.

Thus, there are four principal methods for the analytic study of behavioral development, and these four methods are very much interrelated.

Prenatal development of behavior

Behavior does not begin at birth. All species that have been examined —invertebrate as well as vertebrate—begin to show muscular activity and sensitivity to sensory stimulation in the embryonic, larval, or fetal stages of development. In some species at least, activity precedes reactivity: spontaneous motor (muscular) movements are present before reflexes are evident. Thus, the early behavior or motility in embryos of these species can be said to be truly spontaneous or autogenous (defined as movements which are not dependent on sensory stimulation for their initiation or maintenance). The sensory systems also begin to function early in prenatal development, however, even in cases where they are presumably not yet "hooked-up" to the motor system. In this section we will discuss the motor and sensory aspects of embryonic behavior, especially in bird embryos, where these matters have been best worked out.

Motor primary

Preyer-Tracy hypothesis of autogenous motility. In 1885, W. T. Preyer, the modern-day pioneer of the study of behavioral embryology, observed that the domestic chick embryo begins to move several days in advance of the time when an overt response to stimulation of the skin of the embryo can be elicited. This suggested to Preyer that these early movements are autogenous or spontaneous, possibly arising from autonomous electrical impulses generated in neuromotor cells in the spinal cord of the embryo. In 1926, H. C. Tracy, in studying the behavioral embryology of toadfish, noted that the larvae were active well before hatching, whereas they did not respond overtly to tactile stimulation until after hatching. Tracy concluded that the early swimming movements of toadfish are generated entirely from sources inside the larvae's nervous system, that is, that these movements are of endogeneous motor origin.

The Preyer-Tracy idea of motor primacy is the historical and ontogenetic forerunner to the classical ethological concept of the so-called Fixed Action Pattern. One of the defining characteristics of these highly stereotyped postnatal motor movements (FAPs) is held to be their spontaneity: they are said to occasionally occur in the complete absence of sensory stimulation even in adults. While the spontaneous motor movements of chick embryos are not highly patterned (they appear to be uncoordinated), the autogenous larval movements of the toadfish are highly patterned. They are very regular and co-ordinated in appearance, being part of the pattern of swimming movements which the larvae must engage in immediately upon hatching. A conspicuous feature of the embryonic movements of the chick, toadfish, and many other species which have been investigated is their rhythmicity or periodicity: the embryos are active and inactive in cycles, the length of which is related to the age of the embryo, at least during the early part of embryonic development.

In the 1960s, the neuroembryologist Viktor Hamburger turned his attention to the behavior of the chick embryo. Among his chief interests was the experimental demonstration that the chick embryo is capable of moving without the benefit of sensory stimulation. In an anatomical deprivation (deafferentiation) experiment, involving a very delicate microsurgical technique (see Fig. 37-1), Hamburger and his colleagues removed that part (dorsal half) of the spinal cord from which the sensory nerves innervate the legs of the embryo. If surgically deafferented embryos can move their legs in a rhythmic manner as the intact embryo does, that would provide unequivocal support for the Preyer-Tracy hypothesis of autogenous motility in young embryos. The embryos did just that: the cyclic motility of the deafferented embryos was virtually the same as the embryos which had their sensory nerves completely intact, at least up until about Day 15 of embryonic development. (The domestic chick hatches on Day 20).

During the last few days of incubation, the chick embryo's behavior becomes much more patterned as it goes through the various postures and movements which lead to hatching. During the prehatching stage, various sorts of sensory stimulation (vestibular, tactile, proprioceptive, interoceptive) probably begin to play a more important role in regulating the embryo's movements. Thus, the chick embryo would require intact sensory nerves during the prehatching period. (The deafferented embryos did not hatch, but that could also have been due to degenerative changes which occur in the spinal cord after Day 15 in these particular surgical preparations.)

The picture of developing behavior in the chick embryo suggests that the ill-coordinated, autogenous movements practiced in the early autogenous phase have little, if anything, to do with the more highly coordinated movements in the later prehatching stage. In other words, there would seem to be a discontinuity between the earlier and later stages of the chick embryo at least as far as patterns of motor movement are concerned. From certain other experiments, however, it is clear that the later movements are indirectly affected by the occurrence of the earlier movements. For example, when chick embryos are paralyzed even for short periods of time (1–2 days) by injecting certain drugs into the egg, in the absence of movement their joints, tendons, and muscles become deformed, so that they have difficulty in hatching or do not hatch at all. This kind of dependency of the later movements on the early movements is not the

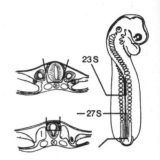

Fig. 37-1. Deafferentation operation. In the deafferentation experiment, the dorsal half of the spinal cord is removed by microsurgery, thereby preventing sensory innervation of the legs of the chick embryo (below segment 27 of the cord in the illustration). In the control group, the dorsal and ventral cord is removed at a higher level (between segments 23–27 in the illustration), thereby preventing sensory or other impulses from higher centers from contributing to leg movements initiated below segment 27. The control group has an intact spinal cord below segment 27, which allows the influence of proprioceptive sensory innervation at the leg level. (Modified from Hamburger, Wenger, and Oppenheim.)

same as demonstrating a continuity of the movements as such, as would be the case if, for example, certain elements of the relatively ill-coordinated early movements could be shown to be the basis for aspects of the later movements. Since the answer to this question is not yet known, we must reserve judgment on the apparently complete discontinuity between early and late embryonic movements in the chick on the present evidence. Before turning to the known continuities that begin to manifest themselves in later embryonic stages, both in motor movement and perception, it is of some interest to compare the quantitative features of motility of the chick to that of other avian species at various stages of embryonic development.

Since the experimental results from studies of the chick embryo serve so widely in our understanding of the early ontogeny of motor movement in birds generally, it is of more than routine interest to compare the chick's motility to that of other avian species. We are in a position to do that (at least quantitatively) for the duck embryo and the pigeon embryo, where investigators in the same laboratory have used the same observational techniques and criteria in counting the movements of embryos of each species on each day of development. The fact that these species are quite divergent ecologically, and in other important ways, makes the comparison all the more interesting and illuminating. For example, pigeons hatch in a rather immature and helpless (altricial) condition after about 17.5 days of incubation, whereas domestic chicks hatch after 20 days of incubation in a rather precocious state of motor development—for example, they can locomote and almost completely fend for themselves—as do domestic Peking ducks, which hatch after 26–27 days of incubation.

As can be seen in Figure 37-2 despite the vast differences in ecology, independence at hatching, and length of the incubation period, the quantitative aspect of motility in pigeon, chick, and duck embryos follows a very similar pattern during embryonic development. Specifically, there is an initial rapid rise in rate of movement which attains peak around the midpoint of the incubation period and then falls off rapidly as the embryos near hatching. While it is interesting that the altricial pigeon embryo shows a lower level of peak activity than the precocial chick and duck embryo, too few species have been compared to be able to conclude that a relatively lower level of activity is characteristic of altricial embryos, in comparison to precocial ones. With respect to the more qualitative aspects of the movements, while there are qualitative as well as the quantitative similarities in the behavior of the embryos of different species—for example, chick and duck embryos turn their head more often to the right than to the left after the midpoint of incubation—these kinds of comparisons have not yet been worked out in detail for chick, duck, and pigeon embryos.

Since the early autonomy of the neuromotor system of the chick embryo has been ably demonstrated by the deafferentation experiment of

Motility in chick, duck and pigeon embryos

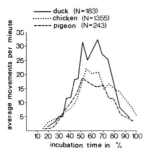

Fig. 37-2. Frequency of movements in pigeon, chicken and duck embryos during embryonic development.

Hamburger and his colleagues, it seems that Preyer speculated correctly in saying that the early movements of the chick embryo arise from neuro-motor cells in the spinal cord. Further evidence for Preyer's hypothesis comes from certain recent and quite elegant neurophysiological studies which correlate bursts of electrical activity in the chick embryo's spinal cord with burst of overt movement in the same animal at the same time! Since the electrical bursts show a fairly regular periodicity up to about thirteen days of embryonic development, as do the embryo's overt movements, it certainly would seem that Preyer's (1885) hypothesis of the spinal origin of these early movements is correct.

While many (or most?) of the chick embryo's early movements are spontaneous and very likely of spinal origin, it should be noted that the chick embryo is also capable of responding overtly to experimentally applied sensory stimulation rather early in its development. In order to avoid confusion, it is necessary to be aware of the dual capability of the chick embryo to move spontaneously as well as in response to various sorts of sensory stimuli. As implied above, the importance of sensory stimulation in instigating, directing, or maintaining chick embryonic movement would appear to increase with age, being relatively unimportant, or of no importance, in the early stage and becoming rather more significant in the late embryonic and early postnatal phase. Motor primacy refers to the fact that autogenous or spontaneous motility predominates at an early age, regardless of whether the sensory systems are hooked-up to the motor side or not. Motor primacy can not be generalized to all vertebrate embryos, however; the situation at the mammalian level has been too little explored to allow any definite conclusions. In the human embryo, for example, sensory function develops very early, about the same time the embryo first begins to move. Of course, this fact in itself does not rule out motor primacy, but in the many instances of the simultaneous onset of sensory and motor function in mammalian and other embryos, it would be difficult to definitely establish motor primacy without resorting to deafferentation.

The only experimental inkling we have of a possible relationship between prehatching and posthatching motor movements comes from one particular experiment which has been repeated several times since 1904. The experiment involves the posthatching swimming behavior of tadpoles which have been paralyzed during the last few stages before hatching, during the time they would ordinarily have exercised their swimming movements. In all cases, the tadpoles could swim after hatching once the effect of the anesthetic wore off. Their swimming movements are quantitatively deficient, however, in that they take longer than normal tadpoles to swim from one end of the tank to the other under continuous probing. Thus, there would seem to be a weak quantitative continuity between the prehatching and posthatching swimming movements in amphibia. If the chick embryo could be paralyzed in such a way (that is,

Neurological correlates of early embryonic motility

▷
Breeding and artificial selection often leads to increased intensity of fighting in the attacker, while at the same time the weaker animal lacks a submission or flight response. Top left: two artificially bred male fighting fish; top right: the corresponding wild species. Below: a cock fight.

Amphibian swimming movements

for short periods) so that it would not become deformed, it would be interesting to determine whether the deprivation of movement early in embryonic development would lead to a lower level of spontaneous or evoked motor activity later in embryonic development, thus demonstrating a quantitative continuity between earlier and later embryonic movements which are not highly patterned and are not primarily controlled by sensory stimulation. (Swimming is a highly patterned activity and it apparently comes under sensory modulation rather quickly in the life of the tadpole.)

Sensory function

Since we have been emphasizing motor movement, particularly the spontaneous movements of chick embryos, it is important to note that the various sensory systems are capable of function in the chick embryo as well as in the embryos of other vertebrate species. The extent to which they do function under normal (non-experimental) conditions of development is not clear, except for later stages when it is known that normally occurring stimulation does affect the embryos' overt activity (see below).

Sequential development of sensory systems

The sequence in which the sensory systems become functional is remarkably similar in birds and mammals (including the human species). The cutaneous or somesthetic system is the first to become functional, with the embryo or fetus showing an "avoidance" reaction to tactile stimulation about the oral region (turning away from the source of stimulation). Next comes the vestibular system, followed by the auditory sense, and finally vision, in that order. In precocial birds (chicks, quail) and precocial mammals (sheep, guinea pig), the capacity for visual function develops prenatally, whereas in highly altricial forms (grackle, rodents, cats), visual sensitivity does not develop until well after hatching or birth.

It has already been noted that sensory stimulation probably begins to play a more prominent role in the overt behavior of the avian embryo around the time of hatching. One of the more dramatic instances of such an influence has to do with the means by which embryos of certain species manage to synchronize their time of hatching within a clutch of eggs. In nature, the highly precocial quail young leave the nest with their parents very soon (perhaps within 5 hours) after hatching. By way of contrast, various duck species leave the nest 24–48 hours after hatching. Thus, in the quail species in which the nest-departure takes place soon after hatching, it is very important that the young hatch around the same time, rather than spread over a several hour period as usually occurs in duck species.

Quail embryos do synchronize their hatching. In order to achieve synchrony in hatching, the eggs must be in physical contact with each other. This has been demonstrated by Margaret Vince in an experiment in which she kept the eggs in some clutches of quail eggs in physical contact with each other and the eggs in other clutches just out of physical contact, both clutches being in an incubator in the laboratory. The embryos in the eggs which were in contact hatched around the same time whereas

◁

Examples of differences between wild and domestic forms. Above: Threatening behavior in a wolf. The expressive behavior of wolves is highly differentiated. Hunting in groups requires living in a group, and this in turn requires social co-ordination. Compared with the ancestral form, wolf, many social behaviors are no longer as greatly differentiated in dogs, most of whose social partners are humans. Such behavior patterns have become simplified or have even disappeared completely. Below: Southeastern Asian pigs display typical features of domestication. In addition to morphological changes (spotted coloration, pug-faced appearance, increased fat deposition), we can also observe behavioral modifications like hypersexualization and the precocious onset of puberty.

the embryos in the eggs which were not in contact showed a dispersion in their hatching time, and this difference held true in nine of the ten experiments in which the eggs were so treated. These findings indicate very clearly that there is some mechanism for inter-embryonic communication by which a clutch of quail embryos can regulate their progress toward hatching, and that this mechanism operates when the eggs are touching each other. Very likely the means of inter-embryonic communication is the peculiar "clicking" sound the embryos make when they begin to breath air during the last day or two of the incubation period. Vince has found that hatching can be accelerated by exposing quail eggs to either auditory or purely vibratory stimuli like the "clicks" late in incubation. Thus, according to this finding, the more advanced embryos in a clutch would accelerate the development of their less advanced siblings by simply beginning to click. In order for hatching synchrony to occur, however, it would seem that the more advanced embryos must slow down their own rate of development. Thus far it has proven much more difficult to decelerate hatching time by artificial stimulation than to accelerate it, so the means by which deceleration (and therefore hatching synchrony) is achieved in quail embryos is not yet completely understood. Other avian species such as chickens and ducks "click," and they also show more synchronized hatching when the eggs are in physical contact rather than out of touch with each other, but the synchrony in these other species is not as precise as in the quail. Since the other species dwell in the nest longer than do the quail, the natural selective pressure toward greater synchrony would not be as great as it is in the quail.

There are hints of sensory modulated, synchronous hatching mechanisms in species other than birds—cockroaches, alligators—but the experimental investigation of these has not yet proceeded to the point where they can be described with certainty. In many species which hatch out of eggs it is very likely that special hormones play an important role in the hatching process.

Whereas it is not entirely clear whether hatching synchrony is mediated primarily by auditory or vibrotactile stimulation, it is known that the auditory system of many animals does become functional in the prenatal or embryonic stage of development. This investigative problem has received the most attention in birds, so our discussion will once again emphasize avian species.

In colonial nesting species, where the nests are very close together, the young show a strong and definite tie to their own parents very soon after birth. This tie assures the orderliness of social life in the nesting community, with each young soliciting care from and being cared for—sheltered, brooded, and fed—by its own parents despite extremely crowded conditions in some instances, such as in penguin rookeries, gulleries, and among the cliff-dwelling guillemots.

An experimental demonstration of the guillemot embryo's capability

Sensory influences on hatching time

Auditory learning in bird embryos

Fig. 37-3. Gulls—as many other birds—nest close together in colonies. Strong bonds exist between parents and young, they know each other individually: For some species this bond is established before birth by vocalizations. In this way newly hatched young will only solicit their own parents and not neighbors for food and parental care.

Species-specific auditory perception: Embryonic background

to learn its parents' calls comes from the work of Beat Tschanz. In this species, as in gulls and penguins, both parents perform incubation duties. In the case of guillemots there is only one egg in the clutch, whereas in the gulls and penguins there are usually two eggs. Tschanz has shown that the guillemot can learn the individual characteristics of the "luring" calls of both of its parents after it has penetrated into the air-space at the large end of the egg and actually pipped a hole in the shell many hours before hatching. After hatching, the young guillemot approaches these calls whether they come from a loudspeaker or its parent. Thus, the initial tie between the young bird and its parent(s) is auditory, and since the hatchling responds selectively to its own parent's calls, it solicits care from its own parents rather than from nearby adults as soon as it hatches.

It is evident from the work of Colin Beer, Monica Impekoven, and Roger Evans that individual auditory recognition of parents by their young is also operative in gull species, but the capacity for selective auditory learning in the embryo has not yet been so clearly demonstrated in the gulls. Impekoven has shown that the laughing gull embryo can respond to sound several days before hatching and that its postnatal behavior (pecking) is influenced by sounds to which it has been exposed in the egg. In the adelie penguins the family bond is also maintained by individual recognition via sound signals.

It would seem likely that visual stimuli would also be important for individual recognition in all these species, but the initial filial bond is clearly of an auditory-tactile-thermal nature. If, for example, young birds—guillemots, domestic chicks, ducklings—are given a choice test involving the call of their own parent and the visual aspect of the parent, they invariably choose the call. As indicated by the study of the behavior and voices of adult birds in a variety of other avian species (including terns and gannets), the capacity for individual recognition by auditory learning persists throughout the lifetime of the individual. The important point is that, in some species at least, the capacity for auditory perception and learning begins to operate when the bird is still in the embryonic state.

Precocial ducklings and domestic chicks are attracted to the assembly or "luring" call of a hen of their own species even in the absence of prior exposure to the call. (The same is true for guillemots. Naturally, learning by direct exposure to the parent's call is required if the bird is to discriminate between, and come to prefer, its own parent's call over the call of another parent of the same species.) The maternal calls of the various precocial species which have been tested—mallard duck, wood duck, mandarin duck, pintail duck, domestic chicken hen—are discriminably different from one another, with much greater variation on several specifiably physical parameters between species and much more similarity on these parameters within the species. The ability of young birds to respond discriminatively (selectively) to the unique characteristics of the assembly call of a hen's call is a hallmark of instinctive behavior, and for this reason poses a particularly interesting problem for developmental analysis. It is highly adaptive, of course, for a young, highly precocial bird to be able to identify the call of a hen of its own species, particularly since this would direct its behavior to the parent. The strong attraction of the assembly call gives the hen a great deal of social control over the movements of young birds which are actually able to leave the nest and feed themselves—and get eaten by a predator—by virtue of their advanced stage of locomotor and perceptual development upon hatching. (Figure 37-4 shows one species of waterfowl before, during, and after the exodus from the nest in nature.)

Naturalistic observations on wild waterfowl indicate that the young ordinarily stay with their mother anywhere from about four to eight weeks of age, partaking of her physical warmth by brooding under her in the evening and perhaps learning to avoid potentially dangerous predatory situations by remaining with her throughout the day. In the marshes and swampy habitat which waterfowl prefer, it happens that members of the brood sometimes become visually separated from their hen, but they are always able to find their way back to her by localizing her call.

While the response of maternally naive ducklings to a hen of their species can be described as species-specific in the auditory modality, it is not so in the visual modality. If maternally naive mallard ducklings are given the opportunity, in a laboratory test situation, of approaching a stuffed model of their own or another species (see Color plate, p. 538), whether the hens are silent or emitting a recording of the species' maternal call, the ducklings do not respond selectively to the visual aspects of a hen of their own species. It appears likely from this result that the ducklings must learn the visual characteristics of a hen of their species based on contact with her, and it seems safe to conclude that this learning occurs in the context of their attraction to her call in the normal course of events in nature. (This matter will be pursued further in the section on visual imprinting.) The lack of specificity in the original visual perception of the hen makes the specificity of the auditory response even more important.

With respect to the mallard ducklings' selective response to the maternal call of its species, behavioral experiments with domestic mallard (Peking) embryos indicate that the selective auditory response is already present on the day before hatching (see Color plate, p. 538). (This finding has been verified in wild ducklings—Marieta Heaton has found that wood duck embryos respond in a specific way to the wood duck maternal call on the day before hatching.) By examining the behavior of many domestic duck embryos at various ages, it has been possible to trace out the prior developmental history of the Peking duck embryo's response to the mallard maternal call, as well as to delineate the conditions necessary to perfect the embryo's auditory perception so that it is truly species-specific.

Fig. 37-4. A wood duck (hole nester) leaves the nest with her young. At first the mother checks out the surroundings, then she jumps into the water and calls her young who follow her.

A mallard hen approaches and is attacked and driven off. The young remain four to six weeks with the mother, who keeps her young together by means of specific call notes.

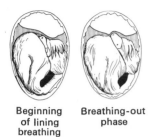

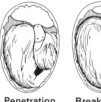

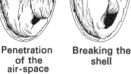

Beginning of lining breathing

Breathing-out phase

Penetration of the air-space

Breaking the shell

Hatching

Test 48 hours later

Fig. 37–5. Some developmental phases of the Peking duck shortly before and after hatching.

The domestic duck embryo first begins to respond consistently to the species' maternal call at five days before hatching, on Day 22 of incubation. At that time, however, the response is not fully differentiated, in the sense that the embryo is not responding to many of the distinctive aspects of the maternal call which set it apart perceptually from many other sounds. For example, the young embryo's response is restricted to the low frequency component of the call (<825 Hz), it is capable of responding to the call at only one repetition rate (4 notes per second), and it also responds to white noise (random, unmodulated sound) if it is pulsed at 4 notes/sec. Several days later, however, when the embryo's bill penetrates into the air-space at the large end of the egg and pulmonary respiration starts so the embryo itself is capable of vocalizing, the embryo's response to the features of the maternal call broadens to include the high frequency components (825–2300 Hz) and it sharpens so it will no longer respond to white noise pulsed at 4/sec. Thus, while the early response to the maternal call is rather generalized and not keyed specifically to the special attributes of the maternal call, the older embryo's response is much more finely "tuned," as one would predict from the general principle stated earlier concerning the gradual development of differentiation of perception. Since the more highly differentiated response occurs only after the embryo has heard its own vocalizations, an investigative question of considerable interest is whether or not the embryo's experience of hearing its own vocalizations plays a role in the differentiation of its auditory perception. As a first test of this proposition, embryos were muted (devocalized) by a simple surgical procedure (Fig. 37-6) and kept in individual isolation so that they were unable to hear either their own vocalizations or those produced by siblings. After hatching, the mute ducklings were tested for the ability to respond to the mallard maternal call in auditory choice tests involving the mallard call and the maternal calls of some other species, a task which is ordinarily quite simple for the (vocal) duckling. The mute ducklings performed rather well in most of these tests. They were deficient, however, in distinguishing the mallard call from the chicken maternal call, a call which has certain low frequency characteristics in common with the mallard call, including an energy peak of about 775 Hz. As one of several control procedures to make certain that it is the lack of exposure to their embryonic vocalizations and not some other factor which is responsible for the mute ducklings' poor performance in the mallard vs. chicken call test, a group of duck embryos was muted 18–24 hours later than usual—well after penetration into the air-space and after they had started vocalizing. This group of muted ducklings had no difficulty in discriminating the mallard call from the chicken call. Thus, in order for the ducklings' species-specific auditory perceptual ability to develop normally, they must hear their own vocalizations. (Further research now in progress will answer the important question of which particular aspects of the embryonic vocalizations must the embryo experience in order to perfect its perception of the key attributes of the mallard call.)

The embryos' auditory perception may still be undergoing differentiation at the time they are devocalized (Fig. 37-6), so it is possible that function (that is, hearing their vocalizations) plays a constructive role in the auditory perceptual differentiation process, not one of merely maintaining peak performance in an already completed perceptual system. Some recent analytic experiments on the postnatal development of visual perception in altricial rodents confirms and extends the view that exposure to normally occurring sensory stimulation may be necessary to bring perception to peak performance or complete differentiation. We will now turn to those studies.

Newly hatched and new born animals (neonates) can and do perform a number of perceptual and motor activities which are admirably suited to their survival. As we have seen in the case of birds and amphibians—where experimental manipulation has been possible—the developmental background for some of these activities begins in the embryo. In this section we will cover some postnatal experimental findings pertaining to the development of perception, motor activities, and problem-solving, with particular emphasis on the outcome of analytic developmental studies, ones where the usual conditions of development have been systematically altered to discover possible functional relationships between normally occurring developmental events and later behavior.

Perceptual learning is an "odd" sort of learning which is not widely studied by experimental psychologists, although it is probably as germane —possibly even more germane—to the understanding of the development of behavior than are the forms of learning which occupy psychologists the most, namely, classical and instrumental conditioning. The essence of the phenomenon of perceptual learning is that perception is improved (becomes more highly differentiated) as a consequence of the animal merely being exposed to certain features or kinds of environmental or self-stimulation. Naturally, the animal must attend to the stimulation. Otherwise no learning will take place, so it is not as passive as the phrase "merely being exposed" implies. This phenomenon differs from the two types of conditions in that there is no necessity for external reinforcement or reward, and the learning is not restricted to a specific motor response.

As we shall see, the improvement of perception (by practice) through exposure to relevant stimulation applies to even the most rudimentary perceptual abilities of young animals, "innate" abilities that have been thought previously to mature completely in the absence of sensory stimulation or "experience." The importance of the phenomenon of perceptual learning is that it is capable of synthesizing the nature-nurture pseudo-dichotomy in a very precise and understandable way; young animals can perform all manner of adaptive behaviors in the apparent absence of prior experience, but they perform them better—in fact, in the fully

Postnatal development of behavior

Fig. 37-6. Embryonic devocalization procedure. After the embryo's head has been pulled out of the shell early on Day 24 (see Fig. 37-5), a topical anesthetic is applied to its chest before the operation. The operation involves 3 further steps. (1) A small incision is made over the embryo's syrinx (voice-box). (2) The syringeal membranes are painted with a non-toxic surgical glue which prevents them from vibrating. Immobilization of these membranes prohibits vocalization. (3) The chest wound is sewed up with two sutures. The embryos hatch on Day 26 and are tested 24–48 hours later to determine the effect of not hearing their own vocalizations on their perception of the maternal assembly call.

species-typical manner—only in the presence of the customary prior experience. The probably biological reason for the improvement of performance of even rudimentary perceptual behaviors in the presence of experience is that evolutionary (natural selection) processes have always and everywhere worked on the phenotype, the phenotype is a product of a developmental manifold, and that manifold usually includes opportunities for experience to perfect phenotypic performance. Thus, in a manner of speaking, animals are "programmed" to perceive and to perform better as a consequence of experience or practice. The only reason for mentioning the probable evolutionary background of the phenomenon of perceptual learning is to make the point that the participation of perceptual learning in the development of innate or instinctive behavior does not make that behavior any less biological or ethological. Heretofore, perceptual learning has been mentioned in the ethological literature only in connection with the rather complicated process of imprinting; recent evidence from comparative and developmental psychology suggests that it may play a role in much more rudimentary behavior as well.

Depth perception

Very young animals show the ability to perceive depth. Day-old precocial birds, for example, tested in a laboratory apparatus called the "visual cliff" (Fig. 37-7) respond to the visually deep side in a manner consistent with their nesting ecology. Groundnesting species step off the center board to the visually shallow side whereas confirmed hole-nesting species, ones which characteristically nest high above the ground, are as likely to step-off to the shallow side as to jump-off to the deep side, indicating that they perceive the visual difference in depth between the two sides. Domestic chicks that have been kept in darkness prior to the test step off to the shallow side, which indicates that patterned visual experience is not a necessary precondition to perceive depth. But a recent ontogenetic analysis by Richard Tees shows that a fine differential discrimination of depth comes about only in the presence of prior visual experience: rats reared in the dark are able to perceive depth but their discriminative ability never reaches that of rats reared in the light. The constructive effect of visual stimulation is shown in that the animal's depth perception becomes optimal only in the presence of light; the maintenance effect of visual stimulation is shown in that the discriminative ability of the rats began to deteriorate after eighty days in the dark and continued to get worse the longer the animals were kept in the dark. While the importance of sensory stimulation (or neuromuscular exercise) for the maintenance of already-developed abilities has always been appreciated, it has been tacitly assumed that the maturation of "innate" abilities such as depth perception could reach its peak in the absence of stimulation. The present study indicates that assumption is incorrect. (The next important step in studies of this kind is to determine which specific features of the prior stimulation are essential for the improvement in performance.)

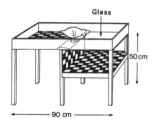

Fig. 37-7. Visual cliff test apparatus. The bird (or other animal) is placed on the center board. It has the option of descending to the visually shallow side, the visually deep side, or not at all.

Before turning to the role of perceptual learning in imprinting, it should be noted that experience probably plays a role in differentiating other aspects of rudimentary perception besides depth perception, but there is relatively little experimental information on this point because few investigations have been addressed to it. Joseph Kovach, for example, has found that dark-reared chicks show a color preference for blue and red over yellow and green. When chicks have been given extensive previous exposure to one of these four colors, the preferences for blue and red are markedly increased whereas the preference for yellow and green are barely affected. This indicates that perceptual learning operates within constraints—the young animal is not merely a "blank slate upon which experience is registered."

As the phenomenon of perceptual learning becomes more widely accepted and appreciated, more experiments will probably be addressed to it. Incidentally, prior exposure need not always have beneficial effects; there are several experiments on record (P. P. G. Bateson) which show that very familiar perceptual patterns are not differentiated as readily as novel ones. There is considerable grist here for the experimental mill.

Imprinting

This term is used to cover the many different developmental situations where a young animal comes to prefer (or forms an attachment to) a certain perceptual pattern—be it odor, a taste, a sight, or a sound, or some combination thereof—based on mere exposure to it during a sensitive or critical period of early life. (Although the experimental condition or operation for producing imprinting is similar to that for perceptual learning, the latter does not necessarily give rise to a positive preference for the experienced pattern, nor has a sensitive or critical period been postulated for perceptual learning.) The best-known example of imprinting has to do with young chicks or ducklings learning the visual characteristics of, and becoming attached to, the first mobile object they come into contact with, be it a hen of their own species or something quite dissimilar. This kind of imprinting has been studied most often using precocial species like quail, chicks, ducklings, geese, and guinea pigs, but altricial species (for example, doves) also show filial imprinting. Imprinting in birds can occur to sound as well as to visual patterns of stimulation, and olfactory imprinting has been demonstrated in some mammalian species, and perhaps gustatory imprinting, in turtles.

The interesting thing about the process of imprinting is that it involves a strong attachment to the familiar perceptual pattern—it is not merely a matter of perceptual learning, that is, improvement in discrimination or perceptual differentiation as a consequence of prior exposure. Whereas perceptual learning can occur in adulthood, imprinting is confined to immature stages of neural, behavioral, and psychological development.

Although the term imprinting conjures up the image of an indelible seal being made on an indifferent piece of wax, it actually occurs within perceptual constraints and it is not as irreversible as once believed. Chicks,

▷
In a zoo we have the opportunity to study details of mimical expression in individual animals, following their entire development and covering all their moods over many years. In this way we can make observations and record these on film far better than was ever possible with animals in the wild. This provides us with important comparisons between and within species. The ethologist must, however, be careful to avoid any anthropomorphization when interpreting expressive behaviors in animals.

Top left: Among the most important differences in behavior between free-living animals and those in zoos is the reduction of the escape tendency. People are no longer considered enemies, and instead are viewed as familiar and trusted fellow creatures, even as conspecifics. Top right: The original threat posture of the pygmy hippopotamus,—the gaping of the mouth and display of the razor-sharp canine teeth—sometimes becomes a begging gesture with zoo inhabitants. Middle left: Like this rhinoceros, most zoo animals set up their runways—here between the bathing site, resting place, and rubbing site—within the enclosure in the same way as in nature. Middle right: The comfort and well-being of a wild animal in the zoo depends less on the size of its living space than on its quality, particularly the presence of important fixed reference points. For the Grevy's zebra, for example, it is important to have a termite hill (artificial, in this case) to rub against. Bottom left: With most wild animals, the most important fixed reference point within their territory is the residence or home, even in a zoo. This is the place of greatest security. The aardwolf, for example, uses this man-made hollow as a den. Bottom right: Modern zoos offer many technical means for displaying the structure of a wild animal's home. This photograph shows a yellow-necked field mouse.

for example, are more readily imprinted to oval visual objects than to intricately shaped objects. Color and size preferences also play an important role in filial imprinting. Irreversibility is a matter of degree and it seems to differ between species; the imprinted animal may always prefer objects with the perceptual attributes of the original imprinting object, but that does not preclude socializing, courting, or otherwise reacting to objects which depart somewhat from the preferred perceptual pattern. Unfortunately, the actual perceptual features to which the young animal becomes imprinted have not been systematically studied: studies of the distinctive features involved in imprinting are sorely needed.

The original formulation of imprinting by Konrad Lorenz held that, in some species at least, mating choice in adulthood is determined by imprinting in infancy. This is called sexual imprinting, and it seems to occur most readily in the males of bird species which are sexually dimorphic (= male and female are colored and marked differently). F. Schutz has found this to be the case in waterfowl, and K. Immelmann has confirmed the effect in several species of finches. It is illuminating that Immelmann found the ease of inter-specific sexual imprinting in finches to be roughly proportional to the degree of overlap in the call repertoire of the two species. This further demonstrates the constraints within which imprinting takes place, as well as the probable facilitation of visual imprinting by species-specific auditory stimulation. The young of many avian species show preferences for species-specific auditory patterns but not for species-specific visual patterns.

Other examples of imprinting or imprinting-like phenomena have been observed in connection with habitat preferences, nest-site preferences, and food preferences. Young tadpoles, for example, show a selective learning of certain substrate patterns and show a preference for these at a later age. Female wood ducks return to their natal pond to nest, even though they have spent only a very brief time there early in their life and have migrated a considerable distance in the meantime. Along the same line, salmon return to their natal stream to spawn when they are two or three years old, after journeying many miles in the open ocean. Presumably they use odor (olfactory imprinting) to detect their home river and tribuary, but how they navigate to that point from the open ocean is another question entirely. On food imprinting, newly hatched snapping turtles come to prefer a certain diet (either horsemeat or worms) after only a single feeding, provided it was their first feeding. In line with the notion that imprinting always takes place within certain constraints or "natural" preferences, horsemeat seems to be more effective than worms in inducing the effect.

Thus, there are numerous examples of imprinting, a form of learning which takes place early in life and markedly influences perceptual preferences exhibited later on. The process of imprinting is akin to perceptual learning in that exposure to stimulation (at the proper age and under the appropriate circumstances) is all that is required. Imprinting is different from perceptual learning in that not only are the characteristics of the object learned, but an attachment or positive preference for the object is also established. Here, as elsewhere, more analytic research is required to determine which features of the stimulating source are critical, both for the learning and for the attachment.

Although it is a commonly used expression, the term motor behavior is a misnomer in the sense that there are very few, if any, cases of behavior which involve solely the motor system to the exclusion of sensory and perceptual factors (including especially proprioceptive and kinesthetic "feedback" or reafference).

Motor behavior

These are very rudimentary patterns of limb movements. Historically speaking, there have been grave disputations on whether these patterns are entirely centrally patterned or not. As we realize now, that is not a developmental approach to the problem; we can fully accept spontaneity or the central control of motor behavior and still want to understand the factors which contribute to its development.

A basic and important developmental question concerns the role of exercise or practice on the rate of development and perfection of motor behavioral patterns. As will be recalled, young amphibians can swim even if they have been prevented from practicing or using the neuromusculature, but they can swim better if they have had the opportunity to use their muscles in the normal course of development. There have been no carefully done manipulative ontogenetic studies of flying behavior, using both the addition and the subtraction methods, although neurophysiologists have studied certain features of the flight patterns such as wing-beat frequency in developing insects—but not in birds. Recent renewed interest in the interest in the ontogeny of walking in human infants has shown that active exercise of the walking (and placing) reflexes in the newborn in the first eight weeks after birth leads to an earlier onset of independent walking (ten months) than passive exercise of the same muscle groups (eleven months).

Although there have been no comprehensive developmental studies of walking, swimming, or flying, the few and fragmentary studies which are available indicate that practice probably contributes to the rate of development and perfection of these motor activities.

Visual-motor
coordination

The act of reaching or placing has a rather clear-cut perceptual and motor component: the animal (or person, for that matter) must orient the far parts of the limb to some visual object or surface. Alan Hein and his colleagues have studied the development of this ability in kittens. The kittens are tested by slowly lowering them toward the prongs of a test board (Fig. 37-8). Upon being lowered toward the board, the kitten reflexively extends its free forelimb, and as it nears the board it places the paw of that leg on a prong. The placing movement would appear to be a very simple behavioral act, but it has a rather "complicated" develop-

Fig. 37-8. Prong test board. A kitten is gently and slowly lowered to the prong board to test its ability to reach out and place its paw on an interrupted surface. The handler clasps the opposite forelimb so that only one leg is tested at a time. The sketch illustrates deficient visually guided placing behavior in a kitten which had been allowed some visual experience, but not the particular kinds and sequences of visual experiences necessary to perfect its reaching response.

mental background. As it turns out, accurate reaching (or placing) is not only dependent upon visual and motor experience: it is dependent on particular visual and motor experiences, and on a particular sequence of visual and motor experiences.

In order to develop skillful visually guided reaching, the young kitten must first fully experience itself moving its body in a visually defined space, such as a room with objects in it. The movement must be self-produced: if the kittens are passively moved about by carrying them, they do not achieve a proper body-centered visual-spatial framework. In addition, to do well in the reaching test, they must be able to perceive their limbs as they locomote about the room. Kittens which locomote about the room with an Elizabethan collar around their neck (so they can't see their forelegs) do not develop skillful reaching; they must have the visual feedback of seeing their forelimbs move in space. Finally, if visually naive kittens are permitted to see one of their forelegs as they move about in an otherwise dark room (applying a special luminescent to the leg makes it visually perceptible in the dark), this is insufficient to establish skillful visually guided reaching with that foreleg. This experience is effective only if the kittens have first been allowed to establish a body-centered visual-spatial framework by locomoting in a lighted room, even wearing the Elizabethan collar. If they are then allowed to view their foreleg in an otherwise dark surrounding, their visually guided reaching behavior becomes perfected. In other words, the development of visually guided reaching with a forelimb takes place only in the context of an already acquired "locomotory space." Kittens which have not been allowed to locomote about the room under their own power do not achieve a body-centered visual spatial framework, as shown by the fact they collide with obstacles in a test enclosure which has only narrow channels of passage (see Color plate, p. 538). They need not have previous experience seeing their forelimbs move to achieve a body-centered framework: kittens allowed to locomote in a lighted room wearing Elizabethan collars later perform well in the obstacle course. A further point of interest, attesting to the specificity of the experiential effects, is that the visually guided reaching which is perfected when the kittens are now allowed to see their luminescent forelimb in an otherwise dark surrounding is restricted to the forelimb viewed! Hein and his collaborators also performed the reciprocal experiment of covering one eye and allowing the kittens to see both their forelimbs as they locomoted in a lighted room. In that case visually guided reaching (with either forelimb) was normal with the "experienced" eye, but deficient with the other eye.

These investigations of visual-motor coordination indicate not only the constructive effects of experience, but also the nature of the requisite experiences. This represents a significant advance in the developmental analysis of behavior. All too few ontogenetic studies attempt to deal with both of these aspects of development.

The production of species-typical or species-specific song in adult songbirds presents a very interesting subject for developmental analysis, and several students (notably Thorpe, Marler, Nottebohm, and Konishi) have delineated some of the main features of the problem.

Under normal conditions, songbirds are cared for by both parents, and the young male is exposed to the songs of its father and perhaps other adult males in the vicinity. (Females sing only in comparatively few species.) The young male usually begins singing in the next season, by which time it has heard itself and sibs make various calls as well as having heard the adult songs of its own and possibly other species. Singing functions to attract potential mates, and it also serves as a spacing and recognition device between males of the same species.

In all songbird species studied thus far, auditory perception must be intact if the bird is to sing normally as an adult. Songbirds which have been deafened as nestlings do not produce species-typical songs in adulthood. They must either be able to hear themselves or be exposed to the song of "models" (adult birds or tape-recordings) if they are to sing in adulthood. In avian species which do not utter the complex sounds defined as song—for example, chickens and ring doves—the call repertoire appears unaffected by early deafening. In fact, deafened songbirds utter sounds, and sometimes songs, in adulthood, but not the intricately phrased and modulated songs of their intact conspecifics.

A very important feature of the learning which seems to be involved in song development is its specificity or selectivity. White-crowned sparrows and chaffinches, for example, learn the song of only their own kind when they have been exposed to conspecific and extra-specific songs. In some species the selectivity of the learning is based on the social bond with the parent or foster-parent. For example, zebra finches and bull-finches learn the song of their "father," be it their actual father or a foster-father (e.g., a canary).

There are critical periods for song learning, but they differ very widely between species. Cardinals, for example, are able to learn conspecific songs up to the time of first breeding (about ten months of age), while the sensitivity of white-crowned sparrows declines after only two months of age.

It seems reasonable that there are properties of the acoustic pattern to which young birds of different species are differently sensitive, but research has not yet been directed toward that question in the vocal learning of birds.

Aside from the ethological studies on imprinting and song-learning, one finds the most striking evidence of the pervasive effects of early experience on later behavior in the experimental psychological literature on learning and "intelligence." These studies offer compelling evidence for the continuity of development, especially when one takes into consideration that the problem-solving feats (or the means of problem-solving)

Bird-song

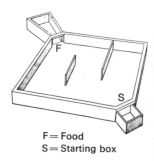

F = Food
S = Starting box

Fig. 37-9. Hebb-Williams closed-field maze. After the animal learns the way from the start to the goal box in this versa-tile maze, it is given a series of changing problems to solve in the same arena. (See Fig. 37-10.)

Learning and intelligence

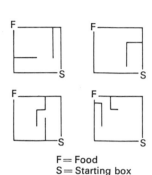

F = Food
S = Starting box

Fig. 37-10. Problems in a Hebb-Williams maze. These are illustrations of the numerous practice and test problems which the animal encounters in this maze. The essential feature is that the path between the start and finish boxes is altered from problem to problem by moving the walls as indicated. Standards of performance have been established for various strains of rodents on these problems. This maze is considerably more difficult than a Y- or T-maze, so it taxes the animal's learning ability to a greater degree. Although, as in other standardized laboratory situations, such a test is rarely tailored to the ecological status of the species tested therein, it does have the advantage of being standardized so that useful information can be obtained from the performance of closely related species in the same situation. (From Rabinovitch and Rosvold.)

may be discontinuous in young and adult animals of the same species. In other words, adult animals may solve problems in ways which younger animals can not (discontinuity), but the early experience of the young animal influences its problem-solving performance in adulthood (continuity). Also, in this area we see once again the operation of optimum periods of experience, periods in which the neurobehavior status of the young animal is maximally set to benefit (or suffer!) from particular kinds of experiences. This critical-period concept carries with it the additional specification that the absence of certain kinds of experiential opportunities circumscribes the development of learning abilities in ways which can be deleterious. We owe these developmental concepts, with their emphasis on the profound effects of early perceptual enrichment and deprivation, to Donald O. Hebb. Although the overwhelming majority of these studies have utilized only rats, the implications have been verified in other mammalian species such as mice, gerbils, and dogs, and they have also been supported by similar studies with rhesus monkeys, the only primate species investigated so far.

A typical experiment employs several groups of young animals reared under certain conditions for certain periods of time and later tested in the rather complicated Hebb-Williams closed-field maze (Figs. 37-9 and 37-10). (This maze is more difficult than a simple T- or Y-maze, which is attested to by the fact that enriched or deprived rearing has no measurable effect on learning the latter). The results show that restricted opportunities for visual exploration hamper mastery of the Hebb-Williams maze while enhanced perceptual experience facilitates it. For example, Hymovitch reared one group of rats in a large, complex pen, another group in a small mesh cage but exposed to the complex pen by moving their cage about in it, a third group in a "stovepipe cage" allowing little diverse visual stimulation or motor activity, and a fourth group in an enclosed activity wheel allowing a lot of motor activity but little variation in visual stimulation. The rats reared in the large pen and those exposed to it in the mesh cages performed equally well in the Hebb-Williams maze, and both groups performed much better than the rats reared under the other two conditions. To confirm the early experience aspect, in another experiment Hymovitch exposed one group to the large complexly designed pen from 30 to 75 days of age while another group remained in the stovepipe cages. At that point, the experiences of the two groups were reversed, with the first group being placed in the stovepipe cages and the second group being given exposure to the complex pen. The animals given the early exposure to the pen performed much better in the Hebb-Williams test than the animals given the late exposure to it, with early exposure being equivalent in its effectiveness to prolonged exposure throughout the experiment, and late exposure being no better than housing in normal cages. As has been found by Rosenzweig and his colleagues, these effects of impoverished and enriched rearing experiences are correlated with increased anatomical weight and neurochemical activity of the rat's brain. In contradistinction to the behavior in the Hebb-Williams test, however, these brain changes can be induced in adulthood and they are reversible; that is, the weight and neurochemical increases are lost if the animal is returned to the impoverished condition.

As has been noted in this chapter, genetic differences circumscribe the phenotype which is developed in any given set of rearing circumstances; this holds true for the present paradigm, too. Genetically different "bright" and "dull" strains of rat were reared in one of three conditions: enriched (cages with objects in them and designs on the walls), normal (regular cages in the rat colony), and impoverished (same cage area as the enriched cages, but with no objects and no designs on the walls). While rats from the bright strain performed as well in the Hebb-Williams test whether they were raised in the normal or the enriched environments, the performance of the dull rats increased significantly under the enriched condition, so much so that they were equal to the bright rats. So these particular enriched conditions helped the dull but not the bright rats. Conversely, the bright rats reared in the impoverished environment showed a much poorer performance than those reared in the normal condition, while the dull rats performed the same (poorly) whether they were reared in the normal or the impoverished condition. So these particular impoverished conditions hurt the performance of the bright rats, but it did not affect the performance of the dull rats. The so-called bright rats performed as poorly as the dull rats in the impoverished condition. In fact, the only condition in which the bright rats excel the dull rats is when both are reared in the normal colony cages!

Thus, "bright" rats can be bright or dull, and "dull" rats can be dull or bright, depending on the developmental background out of which they come—a fitting closing illustration of the multiplicity of developmental pathways between gene and behavior.

38   Behavior-Genetic Analysis

Behavior-genetic
analysis, by J. Hirsch

This discussion describes behavior-genetic analysis, an approach to the study of animals and their behavior. It combines from an evolutionary perspective the concepts and methods of genetic analysis, based on knowledge or control of ancestry, with the concepts and methods of behavioral analysis, based on knowledge or control of individual experience and species habitat.

One difficulty in bridging the gap between behavior study and the study of heredity lies in the very different nature of the concepts applied to the subject matter of the two fields. Students of behavior have been concerned with *types* of events—the activities of animals. But in the geneticist's framework such events are phenotypes, and are to be distinguished from genotypes; this distinction lies at the foundation of genetics. Genetic analysis begins with the study of the variations in the phenotypic expression of specific traits, that is, individual differences, rather than with some typical act, uniform in expression across individuals. The picture that has emerged from a large amount of research, dating back to the rediscovery of genetics in 1900, is that the relationship between heredity (genotype) and behavior (phenotype) is one of neither isomorphism nor independence. Isomorphism might have justified a naive reductionism, and independence, a naive behaviorism; both have been tried but neither approach has yielded an adequate understanding.

We have learned that we can study the behavior of a particular animal, the genetics of a specific population, and individual differences in the expression of some behavior by the members of that particular population. For these reasons we now speak of behavior-genetic analyses, understanding by that expression simultaneously the analysis of well-defined behaviors into their sensory and response components, the reliable and valid measurement of individual differences in the behaviors and in their component responses, and *then* subsequent breeding analysis by the methods of genetics over a specified set of generations in the history of a given population under known ecological conditions. We know full well that both the behavioral and the genetic properties can and will vary (with quite imperfect correlation between them) over time, over ecological conditions, and among populations. That is why their relationship is one of neither isomorphism nor independence. In order to appreciate both genotypic diversity and its relation to those ubiquitous individual differences, which, in so much behavior study, were obscured by too often inappropriate statistical analyses, it became necessary to shift from the typological to the population mode of thought—a major conceptual revolution. As Ernst Mayr has so cogently pointed out, "the philosophical basis in much of early science was typological, going back to the *eidos* of Plato. This implies that the 'typical' aspects of the phenomenon can be described, and that all variation is due to imperfect replicas of type, all variants being, in the terms of Plato's allegory, 'shadows on a cave wall.' Such typological thinking is still prevalent in most branches of physics and chemistry and to a considerable extent in functional biology, where the emphasis is on the performance of a single individual. The typological concept has been completely displaced in evolutionary biology by the population concept. The basis of this concept is the fact that in sexually reproducing species no two individuals are genetically alike, and that every population is therefore to be characterized only by statistical parameters such as means, variances, and frequencies . . . . Genetic variability is universal, a fact which is significant not only for the student of morphology but also for the student of behavior. It is not only wrong to speak of the monkey but even of *the* rhesus monkey. The variability of behavior is evident in the study not only of such a genetically plastic species as man but even of forms with very rigid, stereotyped behaviors such as the hunting wasps . . . . Striking individual differences have been described for predator-prey relations, for the reactions of birds to mimicking or to warning colorations, for child care among primates, and for maternal behavior in rats. It is generally agreed by observers that much of this individual difference is not affected by experience but remains essentially constant throughout the entire lifetime of the individual. Such variability is of the greatest interest to the student of evolution, and it is . . ." now the subject of much research. A more complete account of the genetic rationale for this distinction follows.

The phenotype—appearance, structure, physiology, and behavior—of any organism is determined by the interaction of environment with its genotype—the complete genetic endowment. Each genotype is the end product of many mechanisms which promote genotypic diversity in populations.

Ordinarily, members of a cross-fertilizing, sexually reproducing species possess a diploid, or paired, set of chromosomes. Most species whose behavior we study are sexually dimorphic. The genetic basis of this dimorphism resides in the distribution of the heterosomes, a homologous pair of sex chromosomes (SS) being present in the mammalian female

▷
Our language of facial expressions is the same all over the world. Here, a Peruvian mother is feeding her child (top left), then smiles at him (top right), looks pensive (bottom left), and finally smiles at her husband (bottom right).

▷▷
Above: a Bushman family in front of its hut. The primitive society of hunters and gatherers, too, is built on the basis of family units. No "primitive promiscuous hordes" are to be found, as people believed in an earlier time. Below: A number of bonding rituals, some of them in the context of play, serve to keep the small groups of hunter-gatherer society together. This melon dance game (Bushman) is played only by women and girls. The ball (melon) is thrown from one partner to the next, each following the other as they dance to the beat of singing and hand-clapping. When one of the partners is clumsy and drops the ball, the others laugh at her. Dancing rituals have a cohesive effect because participants engage in a common activity and because the group is closed to others who do not know the rules. Skill is rewarded by group members, whereas clumsiness is ridiculed. This puts pressure on individuals to conform to the group's norms.

Understanding individuality

and an unequal pair of sex chromosomes (SY) in the mammalian male. Sexual dimorphism guarantees that any population will be variable to the extent of at least two classes.

Chromosomes other than sex chromosomes are called autosomes. Every autosome is normally represented by a homologous pair whose members have identical genetic loci. Alternative forms of a gene any of which may occupy a locus are termed alleles. If an individual receives identical alleles from both parents at homologous loci, he is said to be homozygous for that gene. If he receives two alleles that differ, however, he is said to be heterozygous for that gene. The process by which a gene changes from one allelic form to another is called mutation.

When a gene is represented in the population gene pool by two allelic forms, the population will be genotypically polymorphic to the extent of at least three classes. That is, individuals may be homozygous for either of two alleles or heterozygous for their combination.

Study of populations has revealed that often extensive series of alleles exist for a locus. Well-known examples are the three (actually more) alleles at the ABO-blood locus in man and a dozen or more alleles at the white-eye locus in *Drosophila*. In studying the internal genetic architecture of one viral "gene" with a corresponding physical structure of probably less than 2000 nucleotide pairs, Benzer found over 300 distinguishable mutational sites. There is no reason to believe that we shall find less complexity in cellular organisms as further refinement increases the resolving power of our techniques for analyzing them. In general, for each locus having n alleles in the gene pool, a population will contain $n(n + 1)/2$ genotypic classes. Mutation insures variety in the gene itself.

Sexual reproduction involves meiosis—a complex cellular process resulting in a meristic division of the nucleus and formation of gametes (reproductive cells) having single genomes (a haploid chromosome set). One homologue in every chromosome pair in a diploid complement is of paternal origin and the other is of maternal origin. In meiosis, the homologues of a pair segregate and a gamete receives one from each pair. The assortment to gametes of the segregating homologues occurs independently for each pair. This process insures diversity because it maximizes the likelihood that gametes will receive unique genomes. For example, gametogenesis in the fly *Drosophila willistoni* produces eight alternative gametic genomes, which, if we represent the three chromosome pairs of this species by Aa, Bb, and Cc, we designate ABC, ABc, AbC, aBC, Abc, aBc, abC, abc. In general, n pairs of chromosomes produce 2^n genomes (if we ignore the recombination of gene linkages that actually occurs in crossover exchanges between chromosomes). Man, with 23 chromosome pairs, produces gametes with any of 2^{23} alternative genomes. This makes vanishingly small the chances that even siblings (other than monozygotes) will be genetically identical. Since the gamete contributed by each parent is chosen from 2^{23} alternatives, the probability that the

◁◁
Courtship ritual of the Melba (New Guinea): While rolling their head, the partners rub noses according to a prescribed ritual. This is an example of how behavior, here derived from the idea of two people affectionately sniffing at each other, may be greatly simplified or stylized through cultural control.

◁
A Bushman girl shows embarrassment by covering her face with her hand. This movement and pose is a universal one accompanying a feeling of embarrassment.

second offspring born to parents will have exactly the same genotype as their firstborn is $(1/2^{23})^2$, or less than one chance in over 70 trillion! The probability that two unrelated individuals will have the same genotype, then, is effectively zero.

The argument for the genotypic uniqueness of the individuals in diploid species is even more compelling, since other conditions also contribute to diversity. So, it is clear, the animals, whose behavior we study, are intrinsically variable before they undergo differentiating experiences. The mechanisms responsible for this variety are mutation, recombination, and meiosis. Add to these individual experience, and it becomes evident why individuals differ in behavior. In fact, the more reliable our methods of observation become, the more evident this variety will be.

Assumptions about the uniformity and normality of material under investigation are often made in physiology, a science after which, more than any other, so much of behavior study has been patterned. We may, therefore, get a better grasp of the individuality-uniformity distinction by examining the differences between organisms whose behavior is studied by behavioral scientists and systems whose functioning is studied by physiologists.

Since the two disciplines are working at distinctly different levels of biological organization, the meaning of "normality" as operationally determined by them is quite different. Physiologists choose a normal organism to work with—one that looks healthy and does not appear unusual—and study one or more of its systems, such as the adrenals, gonads, or other endocrines, or regions of the nervous system. Either pre-experimentally or post-experimentally, anatomical, histological, or biochemical verification is made of the normality of the material under study, and sometimes of related or adjacent functions as well. In the behavioral sciences we also choose normal-appearing organisms to study. We rarely perform biopsies unless there is a specific physiological interest, in which case we operate as the physiologist does.

Physiological systems are variable, not uniform. Roger Williams amply documented this and pointed out that implicit in our use of "normal" is reference to some region of a statistical distribution arbitrarily designated as not extreme, for example, the median 50%, 95%, or 99%. We choose such a region for every trait. Among n mathematically independent traits—for example, traits influenced by n different chromosomes—the probability that a randomly selected individual will be normal for all n traits is the value for the size of that region raised to the nth power. Where "normal" is the median 50% and $n = 10$, on the average only 1 individual out of 1024 will be normal (for ten traits). When we consider at one time the distributions throughout a population of large numbers of physiological systems, we should expect negative deviates from some distributions to combine with positive deviates from others, both kinds of extreme deviates to combine with centrally located ones, and deviates of similar

The abnormality of the normal

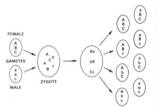

Fig. 38-1. Recombination segregation, and independent assorsment. Maternal components (chromosomes or genes) in capital letters, paternal in lower case.

algebraic sign and magnitude to combine. Each individual's particular balance of physiological endowments will be the developmental result of the genotype he draws in the lotteries of meiosis and the mating ritual. Because of crossing over, most genes assort independently. Hence, we cannot expect high correlations among the systems they generate.

If, underlying every behavior, there were only a single such system —for example, if the male "sexual drive" were mainly dependent on the seminal vesicles, or escape behavior were mainly dependent on the adrenals —then the same kind of distribution might be expected for both the behavior and the underlying system. Whatever uniformity might exist at one level would be reflected at the other. The last few decades of research on the biological correlates of behavior have made it increasingly clear that behavior is the integration of most of these systems rather than the expression of any one of them. Therefore, there is little reason to expect that the many possible combinations and integrations of those systems that go to make up the members of a population will yield a homogeneously normal distribution of responses for many behavioral measures. An organism richly endowed with the components of one subset of systems and poorly endowed with those of another is not to be expected to behave in the same manner as an organism with an entirely different balance of endowments. The obviousness of this fact is well illustrated in the behavior of the various breeds of dogs and horses.

Reductionism

Another conviction, strongly held by some, is that *real* explanations must be reductionistic. Those who hold this view in its most extreme form assert that no behavior can be understood until its physical basis has been unraveled. The search for the physical basis proceeds along physiological, biochemical, biophysical, or genetical lines, depending on the skills and predilections of the investigators.

In laboratory experiments, some rats learn mazes more readily on the basis of visual cues, while others do better with predominantly kinesthetic cues. The kinds of differences in organization that can coexist as alternative forms within a species, as well as some relations between one behavior and the component subsystems that are alternative possibilities, have been further revealed in a series of studies of the effects of domestication. In some domesticated rats, activity in a revolving drum was controlled by the gonads: control rats had daily activity scores as high as 18,000 revolutions, while gonadectomized rats scored only a few hundred revolutions. Cortisone therapy restored a high activity level in the gonadectomized rats. When the same experiment was repeated on wild brown (or Norway) rats, however, the presence or absence of gonads made no detectable difference in measured activity. Further study of differences between these domesticated and wild rats revealed larger adrenals in the wild rats and larger gonads in the domesticated. So it appears that activity may be under the control of adrenal output in one case and gonadal output in the other —that behavior is not a univocal index to an organism's balance of endowments. The fallacy of reductionism lies in assuming the one-to-one relation between different levels of organization. With degeneracy already demonstrated in the genetic code of messenger RNA base triplets for the amino acids of proteins, we should be surprised not to find it at the levels of complexity we are considering.

Thus it becomes clear why individual differences are found in populations of bisexual species: variation occurs because the mechanics of heredity and reproduction generate unique genotypes. Furthermore, reproductively isolated populations develop different gene pools. Even though such populations belong to the same species and share the same genes, the relative frequencies of the different alleles in their gene pools are almost certain to differ. Mutations and recombinations will occur at different places, at different times, and with differing frequencies. Selection pressures will also vary. A consequence of reproductive isolation, therefore, is that each population travels its own unique path of evolutionary divergence.

The population mode of existence has evolved with the bisexual species. While individuals come into existence, pass through developmental stages, and then die, populations endure so long as the supply of individuals is replenished by reproduction. We now realize that populations possess a corporate genotype. Their genetic structure, while clearly a function of the genetic constitution of the component individuals, has its own unique properties distinct from those which characterize the genetics of individuals.

Natural populations occupy a habitat in some region where they exhibit an impressive stability. Their numbers remain relatively constant, being confined within limits by checks on growth, such as food supply, predation, and disease, despite an enormous reproductive potential to which Malthus long ago called attention. The community of interbreeding individuals occupying a habitat is known as a Mendelian population—the group within which gene exchange occurs. If we speak of rabbits and wolves in a forest in Norway and of rabbits in Pennsylvania, we are talking about three separate Mendelian populations, which are reproductively isolated from one another: the first and second because they are different species, the first and third because they occupy different habitats. Reproductive isolation fosters evolutionary divergence by permitting the accumulation of the differences resulting from mutation, recombination, and differing selection pressures.

The habitat adaptation of such populations is interpreted in terms of Darwinian natural selection: from the diversity of genotypes produced in every generation by meiosis and mating combinations, those survive and reproduce whose phenotypes function more effectively in the prevailing environmental conditions.

The behavior of animals is one of the major ways in which they adapt to their habitats. As a habitat changes (for example, changes of season and climate), an animal's behavior changes. When another habitat be-

Populations

comes available and is colonized by a newly immigrant species, reproduction may at first be prolific, population growth can be rapid, and natural selection will have a diversity of types from which to choose in adapting the species to some niche in its new environment.

Behavior-genetic analyses

In certain instances behavior-genetic analyses are now permitting us to understand adaptations and some of the mechanisms responsible for them. Honeybees (*Apis mellifera*) are susceptible to American foul brood disease. Rothenbuhler has shown that two different genes control two distinct components of a highly adaptive hygienic behavior. In an elegant behavior-genetic analysis he has compared the hygienic behavior—when an infected larva dies, worker bees will uncap the cell it occupies, remove, and eject the corpse from the hive—of "resistant" hives with the non-hygienic behavior of "susceptible" hives. By following the segregation of the presence and absence of hygienic behavior through a genetic analysis involving crosses between the two types, he found hygienic behavior to be the expression of two independent genes. One gene controls whether or not the workers uncap the cells containing infected larvae, and the other gene controls their removing and ejecting the corpses. (The independence of these events was shown by a clever manipulation. In a certain genetic combination the cells would be uncapped but the larvae were not ejected. In another genetic combination the cells would not be uncapped but the larvae were ejected when Rothenbuhler uncapped the cells by hand for the bees. Only one genetic combination proved to be appropriate for the appearance of the complete sequence of events comprising hygienic behavior.)

In swordtails (*Xiphophorus*), a tropical fish, the males of two closely related species, *X. montezumae* and *X. helleri*, have somewhat different displays during courtship. Of three movements—sword-bending toward the female from the side, approaching her from a frontal position, and backing toward her from an antiparallel position behind her—the first two are regular features of *X. montezumae* display but not of *X. helleri*, and the third is displayed by *X. helleri* but not *X. montezumae*.

D. Franck has found that the species will cross and produce fertile hybrids. His behavior-genetic analysis showed that all three components could be transmitted genetically to the offspring of the hybrids, but that they assorted completely independently, so that any combination of the presence or absence of the three components would be obtained in a male of the F_2 (the second hybrid) generation.

Flies (Diptera) have daylight vision; they best conserve the moisture needed by their bodies when it is not too hot and dry. The first behavioral act of an adult fly is eclosion, emergence from the pupa case. Since young adults are especially susceptible to desiccation and their wings commonly fail to expand properly in low humidities (for example, in the hot, dry midday), it is most advantageous for eclosion to occur in the light but when there is sufficient moisture, a combination of conditions prevailing

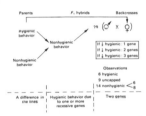

Fig. 38-2. Experiments and their results in diagrammatic form.

at dawn. Clayton and Paietta have provided most impressive evidence for the role of natural selection in "setting the clock" that controls dawn eclosion. They compared flies (*Drosophila melanogaster*) of a population from a natural habitat with others of a population cultured in the laboratory for forty-seven years, and found eclosion time to be much more flexible genetically in the laboratory population than in the natural population.

They selected simultaneously two lines from each population, one bred for early eclosion—well before the dawn optimum—and the other for late eclosion—well after the dawn optimum. The response to selective breeding for eclosion at times both earlier and later than the dawn optimum was pronounced in the laboratory population but much weaker in the natural population. This result is consistent with the evolutionary interpretation that in nature, intense selection pressure has been long and persistently eliminating from the gene pool those mutations and gene combinations which might have generated an array of individual differences with respect to eclosion time. Such mutations and gene combinations however, clearly did accumulate in the protected and unfluctuating environment provided for laboratory cultures, where the selection pressures must have been quite different.

A valuable example illustrating the lack of isomorphism but non-independence between genotype and phenotype and the possibility of uncovering hidden aspects of biological organization can be seen in a behavior-genetic analysis of courtship in several duck species and their hybrid offspring. Working with Konrad Lorenz, Wolfgang von de Wall showed how analysis makes manifest previously latent genetic potentialities and thus throws light on species affinities.

There is a series of postures and sounds displayed by drakes during courtship: A, Grunt-Whistle; B, Head-up-tail-up; C, Down-up; and D, Bridling. The grunt-whistle requires a peculiar bent posture of the drake, which probably stretches the trachea. At the climax of this movement there follows, according to Lorenz, "a loud, sharp whistle, followed by a deep grunting sound, while the head is raised once more and the body sinks back onto the water surface. The grunting sounds as if the air compressed during the whistle were escaping." The head-up, tail-up is one of the most conspicuous movements of the drakes. "The drake first pulls his head back and upward with the chin retracted, while uttering a loud whistle, and simultaneously raises his tail with its strongly erected rump feathers, so that the whole bird becomes peculiarly short and high. The elbows are raised at the same time, so that the projecting curled tail feathers remain visible from the side. This phase lasts for one-twentieth of a second, and then the body resumes its normal posture. Only the head remains raised for a moment, and the beak is then directed at a particular duck present during the social display of the drakes; in paired drakes this is always the mate." The up-and-down movement is characterized by a

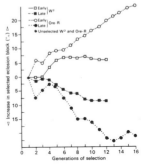

Fig. 38-3. Selection for early and late eclosion strains of *Drosophila melanogaster* wild-caught (W²) and laboratory-reared (Ore-R) populations. The values on the ordinate are the percentages eclosing in block 1 of the early strain and block 3 of the late strain minus the percentages eclosing in these collection blocks of the unselected stocks. The mean number per data point is 362 flies; the range is 153 to 652.

rapid dipping of the beak into the water, followed by a quick raising of the head, during which the breast remains deep in the water. At the highest position of the head, the drake makes a whistle, followed by a rapid "rab-rab" call. In the post-coital display, the drake performs a movement called "bridling" or "jerking up"; immediately after copulating he throws his head and neck backward over the back, sometimes still holding the duck's nape feathers in the beak, so that her head is jerked up, too. Then the birds swim.

Different duck species perform varying arrangements of some or all of these four behavioral elements A, B, C, D (and also other elements not included in this example). The south American teal *Nettion flavirostre* displays elements A, B, and D. The pintail *Dafila acuta* displays just elements A and B. Their hybrid offspring, however, displays all four elements: A, B, C, D.

P   *Nettion flavirostre*   *Dafila acuta*
       A B – D   A B – –
F   A B C D

The appearance of the more complete display including the C element not used by either parent species suggests that they are related to (= share genes in common with) other species, such as the Bahama duck (*Poicilonetta bahamensis*), which displays all four A, B, C, D elements.

The methods of behavior-genetic analysis involve selection, as in the Clayton and Paietta study of *Drosophila* eclosion time, population comparison—comparison of different species, as in the von de Wall study of ducks and the Franck study of swordtail fishes, or different strains within a species, as in Clayton's and Paietta's study of laboratory and wild *Drosophila* flies—and then crossbreeding of groups observed to differ in the expression of the behaviors of interest, as in the Rothenbuhler study of hygienic behavior in bees, where the identification of genetic units seems to have been achieved.

39   Eco-ethology—The Adaptive Value of Behavior

If this is not the best of all possible worlds, as Voltaire's Pangloss would have us believe, it is at least a pretty well-organized one. Excepting where man's presence has intruded, predators don't normally overeat their prey, space and food are equitably shared, individuals are organized into communities characterized by the existence of specific relations between individuals, and there are regularities in the diversity and abundances of individuals on both a local and global scale. Explanations for these patterns and others have customarily been sought in historical processes. Thus, any species whose production of young outstrips its resources will either disappear or shift to using another set of resources (thereby perhaps prodding some other species over the cliff of extinction). On isolated islands, the total number of species represents a balance between the rate at which new colonists have managed to arrive and species of established colonists have emigrated or died out. Even at the level of the individual rather than of the species, historical explanations apply. Two white blossoms that appear identical to us are readily distinguished by bees on the basis of their reflection of ultraviolet light. In the history of bees, a competitive advantage accrued to those bees which could detect UV. In the parlance of the evolutionist, they were more "fit" (by which, incidentally, is merely meant that in the next generation of bees, a larger proportion could trace their ancestry to the "more" fit than to the "less" fit progenitor). In the history of *Homo sapiens* this ability was apparently of insufficient consequence to have favored fitness. Rats can learn certain tasks much more easily when the "reward" is food than when it is water; for instance, a maze-alternation habit, a task that requires the animal to alter its path on successive runs. This presumably has an historical explanation in the fact that in the usual environment of rats, water is found in fixed loci while the location of food varies.

However, these same phenomena, while they must be considered through the eyes of the evolutionist, that is, historically, may be understood at other levels, too. At the other extreme, we could analyze the

By P. H. Klopfer

What is behavioral ecology?

difference between the sensitivity to ultraviolet of man and bee in terms of the photochemistry of the visual pigments of their retinae. This would be a more molecular approach. The level of analysis which will be illustrated below is intermediate and may be characterized as "molar"—above the level of the internal mechanisms that generate responses, below the historical layer of the evolutionist. The behaviorally oriented ecologist seeks his "explanations" in terms of the immediate functional relationships between organisms and their environment (which includes other organisms).

Why don't predators overeat their prey?

Except where man has intervened, predatory species do not altogether eliminate their prey. To have an unrestrained appetite, after all, would not be self-serving. When the last mouse is gone, what will the fox eat? Some mice must be allowed to escape so as to produce the next month's meals. This is not to deny that there may exist cycles in the abundance and scarcity of prey organisms and their predators which occasionally carry a population to extinction. In the far north of the American continent, snowy owls apparently do sometimes exhaust the supply of rodents and must then starve or emigrate. Significantly, these oscillations are local in extent and, in any case, represent exceptional cases found only where the total number of species is relatively small. Where this is the case, as in the Arctic, food chains are simple. There are few alternate sources of food when a primary source becomes rare. Hence, any change in abundance at one end of the chain must necessarily influence abundances all along its length. If one realizes that a predator can actually utilize only about ten percent of the energy contained by its prey, then it will be apparent why even a modest reduction in the availability of prey may cause the starvation of the predator. However, where species diversity is relatively greater, the rise and fall in the numbers of one species will not necessarily be reflected by all other species. Hence, in situations of greater faunal richness, as in the tropics, food chains give way to "food-nets" in which any predator has many alternate sources of energy. Oscillations in numbers are thus easily dampened.

Our interest here lies not so much in the population biology of prey-predator relations alluded to above, but in the devices of individual organisms for escaping predation and capturing prey. Since most organisms are both prey and predators, and since they inhabit an inherently changing environment, we face a situation akin to a dog chasing its tail: what is an advance from one point of view becomes a retreat from another. Still, we may consider the factors that influence the character of the protective adaptations made by prey (and the corresponding stratagems to overcome them made by the predators).

Camouflage

A plant or animal could escape from a predator by taking advantage of the limits to the predator's sensory or motor capacities, or by mimicking unpleasant or dangerous organisms not considered prey-worthy by the predator. Where the prey is simply not perceived by the predator, we speak of it being camouflaged or being cryptically colored.

In England, there are certain species of snails whose predominant color varies with the season of the year. In the spring, there is a higher proportion of snails with yellow or green tints in their shells. This is the time of year when the ground is green and snails of this color are more adequately matched to their substrate than would be the case in winter when the predominant color of the soil is red. The snails are preyed upon by certain birds which, in order to eat the snails, must smash the shells against rocks. This habit affords biologists an opportunity to examine, directly, the proportion of particular color types that are taken by the birds. The proportion of mismatched colors found at the rocks—known as "thrush anvils" —is disproportionately high. It is apparent that thrushes selectively remove those snails whose colors contrast most sharply with the ground. Snails which have a color similar to that of the substrate are less likely to be seen by birds, and, obviously, this improves their likelihood of surviving. It must not be thought from this, however, that every time an organism resembles its background to one degree or another that it is because of the selective advantage of camouflage. It is not necessarily the case that this type of substrate-matching is the result of selection by a predator. For instance, the downy young of certain shore birds have a color that matches that of the particular sand on which their nests are found. The result is that beaches of different sand color have colonies whose downy young also differ in color. It could be that this is because predators have, over the years, removed any bird whose color was mismatched. But it is equally possible that random factors have led to a particular color predominating in a particular beach, with the consistency shown within colonies being merely the result of a pronounced tendency of birds to always return to the site of their hatching when they themselves are of an age to breed. This type of *Ortstreue* or locality imprinting is fairly well known among birds. Consequently, in order to determine whether an apparent resemblance of an organism to its background really does represent camouflage, rather than the outcome of some other process, we need know something about the perceptual capabilities of the predator organisms. A white polar bear on a white ice floe may seem perfectly camouflaged to us, but if we had eyes that were sensitive to ultraviolet, the bear would be very conspicuous, indeed. Nor are perceptual differences between organisms to be attributed solely to differences in sensitivity to particular stimuli. They may be due to radical differences in the central treatment of stimuli. For instance, in the case of the octopus, tactual discriminations are apparently made on the basis of the number of adjacent papillae of the arms which are stimulated. Thus, one can take a series of plastic cylinders and etch different patterns onto these surfaces. The octopus, however, will judge the cylinders as being either "alike" or "different" solely on the basis of how similar or different the ratio of grooves to surface is on the different cylinders, with little regard to pattern. The point to be drawn from such an analysis is that an animal could indeed be cryptic with respect to its predator, and yet be very conspicuous to some other organisms, such as man, or vice versa, because the perceptual and sensory processes of man and the other animals differ so radically. An analysis of the predator's sensory mechanisms and the perceptual processes are a prerequisite for determining whether or not a given organism is truly displaying a cryptic pattern.

There is yet a final complication, even when we have an adequate store of information on the sensory capacity of an animal, and understand the processes involved in perception. We still cannot assume on the basis of laboratory discrimination trials alone whether a given prey organism is camouflaged. The reason is that conditions that we define as re-enforcing, that is, conditions that increase the probability that a response will be repeated or a discrimination learned, vary in their effectiveness according to the context in which they occur. For ourselves this is obvious; icy lemonade is not going to appeal to us on a bitter, cold winter morning nearly to the extent that it would on a hot, summer day. It may be less obvious (but it is no less reasonable) that rats will learn a fixed pathway through a maze readily when a drink of water is provided as a reward. A variable pathway, however, is more easily learned with food as a reward rather than with water. This presumably relates to the fact that in the usual world of rats, the location of water is fixed while that of food varies. This causes differences in the effectiveness of food and water as rewards in different learning contexts. In short, there are constraints that influence the ease with which particular kinds of discriminations are learned. Contrary to the suppositions of Skinnerian psychologists, discriminable stimuli are not all necesarily discriminative, nor are these constraints necessarily all the result of purely exogenous conditions. For instance, a migratory bird, if using a particular star pattern, may use it as a signal for either heading southward or northward, depending on whether the prior photo-period consisted of long days and short nights or short days and long nights. What this amounts to is a serious caution. When we see a prey organism that to us resembles its background, it may, indeed, be camouflaged from its usual predator; but that is not necessarily the case. Nor is it the case that an animal, which to us is conspicuous, is therefore conspicuous to its predators. Conspicuousness, or crypticity, has to be regarded relative to the predator's capacities for recognition of the prey, and those capacities depend on sensory abilities which may vary radically from one animal to another, but which we can hope to study. It varies with the perceptual organization of different animals, and this, too, we can hope to study, though with more difficulty. Finally, the salience of a potential prey organism, the degree to which it "stands out," may be expected to vary with the context in which it occurs, as a result of the constraints on learning ability to which every animal is subject. These are the most difficult of all to study. Thus, the facile suppositions that the green beetle, on the green leaf, possesses that color because it is thereby camouflaged and protected from its predator, while possibly true, could just as likely be false. It will take sophisticated behavioral investigation to determine which is the case, and this demands the insights of both a naturalist and a psychologist.

To be inconspicuous may be a very good ploy for a prey organism, but, for a variety of reasons, it may not be possible. For instance, the life style of the animal may preclude the degree of immobility that successful camouflage requires. An alternate solution to the problem of befuddling or escaping the attention of potential predators is to be found in imitating the appearance or behavior of organisms that are not treated as prey, by virtue of their being unpalatable or especially difficult to capture, or, perhaps, even dangerous. Imagine yourself the potential prey of a particular animal and imagine that your best strategy in escaping predation is to deceive that predator by resembling some other animal that is disregarded by your predator. To what degree must you resemble that model? What degree of difference in the palatability between you and the model must exist? If, for example, a predator, when it makes a mistake and seizes the unpalatable model, suffers greatly as a result of the mistake, then it might be very unwilling to chance the attack. When the difference in the degree of palatability or danger is small, it may take a chance more readily. What must be the relative frequency with which the predator encounters you or your "model" in order for it to learn that the likelihood of a desirable outcome, given an attack, is too low to be worthwhile? If you are much more common than your unpalatable model, and, particularly, if the degree of palatability is not too different, then the predator may not be the least deterred by the resemblance and may be prepared to pay the price of the occasional distasteful mouthful. In short, what are the boundary conditions that allow mimicry to be successful? These are some of the behavioral questions that one encounters in analyzing those conditions where one is interested in knowing whether mimicry might arise, and where one is anxious to discover whether mimicry does, in fact, exist.

A number of workers have attempted to define, experimentally, what the limiting conditions for mimicry are. For instance, in one such study, captive starlings were used as predators and painted meal worms were used as the prey. Certain colored meal worms were rendered unpalatable by being treated with a quinine salt. The birds were confronted with different proportions of the "edible" and "inedible" meal worms, and learning rates could be compared. In general, there was an increased speed in learning to avoid both the inedible (quinine-treated) meal worms and meal worms of the same color as the proportion of the inedible ones increased. But, at the same time, it has to be stated that these results may be expected to vary with changes in the degree of palatability, with the interval between tests, and because of other factors that may effect the memory of the bird for the outcome of a particular encounter. Another complicating factor has been noted by workers who found that in some cases predators are attracted by the most divergent or odd member of an assemblage of prey, even where it is the more cryptic! This kind of "oddity preference"

Mimicry

is of some significance in the removal of mutants or sick or injured animals, but, aside from this, it introduces yet another complication in determining the boundary conditions for the occurrence of mimicry.

Of course, in discussing the issue of mimicry, we have taken for granted that mimicry really does occur, that is, that a vulnerable, tasteful prey organism does attain some protection by resembling a distasteful or dangerous model. The behavioral ecologist has the problem of defining the limiting conditions that must be met for the resemblance to be effective. Some naturalists have claimed that the fundamental question has been begged, for it is not necessarily the case that selective predation is the cause of mimetic resemblances. The doubters have a case. It is possible that what a predator accepts as prey is related to the relative abundance of the prey organisms rather than to appearances. There are a few fairly conclusive field studies of mimicry, but their number is still pitifully small, so the issue of the reality of mimicry is one that persists, even while we are challenged by and seek to answer the question, what are the conditions that might allow mimicry to occur?

How is food shared between species? It has almost become a matter of routine for ecologists to demonstrate that species that live in the same area do not eat the same foods. The theoretical importance of a divergence in food preference lies in the fact that this minimizes or eliminates competition. If the same foods are taken by two kinds of animals, it is expected that they will be taken at different times of the day, or that there will be differences in the size of the items being used, with one species, for instance, taking the larger, and the other the smaller grasshoppers. In this way, they avoid competing for food. The avoidance of competition is necessary for co-existence. If two species are continuously present in the same area, it is inevitable that one will be more efficient, however slightly, than the other in garnering whatever resources limit their numbers. The slightly more efficient species will ultimately displace the other. The fact that two species co-exist in an area over any reasonable period of time, then, implies either that there are no differences in their efficiency, an extremely unlikely situation if they belong to different species, or that they are garnering resources in different ways, or garnering different resources. Those situations which seem to be exceptions to this generalization are ones where the environmental conditions are sufficiently unstable that the advantage in efficiency of exploitation shifts from one species to the next at a frequency that is sufficiently high so as to prevent either of the species from being driven to extinction. This occurs in unstable environments.

For feeding habits to differ and competition to be avoided, the environment must be heterogenous. In a homogenous world, it would not be possible for two species to employ different ways of obtaining their energy. This was demonstrated in a classical experiment. If one takes a jar with water containing a uniform suspension of yeast, it cannot support two different species of protozoans. Only by introducing some grains of rice or bits of grass, which immediately introduce discontinuities in the environment, can we enable more than one species to exist over any period of time. The particular question that the behavioral ecologist addresses himself to doesn't concern itself so much with the theoretical necessity for what is called "niche diversification" or the development of different and non-competing feeding habits, but rather to how the specific feeding habits are maintained through consecutive generations. We know from studies of captive animals that under conditions of captivity, animals are often able to use a wide range of unusual substances for food. The rather narrow preferences shown by some animals are exceptions rather than the rule; that is particularly true of birds. When they are put in non-competitive situations, they are often quite catholic in their tastes. This is not usually the case in nature, where the feeding repertoire appears small. How are these specific differences in feeding preference maintained?

On the one hand, many species have developed a feeding apparatus which is anatomically so specialized that the range of foods that they can utilize or the variety of ways in which they can collect food is severely limited. One has only to regard the peculiar shape of the flamingo's bill, or the parrot's, to realize that neither bird is going to be a very effective insect eater. Even where the design of the bill doesn't itself determine what foods shall or shall not be eaten, other structures may play a determining role. For instance, certain birds have enzyme systems that allow them to digest fats. In the absence of particular enzymes, fat cannot be utilized as a food. In its presence, wax does become a source of energy. One other point that is to be made by this example, of course, is that the distinction between determinants of feeding preferences that we call structural and those that we relate to other phenomena is not altogether straight-forward. The distinction between a bill, a structure, and an enzyme (also a structure) illustrates that structure grades imperceptibly into process. Nonetheless, we can still argue that, in some cases, anatomic features do influence food preferences. These cases, though, are probably not nearly as common as those cases where an anatomic feature, rather than simply allowing or disallowing the use of particular foods, affects the ease or the efficiency with which particular foods can be handled. The cardinal can devour sunflower seeds rather rapidly. Its bill is well suited to the structure of sunflower seeds. The chickadee, a very much smaller bird, can also eat sunflower seeds, but it must hammer these seeds into fragments before its smaller bill can extract the kernel. On the other hand, with millet the chickadee is quite efficient and speedy in extracting the kernel. When eating millet, however, the cardinal must work laboriously for a very small amount of food in return for its effort. It is not surprising that, given both foods, inexperienced cardinals very soon develop a rather strong preference for the one seed and chickadees for the other. In both cases, the preferences that develop are related to those seed types which the anatomy of the bird allows them to utilize with the greatest efficiency.

In some cases, an early experience, or a conditioning to a particular food, appears to establish a long-lasting preference even where there is no obvious anatomical basis for the preference. The most extreme of these situations is seen where the preference is the result of a social convention or tradition. Groups of monkeys that inhabit separate islands off of southern Japan display marked differences in food preferences, as well as in their manner of dealing with food. One group prefers peanuts to peas; another washes its sweet potatoes before eating them. These insular idiosyncrasies are apparently the result of fortuitous events: one animal spontaneously commenced a pattern of behavior, and was gradually copied by its peers until the behavior had spread throughout the colony—but only that colony. Traditions are not solely a human prerogative.

How is space shared?

Interspecific divisions

It is a truism that no two individuals can simultaneously occupy the same space; it is no less true, even if less obvious, that no two (or more) species can co-exist in similar space. In a classical experiment, already cited, individuals of two species of protozoans were placed into a homogeneous culture. Inevitably, only individuals of one species survived. Flour beetles in a uniform bowl of flour also interact in such a manner that only one species persists. The underlying reason is not mysterious: If species represent different forms, then the ability of individuals of different species to utilize any particular environment must differ. Thus, even where the difference in fitness (the proportion of the next generation attributable to an individual) is small, one species will ultimately monopolize the space. The way to assure co-existence is to perturb conditions so that in one generation those of one species are favored, and in another, those of the other are favored. Alternatively, even a stable (unperturbed) environment can allow co-existence of several species if the environment is heterogenous. A few stones in a beaker of water may permit two species of protozoa to co-exist if one is better able to use the crevices between the stones, and the other, the open water. Thus, for space to be shared it should be heterogenous. This is surely the case for the surface of our planet (though of course one must specify the level of the analysis or degree of graininess when one discusses heterogeneity: a uniform sheet of white paper will prove decidedly blotchy if examined under a strong lens). Because the earth's surface has patterns of heterogeneity, there will be patterns of distribution of species over its surface. The question for the behaviorally inclined ecologist is, how are the spatial distributions, the preferences for particular kinds and amounts of space, maintained? What makes one kind of mouse stick to woodlands, another to grasslands? We might begin our inquiry by asking, in a general way, how can space be divided?

If we begin with an imaginary space, we could divide it into horizontal segments. Each layer would then have assigned to it a particular species. This is the kind of situation one finds in many marine environments. The animals stratify themselves horizontally because of differences in their preferences for or tolerance of particular pressures, or salt concentrations.

Space could also be divided vertically, with restriction of one species to one segment and another to a different one. This kind of situation may obtain where distributions are determined by past geological events. The subsistence and then emergence of the Panamanian Isthmus of Central America apparently produced this kind of separation of the rodents of South and North America. Finally, if our space is not uniform, it might be divided into segments of a greater degree of uniformity, that is, into micro-habitats. Translating this into descriptive language, we can distinguish species whose ranges have been determined by structural or physiological adaptations (those for whom space was divided horizontally); species whose ranges are determined by accidents of history and geology (these are the ones whose space is divided vertically) and, finally, those that select particular areas of a heterogeneous environment.

The selction of a micro-habitat within the larger environment could be based on a number of different mechanisms, and it these that we must consider. Figure 39-1 illustrates one way in which species that co-exist in a particular woodland may divide their space into smaller units. In this, their micro-habitat, they function more efficiently than any other species and are able to maintain dominance. The diagram illustrates that different warblers make predominant use of different portions of the trees they use in common. The warblers, however, are not known to have differences in their musculature, or anatomy, that would account for the fact that one is found predominantly on the tips, and the other in the bases of branches. We do know other instances where such anatomic differences do influence the choice of micro-habitat. For instance, amongst the tits of Europe it is clear that the smaller size of the blue tit allows it to feed from the extremities of small twigs upon which the much heavier great tit could not sit. Here the choice of micro-habitat is obviously related to differences in anatomy. There are three species of the deermouse in California which differ in the fine structure of their feet. These differences are related to their ability to climb or to penetrate loose soil, and, apparently, they account in large part for the way in which the mice segregate themselves into different micro-habitats within the same area. But, as intimated, not all such differences in habitat-choice are necessarily attributable to some physical factor. There isn't any obvious necessity, for example, in terms of morphology, for salmon to return to the stream of their youth, and, yet, a number of studies have shown that salmon, as they mature, do return to the home stream. While the size of the tits may determine where in the tree they feed or nest, these size differences don't necessarily dictate a preference for particular kinds of woods, but such preferences are also known to exist. They have been labeled "psychological" factors in habitat selection. What are some of these factors? And why are some species much more flexible in their preferences and others less so? What aspects of a particular habitat does a species perceive, and how? When tits respond preferentially to a particular wood, are they recognizing an individual

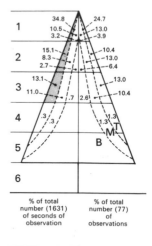

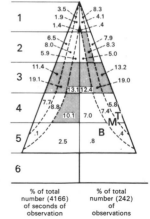

Fig. 39-1. Preferences of two species of *Dendroica* warbler are depicted in these schematic diagrams of spruce trees. The portions of the trees most favored by each species have been shaded until fifty percent of the time (or observations) has been accounted for. One species, it will be noted, is predominately in the outer edge of the upper three layers of the spruces; the other moves uniformly through the third layer. (After MacArthur, 1958.)

species of a tree or a gross configuration of the forest? These kinds of problems, of perception, discrimination, and learning, have scarcely been touched upon in an effort to understand the behavioral basis of habitat.

One study, which attempted to approach some of these problems, was directed to the question of the nature of the stimuli that were perceived by certain birds in making their choice of an appropriate habitat. For instance, in parts of North America there is a small finch, the chipping sparrow, which is generally found in pine foliage rather than oak trees, even where the two trees are side by side. If one tests these animals under artificial conditions, where perching opportunities and the availability of food are equalized, but with perches decorated with simulated pine or simulated oak foliage, it can be shown that the foliage itself is what determines the bird's choice. It is not light intensities, shadow patterns, or any of the other variables which might possibly be correlated with the rather different structure of oak and pine trees. Similar behavior is seen even in birds that have been hand-reared from the time their eyes have opened. However, the white-throated sparrow doesn't make a discrimination between the two types of foliage under these conditions. It habitually selects the darker area of the experimental apparatus. If light intensities are held constant but the size of the blotches of shadow is varied, it becomes apparent that for these birds, a particular size and intensity of shadow pattern seems to be the factor that determines the area in which they choose to pass their time. In nature, incidentally, this species is habitually found in the underbrush. A similar study was performed with mice, with an emphasis on discovering not only those aspects of the environment to which the animal responded, but what role early experience and parental disposition played in fixing the preference of the young. In this instance, the experimental subjects were mice of a variety that habitually lived in the woodland, or, alternatively, in a field. The lab-born offspring displayed the same preferences as their wild-born parents, the preferences corresponding to the usual preference of the adult. A group of grassland (field) variety was then isolated in the lab and bred for over twelve generations totally removed from the normal environment. The offspring of these animals did not show a very pronounced preference for either habitat, as compared with the first-generation animals. If, however, the young were reared for a short time after birth in the grassland, then a very strong subsequent preference for grassland was manifested. On the other hand, if they were reared in the woods, no corresponding preference for the woods habitat would develop. Thus, the mechanisms that are involved in fixing the preference are related both to the lineage of the animal, that is, to events in its phyletic past, as well as to events that transpire after its birth. The system is not infinitely flexible, for the animals can be made to develop a preference of one sort but not of another as a consequence of experience. This is the kind of a situation where learning is constrained, that is, where a behavior pattern does depend upon experience, but where only certain kinds of experiences are effective. It appears to be a much more general phenomenon than has been recognized, and it is of particular importance in situations where a certain degree of stability is desired with, however, some opportunity for change.

In summary, then, we can say that particular species of animals are generally associated with a particular kind of space—their micro-habitat. This association could be the result of extrinsic factors such as selective removal of the more conspicuous animals by predators. This would result in only those whose physical appearance conformed to the micro-habitat being removed. It could also be due to the physical structure of the animals, which makes it able to function effectively only in one and not in another micro-habitat. The structural differences may also be more subtle, reflecting merely differences in the efficiency in which the animal deals with particular micro-habitats. This in itself will eventually lead to restriction to that micro-habitat. Finally, the association may be the result of what are called psychological preferences, the basis of a seeming choice by the animal. Just how these preferences develop is one of the more intriguing and still largely unsolved or unexplored areas of behavioral ethology. It may be based on early experience, on imprintinglike processes that have their effectiveness only at certain stages in the life of the animal, or it may be the result of association with parents and more gradual kinds of learning. It is in these studies of habitat selection that the conventional problems of psychology and ecology come closest together.

Thus far we have considered the division of space between species, but, obviously, some kind of division of space must also occur within one species. We might ask, "What factors determine how densely packed the individuals of a species may become?" that is, "What factors determine the volume of individual space and the manner in which it is apportioned?"

Intraspecific divisions

We must make explicit a distinction that has been implied all along, a distinction between factors which have an immediate, or proximate, effect on an animal and those which may be called the ultimate, or final, causes of behavior. Thus, for example, when birds from the Northern Hemisphere migrate south in the fall, we find, as the ultimate cause, the fact that birds which fail to migrate starve during the ensuing winter. But we know from experimental evidence that the immediate, or proximate, cause of the migration southward is the shortening of the day. Similarly, with respect to the division of space between individuals of the same species, it is obvious that the ultimate cause for there being some finite amount of space below which the animal cannot subsist is the availability of some resource (food or shelter), that is essential to the survival of the animal. We assume that in those species where the population levels are fairly constant, there have evolved responses to factors other than the ones which exert the final or ultimate control, such as to prevent a diminution of the amount of space per individual below that demanded by subsistence.

It is certainly not the case that in all species there have been an anticipation of the effect of an unrestricted increase in numbers and consequent diminution in the amount of space available per individual. A great many insects, for example, show radical changes from one season to the next, or from one year to the next, in their population densities. This is particularly true of insects that live in areas where there are periodic and irregular climatic catastrophies. What appears to happen is that when conditions are favorable, the populations grow and grow until, at some point, they outstrip their resources, at which point there is a rapid diminution in numbers as the result of starvation or other causes related to the exhaustion of the resources. The cycle recurs at some future time. In these situations there is no advantage in developing a mechanism that would limit the rate at which numbers grow so as to avoid exhausting the resources, and there is a consequent instability in the size of the population. The point at which exhaustion of resources occurs is in itself unpredictable, depending on the vagaries of climate. Where the resources, or their availability, however, is constant, it does become possible to anticipate their exhaustion. There is, then, a basis on which natural selection can favor those individuals of a population who gain control of some portion of their resources and use these in a budgeted fashion so as to maintain themselves throughout the year, even during that period when, normally, the pickings would be slim. The question for the evolutionists, of course, is, "What is the basis on which natural selection can develop this kind of anticipatory response?" For the behavioral ecologist the question is "What are the cues to which the animal responds; what are the signs of an impending limit to the number of animals in a population that may be allowed to co-exist?"

The answers to these questions seem to be almost as varied as the number of kinds of animals that show some degree of population regulation. Among certain species of locusts, for instance, rather than there being a restriction of numbers per se as density rises, the animals transform themselves from sedentary to volent forms, simply migrating to other areas, and thereby minimizing the prospect of mass starvation as the result of depletion of local resources. It appears that the signal by which the animals assess the numbers of their conspecifics in the immediate vicinity is in the form of tactile stimulation. Locusts may be artificially transformed from being sedentary to migratory by being repeatedly poked with fine wire brushes. In real populations, one might imagine that as the number of conspecifics grows, the frequency and intensity of elbow-rubbing also grows, and at a certain level this becomes the signal for migration. A somewhat similar (at least by analogy) response is even seen in a mammal. In the black-tailed prairie dog of North America, families inhabit a common territory. As the number of young increases and the frequency of the contact between the young and the adults rises, a point is reached where the adults move out. Prairie dogs are social animals. There is a great deal of mutual grooming and contacts called "kissing." If we may interpret the behavior anthropomorphically, it is as if the adults can tolerate just so many kisses per day after which they move to fresh ground. The effect of this is to maintain the density of the family population, within limits, and also to assure to the forthcoming generation a secure and familiar homesite. This behavior of the prairie-dog family represents a solution to the problem of spacing and anticipatory responses to food shortage. The animals are territorial in the sense that they maintain for the use of a particular group a piece of ground from which other animals are excluded.

Territorial behavior, in one form or another, is one of the most conspicuous ways in which animals divide space and preserve, within limits, a fixed density. A great deal of nonsense has been written about territorial behavior, particularly by individuals who have extrapolated rather freely from the defense of a song-perch by a bird to the defense of his own home by a man. It is perhaps inappropriate to review all of the controversies here. The point surely is that territories can provide a number of different advantages. They may increase the efficiency by which some resource——food or refuges—is used merely because by having a territory, an animal will become more familiar with a particular area and thus can utilize it more efficiently. Having territories may limit the intensity of competition for some resource since the number of territories sets the limit on the total breeding density of a population. Having a fixed home or territory may enhance pair formation or the maintenance of the pair-bond, which, in turn, is an advantage in reproductive behavior. It may reduce predation, again as a result of an enhanced ability to escape on familiar ground as well as by the spacing that territoriality provides. It may inhibit the spread of infectious agents. The variety of functions which territories can conceivably serve is matched by the varieties of territories that have been described. Some territories represent the ground on which mating and nesting and the feeding of young all occur. In others, mating and nesting, but not feeding, may occur on the defended area. In some animals, the territory is merely a mating station; others maintain winter territories in areas distinct from where they maintain summer territories. Some birds have communal territories, which are defended by groups of individuals. This list is far from exclusive. The implication, though, is plain enough. There is a diversity in the various ends that territories may serve, as well as a diversity in the kinds there are. Thus, there is little reason to assume that territorial behavior is based on some kind of simple, or single, neural mechanism. And, if it does not depend on one single neural mechanism, then direct extrapolation from one species to the next is hardly justifiable. This is not to say that one cannot analyze the way in which territorial behavior has evolved, but that one cannot analyze it by listing a sequence through which it is presumed to have evolved, or on the basis of the phyletic relationships of various animals that show different degrees of territoriality.

One way of looking at the evolution of territorial behavior is based on the assumption that what different kinds of territorial defense have in common is an aggressive pattern of behavior that is linked to a particular place in space. To quote J. L. Brown, "This kind of site-dependent aggression may be analyzed on a cost-benefit basis. Too much aggression in the absence of a short supply of a disputed requisite would eventually be detrimental. Consequently, a balance must be achieved between the positive values of acquired food, mate, nesting area, protection of family, etc., and the negative values of loss of time, energy, opportunities, and risk of injuries. Where this balance may lie in any particular species is influenced by a great variety of factors." In short, the kind of territory that will evolve, if any evolves at all, depends upon economic considerations. Many marine birds feed from the sea. These birds can hardly defend a feeding territory. On the other hand, they nest on limited sites on the sea-shore and, for them, defense of a small area to be used for mating and nesting is not only possible, but, given the limitation of suitable terrain, even a necessity. In the case of grouse, which range widely in their search for food, the major cause of death is predation of juveniles. In these animals, too, there seems to be no great value in defending a feeding area. What grouse need to do is to remain inconspicuous while the young are still maturing. Here, too, it is not reasonable to expect the evolution of a feeding territory. What these kinds of ecological considerations suggest is that territorial behavior shows extreme plasticity, appearing or disappearing many times in the history of a species, depending on the ecological conditions that obtain at any particular point. We see an example of this in a pair of closely related North American jays. In one species the bird is very territorial, whereas the second species defends a communal territory in weak and half-hearted fashion. A rather similar situation exists for some of the blackbirds of the western coast of North America. The point is that differences of great magnitude exist in related species, which further supports the view that territorial behavior is a labile trait which is highly responsive to changes in the advantages or selective pressures that exist at any particular point in time. We will return to this question of ecological pressures on the development of behavior, and the use of behavior for constructing evolutionary sequences.

Why are there so many species?

In an earlier section, we commented that where the environment was homogeneous, co-existence of different species became impossible. The likelihood that two species are precisely identical in their efficiencies is too small to be worth considering. Thus, one expects, and, indeed, empirically finds, that wherever two species are in a homogeneous environment, one eventually comes to displace the other. Where some degree of heterogeneity exists, co-existence of species may occur. The Universe which we inhabit is, in fact, a highly heterogeneous one. The possibility for co-existence of many different species is relatively great; but for how many? The degree of graininess is relevant to the question of how many species may co-eixst. To our eyes, at least, a grass field looks more homogeneous than a tropical rain forest. In fact, the latter does have many more species of animals. But, how many more? Is the increase related simply to the increase in environmental heterogeneity or has it other explanations as well? These are difficult questions to answer, not least because of the difficulty in obtaining some kind of quantitative measure of the number of different kinds of species. Simply counting species is likely to provide a list that overlooks those that are rare, and, more seriously, it doesn't allow one to discriminate between species that are casual visitors to an area, or impermanent residents, from those who are truly inhabitants of an area. In short, though it seems absurd that such prosaic difficulties as census-taking should impair our ability to deal with the question, "What determines the number of species in an area?" that is, in fact, the case; and these difficulties have been resolved for only a few organisms. The greatest progress has probably been made in the case of birds, since, at least during the breeding season, birds are relatively easily counted. A census during the breeding season does provide one with a list of more-or-less permanent residents of an area. From such counts, one can get a measure of the diversity in the bird species of an area, although, in general, one treats diversity as something other than merely the absolute number of species. If, for instance, a particular wood has 100 individuals, of which twenty-five belong to each of four species, and a second wood of the same size also has a 100 individuals, with ninety-seven of the birds belonging to one species, and only one each to the remaining three, we would, intuitively, not hesitate to say that the first wood was more diverse. In fact, both have the same number of species. In this instance, our intuition is related to the fact that we are less able to predict the identity of a randomly encountered individual in the first case than in the second. In most discussions of species diversity there is a correction made for the absolute number of different kinds of species represented that takes into account the proportion of the total population that belongs to that species. With such a measure, one can then begin to compare the species diversity of different areas in a way which seems in accord with common-sense assessments. If one does that, one does, indeed, find distinctive patterns of diversity. For instance, as one goes from northern habitats, toward the Tropics in North America, there is an increase in the diversity of bird species, an increase which becomes strikingly large as one enters the tropical zone, the zone in which the mean daily fluctuations in temperature and rainfall are about equal to the annual fluctuation. There are also particular patterns of diversity associated with islands, such that one can predict diversity as a function of the distance of an island from the mainland, or from a large central archipelago and its size. But, let's focus here on the question of why the tropical regions, in particular, seem to have a higher bird-species diversity than regions to the north.

Several explanations have been proposed for the greater diversity of bird species in the Tropics; and, it should be emphasized, many of these are not mutually exclusive. First, it is possible that the Tropics simply had a longer period of time in which there has been a freedom from those sorts of disturbances that cause extinction, such as glaciation. If we assume that in the temperate zones there has been a periodic decimation of species as a result of glacial advance, it is reasonable to expect the undisturbed areas of the Tropics to have more species. Against this explanation is posed the observation that glacial epics haven't really eliminated ecological zones, but merely shifted them latitudinally and, hence, should not have produced the kind of pattern of diversity that one observes as one surveys the globe latitudinally. Another explanation is that the graininess of the Tropics is greater, that is to say, there are more niches—the constellation of factors that determines the *Umwelt* of a species. This is a difficult explanation to evaluate, since, as we noted before, to determine the graininess in an area from the standpoint of an organism demands knowing a great deal about its perceptual apparatus and the way in which it views the world. It is, however, possible to do correlational analysis which allows one to make some general statements about the relationship of the physical structures of an area, and about its suitability to individuals of a given species. For instance, it has been shown that for most terrestrial birds the best single predictor of the presence of a given species is the pattern of foliage diversity at different heights. In a forest or a field, one can make a graph of the diversity of the foliage at different heights above the ground, and express the variation in diversity at different levels in an index which is similar to the index used to express bird-species diversity. In this instance, however, what one is doing is specifying the probability of finding a particular density of foliage at a given randomly chosen level. This foliage-height diversity index is correlated with the total species diversity more closely than any other index yet noted. In other words, by knowing the variations in foliage density at different levels, one can more accurately predict the bird-species diversity than if one actually knows the specific make-up of the plants in that community. What is more, for any particular pattern of foliage diversity, one can predict the particular bird species likely to be present. If one compares a tropical and a temperate forest, one does, in fact, find that the former has a higher foliage-height diversity; and this is associated with an increase in bird-species diversity. But, one can also find tropical forests whose foliage-height diversity is similar to that of particular forests in the north temperate zone, and, in those cases, the tropical forest still has a higher bird-species diversity. It appears that there may, in fact, be more niches, more graininess in tropical environments. A third explanation, that has been advanced to explain the greater diversity of the Tropics, is based on considerations of energetic demands. The argument is that tropical regions being more equitable in temperature, demand less of an organism's energy for maintenance activities, which allows more to be used for growth and reproduction. This, it is premised, should increase the number of individuals, which would allow for greater genetic diversity and, thus, more opportunities for the formation of species. But, at least in birds, the facts of the situation contradict the expectation. Tropical birds seem to take longer to mature than those in the temperate region. They have smaller clutches, and suffer far more losses from predators than is true of their northern counterparts. Thus, increased energy, while it may be relevant to species diversity, seems unlikely to operate as a major cause. Another explanation also focuses on the constancy of tropical climate as a relevant factor, but in a different way. The argument here is that where climatic conditions are unstable, the sources of food, nesting areas, etc., must be variable; that is, the degree to which a bird can specialize in particular ways of maintaining itself must be limited. In other words, a tropical bird can become accustomed to one particular kind of food, or one particular kind of tree, to a much greater extent than an inhabitant of northern regions who has to face different environmental conditions— winter, spring, summer, and fall. The constancy of the conditions of the Tropics, then, allows for a higher degree of specialization, or behavioral stereotypy. Behavioral stereotypy, or specialization, in turn, should allow an increase in the degree of similarity that could be tolerated between co-existing species. A community of specialists, in other words, will have more non-competing members than a community of "jacks-of-all-trades." A testing of this hypothesis depends on determining whether, in fact, birds of the Tropics are more stereotyped, more specialized in the way in which they find and select food and perch sites. The evidence on this point is still rather incomplete and thus far has not given very strong grounds for support, though the hypothesis still appears to be the most reasonable of those projected. One other notion has been advanced recently which brings together a number of the factors that have been alluded to above.

The clutch size, or number of eggs at one set time, is smaller for most tropical birds than that of related species in the temperate zone. Probably this is because the longer spring days in the north allow more time for the daily search for food for hungry nestlings. North of the Arctic Circle, a bird may have nearly twenty-four hours of daylight available to feed nestlings, while no equatorial bird will ever have more than about twelve hours.

In addition to having smaller clutches, tropical birds take longer to develop. (It is not clear why this should be so.)

It is generally believed that most passerine bird species originated in the Tropics. As time passed, and glaciers retreated, colonists were able to migrate northward.

Let us assume that the longer day length (in spring) of higher latitudes produces a higher standing crop of whatever food resources birds depend upon. The briefer the period to which this crop is limited, the greater the pressure for shortening developmental times, irrespective of clutch size. It is thus reasonable that embryonic development and growth should take less time north (or south) of the tropics than within them. At the same time, in the tropics, where a more uniform pattern of abundance prevails, the populations of adult birds have likely expanded to the upper limit supportable by the available resources. When the young birds hatch, therefore, there can be available for them only a small amount of resources beyond what is normally required to support the population of adult birds. The developmental period of the young must as a consequence be stretched out in the Tropics.

As autumn approaches in the temperate latitudes, waning food resources (or low temperatures or short days) require a return to the Tropics. Why don't the migrants breed again while in the Tropics? Probably because the resources available during a single day in the Tropics are not enough to support the high developmental rates of those birds that have become adapted to the seasonally richer north. Hence, these birds and their descendants must continue their annual migration.

So far the argument has been that when resource abundance remains uniform (and low) year-round, development proceeds slowly. But, an inevitable result of a stretched-out developmental period would be an increased predation of nestlings. This will affect the structure of the entire community. If a predator can prevent any one species from monopolizing a limited food supply, the number of species will increase.

Nestling losses in the Tropics are very high; they are probably due largely to snakes, which apparently show little selectivity for particular species of birds or eggs. Hence, snake predation of avian nestlings should increase bird-species diversity. Other cited "causes" of tropical diversity are entirely compatible with this "cause." We should not demand (even though we might prefer) an unitary explanation.

On extrapolations

In the introduction to this section, we asked the question, "What is behavioral ecology?" In this, the concluding section, we would like to return to this question and ask, "Why study behavioral ecology?" As with art, of course, the study may be its own reward. However, the question is still an interesting one to ask because much of the value in the studies of behavioral ecologists relates to other fields and, in particular, to comparative psychology. Since Darwin's time, comparative studies have been based upon an evolutionary model. The limbs of animals, their teeth, and their brains evolve; and, it has been argued, surely their behavior, the things they do, also evolves. We establish evolutionary sequences of animals. Thus, why not establish evolutionary sequences of behavior, sequences which help us to understand the origins of seemingly idiosyncratic behavior patterns? For instance, if we look at the pattern of courtship in a group of dancing flies, we find that there is one species in which the male brings a small ball of silk, which he has secreted, to the female that he is courting. While the female plays with the ball of silk, the male copulates. What a strange pattern of behavior; how do we explain its origin? If we look at related species, we will find that there is another in which copulation proceeds without elaborate courtship, but in this particular species, the female, a voracious individual, often eats the male during the attempted copulation. On those occasions where the female is already engaged in eating the prey organism, the male is not attacked, or at least is not attacked until copulation has been completed. This provides a useful clue as to the significance of the behavior of the species in which a ball of silk is provided. And, if we look at other related species of flies, additional clues fall into line. In some species, the male catches the prey organism and presents it to the female, who eats it, and thus allows copulation to proceed to its conclusion. Yet another species provides the female with prey, but secretes a mass of silky threads around the prey. This has the effect of keeping the female occupied for a much longer period than does unwrapped prey, and the consequence is that the male not only can complete copulation, but he can also escape with his life. In another species, the same behavior occurs, but the wrapping is not as firm, and often the prey falls out. But, so long as the female has been occupied by unwrapping the parcel, though empty, no injury is done to the male. There is even one species where the male wraps objects that are inedible—petals or leaves—but, again, it serves his purpose in that in these cases, too, the female remains occupied with the mass of silk, even though no actual prey is contained therein. So we see that the sequence from a prey organism, to a wrapped prey, to a wrapped substitute, to merely an empty ball of silk—which is the behavior pattern that is characteristic of the first species—makes that latter behavior seem entirely reasonable. We postulate, then, that the behavior of presenting a ball of silk evolved according to this sequence.

Comparative psychologists and ethologists, of course, are interested not merely in establishing the evolutionary sequence of acts, but, also in establishing the lineage of particular organisms where the anatomic evidence does not provide any basis for determining evolutionary relationships. For instance, in spiders, the skeletal remains are not very useful in establishing relationships, but if we assume that behavior evolves in a sequential fashion, it is possible to propose family trees of spiders on the basis of similarities and dissimilarities and complexities in their behavior patterns.

The problem with these two uses of behavior—to understand the evolution or the origin of idiosyncratic patterns and to establish ancestoral lineages—is that they ignore that similarities between organisms may be due to convergences as well as to kinship. For instance, we know that in the tropical rainforests there are many green animals. Their color, though, is based upon many different factors. In one case, it may be refracting scales, or hairs; in another case, particular pigments. Even pigments that reflect similar wavelengths of light may be chemically very different. The animals all are green because green serves a common function in this habitat. It provides for crypticity or camouflage. Thus, a common function, protection from predation, is achieved by very different mechanisms. We call such similarities, that are due to a common function, analogies. Similarities in the wings of birds and bats and insects are considered to be analogous similarities. We distinguish such similarities that arise from common function from similarities which are due to common ancestry. These latter are referred to as homologous. The difficulty in the recognition of a similarity as being homologous or analogous is that it depends in the first instance on prior knowledge of the ancestral relationships and physiology of the animals in question. If we already know that bats are mammals, and that flies are insects, we can classify their wings as analogous structures, a convergent response to a common problem. But, if we don't have some independent basis for deciding upon the relationship of bats and flies, then we have no way of denying that the wings may, in fact, be homologous. This severely limits the usefulness of comparative behavioral studies in deciding upon evolutionary relationships. It means that, at best, the behavioral data can only augment the understanding of relationships arrived at by other means.

There is another even more basic problem that we have to consider. When we treat behavior as some kind of an evolutionary marker, similar in its indication of relationships to the scales of the snake or the teeth of a monkey, we assume that there is some kind of a fixed relationship between a pattern of behavior and the structures that are involved in the elaboration of that act. These structures, which include portions of the brain, are, in turn, assumed to be linked to particular areas of particular chromosomes. There may, in fact, be a fairly tight linkage between particular chromosomal loci (genes) and particular structures. Where that link is rigid, we refer to the attributes in question as heritable; but that expression does not, or should not, imply that the attribute in question is contained wholly within the chromosome. The genetic code, in other words, is not a blueprint of an organism or of an organ, but rather an information-generating device. It can function only when it is operating in an environment that already contains a high degree of order. The problem of how genes act, which is what this is leading up to, is, of course, beyond the scope of this article. But, we can give some indication of what we mean by the gene as an information-generating device. Consider the fact that gull chicks, which prefer particular colors, show this preference, at least partially, because of the presence and distribution of certain oil droplets in their retinal cells. These act as filters that are selective to particular wave lengths of light and that appear to control color preferences. Now we ask the question, "How is the distribution and the appearance of these oil droplets, on which color preferences depend, controlled by the genes?" Color preferences are heritable and, thus, we may reasonably assume the distribution and character of the oil droplets to be so as well. The simplest explanation appears to be that at a particular area of a chromosome there is initiated a process which leads to the production of a nucleic acid that reacts with materials that are outside of the chromosome to produce other substances that influence the rate at which other areas of the chromosome react. In this kind of a feedback fashion, certain oil-soluble pigments are eventually formed. This is obviously a very cursory description of a very complicated process, but the point is that the color preference that has been inherited is not simply the expression of something that is in a region of the chromosome. A chromosome, then, is not a repository of data on what the organism is going to be like when it has been fully developed. As indicated, it is a device that generates information by exploiting the predictable and ordered nature of the environment in which it finds itself. As we move from the structural features of an animal to its behavior, there is generally a further loosening of the coupling between the gene and its end product. By this we mean that there are yet more interactions. This means that behavior does not depend on particularity at the molecular level, or at least that it need not do so—which is not to deny there may not be situations where it does. Thus, no two movements, however stereotyped they are, need involve, for instance, the same muscles, or the same nerve cells. The entire organism may be considered to resemble a system that examines its own output and thereby maintains a functional constancy which does not depend upon structural constancy. This has serious implications for evolutionary and comparative studies of behavior. I think one of the most important is that little can be gained by making extrapolations from one species to another, except for those rare instances where we have some independent evidence as to relationship, and, even then, the value of those extrapolations will be limited by the fact that behavior at the organismic level is too maleable; it is influenced by too many steps beyond the initial action of the gene in early development. Thus, in order to make any kind of useful generalizations that go beyond the animal that we are immediately interested in, we have to generate some "rules of the game." That is, we have to be able to specify all of the factors that provide for constraints, or limits, on the development of particular patterns of behavior. When we do this for an array of species, then we can hope to make generalizations that would have a wider applicability. For instance, students of primate behavior have looked at the relationship between the habitat used by different troops of baboons and monkeys and their social structure and noticed that, quite independent of phyletic relationships, the arboreal habit, the diurnal habit, and the leaf-eating habit seem to be correlated with one particular type of home-range and troop-size; the susceptibility to predation by large cats has produced another common solution in distantly related primate species. In other words, the physical environment imposes restrictions on the kinds of social behavior that can develop. This provides a basis for generalizing about the relationship between structure and behavior.

Consider, for another example, territorial behavior. It is highly varied, as we indicated before, and any attempt to study the evolution of territorial behavior by looking at its manifestations in different groups of animals, even where these relationships are known to us, will produce little but confusion. Even in very closely related species there may be extreme differences in the degree to which territorial behavior is manifested. But, if we approach the study from the standpoint of someone seeking the "rules of the game," we can recognize that the existence of territorial behavior is related to a cost-benefit relationship. This does allow us to specify those limiting conditions under which territorial behavior may or may not occur. If food is available in large packets or clumps, it may pay to defend a feeding territory. If food is diffusely spread over a large area, the time demanded for territorial defense is better spent on searching. Thus, by taking into account ecological factors, a survey of the distribution of territorial behavior can begin to make some sense. In short, given the multiplicity of mechanisms that subserve common ends, even among related species, we have to know what the ecological conditions are that favor or disfavor particular patterns of behavior before we can understand their evolutionary origins.

Since much of human behavior is explained, justified, or even excused on the assumption that it represents a biological heritage that we can no more deny than our parentage, the awareness of the role of ecologic factors is important. In particular, such patterns of behavior as those we label aggressive, or territorial, or loving are often thought to be understood on the basis of extrapolations from some particular species. It is only occasionally admitted that the endpoint of any given extrapolation crucially depends on the species chosen. We know that English robins defend individual territories, but the mockingbirds of the Galapagos Islands have communal territories. Which is a more appropriate starting point for an extrapolation to the behavior of man? Animal studies can enlighten our view of the human condition to the degree that they generate the "rules of the game." To the degree to which they define the functional attributes that particular ecological conditions demand, they are useful. For instance, when a mammal lives in a herd in which there is a good deal of variation in the distances between individuals (as a result of the character of the food the animal depends upon), then there must be an exclusive relationship between the mother and her offspring. Mothers must reject alien young; otherwise some mothers will be nursed too often and others not enough. Where the ecological conditions allow the herd to be stationary and more compact, it is possible for mothers to care for foundlings or aliens. Studies that reveal these kinds of relationships will provide a more reliable basis for extrapolation than purely phyletic comparisons. One of the more intriguing points they have led to is that much behavior is rather more plastic than could be admitted if our (human) foibles are to be blamed on our evolutionary forebears. In any event, the behavioral ecologist can expect to make signal contributions to an understanding of the relationship between the evolutionary past of an organism and the behavior of its present-day descendant.

40   The Evolution of Behavior

The development of organisms is based on developments in their organs, physical structure, and behavior. Since the expressions of life are primarily behavioral, the development of behavior is of special significance. On a very broad level we can say that an organism never develops organs that do not in some way form a part of its behavior. In other words, the structural traits of an animal species are adapted to its particular behavior patterns; they develop within the limits set by behavior. Thus, behavior is viewed as some kind of a "pacemaker" of evolution, since it precedes the development of morphological traits.

Since the behavior of all animal species is more variable than their physical structure, any morphological change that does occur will have been preceded by some corresponding behavior pattern. When a polar bear wants to swim through a body of water, it will hardly wait on shore to grow fins instead of legs; instead, it changes the movements and coordination of its legs for swimming (polar bears swim only with their forelegs, dragging the hindlegs behind). By contrast we have the case of sea lions, which are almost always in the water and constantly use their forelegs for swimming: These limbs later took the shape of fins. This is a simple example to show how behavioral changes occur before the structural ones.

But where is the proof for this general statement? In searching for evidence, we would have to trace the evolution of organs separately from that of behavioral traits and then compare the two. Since we cannot watch evolution take place, we must resort to reconstructing such developments by studying whatever stages have been preserved. This often draws the protest that such a procedure cannot be applied to behavior patterns, since they do not become fossilized and therefore die out with the species that performed them. But what do fossils actually tell us? They certainly do not carry their own labels of what taxonomic category they belong to. It remains our job to classify them by comparisons with other fossils and with living species. Yet this is the same procedure used to establish genetic

The evolution of behavior, by W. Wickler

Methods of reconstruction

relationships between species living today. Hence, fossils on their own do not answer any questions about degree of relatedness, but rather, they pose them. On the other hand, if a fossil is shown to be some kind of intermediate form, it may help to close the gap between degrees of similarity in species living today. Furthermore, fossils found in a geological stratum whose age can be determined will provide a dating method, enabling us to test the validity of evolutionary sequences hypothesized from similarities between present-day species. We would not expect highly developed and recent forms to be found in older strata containing otherwise primitive and more ancient ones.

Behavioral fossils

This validation cannot be done with behavior patterns. It is also not possible to discover fossil links between behavior patterns observed today. On the other hand, behaviors often vary greatly, providing more intermediate forms than do organs. A leg usually grows only once, while a movement occurs many times, frequently with very different degrees of intensity. But we should also note that a few behavior fossils do exist, in a form we can work with: These are tracks made by animals in performing typical movements. Thus we find tracks made by walking in the mud, feeding tracks found in various materials, nests constructed by birds and insects (termites), and so on. However, we can only interpret these tracks by observing how they are made by present-day species. Scientists have in fact evaluated the walking tracks of living as well as fossil salamanders, the walking, feeding, and digging tracks of trilobites, and the complex structures built by termites.

Genetic basis of behavior

Some researchers are sceptical about reconstructing the evolution of behavior patterns. They point to the fact that it is very hard to tell which elements of a behavior are genetically determined, or—to use a common but somewhat misleading term—"innate." Yet any phylogenetic development can only apply to these genetic elements. Since animals can be bred for certain behavioral characteristics—we need only look at fighting bulls, cocks, and fish, or other animals such as pigeons and various breeds of dogs—there is no doubt that behavior must have some genetic base. But we do not know which details of behavior are determined this way (see Chapters 15 and 37).

For instance, an increase in fighting readiness may result from a stronger drive to fight, but it could also come about through a decreased tendency to submit and give up. Similarly, the cause may be a different evaluation of sign stimuli that release fighting. The problem seems easier in cases where hybrids show some behaviors intermediate to those of the two parent species. But even here there are too many morphological and physiological facets involved to establish a clear-cut association between behavior and genome. Hence—according to the sceptics—we may indeed speak of a phylogenetic history of behavior; but at present it seems impossible to trace such a development for any one in particular. One important reason is that groups of behavior patterns are usually named for their biological function; that is, they are interpreted rather than described.

As an example, let us take nest-building behavior. We cannot study its evolutionary development without a precise statement of the behavior elements in question. Birds alone build nests in very different ways: Woodpeckers cut hollow spaces into trees, swiftlets glue sticky drops of saliva together to make flat (edible) nesting bowls. Finches build a nest with grass stalks and pad the inside with feathers. But, according to our usage, nests are also what rat mothers build for their young, or what gorillas and chimpanzees build as a sleeping place for the night. Rats' nests and birds' nests are closely tied to reproduction, whereas those of gorillas and chimpanzees are not. Rats and apes build primarily with their hands, birds, with their beak. In its original sense, "nest" means "a place for sitting down or in"; accordingly, nest-building behavior is any activity involved in preparing such a place. But since this includes a tremendous number of behavior patterns, we know from the start that we cannot look for one evolutionary history of nest-building behavior, nor one evolutionary path for all patterns involved in nest building. This would already be countered by forms of behavior that are more similar in unrelated animals than in related species: Fighting fish and swiftlets build nests with saliva, while both sticklebacks and chaffinches carry plant stalks together using their mouth.

We find a similar case with other behavior units named according to their function: locomotion, food-getting behavior, fighting behavior. We therefore try to define other common characteristics. For example, locomotion is subdivided into swimming, walking, and flying—that is, into locomotory activities on land, in the water, and in the air. Even that is not enough, however, for fishes and jellyfish swim in very different ways: The criterion still lies outside the organism itself. Since we are trying to trace the evolutionary history of organisms only, we must also define the elements of study according to the organism. For instance, in looking at locomotion in the air we are still impeded by the functional definition of "wings." Only once we have clarified whether our object is bird, bat, or insect wings do we provide a basis for studying the phylogenetic development of a certain behavior in detail—here, a particular form of locomotion, namely, flying in birds.

Subdividing behavior into units

Unfortunately, such detailed studies into the phylogenetic history of a behavior pattern are rare. Quite obviously, the patience needed for this kind of research has not kept pace with the desire to make statements of more general validity. Such an impatient kind of interest is usually directed less at questions of detail ("How did the present form of courtship develop in mallard ducks?" "How did elephants come to drink with their trunk?") than at more general issues ("What is the origin and development of monogamy?" "What are the evolutionary laws of innate behavior?") Since these questions can be properly answered only when enough detailed

Detailed investigations and making generalizations

studies have been made, we can well understand the tendency to generalize from isolated findings. However, we must underline the hypothetical nature of such generalizations, whose validity is based entirely on the premise that future investigations will yield the same results as the case we already know.

Phylogenetic development is history. We can basically examine the history of any object or phenomenon. Of course, we can take our investigation only as far as this object or phenomenon remains a unit. But in most cases it does not simply stay as it is—otherwise we could certainly not speak of development. Thus the problem is to recognize a constant theme in its variations, or, in other words, to group the variations according to their related theme and to exclude all commonalities not based on this thematic relationship. I have deliberately made this analogy with music because non-zoologists will understand also what I mean. And in fact, recognizing a theme in all its variations and classifying organisms according to a particular system basically reflect the same methodology as compiling languages into linguistic families or working with other systems of classification: there is a kind of "general comparative science."

For the purposes of comparison, biologists have evolved quite a sophisticated methodology. Unfortunately, this is lacking in other sciences. On the other hand, biologists have paid too little attention to the task of making their methods available to other sciences. Today they consider it almost a methodological error to apply the biological-phylogenetic procedure to human phenomena that have originally been dealt with by other areas.

One example for this is the doctrine of homology. Since, in order to arrive at a natural system of taxonomic classification, phyletic similarities had to be distinguished from similarities due to adaptation, biologists developed special research methods, which we cannot discuss in greater detail at this time. These methods allow us to determine whether congruencies in the traits under study are merely accidental, based on similar adaptations (that is, determined by external circumstances, as would be the fish-shape and dorsal fin of both whales and fishes), or whether they derive from a common ancestor. This latter phyletic similarity is called a homology; and homologizing is one of the most important tools of all comparative researchers. By means of homological studies it was shown, for example, that bats are not birds but mammals, evolving their wings separately from birds, and that our auditory ossicles were originally a hinge in the jaws of reptiles. With this method, scientists test the homology of individual traits, which they interpret as being derived from a common ancestral trait. Biologists have examined a great number of morphological features, and with the aid of the homological method they have gained a good picture of phyletic relationships throughout the entire organismic world.

Linguists have achieved a similar success using the same method,

Classification according to similarity and relatedness

The doctrine of homology

Linguistic research

thereby constructing a good image of how modern languages and those "fossilized" by written records are related to one another. These scientists, too, have examined individual linguistic traits for their homological association, although they normally do not use this term. But while they draw conclusions about the relatedness of languages, they do not concern themselves at all with the peoples that speak them. By contrast, biologists have taken the affinity of organs, skeletal structure, and so on, and have used them to determine the relatedness of organisms carrying these traits. In other words, they deduced the affinity of the trait-bearers from the homology of the traits, thus making further inferences as to the homology of still other traits shown by these organisms. Thus, linguists presume that language traits form a closed system—namely, language itself—and that we can derive the affinity of languages from the affinity of a few traits; but they do not draw similar conclusions for the people that speak them, since everyone knows that language and other, biological traits do not form a closed system in man: Indians, too, are capable of learning Chinese, and interpreters are not necessarily of mixed race. By contrast, biologists presuppose that all the characteristics of an organism do form a closed complex, and that we can indeed derive the relatedness of the entire complex from that of a few traits.

It is well recognized that language is very much a characteristic of humans. But biologists have simply not studied this yet. They have restricted themselves to morphological traits, which are solidly entrenched in the genome (genetic substance) and closely linked to each other. Linguists, on the other hand, have been investigating a typically human behavior pattern, whose form is probably not based on the genome, but is learned. Ethologists deal with both kinds of traits, learned and "innate," in man as well as in animals, whereas the learned traits certainly are not, and the innate traits perhaps are derived from phylogenetic relationships between organisms. The ethologist then recognizes that deriving phyletic relationships of trait carriers from the homology of their traits is a valid procedure only when these characteristics are genetically based. This statement is not a downgrading of homological reasoning, but rather a clarification. After all, homology is no absolute feature of any trait but a relationship between characteristics.

As we mentioned before, we may ask about the history of any trait or phenomenon. However, this does not mean that we can always use the same method to reach our goal. A particularly good illustration of this problem are the popular questions, "innate or learned?" and "homologous or analogous?"

Homologous or analogous?

A few years ago scientists were debating whether the nest-building activities of a rat are innate or learned. Investigations, most prominently those of I. Eibl-Eibesfeldt, have shown that this behavior consists of many elements, such as gathering and carrying the nesting material, splitting tough stalks, making a hollow in the pile, and tamping the wall of the nest tight with the front paws. The form of these individual elements seems to be inborn, although certain fine details may be added through learning; but the sequence of the elements is definitely acquired. Naive rats tamp at a non-existent nest wall, shred single stalks of straw, and so on. If the individual elements are inborn but their sequence learned, we cannot say whether nest-building behavior, on the whole, is innate or acquired. In my opinion, the most important task of scientists dealing with the "nature-nurture" issue is to determine those units of behavior that exclude the possibility of either one or the other (see Chapter 37).

In homological research it is also important to determine precisely which unit is to be studied, so that homology and analogy may be clearly distinguished: The wings of birds and bats are homologous forelimbs, but they developed into wings independently of each other. They are homologous as limbs, but analogous as wings.

Variational sequences

Once we have managed to isolate the relevant traits, and to characterize them sufficiently for easy recognition and delineation from similar but nonhomologous ones, our next step is to organize different variations of the same trait into an evolutionary sequence. For this task, we first create a sequence of variations and then try to determine its historical direction.

Information theory

Our procedure for setting up the variational sequence is based on a very simple principle, first outlined in detail by information theorists and enabling the researcher to correct for errors in transcription. The phrase "tell mim," if corrected, could either read "tell me" or "tell him." However, the amount that "me" and "him" are removed from "mim" is different, measured by the number of steps necessary to change the word: Between "mim" and "him" there is only one modification, but there are two such steps between "mim" and "me." The assumption is that two errors are less probable than one, and so "tell mim" is corrected to "tell him." "Juda," designated as the name of a month, cannot be corrected, because "June" and "July" are equally far removed from it. A sequence containing varying degrees of similarity will have to have at least three units. If all these units are equally far removed from each other, no clear-cut sequence can be established. If the distances (as measured in number of steps toward or away from being the same) are not equal, the the number of steps to one side must be one, to the other side, at least two, making a minimum total of three. In information theory, a communication system with at least three modification steps between each of two relevant symbols is called a *redundant code*.

The same principle is used for the learning matrix developed as a correcting receiver in the field of communications; this matrix serves to a large extent as a model for certain properties of human memory. By learning more about our memory processes, we use this knowledge as a base for studying homologies and evolutionary paths. Our assumption is that nature "strives" to re-create or duplicate any particular trait. It then becomes irrelevant how many mutational steps lie between the trait variations under study; only the number of different stages of similarity that we can discern is important. The more elementary these traits, the harder they are to define and demarcate; and the less someone knows of such characteristics and their variations, the less he will be able to assign a new trait to its proper place in a particular set, where he must find the smallest distance on the similarity scale. Setting up these evolutionary sequences, as with the reconstruction of phylogenetic trees, is basically a problem of information theory. The more one knows about a certain animal group, the more likely he will be able to order its species according to their proper relationships.

When performing this task, of course, the comparative researcher must restrict himself to forms that have actually been found, and not use imaginary ones for comparison. In the case of behavior patterns, even the same individual will show more variations than with organs. This is why ethologists are more likely to set up variational sequences. On the other hand, such behavioral variations appear in far greater abundance, so that frequently an ethologist may set up more sequences but will not have an easier task dealing with them.

The direction of a variational sequence is determined by means of several criteria. What we are looking for is the temporal sequence of forms, and it is logical to assume that the original or basic form appeared earlier than its variations. From an ideal standpoint this need not be the case: In theory it is possible that we are all less than perfect editions of some ideal human type. And in fact, animal breeders or plant cultivators try to make their objects conform to a certain ideal. In the most favorable cases we can read the actual time sequence paleontologically from geological strata. But most of the time—and especially in ethology—we are forced to rely on other criteria.

Criteria for determining temporal sequence

1. The taxonomic system: With the assumption that traits that are more widely distributed within a taxonomic category are older than those shared only within small units, we often find it highly profitable to use the taxonomic system of animals already set up by morphologists to reconstruct the phylogeny of behavior patterns. In making use of this classification system, we also assume that behavioral traits have never developed completely independent of the morphological characteristics already studied, so that we are fully justified in using physical structure as a guide.

The taxonomic system

With this method, K. Lorenz organized the courtship behavior of ducks, N. Tinbergen, the social behavior of gulls, J. Crane, the behavior of jumping (salticid) spiders, R. S. Schmidt, the nest-building activities of termites, and K. G. Noble, the mating properties of amphibia each into their own phylogenetic sequences. We may then show, for example, that the insect family of Bombycidae displays an increasing tendency to use nothing but spinning fibers for its cocoons, or that in fishes the rhythmic and continuous movement of the pectoral fins is older than the arhythmic and situationally dependent movements of the same organs.

**Principles of
modification**

2. Principles of modification: These were established empirically through research in comparative anatomy, and are formulated as follows:

a) Dissimilar organs appearing in numbers together very probably were originally the same. One example is the limbs of crabs and other arthropods, which had evolved very different specializations from the cephalothorax (front) back to the abdomen. Correspondingly, in the study of behavior, this rule enables us to conclude that the singing movements of grasshoppers are derived from walking movements.

b) Within one area of specialization, we may consider the less specialized form as more original or rudimentary, and the more specialized one as derived. From this we conclude, for example, that mouthbreeding (e.g., in cichlids) or pouch breeding (as in sea horses) is more specialized and represents a more recent form of brood care in fishes than spawning the eggs into a sand pit or on top of a stone.

However, even in anatomy the course of development is not always from simple to complex. Neither has it been shown just how well the principles of modification can be applied to behavior. The skull construction of vertebrates becomes more and more simplified, starting with fishes and going up the evolutionary scale, and the same applies to shoulder girdles, starting with monotremes. In developmental physiology we know that individual cells have a very broad potential to begin with, but that during ontogenetic growth these possibilities become increasingly limited. And in the area of behavior, many simple movement patterns are derived from complex ones: The courtship of a number of salticid spiders, fiddler crabs, bower birds, and ducks has undergone a secondary simplification, losing many of its elements. In insects, and especially in many parasites, polyphagy (feeding off a great variety of plants and animals) is generally considered the original method of food consumption, while oligophagy (specializing to a few plants or prey animals) and, finally, monophagy (specializing to only one species of plant or animal) are considered derived forms. We do not know whether the less complex dance of dwarf honey bees (*Apis florea*) is phylogenetically older than the complex dance of honey bees (*A. mellifica*), or whether species of lizards and cichlids which practice injurious forms of fighting are more original in this particular trait than species with ritualized combat forms.

H. E. Evans ordered the various forms of brood care practiced by digger wasps of the genus *Ammophila* into the following phylogenetic sequence, by means of which we want to show how many traits may come together in such a complex: Originally these wasps hunted a caterpillar, stung it, carried it to a protected spot, laid an egg on it, and covered it—as done by the morphologically more ancient *Podalonia* species and other digger wasps. Species belonging to *Ammophila* always build a nest before going out to hunt. More primitive species such as *A. nigricans*, *A. xanthoptera*, and *A. aureonotata* start out by fetching only one large caterpillar, which they hold with their mandibles and drag to the nest by foot.

An important evolutionary step occurred with the wasp closing the nest only temporarily if the first caterpillar—with an egg deposited on top—did not fill out the entire space; the wasp would then fetch a second caterpillar, as practiced by *A. urnaria* and *A. juncea*. Finally, wasp species such as *A. aberti* and *A. placida* have found a new source of food in the small caterpillars which they can carry during flight. While flying, they first hold the caterpillars with their mandibles, then use their middle legs or even a specialized secretion at the abdominal end. Sometimes a wasp needs ten caterpillars to fill its nest. Before it has gathered that many, night may fall or it may start to rain. When it continues after resting for two days (*A. urnaria* and *A. aberti*), the larva has already hatched and may now enjoy fresh caterpillars. Finally, *A. harti* continually fetches food depending on the larval stage of development, then has a "rest period" until the larva hatches. During this time, however, *A. pubescens* creates more nests and packs them with food.

In addition, some species build nests close to the surface and others very deep down; some hunt in trees and others in plants. Species that carry the caterpillars home in flight also fly the soil dug up during nest-building far away. This kind of progressive development from simple to complex behavior, is called "ethokline," according to Evans.

3. Relationships between different courses of development: Traits may develop several times, during different time periods. With behavior patterns, these various courses of development are (a) phylogeny in the succession of generations; (b) ontogeny in the life of an individual; (c) the "awakening" of a behavior pattern that recurs several times (such as courtship behavior, which appears anew each spring); (d) actual genesis, where a behavior undergoes one period of individual development starting with very light intention movements and running through all stages up to the fully intensive behavior. All of these four courses of development do not have to occur with each behavior pattern; some may appear together, for instance, when a behavior pattern occurs only once in animals with a short life span.

Relationships between different courses of development

We can observe the developmental sequences b, c, and d, but a must be deduced. The next obvious step is to look for correspondences between what is observed and what has been inferred. The first three sequences may also be found with morphological features, such as antlers or feathers that are shed; yet comparisons have been made almost exclusively between ontogeny (individual development) and phylogeny (evolution). All too often scientists have tried to translate ontogenetic sequences directly into phylogeny; according to A. Remane, this led to zoologists making wrong conclusions about forty percent of the time. As far as behavior patterns are concerned, we cannot even suggest an error probability. In any case, we do not know of any basic mechanisms whereby these various developmental courses would have to apply all at the same time. Thus we can rely neither on the biogenetic law ("ontogeny recapitulates phylogeny"; see Chapter 37), nor on other, similar inferences. Whatever parallels we may find will depend in part on the behavior system. The autumn song of many birds is similar to their youth song, and in both cases develops into the finished product, as we suspect also with the song's phylogenetic history. Larks, which we know are descended from hopping birds, hop when they are young, before starting to walk. Similarly, the bearded titmouse goes so far as crawling with all four limbs moved in alternating diagonal pairs before achieving its normal method of locomotion.

More frequently, scientists try to compare the development of functional behavior systems during phylogeny with individual development, or ontogeny. In morphology, this would roughly correspond to examining in what order mouth opening, teeth, intestinal canal, stomach, digestive enzymes, and so on appeared during development. One obstruction to using this method is the fact that, during ontogeny, the development of the sense organs and that of the musculature does not always occur hand in hand; similarly, behavior patterns often mature at different times from their effector organs. For example, male crickets have a fully developed song apparatus from the time they hatch, yet their song matures later; conversely, many songbird nestlings groom their "feathers" before these have even developed. We have observed some cases, such as the courtship of various hummingbird species, where ontogeny does follow the same course as phylogeny; in other cases it does not, such as with ravens and gulls, which simplify their courtship behavior as they gain more experience at mating. Similar variations have been found in all other comparisons to date between the four courses of development. This means that, as yet, they are of little help in reconstructing evolutionary history.

Every organism displays characteristics that changed more rapidly than others during both phylogeny and ontogeny. All in all, this has resulted in a mosaic-like distribution of derived and original traits, both organic and behavioral within themselves and among each other and among those characteristics forming functional associations.

Mosaiclike distribution of derived and original traits

In this way, for example, the tropical cichlids have evolved from substrate breeding to mouthbreeding. Substrate breeders typically form stable pairs, do not show any sexual dimorphism in body structure or behavior, and produce a large number of small and very sticky eggs. In contrast, highly specialized mouthbreeders do not form pairs, show a prominent sexual dimorphism in body structure and behavior, and produce a few large and nonsticky eggs. However, when we order the known mouthbreeding species according to their degree of specialization in these traits, we find both pair-forming but already dimorphic species whose eggs are still sticky, on the one hand, as well as species with a nondimorphic body structure but highly dimorphic behavior and whose giant eggs do not stick, and so on.

Historical remnants

Features that change very slowly can still be seen in particularly rudi-

mentary form. Being at a primitive stage of development, such traits are
the only means of determining the natural relationships between present-
day species where other characteristics have diverged. Such "historical
remnants" (or rudiments) are particularly noticeable in species that developed different specializations in response to ecological pressure—that is,
that changed their original habitat but still show "adaptations of the past."
A very good example is the egg-rolling movement of many birds—
examined by H. Poulsen—with which they roll an egg back into their
nest (Fig. 1-10). This behavior is typical for ground breeders but is lacking
in most tree breeders, since in the latter case any egg that had somehow
been shifted outside the nest wall would be lost anyway. Nonetheless,
there are exceptions: The meadow lark and meadow pipit are ground
breeders, but they do not roll their eggs back into the nest; on the other
hand, tree-nesting pigeons and some tree-nesting rails do perform this
movement if we build a collar around their nest to hold eggs on the outside. In cases like this, we know that the species changed from another
method of breeding to their present one relatively recently: Secondary
tree breeders "still" perform the egg-rolling movement of their ground-
breeding ancestors, while secondary ground breeders do not do this any
more than their tree-breeding forebears. Many historical remnants in
behavior should be seen not so much as old adaptations but rather as
peculiarities that were retained by a species and whose adaptive value is
no longer clear. Occasionally a behavior pattern survives longer than its
effector organ: Deer of the subgenus *Rusa* retract their lips upward when
threatening, thus baring the long-lost canine teeth, while the primitive
muntjac uses the same movement to display its long and sharp canines.
Conversely, crickets possess a complete flying apparatus, yet never fly on
their own, although we can trigger flying movements by electrical
stimulation of their "brain."

The egg-rolling movement has a clear adaptive advantage (see Chapters
15 and 17). By means of such an adaptation—which enables the bearer to
interact with its environment and which can be transmitted to offspring,
either by genes or by tradition—the average number of offspring is increased. Speaking very generally, behavior is superbly suited for adaptation
due to its flexibility. This is why behavioral traits can easily change their
function. The blood circulation of the skin, originally an aid in respiration,
now helps mammals to regulate their temperature and finally also serves
as a signal (blushing, growing pale). The origin of signals has been studied
in the most detail within this context, with a good description provided
in Chapter 18. On the whole we might say that many behavior patterns
used by the animal to defend itself against predators and rivals originally
derived from food-getting or feeding behavior, and that most of the
appeasing behavior patterns (those intended to inhibit aggression) are
derived from preliminary mating movements or from parental behavior.

Ritualization is a process of modification whereby a behavior pattern

Adaptation and change in function

Semanticization and ritualization

attains signal function. In some species of weaverfinches, the females build the nest while the males bring the necessary materials. From this pattern of fetching material, the males have developed a "courtship with stalks," making modified nest-building movements while holding a blade of grass in the beak. Some species cannot court or copulate successfully without a stalk.

Some movement patterns attain signal value simply through the fact that some other individual learns to understand them without in turn affecting their form or other characteristics; this is another way in which behaviors can serve a new function, that is, of communication. But the adaptation may also occur only with the receiver, who interprets the behavior in a certain way, giving it a particular meaning. I have therefore suggested the term "semanticization" for all those processes by which a signal attains its meaning, whereas ritualization denotes only changes occurring in the sender.

We may compare not only species and traits but also consequences. Ethologists in particular make use of this method. It entails labeling behavior patterns according to their function—irrespective of whether they are phylogenetically related or not. Examples then repeatedly show that animal groups at very different phyletic levels use different means to attain the same ends: Jellyfish are already viviparous (bear live young), almost all classes of animals have members where the mother provides for her brood and feeds them, crabs already recognize their conspecifics individually, stable monogamous pair-bonds are found pretty well at all levels, and so on. We can distinguish different levels of evolutionary development within these functional contexts, as well as within individual traits; but we cannot use such functional behavior systems to determine the phyletic level of the species itself. This evolutionary stage can only be decided by how many cues or data a species can receive and utilize from its environment.

41 The Behavior of Domestic Animals

Man first domesticated animals by isolating small groups of the original wild species and then preventing further interbreeding with wild individuals. For these animals, which then reproduced in captivity, man created a special environment deviating more and more from the living conditions of the wild form. Domestic animals adapted to this new environment and to man's artificial selection pressures. In this way we have an example of natural development under conditions similar to controlled experiments, providing us with insight into the developmental capacities of animal species. Under the more uniform conditions of nature, only a small part of this potential may usually be observed. The same applies to special features of behavior.

The behavior of domestic animals, by W. Herre and M. Röhrs

Since man breeds domestic animals for his own purposes, these creatures must adapt to his needs and preferences. They must become less aggressive, for man wants docile animals that are easy to handle. A high rate of reproduction is also desired. Associated with these special environmental features, which are brought about by man, a variety of unique behavioral traits are found in domestic animals. In order to learn whether these are special characteristics determined by environmental factors or whether they result from the selection of hereditary traits, we must first examine the behavioral differences between domestic animals and their ancestral species. We should note, however, that studies in this area are still very scant. With some domestic animals—such as cattle, camel, horse, and donkey—the ancestral forms are either extinct or no longer roam the wilds in sufficiently large numbers. In these cases a comparison between wild and domesticated animals is practically impossible. Furthermore, even in ancestral forms that still exist, behavior in the wilds has not been studied enough to provide a clear picture of what changes have occurred from wild to domesticated species. For that reason we must look for any aids or methods providing us with possible clues.

The environment of domestic animals

Behavior is most certainly related to the development of the brain. Hence it is important to know that the brain of domestic animals is more

Studies of the brain

than thirty percent lighter than that of comparable wild species. This reduction by about roughly a third indicates that demands on domestic animals are considerably less than on the wild species. The various parts of the brain are reduced in different amounts, indicating that not all the demands have changed in the same way. In domestic animals, the smallest development takes place in brain regions dealing with the visual and olfactory senses, emotional behavior, attention, and general activity level. On the other hand, areas controlling reproductive behavior have changed but little.

From wolf to poodle

We can give some examples: A comparison of the behavior of wolves and poodles was published by E. Zimen in 1970. Wolves—that is, the wild species—typically show vigorous and well-coordinated movements, and are far more flexible than domestic dogs. The locomotory behavior of poodles has lost the flowing movements of wolves, especially in trotting and galloping. Poodles also do not show the very tense orientation posture of wolves. Furthermore, they lack the reconnaisance jump, which requires a great deal of body control. The particular posturing of individual body parts used by wolves in social communication have disappeared or at least been reduced in dogs. Poodles show less than half and even these are far less intense.

The same applies to play behavior. The highly expressive running games of wolves are barely performed by poodles. When wolves play, the drive elements of their hunting behavior mature independently of each other and are not coordinated until later. The releasing mechanisms of some of these behavior patterns mature very early. In poodles, they start to develop at about the same time as in wolves, but then they do not become differentiated during later development. Adult poodles basically behave somewhat like young wolves. Wolves also attack with much greater force and intensity than poodles. Whereas poodles attack strange dogs only in their own territory, wolves show a far greater aversion to outsiders, especially during the breeding season. Poodles do not seem to show increased aggression at certain times of the year or when bitches are in heat; again, this is similar to the behavior of immature wolves.

Prominent differences also appear in feeding behavior. At their feeding site, wolves are extremely social; young animals and nursing mothers have priority. The males share in raising the young, by feeding them. Male poodles do not show this behavior. They are very aggressive at the feeding site. When adult wolves are in a group of juveniles, they seem to move with special care, while poodles do not show any such increase in orienting behaviors. The dominance hierarchy in a wolf pack is not fixed; ranks may be challenged and are in constant flux. By contrast, poodles in a pack show a very rigid dominance hierarchy. When sleeping, adult wolves keep a constant watch on their individual distances. Poodles maintain little or no distance between each other when sleeping, and also stay very close together when running in a pack. This behavior, too, is similar to that of young wolves.

Great changes have taken place in vocal behavior. When orienting, wolves usually do not make any sounds, while poodles bark. In domestic dogs this barking has evolved into a versatile form of expression. On the other hand, the "coughing" of wolves—which is lacking in poodles—plays an important part in expression. Also important in wolves is their howling behavior, while this is no longer very significant in poodles. This comparison shows that the behavior patterns of poodles, a breed of domestic dog, have changed from those of wolves, their ancestral form. On the other hand, these modifications need not appear in all individuals; as far as domestic dogs are concerned, behavior patterns are extremely variable.

Large behavioral differences also occur between domestic pigs and their wild counterparts. In domestic pigs, the group structure typical of ancestral forms has disappeared, the activity level has decreased, escape behavior is weaker, and the escape distance is much smaller. Wild pigs are extremely aggressive and will even attack enemies in groups. Dampening this aggressive drive was an important requirement for domestication. Similarly, domestic pigs lack some essential elements of food-getting behavior; predation has become especially rare. Wild pigs hunt insect larvae as well as frogs, reptiles, rodents, hares, and even fawns. In the context of reproductive behavior, domestic boars do not fight their rivals as much as wild ones do; correspondingly, domestic sows do not show the varied nest-building behaviors of their wild counterparts.

From wild to domestic pig

Among wild chickens, the jungle fowl (*Gallus*) is considered to be ancestral to domestic chickens. Jungle cocks on the whole are more temperamental than domestic cocks and show far more intensive courtship and fighting behavior. Before treading the hen, a red jungle cock (*G. gallus*) will court by displaying close in front of her, then running with fast tripping steps toward her, his body raised high and his neck ruff spread wide. The domestic cock does not display this way; he approaches the hen very casually, with his body less raised and his neck ruff in normal position. After treading, the jungle cock performs a prominent courtship loop, courting back and forth in front of the hen with fast tripping steps. In the domestic cock this courtship loop is incomplete, performed with slow steps and frequently reduced to a "waltz" or a "bow."

The effects of domestication on chickens

In the three other species of wild fowl (Sonnerat's, Lafayette's, the green jungle fowl), the cocks also fight a great deal and very intensively. They defend their territories even against members of other species. The fighting behavior of domestic cocks is quite variable; differences occur not only between breeds but also between individual animals. In some, the fighting drive may even be totally eliminated; these cocks avoid encounters or terminate them quickly. For domestication, such animals are no doubt an advantage. But even the oriental domestic cocks that have been bred for fighting do not show more bellicosity than their wild counterpart.

In the following paragraphs we shall try to put some order to the changes that occurred in domestic animals. We shall begin by describing

Organization of behavioral changes

some behavior patterns that have either decreased or completely disappeared in domestic animals. Laboratory rats already display far less general activity than the brown (or Norway) rats, also less warning, escape, and defensive behavior. While brown rats emit a characteristic vocalization when in danger, this call does not appear in laboratory rats.

In the wild, guanaco stallions use a shrill neighing to warn their family group of danger. The others then flee while the stallion follows and often tries to divert the enemy. We never observed such behavior in the llamas and alpacas of South America, which are descended from the guanaco and which are kept under relatively free conditions. On a very general level we can say that in domestic animals the escape distances are greatly reduced, with flight behavior also being weaker. In some of these creatures, escape behavior has even been almost totally eliminated through artificial selection.

Reduction of aggressiveness in domestic animals

Aggressiveness can also be greatly reduced in domestic animals. Caged brown rats almost always kill mice that are placed in their enclosure, while laboratory rats do this very rarely. If two wild rats are placed in a cage together and given electric shocks, they will fight each other with unusual ferocity; by contrast, laboratory rats in the same situation try to flee. It is obvious that from the start, man placed great emphasis on curbing the aggressive drive of his animals, since large domestic animals would otherwise be very hard to control.

Decrease in prey-catching behaviors

Other behavior patterns that may be reduced and in part eliminated in domestic animals are those of food getting. Ferrets, dogs, and cats (all domestic carnivores), commonly perform prey-catching behaviors less frequently than their wild forms. This could result merely from the simple fact that man feeds these animals. Some domestic dogs and cats no longer catch, kill, and eat prey animals, even if they are offered to them; in others, parts of this behavior are omitted. Thus we have house cats that catch and kill prey animals but no longer eat them. Others no longer catch and kill animals but eat them if they are offered dead. In some working dogs, the final parts of prey-catching behavior are eliminated, so that the chain of action literally comes to a halt in "pointing."

In pigs, the ability to catch prey need not always be lost. There are isolated examples that show how certain behavior patterns may be bred independently of others. For instance, from the 11th to the 15th Century, people in Hampshire, England, kept hunting pigs; one of these, called "Slut," became very famous. This pig was better than many hunting dogs. It could track all kinds of game, point, and even retrieve.

Domestic animals that live only on grass and other plants no longer need to search for feeding places or defend their grazing territories when they form part of the human household. However, recent studies have shown that even these animals underwent considerable changes in food-selection behavior. While many wild species require certain food plants and are able to pick them out from an abundance of other plants, domestic animals cannot choose anymore even when given the opportunity.

Such reduction and elimination of behavior patterns occurs most frequently in the area of reproduction. Swan goose ganders generally do not court any more, instead rushing at the geese without any further preparation. According to K. Immelmann, domesticated Australian weaverfinches show a reduced incubation drive. Many domestic chickens do not incubate at all. Today this deficiency is encouraged even further by artificial selection with intensive methods of breeding. Ignoring a few exceptions, even the defense of young is far weaker in most domestic animals than in wild ones. Domestic geese no longer form stable monogamous pairs, as does their ancestral species, the greylag goose; instead, ganders are polygamous. We have already mentioned deficiencies in nest-building behavior and defense of young.

Conversely, however, domestic animals may also show excessive development, or hypertrophy, of behavior patterns. The drive to eat, for example, may be greatly increased. In the prey-catching movements of carnivores, single elements may be exaggerated. For instance, some house cats catch the same mouse over and over again without killing it; others repeatedly carry killed prey animals into their nest or to the house. Dogs, too, may be pronounced "rat biters." In some domestic dogs, aggression is not reduced but rather accentuated; they attack animals that a wolf would not.

Primarily in the area of reproduction, domestic animals show hypertrophied behavior. These animals, for example, generally reach sexual maturity at an earlier age than wild ones. The number of births and offspring is increased. Domesticated zebra finches may undergo a substantial increase in nest building, and they often lay one clutch of eggs after another without incubating them all. During the display flight, the male rock pigeon claps its wings above the back. This behavior pattern has been developed further during domestication and increased through artificial selection. In several breeds of pigeon the clapping movement is much stronger and more extensive than in the rock pigeon, so that the tips of the primary feathers wear out increasingly from spring onward. The position of the wings during the subsequent glide, too, is hypertrophied. While the rock pigeon lifts its wings only slightly above horizontal, pouters touch the tips of their wings and quickly lose altitude. In the various breeds of roller pigeons, this gliding display has even become a repeated backward somersaulting; occasionally a pigeon will make such rapid somersaults that it shoots down in a swirling movement from several hundred meters, breaking the fall only when just above the ground.

The hypertrophied behavior patterns of domestic animals may even lead to a change in the way they hold their bodies. When displaying, a wolf raises its normally hanging tail and curls it slightly over the back. In some dogs, this display posture has become a permanent feature. A very inhibited or intimidated wolf carries its tail between the legs, and this may again be a fixed posture in some domestic dogs. Display postures of the

Deficiencies in reproductive behavior

Hypertrophy of behavior patterns

wild forms have also become fixed in swan geese and Japanese Chabo chickens. However, we should not assume from these permanent body features that such domestic animals are constantly displaying or are always intimidated.

According to K. Lorenz, behavior patterns originally belonging together may become dissociated in a domestic setting. Swan geese ganders that copulate with females without courting, and domestic dogs that cut their prey-catching activity short at pointing already show this kind of dissociation. In domestic geese, too, Lorenz tells us that the instinctive actions of "falling in love," which lead to stable monogamous pair-bonding, have become dissociated. As a result, domestic geese are no longer at all selective in who they mate with. Immelmann reports that in Australian weaverfinches, disintegration takes place most frequently in pair-bonding and copulation as well as in pair-bonding and social grooming. The processes of pair-formation and monogamous pair-maintenance described in wolves may no longer be found in domestic dogs.

Changes in releasing mechanisms

According to Lorenz, a second set of alterations in the behavior of domestic animals lies in the area of releasing mechanisms, which allow the discharge of innate behavior patterns in response to sign stimuli. Thus, domestic animals may show raised or lowered thresholds for sign stimuli and their innate releasing mechanisms (IRM; see Chapter 1). Jungle fowl, for example, respond with parental behavior only to chicks sporting a certain pattern on the top of their head and on their back; they kill all chicks with different colorations. Most domestic chickens, however, accept young with any coloration; many will even raise chicks of other species, such as ducklings, goslings, and young turkeys. Domesticated zebra finches will feed even nestlings without the species-specific throat coloration, and will court very crude models.

In many domestic animals certain behaviors may also be released with substitute stimuli; that is, they perform actions toward completely wrong objects. Substitute objects such as balls of wool, stones, and twigs are used by dogs and cats, first during play when they are young, then in adulthood for making prey-catching movements. Very crude sign stimuli may also release sexual behavior in domestic animals. This is why breeders are able to induce domestic boars and bulls to copulate with very simple models—a useful action in artificial insemination.

On the subject of behavior changes in domestic animals, we should mention one other fact: In many of these creatures the body structure has been modified to such an extent that they are no longer even capable of performing the behavior patterns of their wild counterparts in the proper way. The considerable changes in body size alone may play a part here. Domestic cattle, goats, and sheep with their greatly modified horns can no longer fight rivals as their wild ancestors did. Dogs with drooping ears cannot straighten them up anymore, although they try. In many breeds of pigeon the beak is so short that the hatching young are no longer able to crack open their egg shell unaided. Indian runner ducks are hardly capable of behaving like mallard ducks; when feeding, for example, they sometimes have to lie on their stomach.

The behavioral changes we have described under conditions of domestication involve largely innate behavior patterns. Many of the modifications observed in domestic animals are surely hereditary. This is substantiated by the fact that breeders successfully select for certain behavior traits. By crossbreeding experiments in dogs it was shown that behavior traits resulting from domestication can indeed be hereditary. Similar statements can be made from findings of hybridization between wolves and poodles in Kiel, Germany. Most certainly, even at the beginnings of domestication some kind of selection had taken place unconsciously in dealing with wild animals. To this point, interesting observations were made on caged laboratory rats: Especially the more tame and docile animals reproduce under human care, while minimal success is attained in breeding wild and aggressive ones. On the other hand, it is obvious that not all behavior changes stemming from domestication are hereditary. Thus, innate behavior patterns may be performed differently in a domestic setting—depending on the situation or environment—than is the case for wild animals.

Because of the special conditions of domestication, these animals have gained certain freedoms over wild ones. Man takes over the tasks of providing food, protecting against natural enemies, and so on. The almost constant tension in wild animals for keeping alert is no longer to be found in domestic species. Domestic animals live in a "destressed" field, as K. Lorenz puts it; they lack the manifold demands that animals in the wild are almost constantly subjected to. In connection with this, domestic animals display a remarkable variety of acquired and learned behavior patterns. In particular this is shown by the various behaviors to which domestic animals are trained—for example, dressage—or conditioned. We need only look at such training in dogs, pigs, or horses. These accomplishments, however, do not represent truly novel acquisitions in the course of domestication. Even wild animals may, after all, be trained quite well in captivity, learning behaviors they would never perform in nature. Often enough such conditioned behaviors are easy to achieve in domestic animals because they have a need for activities. Furthermore, imprinting through man may have a number of effects.

Environment and behavior

Compared with their wild counterparts, domestic animals show a remarkable variety of behavior patterns. Yet the sequences of these patterns have not changed in any basic way. No new behaviors have evolved. Changes do not appear equally in all domestic animals. Some still bear a great resemblance to wild animals, while others are quite different. This is where the variety is great. On the other hand, it often only appears that way because certain individual elements change in different ways, so that behavior patterns forming part of one biological group may become dissociated from each other. These observations indicate that in a wild species

Increased variety of behavior patterns in a domestic setting

a sequence of co-ordinated behavior patterns is not controlled by the same genetic factor; rather, a number of independent factors join into a biologically co-ordinated unit. Under conditions of natural selection, this fusion of elements has advantages, while in a domestic setting the elements may separate from each other. In this way certain individuals can develop one-sided specializations, such as pointers, which may even be encouraged by man. In contrast to wild species, which are adapted to a particular environment, such domestic "specialists" show a "disharmony." The behavioral changes of domestic animals may therefore be taken only as a very limited model for natural evolution. But they do demonstrate how greatly expressions may vary when animals live in conditions no longer corresponding to normal biological demands.

Degeneration or adaptation?

These modifications in behavior may give us the impression that domestic animals are "degenerate wild animals." But this attitude is unjustified from a biological point of view. The changes we have described are adaptations to environmental conditions of domestication created by man. The question remains, how much farther can these conditions be changed; and what are the limits of adaptation? Here it is interesting to examine the modern methods of animal husbandry. The massed housing of pigs and chickens has led to a great deal of controversy. We can mention only briefly the ethical viewpoint that such methods are often extremely cruel to the animals.

Limits of potential for change

All in all, domestic animals are capable of adapting to the peculiar conditions of domestication even in their genetic base, and in a variety of ways. We cannot yet determine whether this potential for development has limits that make certain forms of animal husbandry simply impossible. Thus we can imagine that further breeding may enable even more extreme forms of husbandry than is common today. But this is an uncertain issue. By far the greatest number of behavioral changes that occurred from wild species to domestic animal are genetically determined. They can be linked to the special selection pressures of the ever-changing environment created by man. Further studies in greater detail, dealing with the behavioral changes from wild species to domestic ones and examining all the causal factors, will no doubt contribute to a better understanding of the basic questions involved in animal behavior. Knowledge of this kind will also help to explain certain peculiarities of human behavior, because man is changing his own environment in ways sometimes reminiscent of the living conditions of domestic animals.

42 Ethology and the Study of Animals in Captivity

Ethologists may study basically two large categories of animals: those living in nature under their original conditions, and those living under the care of humans. Fifty years ago this crude division may have been relevant, since there were still a few places on this earth with truly untouched areas. But today barely any biotopes remain that have not felt the direct or indirect influence of man; and these human influences all too often consist of drastic interference with the habitat or even of brutal destruction.

Even national parks and similar sanctuaries, where modern field workers do almost all of their research (especially on large animals), do not represent the remnants of an original habitat but are all more or less affected by man. For example, animals can no longer go on their long migrations, and the population density increases due to hunting by man in the surrounding areas. Furthermore, these animals have to a large extent grown used to motorized vehicles, their escape distances have grown smaller, and so on. Nonetheless, these sanctuaries still provide wild animals with a fairly natural form of existence, especially in regard to space, landscape (biotope), climate, vegetation, and food. Thus, in a certain way they represent a middle path between the original environment and zoological gardens; the latter, in turn, now contrasts with the old-fashioned menageries common at the turn of the century by also displaying a certain naturalness, at least far more so than do laboratories.

The main differences between zoological gardens and sanctuaries are that zoo animals receive all their necessary food from humans. This means that living space or even territory may be a thousand or even ten thousand times smaller than in the wilds. Antelopes in a zoo do not require huge areas of steppe, and lions do not need their many square kilometers of savanna with its hundreds or thousands of grazing prey animals (antelopes, zebras, wart hogs, etc.). In a zoo these animals are protected from hunger and thirst, and also from predators and to a large extent even from their conspecific rivals. They enjoy a security that is in some way unnatural, and which prevents certain behavior patterns from really developing.

Ethology and the study of animals in captivity, by H. Hediger

Differences between zoos and natural parks

Fig. 42-1. For all antelope-like animals with well-developed preorbital glands (G), ramified branches for depositing the secretion are important parts of their territory.

Fig. 42-2. Only with very tame and diurnal zoo beavers could we observe and film in every detail how the young are carried about and "spoiled" by their parents and older siblings.

Migrations are impossible in a zoo; in most cases the food consists of substitutes more or less differing from the natural kind. Furthermore, humans, which the animals avoid in the wild, are almost always present in a zoo. And yet, although all these conditions are to some degree unnatural, they are still far closer to nature than if the animals were kept alone and in totally bare and tiny spaces, typical both of the old menageries and of modern laboratory conditions. In 1942, I founded a branch of biology dealing specifically with animals in zoological gardens. I have continued to expand this area of study, with the goal of providing animals in human captivity in zoos with artificial territories that, despite severe restrictions in space, would display all the essential elements of their natural territories —not only in physical layout, but also in a social sense.

This crucial demand has led to the condition that zoo animals can now exercise at least those behavior patterns associated with their territoriality and social life. Observing these two areas of behavior from a close vantage point alone has opened a number of doors to ethologists, with considerable opportunities made possible by zoo biology. This has also been openly acknowledged by many prominent ethologists, such as N. Tinbergen in 1953, when he published his work on animal sociology.

Hence, this is a major area where zoo biology, or rather the keeping of animals in zoos according to these principles, has made a highly significant contribution to ethology. For example, only in a zoo setting was it possible to distinguish the finely-detailed expressions of wolves or monkeys, and to recognize their meaning in the social interactions of these animals. Only the precise research done in zoos made it possible to recognize the mimicry, gesticulation, behavioral groupings, and other factors in the field and then to examine them further in their natural setting. In this way, for instance, H. Spivak observed the group of gelada baboons in the Zurich zoo for many years and was able to record no less than twenty-one clearly defined expressions—without accompanying sounds— that included raising the brows, retracting the upper lip, curling the upper lip, lip and tongue movements, sudden opening and closing of the mouth, pulling hairs, dipping the tail in water, and many more. Such detailed studies would never have been possible in the wild. Now that the individual expressions have been clearly defined and their meaning understood, however, it is likely that they can be recognized as well under the more difficult working conditions of nature, and to utilize them in further research.

Another ethologist to emphasize the value of zoo observations for subsequent field work, and the necessity to combine these two kinds of studies especially with primates, was H. Kummer, who started out in 1957 by making detailed investigations on one group of hamadryas baboons in a zoo; then, in 1965, he began observing another group in Ethiopia. Naturally, the requirement for this research combination is the keeping of monkeys according to the principles of zoo biology, that is, within a larger social unit and in a sufficiently large area physically designed to suit the purpose. Keeping animals in isolation or in tiny groups within tight cages, as used to be the common practice, naturally provides only a distorted picture of normal social behavior.

Kummer found that, compared to the natural habitat, a zoo environment offers certain advantages to the sociological researcher. Apart from the "tameness" or calm of the animals, which allows the observer to come much closer, there are three main benefits: 1. Animals in a zoo spend more time on social interactions. 2. The number of social interactions is higher in a zoo. 3. Some behavior patterns that in the wild occur only in very special situations and with only a few individuals may be observed more frequently in a zoo.

On the other hand, by observing baboons in the wild, Kummer regularly found one phenomenon that is very important for zoo biology, and that scientists had also confirmed with other species: As soon as food was offered to the free-living animals and individuals came closer together, the social interactions increased. In other words: Less space increases aggression. This principle corresponds to an observation made long ago by zoo keepers—and also by human psychologists, for example, when observing people in slums; but the surprising factor is that merely concentrating the food leads to increased aggression, even though the animals have otherwise unlimited space.

Putting food all in one place creates a very artificial situation for wild monkeys. By contrast, a family or group of lions crowd around a single source of food even in nature, for example when feeding off a slain zebra. We know that this often leads to abusive behavior even among members of the closest family. This is why in a zoo or a circus, attendants often isolate the large carnivores in particular for feeding—a precaution no doubt appreciated by these animals, too. In other cases, attempts are made to divide monkey enclosures into different visual and therefore social sections, using partial walls or partitions. In a similar way to sociological research, zoo biology also played an important part in stimulating the study of territories. For example, observers first noticed the way mammals mark their territories with odor (scent marking) in a zoo.

One of the main tasks of zoo biology is to find the best possible conditions for keeping wild animals under human care. Starting with the most natural habitats, scientists soon discovered that so-called "free-living" animals did not actually have that much freedom—neither in a spatial, temporal, nor personal (social) sense. We already mentioned that sociology deals with the personal kinds of restrictions, that is, with social-status relationships and with the great variety of intraspecific interactions and bonds (see Chapter 31). Research into rhythmic phases or cycles deals with temporal restrictions, such as the periodic migrations, hibernation, breeding seasons, and activity and rest cycles (see Chapter 12); again, much of the initiative has come from zoo biology. Zoo keepers have al-

Fig. 42-3. This young elk cow has zoomorphized the person—that is, has assimilated her as a conspecific. It now treats her as a playmate and invites her to join in the favorite game of "castle."

Fig. 42-4. Now the elk cow wants to defend its castle—a sand hill. The playmate must try to push it off.

Fig. 42-5. As with a human child, the mood of a young animal can easily change from play to earnest. Here, we can see this reversal by the mimicry (position of ears) and the raising of one foreleg.

ways tried to adapt the natural actograms (activities and activity periods) of their animals to the human actogram—the visiting hours—because active animals are more attractive than resting or sleeping ones.

Night houses

Fig. 42-6. Without the addition of fruit trees in the animal's enclosure, we would never have observed the amazing "fruit-picking" behavior of red deer.

Because of this, many zoos have recently started to build "night houses," where the animals are subjected to very bright illumination at night, while during the day, when people visit, they are displayed in a dull blue or red light to make them more active. To be sure, in some species this attempt to display individuals in an "artificial night" did not work: In addition to animals that are active only during the day or only during the night (biphasic animals), there are also species whose activity and rest cycles occur partly in daylight and partly at night (polyphasic animals). Some animals that have been thought of as strictly nocturnal actually crave sunlight. Even bats and owls do not fly all night.

The spatial restrictions of so-called "free-living" animals are still widely misunderstood, and in this area attitudes have been particularly hard to change. The naive assumption that animals in the wild have unlimited space, that they are totally free to roam wherever they wish, has at least led to intense and successful protests against the old-fashioned menagerie method of keeping individuals in dungeonlike cages. As a result, the method used in the Hagenbeck zoo of providing the animals with large and attractive open enclosures has met with an enthusiastic response throughout the world.

In the meantime, however, this new style—pleasing to both people and animals—became somewhat misunderstood. People believed the more room they had, the happier animals kept by humans would be. But zoo biology has shown that—within certain limits—it is less the size of the enclosure than the landscaping and design, which is important. After all, even in nature most animals, or their social units (individual animal, pair, family, herd, etc.), are restricted to a certain area whose size can be defined and measured. This is called a home range or territory (see Chapter 25). Small animals have small territories and large animals have large ones, because the amount of space involved is determined, first, by the nutritional needs of its inhabitants and, second, by the food naturally available within this area. While a mouse needs no more than a few square meters, a lion must cover a territory of several dozen square kilometers. We can understand that a predatory animal requires far more space than herbivores (plant-eaters) of the same size, because the carnivore has to "keep" many of its prey animals—which are self-sufficient—within the territory.

Fig. 42-7. The needs of many animals in captivity may include certain elements of their natural habitat. The remora (Echeneis) requires a marine turtle or a large fish, to which it attaches itself to be constantly carried about.

The simplest way to define a territory is as the space occupied by an individual or a social unit. In most cases this area has to be taken by force and constantly defended—especially against conspecifics. But a territory is not merely some abstract piece of space, the same from one end to the other; it also has an internal arrangement that is extremely important to the animal. Basically, this arrangement consists of fixed reference points and the lines connecting them. Such reference points, for example, are the residence (den, shelter), food-storage places, and sites for defecating and urinating, wallowing in the mud, drinking, marking, and so on. The connecting lines are paths made by the animal itself, the haunts or runways.

This structuring of space is basically found in humans as well, and has always had an important influence on our behavior. Thus, for example, the loss of one's home, the fixed reference point offering the greatest security, can lead to severe psychological damage in both animals and man. The great significance of a familiar path for both animals and humans was closely studied by the psychotherapist F. Fischer, in 1972.

One of the most important demands of zoo biology is that animals in captivity are given all the fixed reference points otherwise found in their natural territories. Deer, for example, need neither wallowing places nor storage sites, whereas squirrels do need places to store their food. Beavers must have a defecation site in the water; without it, they can suffer lethal constipation. Lions and tigers require trees or poles to sharpen and groom their claws. In some zoological gardens we may find enclosures that have plenty of space—sometimes even more than enough, so that their inhabitants use only small parts of it—but that these areas have only very inadequate landscaping. Thus, for example, an enclosure for capybaras may lack a bathing spot, and the area for black bucks may not have trees and branches suitable for scent marking.

Fig. 42-8. Among the most important fixed reference points in the territory of many animals are the marking sites. Here, a brown bear has chosen a particular rough spot on the wall of its pit for marking with the back. Tree trunks are often also used for this purpose.

If all the needs of an animal are adequately met, the zoo enclosure offers its inhabitant a man-made, miniature territory with all the properties of a natural one. The animal will then consider the territory its own: It marks and defends the area and does not feel imprisoned. This territory is defended mainly against conspecifics. People themselves may be viewed as rivals of this kind, which is explained by the "assimilation tendency" (zoomorphism) to be discussed later.

We have to imagine all those habitable parts of the earth's surface as if they were a multiple layer of overlapping mosaics the size of different stones. Each pebble represents a territory, which must be taken by force, occupied, and defended by the owner. Hence, we might say that there is considerable pressure on the occupant, caused by a constant threat of intrusion by conspecifics. Violations of strange territories lead to fights, very often a matter of life and death. This applies to rats as much as to lions and nightingales. Identification (marking) of the territory occurs through singing in the nightingale, in mammals with acute olfactory ability (macrosmats), primarily through scent marks like gland secretions (many antelopes) or feces and urine (e.g., hippopotamus). These marks undoubtedly help to keep away intruders and thus to avoid fights.

We should not underestimate the effects of such scent marks. They are not aimed at the weak olfactory sense of humans, but rather at the highly sensitive noses of macrosmats, whose olfactory capacities are a hundred thousand times—or even a multiple of that—superior to man's. In the environment of such macrosmats, one marking site virtually shines like a

Fig. 42-9. The erect posture of the "zoo-morphized" attendant is misinterpreted by the male giant kangaroo as a fighting pose. It responds with a dangerous attack, attempting to defend its territory and drive off the rival.

Fig. 42-10. As soon as the attendant bows down, eliminating the posture that in a kangaroo's eyes represents a challenge to fight, peace is restored.

Fig. 42-11. This shoebill (*Balaeniceps rex*) is imprinted to a person (an attendant), and greets him with head-shaking and bill-clacking movements, normally displayed toward the mate.

"scent-watchtower" far into the countryside. This explains the behavior occasionally observed in a zoo, when two macrosmats that do not know each other are put together for the first time in neighboring enclosures. Sometimes they will march straight past each other and begin by exploring the new area with their noses, smelling it out, so to speak. The visual encounter may occur much later. We know this can happen with carnivores as well as artiodactyls (even-toed animals).

In earlier times, the depositing of gland secretions on parts of the enclosure or the spraying of urine and distribution of feces at various spots was occasionally seen simply as "bad manners." Today, we know that these scent marks are used not only as warning signals but frequently also as orientation aids by the animal itself and as information signals concerning the rut, incidence of predators, and so on. The study of such odor marks (called pheromones, substances with informational value; see Chapter 32) has become increasingly important in ethology.

Because of its importance, we may note that the first examples of such functional scent marking were described by zoo biologists, at first probably in Indian black buck, whose bucks use large preorbital glands to deposit whole clumps of a resinous secretion on projecting objects, when necessary even on a wire or lock if no suitable branches are available. Only some time later did I discover a similar behavior in the small Oribi antelopes in the Congo region. Bucks of this species deposited a blackish secretion from their preorbital glands onto bare branches or single jutting stubbles. In the meantime we have learned, especially through the studies of F. Walther, that this kind of behavior is quite common among antelopes.

As with their social interactions, animals in a zoo mark more frequently and more noticeably than in the wild. No doubt this occurs because these artificial territories are much smaller than natural ones. The fixed reference points are closer to each other, and every encounter with such a point where the animal deposits its habitual marks will induce it to make a fresh one.

The narrowing of fixed reference points and the resulting shorter runways may lead to stereotypies—to movements constantly repeated in the same way. The occasional behavior of looking out for the attendant who brings the food may initially be quite natural and functional at the usual feeding time. If, however, the animal lacks other distractions and activities, such as conspecifics, this walk to the look-out spot may become far too frequent, perhaps occurring several times a minute; or the animal may pass and re-mark the same marking point every few minutes on a short runway. In this way the originally functional behavior loses its purpose, and the repetition becomes obsessive, or stereotypical.

This is only one of the many ways behavioral stereotypies can arise (see Chapter 24), but it demonstrates that ethology has profited much from zoo biology in the area of abnormal and pathological behavior. M. Meyer-

Holzapfel made some highly detailed studies of movement stereotypies in zoos, and in 1964 even published a thorough investigation on the relationships between animal psychology, ethology, and psychiatry.

These three key words show important relationships and cross-connections largely overlapping with the term "zoo biology." As with scientists in these other areas, zoo biologists do not just restrict themselves to animals, but also deal with various basic views on human behavior. On the one hand, zoo biology provides us with the scientific background information needed to create the best and most functional environment possible for wild animals in a zoo, and on the other hand, it investigates and formulates the special biological laws that may be obtained for both animals and man from this method of keeping animals.

Zoo biology also includes animal psychology, which in turn forms part of comparative psychology comprising the study of animals and humans. We are making this explicit, because the so-called "objective" study of behavior, that is, ethology, founded by K. Lorenz and N. Tinbergen, often excludes all psychological aspects. It seems plausible that zoo biologists cannot ignore all the psychological, subjective, even individual elements of the animal. In a zoo we are dealing primarily with large animals that often have a unique life history, a special kind of character, and their own personal peculiarities. More than that, it is one of the tasks of a zoo biologist to deal with an animal's subjective and psychological manifestations. In no way does this embody the unrestrained anthropomorphism fairly common during the early years of animal psychology. Instead, zoo biology—or rather the kind of animal psychology employed here—is basically concerned with what may be called a critical, controlled empathy.

The animal psychologist working in a zoo must use his zoological, ecological, physiological, and—not the least—ethological knowledge of the animal to project himself into that individual, to such an extent that he not only understands and sympathizes with its behavior in certain situations but also learns to predict and to test his predictions.

When animals are trained for the circus, many of these factors become even more obvious, because with each performance the trainer must be acutely aware of the situation and of the animal's capacities, personal traits, and mood. No work can be done properly without a great deal of emphatic understanding: Often enough, predictions of an expected behavior must be accurate within fractions of seconds and centimeters.

In this context we shall mention the square-lipped rhinoceros of the Knie Circus, which was captured in the South African wilds; after a while it could not only be shown in the ring but also led through busy streets completely on its own. Such a tame rhino, of course, reacts differently and must be understood in a different way from its conspecifics still living in the African bush, although it should hardly diverge in its physiological make-up, especially its nervous capacities. A prominent, even dominant

Fig. 42-12.

a. A hand-reared emu cock courts its zoomorphized attendant.

b. As the attendant squats down to do something, the emu approaches him with an unmistakeable intention to copulate.

c. The emu cock approaches its caretaker, mistaken as a conspecific female, in typical fashion.

d. The copulatory act performed on an unsuitable object—the zoomorphized person—is nearly complete.

factor in this rhino's behavior is its tameness; yet it is just this term, so important in the vocabulary of zoo biology, that is lacking in ethological literature. In our case, "tameness" means a reduction in escape tendency pretty well to zero, and furthermore implies a complete reversal of the meaning humans hold for these animals: From an initial enemy (a threat to the animal's freedom, of being captured), man has become a familiar and trusted fellow creature.

Zoo biologists often come across this complete change of meaning or attitude. The "assimilation tendency" can lead animals and humans not only to view members of other species as conspecifics, but even to assimilate them into their own social structure, for example, as social superiors or inferiors, as opponents in a fight, and even as sexual partners. When a person gives his dog a human name and talks with it like a human, we speak of anthropomorphism. In the case of a hand-reared roebuck that regularly attacks its foster parents after puberty—as roebucks normally do among each other—we speak of zoomorphism.

A human being can be assimilated slowly (over months or years) or very suddenly (within hours or minutes) as a conspecific into the personal *Umwelt* of an animal. In 1941, O. Heinroth pointed to these possibly very grave consequences in birds and mammals, designating the sudden process as *Prägung* (imprinting) (see Chapters 1 and 24). The phenomenon of imprinting—especially in greylag geese—was studied by K. Lorenz in his classical research. Not only does it play a significant part in the life of many animals, however; it also helps to influence the fate of human beings in, for example, their choice of profession or sexual partner (see Chapter 24). Hence the findings of this kind of research are extremely important not only in zoo biology but also in ethology—including the study of man.

The valuable contributions and impulses that ethological research gained from zoological gardens did not simply start in the "official" founding year of zoo biology (1942). Competent and ingenious zoo workers have always provided ethology with invaluable observational data. We need only mention K. M. Schneider, who directed the Leipzig zoo, and O. Heinroth, who performed his work in the Berlin zoo and whom Lorenz described as his teacher and the "father of ethology."

43 Phylogenetic Adaptations in Human Behavior

The facts and ideas discussed by ethologists in the preceding chapters also have some bearing on the study of man. We know that humans, too, have an evolutionary history. We may therefore reasonably inquire whether certain aspects of human behavior have become more ingrained (preprogramed) through phylogenetic adaptations, similar to the evolutionary development of other animals. In recent years, this question has ignited a good number of lively and even heated discussions. In many cases people would assert that an individual's development is determined only by the way he is raised—in other words, by what he experiences and learns. The fact that ethological findings are often ignored or rejected is probably based, among other things, on a fear that anything innate cannot be altered. But this is surely an unnecessary assumption.

From the day of birth, every infant is already equipped with a number of movement patterns essential for survival. The infant can suck, breathe, hold on tight with his hands, cry, and a great deal more. In general no one doubts any longer that such movements are inborn. However, many other behavior patterns develop only gradually in the course of an individual's growth. It then becomes harder to determine whether they are innate. But we may find such evidence in persons who have grown up under conditions of deprivation, for example, individuals born both blind and deaf. Tragically, they grew up in eternal darkness and silence, never to see or hear how other people behave when they are happy, angry, or sad. If man would indeed have to learn the appropriate behavior to express a certain feeling, as claimed by some, then deaf-blind children would behave very differently from normal persons. Interestingly enough, this is not the case: Children born deaf and blind laugh and cry like everyone else.

On the other hand, the study of deaf-blind people tells us little about how phylogenetic preprograming influences the complex social behavior patterns of man. The reason is that many of these behavior patterns are released primarily through visual and auditory signals. A single friendly smile may initiate a wealth of behavior! If we want to know how much

Phylogenetic adaptations in human social behavior, by I. Eibl-Eibesfeldt

Innate movement patterns

of this is innate, we can start with certain experiments on individuals who were blind from birth. When I complimented a ten-year-old girl on her good piano-playing, she turned toward me with her sightless eyes, ceasing her otherwise restless eye movements; then she lowered her head, blushed, smiled "embarrassedly," and looked at me again, just like sighted girls when they feel shy.

Cross-cultural comparison

Another step in our inquiry is to make cross-cultural comparisons. We may start with the assumption that human beings tend to create cultural variations wherever possible. Should we then find that, in the most diverse cultures, people display the same behavior patterns in certain situations, right down to the last detail, the probability is very high that these behaviors are basically innate.

In the last several years we have been filming social interactions in primitive and advanced societies, and in the most diverse regions of the world. Our technique is to use mirror lenses so that we can film to the side of the camera without the subjects noticing it. Furthermore, we made notes of what these persons were doing before and after being filmed, and the social context in which they performed the behavior in question. Our analysis of the films showed that human beings are basically alike in their social behavior patterns. For example, mothers everywhere fondle their children with the same movements: They rub their faces on their children, kiss them, fondle them, nod at them while performing quick upward eyebrow movements, and talk to the children in a high-pitched voice. Some expressive movements of threat, too, are found all over the world. Staring at someone, for example, is considered threatening behavior everywhere. Pouting, on the other hand, is a wide-spread form of appeasement.

These expressions and their meaning correspond in great detail. Another example is our observation that people nod at each other all over the world when greeting at a certain distance. In addition, they will smile and, while starting to nod, raise their eyebrows quickly, for a fraction of a second. This eyebrow flash generally denotes a willingness to take up social contact. It may have derived from an expression of joyful surprise.

An old primate legacy

The legacy of our primate ancestors may be recognized in many of our behavior patterns expressing affection and tenderness. Chimpanzees hug, fondle, and kiss their young as we do. Adults also display these behaviors with each other. The kiss may have its origin in the movement of mouth-to-mouth feeding. Orangutan, chimpanzee, and human mothers supplement their infant's diet with prechewed food that they pass on from mouth to mouth (kiss-feeding). When people are very angry, they will show their teeth, primarily opening and pulling down the corners of the mouth. This is the way that some monkeys bare their long upper canine teeth. It seems that humans have retained this movement of threat although their own canine teeth, which would originally have been displayed in this way, have become involuted.

Some of the behavior patterns encountered everywhere may develop from similar experiences that all individuals would be subjected to in their earliest childhood. One example is the fully fed infant who, as Darwin had already observed, will turn his head away from the breast in a sideways movement. If the mother keeps offering him the breast, the infant will likely repeat this sideway movement and develop a kind of head-shaking to indicate rejection. Another point is that children push away objects they don't want. In this way, rejecting movements such as head-shaking ("no") or putting out the upright hand with open palm—both of which are wide-spread expressive movements—may have been acquired through similar experiences. Whether this interpretation is true, however, is in no way certain.

Numerous experiments have shown that animals respond in a certain way to particular sign stimuli, without first having to learn this complex of stimuli and responses. For example, immediately after completing its metamorphosis, a frog will use its tongue to hit small moving objects. It "knows" its prey before any relevant experience was possible. Similarly, a male stickleback recognizes a red-bellied conspecific as a rival and a silver-bellied female as its mate. Male sticklebacks raised in isolation will attack simple red-bellied models, and will perform courtship movements when presented with a silver-bellied one (see Chapter 17). Young hares stop short at the top of an abyss covered with a glass plate—experimenters call this arrangement the visual cliff—even if they never had the experience of falling down a precipice.

Innate recognition of sign stimuli

Results of research done in recent years has indicated that human beings, too, are equipped with some kind of "innate releasing mechanisms" (IRM's, see Chapters 1 and 15). If two-week-old infants are fastened into a small chair, and we then project onto a screen in front of them a dark patch expanding evenly, they will act as if an object is coming right toward them: They make defensive movements with their hands, turn their face away, and show an increase in pulse rate. Yet these infants had never collided with any such object. On the other hand, they will give no sign of fear if the patch expands over the same area but asymmetrically, or if it becomes smaller. Certain visual impressions seem to make the infants expect a particular tactile impression, again prior to any relevant experience. Other examples may be given. With a special projection technique, we can give the infant the impression of an object right in front of him. The subject will then reach for it, showing all indications of arousal when he does not meet with the expected tactile stimulus: Again the pulse rate increases, and many babies start to cry. If, on the other hand, they are allowed to touch an object that they see, their pulse rate stays even and they remain cheerful, everything being right in the world, so to speak. Another test is to cover an object lying in front of them with a screen. When we remove the screen after a short time, the infants expect to see the object again. But if we remove the object before moving the screen away, even twenty-day-old youngsters will react with an increased pulse rate, indicating arousal—provided that we did not wait too long between the two events.

These experiments are very important because they indicate the presence of innate mechanisms for processing data. It seems that in adult life, too, these mechanisms play a greater role than is generally assumed. For example, adults display a marked reaction to certain traits in small children, even when these traits are combined in a greatly simplified model or dummy. We call this the baby schema (or kewpie-doll schema), and it includes a strongly pronounced forehead in relation to a small face, relatively large eyes, rounded body contours created by special fat deposits, and chubby cheeks, all of which probably developed in the course of evolution as social signals. If an object such as a toy animal displays some of these traits, we find it cute and are tempted to hug it. The toy induces a basically friendly attitude in us, just like a small child would (see Fig. 1-12).

It is in fact the doll industry that offers numerous examples for this phenomenon. The heads of these cute little artificial creatures are often larger than their bodies, and the protrusion of the forehead is also exaggerated. Lap dogs and other such pets, which serve people as substitute objects for expressing their nurturing needs, generally display infantile traits. Exaggerating these traits raises their effectiveness as sign stimuli. If we want someone to feel friendly toward us, we often employ baby traits and baby signals; this is very prominent in the advertising industry.

There is a strong indication that we may even recognize the mimicry of our fellow humans innately on the basis of very simple relational signals. We have no conclusive evidence for this, yet we may point to the fact that, not only do we respond to simple outline sketches of human facial expressions, but we even interpret and classify animal faces despite our knowledge to the better, as if they expressed feelings and attitudes like those of people. We say the eagle looks "bold" because his supraorbital bony ridges remind us of a person gazing resolutely into the distance, and this expression of "proud determination" is further underlined by the severe cleft of the beak. The camel conveys an expression of arrogance because we misinterpret the relative elevation level of eye to nose (see also Fig. 1-2) —in humans often signifying contempt. The investigations of P. Ekman and W. Friesen have shown that Papuans, a people with Stone Age technology, were able to give correct interpretations of videotaped expressive movements performed by Japanese.

We judge others according to standards that again are perhaps phylogenetically determined. One indication of this is the wide-spread agreement on certain ideals of masculine and feminine attractiveness. In a great many cultures people value narrow hips and broad shoulders in a man, and often enough the shoulders are emphasized by ornaments or clothing. P. Leyhausen pointed out that our hairline on the back and front runs in such a way that with a fully developed coat of hair (like our ancestors probably possessed) we would have tufts standing up from the shoulders,

Baby schema

Facial expressions

resulting in a much greater body outline when a person is standing. This adaptation must have evolved in association with the development of upright walking, and even when man's coat of hair receded, he retained his tendency to emphasize the shoulders. For women there seems to be two basic kinds of beauty ideal, represented on the one hand by the classical Venus and on the other by the prehistoric Venus of Willendorf.

People seem to recognize signals innately in other sense modalities as well. Studies have shown that women are able to smell odors belonging to the group of musk substances especially well during their entire reproductive period, from the start of the menstrual cycle to its end. Furthermore, this olfactory threshold undergoes fluctuations within the estrus cycle. It is very low at the time of ovulation. By contrast, men are able to smell these substances only in a strong concentration. Even in males, the olfactory threshold can be lowered if they are injected with female sex hormones (estrogens).

There are probably predetermined propensities to recognize signals in the auditory sense modality, too. Studies have shown that particular rhythms and melodies tend to have certain physiological effects. A German lullaby, for example, has a soothing effect on the listener just as much as a Chinese one. Then there are other melodies and rhythms that have an arousing effect.

A number of chapters in this volume have shown that the behavior of animals is often motivated by strong drives. All these drives lead to so-called appetitive behavior, whereby the animal seeks out situations that allow it to reduce this drive by means of certain movements. We know today that individual drives may be based on very different mechanisms.

It is generally agreed that humans are also subject to drives based on similar physiological processes. A great deal of research has been done on the processes underlying hunger and thirst, as well as on the sex drive. On the other hand, there is considerable controversy on whether humans have an aggression drive (see Chapter 24), like some animals are definitely known to display. A series of experiments have demonstrated the build-up of a measurable aggression stow which can be discharged, or redirected. In one set-up, for example, experimenters deliberately made a group of students angry. These subjects exhibited a marked increase in blood pressure. Next, the experimenter who had angered the subjects divided them into two groups and told each that he himself would now be the subject. He would undergo a series of tasks and the students were to press a button every time he made a mistake. One group was told that a blue light would flash on; the other group was told that the subject would receive an electric shock when they pressed the button. In the latter group, who thought they could punish the experimenter for making mistakes, blood pressure quickly decreased, indicating that they discharged their aggression. But the others, who thought they would merely turn on a blue light when pressing the button remained angry.

Few people still doubt that humans have an aggressive drive, but there is much disagreement on whether this is acquired through learning or whether aggression has an innate basis. The universal occurrence of aggressive behavior would indicate that these drives are inborn; those cultures striving for peaceful ideals have evolved customs and practices that allow their members to discharge aggressive tendencies in a harmless manner (safety-valve customs). Frustrating experiences in early childhood, as well as the fact that aggressive acts can lead to success, will certainly favor the learning of aggressive attitudes and dispositions. We might well accept this as an entirely satisfactory explanation, were it not for certain research findings in neurophysiology that give evidence to the contrary.

In addition, we have discovered that human aggressive behavior has a physiological foundation in those ancient parts of the brain whose corresponding structures in other vertebrates generate the same kind of behavior. Electrical stimulation of these parts will trigger aggressive behavior in both animals and men. Not only that; we also know of spontaneous, rage reactions—that is, independent of external stimulation—which can be traced to the firing of nerve cells in the temporal lobe and the brain stem. Individual patients experienced anger and behaved aggressively during these experiments. This leads us to assume that there are neuron chains, or loops, that underlie aggressive behavior, and we may say that the hypothesis of an innate aggressive drive, as K. Lorenz suggested in his book *On Aggression*, has much to support it. Although many others repeatedly argue that aggression is very much influenced by learning, this does not automatically refute the idea of an innate basis. In any case, ethologists do not deny that learning plays some part in the development and expression of aggressive tendencies.

The learning capacity of individual animal species is not unlimited, nor do animals learn everything equally well at different times of life. In many cases, different things are learned either better or exclusively within certain "sensitive periods" (Chapter 24). Animal species also vary in their particular learning dispositions, since each species must adapt its behavior to the changing demands of its environment through learning. Phylogenetic adaptations ensure that animals learn the right things at the right time. In particular, such innate learning propensities have been closely studied in birds. Some have to learn the songs typical of their species, but a number of these animals seem to recognize their species-specific calls without direct experience, "knowing" immediately what they will be required to learn: If we let these birds listen to several different songs on tape, they will always choose the species-typical one to imitate. Some birds also learn the object of their future sexual behaviors in early childhood, long before puberty, during a particular sensitive phase. As a result, for example, jackdaws, turkeys, and many other species will end up courting humans if these birds are raised by hand, even when they are put together with conspecifics during some interim period. This mating preference does not change even when our hand-raised birds are forced to cohabit with conspecifics of the opposite sex.

Humans possess quite an array of innate learning dispositions, and one of the more prominent ones is our marked propensity for learning to speak. Thus, a child of only eighteen months is able to repeat words, that is, to transpose sounds into muscular movement. Yet at the same age he or she is not capable of copying a circle on paper, although this task requires a much simpler coordination of movement. Furthermore, psychological and psychoanalytic research has shown that human ontogenetic development, too, involves sensitive periods during which the individual acquires certain basic attitudes, a process much like imprinting. In this way, for example, a child learns to hold what we may call a "basic trust" in others during its first year of life, and this primary attitude of trust is a cornerstone of any healthy personality, a basic requirement if people are to live together in an orderly society. After all, in everyday life we are continually forced to trust even total strangers. It was not until some researchers undertook the study of children raised in institutions, that we realized how strongly the development of this basic trust could be curtailed or even entirely suppressed.

We can already see the influence of phylogenetic adaptations on our social behavior by the fact that we use innate expressive movements in nonverbal communication. Many friendly as well as hostile behavior patterns and rituals have some genetic basis. Yet our social traits are preprogrammed to an even far greater extent than that. Let us start with the relationship between a mother and a child, who are bonded by means of a variety of responses. The child possesses baby traits and displays certain behavior patterns such as smiling and crying which trigger or reinforce caretaking responses in the mother. We can no longer accept the idea that a child becomes attached to its mother only through conditioning, so to speak, where food is the reward. Starting out with only a few signals, this mutual mother-child relationship rapidly intensifies—not only because the mother forms this personal attachment to her child, but also because the infant itself enters into a strong emotional bond with the mother during the second part of its first year of life. Every child has the need to become attached to some reference person; if it does not have this opportunity, severe developmental problems will result. Such personal bonding must have evolved in the service of brood-care functioning. In this way, the parent would always recognize its own young, ensuring their proper care.

Pre-adaptations in human social behavior

At the same time that the child develops a personal bond with its mother or other reference persons, it begins to behave adversely toward strangers. This attitude starts out as a fear of strangers and later even develops into rejection. Humans have an "enemy schema" with the formula "familiar = friend, and strange = enemy" which seems to have an innate basis. My own observations on children born deaf and blind indicate that the small child does not acquire this attitude on the basis of experience:

Distinguishing between "friend" and "foe"

Even these deaf-blind youngsters, who have never had bad experiences with strangers would develop both a fear and a rejection of strangers. Furthermore, such behavior can be observed in children of all cultures.

The development of this personal bond and of mutual mother–child signals in the service of brood-care functioning provide us with the basic pre-adaptations necessary to live in society. The binding relationship of married couples partly develops through behavior patterns resembling mother–child interactions, and these continue to reinforce the conjugal bond. In addition, there are a number of physiological characteristics peculiar to human sexuality, and these help to ensure the stability of marriage. While most mammals tend to pair off only during the female's brief reproduction periods, in higher primates, and notably in man, sexual bonding is no longer dependent on the reproductive cycles. Most of the time, a woman is physiologically prepared to comply with the man's sexual demands, even outside of her reproductively receptive days. By rewarding the man sexually, she can bind him into this permanent relationship. But this pair-bonding is also facilitated by the ability of the woman to experience a climax comparable to the male orgasm, which on her part strengthens the emotional bond to her partner and raises her willingness to meet his needs. In this way, the sex act in humans has become meaningful beyond mere reproduction.

The family vs. mass society

Aside from this programing which leads to a permanent bonding between sexual partners, there seems to be no single set pattern for conjugality. But whatever the individual structure of this partnership—whether monogamous or polygamous—there is one characteristic common to all families: their closed nature or exclusivity. Families delineate themselves as a group and even draw spatial boundaries to set themselves off from others. Human family groups are usually embedded within larger groups. In the beginning, members of such a group were bonded to one another through personal recognition. This was the case, for example, in the bands of hunters and gatherers or in the small village communities of horticultural peoples. Anonymous associations did not come about until the advent of mass society, where people are held together by means of a common symbol (symbolic identification). Yet a group where members know one another personally still embodies the original state of human social life. This "individualized group" is to some degree closed and territorial. Intruders are met with hostility and aggression.

Some people have proposed that territoriality, defined as intolerance toward strangers within certain spatial boundaries, evolved only with man's transition to agriculture. In other words, hunters and gatherers such as the African Bushmen were said not to defend any areas of their own against outsiders. When we examine these claims more critically, they turn out to be false. More recent investigations have shown that Bushmen are indeed territorial.

Another fundamental factor in social life is the genetically determined differences between men and women. Males in all cultures have a tendency to form groups that exclude women, all-male associations which may represent an adaptation to hunting and fighting behavior (see Chapter 35). In most cases, members of a group form rank orders, or dominance hierarchies. This would mean, on the one hand, that individuals strive for dominance, on the other, that those who end up on a lower rank subordinate themselves and follow their leaders. A brief note may be added about the dangers of blind obedience inherent in such a system.

At this point we cannot make any precise statement on the extent to which individual phylogenetic adaptations influence our social behavior. There is a great deal to suggest that striving for dominance or status, willingness to subordinate oneself and obey a higher authority, intolerance against outsiders, aggression, but also our tendencies to be altruistic and the drive to establish bonds with others—in short, to feel and express love in a general sense—are all behavior traits preprogramed by evolutionary adaptations.

Man the "cultural animal"

Should this be the case, there is no need to assume that we must yield to all our drives, that we are powerless to do anything about them. Some people claim that ethologists, because they study innate behavior patterns, provide support for conservative ideas such as the doctrine of society's immutability. To be sure, we recognize a real danger that the findings of ethological research will be misused. But ethologists have tried to prevent and counteract such abuse by repeatedly emphasizing that many phylogenetic adaptations are no longer to our advantage. Much like our vermiform appendix, which has lost its adaptive value and today remains only an evolutionary burden, many of our behavior tendencies could also prove to be "appendixes." We have to cope with such historical ballast, and, as "cultural creatures by nature" (in the words of A. Gehlen), we should well be able to do so.

While many animals possess not only innate drives, but also the appropriate movement sequences or action patterns which for the most part are genetically fixed, this is by no means the case in humans. Although we do have drives, display certain movement patterns in the form of *Erbkoordinationen*, and even respond innately to certain unconditioned stimuli, there are no fixed controls determining the general course of human behavior. Instead, we perform our overall patterns of behavior within much broader boundaries; these are not totally flexible but certainly capable of modification.

Cultural patterns of behavior control

Only the cultural patterns of control restrict this behavioral variability to a larger extent. Since these control patterns in turn vary from one place to another, people could adapt very quickly to different environmental demands. After all, Eskimos need different discharge controls for their aggressive or sexual drives than, say, the Masai in Africa or modern European city dwellers. Furthermore, we can change cultural patterns of behavior control, when change becomes necessary: The present time is one of upheaval. Now some people are suggesting that children should not be subjected to any guidelines at all, since they are well able to grow and mature on their own. But what resources are available? Our own innate propensities? These are, for the most part, based on drives. Furthermore, these phylogenetic preadaptations are not in themselves enough to guarantee a social life without conflict. Man must depend on cultural patterns of control to teach him how to live in peaceful society with others.

On the one hand, our cultural ground rules must always remain flexible. Yet from a biological point of view any process of evolution is a gradual matter. This applies to cultural evolution as well. As K. Lorenz emphasized, there is in fact a danger that some radical upheaval overthrowing all traditions would inevitably result not in continuing growth but in destruction. If our cultural evolution should not stagger onward blindly through trial and error, we must continue our research into the nature of man. We have valuable knowledge to gain of basic cause-and-effect relationships and especially of biological pre-adaptations influencing our behavior. This should help us considerably in finding solutions to an ailing human society.

Supplementary Readings

Scientific Journal Articles and Chapters

Ader, R. (1971): Experimentally induced gastric lesions. Results and implications of studies in animals. *Adv. Psychosom. Med.*: 6, 1–39.

Ader, R. and Plaut, S. M. (1968): Effects of prenatal maternal handling and differential housing on offspring emotionality, plasma corticosterone levels, and susceptibility to gastric erosions. *Psychosom. Med.*: 30, 277–286.

Archer, J. (1970): Effects of population density on behaviour in rodents. In: J. H. Crook (Ed.) *Social Behaviour in Birds and Mammals.* Academic Press, New York, p. 169–210.

Baerends, G. P. and Baerends, J. M. (1939): An Introduction to the Ethology of Cichlid Fishes. *Behavior*: Supplement I, 1–242.

Barnett, S. A. (1964): Social stress. The concept of stress. In: J. C. Carthy, C. L. Duddington (Eds.) *Viewpoints in Biology*, Vol. 3. Butterworths, London. P. 170–218.

Bartholomew, G. A. (1970): A model for the evolution of pinniped polygyny. *Evolution*: 24, 546–559.

Beach, F. A. (1950): The snark was a boojum. *American Psychologist*; 5, 115–124.

Berry, R. J. (1969): The genetical implication of domestication in animals. In:P. J. Ucko and G. W. Dimbleby (Eds.) *The Domestication and Exploitation of Plants and Animals.* London, P. 207–217.

Blaine, E. H. (1973): Elevated arterial blood pressure in an asymptotic population of meadow voles (*Microtus pennsylvanicus*). *Nature*: 242, p. 135.

Bourne, P. G. (1971): Altered adrenal function in two combat situations in Viet Nam. In: B. E. Eleftheriou, J. P. Scott, (Eds.) *The Physiology of Aggression and Defeat.* Plenum Press: New York, London, p. 265–290.

Brady, J. V. (1958): Ulcers in "excutive" monkeys. *Scient. Amer.*: 10.

Burghardt, G. M. (1973): Instinct and innate behavior: towards an ethological psychology. In: J. A. Nevin, (Ed.) *The Study of Behavior.* Scott, Foresman, Glenview, IL, p. 322–400.

Calhoun, J. B. (1962): Population density and social pathology. *Scient. Amer.*: 206, p. 139–148.

Calhoun, J. B. (1971): Space and the strategy of life. In: A. H. Esser (Ed.) *Behavior and Environment.* Plenum Press: New York, London, p. 329–387.

Christian, J. J. (1963a): Endocrine adaptive mechanisms and the physiological regulation of population growth. In: W. V. Mayer, R. G. van Gelder (Eds.) *Mammalian Populations*, Vol. 1. Academic Press: New York, London, p. 189–353.

Christian, J. J., Lloyd, J. A., Davis, D. E. (1965): The role of endocrines in the self-regulation of mammalian populations. *Recent Progr. Hormone Res*: 21, 501–578.

Christian, J. J. and Lemunyan, C. D. (1958): Adverse effects of crowding on lactation and reproduction of mice and two generations of their progeny. *Endocr.*: 63, 517–529.

Christian, J. J. (1963b): The pathology of overpopulation. *Military Medicine*: 128, 571–603.

Christian, J. J. (1950): The adrenopituitary system and population cycles in mammals. *J. Mamml.*: 31, 247–259.

Clayton, D. C. and Paietta, J. V. (1972): Selection for circadian eclosion time in *Drosophila melanogaster. Science*: 178, 994–995.

Coulson, J. C. (1966): The influence of the pair-bond and age on the breeding biology of the kittiwake gull *Rissa tridactyla. J. Anim. Ecol*: 35, 269–279.

Cowley, J. J. and Wise, D. R. (1970): Pheromones, growth and behaviour. In: *Chemical Influences on Behaviour*: Ciba Foundation Study Group. (R. Porter and J. Birch, eds.) Churchill; London, p. 144–170.

Crook, J. H. (1965): The adaptive significance of avian social organisations. *Symp. Zool. Soc. Lond.*: 14, 181–218.

Crook, J. H. (1964): The evolution of social organization and visual communication in the weaver bird (Ploceinae). *Behaviour*: Suppl. 10.

Crook, J. H. (1970): Social organization and the environment: Aspects of contemporary social ethology. *Animal Behavior*: 18, 197–209.

Crook, J. H. (1970): The Socio-ecology of Primates. In: J. H. Crook, Ed., *Social Behavior in Birds and Mammals.* Academic Press: London.

Denenberg, V. H., and Rosenberg, K. M. (1967): Nongenetic transmission of information. *Nature* 216: 549–550.

Denenberg, V. H. (1967): Stimulation in infancy, emotional reactivity, and exploratory behavior. In: D. C. Glass, (Ed.) *Neurophysiology and Emotion.* Rockefeller University Press, New York, 161–190.

Diamond, S. (1971): Gestation of the instinct concept. *Journal of the History of the Behavioral Sciences* 7: 323–336.

Eisenberg, J. F. (1966): The social organization of animals. *Handbuch der Zoology*, Band 8, Lieferung 39 (10) 7, 1–97.

Eleftheriou, B. E. (1971): Effects of aggression and defeat on brain macromolecules. In: B. E. Eleftherious and J. P. Scott (Eds.) *The Physiology of Aggression and Defeat.* Plenum Press, New York and London, p. 65–90.

Erickson, C. J. and Morris, R. L. (1972): Effects of mate familiarity on the courtship and reproductive success of the ring dove (*Steptopelia risoria*). *Anim. Behav.* 20: 341–344.

Frank, F.: (1935): Untersuchungen über den Zussamenbruch von Feldmausplagen (*Microtus arvalis Pallas*) *Zool. Jahrb.* 82: 95–136.

Galle, O. R., Gove, W. R., McPherson, J. M. (1972): Population density and pathology. What are the relations for man? *Science* 176: 23–29.

Gleason, K. K. and Reynierse, J. H. (1969): The behavioral significance of pheromones in vertebrates. *Psychol. Bull.*: 71, 58–73.

Goldman, P. S. (1969): The relationship between amount of stimulation in infancy and subsequent emotionality. *Ann. N.Y. Acad. Sci.* 159: 640–650.

Henry, J. P. Ely, D. L., Stephens, P. J., Ratcliffe, J. L. Santisteban, G. A., Shapiro, A. P. (1971): The role of psychosocial factors in the development of arteriosclerosis in CBA mice: Observations on the heart, kidney and aorta. *Atherosclerosis* 14: 203–218.

Henry, J. P. and Cassel, J. C. (1969): Psychosocial factors in essential hypertension. Recent epidemiologic and animal experimental evidence. *Amer. J. Epidemiol* 90: 171–200.

Holst, D. v. (1969): Sozialer Stress bei Tupajas (*Tupaia belangeri*). Die Aktiverung des sympathetischen Nervensystems und ihre Beziehung zu hormonal ausgelösten ethologischen und physiologischen Veränderungen. *Z. Vergl. Physiol*: 63, 1–58.

Holst, D. v. (1972a): Renal failure as the cause of death in *Tupaia belangeri* exposed to persistent social stress. *J. Comp. Physiol.* 28: 274–288.

Holst, D. v. (1972b) Die Nebenniere von *Tupaia belangeri. J. Comp. Physiol* 28: p. 274–288.

Huttunen, M. O. (1971): Persistent alteration of turnover of brain noradrenaline in the offspring of rats subjected to stress during pregnancy. *Nature* 230: 53–55.

Huxley, J. (1914): The courtship habits of the great crested grebe *Podiceps cristatus*: with an addition to the theory of sexual selection. *Proc. Zool. Soc. Lond.*, p. 492–562.

Keeler, K. (1962): Prenatal influence on behavior of offspring of crowded mice. *Science* 135; 44–45.

Kummer, H. and Kurt, F. (1963): Social units of a free-living population of Hamadryas baboons. *Folia Primat.* 1: 4–19.

Kummer, H. and F. Kurt (1965): A comparison of social behavior in captive and wild Hamadryas baboons. In: H. Vagtborg (Ed.) *The Baboon in Medical Research.* University of Texas Press, Austin.

Lack, D. (1940): Pair-formation in birds. *Condor 42*: 769–786.

Lapin, B. A. and Cherkovich, G. M. (1971): Environmental changes causing the development of neuroses and corticovisceral pathology in monkeys. In: L. Levi (Ed.) *Society, Stress and Disease*. Vol. 1. *The Psychosocial Environment and Psychosomatic Diseases*. Oxford University Press, New York, London, p. 266–279.

Laws, R.m. and Parker, L. S. C. (1968): Recent studies on elephant populations in East Africa. *Symp. Zool. Soc. Lond. 21*: 319–359.

Lehrman, D. S. (1953): A critique of Konrad Lorenz's theory of instinctive behavior. *Quarterly Review of Biology 28*: 337–363.

LeResche, R. B. and Sladen, W. J. L. (1970): The establishment of pair and breeding site bonds by young known-age Adelie penguins (*Pygoscelis adeliae*). *Anim. Behav. 18*: 517–526.

Levine, S. and Thoman, E. B. (1969): Physiological and behavior consequences of postnatal maternal stress in rats. *Physiol. Behav. 4*: 139–142.

Mason, J. W. (1972): Organization of Psychoendocrine mechanisms. In: N. S. Greenfield and R. A. Sternbach (Eds.) *Handbook of Psychophysiology*. Holt, Rinehart & Winston, New York, p. 3–91.

Morris, R. L. and Erickson, C. J. (1971): Pair bond maintenance in the ring dove (*Streptopelia risoria*). *Anim. Behav. 19*: 398–406.

Morrison, B. J., and Thatcher, K.: (1969): Overpopulation effects on social reduction of emotionality in the albino rat. *J. Comp. Physiol. Psych. 69*: 658–662.

Mundinger, P. C. (1970): Vocal imitation and individual recognition of finch calls. *Science 168*: 480–482.

Myers, K. (1966): The effect of density on sociality and health in mammals. *Proc. Ecol. Soc. Aust. 1*: 40–64.

Nicolai, J. (1968): Die isolierte Frühmauser der Farbmerkmale des Kopfgefieders. *Z. Tierpsychol. 25*: 854–861.

Nicolai, J. (1956): Zur Biologie und Ethologie des Gimpels (*Pyrrula pyrrula L.*). *Z. Tierspsychol. 13*: 93–132.

Plaut, S. M., Graham, D. W., III, and Leiner, K. Y. (1972): Effects of prenatal maternal handling and rearing with aunts on behavior, brain weight, and whole-brain serotonin levels. *Developmental Psychobiol. 5*: 215–221.

Ratcliffe, H. L. (1968): Environment, behavior and disease. In: E. Stellar and J. M. Sprague (Eds.) *Progress in Physiological Psychology*. Academic Press, London and New York, 161–228.

Ritcher, C. P. (1954): The effects of domestication and selection on the behavior of the Norway rat. In: *J. of the Nat. Cancer Inst. 15*.

Rose, R. M., Gordon, T. P., Bernstein, I. S. (1972): Plasma testosterone levels in the male rhesus: Influences of sexual and social stimuli. *Science 178*: 643–645.

Roth, H. H. et al (1970–71): Studies on the agricultural utilisation of semi-domesticated eland (*Taurotragus oryx*) in Rhodesia. *I. Introduction Rhod. J. Agric. Res. 8*: 67–70.

Sassenrath, E. N. (1970): Increased adrenal responsiveness related to social stress in rhesus monkeys. *Hormones Behav. 1*: 283–298.

Schneirla, T. C. (1946): Problems in the biopsychology of social organization. *J. Abnorm. Soc. Psychol. 41*: 385–402.

Schneirla, T. C.: Aspects of stimulation and organization in approach/withdrawal processes underlying vertebrate behavioral development. In: D. L. Lehrman, R. A. Hinde, and E. Shaw (Eds) 1965: *Advances in the Study of Behavior*. Vol. 1. Academic Press, New York, p. 1–74.

Scott, J. P. (1954): The effects of selection and domestication upon the behavior of the dog. *J. Nat. Cancer Inst 13*.

Selye, H. (1950): The physiology and pathology of exposure to stress. Acta, Montreal.

Snyder, R. L. (1968): Reproduction and population pressures. In: E. Stellar, J. M. Sprague (Eds.) *Progress in Physiological Psychology*. Academic Press, New York and London, p. 119–160.

Thant, U. (1965): New Ideas for a New World. *Saturday Review*: July 24. P. 24–25.

Thorpe, W. H. and North, N. E. W. (1965): Origin and significance of the power of vocal imitation: With special reference to the antiphonal singing of birds. *Nature 208*: 219–222.

Tinbergen, N. (1968): On war and peace in animals and man. *Science 160;* 1411–1418.

Verner, J. and Willson, M. D. (1966): The influence of habitats on mating systems of North American passerine birds. *Ecology 47*: 143–147.

Wall, W. van de (1968): Le comportement de canards hybrides. *Annales de la Societe Royale Zoologique de Belgique 98*: 125–137.

Welch, B. L. (1964): Psychophysiological response to the mean level of environmental integration. *Symp. of Medical Aspects of Stress in Military Climate*: 39–99.

Wehmer, R., Porter, R. H., Scales, R. (1970): Premating stress and pregnancy stress in rats affects behaviour of grand pups. *Nature 227*: p. 622.

Wickler, W. and Seibt, U. (1970): Das Verhalten von *Hymenocera picta Dana*, einer Seesterne fressender Garnele (*Decapods, Natantia, Gnatophyllidae*). *Z. Tierpsychol. 27*: 352–368.

Wynne-Edwards, V. C. (1963): Intergroup selection in the evolution of social systems. *Nature: 200*, 623–626.

Zeuner, F. E. (1973): Summary of the Symposium: Man and Cattle. In: Mourant and Zeuner: *Man and Cattle*. Proceedings of a symposium on domestication at the Royal Anthropological Institute, London. P. 158–166

Books

Allee, W. et al (1955): *Principles of Animal Ecology* Saunders, Philadelphia.

Ardrey, R. (1966): *The Territorial Imperative*. Antheneum Publishers, New York.

Ardrey, R. (1970): *The Social Contract*. Antheneum Publishers, New York.

Aristotle. History of Animals. In: Robert M. Hutchins, (Ed.) *Great Books of the Western World*. Vol 9, Aristotle II. William Benton, Chicago.

Armstrong, E. A. (1955): *The Wren*. Collins, London.

Aronson, L., et al (Eds.) (1970): *Development and Evolution of Behaviour*. W. H. Freeman, San Francisco.

Bajusz, E. (1969): *Physiology and Pathology of Adaptation Mechanisms—Neural, Neuroendocrine, and Humoral*. Pergamon Press, Oxford, London.

Barnett, S. A. (1963): *The Rat: A Study in Behaviour* Methuen, London.

Bastock, M. (1967): *Courtship: An Ethological Study* Heinemann Educational Books, Ltd., London.

Bateson, P. P. G. and Klopfer, P. (Eds.) (1973): *Perspectives in Ethology I*. Plenum Press, New York.

Bateson, P. P. G. and Klopfer, P. H. (Eds.) (1975): *Perspectives in Ethology II*. Plenum Press, New York.

Bourlière, F. (1955): *The Natural History of Mammals* Harrap, London.

Brown, Jerram L. (1975): *The Evolution of Behavior* W. W. Norton, Co., Inc. New York.

Cannon, W. B. (1915): *Bodily Changes in Pain, Fear, and Rage*. Appleton, New York and London.

Carlestam, G. and Levi, L. (1971): *Urban Conglomerates as Psychosocial Human Stressors. General Aspects, Swedish Trends, and Psychosocial and Medical Implications*. Kungl. Boktryckeriet, Stockholm.

Carthy, J. D. and Ebling, F. J. (1964): *The Natural History of Aggression*. Academic Press, London and New York.

Chapple, Elliot, D. (1970): *Culture and Biological Man*. Holt, Rinehart and Winston, Inc., New York.

Cody, M. L. (1974): *Competition and the Structure of Bird Communities*. Princeton Univ. Press.

Coon, Carleton S. (1971): *The Hunting Peoples*. Little, Brown and Co. Boston and Toronto.

Count, E. (1973): *Being. and Becoming Human*. Van Nostrand Reinhold Co., New York, N.Y

Crowcroft, Peter (1966): *Mice All Over*. Foulis, London.

Darwin, C. (1871): *The Descent of Man and Selection in Relation to Sex*. Murray, London. 2 Volumes.

Darwin, C. (1872): *The Expression of the Emotions in Man and Animals*. Murray, London. (Republished: Chicago, Univ. of Chicago Press, 1965).

Davis, David E. (1966): *Integral Animal Behavior*. Macmillan.

DeVore, I. (1965): *Primate Behavior*. Holt, Rinehart and Winston, New York.

Dobzhansky, Theodosius. *The Biological Basis of Human Freedom*. Columbia University Press, New York.

Dobzhansky, Theodosius (1963): *Evolution, Genetics and Man*. Wiley & Sons, New York.

Dreiack, M. J. (1970): Zum Verhalten von Hähnen in der intensiven Bodenhaltung. Dis.Tierärztl. Hochschule Hannover.

Duha, J. (1970) Veränderungen in der Nest-Biologie der Art *Anser anser* L. durch den Einfluss der Domestikation. Sarus: Vorträge auf dem internationalen zooligischen Symposium Bratislava, 101–109.

Eibl-Eibesfeldt, I. (1961): *Galapagos*. Doubleday, Garden City, N.Y.

Eibl-Eibesfeldt, I. (1971): *Love and Hate*. The Natural History of Behavior Patterns. Holt, Rinehart and Winston, New York.

Eibl-Eibesfeldt, I. (1975): *Ethology—The Biology of Behavior*. 2nd Edition. Holt, Rinehart and Winston, New York.

Ekman, Paul (Ed.) (1973): *Darwin and Facial Expression* Academic Press, Inc., New York and London.

Ekman, Paul and Friesen, W. C. (1975): *Unmasking the Face*. Prentice-Hall, Inc., Englewood Cliffs, N.J.

Ekman, P., Friesen, W. C., Ellsworth, P. (1972): *Emotion in the Human Face*. Pergamon Press, Inc., New York.

Etkin, W. (Ed.) (1964): *Social Behavior and Organization Among Vertebrates*. The University of Chicago Press, Chicago.

Evans, Richard I. : *Konrad Lorenz: The Man and His Ideas*. Harcourt, Brace and Jovanovich.

Ewer, R. F. (1973): *The Carnivores*. Cornell University Press, Ithaca, N.Y.

Ewer, R. F. (1968): *Ethology of Mammals*. Plenum Press, New York.

Fischer, F. (1972): *Der Animale Weg*. Verlag Artemis Zurich.

Fletcher, Ronald (1957): *Instinct in Man*. International Universities, Inc.

Fraenkel, F. S. and Gunn, D. L. (1961): *The Orientation of Animals*. Dover Publications, New York.

Fraser, Andrew F. (1974): *Farm Animal Behavior* William Wilkinson Co., Baltimore, Md.

Frisch, K. v. (1955): *The Dancing Bees*. Harcourt, Brace & Co., New York.

Gibson, E. J. (1969): *Principles of Perceptual Learning and Development*. Appleton-Century-Crofts, New York.

Geist, V. (1971): *Mountain Sheep*. Chicago University Press, Chicago.

Goffman, Irving (1963): *Behavior in Public Places* Free Press of Glencoe.

Gottlieb, G. (Ed.) (1973): *Studies on the Development of Behavior and the Nervous System* Vol. 1. *Behavioral Embryology*. Academic Press, New York.

Hafez, E. S. E. (Ed.) (1962): *The Behavior of Domestic Animals*. Williams and Wilkins, Baltimore, Md.

Hall, E. T. (1966): *The Hidden Dimension*. Doubleday, Garden City, N.Y.

Hall, E. T. (1959): *The Silent Language*. Doubleday, Garden City, New York.

Hass, H. (1970): *The Human Animal*. G. P. Putnam's Sons, New York.

Hediger, H. (1971): Ob Tiere Träumen? In: H. J. Schultz (Ed.) *Was Weiss Man von den Träumen?* Kreuz Verlag Stuttgart, Berlin.

Hediger, H. (1950): *Wild Animals in Captivity*. Dover Publications, New York.

Hediger, H. (1955): *The Psychology and Behavior of Animals in Zoos and Circuses*. Butterworths Scientific Publications, London.

Hediger, H. (1968): *Man and Animal in the Zoo*. Delacorte Press, N.Y.

Hendrichs, H. and H., (1971): *Dikdik und Elefanten* Piper Verlag, Munich.

Heinroth, O. and M. (1924–1933) *Die Vögel Mittel-europas*, Vol. I-IV, Verlag Harri Deutsch, Frankfurt a.m. und Zurich.

Heinroth, O. and K. (1958): *The Birds*. University of Michigan press, Ann Arbor.

Heinroth, K. (1971): *Oskar Heinroth: Vater der Verhaltensforschung*. Wiss. Verlagsges. Stuttgart.

Henry, J. P., Stephen, P. M., Ciaranello, R. D., Santisteba, G. A.: The induction of fixed hypertension and arteriosclerosis by prolonged social environmental stimulation. (In Prep.)

Henry, M. P., Ely, D. L., Watson, F. M. C., Stephen, P. M.: Ethological methods as applied to the measurement of emotion. (In. prep.)

Hess, E. H. (1973): *Imprinting*. Van Nostrand Reinhold Co., New York.

Hinde, R. A. (1970): *Animal Behavior*. McGraw-Hill Book Co., New York, 2nd edition.

Hinde, R. A. (1974): *Biological Bases of Human Social Behavior* McGraw-Hill Book Co., New York.

Hirsch, J. (1967): *Behavior-Genetic Analysis*. McGraw-Hill Book Co., New York.

Holst, D. v. (1973): Sozialverhalten und sozial Stress bei Tupajas. *Umschau 1*: 8–12.

Holst, Erich v. (1973): *The Behavioral Physiology of Animals and Man*. Vol. I and II. University of Miami Press, Coral Gables., Fla.

Hutt, S. J. and Hutt, C., (1970): *Direct Observation and Measurement of Behavior*. Charles C. Thomas, Springfield, Il.

Johnson, Roger N. (1972): *Aggression in Man and Animals*. W. B. Saunders Co. Philadelphia.

Klein, R. D. (1970): Food selection by North American deer and their response to over-utilization of preferred plant species. In: A. Watson (Ed.) *Animal Populations in Relation to their Food Resources*. British Ecological Society Symposium Number Ten.

Klinghammer, E. (1969): Factors Influencing Choice of Mate in Altricial Birds. In: H. W. Stevenson, et al (Eds.) *Early Behavior*. Wiley & Sons, New York.

Klopfer, P. H. and Hailman, M. P. (Eds.) (1972): *Control and Development of Behavior*. Addison-Wesley Publishing Co., Reading, Mass.

Klopfer, P. H. and Hailman, M. P. (Eds.) (1972): *Function and Evolution of Behavior*. Addison-Wesley Publishing Co., Reading, Mass.

Klopfer, P. H. (1973): *Behavioral Aspects of Ecology* Prentice-Hall, Englewood Cliffs, N.J.

Kummer, H. (1971): *Primate Societies: Group Techniques of Ecological Adaptation*. Aldine-Atherton, Chicago.

Kuo, Z. Y. (1967): *The Dynamics of Behavior Development* Random House, New York.

Kruuk, H. (1973): *The Spotted Hyena*. University of Chicago Press, Chicago.

Lack, D. (1968): *Ecological Adaptations for Breeding in Birds*. Methuen, London.

Levi, L. (Ed.) (1971): *Society, Stress and Disease* Vol. 1: *The Psychosocial Environment and Psychosomatic Diseases*. Oxford University Press, London and New York.

Leyhausen, P. (1960): *Verhaltensstudien an Katzen* 2nd Ed., Verlag Parey, Berlin and Hamburg.

Lindauer, M. (1961): *Communication Among Social Bees* Harvard University Press, Cambridge, Mass.

Lorenz, K. (1952): *King Solomon's Ring*. Crowell, New York.

Lorenz, K. (1955): *Man Meets Dog*. Crowell, New York.

Lorenz, K. (1965): *Evolution and Modification of Behavior*. Phoenix Science Series. The University of Chicago Press, Chicago.

Lorenz, K. (1967): *On Aggression*. Bantam Books, New York.

Lorenz, K. (1971): *Studies in Animal and Human Behavior*. Vol. (1970) and Vol. II (1971). Harvard University Press, Cambridge, Mass.

Lorenz, K. and Leyhausen, P. (1973): *Motivation of Human and Animal Behavior*. Van Nostrand Reinhold Co., New York.

Manning, A. (1971): *An Introduction to Animal Behavior* Addison-Wesley Publishing Co., Reading, Mass.

Marler, P. and Hamilton, W. J. (1966): *Mechanisms of Animal Behavior*. John Wiley and Sons, Inc., New York.

Mayr, Ernst (1963): *Animal Species and Evolution*. Belknap Press of Harvard University Press, Cambridge, Mass.

McGill, T. E., (Ed.) (1965): *Readings in Animal Behavior* Holt, Rinehart and Winston, New York.

Meyers, J. S. and Bowman, R. E. (1972): Rearing experiences, stress, and adrenocorticosteroids in the rhesus monkey. *Physiol. Behav.* 8: 339–344.

Morris, D. (1969): *The Human Zoo.* McGraw-Hill, New York.

Murie, Adolph (1944): *The Wolves of Mount McKinley* United States Gov. Printing Office, Washington, D.C. U.S. Dept. of Interior, National Park Service, Fauna Series 5.

Portmann, A. (1961): *Animals as Social Beings* Viking Press, New York.

Primbram, K. H. (Ed.) (1969): *On the Biology of Learning.* Harcourt, Brace and World, Inc., New York.

Roe, A. and Simpson, G. G. (1958): *Behavior and Evolution.* Yale University Press, New Haven.

Roeder, K. D. (1963): *Nerve Cells and Insect Behavior* Harvard University Press, Cambridge, Mass.

Ruesch, Jurgen, and Kees, Weldon (1956): *Nonverbal Communication.* University of California Press, Berkeley and Los Angeles.

Rutter, R. T. and Pimlott, D. H. (1968): *The World of the Wolf.* J. B. Lippincott & Co., Philadelphia and New York.

Rutzler, W. (1972): *Rotwild.* Bayerischer Landwirtschaftsverlag.

Schaller, G. B. (1972): *The Serengeti Lion.* Chicago University Press, Chicago.

Scientific American (1967): *Psychobiology.* W. H. Freeman & Co., San Francisco.

Scott, John Paul and Fuller, John L. (1965): *Genetics and the Social Behavior of the Dog.* University of Chicago Press, Chicago.

Sluckin, W. (1965): *Imprinting and Early Learning* Aldine, Chicago.

Sommer, Robert (1969): *Personal Space.* Prentice-Hall, Inc. Englewood Cliffs, New Jersey.

Southwick, Charles H. (1963): *Primate Social Behavior* Van Nostrand Reinhold Co., New York.

Spivak, H. (1968): Ausdrucksformen und soziale Beziehungen in einer Dschelada-Gruppe (*Theropithecus gelada*) im Zoo. Juris Verlag, Zurich.

Stauffer, R. C. (1975): *Charles Darwin's Natural Selection.* Cambridge University Press, Cambridge.

Stebbins, G. Ledyard (1966): *Processes of Organic Evolution.* Prentice-Hall, Inc., Englewood Cliffs, New Jersey.

Stieve, H. (1952): *Der Einfluss des Nervensystems auf Bau und Tätigkeit der Geschlechtsorgane des Menschen.* Thieme Verlag, Stuttgart.

Stokes, Allen W. (1974): *Territory.* Dowden, Hutchison & Ross, Inc., Pa.

Stumpfe, K. (1973): *Der Psychogene Tod.* Hippokrates Verlag, Stuttgart.

Thorpe, W. H. (1963): *Learning and Instinct in Animals* 2nd Ed., Methuen, London.

Thorpe, W. H. (1974): *Animal Nature and Human Nature* Anchor Press/Doubleday, Garden City, N.Y.

Tinbergen, N. (1951): *The Study of Instinct.* Oxford University Press, Oxford.

Tinbergen, N. (1953): *Sosial Behaviour in Animals* Methuen & Co., Ltd., London, and John Wiley & Sons, Inc., New York.

Tinbergen, N. (1960): *The Herring Gull's World* Basic Books, New York, N.Y.

Tinbergen, N. (1965): *Animal Behavior.* Life Nature Library, Time, Inc., New York.

Tinbergen, N. (1968): *Curious Naturalists.* Basic Books, New York.

Tinbergen, N. (1973): *The Animal in its World.* Vol. 1 (1972) and Vol. 2 (1973): Harvard University Press, Cambridge, Mass.

Tobach, E., et al (Eds.) (1971): *The Biopsychology of Development.* Academic Press, New York.

Van Lawick-Goodall, J. and Van Lawick-Goodall, H. (1971): *Innocent Killers.* Houghton Mifffin Co., Boston, Mass.

Von Buddenbrock W. (1958): *The Love-Life of Animals.* Crowell, New York.

Walther, F. (1968): *Verhalten der Gazellen.* Die Neue Brehmbücherei Nr. 373.

Walther, F. (1966): *Mit Horn und Huf. Vom Verhalten der Horntiere.* Paul Parey, Berlin-Hamburg.

Waters, R. H., et al (1960): *Principles of Comparative Psychology.* McGraw Hill Co., Inc., New York, Toronto, and London.

White, T. H. (1960): *The Bestiary: A Book of Beasts.* Putnam, New York.

Whitten, W. K. and Bronson, F. H. (1970): The role of pheromones in mammalian reproduction. In: J. W. Johnston, Jr., D.C. Moulton, A. Turk (Eds.) *Communication by Chemical Signals.* Appleton-Century-Crofts, New York.

Wickler, Wolfgang (1968): *Mimicry.* World University Library, McGraw-Hill Book Co., New York and Toronto.

Wickler, Wolfgang (1973): *The Sexual Code.* Anchor Books, Anchor Press/Doubleday, Garden City, N.Y.

Wilson, E. O. (1975): *SocioBiology.* Harvard University Press, Cambridge, Mass.

Wynne-Edwards, V. (1962): *Animal Dispersion in Relation to Social Behavior.* Hafner, New York.

Dictionary of Ethological Terms

Within a few decades, ethology has created a vocabulary of its own, just like other areas of biology and other sciences have done before. But it cannot be ignored that ethology in particular has suffered great difficulties in creating this common ground for communication, especially where it has been necessary to find truly objective and descriptive terms for certain processes, terms which do not already imply a certain interpretation. This is particularly important in ethology, because even laymen naturally tend to interpret behavioral traits to a far greater extent than they do other processes of life. We need only consider how easily and carelessly people interpret animal behavior as showing maternal love, courage, stupidity, wickedness, or intelligence; in other words, characteristics which an animal—at least in the human sense—certainly does not have.

In the context of historical development, it is easy to see the difficulties in developing a terminology for ethology. In the beginning of ethological research it seemed most convenient to use the two "sources" already present, the realm of everyday language and the technical jargon of psychology. Different terms and concepts were thus taken from both sources and were quickly integrated into the field. As ethology progressed, it became evident that many of these terms could not be applied to animal and human behavior in the same way, after all, or that if they were indeed used to mean the same thing it immediately led to the problem mentioned above, the danger of inaccurate interpretation. Despite these substantial differences and problems, it was inevitable that, once accepted, such terms remained as part of the ethological language.

As a result, many words in ethology today carry a different meaning from that of everyday speech. Striking examples for this are words such as "dummy," "learning," "social," and even the controversial term "instinctive". Similar differences exist between ethology and psychology.

But even within ethology many technical terms —partly for historical reasons i.e. depending on the "school" that a scientist was trained in—are occasionally used in different ways. If one reads carefully such variations in meaning can even be detected in a few of these chapters.

Considering the many books on the market today dealing with ethological topics this state of affairs has led to considerable differences in the way certain technical terms are defined and explained. As a result any reader who has not had some training in ethology may experience a certain confusion.

The following is an attempt to give brief explanations of the main ethological terms. There are actually two aims to this "dictionary". For one, it is intended as a "reference section" so that the reader of this volume has a chance to learn certain terms relatively quickly that he may not be fully familiar with. Secondly, it is designed as a summary of ethological vocabulary, since we may assume that any interested reader will also want to gain a general picture of the "technical language" of this field.

Because of the above-mentioned variations in the usage of certain terms, our dictionary is o course also subject to personal interpretations or preferences, and we are certain that many of our colleagues in the field will occasionally disagree with the definitions. This may —and we are saying this as an "excuse"—be partly due to the fact that our space is necessarily limited, so that we could not delve into all the details and had to dispense with many additional comments that would otherwise have been important for understanding. I have tried, however, to give a brief indication whenever a term is subject to disagreement within ethology or when it is not always used in the same manner.

Furthermore, many technical terms are explained and discussed in detail in individual chapters. More specifically, this dictionary is drawn up in the following way: For each term, we first offer —where applicable—other words with the same or similar meaning. These may already provide a preliminary, catchword type of explanation, especially where the accepted ethological term is fairly technical and where there are other, more common terms with related meanings. Next, we present a brief explanation of the concept, and finally—if necessary for clarification—we give one or several examples.

Certain technical terms from the realm of sensory physiology or neurophysiology, which appeared in the first part of this volume, were omitted from the dictionary for reasons of space because they were used in only one chapter respectively and have therefore already been explained. On the other hand we are presenting terms not directly stemming from the realm of ethology but which are used on several occasions in this volume, and where we have reason to assume that not all readers are immediately familiar with them.

action readiness—motivation
adaptation
The development of traits by—*natural selection*, making the organism more suited to its environment, i.e., increasing its own chances of survival and those of its descendants.
adequate stimulus—receptor
adult
sexually mature, grown up
afference
The total number of nervous excitations leading from the sense organs (including the—*proprioceptors*) to the cultural nervous system (spinal cord and brain). The nervous pathways conducting these impulses are called afferent or sensory nerves.
agression
Attack and threat behavior. A distinction is made between intraspecific and interspecific aggression, depending on whether the conflict takes place between conspecifics or between members of different species (in the latter case, for example, between predator and prey or between competing species). Behavior patterns may be highly diverse for both forms of aggression (—*damaging fights*, —*ritualized fighting*).
aggressiveness
Degree of readiness to attack in an individual or species.
agonistic behavior
General term for all behavior patterns associated with conflict between individuals. This concept includes attack (—*aggression*) and escape.
altruism
Selfless behavior. Opposite of egoism. Behavior by which the individual disregards its own well-being and serves that of another conspecific. Altruism usually occurs in the area of—*brood care*, or parental behavior. Between adult animals, it is commonly seen among state-forming insects, while among vertebrates true altruism is quite rare. Clearcut evidence for this has been obtained only for African wild dogs, dolphins, and some primates. We may assume, however, that altruistic behavior also occurs in the individualized

groups of several other mammals, as well as in birds with a high level of social development.
ambivalent behavior
Behavior patterns consisting of parts belonging to different and often opposing kinds of—*motivation*, with the parts occurring simultaneously or in rapid succession. Thus, for example, many threatening movements contain elements of both attack and escape behavior.
analogy
Similarity of adaptations. Correspondences, as in morphology or behavior, appearing in organisms that are not closely related, but instead deriving from similar adaptations to certain conditions in the environment and with an independent evolutionary history. Examples are the wings of birds and insects, in the area of behavior the suction technique of drinking used by the otherwise diverse groups of pigeons, sand grouses, and weaver-finches.
androgens
General term for the male—*sex hormones*, produced mainly in the testes. The most important androgen is testosterone.
animal psychology
An older term for ethology, which has become somewhat misleading due to the increasingly physiological orientation of ethology. Hence it is no longer used this way. Today, animal psychology is considered a subarea of ethology that deals less with the "normal behavior" of an animal species than with the more individual, subjective (as far as they can be studied at all) phenomena and even the pathological behaviors of animals, in this sense being more a "comparative psychology." Important results in the field of animal psychology have been obtained with zoo and circus animals.
animal sociology
A subarea of ethology that deals with the—*social systems* of individual animal species, and with the behavior patterns used to establish and maintain these social systems.
anonymous association
An animal group in which – as opposed to an—

individualized group – the members do not know each other individually. Examples: Schools of fish. flocks of birds, and migrating herds of mammals.

antagonism
Opposing force. Opposing effects of muscles (for example, flexor and extensor of limbs), hormones (e.g., male and female sex hormones), or nerves (e.g., facilitating and inhibiting effect on the activity of internal organs). The muscles, hormones, or nerves involved are called antagonists. In behavior, too, there are antagonistic elements, such as attack and flight.

appeasement posture(s)
Behavior which inhibits attack. General term for all behavior patterns that may inhibit intraspecific aggression. In—*social animals* they occur during "greeting" between two members of a group or partners of a pair, and facilitate approach. To a large extent their form is often the opposite of a—*threatening movement*. A large number of appeasement postures contain elements of juvenile behavior (e.g., begging postures and movements) or from the sexual realm: In many monkeys, for example, the female invitation to copulate functions as an appeasement posture. A particularly prominent appeasement behavior is the—*submission posture*.

appetitive behavior
Searching behavior. Active striving for a releasing stimulus situation. Here we are dealing with a goal-directed, oriented behavior whose aim is the discharge of a—*consummatory behavior*. In the most simple case, appetitive behavior consists only of a—*axis*. But most of the time it is a variable sequence of movements and orienting behaviors or adjustments which may last for a considerable period. Examples of particularly longterm appetitive behavior are the migrations of various fish and bird species in search of their spawning or breeding grounds.

ARM—releasing mechanism
behavior
The behavior of an animal is defined as the totality of its movements, sound emissions, and body postures; also the externally noticeable changes that serve bilateral communication and can therefore release other behaviors in the partner (changes in color, secretion of odorous substances, etc.).

behaviorism
A branch of psychological research concerned mainly with the relationships between behavior patterns and the conditions under which they occur. For example, it studies the question of how events following a certain behavior (reward, punishment, etc.) reinforce or decrease it. However, the term "behaviorism" is presently applied, not only to the research approach itself, but also to the basic attitude that has emerged from the behavioristic "school" to animal behavior. This approach holds that all behavior is the product of learning, i.e., there are no—*innate* elements. An important criticism sometimes directed at behavioristic methods is that experimenters in this field ignore the—*spontaneous* elements of behavior.

behavior genetics
The study of how innate behavior patterns are inherited, using the methods of genetics. Cross-breeding experiments have been used to demonstrate that the behavior patterns are passed on from one generation to the next in the same way as other body traits or functions.

behavior rudiment—rudiment
behavior system—functional category
behavioral physiology
A subarea of ethology that studies the behavior of an animal "from the inside," investigating the processes within the sense organs and in the central nervous system (spinal cord and brain) which form the basis for a certain behavior.

biological clock
Also internal clock. Sense of time. The ability to determine time independent (endogenous) of external stimuli (e.g., position of the sun). This capacity is based on the fact that certain phenomena of life show regular and recurrent fluctuations, or phases (—*periodicity*). The "seat" and the exact physiological bases of biological clocks are still unknown. The biological clock plays an important role in the "language" of bees and in the orientation of migrating birds and other animals (—*biorhythmicity*).

bio-rhythmicity
An area of biology investigating the rhythms (daily, annual periods etc.) of an organism and their underlying physiological processes. Since this kind of—*periodicity* has a strong influence on behavior as well, the study of biorhythms has considerable bearing on ethology (—*biological clock*).

biotope
Living space. A sphere offering certin living conditions essential for the existence of plant and animal species adapted to this particular environment. Each biotope is inhabited by a number of organisms typical for this environment, which together are called a biological community or biocene.

brood care
Behavior patterns to protect, feed, and nurture the eggs or young. These include transfer of food, skin grooming, removing feces and other impurities, camouflage (e.g., transporting the young to a protected place), as well as defending and leading the young. In egg-laying animals, brood care activities also include the incubation of eggs in birds and the active provision of oxygen through fanning movements in some fishes. Aside from the parents' direct actions, brood care offers the young – especially with long periods of dependence – the very important opportunity to learn from their parents. In a very general way scientists speak of brood care only if there is a direct contact between parents and young (—*provision for young*).

brood parasites
Animals that find other host species to raise their young (*brood care*) entirely or partly. Brood parasitism requires a high level of adaptation, which may involve the appearance of the eggs and young or even the behavior

of parents and young animals. Thus it is in no way more "primitive" or "simple" than raising one's own offspring. The most familiar example of a brood parasite is the cuckoo.

Caspar Hauser (animal) experiment
An animal raised under conditions of deprivation so that it could not have the experiences necessary for normal development of behavior. Complete deprivation conditions can never be attained, since even with total isolation in the dark the animal can still make certain experiences, at least with its own body. Depending on the purpose of the study, the young animal is often deprived only of certain stimuli (e.g., social, i.e., originating with conspecifics). Deprivation (or isolation) experiments are used for investigating innate behavior elements.

caste
Group of individuals of a species that have certain tasks and are adapted to these in morphology and behavior. True castes are found primarily in state-forming insects (e.g., reproductives, workers, and soldiers). Formation of castes is associated with the—*division of labor* within the state.

castration
Removal of the—*gonads*. In ethological studies, castration is an important method for examining the effects of sex hormones on behavior.

cell membrane
Skin of the cell. An enclosing layer that separates the cell from its surroundings. It displays the same basic structure throughout the entire animal kingdom.

chemoreceptor
Sensory cell responding to chemical stimuli. The chemical senses include, among others, those of taste and smell (olfaction).

circadian rhythmicity
Approximate daily rhythmicity, day-night rhythm, 24-hour rhythm. The daily change in behavior and in other body functions. In many cases this comes about by way of the organism's own, internal (endogenous) —*periodicity* (—*biological clock*), which is temporally synchronized with the environment's periodicity by way of a—*Zeitgeber*.

cocoon
Protective covering for eggs, larvae, or resting stages. Egg cocoons are constructed by the mother, larva cocoons by the larva itself (e.g., by insect larvae for their pupal resting stage).

coition—copulation

comfort behavior
In the ethological literature, this term was used for two different groups of behavior patterns: body grooming (grooming and scratching, movements, shaking oneself, scrubbing oneself), and movements associated with body metabolism, especially with the intake of oxygen (stretching movements, yawning). But there is no unanimous agreement on the meaning of this expression. Some ethologists use it in the broadest sense, others restrict the term to only a few of these behavior patterns.

communication

Transfer of information. General term for signals exchanged between animals. Depending on the sense modality involved, we speak of visual, auditory, or chemical communication. Communication between members of the same species is called intraspecific, between members of different species (e.g., between partners in a symbiosis) interspecific.

conditioning
Attaining a training effect or a conditioned—*reflex* with repeated reward or punishment.

conditioned reflex—reflex

conflict behavior
Behavior patterns occurring when two mutually exclusive behavior tendencies (e.g., attack and escape) are activated simultaneously. These include —*ambivalent behavior*, —*displacement activities*, and—*redirection activities*.

consummatory behavior
A—*fixed action pattern* occurring at the end of a series of—*appetitive behaviors* and resulting in a shorter or longer interruption of the appetitive behavior. In other words, it "consumes" the corresponding energy and therefore allows the "satisfaction" of a "drive." Examples: feeding movements (breaking down food and swallowing it) are consummatory behaviors in the functional category of seeking food, while the discharge of copulatory movements is a consummatory behavior in the area of sex. These consummatory acts may be preceded by long series of appetitive behaviors aimed at finding and seizing the prey, or at—*synchronization* of activities with those of a mate.

contact animals
Animal species lacking an—*individual distance*.

contact behavior
Conspecifics moving closely together. Many contact animals strive to gain as much physical contact with conspecifics as possible when resting. The original purpose of this behavior was probably too keep warm, but in many animal species it has acquired an additional or complete social function (pair-bonding and group bonding).

copulation
Coition. Joining of the male and female sex organs. In mammals and some members of other classes of vertebrates, the male copulatory organ (penis) introduced into the female genital opening (vagina). In most birds, which have no penis, the genital openings are merely pressed together. Many lower marine animals do not practice copulation, instead depositing their sex products (egg and sperm cells) into the sea. In these cases fertilization take place outside of the female's body.

courtship
Precopulatory or precoital behavior. More specifically, courtship is a general term for all behavior patterns leading to or possibly leading to mating. However, such behavior patterns do not always result directly in copulation, instead often serving the process of—*pair-forming* or simply reinforcing a pair bond. Furthermore, the

boundaries between precoital behavior and that serving only the latter function are often unclear. For these reasons the term courtship is often used in a broader sense to include all behavior patterns of precopulation pair-formation, and—*pair-bonding*. On the other hand, this term in particular is sometimes applied with very different meanings by various scientists. This is why an author writing on courtship often indicates his own meaning at the beginning of his publication.

courtship feeding
The (real or pretended) transfer of food from one partner to the other as a courtship or pair-bonding behavior.

critical period—sensitive phase—optimal period
cuticula
A protective surface layer secreted by the epidermis. This layer can be strongly developed especially in crabs and insects (e.g., beetles) where it also serves as an exoskeleton.

daily rhythms—circadian rhythmicity
damaging fight
A form of combat where – in contrast to—*ritualized fighting* – the opponents do not simply aim at driving the other off but also try to injure or even kill each other. As a rule, damaging fights occur only between members of different species, e.g., when two competing species of roughly similar "strength" fight for a certain piece of food, or when an (armipotent) prey animal defends itself against a predator. Damaging fights are very rare among members of the same species. Here, they occur mainly in social species (—*social animals*) between members of different groups.

decrement
The decrease in nervous excitation with increasing distance from the locus of stimulation, i.e., with the length of nervous conduction.

deprivation experiments—Caspar Hauser (animal) experiment
directional specificity
The property of a receptor cell to respond to stimuli from a certain direction only.

displacement activity
An "unexpected" movement occurring outside of the behavior sequence for which it was initially developed. This kind of activity occurs in conflict situations, and is usually "ineffective" in the sense that it does not fulfill the function to which it is actually adapted. Examples: displacement pecking in fighting domestic cocks, displacement fanning in courting sticklebacks. The German term "Übersprung" (literally "spark-over") was based on the original assumption that two opposing and mutually exclusive behavior tendencies (e.g., attack and escape) inhibit each other so that the "behavioral energy" sparks over to a third behavior system.

display behavior
Intermixed with—*courtship* movements, this is a —*threatening behavior* that repels same-sexed rivals and attracts members of the opposite sex. It frequently plays a role in—*pair-formation*.

distance-maintaining animals
Species that do not normally tolerate physical contact with conspecifics but always keep a certain distance from each other, i.e., maintain an—*individual distance*. Intrusion is permitted only for copulation and brood care.

division of labor
Assigning tasks. Different tasks. Differential distribution of activities within a social unit. This can be observed most prominently in state-forming insects. In many cases the activity of an animal is based on its morphological structure (—*caste differentiation* in social insects), its sex, or its age. Where the latter applies, the same animal may fill various roles in the course of life (e.g., in honeybees). In vertebrates, division of labor occurs primarily in social mammals. In some brood-caring vertebrates there is also a certain division of labor between the sexes (e.g., between guarding and feeding the young, or between gathering the food and distributing it).

domestication
Change into a domestic animal. The breeding of animals under human care, thus replacing—*natural selection* with an artificial selection process which in some instances may lead in a completely different direction. As a rule, animal species that have been subject to this altered selection pressure for several generations differ from the wild ancestral form in a number of traits.

domestication, results of
Hereditary changes in the behavior and morphology of a domestic animal as compared with its wild ancestral form. Observations on very "young" domestic animals, such as the budgerigar which has been bred as a house bird for several decades, have shown that such features of domestication can sometimes appear after only a few generations of in captivity.

dominance
Superior status in the social order, or—*dominance hierarchy*. A high-ranking animal is called dominant.

dominance hierarchy
Rank order, social hierarchy, status system. The ordered distribution of rights and duties within a group. Strong and experienced animals are at the top of the dominance hierarchy and have certain privileges (priority at the food or drinking site, access to preferred sleeping sites mating priorities), but may also have to perform certain duties (leadership role, guard duties, defense). The highest-ranking group member is called the alpha animal, the lowest one the omega animal. Individual superior to others in the group with respect to status are considered dominant to these others. Dominance relationships occur in many social species, and contribute to the stability of relations within the group by limiting conflicts essentially to the formation or alteration of a status hierarchy (e.g., with the integration of maturing young), while differential privileges and rights are respected without combat at other times.

drive
An older expression for the specific readiness of an

animal to perform a certain behavior. However, because of its highly varied usage in everyday language, this term is imbued with considerable connotation and is hard to define. Hence it is rarely used in ethology today, having been largely replaced by the term—*motivation*, although some ethologists hold that the two concepts do not mean exactly the same. Occasionally, a distinction is also made between "drive" and "impulse": In this case, drive is used to denote the latent condition (resting state), while the impulse (or urge) is the actual state of arousal.

dummy (model)

Ethologists use this term somewhat differently from common language, where it means to copy some object and make it look as realistic as possible. In ethological studies we may also find completely "unnatural" models. Here, a dummy is any kind of imitation or reproduction used to test the behavioral response of an animal, with the aim of discovering the characters of whatever—*sign stimuli* and—*releasers* are important for that particular behavior. For example, often the experimenter will offer only reproduced parts of another animal (bill dummies, head dummies), or he will alter various features of the object presented (increasing or reducing size, changing shape or color). Even completely unnatural objects (e.g., colored wooden balls or cubes) may at times be used as dummies. Furthermore, the experimenter may use audiotape recordings of relevant sounds, or artificial odors as auditory or olfactory dummies. Agains we can alter individual traits of the natural model in order to test the effective characters of the presented stimuli. Dummy experiments have played an important part in the methods of ethology ever since they were first used.

echolocation

Echosounding; gauging, taking bearing, or taking direction by means of echoes. The ability to perceive obstacles or prey objects with echoes produced by the emission of locating sounds. Echolocation is highly developed especially in bats. It has also been demonstrated in various insectivores (e.g., shrews) and aquatic mammals (especially dolphins), as well as in some birds, specifically the South American oil bird and the Asiatic ice fishes. The locating sounds of most species lie in the supersonic range of frequencies, i.e., above the range of human hearing.

eco-ethology

A recent area of ethology that studies the relationships between the behavior of an animal species and the conditions of its animate and inanimate environment. The main focus is on parallel behavioral adaptations occurring within certain biotopes even when the species concerned are not closely related to each other (—*analogy*).

ecology

Study of the environment. The study of the interrelationships between organisms and their animate and inanimate environment.

effector organ

Executive organ. General term for muscles, glands, and pigment cells. Nerves leading from the central nervous system to the effector organs are called efferent nerves.

efference

Opposite of—*afference*. The totality of nervous excitations leading from the central nervous system (spinal cord and brain) to the—*effector organs*. The nervous pathways conducting these impulses are called efferent or motoric nerves.

egg follicle—ovulation
embryonal development—ontogeny
endogenous

Originating inside the organism; intrasystemic, internal.

entraining agent—Zeitberger
enzyme

An effective substance that facilitates certain chemical transformations in the body, without being consumed in the process. There are a large number of enzymes in the body, playing an essential role in all metabolic processes (—*metabolism*) (breathing, digestion, excretion, etc.).

epidermis

The external skin. The uppermost layer(s) of cells on the body. In invertebrates the epidermis consists of one layer only, while all vertebrates have several layers of epidermal cells. Underneath the epidermis is the underskin, or hypodermis, which consists of several layers in vertebrates as well as in most invertebrates.

escape distance (flight distance)

Also flight distance. The limit beyond which, if intruded upon by a certain object such as a predator or rival, the animal will flee. It may vary considerably from one animal species to another, as well as between members of one species depending on previous experience with this object.

estrogens

A group of female sex hormones produced mainly in the ovary. They control the growth of the egg, as well as providing for the development and retention of female sex characteristics and of female reproductive behavior. The most important estrogen is the follicle hormone (estradiol).

ethogram

Inventory of behavior, action catalog. Includes, as much as possible, all the behavior patterns displayed by a certain species. The ethogram forms the basis for all scientific study of behavior.

evolution

Phylogenetic development. The development of more specialized forms to higher levels of organizations from earlier and more rudimentary stages. The most important "engines" of evolution are—*mutation* and—*natural selection*.

expressive behavior

Behavior patterns acting as a—*releaser* for communication between members of a species (and occasionally between different species). Examples: movements of threat and appeasement, begging movements of young animals.

external stimulus

Sensory stimulus originating outside of the body. It is perceived by way of—*exteroreceptors*.

exteroceptor

A sensory cell that – in contrast to a—*proprioceptor* – responds to sensory stimuli coming from outside the body.

exogenous

Originating outside the organism; extrasystemic, external; environmental.

facilitation

The mutual increase of nervous excitation. Scientists differentiate between temporal and spatial facilitation. In temporal facilitation, a prior stimulus (or stimuli) increases the capacity of a nerve cell to respond to a stimulus (i.e., facilitates the production of impulses); in spatial facilitation, simultaneously incoming excitations that would have remained ineffective individually may together produce an effect on some—*effector organ*. Also—where the presence of another animal effects an increase in a behavior, e.g., Seeing another eating stimulates eating.

fertilization

Union of nuclei of egg and sperm cells.

filtering

Out of all the stimuli impinging from the environment, this process involves selecting those (—*sign stimuli*) that are essential for a species' survival, and where the animal must respond with appropriate behavior patterns. Filtering takes place in the sense organs (peripheral filtering) as well as in the central nervous system (central filtering). As a rule, this process occurs in a number of succeeding steps.

fixed-action-pattern

Movement normative of a species. *Species-specific* movement sequence. Part of an—*instinctive behavior*. A relatively rigid or fixed series of movements ordered in time and space (constant in form). It is triggered by external stimuli, which also affect intensity and orientation of the behavior (—*taxis component*). By contrast, its discharge pattern or sequence, i.e., the type of movement, is independent of external stimuli and is fixed in a species-typical way. For that reason, fixed action patterns may be used to identify and classify species in the same way as morphological traits. Fixed action patterns are particularly prominent in grooming and courtship activities. (See also model action pattern).

follicle hormone—estrogens

frustration

There are considerable differences in the usage of this term between ethology and psychology, and even within these scientific areas. On the one hand, it refers to the prevention or obstruction of a goal-directed activity, on the other hand to the motivational or emotional state (—*motivation*) resulting from such failures. In other words, the concept of frustration is used to mean the process as well as the resulting condition. However, its range of application is narrower in ethology than in psychology: Ethologists will usually speak of frustra-

tion only when external factors prevent an animal from completing a behavior pattern, i.e., when the goal of an action already begun becomes unattainable. By contrast, the human psychologist uses this term for every kind of obstruction, i.e., for actions already in process as well as those not yet begun, and refers to tangible as well as imaginary goals.

functional category

Behavior system, functional group, functional classification of behavior, or functionally related activities. Blanket term for behavior patterns with similar or identical function and effect, e.g., locomotion, consumption of food, courtship, brood care, or aggression.

functional group—functional category

gamete

Germ cell, reproductive cell. In most animals there are two types of gametes: motile sperm cells (*spermatozoa*) in males and sedentary stationary egg cells in females. During fertilization, the nuclei of one sperm and egg cell fuse to form the fertilized egg cell, called a zygote.

gene pool

All the genomes of a—*population*, i.e., all the genetic traits present within a population unit.

genome

genetic substance, genetic material. All the hereditary information stored in the cell nucleus in form of genome molecules.

germ cell—gamete

germ gland—gonad

gestagens

Group of female sex hormones formed in the ovary by the corpus luteum, in mammals also in the—*placenta*. They play a part in the regulation of sexual processes, but as a rule their effects come about later than those of estrogens and thus they also influence reproductive processes following copulation (preparing the uterus for attachment of the egg, brooding activities, feeding behavior). The most important gestagen is progesterone.

gland—secretion

gonadotrophins

Hormones produced by the hypophysis, influencing the activity of the—*gonads* and thus the secretion of—*sex hormones*.

gonads

Germ gland. The male gonads are called testes, the female ones ovaries. Gonads have two main functions: They produce the reproductive cells (sperm and eggs) and are also the main production centers for the *sex hormones*.

habituation

The gradually decreasing tendency of an animal to respond to repetitive stimuli that have proved meaningless. Habituation can be tested particularly well with dummy experiments. It may be considered as one of the simplest forms of learning.

harem

One-male or unimale group. The stable and permanent

association of one male with several females. Harems are one of several forms of —*polygamy*.

home range—range

homology

Evolutionary congruency. Correspondences in morphology and behavior of organisms that are based on descendance from the same ancestor. While in many cases homology of organs can be recognized fairly readily through the fossilized presence of intermediate forms, behavioral homologies must be inferred with a great deal of difficulty by means of a number of characters and regularities.

hormone

Organic substances formed in the endocrine glands, secreted into the blood stream, and transported to other parts of the body where they have different effects (controlling metabolic and growth processes). The behavior of animals, too, especially those patterns associated with reproduction, are influenced by hormones in a number of ways.

hormone treatment

Artificial introduction of hormones. This may be done in different ways: oral (i.e., by adding to food or drinking water), through injection into the bloodstream (intravenous) or into the muscles (intramuscular), or through implantation of hormone crystals underneath the skin or into the brain.

hospitalism

General term for all developmental problems in the realm of behavior arising through social isolation in early childhood, e.g., apathy, restless movements, movement stereotypies, or inability for normal social behavior. The most extreme hospitalism occurs in animals raised without a mother. This term is used by both ethologists and human psychologists in the same way.

human ethology

A recent branch of ethology that attempts to study human behavior with ethological methods. Its main focus is the phylogenetic and genetically predetermined regularities and ranges of variation in human behavior.

hypophysis

Pituitary gland. The body's most important control organ, located at the underside of the diencephalon. It produces two kinds of hormones: body hormones, which travel directly to their effective sites, and "glandotropic" hormones, which influence the activity of other hormone glands (e.g.,—*gonads*), thus having only an indirect effect on the rest of the body.

imitation

Adopting movements or sounds from another animal by observation or listening.

imprinting

A relatively rapid learning process that takes place during a short, sensitive period in early youth. It has a prominent—*sensitive phase* and a stable, often—*irreversible* effect. Young animals learn their own species (or other species or models if exposed to them) with whom they will mate as adults. Also applies to habitat imprinting.

incidence threshold—threshold

individual distance

In—*distance-maintaining animals*, this is the maximum distance at which conspecifics will approach each other. Intrusion into this space results in aggression. The individual distance is species-specific, but may vary even within a species depending on season and time of day. Furthermore, there may be differences between mated and unmated animals, between males and females, or between young and old animals. In some situations (copulation, brood care, fighting) such a distance may be reduced to zero.

individualized group

An animal group where the members know each other individually (by smell, appearance, or voice). Such individual recognition is prerequisite for the formation of a—*dominance hierarchy*. Individualized associations are known only for the highest level of vertebrates, especially for many species of birds and mammals.

infantilism

Occurrence in adults of behavior patterns characteristic of young animals. Infantile behaviors are displayed most often in the functional category of—*courtship*, e.g., consisting of begging postures and movements usually seen in young, or of invitation to mate in adult females. Because they inhibit aggression, infantilisms are also a common element of—*appeasement postures*.

inhibition

Blocking a behavior pattern by means of certain external or internal stimuli.

innate

One of the most controversial terms in ethology. It states that a certain behavioral trait (e.g., the discharge of a movement or the recognition of an object, such as the proper food or a suitable mate) is basically determined by genes. It does not say that this trait must be present at birth and that the environment has no influence at all on its development. Instead – as with the ontogenetic growth (—*ontogeny*) of all other traits and properties in an organism – there is a constant interplay between the genes and environment. Thus, the term "innate" or "inborn" should be interpreted in the same way as with, for example, a "congenital heart defect." This characteristic, too, often appears at a later stage or perhaps only under certain circumstances.

innate releasing mechanism (IRM)—releasing mechanism

innate schema

An older expression for "innate releasing mechanism" (IRM). This term, however, may give the mistaken impression that we are dealing with a purely "schematic" behavioral response. For that reason it is no longer in common usage.

instinct

By far the most controversial term in ethology. As in the past, its usage varies tremendously with different scientists who each have their own interpretation. Since it also appears in everyday language (again with different meanings) and may here even be associated with a certain value judgment, most ethologists now avoid even using this hard-to-define concept. In ethology, instinct is most frequently understood as an—*innate* behavior mechanism expressed in ordered movement sequences (so called—*fixed action patterns*) and often triggered with certain stimuli by way of a—hypothetical *releasing mechanism* (RM). Within this context, an "instinctive act" (or instinctive behavior) is an ordered sequence of fixed action patterns. G. W. Barlow has proposed to replace the term instinctive act, conflictual in so many ways, with the neutral expression "modal action pattern."

instinctive
This term is almost as controversial and hard to define as —*instinct*. Nonetheless it is used far more often. In general, instinctive behavior patterns are defined as those based on genetic factors. In this sense, then, the word means the same as —*innate*.

internal clock—biological clock

insemination
The process of placing sperm into the female genital tract. This leads to fertilization—the union of egg and sperm cells.

intention movement
Indicative movement, mood behavior. An incomplete act preliminary to a behavior pattern (e.g., repeatedly pecking at nesting material before starting the actual nestbuilding activity). It expresses the momentary mood of an animal and may therefore serve bilateral communication between conspecifics by indicating readiness for a certain behavior. Thus, in many bird species members of a group will make intention movements prior to flying, raising the flying mood of the whole group to a higher "pitch" and perhaps causing the entire flock to fly up at the same time.

internal stimulus
Sensory stimulus originating inside the animal's own body. It is perceived by—*proprioceptors*.

interoceptor
Sensory (or—*receptor*) cell responding to the conditions or changes within the animal's own body (—*internal stimulus*). Among other things, interoceptors convey information about changes in blood pressure and body temperature, degree of stomach filling, or oxygen content of the blood. Interoceptors associated with the muscles are called—*proprioceptors*.

interspecific behavior
Interactions between members of different species.

intraspecific behavior
Interactions between members of the same species. This term is almost but not completely indentical with—*social behavior*.

isolation experiments—Caspar Hauser (animal) experiment—deprivation experiment

irreversibility
Cannot be extinguished. The fixed and permanent nature of learning in some cases of imprinting.

juvenile
Youthful, prepubertal, not yet sexually mature.

latency period
The time interval between signal and response, i.e., between the onset of a stimulus and the appearance of the response released by it (e.g., a behavior pattern or a nervous impulse).

learning
General term for all behavioral changes affected by experience, e.g., reward or punishment. In ethology as well as human psychology, this term may be used in a much broader sense than in common language. Thus, in addition to actual learning, i.e., associating a stimulus with a response, this term may also include general environmental influences such as the stimulating effects of varied surroundings on a certain behavior.

learning disposition
Genetically predetermined propensity for learning. Many animals learn certain things from their environment that are of special importance (e.g., sign stimuli belonging to food or predators) particularly fast, while learning less significant things only slowly or not at all. —*Social species* often have a special capacity for learning to recognize members of their own group individually ("social intelligence,"—*individualized group*).

libido
Sexual arousal.

liquormone
Hormone-like substance produced by cells in the wall of the cerebral ventricle and secreted into the ventricular fluid, the liquor. These substances are significant for behavior in that they can be assimilated by nervous receptor cells at other points in the ventricular wall. In this way liquormones influence the activity of these cells and thus also behaviors under their control.

living space—biotope

localization
Determining the site of important reference points (sources of food, conspecifics, predators).

lousing—social grooming

lowering of threshold
The increasingly greater ease with which a behavior pattern may be released, growing with the length of time since it last occurred. This change is based on lowering of the—*threshold* for the corresponding stimuli.

macrosmatic animals
Animals with well-developed capacity for smell. The classification into macrosmats and microsmats is applied particularly to mammals, where – in contrast to, e.g., birds – most species have a good sense of smell. Among mammals it is mainly the primates (including man) that are microsmats.

marking behavior
Behavior patterns used to indentify a—*territory*. These include conspicuous display in one's own territory (visual marking), singing and other sounds made by frogs, birds, or certain monkeys (auditory marking), and depositing —*scent marks* (olfactory marking). In addition to marking their territory, many mammals also deposit marks on conspecifics, e.g., their offspring or mate.

maturation

The development of a behavior pattern without practice. A behavior matures if during—*ontogeny* its performance improves even when the animal has no opportunity to practice it. We can demonstrate maturation processes if we prevent young animals from performing a behavior (e.g., flying) for a while, then comparing them at a certain age with conspecifics raised under normal conditions.

maze

Labyrinth. A system of passages where only one leads to the goal (with a reward) while the others come to dead ends. Maze experiments have played a very important role in the psychology of learning, namely in the "school" of—*behaviorism*. Here, an important measure for learning is the number of trial runs that an animal needs to find the goal without errors, i.e., without entering the blind alleys. The degree of difficulty in running the maze is determined by the number of alleys branching off. The most simple kind of "maze" is a T-shaped alley requiring only one decision. Most maze experiments were carried out with rats.

mechanoreceptors

Sensory cells responding primarily to mechanical stimuli, i.e., receptors that are mechanosensitive. Mechanical stimuli include, for example, pressure, drafts, sound waves, gravity, and acceleration.

membrane—cell membrane

metabolism

General term for all chemical conversions (e.g., digestion, respiration, excretion) taking place inside an organism.

microsmatic animals

Animals with poor capacity for smell (—*macrosmatic animals*).

mimesis

Camouflaged appearance. Resemblance to single objects in the environment (stones, leaves, bird feces, flowers, branches) that are ignored by other animals. Best-known example: the "walking sticks".

mimicry

Imitating signals. The imitation of an animal or parts of an animal to gain a biological advantage. The most familiar kind is Batesian mimicry (false warning coloration), where a protected (armipotent or bad-tasting) species is imitated by an unprotected one, such as the wasp mimicry of various insects. In Peckhamsian mimicry (attack mimicry) it is the predator species that imitate the animals normally eaten, in turn, by their prey (e.g., through worm-shaped body appendages), luring the prey to approach them (e.g. anglerfish).

modal action pattern—fixed-action-pattern

monogamy

Marriage to one partner only. True monogamy, with the mating bond sometimes lasting years or throughout the animals' entire life, is found among members of all classes of vertebrates and occasionally among invertebrates (e.g., some crabs). Monogamy is particularly common among birds. By contrast, it is very rare among mammals. Clear-cut evidence has been found only for a few canid carnivores (e.g., black-backed jackal), some rodents (e.g., agoutis), some hoofed animals (e.g., dikdik and klippspringer), and various primates (e.g., marmosets), titi monkeys and gibbons).

mood—drive—motivation

mood transfer

"Contagion." The tendency common among social animal species to do the same things at the same time. Example: a satiated hen is "carried along" by food-pecking conspecifics to start pecking at grains again. Transfer of mood is an important means of synchronizing behavior within a group (—*synchronization*).

motivation

Mood, readiness for behavior, "Drive." The readiness or tendency of an animal for a certain behavior. This is determined by a number of factors, which include—*external stimuli*, —*internal stimuli*, —*hormones*, and the history of the behavior in question (i.e., the time elapsed since it last occurred). Every behavior pattern has a certain motivational value at all times. This value declines when the act is performed (at times erratically), then rises again. Externally, a change in mood can be recognized by the fact that the animal responds to the same stimulus differently at different times. However, the concept of motivation, which has been the subject of countless lively discussions, is not always used in the same way by ethologists. Many scientists also include maturation processes, habituation, learning processes, and fatigue effects, because these factors, too, may influence a behavior readiness. Other ethologists reject this broader interpretation.

motivational analysis

The investigation of an animal's behavior readiness, i.e., the external and internal factors that cause an animal to behave in a certain way. Motivation cannot be measured and so it must be determined from the behavior itself (strength and frequency of an act). The most important requirement for a precise motivational analysis is that the external conditions under which the animal is tested remain constant, or that each time only one external factor is changed in a precisely predetermined manner.

motor nerves—efference

movement stereotypy—stereotypy

mutation

Sudden change in the genetic substance of a cell. Under natural conditions, mutations occur "spontaneously," i.e., without any recognizable, external influence. In an experiment they may also be induced artificially, e.g., through short-wave radiation or certain chemical substances. Mutations provide the "raw material" for the process of—*natural selection*, and are therefore an essential requirement for—*evolution*.

natural selection

A process by which the environment provides the greatest chance for survival and reproduction to those members of a—*population* whose genes make them most highly adapted to prevailing conditions of life. It is an important factor in—*evolution*.

navigation
Setting a course. The ability to find a goal without using familiar landmarks. Many migrating animals are known to use true navigation, e.g., migratory birds, wales, and sea turtles. Navigation also plays a part in the honeybee's indication of direction. Orienting aids are the apparent movement of sun and stars (used in conjunction with a —*biological clock*), and occasionally also the earth's magnetic field.

neurohormone
A hormone not produced by the endocrine glands but rather by cells of the central nervous system (e.g., in the brain). These cells are called neurosecretory cells, and the process is known as neurosecretion.

neuron
Nerve cell. A cell that produces electrical signals, processes them, and conducts them to other cells, thus providing for the transmission of information within the body.

neurosecretion—neurohormone

one-male group—harem

ontogeny
Development of an organism from the fertilized egg until death. It comprises prenatal development in the egg or uterus (embryonal development), juvenile development, the condition of the adult, sexually mature organism, and the processes of aging and decline. Behavior also undergoes ontogeny.

orientation
The ability of an organism to direct (orient) its position and movements in space and time according to certain conditions and events.

ovulation
The release of eggs ready to be fertilized. The eggs are released from a shell known as the egg follicle, whose primary function is supplying the maturing egg cell with nutrients. In mammals, the follicle also secretes—*estrogens*.

pair-bonding
A union between a male and a female lasting for a certain period of time (e.g., over one reproductive period). Such a bond may occur either between one male and one female (*monogamy*), or between one animal and several others of the opposite sex (—*polygamy*).

pair-bonding behavior
Behavior patterns that reinforce the bond between partners of a pair. These may involve true—*courtship behaviors* and copulatory acts as well as special behaviors (in birds frequently ruffling each other's feathers). However, in many cases the latter behavior patterns would also have originated from courtship acts, then undergone a separate phylogenetic development. Joint —*brood care* behaviors may also reinforce the bond.

pair-formation
The creation of a—*pair-bond*. The corresponding behavior patterns originate to a large extent from the area of courtship, but elements of other behavior systems (e.g.,—*display behavior*) may also be included in pair-formation.

peck order
Coined in the early stages of ethology, this term has generally been replaced by—*dominance order*, which is commonly used today. Historically, the concept orginated from the fact that differences in rank were first noted in the social order of domestic chickens, where they are actually fought for by pecking with the beak. It was then generalized to other social animal groups.

periodicity
Rhythmicity. The regularly recurring fluctuations of certain phenomena of life (body temperature, liver activity, readiness to breed, general activity, etc.). These cyclical changes are based on internal (endogenous) physiological processes that obviously take place in the individual cells themselves. However, their exact nature is not yet known. The periodic changes are adjusted to corresponding changes in environmental conditions by means of—*Zeitgeber*.

peripheral
To the outside. Located at the surface of the body or in that vicinity. Also refers to animals that live at the periphery of social groups, e.g., lower-ranking, not participating in reproductive activities, or not having access to choice feeding places or nest sites.

phagocyte
A cell that assimilates small particles, such as bacteria or pieces of other cells, from the surroundings into its inner part by "flowing around" them. This process is known as phagocytosis.

phenotype
The externally visible manifestations of an organism. This is the product of an interaction between the genes (genotype) and environmental conditions acting upon the organism during ontogeny.

pheromone
Substance that conveys information. Hormone-like substance formed in certain glands and – in contrast to the actual hormones – secreted externally. It serves as a "chemical releaser" for olfactory communication between conspecifics, and is used, for example, in territorial marking or in attracting a mate.

phoresia
Temporarily using another animal as a "means of transportation." Example: dung-eating mites attach themselves to a dung beetle and in this way are carried from one dung heap to another. No parasitism is involved.

photoreceptors
Visual sensory (or receptor) cells. Sensory cells that respond to light (or visual) stimuli. In most animal species they are located in special sense organs, the eyes. In some lower animals, however, visual receptor cells may also be distributed over large areas of the body's surface.

phylogeny
Evolutionary development. Sequence of generations. The phylogenetic development of behavior is subject to the same rules as the phylogeny of all other organismic traits. The main factors are—*mutation* and—*natural selection*.

physiology
The study of the internal functions and activities of an organism, e.g. organs, cells etc.

placenta
An organ found in viviparous animals (especially in the great majority of mammals), which surrounds the embryo and supplies it with nutrients. It is constructed of both embryonal tissues and those of the mother (uterine wall).

play behavior
Behavior patterns that are not "in earnest." Play behavior, which occurs only in the higher vertebrates, has an important biological function: It helps the animal to become familiarized "with itself and with the world," i.e., to learn about its own body and behavioral capacities and to gather experiences with elements of the animate and inanimate environment through trial and error and through imitation.

polyandry—polygamy

polygamy
Sexual relationships between one individual and several others of the opposite sex. We distinguish between polygyny (many wives) and polyandry (many husbands). With respect to timing, there are two possibilities: In some species a member of one sex will mate with several of the opposite sex in succession, e.g., the males of many gallinaceous birds (peacock, heath cock). In these species there is no bonding between partners, and after successful copulation the females alone care for their offspring (thus we cannot speak of "marriage" here.) In other species, a member of one sex enters into a stable bond with several conspecifics of the opposite sex, and this relationship may last years or even until death. If one male is mated with several females, we speak of a—*harem*. The best-known examples are plains and mountain zebras and some monkeys (red hussars, gelda baboons, hamdryas baboons). The opposite case (for which there is no special name) can be found in the Tasmanian water hen, where two males are always mated with one female.

polygyny—polygamy

population
The unit of all animals of one species within a certain area. Within a population there is a constant blending of genes (gene flow). This makes a population the locus of evolutionary processes (—*mutation*, —*natural selection*).

prenatal learning
Learning processes taking place within the uterus or in the egg. Example: The chicks of some bird species (e.g., the guillemet or murre, which breeds on ledges of cliffs) learn to recognize the call of their parents and to distinguish it from that of neighboring pairs even before hatching.

primates
An Order of mammals that includes prosimians, monkeys, apea, and man. The branch of science studying the primates is called primatology.

progesterone—gestagens

promiscuity
In promiscuous animal species, males and females do not enter into a real—*pair-bond*, instead coming together only for copulation. Thus, a male can copulate with many females in succession, and vice versa.

proprioceptor
An internal sensory (or receptor) cell (—*interoceptor*) that registers the position and movement of body parts relative to each other, thus providing information about the (active and passive) movements of the body. Proprioceptors are particularly common in the musculature of vertebrates (muscle and tendon spindles) and in arthropods (extensor receptors).

provision for young
General term for all behavior patterns by which parent animals create favorable conditions for the young prior to the latter's arrival: building and maintaining protective structures or nests, gathering food to create a single storage supply, or laying the eggs near a suitable source of food for the young. On the whole, this term refers only to the period up to the time the eggs are fully disposed of and provided for. If the parents continue to nurture the young after hatching, and if direct contact develops between parents and young, we speak of—*brood care*.

range
Also home range. Spatial boundaries of an animal's activities. The area regularly inhibited by an individual or group, but where – as opposed to a—*territory* – conspecifics are not driven away. Frequently, the range is a "neutral" area between two or more territories.

rank order—dominance hierarchy

reafference
Feedback. Sensory stimuli produced by the body's own active movements.

receptor
Sensory cell. A cell that registers events within and outside the body, i.e., records stimuli and transduces (transforms) them into the informational code of the body, which is in the form of electrical signals. The various receptor cells each respond to very specific (adequate) stimuli only (e.g., to light, to sound waves, or to odorous substances). Receptor cells either stand on their own or are joined into —*sense organs*.

recognition stimulus—sign stimulus

redirected activity
Substitute behavior. A movement that is not directed at its "normal" object. Redirected activities occur most frequently in fighting encounters: If an animal is threatened or attacked by a high-ranking conspecific, its response will often not be discharged at the aggressor itself but rather at a lower-ranking conspecific or at a neutral object, such as tuft of grass. Like displacement activity, redirection behavior is typical for conflict situations.

reinforcement
A hypothetical construct invoked to explain the increase in performance of a behavior as a result of reward (positive r.) or punishment (negative r.) which results in a decrease.—Learning.

reflex

The certain response of an—*effector* to a sensory stimulus. Reflexes are characterized by a particularly rigid—*stimulus-response association*. If this connection is based on genetic factors, we speak of an "unconditioned reflex" (pupillary-, knee-jerk-, and salivary reflex). If it results from training, we speak of a "conditioned reflex." Conditioned reflexes were studied primarily by the Russian physiologist I. Pavlov.

releaser

General term for all of an animal's physical and behavioral manifestations aimed at releasing a response in another animal (sexual partner, parent or young, member of the group, symbiotic partner). Releasers include characters of color and shape, sounds, odors (such as sexual attractants or pheromones), as well as various movement patterns (e.g., courtship and threatening movements) which in general are called—*expressive behavior*. The terms releaser and—*sign stimulus* are often confused with each other, distinguished only vaguely, or even applied in the same way, to mean the same thing. There is a distinction, however, because in the course of evolution the primary factor for releasers was bilateral communication. The main purpose of a releaser is to trigger a response, and it is particularly adapted to this function. By contrast, in a sign stimulus the transmission of information is an "accidental" byproduct: A tree never grows colored branches just so that certain bird species may recognize it more easily as a nesting site, and in a predator-prey relationship any conspicuous traits that could aid the other's recognition are even avoided as far as possible. In other words, with a releaser both parties are "interested" in the information process, while with a sign stimulus only the receiver is actively concerned. Most releasers occur with intraspecific communication, but there are also relationships between different species (e.g., in—*symbiosis*) that require some kind of mutual communication. In the latter case, the effective stimuli are called interspecific releasers. In order to make a further distinction, stimuli serving intraspecific communication are sometimes called "social releasers."

releasing mechanism (RM)

Stimulus filter, filtering mechanism. General term for all the nervous system activities leading to a *filtering* of incoming stimuli so that only the "proper" stimuli (*sign stimuli*) release a behavior pattern. The process involved in a releasing mechanism is not homogeneous, but consists of a series of individual steps all linked in a certain sequence. If the animal responds to a stimulus pattern appropriately without ever having encountered it before, so that it had no opportunity to learn, it, ethologists refer to an "innate releasing mechanism" (IRM). This means that the animal forms an association between an existing stimulus and a corresponding response without prior experience. When the characters of the releasing stimuli have to be learned, we speak of an "acquired releasing mechanism" (ARM).

response chain

Also known as an action chain. Sequence of behaviors that is genetically programmed. Response chains occur primarily in social behavior (e.g., during courtship or ritualized fighting), where many behavior sequences consist of a fixed chain of single acts always discharging in the same way.

rhythmicity—periodicity

ritualization

An evolutionary process whereby a behavior pattern functioning as a—*releaser* becomes more and more effective in bilateral communication, i.e., grows increasingly prominent. Changes leading to this goal are, among others, the development of additional conspicuous traits, simplification or rhythmical repetition of the movement pattern concerned, or the special emphasis of single elements within this pattern. A behavior that has become adapted to its signal function in this manner is called ritualized. This applies above all to many behavior patterns of courtship and aggression.

ritualized fighting

Also ritualized combat. As opposed to—*damaging fights*, this is a relatively "harmless" type of combat enacted according to fixed "rules." Ritualized fights are used to drive off a rival (e.g., from the animal's—*territory*, from a female, or from a source of food) or to determine the stronger individual in a—*dominance hierarchy*. Animals with "dangerous" weapons (horns, hooves, teeth) do not use them at all or only in such a way that serious injury is largely avoided. Most ritualized fights are preceded by intensive—*threatening behavior*.

rudiment

Evolutionary remnant. General term for all organs or behavior patterns that have lost their initial function in the course of evolution (e.g., remnants of limbs in some giant snakes). In many cases, rudiments appear only temporarily during embryonal and juvenile development (e.g., gill openings in various terrestrial vertebrates). They may offer clues to the relative position of an animal species and therefore play an important role in taxonomic classification.

scent marks

Substances deposited for—*marking*, e.g., a *territory*. Animals use their urine, feces, or secretions of certain glands (called scent glands) for marking. The latter kinds of scent marks belong to the—*pheromones*.

search image

An ARM (—*releasing mechanism*) with a very short-lived effect and which therefore requires constant—*reinforcement*. To date, this term has been used primarily for the functional category of searching for food.

secretion

Release of a substance by a cell or gland. This substance is known as a secretion, and the cell's or gland's activity as secretory. Cells with the primary function of secretion are gland cells. Two groups of glandular secretions have a particularly high bearing on the behavior of an animal: the—*hormones* and the—*pheromones*.

sense organ

Organ that receives stimuli. A collection of sensory cells (—*receptors*), nerve cells, and so-called auxiliary cells (protective and supporting cells, screening cells or cell groups).

sensitive phase

Also known as critical period. That period of life in which an organism is particularly receptive to certain experiences. The timing of such a phase may vary for different learning processes. Sensitive phases are most noticeable with—*imprinting* processes, where they sometimes last only a few minutes or hours, as with imprinting of the following response in young precocial animals. Also optimal period.

sensory cell—receptor

sensory nerve—afference

sex hormones

Produced mainly in the—*gonads*, the primary function of these hormones is to control the developments and processes concerned with reproduction. These include the growth of many body organs known as "secondary sex characteristics" (e.g., antlers, special display coloration in birds, or colored patterns in many fishes), many of which function as releasers, and also behavior patterns associated with reproduction (e.g., courtship, nestbuilding, brood care). Male sex hormones are known as androgens, female ones as estrogens and gestagens.

sexual attractants

—*Pheromones* whose function is to attract a mate.

sexual dimorphism

Differences between males and females in external appearance and behavior. Examples: the "display coloration" of many male birds as compared to the plain feathers of females, or the occurrence of certain behavior patterns in one sex and not the other, such as the song of males in many bird species or the brood care behavior of females in many mammals. Animal species where the sexes are similar in appearance and behavior are called monomorphic.

sign stimulus

Sometimes also called a key-stimulus. An external stimulus that can trigger a certain behavior. If this stimulus evolved specifically for bilateral communication between animals (e.g., in the form of an odorous substance, a particular sound, or a color trait), we speak of a—*releaser*. Aside from their releasing effect, sign stimuli may also influence the orientation of a behavior pattern or the mood of an animal (orienting sign stimuli, mood-inducing sign stimuli;—*motivation*). In German, the term *Schlüsselreiz* (literally "key stimulus") was coined in conjunction with the phenomenon of—*filtering*. External stimuli were regarded as "keys" that could "unlock" the filters, in turn viewed as "door locks" (*releasing mechanism*).

Skinner box

Named after its inventor, the psychologist B. F. Skinner, this enclosed apparatus is used for investigating the learning capacity of animals. It works according to the reward principle: A certain behavior pattern, usually the operation of a lever bar or some other "manipulandum," is reinforced with a reward and thus changes in the nature and frequency of its occurrence. The study of such behavioral changes provides information on the effects of certain stimuli (light signals, sounds, shapes, colors) and of various conditions of reinforcement (time, amount, kind, or frequency of—*reinforcement*) on behavior. The results of these experiments may tell us, among other things, about an animal's speed of learning i.e., how quickly it "understands" the connection between behavior and reward. Most experiments according to the Skinnerian method have been carried out on rats, mice, and pigeons.

social animals

Species living in a stable—*pair-bond* or in families, groups, or states, and where—*social behavior* represents a large proportion of activities in general. By contrast, animals living alone are called "solitary."

social behavior

Behavior directed at a conspecific. In ethology, this term is used without any value connotation at all, and therefore has a different meaning as when applied to human society. It merely states that the behavior pattern in question serves the function of intraspecific communication. In this sense, even aggressive encounters are classified as social behavior. While most ethologists define all behaviors which occur between members of a species as social, thus also including courtship, brood care, or encounters at territorial boundaries, others use the term in a narrower sense and equate it with something like "group behavior." In other words, they restrict the concept to animals that are truly "social" and live in stable association with others of their species. In certain cases, as when there is a—*symbiosis* between two animal species which involve certain behavior patterns for bilateral communication, we may also speak of social behavior on an interspecific level.

social grooming (preening in birds)

Manipulating the skin, hair, or feathers of a conspecific. This is done mostly (but not exclusively) on parts of the body that the passive partner cannot reach for himself. Its initial function was probably mutual cleaning and the removal of parasites. In addition, however, social grooming may also aid in pair-bonding and group cohesion, a purpose that in many cases has become far more important than the original one. Birds use their bill for this behavior (ruffling the feathers), primates their hands ("lousing" in monkeys), and many other mammals their tongue.

social inhibition

A blocking of behavior occurring in animal species with a strong—*dominance hierarchy*, and mainly in low-ranking individuals: They yield to higher-ranking members of the group who claim priority at a certain behavior (e.g., copulation) or its object (e.g., a receptive female).

social system

The group organization of animals of one species, for example, a—*pair-bond* or a—*harem*, and within a—*territory* or a—*range*.

solitary—social animals
species-specific, species-typical
species-characteristic, typical of a certain animal species. Behavioral elements belonging to this category include primarily innate movement sequences (—*fixed action patterns*) and—*learning dispositions*.

spermatophore
Sperm packet, sperm carrier. A device that encloses a certain amount of male spermatozoa for a certain period of time. Spermatophores are either transferred directly to the female, or deposited on the ground or at some other spot and then taken in by the female on her own. In some species they are stored for a while within the female's body before the case dissolves to release the spermatozoa.

spermatozoa
Male sperm cells. These are formed in the male germ glands (testes) and usually released together with the secretions of various appendicular glands (in vertebrates, for example, the prostate glands). The mass of spermatozoa and secretions is called sperm. In most animal species the spermatozoa are actively mobile through thread-like appendages (flagella).

spontaneous behavior
Behavior originating autonomously, i.e., from "within," and which is therefore not released or maintained by external stimuli. In practice it is extremely difficult to determine whether a behavior is truly "spontaneous," since many external stimuli are impossible to recognize, or are noticed by the human observer only under certain provisions. Furthermore, even constant environmental conditions may ultimately function as external stimuli, e.g., after a certain period of influence. The most familiar example of spontaneous behavior are—*vacuum activities*.

statocyst
Organ of equilibrium, sense organ of gravity. A sense organ that provides information about the body's position in relation to the earth's gravity.

stereotypy
Constant, uniform repetition of behavior patterns or sound emissions, e.g., the monotonously repetitive cooing of a male pigeon.

stimulus-response association
The (temporal and quantitative) connection between the presentation of a certain stimulus pattern, and the occurrence of its "corresponding" behavior pattern. In simple—*reflexes* this connection is rigid and unmodifiable, but in other behavior patterns it is determined by momentary environmental conditions as well as the animal's internal state (—*motivation*). S-R learning.

stimulus specificity
The property of a receptor cell to respond to very specific stimuli only. These stimuli are termed as adequate.

stimulus summation
Mutual increase in stimulus intensity.—*Sign stimuli* may facilitate each other in their effects. In some cases, this facilitation has the nature of true summation. The joint and total effect of sign stimuli corresponds precisely to the sum of their individual effects (law of stimulus summation or law of heterogeneous summation).

submissive posture
Extreme—*appeasement posture*. In a fight between conspecifics, the loser may take up a submission posture, which can inhibit aggression to such an extent that the fight is broken off. In this way injuries can be avoided even when the combatants are very different in strength.

subthreshold stimulus—threshold
supernormal stimulus
A stimulus that releases a certain behavior pattern better than the normally "appropriate"—*releaser*. The fact that the latter may not necessarily represent the optimum stimulus was first noted in dummy experiments.—*Dummies* (models) that "exaggerate" certain characters serving as releasers often have a stronger effect than realistic imitations of these traits. However, supernormal releasers may also occur under natural conditions: A familiar example is the gaping throat of young cuckoos, which is far more conspicuous than that of their hosts' own nestlings and therefore also has a far greater releasing value.

symbiosis
Organisms of differential species systematically living together for their mutual benefit. Symbioses occur between plants and plants (fungi and algae as lichen), plants and animals (pollination of flowers), and animals and animals (crabs and hydrozoons, cleaner fishes).

synapse
Point of contact transmitting excitation between two nerve cells, between a nerve cell and a receptor cell, or between a nerve cell and a cell belonging to its—*effector organ* (muscle, gland).

synchronization
Temporal coordination between two different events or processes. Ethologists focus mainly on the synchronization phenomena occurring between two or more individuals (e.g., during courtship or group hunting) and between the behavior of an animal and the changes in its environment (e.g., day-night fluctuations, seasons).

synchronizer—Zeitgeber
synergism
Opposite of—*antagonism*. Simultaneous activity or functioning of muscles, nerves, or hormones.

taxis
Orienting response. Ethologists do not always use this term in the same way. It always refers to the spatial orientation of an animal to a source of stimulation (toward the stimulus, away from it, or at a certain angle to it). While some scientists also define the animal's corresponding movements in relation to the stimulus source as taxis, the most common definition today refers only to the orientation of the body, i.e., the positioning of the body axis in relation to the source of stimulation.

taxis component
Stimulus-dependent orienting part of a behavior pattern. This may occur separate from the—*fixed action pattern*, or interwoven with it in various ways.

testosterone

The most important male sex hormone (—*androgens*).

threatening behavior

Behavior patterns aimed at intimidating a rival and causing him to turn away before an actual fight ensues. Frequently, the outline of the body is enlarged by erecting hairs, feathers, folds of the skin, or combs. "Weapons" (teeth, horns) may also be prominently displayed, and this can be combined with—*intention movements* of real fighting behavior. On the other hand we may also observe elements of escape behavior, indicating that both flight and attack tendencies are involved in threat.

territory

A habitational area defended against others. An area in which the presence of its occupant excludes the presence of same-sexed conspecifics or all intraspecific (occasionally also interspecific) rivals.

territorial behavior

Behavior patterns used to identify and defend a territory. These include—*marking behavior,*—*threatening behavior*, and fighting.

threshold

The intensity of a stimulus necessary to just barely release a certain response (e.g., a behavior pattern or a nervous impulse; for example, auditory threshold: the lowest just barely audible sound intensity). Stimuli that do not attain this value and therefore do not elicit a noticeable response are called subthreshold or subliminal. Threshold values are not absolute, but may be influenced by various internal and external factors.

threshold decline—lowering of threshold

tradition

The transmission of acquired (i.e., learned) pieces of information to the next generation. In animals there are two kinds of tradition: In the "direct" form, the young animal learns straight from the parents (e.g., by imitation) and in this way can acquire their skills. In the "indirect" form, such as found in plant-parasitic insects, the transmission of information consists merely of the female depositing its eggs onto a certain host plant, so that the hatching larva becomes very much fixed on this particular plant species and later returns as an adult to lay its own eggs on it. In this manner, the preference for a plant species is transmitted from one generation to the next without members of two successive generations ever meeting each other. Prerequisite for this kind of tradition is merely a certain—*provision for young*. By contrast, the direct form of tradition is based on a definite and fairly long-term —*brood care*.

The transfer of chemical information (—*pheromones*) associated with the transfer of food, occurring in insects that feed each other.

unconditioned reflex—reflex

unimale group—harem

vaccuum activity

The—*spontaneous* occurrence of a behavior pattern without the external stimuli that normally release it. Vaccuum activities result from a severe—*lowering of threshold*.

visual receptor cells—photoreceptors

Zeitgeber

Also synchronizer, entraining agent. An external stimulus that temporally synchronizes an inner (endogenous)—*periodicity* (e.g., a daily or annual rhythm) of an animal with the periodic (cyclical) changes in its environment. The most important Zeitgeber for the daily periodicity of most organisms in the alternation of light and dark. In the realm of annual periods, a common Zeitgeber is the lengthening of days in the spring.

zygote

Fertilized egg cell.

Picture Credits

Photographs

Almasy/Bavaria (p. 334 bottom). Prof. Dr. H. Altner, Regensburg (p. 94 bottom left; p. 95 bottom left and right). T. Angermayer, München (p. 308 top and center left); p. 441 bottom; p. 599 all photographs except bottom right). H. Arndt, Lübeck (p. 9). O. Attiger, Windisch/Schweiz (p. 237 top). Bannister/A. G. E., Spanien (p. 477 center left). Bavaria, Gauting (p. 334 top; p. 364 top). D. Blum, Eßlingen (p. 456 all photographs; p. 509 bottom right; p. 510 bottom left). Prof. Dr. J. Boeckh, Regensburg (p. 94 top left and right; p. 95 top). Dr. G. Bretfeld, Kiel (p. 414 and p. 415 all bottom photographs). A. Bhutz, Heidelberg (p. 413 bottom right). Prof. Dr. D. Burkhardt, Regensburg (p. 20 bottom; p. 80 bottom left and center). Burton/Photo Researchers (p. 387 top; p. 478 top left and center). Campbell/Photo Researchers (p. 478 bottom). Chaumeton/Jacana (p. 186 bottom). I. Czimmek, Schalksmühle (p. 307 bottom left and right). Dr. M. Dambach, Köln (p. 248 all bottom photographs). H. Dossenbach, Obserschlatt/Schweiz (p. 432 bottom right). dpa, Frankfurt (p. 464 top left). Dubois/Jacana (p. 369 bottom). Dupont/Explorer (p. 333). Dupont/Jacana (p. 370 bottom). Edouard/Explorer (p. 474 bottom left). Prof. Dr. I. Eibl-Eibesfeldt (p. 609, 610, 611, 612 all photographs). Eliott/Jacana (p. 422 top left and right). Dr. W. Foersch, München (p. 589 top left and right). Dr. W. Götz, Widen/Schweiz (p. 531 top left; p. 532 all photographs, except bottom left). Dr. G. Gottlieb, Raleigh/USA (p. 19 bottom right; p. 538 top left and right; bottom left). M. Gruber, München (p. 186 top). A. Hein (p. 538 bottom right). Dr. K. Heinroth, Berlin (p. 10 bottom). U. Hirsch, Köln (p. 247 bottom right). T. Hölldobler, Cambridge/USA (p. 499 bottom). Dr. D. v. Holst, München (p. 537 all photographs). E. Hosking, London (p. 441 top; p. 463 bottom left). Prof. Dr. F. Huber, Köln (p. 248 top). Prof. Dr. K. Immelmann, Bielefeld (p. 343; p. 344 top left, center left and right; p. 422 bottom; p. 464 bottom left; p. 484 top and bottom left; p. 515 bottom right). Jacana, Paris (p. 364 bottom). H. Kacher, Seewiesen (p. 10 bottom). Prof. Dr. H. Klingel, Braunschweig (p. 509 top; p. 510 top left and right, center right). Prof. Dr. M. Konishi, Princeton/USA (p. 128). G. Konrad, Heidelberg (p. 431 all photographs). Prof. Dr. E. Kullmann, Kiel (p. 416 center left; p. 457 top right; p. 477 top left). Prof. Dr. H. Kummer, Zürich (p. 531 all photographs, except top left; p. 532 bottom left). W. Layer, Mannheim (p. 80 bottom right; p. 96 top; p. 237 center left; p. 463 top right; p. 499 top; p. 600 bottom right). Prof. Dr. M. Lindauer, Würzburg (p. 500 all photographs). Prof. Dr. K. E. Linsenmair, Regensburg (p. 117 all photographs; p. 118 center and bottom; p. 364 center right; p. 370 center right; p. 477 center and bottom right). K. H. Löhr, Koblenz (p. 416 bottom left and right). Lorenz/roebild (p. 20 top left and right). Dr. J. Lamprecht, Seronera/Ostafrika (p. 19 all photographs, except bottom right). Prof. Dr. D. B. E. Magnus, Darmstadt (p. 369 top left; p. 473 top left and right; p. 474 top left and right). Micha/Mauritius (p. 297 top). Mohn/Zefa (p. 237 center right). Müller/roebild (p. 455; p. 458 bottom left; p. 473 bottom; p. 510 center left; p. 515 top right; p. 599 bottom right). Dr. Müller-Schwarze/Okapia (p. 422 bottom). Prof. Dr. G. Neuweiler, Frankfurt (p. 195 all photographs). A. v. d. Nieuwenhuizen, Heemstede/Holland (p. 457 bottom right). Okapia, Frankfurt (p. 388 all photographs; p. 464 bottom right; p. 484 bottom right; p. 515 top left; p. 589 bottom; p. 600 center right and bottom left). K. Paysan, Stuttgart (p. 237 bottom; p. 298 all photographs, except top left and right; p. 307 top left and right; p. 386 bottom; p. 387 bottom; p. 416 top right; p. 432 all photographs, except bottom right; p. 516 all photographs, except top right). W. S. Peckover, Port Moresby/Neuguinea (p. 422 center left). Pfletschinger/Bavaria (p. 386 top; p. 483 center left). Pfletschinger, Ebersbach (p. 238 center right). Photo Researchers, Uxbridge/England (p. 369 center; p. 370 top right). G. Quedens, Norddorf/Amrum (p. 80 top). Dipl.-Biol. H. U. Reyer, Seewiesen (p. 370 top left, p. 458 right, all 3 photographs; top right). W. Rohdich, Münster (p. 308 top and bottom right). A. Root/Okapia (p. 364 center left). Rouxaime/Jacana (p. 421 top). Dr. G. Rüppell, Erlangen (p. 363; p. 369 top right; p. 516 top right). H. Sass, Regensburg (p. 94 bottom right). Dr. F. Sauer, München (p. 118 top; p. 238 bottom left, top right; p. 247 bottom left; p. 413 all photographs, except bottom right; p. 416 center right; p. 442 top left; p. 474 bottom right; p. 483 top; p. 499 center). Dr. F. Sauer/Bavaria (p. 385). F. Sieber, Krumbach (p. 463 top left). F. Siedel, Sande (p. 510 bottom right). R. Siegel, Breckerfeld (p. 185; p. 464 top right). Dr. R. Sossinka, Bielefeld (p. 457 left, all 3 photographs; p. 590 bottom). D. Schmidl, Seewiesen (p. 422 center right). W. Schmidt, Berlin (p. 509 bottom left). Prof. Dr. H. U. Schnitzler, Frankfurt (p. 196 all photographs). H. Schrempp. Oberrimsingen (p. 96 bottom; p. 238 top and center left, bottom right; p. 247 top left; p. 308 bottom left; p. 477 bottom left; p. 478 top right; p. 483 center right). F. Schutz, Seewiesen (p. 344 top right, bottom left and right). Stonehouse/Ardea (p. 474 center). Prof. Dr. N. Tinbergen, Oxford/England (p. 10 top and center right). Urban, München (p. 247 top right; p. 421 bottom left and right). V-Dia, Heidelberg (p. 297 bottom; p. 483 bottom). Dr. U. Waldow, Regensburg (p. 95 center). Prof. Dr. P. Weygoldt, Freiburg (p. 414 and p. 415 all top photographs; p. 458 top left; p. 463 center and bottom right). Dr. A. Wünschmann, München (p. 600 top left and right, center left). Zefa, Düsseldorf (p. 416 top left). G. Ziesler, München (p. 298 top left and right; p. 442 center; p. 515 bottom left). Dr. E. Zimen, Spiegelau

(p. 590 top). D. Zingel, Wiesbaden (p. 442 top right). Color Plates: K. Großmann, Frankfurt (p. 45; Consultant Prof. Dr. G. Neuweiler, Frankfurt). J. Kühn, Heidelberg (p. 46; Consultant Prof. Dr. G. Neuweiler, Frankfurt. p. 79; Consultant Prof. Dr. D. Burkhardt, Regensburg. p. 93; Consultant Prof. Dr. H. Altner, Regensburg, p. 127).

Sources of portraits in ch. 1: Prof. Dr. K. v. Frisch, München (p. 17). Dr. K. Heinroth, Berlin (p. 6). Dr. D. v. Holst, München (p. 14). Prof. Dr. O. Koehler, Freiburg (p. 15).

Line drawings prepared by: Prof. Dr. H. Altner, Regensburg (ch. 6). A. Borowski, Köln, and I. Litcoff, Hamburg (ch. 20). K. Großmann, Frankfurt (p. 40, 42, 48, 51). T. Hölldobler, Cambridge/Mass. (ch. 32). Klein, Tübingen (p. 206, 207, 208, 211). J. Kühn, Heidelberg (p. 50, 54, 56, 59, 64; ch. 4 and 5; p. 130). Ch. Linsenmair, Regensburg (ch. 7). D. Schmidl, Seewiesen (p. 170, 172, 173, 174, 183, 184, 326 bottom, 327, 336, 338 bottom, 339, 461 bottom). Prof. Dr. H. Schöne, Seewiesen (p. 165, 166 top left, 168, 169, 176, 181). Dr. R. Sossinka, Bielefeld (p. 314, 315, 320; ch. 30). Dr. T. Szabo, Gif sur Yvette (p. 139, 140, 142, 143, 144). H. Tabel, Göttingen (ch. 13). All other drawings by K.-H. Steffel, München-Garching.

Source of line drawings other than those provided by the authors (from top to bottom): p. 24: Schleidt; Lorenz. p. 29: Lorenz. p. 31: Tinbergen, in *The Study of Instinct* (The Clarendon Press, Oxford), as most of the others after Tinburgen. p. 32: Tinbergen. p. 33: in Marler/Hamilton. p. 36 bottom, p. 37 top: in Wickler. p. 37 center and bottom, p. 38 top: Tinbergen. p. 81: Hesse. p. 82, Goldsmith/Philpott; Kühn. p. 83 bottom: Müller-Limroth. p. 84: Cohen. p. 85: Boycott/Dowling. p. 91: Seitz. p. 101: Boeckh/Waldow. p. 102, 103 center: Hansen/Heumann; Atema. p. 105, 106: Sass; Teichmann; Popesko. p. 130: Roeder. p. 139 top: Belbenoit. p. 168, 169: Braemer/Schwassmann. p. 172: v. Holst/Mittelstaedt. p. 180: Bullock/Diecke. p. 188: Beier; Enright; Hauenschild. p. 189: Pengelley/Asmundson. p. 190 top/192/193: Aschoff/Wever. p. 190 bottom: Hashimoto/Remmert. p. 200: Serventy; Cochran. p. 201: Emlen. p. 203: Perdeck. p. 213: Griffin/Webster. p. 219: Wilson/Gettrup. p. 222, 224: Kühme; Waldon. p. 225 bottom, p. 226: Sauer/Sauer; Schenkel; Bower. p. 231: Echelle. p. 232 bottom: Lorenz. p. 241: Hess.

p. 242: Heinroth; Kruijt. p. 243: Tinbergen. p. 251: after photograph by O. Paris. p. 252: Feekes. p. 259, 260 top, 260 bottom, 261, 262 top: p. 262 bottom, 263: in Wickler. p. 264, 265: Tinbergen. p. 266: Baerends. p. 267: Prechtl. p. 282: Harlow/Harlow. p. 283: De-Long; Jouvet, in *The States of Sleep*. Copyright 1967 by Scientific American, Inc. p. 284: Karten. p. 285 top: Instrument der Stoelting Co. 288 top: v. St. Paul. p. 290, 291: Åkerman. p. 292 top: Flynn. p. 292 bottom: Fisher. p. 293: Olds, in *Pleasure Centers in the Brain*. Copyright 1956 by Scientific American, Inc. p. 316: Salzen. p. 321: Hutchison. p. 322: Hinde. p. 326 top and center: in Hediger. p. 326 bottom, 327: Pawlow; Eibl-Eibesfeldt. p. 331: Yerkes. p. 337: in Buchholtz. p. 338 top: Baerends. p. 339 bottom: Eibl-Eibesfeldt. p. 349: in Pringle. p. 350: Koehler/Dinger. p. 353 bottom: Ferster. p. 355: Rensch/Dücker. p. 356: after photographs by Menzel. p. 358: Döhl. p. 359: Jantschke. p. 360: Premack. p. 366, 367 top: Portmann; in Eibl-Eibesfeldt; Hess. p. 367 bottom: after photograph by Klages. p. 372: Schutz. p. 376, 377, 378: in Eibl-Eibesfeldt. p. 381 top: Eibl-Eibesfeldt. p. 382: in Heymer; in Hediger. p. 384, 390: Tinbergen; v. d. Assem. p. 391: Marler/Hamilton. p. 392 bottom, 393 top: in Wickler; Schoener. p. 393 bottom, 394: Lack; Southwick. p. 395, 396: Tinbergen; Wilson. p. 397, 398 top: Hediger; Eibl-Eibesfeldt. p. 398 bottom, 399 top and center in Wickler (*Sind wir Sünder* München, Droemer, 1969); in Wickler (*Sünder*); Eibl-Eibesfeldt. p. 400: Moynihan. p. 401 top: Tembrock. p. 402: Geist; after photograph; Tinbergen. p. 406: in Heymer; Walther. p. 407 bottom, 408: Lagerspetz; Rasa. p. 409 top: Courchesne/Barlow. p. 410: in Thomson; in Suchowski. p. 419, 420: Kleinschmidt. p. 426: v. Graff; in Buchner. p. 427: Enriques/Belar. p. 428: Baltzer. p. 429, 430: Wells/Huxley; in Buchner; Bridge. p. 433: Zawadowsky. p. 434: Cleveland. p. 435: in Houillon. p. 450, 451: Walther. p. 452 bottom: Nicolai. p. 453: Nicolai; Wickler; in Wickler. p. 460: Tinbergen. p. 462, 465, 466 top: Tinbergen; in Brehms Tierleben; Tinbergen. p. 466 center and bottom: Walther; Tinbergen. p. 467: in Lorenz; Berger. p. 468: Armstrong. p. 504, 505: in Wickler (*Sünder*). p. 586: Hamburger/Wenger/Oppenheim. p. 587: Oppenheim/Harth. p. 593: after photograph by Beer. p. 597: changed according to drawing by Kear. p. 603: Hein, in Science 158, 1967, p. 390–392, Copyright American Association for the Advancement of Science. p. 604, 605: Rabinovitch/Rosvold. p. 663, 665, 667 bottom; p. 668, 669: after photographs by Klages.

Index

Abbot of Condillae, 2
Abnormality, 614
Acuity, visual, 82, 83★
Adaptation, 27, 404, 515★, 516★, 621, 652
Aelianus, 1
Afferent, 71
Aggregations of animals, 473★, 474★, 495, 511, 540
Aggression, 497
Aggressive behavior, 37, 374ff, 380ff, 504
Aggressiveness, 380, 403
Agonistic behavior, 381
Akerman, B., 290
Alarming, 465
Alpha animal, 389
Altmann, J., 588
—, S., 558
Altricial, 365, 366★, 488
Altruism, 498
Altrum, Bernhard, 4
Analogy or convergence, 28f
Andrew, B. R. J., 253
Appeasement behavior, 226, 398★, 399★, 399, 466
Appetitive behavior, 5, 13, 222
Aristotle, 1, 1★, 219, 469
Armstrong, Edward, 475
Aschoff, Jürgen, 16
Assem, J.v.d., 384, 393
Attraction, 498
Auditory perception, 593ff

Baby schema, 673
Baerends, G. P., XVII, 12, 14, 17, 227, 265f
Barlow, G. W., 408
Barraud, F. M., 261
Bartholomew, George, 476
Bateson, P. P. G., 598
Beach, F. A., 18, 321
Becker, E., 355
Beer, Colin, 593
Behavior, concept of, 23f
— development, 579ff
— evolution of, 642ff
— genetics, 30, 607ff
Behavior patterns, 23, 657
Bentley, D. R., 276
Benzer, 613
Biogenetic law, 580
Biological clock, 187
Biorhythm or circadian rhythm, 187f, 188★
Birth, 456★
Bisexuality, 427
Blakeslee, B. P., 354
Blest, A. D., 264
Bol, A., 252, 255
Bossert, W. H., 277f
Bower, T. G. R., 227, 266
Box-camera eye, 77★, 77
Brain stimulation, 284ff
Brehm, A. E., XIV, 3
Breuer, 231
Bristles, 99
Broodcare, 457★, 458★, 462, 464★, 487ff
Brood parasitism, 463★, 505f
Brown, J. L., 292, 633
Bubble eye, 77★, 77
Buddenbrock, von, 479
Bühler, K., 366
Bullock, Theodore, XVII
Butenandt, A., 102

Calhoun, J., 575

Camouflage, 621
Cannibalism, 568, 568★
Capranica, R. R., 132f
Captivity, 662
Carey, R., 291
Carpenter, C. C., 21
Castes, 522
Celestial compass orientation, 162
Cephalopoda, 295
Chase, J., 213
Chemical communication, 519ff
— sense, 180
Chemotactic, 175
Christian, J., 566, 574
Chromosomes, 613, 639f
Circadian rhythm, see biorhythm
Clan, 512
Clayton, 618f
Colony, 512
Color vision, 85, 92
Communes, 484★
Communication, 527
Compass orientation, 165
Compound eye, 78, 79★, 80★, 81★, 87
Conditioned aversion, 331
— inhibition, 332
— reflex, 336
Conditioning, 31, 31★, 327
Conjugation, 420, 420★, 430, 451
Consummatory action or behavior, 13, 222
— situation, 223
Copulation, 461
Coulson, John, 471, 482
Courchesne, Eric, 233, 408
Course orientation, 165
Courtship behavior, 422★, 448ff, 448★, 481
Courtship feeding, 465, 485
Craig, Wallace, 5f, 13, 222
Crane, J., 648
Crick, I., 419
Critical period, 598
— phase, 362
Crook, John, 471, 475
Croze, H., 354
Cup eye, 77★, 77
Curio, Eberhard, 17, 244, 266f, 357
Curiosity behavior, 339, 340★
Cuvier, Georges, 205

Darwin, Charles, XIV, 3★, 3f, 219, 231, 235, 275, 391, 637, 672
Davenport, R. K., 354
Dawkins, C. Richard, 241
deBeer, G., 580
Delius, J., 252
Deprivation experiment, see Kasper Hauser, 347
— syndrome, 378
Descartes, XIV, 2
Digeser-Knoll, M., 350
Dijkgraaf, S., 206
Diopter, 82
Disinhibition, 254
Displacement activity or behavior, 235, 246ff, 257★
Division of labor, 506, 521
Dollard, J., 405
Domestic animals, 654ff
Dominance, 541, 573
Dominance hierarchy, 389
Drive, 405
Dücker, G., 355, 357

Echo-location, 205ff
Echo-orientation, 182

Ecology, behavioral, 620ff
Efferent, 70
Eibl-Eibesfeldt, Irenäus, 8, 16, 329, 357, 399, 405, 646
Eisenberg, John, 471
Eisner, T., 280
Ekman, P., 673
Electrical stimulation experiments, 284f
Epicurus, 1
Erickson, Carl, 471
Escherich, 6
Estes, R. D., 543
Ethogram, 6, 24, 231
Ethology, 6, 219
Ettlinger, G., 354
Evans, H. E., 649
— Roger, 593
Evolution, 3
— of behavior, 28, 638ff
Ewer, R. F., 17
Ewert, J. P., 63
Excitation, 66
Experience, 581
Exploratory behavior, 339
Expressive movements (Ausdrucks-bewegungen), 448
Exteroreceptors, 109
Eye, 77, 77★, 78, 79★, 81f, 82★, 83★

Fabré, H., 17
Fabricius, E., 15, 17
Facet eye, 78, 87
Facial expression, 609★
Facilitation, 299
Family, 462
Fay, 132
Feedback control system, 170
Feeding, 490, 499★
Feekes, F., 252, 252★, 255ff
"Female," 420
Fentress, John, 252
Ferster, C. B., 355
Fertilization, 452
Fighting, 589★, 644
Fischer, F., 666
Flynn, J., 291
Fixed action patterns, 6, 223, 329, 585
Food imprinting, 374
— getting, 644
— preference, 625
Forel, 6
Fossils, 643
Fouts, R., 359
Fraenkel, G., 175f
Franck, D., 617
Freud, Sigmund, XIX, 231, 372, 579
Friesen, W., 673
Frish, Karl von, 16, 17★, 326, 328, 523, 528f, 534
— Otto von, 243
Frishkopf, L. S., 132
Frustration, 405
Function, 583

Galambos, R., 206
Gamete, 423
Gardner, 359
Gehlen, A., 678
Geist, V., 391, 402
Genetic, 581
Genotype, 608
Geographic orientation, 149, 162f
Geotactic, 175
Gestalt, perception, 227
Glueck, E., 375
— S., 375

Goldstein, M. H., 132
Gonopodia, 439
Gottlieb, Gilbert, 135, 239
Gould, E., 211
Grabowski, U., 350
Gravity receptors, 121ff
Grégoive, A., 350
Griffin, D. R., 18, 206f, 211ff, 217
Grinnell, A. D., 216
Grossman, S., 292
Grossmann, K., XIII, 328
Group behavior, 496, 531★, 532★
— mechanisms, 38
— structure, 553
Grzimek, Bernhard, XIII
Gunn, D. L., 175f
Gunter, R., 354
Gustatory sense, or sense of taste, 98, 104
Gwinner, E., 349

Haan, Bierens de, 16
Habitat choice, 628
Habituation, 228, 268ff
Haeckel, E., 580
Hahn, W. L., 206
Hailman, Jack P., 228, 263
Hair, 109★, 110, 110★
Hall, G. Stanley, 580
Hamburger, Viktor, 586, 586★, 588
Hamilton, W. D., 498, 508
—, W. J., 405
Harlow, Harry F., 15 241, 376★, 377★, 378f, 378★, 378
—, M., 15
Hartmann, M., 420, 425
Hartridge, H., 206, 216
Harvey, William, 2
Harwood, D., 288
Heaton, Marieta, 594
Hebb, Donald O., 605
Hediger, H., 17, 389
Heiligenberg, Walter, XVIII, 233, 407f
Hein, Alan, 602f
Heinroth, Katharina, XIV
—, Magdalena (née Wiebe), 8, 15
Heinroth, Oskar, XVIf, 6ff, 6★, 10★, 14, 669
—, Otto, 241
Henry, J. P., 577
Henty, Clifford, 244
Herd, 541, 554ff
Heredity, 346, 660
Hermaphroditism, 416★, 425, 426★, 427★
Herrick, 5
Hertwig, O., 296
—, R., 296
Hess, Eckard H., 15, 17, 241, 365, 367★
—, J., 341
—, W. R., 283
Heterogenous summation, 226, 265, 326
Hierarchy, 38, 560ff
Hinde, Robert, 17, 223, 234, 261, 322★
Hofer, H., 357f
Hoffman, K., 193
Höhl, J., 358
Holistic psychologists, 5
Hologamy, 423
Holst, Erich von, XVff, XX, 13, 14★, 170f, 268, 287, 292, 409
Home range orientation, 149, 162
Homology, 28f, 645

Hormones, influence on behavior, 314ff
Howard, H. E., 6
Hubel, D. H., 261
Huber, F., 304
Human behavior, 670f
Hunsperger, R., 290
Huxley, Julian, 6, 9*, 482
Hymovitch, 605

Iersel, J. J. A. van, 252, 255
Imitating behavior, 341
Immelmann, K., 15, 331, 601, 658f
Impekoven, Monica, 593
Imprinting, 10*, 14, 243, 344*, 362ff, 372*, 373*, 598
Individual distance, 384
Information processing, 41
— theory, 647
Inhibition, 332
Innate movement patterns, 670
— releasing mechanism, 12, 261, 263, 346, 672
Insectivorous bats (Microchiroptera), 206
Instinct, 2ff, 236, 581
Instinctive behavior, 18, 236
Instinct-training intercalation, 11
Intelligence, 604
Intention movements, 7
Internal clock, 168
Intersexes, 426f
Interspecific division, 627
Intraspecific communication, 137, 509*
— divisions, 630f
Irreversibility, 368
Isolation experiment, 347

Jakobson, R., 350
James, William, 236
Jantschke, F., 359
Josephson, R. K., 300
Jurine, Ludwig, 205, 213

Kasper Hauser experiment, a.k.a. deprivation experiment, 8, 35, 348
Keller, Helen, 360
Key stimulus, 260
Kinesis, 176
Kipp, F. A., 349
Kirkman, 6
Klatt, D. H., 349
Klinghammer, E., XIII, 15
Klinokinesis, 176
Klopfer, Peter, 227
Klüver, H., 355
Knecht, S., 355
Koehler, Otto, 12, 15f, 15*
Koenig, Otto, 7
Köhler, W., 352, 356, 358
Kohts, N., 354, 358
Konishi, M., 133, 349, 604
Konstantinov, A., 214
Kortlandt, A., 17, 249, 251, 357
Kovach, Joseph, 598
Kramer, Gustav, 242
—, P., 357
Kruijt, Jaap, 241, 252, 255
Kuehen, 12
Kühn, A., 175f
—, E., 355
Kulzer, E., 206
Kummer, Hans, 481, 663f
Kuo, Zing-Yang, 581
Kutsch, W., 276

Lack, David, 400, 471
Lamarck, 3
Lamprecht, J., 397
Language, 527
Lawick-Goodall, Jane van, 357, 375, 552

Le Resche, 482
Learning, 236, 325ff, 581
Lehrman, Daniel S., 18, 239, 582
Lemmon, W. B., 359
Lemunyan, D. C., 574
Leorig, C.-Y., 271
—, Daisy, 227
Leyhausen, P., 16, 383, 389, 405, 673
Liebermann, P., 349
Lilly, J. C., 355
Lind, 256
Lindauer, 535
Lissmann, H. W., 139
Littlejohn, M. T., 132
Localization, 177, 211
Locomotion, 221, 644
Loeb, Jacques, 5
Loftus-Hills, J. J., 133
Lögler, 354
Löhrl, H., 374
Lorenz, Konrad, 6ff, 10*, 11ff, 12*, 12ff, 16, 219f, 222, 225, 231f, 242, 258, 260f, 263, 275, 325, 362, 368, 374, 390, 393, 399, 403, 405, 407, 409, 601, 618, 648, 659f, 668f, 675, 679
—, Margarete, 11

MacArthur, 628*
Mainardi, D., 17
"Male," 420
Malthus, 616
Mammalian ear, 127*, 135ff
Man, 577
Manning, Aubrey, 17, 240
Marking, 398
Marler, P., 18, 276, 279, 349, 384, 394, 405, 604
Mason, E. A., 561
Mating, 470
Maturation, 581
Mayr, Ernst, 3, 608
McBride, G., 17
McCue, J. J. G., 213
McDougall, William, XV, 5
McFarland, D., 17
Mechanists, 5
Mechanoreceptors, 108
Memory, 325
Mendel, 3
Mendelson, H., 17
Menotaxis, 176
Menzel, E. W., 358, 561
Messmer, E., 347ff
—, I., 347ff
Meyer-Holzapfel, M., 17, 357, 667
Michel, R., 17
Michelsen, A., 132
Miescher, J. F., 419
Migration, 199, 543
Milner, P., 293
Mimicry, 36, 37*, 238* 281, 624f
Mitchell, R. P., 289
Mittelstaedt, H., 170f
Modal action patterns, 223, 237*
Modification, 649
Möhres, F. P., 211
Molina, F. de, 290
Mood transfer, 559
Morgan, Lloyd, 4
Morris, D., 256
—, Robert, 471
Mortor, E. S., 278f
Motivation, 33, 230
Müller, J. von, 70, 89
Mundinger, Paul, 485
Mutation, 26, 613
Mutual grooming, 485, 504, 505*

Natural selection, 391, 433
Nature-nurture, 235ff
Neuro-ethology, 282ff

Neuweiler, G., 215
Nicolai, Jürgen, 12, 264, 482
Noble, G. K., 225, 648
Nolte, A., 357
Nonverbal thinking, 345ff
North, 485
Nottebohm, 604
Novick, 211
Nystagmus, 83

Object orientation, 149, 152f
Oken, 3
Olds, J., 293
Olfactory bristles (Sensilla basiconica), 93*, 95*
Olfactory sense, or sense of smell, 98, 106
Omega animal, 389
Ontogeny, 18, 34, 579
Operant conditioning, 329
Oppenheim, 586*
Oppenheimer, J., 580
Orientation, 32, 66, 146ff, 351, 534
Orthokinesis, 176
Overpopulation, 574

Paietta, 618f
Pair bond, 392, 462, 469ff, 677
Pairs, 540
Parental behavior, 487ff
Parent-child interaction, 504
Parthenogenesis, 429
Pastore, N., 355
Paulsen, H., 652
Pavlov, Ivan, P., 5, 5*, 32, 326*, 327
Payne, R. S., 133
Peckham, E., 6
—, G., 6
Perception, 596
Periodicity, 187ff
Pernau, Ferdinand Adam von, 2, 348
Perrault, Claude, 2
Personal bonding, 676
Personal space, 384
Perversions, 373
Phenotype, 608
Pheromones, 101f, 319, 427, 438, 461, 519
Phobotaxis, 175
Photoreceptors, 75, 84*, 85*
Phototactic, 175
Phototaxis, 175
Phylogeny, 580
Physiological clock, 158
Pierce, G. W., 206
Pit cones, 94*
Place loyalty, 502
Play, 325ff
"Play dead," 112, 118*
Ploog, D., 285
Plutarch, 1
Polarized vision, 91
Polygyny, 472, 475
Population control, 565
— genetics, 498
— regulation, 631
Populations, 616f
Portielje, A., 246
Positional orientation, 149ff
Prechtl, H. F. P., 268
Precocial, 365, 366*, 488
Premack, D., 360
Preyer, W. T., 585, 588
Priesner, E., 102
Primate groups, 550ff
Proprioceptors, 109, 169
Protective coloration, 308*
Pseudohermaphrodite, 428
Psychophylogeny, 373
Punishment, 335
Pye, D., 211

Rabinovitch, 605*
Ramsey, O., 15
Rank order, 389, 560
Rasa, Anne, 232, 395
Reafference principle, 170f, 170* 172*
Receptors, 67
Reese, Ernst, 221*, 222
Regen, J., 306
Reimarus, Hermann S., 2
Reinert, J., 356
Releaser, 12, 36, 36*, 225
Releasing mechanisms, 228, 261, 659
Releasing stimuli, 396f
Remane, Adolf, 29
Rensch, B., 355, 357, 373
Reproduction, asexual, 413*
Reward, 335
Reyer, H. U., XIII
Rheotactic response, 119
Rhythms, internal, 34
Riddle, Oscar, 5
Ritualization, 9*, 224
Ritualized fight, 401
Rituals, 610*, 611*
Roberts, W., 291
Roeder, Kenneth D., XVIIf, 130, 229
Roger, C. M., 354
Rollinat, R., 206
Romanes, J. J., 299
Rose, W., 451
Rosenzweig, 605
Rossmann, A., 355
Rosvold, 605*
Rotary orientation, 165
Rothenbuhler, 617, 619
Rowell, C., 253
Roy, Charles de, 3
Russell, Bertrand, 236

Sackett, G. P., 374
Sade, D. S., 562
Saint Paul, U. von, 268, 287, 292, 409
Sauer, I., 347, 357
Scent marking, 391, 398, 502, 570, 667
Schäfer, E. A., 296
Schein, 15
Scheitlin, 3
Schenkel, R., 398
Schief, 129
Schjelderup-Ebbe, T., 390
Schleidt, Wolfgang M., 228, 260, 262f
Schmidt, R. S., 648
Schneider, D., 102
—, G., 282
—, K. M., 669
Schneirla, Theodore C., 18, 480, 582
Schoener, T. W., 392
Schutz, F., 15, 367, 372*, 601
Scott, J. Paul, 18, 233, 406f
Search image, 229, 354
Seibt, Uta, 480
Seiz, Alfred, 12, 226f, 264
Selection, 27, 391
Selous, E., 6
Selye, H., 565
Sem-Jacobson, C., 294
Sensilla, 99f
Sensitive period, 243, 362
Sensory function, 591
— physiology, 67ff
Sex characteristics, 424
Sexual dimorphism, 416*
Sexuality, 419
Sexual reproduction, 434
Shaw, Evelyn, 241
Sherrington, XVf
Shock-vibration, perception of, 309

Signals, 277, 653
Sign stimuli, 225, 261, 346
Simmons, J. A., 214
Simonds, P. E., 560
Skinner, B. F., 328f
Sladen, 482
Social deprivation, 366
Social groups, 477★, 478★, 540
Social hierarchy, 389
— inhibition, 558
— life, 496f, 507, 511
— organization, 540f
Social stress, 564
Socialization, 376, 561
Solitary animals, 540
Sollwert, 168
Sorelli, 2
Sound intensity, 131
Spacing out, 394, 632
Spalding, 14
Spallanzani, Lazzro, 205, 213
Spatial orientation, 164
Species-specific action patterns, 13
Species-specific instinctive actions, 6
Species-specific signals, 275ff
Specific action potential, 231
Spencer, Herbert, 4
Spermatophores, 437ff
Sperry, R. W., 170f
Spivak, H., 663
Statocyst, 121★, 122★, 123
Statoliths, 121f, 122★
Stereotypy, behavioral, 636
Stimulus and response, 30
— filtering, 230

Stimulus patterns, 258f
Stimulus summation, 264f
Strato-orientation, 149, 156f
Stress, 565
Submission, 402
Suga, N., 213
Sun orientation, 163
Supernormal releaser, 228, 433
Suthers, R., 212
Swarming, 521
Swimbladder, 125, 125★
Symbolization, 465
Synchronization, 467, 591

Tactile sense, 109
Taste buds, 100
Tavolga, 132
Taxonomy, 29, 29★, 648
Tees, Richard, 597
Teleology, 2f
Telotaxis, 175f
Territory, 39, 39★, 369★, 370★, 382ff, 466, 541f, 632
Thorndike, E. L., 5
Thorpe, W. H., 15, 18, 348, 356, 485, 604
Threat, 363★, 364★, 399
Threshold, 72, 110, 286
Tinbergen, Luke, 229
—, Niko, XVII, 10★, 12, 14, 16, 17★, 167, 219f, 234, 243, 249, 251, 256f, 259ff, 357, 384, 394, 397, 405, 648, 663, 668
Tischner, 129
Tolerance, 403

Tool-using, 357
Topographic orientation, 149, 161f
Tracking, 181
Tracy, H. C., 585
Transference, 372
Tretzel, E., 355★, 356
Tropotaxis, 175f
Trouessart, E., 206
Tschanz, Beat, 135, 593
Turning tendency, 172ff

Übersprung, 250f
Uexküll, Jakob J. von, 12, 261
Ursin, H., 17

Vacuum activity or behavior, 232, 232★
Valenstein, E., 292
Van de Wall, Wolfgang, 618
Vector orientation, 163, 165, 202
Verner, Jared, 475f
Verplanck, W. S., 17
Vibratory sense, 113ff
Vibrissae, 110f
Vince, Margaret, 591f
Visual sense, 75ff, 179
Vitalists, 5
Vocalization, 485
Vowles, D., 288

Waggle dance, 351, 528ff
Wallace, Alfred Russel, 4
Walther, F., 401, 667
Wasman, E., 6

Watson, J., 419
—, R. M., 543
Webster, F. A., 312f
Weidmann, Ulli, 17, 227
Weisman, 4
Weiss, Paul, XVII, 220★, 221
Wendt, Herbert, XIII
Wenger, 586★
Whitman, C. O., XVIf, 4ff, 13
Wickler, Wolfgang, 16, 403f, 417, 480f
Wiesel, T. N., 261
Williams, J., 214
—, Roger, 614
—, T. C., 214
Willows, A. O. D., 311
Willson, Mary, 475f
Wilson, Donald, 219★, 221
—, D. M., 302
—, E. O., 18, 21, 277f
—, W. H., 349
Wilz, K. J., 257
Wolfe, J. B., 330
Wynne-Edwards, V. C., 393f, 566

Zahavi, A., 17
Zeier, H., 353
Ziegler, H. E., 6
Zimen, E., 655
Zonal orientation, 149, 159f
Zoomorphism, 666f
Zoos, 662ff
Zorn, Johann Friedrich, 2
Zygote, 423

77767